From Newton to Chaos

Modern Techniques for Understanding and Coping with Chaos in N-Body Dynamical Systems

NATO ASI Series

Advanced Science Institutes Series

A series presenting the results of activities sponsored by the NATO Science Committee, which aims at the dissemination of advanced scientific and technological knowledge, with a view to strengthening links between scientific communities.

The series is published by an international board of publishers in conjunction with the NATO Scientific Affairs Division

A	**Life Sciences**	Plenum Publishing Corporation
B	**Physics**	New York and London
C	**Mathematical and Physical Sciences**	Kluwer Academic Publishers
D	**Behavioral and Social Sciences**	Dordrecht, Boston, and London
E	**Applied Sciences**	
F	**Computer and Systems Sciences**	Springer-Verlag
G	**Ecological Sciences**	Berlin, Heidelberg, New York, London,
H	**Cell Biology**	Paris, Tokyo, Hong Kong, and Barcelona
I	**Global Environmental Change**	

Recent Volumes in this Series

Volume 333—Perspectives in the Structure of Hadronic Systems
edited by M. N. Harakeh, J. H. Kock, and O. Scholten

Volume 334—Frontier Topics in Nuclear Physics
edited by Werner Scheid and Aurel Sandulescu

Volume 335—Hot and Dense Nuclear Matter
edited by Walter Greiner, Horst Stöcker, and André Gallmann

Volume 336—From Newton to Chaos: Modern Techniques for Understanding and Coping with Chaos in N-Body Dynamical Systems
edited by Archie E. Roy and Bonnie A. Steves

Volume 337—Density Functional Theory
edited by Eberhard K. U. Gross and Reiner M. Dreizler

Volume 338—Electroweak Physics and the Early Universe
edited by Jorge C. Romão and Filipe Freire

Volume 339—Nonlinear Spectroscopy of Solids: Advances and Applications
edited by Baldassare Di Bartolo

Volume 340—Confined Electrons and Photons: New Physics and Applications
edited by Elias Burstein and Claude Weisbuch

Volume 341—Turbulence: A Tentative Dictionary
edited by P. Tabeling and O. Cardoso

Series B: Physics

From Newton to Chaos

Modern Techniques for Understanding and Coping with Chaos in N-Body Dynamical Systems

Edited by

Archie E. Roy

University of Glasgow
Glasgow, Scotland

and

Bonnie A. Steves

Glasgow Caledonian University
Glasgow, Scotland

Plenum Press
New York and London
Published in cooperation with NATO Scientific Affairs Division

Proceedings of a NATO Advanced Study Institute on
From Newton to Chaos: Modern Techniques for Understanding and Coping with Chaos in N-Body Dynamical Systems,
held July 25–August 6, 1993,
in Cortina D'Ampezzo, Italy

NATO-PCO-DATA BASE

The electronic index to the NATO ASI Series provides full bibliographical references (with keywords and/or abstracts) to more than 30,000 contributions from international scientists published in all sections of the NATO ASI Series. Access to the NATO-PCO-DATA BASE is possible in two ways:

—via online FILE 128 (NATO-PCO-DATA BASE) hosted by ESRIN, Via Galileo Galilei, I-00044 Frascati, Italy

—via CD-ROM "NATO Science and Technology Disk" with user-friendly retrieval software in English, French, and German (©WTV GmbH and DATAWARE Technologies, Inc. 1989). The CD-ROM also contains the AGARD Aerospace Database.

The CD-ROM can be ordered through any member of the Board of Publishers or through NATO-PCO, Overijse, Belgium.

Library of Congress Cataloging-in-Publication Data

From Newton to chaos : modern techniques for understanding and coping with chaos in n-body dynamical systems / Archie E. Roy and Bonnie A. Steves.
p. cm. -- (NATO ASI series. Series B, Physics ; vol. 336)
"Published in cooperation with NATO Scientific Affairs Division."
"Proceedings of a NATO Advanced Study Institute on From Newton to Chaos: Modern Techniques for Understanding and Coping with Chaos in N-Body Dynamical Systems, held July 25-August 6, 1993, in Cortina d'Ampezzo, Italy"--T.p. verso.
Includes bibliographical references and index.
ISBN 0-306-44904-8
1. Many-body problem--Congresses. 2. Celestial mechanics--Congresses. 3. Dynamics--Congresses. 4. Planetary theory--Congresses. I. Roy, A. E. (Archie E.), 1924- . II. Steves, Bonnie A. III. North Atlantic Treaty Organization. Scientific Affairs Division. IV. NATO Advanced Study Institute on From Newton to Chaos: Modern Techniques for Understanding and Coping with Chaos in N-Body Dynamical Systems (1993 : Cortina d'Ampezzo, Italy) V. Series: NATO ASI series. Series B, Physics ; v. 336.
QB362.M3F76 1995
521--dc20 95-1181
CIP

ISBN 0-306-44904-8

A Division of Plenum Publishing Corporation
233 Spring Street, New York, N. Y. 10013

10 9 8 7 6 5 4 3 2 1

Printed in the United States of America

SPECIAL PROGRAM ON CHAOS, ORDER, AND PATTERNS

This book contains the proceedings of a NATO Advanced Study Institute held within the program of activities of the NATO Special Program on Chaos, Order, and Patterns.

SPECIAL PROGRAM ON CHAOS, ORDER, AND PATTERNS

Volume 270 — COMPLEXITY, CHAOS, AND BIOLOGICAL EVOLUTION
edited by Erik Mosekilde and Lis Mosekilde

Volume 272 — PREDICTABILITY, STABILITY, AND CHAOS IN N-BODY DYNAMICAL SYSTEMS
edited by Archie E. Roy

Volume 276 — GROWTH AND FORM: Nonlinear Aspects
edited by M. Ben Amar, P. Pelcé, and P. Tabeling

Volume 278 — PAINLEVÉ TRANSCENDENTS: Their Asymptotics and Physical Applications
edited by Decio Levi and Pavel Winternitz

Volume 280 — CHAOS, ORDER, AND PATTERNS
edited by Roberto Artuso, Predrag Cvitanović, and Giulio Casati

Volume 284 — ASYMPTOTICS BEYOND ALL ORDERS
edited by Harvey Segur, Saleh Tanveer, and Herbert Levine

Volume 291 — PROTEIN TRANSFER IN HYDROGEN-BONDED SYSTEMS
edited by T. Bountis

Volume 292 — MICROSCOPIC SIMULATIONS OF COMPLEX HYDRODYNAMIC PHENOMENA
edited by Michel Mareschal and Brad Lee Holian

Volume 298 — CHAOTIC DYNAMICS: Theory and Practice
edited by T. Bountis

Volume 312 — FUTURE DIRECTIONS OF NONLINEAR DYNAMICS IN PHYSICAL AND BIOLOGICAL SYSTEMS
edited by P. L. Christiansen, J. C. Eilbeck, and R. D. Parmentier

Volume 320 — SINGULAR LIMITS OF DISPERSIVE WAVES
edited by N. M. Ercolani, I. R. Gabitov, C. D. Levermore, and D. Serre

Volume 323 — SOFT ORDER IN PHYSICAL SYSTEMS
edited by Y. Rabin and R. Bruinsma

Volume 329 — NONLINEAR COHERENT STRUCTURES IN PHYSICS AND BIOLOGY
edited by K. H. Spatschek and F. G. Mertens

Volume 331 — HAMILTONIAN MECHANICS: Integrability and Chaotic Behavior
edited by John Seimenis

SPECIAL PROGRAM ON CHAOS, ORDER, AND PATTERNS

Volume 336 — FROM NEWTON TO CHAOS: Modern Techniques for Understanding and Coping with Chaos in N-Body Dynamical Systems
edited by Archie E. Roy and Bonnie A. Steves

Volume 341 — TURBULENCE: A Tentative Dictionary
edited by P. Tabeling and O. Cardoso

These Proceedings of the 1993 NATO Advanced Study
Institute held in Cortina d'Ampezzo, Italy
between July 25 and August 6
are dedicated to the
memory of

Henri Poincaré (1854 - 1912)

distinguished French mathematician and theoretical astronomer,
pioneer in the definition and study of chaos.

PREFACE

The reader will find in this volume the Proceedings of the NATO Advanced Study Institute held in Cortina d'Ampezzo, Italy, between July 25 and August 6, 1993, under the title From Newton to Chaos: Modern Techniques for Understanding and Coping With Chaos in N-Body Dynamical Systems.

This institute was the latest in a series of meetings held every three years from 1972 to 1990 in dynamical astronomy, theoretical mechanics and celestial mechanics. The proceedings from these institutes have been well-received in the international community of research workers in these disciplines. The present institute was well attended with 15 series of lectures being given by invited speakers: in addition some 40 presentations were made by the other participants. The majority of these contributions are included in these proceedings.

The all-pervading influence of chaos in dynamical systems (of even a few variables) has now been universally recognised by researchers, a recognition forced on us by our ability, using powerful computer hardware and software, to tackle dynamical problems that until twenty-five years ago were intractable. Doubtless it was felt by many that these new techniques provided a break-through in celestial mechanics and its related disciplines. And so they were. It was wonderful to be able to integrate the orbits of a star cluster and obtain some of the properties of the cluster's evolution; or to study the orbits of the outer planets over periods of 10^8 years or the inner planets for some 10^7 years; or to follow the pinball progress of a comet through the solar system.

Nevertheless the part played by chaos in all these examples - and many more - is clear. The existence of the predictability horizon of the given problem, a time limit beyond which no confidence can be placed in the calculations, has been recognised. Within such predictability horizons our armoury of analytical and computer techniques is able to produce results accurate enough for all practical purposes. Nevertheless there remain many dynamical problems that still escape our understanding because of their essentially chaotic nature.

It is a measure of the genius of the great French mathematician and theoretical astronomer Henri Poincaré, that almost a century ago, he grasped clearly the concept of chaos and its prevailing influence not only in dynamical systems but in human life. Some of his contributions to our understanding of the part chaos plays will be found in this volume. Lesser mortals such as ourselves can only marvel at his brilliant farsightedness and acknowledge it in some small way by respectfully dedicating this volume to his memory. In it, under five sections, namely: Planetary Theory; Dynamics of Asteroids, Comets and Meteors; Dynamics of Natural and Artificial Satellites; The Three Body Problem; and Selected Topics in Dynamics, will be found the contributions of this NATO

ASI. There is no section devoted exclusively to chaos: there cannot be for chaos is universal.

Many more researchers applied to participate in this NATO ASI than could be accommodated. Nevertheless, the Organising Committee was particularly happy to be able, with NATO's help, to invite some of its colleagues from a number of eastern European countries. The warm international friendship demonstrated by the participants during the duration of the NATO ASI was a most valuable feature of the meeting.

The Organising Committee (Professor A. E. Roy (Director), Professor V. Szebehely (Associate Director), Dr P. J. Message, Professor A. Milani and Dr B. A. Steves) would like to take this opportunity to thank all those whose contributions made the ASI a success and such a very happy occasion. In particular the Committee would like to thank the staff of the Antonelli Institute where the ASI was held. As always they were unstintingly helpful and friendly.

The Committee is also sincerely grateful to the Scientific Affairs Division for their guidance, counsel and support.

The editors of this volume would like to take the opportunity of thanking Andrew Conway, Sarah Matthew, Elaine Rowan and David Ramsay of the Department of Physics and Astronomy of Glasgow University for their valuable help in preparing some of the manuscripts. The editors would also like to thank all those who came to Cortina and who supplied papers for this volume. Their cooperation and friendship are very much appreciated.

Archie E. Roy
Dept. of Physics and Astronomy
Glasgow University,
Glasgow G12 8QQ
UNITED KINGDOM

Bonnie A. Steves
Dept. of Mathematics
Glasgow Caledonian University
Glasgow G4 0BA
UNITED KINGDOM

CONTENTS

PART TWO: Dynamics of Asteroids, Comets and Meteors

PART THREE: Dynamics of Natural and Artificial Satellites

PART FIVE: Selected Topics in Dynamics

PRELUDE

POINCARÉ'S CONTRIBUTIONS TO CHANCE AND CHAOS

V. Szebehely

The Department of Aerospace Engineering
and Engineering Mechanics
The University of Texas at Austin
Austin, Texas 78712-1085

ABSTRACT

In this paper a short and condensed biography of Henri Poincaré is presented with detailed information concerning several biographical references. This is followed by a review of his publications emphasizing his work in celestial mechanics and on the problem of three bodies. His article "Le Hasard" is reviewed in detail discussing his contributions to chaos.

BIOGRAPHY OF HENRI POINCARÉ

He was born on April 29, 1854 in Nancy (Lorrain) and died in Paris on July 17, 1912. He was married and had three daughters and a son.

He enjoyed some variations of his name, such as Pontcaré (square bridge) and Pointcaré (square point), but the original family name, traced back to the fifteenth century, turned out to be Poingquarré (clenched fist).

His father Leon Poincaré was a Professor of Medicine and his cousin, Raymond Poincaré, was one of the Presidents of the French Republic. He spent his childhood in the circle of savants and university people. The attack of diphtheria at five resulted in illness and in frailty, feebleness and general timidity. In order to avoid the repeated brutalities of his childhood friends, he spent considerable time reading. After finishing his high school (Lycée) in Nancy be attended the École Polytechnique (1879). Apparently there was a problem concerning his entrance examination at the École Polytechnique because of his poor drawing and sketching ability, but his high competence in other fields compensated for this problem and he was admitted.

In 1879 he became a member of the faculty at Caen and in 1886 a professor of mathematical physics and calculus of probabilities at the University of Paris. He became a member of the Academy of Science in 1887, its president in 1906 and a member of the Academie Française in 1908. Among the many honors and prizes received two are of special interest concerning our field; in 1889 he received King Oscar's Gold Medal for his contribution to the problem of three bodies and in 1909 the Royal Astronomical Society's Gold Medal.

Several excellent biographical references are available concerning the life and scientific activities of Poincaré. These are listed at the end of this paper and at this point only a few explanatory remarks are offered. One short and excellent sketch is by J. R. Newman published in the World of Mathematics. This article precedes the translated version of Poincaré's paper "La Hasard".

Another, highly readable biographical sketch of Poincaré was written by E. T. Bell

From Newton to Chaos, Edited by A.E. Roy
and B.A. Steves, Plenum Press, New York, 1995

(1937) in Men of Mathematics where he is referred to as the "last universalist" and is compared to K. F. Gauss. The list of biographies (in addition to Bell and Newman) contain Bellivier (1956), Dantzig (1954), Darboux (1913), Hadamard (1922 and 1933), Jourdain (1912), Prasad (1934) and Slosson (1914). A remarkable quotation from Jourdain is as follows: "One of the reasons for which Poincaré will live is because he made it possible for us to understand him as well as to admire him".

PUBLICATIONS BY POINCARÉ IN THE FIELD OF CELESTIAL MECHANICS

He is the author of 30 books on mathematical physics and celestial mechanics, four books on the philosophy of science and two books of popular essays. In addition he published about 500 papers. A good collection of his works was published in eleven volumes, see References. His major books on celestial mechanics are "The New Methods" and "Lectures on Celestial Mechanics". The first has been translated in English and it is available either from NASA (1952) or from Am. Inst. Phys. NY (1993). The famous three volumes of "The New Methods" are rather concise, contain several misprints and are hard to follow without a solid and advanced background in celestial mechanics. According to an unverified story, Poincaré was proofreading his book while using it during his lectures. This combination certainly did not help either his lectures or the proofreading as the reader can easily observe. His books on the philosophy of science are much easier to read and contain many original and fascinating ideas.

His interest in the problem of three bodies resulted in several articles, seven of which were published in the Comptes Rendus of the Academy of Sciences (1883, 1895, 1897), in the Acta Mathematica (1890) and in the Bulletin Astronomique (1884, 1891, 1897). "The New Methods" deals also with the problem of three bodies, see Volumes 1 and 3. One of his major contributions is the concept of non-integrability in dynamics and of the problem of three bodies. The non-integrability of the restricted problem of three bodies is treated in his article in the Acta Mathematica, Vol. XIII, p 259, 1890 and in "The New Methods" Vol. I, p 233, 1892. The general problem of three bodies is treated in "The New Methods" Vol. I, p 253, 1892. The references regarding his work on periodic orbits are Acta Mathematica, Vol. XIII, pp. 67-72, 1890 and "The New Methods", Vol. III, pp. 140-157, 1899.

POINCARÉ ON CHAOS

Poincaré's article "Chance" was published in 1907 and its translation in English appeared in 1956. In this paper he uses the word chaos and associates it with large effects in the final outcome produced by slight differences in the initial conditions, with unpredictability and with irreversibility. Not knowing the exact laws of nature and not knowing the exact initial conditions, precise predictions become impossible. He also points out that simple integrable systems show no chaos, the solution is well defined by the uniform integrals (by the constants of the motion) and the final situation is not independent of the initial conditions.

He offers several explanations and examples and discusses them in considerable detail. He explains that very large effects in history can be the consequences of very small changes. He points out that the greatest change in history can come from the birth of a great man. The meeting of cells of a male and of a female might result in the birth of an idiot or a genius. Or, they might be deflected by a tenth of a millimeter and the great man influencing history (such as Napoleon) would not have been born. Poincaré says that "no example can make us better understand the veritable characteristics of chance".

Another example is related to our limitation which makes us to cut the universe into pieces since we cannot understand the whole. When these pieces interact we often refer to chance. When a man goes to work one morning walking on a street we can form laws of his motion, behavior and process. Quite independently another man is hired to wok on the roof of a building on the same street. The two phenomena are completely separate and no interaction is being considered between these two worlds. Now the man on the roof drops a tile which hits and kills the man on the street. We speak about chance, once again.

Poincaré points out the effect of unstable equilibrium in the atmosphere as the reason for the long time unpredictability of weather: "people pray for rain and at the same time it is considered ridiculous to pray for an eclipse".

His article contains many examples; roulette, cardshaffling, fluids mixing, distribution and size of rain drops, etc.

His conclusions are that unpredictability is as natural and general as irreversibility of dynamical systems and that chance is only a measure of our ignorance.

REFERENCES

Bell, E. T., "Men of Mathematics". pp. 526-554, Simon and Schuster Publ., New York, 1937.

Bellivier, A., "Henri Poincaré ou la Vocation Souveraine", Gillimard Publ., Paris, 1956.

Dantzig, T., "Henri Poincaré, Critic of Crisis", Charles Scribner Publ., NY, 1954.

Darboux, G., "Éloge Historique d'Henri Poincaré", Le Temps, December 15, 1913.

Hadamard, J. S., "The Early Scientific Works of Henri Poincaré", 1922.
"The Later Scientific works of Henri Poincaré", 1933.

Jourdain, P. E. B., "Obituary of Poincaré", Monist, Vol. 22, 1912.

Newman, J. R., "Commentary on an Absent-minded Genius and the Laws of Chance", The World of Mathematics, Vol. II, pp. 1353-1356, Tempus Books, Redmond, Washington, 1956.

Poincaré, H. "Oeuvres", Eleven Volumes, Gauthier- Villars, Paris, 1916-1956.
"Les Methodes Nouvelles de la Mécanique Céleste", Gauthier-Villars, Paris, 1892, 1893, 1899.
"Leçons de Mécanique Céleste", Gauthier-Villars, Paris, 1905, 1907, 1909, 1910.
"Le Hasard", La Revue du Mois, Vol. III, pp. 257-276, Paris, Mars 19, 1907. Editor Emile Borel. English Translation: "Chance", The World of Mathematics, Vol. 2, pp. 1359-1372, Tempus Books, Redmond, Wash. Editor, James R. Newman, 1956.
"La Science et l'hyopthèse", (1902),
"La Valeur de la Science", (1905),
"Science et Méthode", (1908),
"Derniéres Pensées", (1913)
The last four items were published by Flammarion Publ., Paris.

Prasad, Ganesh, "Some Great Mathematicians of the Nineteenth Century", Vol. II, p 279, Benares City, 1934.

Slosson, E. E., "Major Prophets of Today", p 131, Boston, 1914.

PART ONE

PLANETARY THEORY

PERTURBATION THEORY: TECHNIQUES AND LIMITATIONS

P.J. Message

Department of Applied Mathematics and Theoretical Physics
University of Liverpool
United Kingdom

1. INTRODUCTION

The aim of these lectures is to give a very brief review of the development of some aspects of solar system perturbation theory, indicating how formal expressions may be constructed for the mutual perturbations of the planets on each other's orbital parameters. Two remarks are in order here. The processes show how, if the instantaneous orbits of the planets are specified by an appropriate choice of parameters, the expressions derived by perturbation methods for these parameters as functions of time may be expressed entirely in terms of periodic terms, indicating stability of the system, and completely regular behaviour of the motions. Also, predictions of planetary positions based on these methods have always been confirmed by observation, and by comparison with numerical integrations of the equations of motion, over all but the very longest of time intervals, up to many millions of years. However, indicators of quasi-ergodic types of motion, "wildness", or "chaos", which had been found since 1964 in numerical explorations of dynamical systems, chosen to model situations in stellar dynamics, in rather different circumstances from the planetary-type motions, have now been identified in the results of recent numerical integrations, which necessarily lead to qualification of any conclusion as to the very long-term regularity of the system, or possibility even as to its very long-term stability. There nevertheless remains the undoubted success of classical perturbation theory in predicting the positions of the planets over centuries, and in reliably modelling their motions over at least many hundreds of thousands of years. This success is all the more remarkable in the face of the obstacles to any uniform convergence of the asymptotic series employed which are presented by the everywhere-dense commensurabilities of mean motions, which prevents validation of the series by the usual methods of rigourous mathematical analysis. For further discussion of these relationships, and a brief review of earlier explorations finding "wildness", &ce, see the author's lectures given at Advanced Study Institute, "Predictability, Stability, and Chaos in N-Body Dynamical Systems", in 1990 (Message, 1991).

The presentation here will be in the framework of Hamiltonian dynamics, beginning with a very brief review of the use of contact transformations and the Hamilton-Jacobi equation, and of the use of the

From Newton to Chaos, Edited by A.E. Roy
and B.A. Steves, Plenum Press, New York, 1995

method of separation of variables to derive a canonical set of parameters for the Keplerian elliptic orbit. This is used to set up the equations for the mutual perturbations in the N-planet problem. (A non-canonical approach was used in the earlier, 1990, lectures mentioned above.) A very brief review is given of a process by which formal series expressions may be derived for the perturbations, of short-periodic and secular type, to any order of perturbation, showing how these, for an appropriate choice of parameters, may always be expressed entirely in terms of periodic functions, which would indicate the stability of the system. However, these expressions are subject to the limitatioms imposed by near-resonances between the frequencies of the long-period motion (the *secular resonances*) The analysis by Laskar (1991) of his numerical integrations of the secular system shows some features less regular than we might expect from the form of our expressions.

2. HAMILTON'S EQUATIONS, CONTACT TRANSFORMATIONS, AND THE HAMILTON-JACOBI EQUATION

We suppose the state of motion (positions and velocities) of a dynamical system at a given time to be specified by the values then of a set of $2n$ *canonical variables*, n of them designated as *co-ordinates*, $q = (q_1, q_2, q_3, \ldots, q_n)$, and n designated as *momenta*, $p = (p_1, p_2, p_3, \ldots, p_n)$. We also suppose that the equations of motion may be put into the form of *Hamilton's equations*

$$\dot{q}_i = \frac{\partial \mathcal{H}}{\partial p_i}, \text{ and } \dot{p}_i = -\frac{\partial \mathcal{H}}{\partial q_i}, \quad (\text{for } i = 1, 2, 3, \ldots, n), \tag{2.1}$$

where $\mathcal{H}$, the *Hamiltonian function*, is a function of the q_i and the p_i and perhaps also of the time, t. According to *Hamilton's Principle*, these equations are equivalent to the statement that the dynamical paths are the extremal curves of the integral

$$\int (p.\dot{q} - \mathcal{H})\, dt$$

(with the usual inner product $a.b = \sum_{i=1}^{n} a_i b_i$).

A non-singular transformation from one set of canonical variables, (q, p), to another set, (Q, P), is a *contact transformation* if there is a function W of either set satisfying the identity in differential forms

$$p.dq - P.dQ = dW. \tag{2.2}$$

If the *generating function* of the transformation, $S = P.Q - W$, may be expressed in terms of the q_i and the P_i, then the equations giving the transformation are

$$p_i = \frac{\partial S}{\partial q_i} \tag{2.3a}$$

$$\text{and } Q_i = \frac{\partial S}{\partial P_i}, \quad (\text{both for } i = 1, 2, 3, \ldots, n), \tag{2.3b}$$

and the Hamiltonian function for the set (Q, P) is $\tilde{\mathcal{H}}(Q, P) \equiv \mathcal{H}(q, p)$.

Suppose now we seek S so that $\tilde{\mathcal{H}} = P_1$, and replace the p_i in $\mathcal{H}$ using (2.3a), we obtain the partial differential equation for S:

$$\mathcal{H}\left(q, \frac{\partial S}{\partial q}\right) = P_1, \tag{2.4}$$

which is the form in which we shall use the *Hamilton-Jacobi Equation*. If a solution of this equation for S is used as generating function to define the transformation $(q, p) \to (Q, P)$, then the motion is described by

$$Q_1 = t + constant, \tag{2.5}$$

all other Q_i's, and all P_i's, being constants of the motion. The set of constants so derived is a canonical set, and is available for use if a perturbed problem is tackled using variation of parameters. Thus, if the Hamiltonian function of the system actually to be studied is in fact

$$\mathcal{H}(q, p) = \mathcal{H}_0(q, p) - R, \tag{2.6}$$

and we are able to solve the Hamilton-Jacobi equation

$$\mathcal{H}_0\left(q, \frac{\partial S}{\partial q}\right) = P_1, \tag{2.7}$$

and then we use the solution S to generate the transformation $(q, p) \to (Q, P)$, then the Hamiltonian function for the transformed system is

$$\tilde{\mathcal{H}}(Q, P) = P_1 - R. \tag{2.8}$$

3. CANONICAL PARAMETERS FOR THE KEPLER PROBLEM, FROM THE HAMILTON-JACOBI EQUATION

In the *Kepler Problem*, a particle, P, of mass m is subject to a force $\mu m/r^2$ towards a fixed point O, r being the distance OP. As co-ordinates, choose r, and ψ, the *latitude*, being the angle which OP makes with a reference plane (with a convention of sign to distinguish the two sides of the plane), and also ϕ, the polar or azimuthal angle, in the reference plane, of the projection of P onto the plane, measured from a reference direction in the plane. Then the kinetic energy of P is

$$T = \frac{1}{2}m\left(\dot{r}^2 + r^2\dot{\psi}^2 + r^2\cos^2\psi\ \dot{\phi}^2\right), \tag{3.1}$$

from which we deduce that the momenta conjugate to r, ψ, and ϕ are, respectively,

$$p_r = \frac{\partial T}{\partial \dot{r}} = m\dot{r}, \quad p_\psi = \frac{\partial T}{\partial \dot{\psi}} = mr^2\dot{\psi}, \text{ and } p_\phi = \frac{\partial T}{\partial \dot{\phi}} = mr^2\cos^2\psi\ \dot{\phi}, \tag{3.2}$$

and, in turn, that the Hamiltonian function is

$$\mathcal{H} = \frac{1}{2m}\left(p_r^{\,2} + \frac{1}{r^2}\,p_\psi^{\,2} + \frac{1}{r^2\cos^2\psi}\,p_\phi^{\,2}\right) - \frac{\mu m}{r} \tag{3.3}$$

(noting that the potential energy is $-\mu m/r$). Thus, from (2.4), above, we see that the Hamilton-Jacobi equation is

$$\frac{1}{2m}\left(\left(\frac{\partial S}{\partial r}\right)^2 + \frac{1}{r^2}\left(\frac{\partial S}{\partial \psi}\right)^2 + \frac{1}{r^2\cos^2\psi}\left(\frac{\partial S}{\partial \phi}\right)^2\right) - \frac{\mu m}{r} = P_1. \tag{3.4}$$

We seek a solution of this equation (with enough disposable parameters to identify with all of the P_i) to use to generate a transformation $(r, \psi, \phi;\ p_r, p_\psi, p_\phi) \to (Q_1, Q_2, Q_3;\ P_1, P_2, P_3)$ given (as from (2.3)) by

$$p_r = \frac{\partial S}{\partial r}, \quad p_\psi = \frac{\partial S}{\partial \psi}, \quad p_\phi = \frac{\partial S}{\partial \phi}, \text{ and } Q_i = \frac{\partial S}{\partial P_i}, \quad (i = 1, 2, 3,). \tag{3.5}$$

Then the solution of the equations of motion for the Keplerian problem is described (as by equation (2.5)) with $Q_1 = t$ + constant, and with Q_2 and Q_3, and all three P_i's, as constants. Suppose we seek a solution

$S = \Phi(\phi, P) + W(r, \psi, P)$. Then (3.4) gives

$$\{\Phi'(\phi)\}^2 = r^2\cos^2\psi \left(2mP_1 + \frac{2\mu m^2}{r} - \left(\frac{\partial W}{\partial r}\right)^2 - \frac{1}{r^2}\left(\frac{\partial W}{\partial \psi}\right)^2\right), \tag{3.6}$$

where the right-hand side is independent of ϕ, and so $\Phi'(\phi)$ must be, and so must be independent of r, ψ, and of ϕ. We may choose it to be P_3, and so $P_3 = \frac{\partial S}{\partial \phi} = p_\phi = mr^2\cos^2\psi\dot{\phi}$, the angular momentum about the normal, through O, to the reference plane. Next set $W(r, \psi, P) = U_1(r, P) + U_2(\psi, P)$, and note from (3.6) that

$$\left(\frac{\partial U_2}{\partial \psi}\right)^2 + \frac{P_3^{\ 2}}{\cos^2\psi} = -\left(\frac{\partial U_1}{\partial r}\right)^2 r^2 + 2mP_1r^2 + 2\mu m^2 r, \tag{3.7}$$

in which the left-hand side is independent of r, and the right-hand side of ψ, so we may choose each to equal $P_2^{\ 2}$. Since $\frac{\partial U_2}{\partial \psi} = p_\psi = mr^2\dot{\psi}$, we have $P_2^{\ 2} = m^2r^4(\dot{\psi}^2 + \cos^2\psi\dot{\phi}^2)$, so that P_2 is the magnitude of the total angular momentum about O. To complete the identifications of the P_i, we note that P_1 is equal to the Hamiltonian, and so is equal to the total Keplerian energy of the motion of the particle P. We now have

$$S = P_3\,\phi + \int \Psi d\psi + \int (\mathcal{R}/r)dr, \tag{3.8}$$

with $\Psi^2 = P_2^{\ 2} - P_3^{\ 2}\sec^2\psi$ and $\mathcal{R}^2 = 2mP_1r^2 + 2\mu m^2 r - P_2^{\ 2}$. We also have the equations $Q_i = \frac{\partial S}{\partial P_i}$, (for $i = 1, 2$, and 3). (3.9)

For $i = 3$ this leads to $\tan\psi = \tan I \sin(\phi - Q_3)$, (3.10)

where $I = \arccos(P_3/P_2)$. Equation (3.10) describes a plane, the *orbit plane*, inclined at an angle I to the reference plane, intersecting that plane in that line, the *line of nodes*, one half of which is given by $\phi = Q_3$, and the other half by $\phi = Q_3 + \pi$. Thus the angle Q_3, the *longitude of the ascending node*, denoted sometimes by h, or, sometimes, by Ω, is the azimuthal angle of that half of the intersection of the orbit plane with the reference plane on which ψ and ϕ increase together. So (3.10) may be written

$$\tan\psi = \tan I \sin(\phi - \Omega). \tag{3.10a}$$

Now suppose that $P_1 < 0$, that is, confine discussion to bound motions. We have $m\dot{r} = p_r = \frac{\partial S}{\partial r} = \mathcal{R}/r = (1/r)\sqrt{\{2mP_1r^2 + 2\mu m^2 r - P_2^{\ 2}\}}$

$$= (1/r)\sqrt{\{2mP_1(r - r_1)(r - r_2)\}},$$

say, where r_1 and r_2 are introduced so that

$$\mu m = -P_1(r_1 + r_2), \text{ and } P_2^2 = -2mP_1r_1r_2. \tag{3.11}$$

Now $\dot{r}$ is real only when r is between r_1 and r_2, and is zero only at these values, so r oscillates between them. We define a, the *major semi-axis*, and e, the *eccentricity*, by

$$a = (r_1 + r_2)/2, \text{ and } e = (r_2 - r_1)/(r_2 + r_1). \tag{3.12}$$

Thus $r_1 = a(1 - e)$ and $r_2 = a(1 + e)$, (3.13)

and the equations (3.11) give

$$P_1 = -\mu m/(2a) \text{ and } P_2^2 = m^2\mu a(1 - e^2). \tag{3.14}$$

We define also an auxiliary angle, E, the *eccentric anomaly*, defined by

$$r = a(1 - e\cos E), \tag{3.15}$$

which increases monotonically, completing one revolution for each complete oscillation of r. Equation (3.9), for $i = 2$, gives

$$Q_2 = \arcsin\left(\frac{\sin\psi}{\sin I}\right) - \sqrt{(1 - e^2)}\int \frac{dE}{1 - e\cos E}$$

$$= u - 2\arctan\left(\sqrt{\left(\frac{1 + e}{1 - e}\right)}\tan\left(\frac{E}{2}\right)\right), \tag{3.16}$$

where u, the *argument of latitude*, is the polar angle, in the orbit plane, from the direction of the ascending node to the radius vector. So Q_2 is the value of u when $E = 0$, that is, at the closest point to O, the centre of force. This point is the *pericentre*, or, in the case of an orbit about the Sun, the *perihelion*. The diametrically opposite point in the orbit is the *apocentre*, or *aphelion*, and these two points are collectively the *apses*. This value of u, equal to Q_2, is the *argument of pericentre*, denoted by g, or, sometimes, ω. Put $u = g + f$, then f is the *true anomaly*. Equation (3.16) may be written

$$\tan\left(\frac{f}{2}\right) = \sqrt{\left(\frac{1 + e}{1 - e}\right)}\tan\left(\frac{E}{2}\right). \tag{3.16a}$$

Substituting this into (3.15) gives, after some algebra,

$$r = \frac{p}{1 + e\cos f}, \tag{3.17}$$

where $p = a(1 - e^2)$ is the *semi-latus rectum*. Equation (3.17) is the equation of an ellipse with a focus at O, the centre of force. Equation (3.9), for $i = 3$, gives

$$Q_1 = (E - e\sin E)/n,$$

where $n = \sqrt{(a^3/\mu)}$ is the *mean motion in longitude*. Put $\ell = n\,Q_1$, the *mean anomaly*, which, like E, increases through a revolution for each complete oscillation of r, that is, for each complete orbit, but, in unperturbed Keplerian motion, increases uniformly with time. This gives *Kepler's Equation*,

$$\ell = E - e\sin E, \tag{3.18}$$

giving the relationship between position in the orbit and time.

From this solution of the Hamilton-Jacobi equation we have derived the canonical set of co-ordinates:

$$Q_1 = \ell/n,\quad Q_2 = g = \omega,\quad \text{and}\quad Q_3 = h = \Omega, \tag{3.19a}$$

and their conjugate momenta:

$$P_1 = -\mu m/(2a),\quad P_2 = m\sqrt{\{\mu a(1 - e^2)\}},\quad \text{and } P_3 = P_2\cos I. \tag{3.19b}$$

Sometimes it is convenient to use as a canonical co-ordinate the mean anomaly, ℓ. Then the appropriate conjugate momentum, L, say, to ℓ, must be such that $P_1 dQ_1 - L d\ell$ is a total differential, which is found to require that

$$L = m\sqrt{(\mu a)}. \tag{3.20}$$

This leads to the set of canonical parameters for the Keplerian orbit often called the *Delaunay set*, the co-ordinates being

$$\ell,\ g = \omega,\ \text{and } h = \Omega, \tag{3.21a}$$

their conjugate momenta being, respectively, in Delaunay's notation,

$$L = m\sqrt{(\mu a)},\quad G = P_2 = m\sqrt{\{\mu a(1 - e^2)\}},\quad \text{and } H = P_3 = G\cos I. \tag{3.21b}$$

When $I = 0$, then the line of nodes is undefined, and so g is undefined, and when $e = 0$, the line of apses is undefined, and so ℓ is undefined. To avoid these singularities in the co-ordinate system, we use new co-ordinates $\varpi = g + h$, the *longitude of pericentre*, or of *perihelion*, and $\lambda = \ell + \varpi$, the *mean longitude*. The momenta conjugate to

$$\lambda,\ \varpi,\ \text{and}\ \Omega \tag{3.22a}$$

are found to be, respectively,

$$\Lambda = L = m\sqrt{(\mu a)},$$

$$\Pi = G - L = m\sqrt{(\mu a)}\left(\sqrt{(1 - e^2)} - 1\right),$$

$$\text{and}\quad N = H - G = m\sqrt{(\mu a)}\sqrt{(1 - e^2)}(\cos I - 1). \tag{3.22b}$$

We note that $P_1 = -\mu^2 m^3/(2\Lambda^2)$, (3.23)

this being the Hamiltonian function corresponding to unperturbed Keplerian motion for this last set of canonical variables.

4. SYSTEM OF SUN AND *N* PLANETS: THE EQUATIONS FOR THE PERTURBATIONS.

Consider a system consisting of $N+1$ particles moving under their mutual gravitational attractions: a primary particle, S (the *Sun*), of mass m_0, and N particles, $P_1, P_2, \ldots, P_N$, (the *planets*), the mass of P_i being m_i. We use Jacobi's system of relative position vectors, in which the position of each body is referred to the mass-centre of all those before it in this enumeration, that is, we use the vectors

$$\mathbf{r}_1 = SP_1, \quad \text{and, for } j = 2, 3, \ldots, N, \quad \mathbf{r}_j = G_{j-1}P_j, \tag{4.1}$$

where G_j is the mass-centre of $S, P_1, P_2, \ldots, P_j$. In terms of these, the kinetic energy, T, of the system can be expressed as a sum of squares:

$$2T = M_N\dot{\rho}^2 + \sum_{j=1}^{N} \varepsilon\beta_j\dot{\mathbf{r}}_j^2 \tag{4.2}$$

in which ρ is the position vector of the mass-centre of the whole system, referred to an inertial origin, $M_j = \sum_{k=0}^{j} m_k$ is the mass of the system out to P_j (so that M_N is the mass of the whole system), and also $\varepsilon\beta_j = m_jM_{j-1}/M_j$, where ε is the greatest of the ratios m_j/m_0 for $j = 1, 2, \ldots, N$, and so is small when also are the ratios of the masses of the planets P_j to that of the primary, S. Also the potential energy, $V = -G\sum_{j=0}^{N-1}\sum_{k=j+1}^{N}\frac{m_jm_k}{P_jP_k}$, in which G is the gravitational constant, may be expressed as

$$V = -\varepsilon\sum_{j=1}^{N}\frac{\mu_j\beta_j}{|\mathbf{r}_j|} - \varepsilon R \tag{4.3}$$

where μ_j is chosen so that $\varepsilon\mu_j\beta_j = Gm_0m_j$, and

$$R = \varepsilon\sum_{j=1}^{N-1}\sum_{k=j+1}^{N}\beta_j\beta_k\left(\frac{1}{|\mathbf{r}_j-\mathbf{r}_k|} - \frac{\mathbf{r}_j\cdot\mathbf{r}_k}{|\mathbf{r}_j|^3}\right) + O(\varepsilon^2), \tag{4.4}$$

and it will be important later that R is entirely expressible in terms of the quantities $|\mathbf{r}_j|$, $|\mathbf{r}_j-\mathbf{r}_k|$, the β_j and ε. From equation (4.3) we see that the equations governing the relative motion are (after cancelling the factor ε)

$$\beta_j\frac{d^2\mathbf{r}_j}{dt^2} = -\mu_j\beta_j\mathbf{r}_j/|\mathbf{r}_j|^3 + \frac{\partial R}{\partial\mathbf{r}_j}, \tag{4.5}$$

In the limit $\varepsilon \to 0$, we have $R \to 0$, and we have N independent Kepler inverse square central force problems. Let us regard this as the "unperturbed" situation, and develop perturbation methods to study the actual, $\varepsilon \neq 0$, situation.

The equations may be put into Hamiltonian form by taking $\mathbf{p}_j = \beta_j\dot{\mathbf{r}}_j$ as the momentum conjugate to $\mathbf{r}_j$, and then the Hamiltonian function is

$$\mathcal{H}(\mathbf{r}, \mathbf{p}) = \sum_{j=1}^{N}\left\{\mathbf{p}_j^2/(2\beta_j) - \mu\beta_j/|\mathbf{r}_j|\right\} - R, \tag{4.6}$$

in which $\mathbf{r}$ is the set of N 3-vectors $(\mathbf{r}_1, \mathbf{r}_2, \ldots, \mathbf{r}_N)$, and $\mathbf{p}$ is the set of N 3-vectors $(\mathbf{p}_1, \mathbf{p}_2, \ldots, \mathbf{p}_N)$. Associated with each planet, P_j, define three canonical co-ordinates, λ_j, ϖ_j, and Ω_j, and three Keplerian elements a_j, e_j, and I_j, these six quantities being related to $\mathbf{r}_j$ and $\dot{\mathbf{r}}_j$ exactly as λ, ϖ, Ω, a, e, and I are related to the position and velocity relative to the centre of force in Keplerian motion, as described in section 3. Conjugate to λ_j, ϖ_j, and Ω_j, are their conjugate momenta, Λ_j, Π_j, and N_j, respectively, which are related to a_j, e_j, and I_j by the following equations, corresponding to (3.22b),

$$\Lambda_j = \beta_j\sqrt{(\mu_j a_j)},$$

$$\Pi_j = \beta_j\sqrt{(\mu_j a_j)}\left(\sqrt{(1 - e_j^2)} - 1\right),$$

and
$$N_j = \beta_j\sqrt{(\mu_j a_j)}\sqrt{(1 - e^2)}(\cos I_j - 1). \tag{4.7}$$

Then a set of $3N$ canonical co-ordinates for the whole relative motion is $(\lambda, \varpi, \Omega)$, where λ denotes $(\lambda_1, \lambda_2, \ldots, \lambda_N)$, ϖ denotes $(\varpi_1, \varpi_2, \ldots, \varpi_N)$, and Ω denotes $(\Omega_1, \Omega_2, \ldots, \Omega_N)$. The conjugate momenta are, respectively, (Λ, Π, N), where Λ denotes $(\Lambda_1, \Lambda_2, \ldots, \Lambda_N)$, Π denotes $(\Pi_1, \Pi_2, \ldots, \Pi_N)$, and N denotes $(N_1, N_2, \ldots, N_N)$. The Hamiltonian function giving the equations of motion in terms of these is, using (2.8) and (3.23),

$$\mathcal{H}'(\lambda, \varpi, \Omega, \Lambda, \Pi, N) = \mathcal{H}_0 - R, \tag{4.8}$$

where
$$\mathcal{H}_0 = -\sum_{j=1}^{N}\beta_j^3\mu_j^2/(2\Lambda_j^2). \tag{4.8a}$$

In the unperturbed case, with R identically zero, it is clear that all the ϖ_j, Ω_j, Λ_j, Π_j, and N_j are constants, and

$$\lambda_j = n_j t + \varepsilon_j, \tag{4.9}$$

where the ε_j are constants, as are the mean motions

$$n_j = \beta_j^3\mu_j^2/\Lambda_j^3 = \sqrt{(\mu_j a_j^{-3})}. \tag{4.10}$$

Since R is a periodic function of each of the co-ordinates, it may be written

$$R = \sum_{\nu} K_\nu \cos P_\nu, \tag{4.11}$$

where ν denotes the $3N$-vector with integer components:

$$\nu = (\nu_1, \nu_2, \ldots, \nu_N, \nu_1', \nu_2', \ldots, \nu_N', \nu_1'', \nu_2'', \ldots, \nu_N''),$$

and
$$P_j = \sum_{j=1}^{N}(\nu_j\lambda_j + \nu_j'\varpi_j + \nu_j''\Omega_j), \tag{4.12}$$

the summation in (4.11) being over all sets ν with $\nu_1 \geq 0$, and, since R depends only on the relative positions of the bodies, which is not affected by a change of origin of the longitudes, with only those terms for which

$$\sum_{j=1}^{N} (\nu_j + \nu'_j + \nu''_j) = 0. \qquad (4.13)$$

Sine terms do not appear because the lengths and scalar products of the $\mathbf{r}_j$ are not altered if every one of the co-ordinates λ_j, ϖ_j, Ω_j, is changed in sign, since this simply reflects the figure formed by S and the P_j in the line of origin of longitudes. Part of the process of forming these expansions involves the expression of the position of each P_j in terms of Fourier series in the mean anomaly, ℓ_j, using Kepler's equation, (3.18). From this equation we find that (with $p \neq 0$)

$$\exp\{\sqrt{(-1)}pE\} = \sum_{j=-\infty}^{\infty} c_{p,s} \exp\{\sqrt{(-1)}s\ell\} \qquad (4.14)$$

in which

$$c_{p,s} = \begin{cases} (p/s)\,J_{s-p}(se), & \text{for } s \neq 0, \\ 0, & \text{if } s = 0 \text{ and } p \neq \pm 1, \\ -\,e/2, & \text{if } s = 0 \text{ and } p = \pm 1 \end{cases} \qquad (4.14a)$$

where $J_n(x)$ is the *Bessel function* (whose first appearance was in this context), given by

$$J_n(x) = \sum_{r=0}^{\infty} \frac{(-1)^r (x/2)^{n+2r}}{r!\,(n + r)!} \text{ if } n \geq 0, \qquad (4.15a)$$

and $$J_{-n}(x) = (-1)^n J_n(x). \qquad (4.15b)$$

Note that $J_n(x)$ has the factor $x^{|n|}$, this being multiplied by a power series in x^2. Now from equation (3.15) there follows (using $J'_n = (J_{n-1} - J_{n+1})/2$)

$$r/a = 1 + \tfrac{1}{2}e^2 - 2e \sum_{s=1}^{\infty} \tfrac{1}{s} J'_s(se)\cos(s\ell), \qquad (4.16)$$

so that, in this Fourier series, the co-efficient of $\cos(s\ell)$ has the factor e^s multiplied by a power series in e^2. This is the *d'Alembert* property, which may also be described by saying that r/a may be expressed as a power series in $e \cos\ell$ and $e \sin\ell$. In the final expansion of R, the d'Alembert property manifests itself in that R may be expressed as a power series in the quantities $e_j\cos\varpi_j$, $e_j\sin\varpi_j$, $\sin(I_j/2)\cos\Omega_j$, and $\sin(I_j/2)\sin\Omega_j$. Alternatively, it may be expressed by saying that K_ν has each of the N factors $e_j^{|\nu'_j|}.\sin^{|\nu''_j|}(I_j/2)$, multiplied by a power series in the quantities e_j^2 and $\sin^2(I_j/2)$. Since the Hamiltonian function $\mathcal{H}'$ is as given by (4.8), the equations for the perturbations take the forms

$$\dot{\lambda}_j = \frac{\partial \mathcal{H}'}{\partial \Lambda_j} = n_j - \frac{\partial R}{\partial \Lambda_j} = n_j - \sum_{\nu} \frac{\partial K_\nu}{\partial \Lambda_j} \cos P_\nu,$$

$$\dot{\varpi}_j = \frac{\partial \mathcal{H}'}{\partial \Pi_j} = - \frac{\partial R}{\partial \Pi_j} = - \sum_{\nu} \frac{\partial K_\nu}{\partial \Pi_j} \cos P_\nu,$$

$$\dot{\Omega}_j = \frac{\partial \mathcal{H}'}{\partial \mathcal{N}_j} = -\frac{\partial R}{\partial \mathcal{N}_j} = -\sum_\nu \frac{\partial K_\nu}{\partial \mathcal{N}_j} \cos P_\nu,$$

$$\dot{\Lambda}_j = -\frac{\partial \mathcal{H}'}{\partial \lambda_j} = \frac{\partial R}{\partial \lambda_j} = -\sum_\nu \nu_j K_\nu \sin P_\nu,$$

$$\dot{\Pi}_j = -\frac{\partial \mathcal{H}'}{\partial \varpi_j} = \frac{\partial R}{\partial \varpi_j} = -\sum_\nu \nu'_j K_\nu \sin P_\nu,$$

$$\dot{\mathcal{N}}_j = -\frac{\partial \mathcal{H}'}{\partial \Omega_j} = \frac{\partial R}{\partial \Omega_j} = -\sum_\nu \nu''_j K_\nu \sin P_\nu, \qquad (4.17)$$

in which n_j is still given as by (4.10), but of course will not now be exactly constant. If these equations are expressed in terms of the elements a_j, e_j, and I_j, in place of the momenta Λ_j, Π_j, and $\mathcal{N}_j$, we obtain *Lagrange's planetary equations*

$$\dot{a}_j = \frac{2}{n_j a_j} \frac{\partial R}{\partial \lambda_j} = -\frac{2}{n_j a_j} \sum_\nu \nu_j K_\nu \sin P_\nu,$$

$$\dot{e}_j = \frac{b_j(b_j - a_j)}{n_j a_j^4 e_j} \frac{\partial R}{\partial \lambda_j} - \frac{b_j}{n_j a_j^3 e_j} \frac{\partial R}{\partial \varpi_j} = \sum_\nu \frac{b_j(a_j - b_j)\nu_j + a_j b_j \nu'_j}{n_j a_j^4 e_j} K_\nu \sin P_\nu,$$

$$\dot{I}_j = -\frac{1}{n_j a_j b_j \sin I_j} \frac{\partial R}{\partial \Omega_j} - \frac{\tan(I_j/2)}{n_j a_j b_j} \left(\frac{\partial R}{\partial \lambda_j} + \frac{\partial R}{\partial \varpi_j} \right)$$

$$= \frac{1}{n_j a_j b_j} \sum_\nu \left\{ (\nu_j + \nu'_j)\tan(I_j/2) + \nu''_j \operatorname{cosec} I_j \right\} K_\nu \sin P_\nu,$$

$$\dot{\lambda}_j = n_j - \frac{2}{n_j a_j} \frac{\partial R}{\partial a_j} + \frac{b_j(a_j - b_j)}{n_j a_j^4 e_j} \frac{\partial R}{\partial e_j} + \frac{\tan(I_j/2)}{n_j a_j b_j} \frac{\partial R}{\partial I_j}$$

$$= n_j + \frac{1}{n_j a_j^2} \sum_\nu \left\{ -2a_j \frac{\partial K_\nu}{\partial a_j} + \frac{b_j(a_j - b_j)}{n_j a_j^2} \frac{\partial K_\nu}{\partial e_j} + \frac{a_j}{b_j} \tan(I_j/2) \frac{\partial K_\nu}{\partial I_j} \right\} \cos P_\nu,$$

$$\dot{\varpi}_j = \frac{b_j}{n_j a_j^3 e_j} \frac{\partial R}{\partial e_j} + \frac{\tan(I_j/2)}{n_j a_j b_j} \frac{\partial R}{\partial I_j}$$

$$= \frac{1}{n_j a_j^2} \sum_\nu \left\{ \frac{b_j}{a_j e_j} \frac{\partial K_\nu}{\partial e_j} + \frac{a_j}{b_j} \tan(I_j/2) \frac{\partial K_\nu}{\partial I_j} \right\} \cos P_\nu,$$

and $$\dot{\Omega}_j = \frac{1}{n_j a_j b_j \sin I_j} \frac{\partial R}{\partial I_j} = \frac{1}{n_j a_j b_j \sin I_j} \sum_\nu \frac{\partial K_\nu}{\partial I_j} \cos P_\nu. \qquad (4.18)$$

5. Some Techniques for the solution of the equations for the Perturbations, and a complete formal solution

If, on the right-hand sides of equations (4.17) or (4.18), constant values are used for the elements ϖ_j, Ω_j, a_j, e_j, and I_j, and expressions linear in time used for the λ_j, then the P_ν are represented by linear functions of time, and the equations may be integrated, to give approximate expressions for the perturbations in the elements. These approximating expressions are known as the *first-order perturbations*. Before integrating each equation for one of the λ_j, its first term n_j is replaced by $n_{j0} - \frac{3n_{j0}}{2a_{j0}}\delta a_j$, where δa_j is the first-order perturbation in a_j, a_{j0} is the constant approximation used for a_j, and n_{j0} is the corresponding value of n_j). Those terms in the expansion in which the ν_j are not all zero give rise to terms which are periodic, and have denominator $\sum_\nu \nu_j n_j$, (or its square in the case of some terms in λ_j). Those terms in R in which the ν_j are all zero, called *secular*, give rise to contributions which are linear in t. The right-hand sides of the equations are then extended to include the contributions of the first-order perturbations to the elements, following which the equations can be again integrated, giving rise to improved approximations to the perturbations, the changes being called *second-order* perturbations. This process may clearly be continued to provide perturbations of any order. Computing expressions for the perturbations by proceeding in this way has come to be called *Poisson's method*.

A more formal way of developing expressions for the perturbations makes use of *Lie series transformations*. In this we use the operator $\mathcal{L}_W$, associated with a funcion W, (the *generating function* of the Lie series transformation), defined through its action on any function ℓ by

$$\mathcal{L}_W(\ell) = \{\ell, W\} = \sum\left(\frac{\partial \ell}{\partial q_i}\frac{\partial W}{\partial p_i} - \frac{\partial \ell}{\partial p_i}\frac{\partial W}{\partial q_i}\right), \tag{5.1}$$

where (q, p) is any set of co-ordinates and their conjugate momenta for the system. That is, $\mathcal{L}_W(\ell)$ is the Poisson bracket of ℓ and W. In turn we define $\mathcal{L}_W^{\,2}(\ell) = \mathcal{L}_W\{\mathcal{L}_W(\ell)\}$, $\mathcal{L}_W^{\,3}(\ell) = \mathcal{L}_W\{\mathcal{L}_W^{\,2}(\ell)\}$, &ce., and in general $\mathcal{L}_W^{\,n+1}(\ell) = \mathcal{L}_W\{\mathcal{L}_W^{\,n}(\ell)\}$ for $n = 1, 2, 3, \ldots\ldots$ (5.2)

Then the transformation from (q, p) to another set (Q, P) is a Lie series transformation if it is given by

$$q_i = Q_i + \sum_{n=1}^{\infty}\frac{1}{n!}\mathcal{L}_W^{\,n}(Q_i) \quad \text{and} \quad p_i = P_i + \sum_{n=1}^{\infty}\frac{1}{n!}\mathcal{L}_W^{\,n}(P_i). \tag{5.3}$$

(Note that the transition from (Q, P) to (q, p) corresponds to the change over unit time in a fictitious dynamical system which has W as its Hamiltonian function.) Now, in fact,

$$\mathcal{L}_W(Q_i) = \frac{\partial W}{\partial P_i}, \quad \text{and} \quad \mathcal{L}_W(P_i) = -\frac{\partial W}{\partial Q_i},$$

so that (5.3) become

$$q_i = Q_i + \frac{\partial W}{\partial P_i} + \sum_{n=1}^{\infty}\frac{1}{(n+1)!}\mathcal{L}_W^{\,n}\left(\frac{\partial W}{\partial P_i}\right)$$

and

$$p_i = P_i - \frac{\partial W}{\partial Q_i} - \sum_{n=1}^{\infty}\frac{1}{(n+1)!}\mathcal{L}_W^{\,n}\left(\frac{\partial W}{\partial Q_i}\right). \tag{5.3a}$$

For any function $f(q, p)$,

$$f(q, p) \equiv f(Q, P) + \sum_{n=1}^{\infty} \frac{1}{n!} \mathcal{L}_W^{\,n}\{f(Q, P)\} \qquad (5.4)$$

In particular if $\mathcal{H}(q, p)$ is the Hamiltonian function before the Lie series transformation, and $\tilde{\mathcal{H}}(Q, P)$ is the Hamiltonian function after, then, if the generating function W, and therefore the transformation, is time-independent, we have

$$\begin{aligned} \tilde{\mathcal{H}}(Q, P) &\equiv \mathcal{H}(q, p) \\ &\equiv \mathcal{H}(Q, P) + \sum_{n=1}^{\infty} \frac{1}{n!} \mathcal{L}_W^{\,n}\{\mathcal{H}(Q, P)\}, \end{aligned} \qquad (5.5)$$

which gives a functional identity which must be satisfied by the generating function $W(Q, P)$ and the new Hamiltonian function $\tilde{\mathcal{H}}(Q, P)$. It will be noticed that, in the case where the generating function is identically zero, the transformation becomes the identity. In the application to perturbation theory, the choice of $W(Q, P)$ is made so that there are no short-periodic terms in $\tilde{\mathcal{H}}(Q, P)$. Thus, if one were to carry out such a transformation, for this purpose, on the equations for perturbations in a planetary system, using the set of co-ordinates and momenta set up in section 4 above, the generating function W would be sought as an infinite series of terms of increasing order in the perturbation parameter, ε:

$$W = W_1 + W_2 + W_3 + \ldots\ldots + W_n + \ldots\ldots, \qquad (5.6)$$

with a similar expansion for the new Hamiltonian function:

$$\tilde{\mathcal{H}} = \tilde{\mathcal{H}}_0 + \tilde{\mathcal{H}}_1 + \tilde{\mathcal{H}}_2 + \tilde{\mathcal{H}}_3 + \ldots + \tilde{\mathcal{H}}_n + \ldots\ldots, \qquad (5.7)$$

in which the first term, $\tilde{\mathcal{H}}_0$, is chosen to be the same function of the new co-ordinates and momenta that $\mathcal{H}_0$ is of the old, and, in order that $\tilde{\mathcal{H}}_1$ shall contain no short-period terms, we choose (recalling that $\mathcal{H} = \mathcal{H}_0 - R$, as given in (4.8), and the expansion of R is $\sum_{\nu} K_{\nu} \cos P_{\nu}$, as given in (4.11)),

$$W_1 = - \sum_{\nu} (K_{\nu}/\rho) \sin P_{\nu}, \qquad (5.8)$$

where $\rho = \sum_{j=1}^{N} \nu_j n_j$, the summation in (5.8) being over all sets ν in which the ν_j are not all zero. Then W_n and $\tilde{\mathcal{H}}_n$, for successive values of n, are chosen in similar manner, by repeated use of (5.5), as expanded using (5.6) and (5.7). Then the expressions (5.3) give formal expressions for the short-periodic perturbations to all orders. For fuller details see, for example, lectures given to the 1987 Advanced Study Institute, "Long-Term Dynamical Behaviour of Natural and Artificial N-Body Systems" (Message, 1988a). For a discussion of the impact of near resonances on perturbations, and how the d'Alembert property shows the effect is tempered when the orbits are nearly circular and nearly coplanar, unless the resonance corresponds to small integers, see, for example, the author's lectures given at the 1990 Advanced Study Institute, "Predictability, Stability, and Chaos in N-Body Dynamical Systems" (Message, 1991).

The lectures continued with a brief description of a succession of transformations leading to a complete formal solution of the equations of motion. First a change to rectangular-type co-ordinates, as introduced by Poincaré:

$$\xi_j = \sqrt{(-2\Pi_j)} \cos \varpi_j \quad \text{and} \quad q_j = \sqrt{(-2N_j)} \cos \Omega_j \qquad (5.9a)$$

and their conjugate momenta:

$$\eta_j = \sqrt{(-2\Pi_j)} \sin \varpi_j \quad \text{and} \quad p_j = \sqrt{(-2N_j)} \sin \Omega_j \qquad (5.9b)$$

where we notice that

$$\surd(-2\Pi_j) = \surd(\Lambda_j)\varepsilon_j, \text{ where } \varepsilon_j = \surd\left\{2\left(1 - \surd(1 - e_j^2)\right)\right\}$$
$$= e_j + (e_j^3)/8 + \ldots\ldots,$$

and that

$$\surd(-2N_j) = \surd(\Lambda_j+\Pi_j)\sin(I_j/2),$$

so that use of these variables removes the (purely notational) singularities at $e_j = 0$ and $I_j = 0$ of the previous set (just as the change from plane polar, to rectangular, co-ordinates removes the singularity arising because the polar angle is not defined at the origin). This change of variables is time-independent, so the Hamiltonian function for the $(\lambda, \xi, q; \Lambda, \eta, p)$ is obtained by substituting for the ϖ_j, Ω_j, Π_j, and N_j in favour of the ξ_j, q_j, η_j, and p_j as given by the equation (5.9). Then we use the Lie series transformation to remove the short-period terms, that is, the terms dependent on the λ_j. This gives the transformation

$$(\lambda, \xi, q; \Lambda, \eta, p) \mapsto (\lambda', \xi', q'; \Lambda', \eta', p'), \tag{5.10}$$

with the new Hamiltonian function

$$\mathcal{H}'(\lambda', \xi', q', \Lambda', \eta', p') = -\sum_{j=1}^{N} \beta_j^3\mu_j^2/(2\Lambda_j'^2) - R', \tag{5.11}$$

and recall that the transformation is chosen so that the co-ordinates λ' are ignorable, that is, $\mathcal{H}'$ is independent of the λ'. (For more details of this transformation see Message (1982), pages 30 to 33, and for a modification of the method to deal with a single small-integer near-commensurability, see Message (1988b).) We note that, in this transformed system, the d'Alembert property implies that $\mathcal{H}'$ is expressible in terms of quantities of the type $\xi_j'\xi_k' + \eta_j'\eta_k'$, $\xi_j'^2 + \eta_j'^2$, $q_j'q_k' + p_j'p_k'$, $q_j'^2 + p_j'^2$, and $\xi_j'q_k' + \eta_j'p_k'$, as well as of Λ_j', and so shows that the terms in R' of lowest degree in the ξ_j', η_j', q_j', and p_j' are of degree two, and so of the form

$$\sum\left\{K_{jk}(\xi_j'\xi_k' + \eta_j'\eta_k') + K^*_{jk}(q_j'q_k' + p_j'p_k')\right\}, \tag{5.12}$$

the summation being over all pairs (j,k) with $1 \le j \le k \le N$. The *secular variations* theory of Laplace and Lagrange results from the approximation in which terms of higher order than these are omitted, so that the ξ_j', q_j', η_j', and the p_j' are governed by a system of linear differential equations, whose solution is a superposition of oscillations of periods between 47,000 and 2,000,000 years.

Now consider a linear transformation of the co-ordinates ξ_j' and q_j' to a new set σ_1, σ_2, ..., σ_{2N}, given by

$$\sigma_r = \sum_{j=1}^{N} A_{rj}\xi_j' \text{ and } \sigma_{N+r} = \sum_{j=1}^{N} B_{rj}q_j', \text{ for } r = 1, 2, 3, \ldots, N. \tag{5.13}$$

This defines a contact transformation if we also introduce a new set of momenta, τ_1, τ_2, ..., τ_{2N}, conjugate to the σ_r, and given by

$$\tau_r = \sum_{j=1}^{N} A_{rj}\eta_j' \text{ and } \tau_{N+r} = \sum_{j=1}^{N} B_{rj}p_j', \text{ for } r = 1, 2, 3, \ldots, N. \tag{5.14}$$

provided that the matrices (A_{rj}) and (B_{rj}) are orthogonal, and that the

λ'_j and Λ'_j are also replaced by λ''_j and Λ''_j, respectively, where for each j

$$\Lambda'_j = \Lambda''_j$$

$$\text{and} \quad \lambda'_j = \lambda''_j - \sum_{r=1}^{N} \sum_{s=1}^{N} \sum_{k=1}^{N} \left(A_{rk} \frac{\partial A_{sk}}{\partial \Lambda''_j} \sigma_r \tau_s + B_{rk} \frac{\partial B_{sk}}{\partial \Lambda''_j} \sigma_{N+r} \tau_{N+s} \right). \qquad (5.15)$$

The matrices (A_{rj}) and (B_{rj}) may still be chosen to transform the quadratic form (5.12) to a sum of squares, that is, to reduce the Laplace and Lagrange secular variation theory to normal co-ordinates, if the quantities γ_r $(1 \le r \le 2N)$ are all different, where $\gamma_r/2$ (for $1 \le r \le N$) are the eigenvalues of the matrix (K_{jk}), and the $\gamma_{N+r}/2$ (for $1 \le r \le N$) are the eigenvalues of the matrix (K^*_{jk}). The Hamiltonian function is now of the form

$$\mathcal{H}''(\lambda'', \sigma, \Lambda'', \tau) = - \sum_{j=1}^{N} \beta_j^3 \mu_j^2 / (2\Lambda_j''^2) - R'' \qquad (5.16)$$

(in which σ represents $(\sigma_1, \sigma_2, \ldots, \sigma_{2N})$ and τ, $(\tau_1, \tau_2, \ldots, \tau_{2N})$), where λ'' is ignorable, the terms in R'' of degree two in the σ_r and τ_r (that is, the terms retained in the Laplace and Lagrange theory) are $\frac{1}{2} \sum_{r=1}^{2N} \gamma_r (\sigma_r^2 + \tau_r^2)$, and the further terms are power series in the quantities of the types $\sigma_r^2 + \tau_r^2$ and $\sigma_r \sigma_s + \tau_r \tau_s$. In the Laplace and Lagrange theory, the γ_r, for $1 \le r \le N$, are the frequencies of the oscillations in the ξ_j and η_j, that is, in the eccentricities and apses, and the γ_{N+r} (for $1 \le r \le N$) are the frequencies in the q_j and p_j, that is in the motion of the orbit planes. One of the latter is zero, corresponding to the possible motion in which the relative motion of the primary and planets is confined to a fixed plane, so that all the values of I_j are the same, and all the Ω_j are the same. (In the unapproximated problem, representing the actual motion, the other γ_r are the limiting values of the frequencies in the limit of zero amplitudes of the oscillations. In fact the γ_r, for $1 \le r \le N$, are the limiting values of what are often denoted by "g_r", and the γ_{n+r}, for $1 \le r \le N$, are the limiting values of the "s_r", in the secular variation theory.)

Next, return to action and angle form, introducing new co-ordinates $\omega = (\omega_1, \omega_2, \ldots, \omega_{2N})$, and new momenta $\Omega = (\Omega_1, \Omega_2, \ldots, \Omega_{2N})$, by

$$\sigma_r = \sqrt{(-2\Omega_r)} \cos \omega_r \text{ and } \tau_r = \sqrt{(-2\Omega_r)} \sin \omega_r . \qquad (5.17)$$

The Hamiltonian function now becomes

$$\mathcal{H}^\dagger(\lambda'', \omega, \Lambda'', \Omega) = - \sum_{j=1}^{N} \beta_j^3 \mu_j^2 / (2\Lambda_j''^2) - R^\dagger \qquad (5.18)$$

where

$$R^\dagger = - \sum_{r=1}^{2N} \gamma_r \Omega_r + \sum_{\nu} K_\nu^\dagger \cos P_\nu , \qquad (5.19)$$

with

$$P_\nu = \sum_{r=1}^{2N} \nu_r \omega_r , \qquad (5.20)$$

the summation in (5.19) being over those sets $\nu = (\nu_1, \nu_2, \ldots, \nu_{2N})$ with $\sum_{r=1}^{2N} \nu_r = 0$. The co-efficients $K_\nu^\dagger$ are expressible as power series in the Ω_r, with terms of order two or higher, and the d'Alembert property ensures that $K^{\dagger}_{\nu}$ has each of the factors $\Omega_r^{|\nu_r/2|}$. The formal solution is then completed by carrying out a Lie series transformation to remove the terms dependent on the ω_r, in a manner similar to that used earlier to remove the λ_j, resulting finally in a system of equations with co-ordinates $\lambda^* = (\lambda^*_1, \lambda^*_2, \ldots, \lambda^*_{2N})$, and $\omega^* = (\omega^*_1, \omega^*_2, \ldots, \omega^*_{2N})$, and momenta $\Lambda^* = (\Lambda^*_1, \Lambda^*_2, \ldots, \Lambda^*_{2N})$, and $\Omega^* = (\Omega^*_1, \Omega^*_2, \ldots, \Omega^*_{2N})$, with the Hamiltonian function

$$\mathcal{H}^*(\lambda^*, \omega^*, \Lambda^*, \Omega^*) = - \sum_{j=1}^{N} \beta_j\mu_j/(2\Lambda_j^{*2}) - R^*(\Lambda^*, \Omega^*). \tag{5.21}$$

(For more details of this transformation see Message (1982), pages 36 to 39.) In this formal system, all the co-ordinates, λ^*_r and ω^*_r are ignorable, and their conjugate momenta, Λ^*_r and Ω^*_r, are therefore constants of the motion, so that, in the formal solution, the λ^*_r and ω^*_r are linear functions of the time. In fact

$$\dot{\omega}_r = - \frac{\partial R^*}{\partial \Omega^*_r} = \gamma^*_r, \text{ say, for } r = 1, 2, \ldots, 2N, \tag{5.22}$$

which is then constant in this formal solution, and equal to g_r for $r = 1, 2, \ldots, N$, and to s_{r-N} for $r = N+1, N+2, \ldots, 2N$). Substituting back through the sequence of transformations, we see, in this formal solution, the complete expressions for the perturbations in the original parameters of the orbits represented entirely as superpositions of series of periodic terms, some of the periods being those of short period, that is, with frequencies which are linear combinations of the orbital mean motions, and others of course being of the order of magnitude of the periods of the secular variation oscillations, with frequencies which are linear combinations of the γ^*_r, and of course beat frequencies of all of these. In particular we note that $\Lambda'_r = \Lambda''_r = \Lambda^*_r$, which is constant in this solution, in which therefore the Λ_r, and therefore also the major semi-axes a_r, have only short-period perturbations.

6. Limitations of this formal solution

We note that the equations giving this last transformation will involve expressions containing denominators which are linear combinations of two or more of the frequencies γ^*_r, and so, if there are near-commensurabilities between these frequencies (*secular resonances*), similar considerations will apply in this transformation to those which qualify the use of the first Lie series transformation in the presence of near-commensurabilities between the orbital mean motions. (Note, however, that the fact that one of the γ^*_r is zero, corresponding to the planar solution, does not lead to a difficulty of this sort, since in no term of the generating function of this last Lie series transformation does a single frequency γ^*_r appear alone as a denominator.) In fact Laskar, in his analysis of the results of his numerical integration of the equations for the secular variations of the eight major planets (Laskar 1991), identifies a critical argument of the secular motion, $2\omega_4 - 2\omega_3 + \Omega_3 - \Omega_4$ in the notation we are using here, associated with the apses and nodes of the orbits of Earth and Mars, which makes transitions, from an interval during which it is in libration about zero, through one during which it is circulating, i.e., is changing monotonically through complete revolutions, in turn to an interval of libration about 180°, all during the *200* million years of his numerical integration.

REFERENCES

Laskar, J., 1991, *Icarus*, **88**, *266-291*.

Message, P.J., 1982, *Celestial Mechanics*, **26**, *25-39*.

Message, P.J., 1988a, "Planetary Perturbations from Lie Series, including Resonance", in "Long-Term Dynamical Behaviour of Natural and Artificial N-Body Systems" (ed. A.E. Roy, Kluwer), pp.47-72.

Message, P.J., 1988b, *Celestial Mechanics*, **43**, 119-125.

Message, P.J., 1991, "Perturbation theory, resonance, librations, chaos, and Halley's comet", in "Predictability, Stability, and Chaos in N-Body Dynamical Systems", (ed. A.E. Roy, Plenum), pp.239-247.

QUANTITATIVE METHODS IN CLASSICAL PERTURBATION THEORY

Antonio Giorgilli

Dipartimento di Matematica dell' Università di Milano
Via Saldini 50, 20133 Milano, Italy

OVERVIEW

At the beginning of the second volume of his *Méthodes nouvelles de la Mécanique Céleste* Poincaré devoted the chapter VIII to the problem of the reliability of the formal expansions of perturbation theory. He proved that the series commonly used in Celestial mechanics are typically non convergent, although their usefulness is generally evident. In particular, he pointed out that these series could have the same character of the Stirling's series. Recent work in perturbation theory has enlighten this conjecture of Poincaré, bringing into evidence that the series of perturbation theory, although non convergent in general, furnish nevertheless valuable approximations to the true orbits for a very large time, which in some practical cases could be comparable with the age of the universe.

The aim of my lectures is to introduce the quantitative methods of perturbation theory which allow to obtain such powerful results.

It is well known that the hardest problem in dealing with the possible convergence of the series is the appearance of the so called *small denominators.* It is also known that Kolmogorov (1954) proved that in some special cases these series do converge, but only for a strange set of initial conditions which has large measure, but is nowhere dense in phase space. I refer here to the celebrated theorem on the persistence under small perturbation of conditionally periodic motions on invariant tori, usually referred to as KAM theorem.

On the other hand, the problem of establishing the asymptotic character of the formal expansions seems to have been less considered by mathematicians: as far as I know, it has been discussed by Moser (1955) and Littlewood (1959), but these works have essentially been forgotten. More recently, a general theorem in this direction was proven by Nekhoroshev (1977). This will be the subject of part of my lectures.

The theorem of Nekhoroshev can be considered as a quantitative approach to the normal form methods in perturbation theory. The normal form theory is a rather well known topic, at least as far the formal aspect is concerned (an application to the

From Newton to Chaos, Edited by A.E. Roy
and B.A. Steves, Plenum Press, New York, 1995

problem of the solar system has been given in the lectures of Message). The problem can be stated as follows: given an Hamiltonian of the form $H(p,q) = h(p) + \varepsilon f(p,q)$, where p, q are action angle variables and ε a small parameter, to find a canonical transformation which gives the new Hamiltonian, Z say, a "simple" form (the so called *normal form*). A rather detailed discussion is required to explain what is meant by "simple". In the most naive approach one tries to remove the dependence on the angle variables, so that Z depends only on the action variables, and the system turns out to be integrable. But such an approach turns out to be too naive: the existence of resonances, which manifests itself through the appearance of small denominators, makes the procedure inconsistent, in general. This was first stated by Poincaré in his celebrated theorem on the nonexistence of uniform integrals (see Poincaré, 1892, chapt. V).

However, it is a common procedure in Celestial mechanics (and more generally in perturbation theory) to perform a partial normalization of the system. The naive argument is more or less the following. Let us consider the Hamiltonian above, and truncate the Fourier expansion of $f(p,q)$ in the angles to some (not too high) order. Then we can find a canonical transformation which gives the Hamiltonian a normal form up to terms of order ε^2, i.e., (denoting by p', q' the new variables) the form $H'(p',q') = h(p') + \varepsilon \overline{f}(p') + \varepsilon^2 f'(p,q,\varepsilon)$. Here, $\overline{f}$ is in fact the average of f with respect to the angles. Forget now the terms of order ε^2; then we have an integrable system, the solutions of which can be considered as reliable up to a time of the order of ε^{-2}. If we iterate the same procedure a finite number of times, r say, then we obtain an Hamiltonian which is integrable up to remainder terms of order ε^r. This gives results which are reliable up to a time of the order of ε^{-r}.

As is, this argument is wrong, for two reasons. The first reason is that the integrable normal form can be obtained only in regions of the phase space which are far from resonances among the frequencies. This fact is well known, and leads to theories adapted to resonant situations. The second, and more hidden, reason is that when one says "a remainder of order ε^r" one actually means "a term of the form $\varepsilon^r f^{(r)}(p,q,\varepsilon)$, with some function $f^{(r)}$". But nothing is said about the *size* of the function $f^{(r)}$. In fact, according to quantitative estimates, the size of $f^{(r)}$ grows typically as fast as some power of $r!$, which makes the series to diverge.

Now, the underlying idea of Nekhoroshev's theorem is to use a standard argument in the theory of asymptotic expansions. In its simplest formulation the argument is the following. Forget for a moment the problems due to the resonances; let the size of the remainder after r normalization steps be $r!\,\varepsilon^r$, and compare it with the size $(r-1)!\varepsilon^{r-1}$ after $r-1$ steps; remark that $r!\varepsilon^r = r\varepsilon \cdot (r-1)!\varepsilon^{r-1}$. Thus, performing the step r is useful (i.e., the size of the remainder actually decreases) only if $r\varepsilon < 1$. This means on the one hand that the normalization procedure is useful only if ε is small enough, and on the other hand that there is an optimal choice of the number r of normalization steps as a function of ε, namely $r \simeq 1/\varepsilon$. Replace this value in $r!\varepsilon^r$, also using the Stirling's formula $r! \simeq r^r e^{-r}$; this gives $r!\varepsilon^r \simeq e^{-r} \simeq \exp(-1/\varepsilon)$. The final conclusion is that one has results which are reliable up to a time of the order of $\exp(1/\varepsilon)$. In the words of Littlewood, "if not eternity, this is a considerable slice of it".

Taking into account the resonances is in fact, in my opinion, the main contribution of Nekhoroshev. This requires the construction of a kind of geography of the phase space, with some suitable properties. This part is left for the detailed discussion in section 3. The final result can be roughly formulated as follows: the action variables satisfy the bound

$$|p(t) - p(0)| < C\varepsilon^a \quad for \quad |t| < T_* \exp\left[\left(\frac{\varepsilon_*}{\varepsilon}\right)^a\right] ,$$

with constants C, T_* and ε_*, and with typically $a \sim 1/n$, where n is the number of degrees of freedom.

The lectures are organized in two parts. The first part has a tutorial character. Its goal is to introduce the technical tools of quantitative perturbation theory. I will first recall, at a formal level, the definition of Lie series and of Lie transform. Next, I will introduce some elements of the theory of analytic functions, with particular attention to the Cauchy estimates. Finally, I will show how all these things work together by giving the conditions for the convergence of the Lie series. The second part will be devoted to a general discussion of the theorem of Nekhoroshev. In order to avoid unnecessary complications of a purely technical character, I will give the proof for a particular, but relevant, system, which contains almost all the essential elements of the theorem. This part is based on a recent paper by Zehnder and myself (see Giorgilli and Zehnder, 1992); thus, I will stress the relevant points of the proof, without insisting on technical details that can be found in that paper.

LIE SERIES AND LIE TRANSFORMS

The use of Lie series and Lie transforms is quite recent in Celestial Mechanics, and is generally known at a formal level. I give here a quantitative formulation of the theory.

Formal definitions

On a $2n$–dimensional phase space endowed with canonical coordinates p, q, consider an analytic function $\chi(p,q)$, which will be called a *generating function*. The *Lie derivative* L_χ is a linear differential operator acting on functions on the phase space, defined as

$$L_\chi \cdot = \{\chi, \cdot\}. \tag{1}$$

This is nothing but the derivative along the Hamiltonian field generated by χ. The *Lie series* operator is then defined as the exponential of L_χ, namely

$$\exp(L_\chi) = \sum_{s\geq 0} \frac{1}{s!} L_\chi^s \tag{2}$$

(see for example Gröbner, 1960). It is known that this operator represents the time one evolution of the canonical flow generated by the autonomous Hamiltonian χ.

As an example, consider the case of action–angle variables $p \in R^n$ and $q \in T^n$. The function $\chi = \sum_l \xi_l q_l$, with $\xi \in R^n$ generates through the Lie series operator the canonical transformation $p' = \exp(L_\chi)p = p + \xi$, $q' = \exp(L_\chi)q = q$, namely a translation in the action space. Similarly, a generating function $\chi = \chi(q)$ (independent of p) generates the transformation $p' = p + \frac{\partial \chi}{\partial q}$, $q' = q$, namely a deformation of the action variables.

The *Lie transform* differs from the Lie series in that it is connected with the flow of a *nonautonomous* (i.e., time dependent) Hamiltonian. In such a form, it has been introduced by Hori (1966) and Deprit (1969). In fact, several algorithms have been devised in order to give the Lie transform an algorithmic recursive form: everybody, of course, has his favourite one. To make a definite choice, I will make use here of my favourite one, which is related to the "algorithm of the inverse," introduced by Henrard.

Consider a *generating sequence* $\chi = \{\chi_s\}_{s\geq 1}$ of analytic functions on the phase space. The Lie transform operator T_χ is defined as

$$T_\chi = \sum_{s\geq 0} E_s \ , \tag{3}$$

where the sequence $\{E_s\}_{s\geq 0}$ of operators is recursively defined as

$$E_0 = \mathrm{Id} \ , \quad E_s = \sum_{j=1}^{s} \frac{j}{s} L_{\chi_j} E_{s-j} \ . \tag{4}$$

As a simple example, consider the case $\chi = \{\chi_1, 0, 0, \ldots\}$, namely a generating sequence containing only the first term. Then the Lie transform generated by χ coincides with the Lie series generated by χ_1, i.e.,

$$T_\chi = \exp(L_{\chi_1}) \ .$$

¿From now on, it will be useful to look at the algebraic aspect of the definitions of Lie series and Lie transform, rather than to their property of being related to the canonical flow of some Hamiltonian. In particular, all the properties stated here can be proved on a purely algebraic basis, without any reference to the canonical flow (see for example Giorgilli and Galgani, 1978). The relevant properties are the following: both the Lie series and the Lie transform operators are linear, and preserve products and Poisson brackets, namely

$$T_\chi(fg) = (T_\chi f)(T_\chi g) \ , \quad T_\chi \{f, g\} = \{T_\chi f, T_\chi g\} \ ,$$

with analogous formulæ for the Lie series. Moreover, both the operators are invertible. Finding the inverse of the Lie series is an easy matter: recalling that it represents the flow of an autonomous canonical system, one immediately concludes that the inverse of $\exp(L_\chi)$ is $\exp(L_{-\chi}) = \exp(-L_\chi)$, i.e., the generating function of the inverse is $-\chi$. Concerning the Lie transform, a careful analysis of the algorithm3–4 shows that finding the inverse transformation is an easy matter, although finding the generating sequence of the inverse is not as easy. Indeed, assume we are given a function $f = f_0 + f_1 + \ldots$, and denote by $g = g_0 + g_1 + \ldots$ its transformed function $g = T_\chi f$; using the linearity of the Lie transform, we can apply T_χ separately to every term of f. It is useful to rearrange the terms according to the triangular diagram

$$\begin{array}{cccccc}
g_0 & f_0 & & & & \\
 & \downarrow & & & & \\
g_1 & E_1 f_0 & f_1 & & & \\
 & \downarrow & \downarrow & & & \\
g_2 & E_2 f_0 & E_1 f_1 & f_2 & & \\
 & \downarrow & \downarrow & \downarrow & & \\
g_3 & E_3 f_0 & E_2 f_1 & E_1 f_2 & f_3 & \\
 & \downarrow & \downarrow & \downarrow & \downarrow & \\
\vdots & \vdots & \vdots & \vdots & \vdots & \ddots
\end{array} \tag{5}$$

where terms of the same order appear on the same line. Remark that the operator T_χ acts *by columns*, as indicated by the arrows: the knowledge of f_j and of the generating sequence allows one to construct the whole column below f_j. Thus, the first line gives

g_0, the second line gives $g_1 = E_1 f_0 + f_1$, and so on. This shows how to practically perform the transformation. Concerning the inverse, assume that g is given, and f is unknown. Then, the first line gives $f_0 = g_0$; having determined f_0, all the column below f_0 can be constructed, and the second line gives immediately $f_1 = g_1 - E_1 f_0$; having determined f_1, all the corresponding column can be constructed, so that f_2 can be determined from the third line as $f_2 = g_2 - E_2 f_0 - E_1 f_1$, and so on. In fact, there is also an explicit formula for the inverse, namely

$$T_\chi^{-1} = \sum_{s \geq 0} D_s \ ,$$

where

$$D_0 = \mathrm{Id} \ , \quad D_s = -\sum_{j=1}^{s} \frac{j}{s} D_{s-j} L_{\chi_j} \ .$$

This expression is actually useless for a practical computation: the algorithm described above is much more efficient. Nevertheless, the explicit recursive formula is useful for quantitative estimates.

The property of preserving the Poisson brackets is relevant, as it means that the coordinate transformation

$$p = T_\chi p' \ , \quad q = T_\chi q' \ ,$$

where the generating sequence χ depends on the "new" variables p', q', is canonical (I omit here the analogous statement concerning the Lie series). Now, if we consider a function $f(p, q)$ on the phase space, we should in principle compute the transformed function, $f'(p', q')$ say, as

$$f'(p', q') = f(p, q)|_{p=T_\chi p', q=T_\chi q'} \ ,$$

or, equivalently,

$$f'(p', q') = f(T_\chi p', T_\chi q') \ ,$$

namely by substitution of the coordinate transformation. Now, the properties of being linear and of preserving products can be used to prove that in fact one has

$$f(T_\chi p', T_\chi q') = (T_\chi f)(p', q') \ . \tag{6}$$

Briefly, "don't try to make substitutions: just transform the function, and change the name of the variables".

Analytic framework

The purpose of this section is to introduce the basic elements of the quantitative perturbation theory. Most of the technical elements are concerned with the theory of analytic functions, and so with functions defined on complex domains.

First of all, to avoid unnecessary complications of mathematical character, it is useful to have some precise characterization of the phase space. Thus, I shall consider the common case of action–angle variables $q \in \mathbf{T}^n$ and $p \in \mathcal{G} \in \mathbf{R}^n$, where $\mathcal{G}$ is a subset of $\mathbf{R}^n$.

Next, I will introduce a *complexification* of the domains. For $p \in \mathcal{G}$ consider the complex polydisk $\Delta_\varrho(p)$ of radius $\varrho > 0$ with center p defined as

$$\Delta_\varrho(p) = \{p' \in \mathbf{C}^n \ : \ |p' - p| < \varrho\} \ , \tag{7}$$

where

$$|p' - p| = \max_j \left|p'_j - p_j\right| \;;$$

this is nothing but the cartesian product of complex disks of radius ϱ in the complex plane. The complexification $\mathcal{G}_\varrho$ of the real domain $\mathcal{G}$ is defined as

$$\mathcal{G}_\varrho = \bigcup_{p \in \mathcal{G}} \Delta_\varrho(p) \;. \tag{8}$$

Similarly, the complexification $\mathbf{T}^n_\sigma$ for $\sigma > 0$ of the n–torus is defined as

$$\mathbf{T}^n_\sigma = \{q \in \mathbf{C}^n \;:\; |\,Im\, q| < \sigma\} \;, \tag{9}$$

namely the cartesian product of strips of width σ around the real axis on the complex plane. From now on, the phase space will be the domain

$$\mathcal{D}_{(\varrho,\sigma)} = \mathcal{G}_\varrho \times \mathbf{T}^n_\sigma \;.$$

The next step is concerned with the introduction of *norms* on the space of analytic functions on the domain $\mathcal{D}_{(\varrho,\sigma)}$. The most natural norm is the *supremum norm.* For a function f analytic on $\mathcal{G}_\varrho$ the supremum norm $|f|_\varrho$ of f is defined as

$$|f|_\varrho = \sup_{p \in \mathcal{G}_\varrho} |f(p)| \;. \tag{10}$$

Similarly, for a function g analytic on $\mathcal{D}_{(\varrho,\sigma)}$ the supremum norm $|g|_{(\varrho,\sigma)}$ is

$$|g|_{(\varrho,\sigma)} = \sup_{(p,q)\in\mathcal{D}_{(\varrho,\sigma)}} |g(p,q)| \;. \tag{11}$$

As a matter of fact, the supremum norm is not the best one, to my knowledge, when dealing with small denominators. A more useful norm is the so called *weighted Fourier norm,* which can be introduced on the basis of the following considerations. It is known that an analytic function f on the domain $\mathcal{D}_{(\varrho,\sigma')}$ admits the Fourier expansion

$$f(p,q) = \sum_{k\in\mathbf{Z}^n} f_k(p) e^{ik\cdot q} \;.$$

It is also known that the coefficients $f_k(p)$ of the Fourier expansion decay exponentially with k. More precisely, one has

$$|f_k|_\varrho \le |f|_{(\varrho,\sigma')} e^{-|k|\sigma'} \;,$$

where $|k| = |k_1| + \cdots + |k_n|$. Thus, it is natural to define the weighted Fourier norm $\|f\|_{(\varrho,\sigma)}$ for $\sigma < \sigma'$ as

$$\|f\|_{(\varrho,\sigma)} = \sum_{k\in\mathbf{Z}^n} |f_k|_\varrho e^{|k|\sigma} \;. \tag{12}$$

The exponential decay of the coefficients and the condition $\sigma < \sigma'$ ensure that the series defining the norm converges. It is also an easy matter to check that the supremum norm is bounded by the weighted Fourier norm, i.e.,

$$|f|_{(\varrho,\sigma)} \le \|f\|_{(\varrho,\sigma)} \;.$$

Cauchy Estimates

The need for analytic functions on complex domains is due to the availability of the Cauchy estimates for the derivatives of analytic functions. Let me briefly recall the essentials. For simplicity, consider a disk $\Delta_\varrho(0)$ centered at the origin of the complex plane $\mathbf{C}$. Consider a function f analytic and bounded on the disk $\Delta_\varrho(0)$. The Cauchy estimate for the derivative f' of f at the origin states that

$$|f'(0)| \le \frac{1}{\varrho}|f|_\varrho \, .$$

More generally, for the s–th derivative $f^{(s)}$ one has the estimate

$$|f^{(s)}(0)| \le \frac{s!}{\varrho^s}|f|_\varrho \, .$$

For instance, let $\varrho = 1$, and consider the function $f(z) = z^s$; it is an easy matter to check that $|f|_1 = 1$, so that the Cauchy estimate gives $|f^{(s)}(0)| \le s!$. This shows that the estimate cannot be improved in general. Similarly, consider a polydisk $\Delta_\varrho(0)$ centered at the origin of $\mathbf{C}^n$. Then one has the estimate

$$\left|\frac{\partial f}{\partial z_j}(0)\right| \le \frac{1}{\varrho}|f|_\varrho \, , \quad 1 \le j \le n \, .$$

I omit here the estimates for higher order derivatives, since we shall not need them.

Let us now come back to our analytic setting. Consider a function f analytic and bounded on the complex domain $\mathcal{G}_\varrho$. Consider now a point $p \in \mathcal{G}_{\varrho-\delta}$ for $0 < \delta \le \varrho$ (that is, the union of polydisks of radius $\varrho - \delta$ centered at every point of $\mathcal{G}$). Remark that the polydisk $\Delta_\delta(p)$ is a subset of $\mathcal{G}_\varrho$, so that f is analytic and bounded on $\Delta_\delta(p)$, and moreover we have the estimate $|f(p')| \le |f|_\varrho$ for all $p' \in \Delta_\delta(p)$. By the Cauchy estimate, we immediately get

$$\left|\frac{\partial f}{\partial p_j}(p)\right| \le \frac{1}{\delta}|f|_\varrho \, , \quad 1 \le j \le n \, .$$

Since this is true for every point $p \in \mathcal{G}_{\varrho-\delta}$, we immediately get

$$\left|\frac{\partial f}{\partial p_j}\right|_{\varrho-\delta} \le \frac{1}{\delta}|f|_\varrho \, , \quad 1 \le j \le n \, .$$

The relevant fact here is that we have an explicit bound on the derivative of a function, but we must pay this information with a restriction of the complex domain.

Generalized Cauchy Estimates

The next step is concerned with the generalization of the Cauchy estimates, taking into account that we are interested in Lie derivatives (or Poisson brackets), and that we are going to use the weighted Fourier norm instead of the supremum one.

A first result concerns the Poisson bracket between two functions, and the derivative of a function with respect to one of the canonical coordinates (i.e., the Poisson bracket with the conjugate one).

Lemma 1: Let f be analytic on the domain $\mathcal{D}_{(\varrho,\sigma)}$, and g be analytic in $\mathcal{D}_{(1-d')(\varrho,\sigma)}$ for some $0 \le d' < 1$; assume moreover that $\|f\|_{(\varrho,\sigma)}$ and $\|g\|_{(1-d')(\varrho,\sigma)}$ are finite. Then: i. for $0 < d < 1$ and for $1 \le j \le n$ one has

$$\left\|\frac{\partial f}{\partial p_j}\right\|_{(1-d)(\varrho,\sigma)} \le \frac{1}{d\varrho}\|f\|_{(\varrho,\sigma)} \, , \qquad \left\|\frac{\partial f}{\partial q_j}\right\|_{(1-d)(\varrho,\sigma)} \le \frac{1}{ed\sigma}\|f\|_{(\varrho,\sigma)} \, ; \tag{13}$$

ii. for $0 < d < 1 - d'$ one has

$$\| \{f, g\} \|_{(1-d'-d)(\varrho,\sigma)} \leq \frac{2}{ed(d+d')\varrho\sigma} \|f\|_{(\varrho,\sigma)} \|g\|_{(1-d')(\varrho,\sigma)} \cdot \tag{14}$$

Proof. Using the Fourier expansion, compute

$$\frac{\partial f}{\partial p_j} = \sum_k \frac{\partial f_k}{\partial p_j}(p) e^{ik\cdot q} .$$

According to the definition of the norm, we have

$$\left\| \frac{\partial f}{\partial p_j} \right\|_{(1-d)(\varrho,\sigma)} = \sum_k \left| \frac{\partial f_k}{\partial p_j} \right|_{(1-d)\varrho} e^{(1-d)|k|\sigma} ;$$

by the Cauchy estimate, and taking out d from the exponent, the right hand side of the latter expression is bounded by

$$\frac{1}{d\varrho} \sum_k |f_k|_\varrho e^{|k|\sigma} ,$$

and the first of 13 follows in view of the definition of the norm. Coming to the second one, compute

$$\frac{\partial f}{\partial q_j} = i \sum_k k_j f_k(p) e^{ik\cdot q} ,$$

and use again the definition of the norm to estimate

$$\begin{aligned} \left\| \frac{\partial f}{\partial q_j} \right\|_{(1-d)(\varrho,\sigma)} &= \textstyle\sum_k |k_j| \, |f_k|_{(1-d)\varrho} e^{(1-d)|k|\sigma} \\ &\leq \textstyle\sum_k |f_k|_\varrho e^{|k|\sigma} |k| e^{-d|k|\sigma} . \end{aligned}$$

By the general inequality

$$x^\alpha e^{-\delta x} \leq \left(\frac{\alpha}{e\delta}\right)^\alpha \qquad \textit{for positive } \alpha,\ x,\ \delta \tag{15}$$

(just compute the supremum over x of $x^\alpha e^{-\delta x}$), with $\alpha = 1$ and $|k|$ and $d\sigma$ in place of x and δ respectively, one gets $|k| e^{-d|k|\sigma} \leq 1/(ed\sigma)$; the second of 13 then follows in view of the definition of the norm. The proof of 14 requires some more work. Compute

$$\{f, g\} = i \sum_{k,k'} \left[\sum_{l=1}^{n} \left(k_l \frac{\partial g_{k'}}{\partial p_l} f_k + k'_l \frac{\partial f_k}{\partial p_l} g_{k'} \right) \right] e^{ik\cdot q} e^{ik'\cdot q} ,$$

and use the definition of the norm to estimate

$$\begin{aligned} \| \{f, g\} \|_{(1-d'-d)(\varrho,\sigma)} < \textstyle\sum_{k,k'} \Bigg[\sum_{l=1}^{n} \Bigg(|k_l| \left| \frac{\partial g_{k'}}{\partial p_l} \right|_{(1-d'-d)\varrho} |f_k|_\varrho \\ + |k'_l| \left| \frac{\partial f_{k'}}{\partial p_l} \right|_{(1-d'-d)\varrho} |g_k|_{(1-d')\varrho} \Bigg) \Bigg] e^{(1-d'-d)|k+k'|\sigma} . \end{aligned}$$

Using now the Cauchy estimates, the right hand side of the latter expression is found to be smaller than

$$\begin{aligned}&\tfrac{1}{d\varrho}\textstyle\sum_{k,k'}\ |g'_k|_{(1-d')\varrho}e^{(1-d')|k'|\sigma}|f_k|_\varrho e^{|k|\sigma}\sum_{l=1}^n |k_l|e^{-(d'+d)|k|\sigma}\\ &+\tfrac{1}{(d'+d)\varrho}\textstyle\sum_{k,k'}|g_{k'}|_{(1-d')\varrho}e^{(1-d')|k'|\sigma}|f_k|_\varrho e^{|k|\sigma}\sum_{l=1}^n |k'_l|e^{-d|k'|\sigma}\ .\end{aligned} \tag{16}$$

Remark now that the only terms depending on the index l are $\sum_l |k_l| = |k|$ and $\sum_l |k'_l| = |k'|$; moreover, using the inequality 15 one concludes

$$|k|e^{-(d'+d)|k|\sigma} \leq \frac{1}{(d'+d)e\sigma}\ ,\qquad |k'|e^{-d|k'|\sigma}\leq \frac{1}{de\sigma}\ .$$

Reordering the terms, the expression in 16 is found to be smaller than

$$\frac{2}{ed(d'+d)\varrho\sigma}\sum_{k'}|g'_k|_{(1-d')\varrho}e^{(1-d')|k'|\sigma}\cdot\sum_k |f_k|_\varrho e^{|k|\sigma}\ ,$$

and the conclusion follows in view of the definition of the norms $\|f\|_{(\varrho,\sigma)}$ and $\|g\|_{(1-d')(\varrho,\sigma)}$.

The next step concerns the estimate of multiple Poisson brackets, namely of an expression of the form $L_\chi^s f$. Since $L_\chi^s = L_\chi \circ L_\chi^{s-1}$, one can iterate s times the estimate of lemma 2.1. However, one must take into account the restriction of the domains. To this end, having fixed the final restriction d, we choose s positive quantities $\delta_1,\ldots,\delta_s$ the sum of which is d, and estimate $L_\chi^j f$ on the domain restricted by the factor $1-\delta_1-\ldots-\delta_j$. The simplest choice is $\delta_1=\ldots=\delta_s=d/s$. The final result is given by

Lemma 2: Let f and χ be analytic on the domain $\mathcal{D}_{(\varrho,\sigma)}$, and assume that $\|f\|_{(\varrho,\sigma)}$ and $\|\chi\|_{(\varrho,\sigma)}$ are finite. Then for every positive $d<1$ one has

$$\left\|L_\chi^s f\right\|_{(1-d)(\varrho,\sigma)} \leq \frac{s!}{e^2}\left(\frac{2e}{d^2\varrho\sigma}\right)^s \|\chi\|_{(\varrho,\sigma)}^s\|f\|_{(\varrho,\sigma)} \tag{17}$$

Proof: For $s=1$ this is nothing but lemma 1. For $s>1$ let $\delta=d/s$. A straightforward application of lemma 1 gives

$$\|L_\chi f\|_{(1-\delta)(\varrho,\sigma)} \leq \frac{2}{e\delta^2\varrho\sigma}\|\chi\|_{(\varrho,\sigma)}\|f\|_{(\varrho,\sigma)}\ .$$

For $j=2,\ldots,s$, recalling that $L_\chi^j f$ can be estimated on the domain restricted by the factor $1-j\delta$, use again lemma 2 with δ in place of d and $(j-1)\delta$ in place of d', and get the recursive estimate

$$\left\|L_\chi^j f\right\|_{(1-j\delta)(\varrho,\sigma)} \leq \frac{2}{ej\delta^2\varrho\sigma}\|\chi\|_{(\varrho,\sigma)}\|L_\chi^{j-1}f\|_{(1-(j-1)\delta)(\varrho,\sigma)}\ .$$

By recursive application of this formula up to s, and recalling $\delta=d/s$ we get

$$\left\|L_\chi^s f\right\|_{(1-d)(\varrho,\sigma)} \leq \left(\frac{2}{ed^2\varrho\sigma}\right)^s\frac{s^{2s}}{s!}\|\chi\|_{(\varrho,\sigma)}^s\|f\|_{(\varrho,\sigma)}\ .$$

Then, 17 follows in view of the trivial inequality $s^s\leq e^{s-1}s!$ for $s\geq 1$ (prove it by induction).

Analyticity of the Lie Series and of the Lie Transform

The Lie series and the Lie transform have been defined as series of analytic functions. It is natural to ask whether these series are convergent, or, more precisely, whether the sums of these series are analytic functions on some domain. Sufficient conditions can be found on the basis of the results of the previous section. This is stated by

Proposition 1: Let f and χ be analytic on the domain $\mathcal{D}_{(\varrho,\sigma)}$, and assume that $\|f\|_{(\varrho,\sigma)}$ and $\|\chi\|_{(\varrho,\sigma)}$ are finite. Then for every positive $d < 1/2$ the following statement holds true: if the condition

$$\frac{2e\|\chi\|_{(\varrho,\sigma)}}{d^2\varrho\sigma} \leq \frac{1}{2} \tag{18}$$

is satisfied, then the operator $\exp(L_\chi)$ and its inverse $\exp(-L_\chi)$ define an analytic canonical transformation on the domain $\mathcal{D}_{(1-d)(\varrho,\sigma)}$ with the properties

$$\begin{aligned}
\mathcal{D}_{(1-2d)(\varrho,\sigma)} &\subset \exp(L_\chi)\mathcal{D}_{(1-d)(\varrho,\sigma)} \subset \mathcal{D}_{(\varrho,\sigma)} ,\\
\mathcal{D}_{(1-2d)(\varrho,\sigma)} &\subset \exp(-L_\chi)\mathcal{D}_{(1-d)(\varrho,\sigma)} \subset \mathcal{D}_{(\varrho,\sigma)} .
\end{aligned}$$

The proof of the proposition is based on the following

Lemma 3: With the hypotheses of proposition 1 the series $\exp(\pm L_\chi)f$, $\exp(\pm L_\chi)p$ and $\exp(\pm L_\chi)q$ are absolutely convergent on $\mathcal{D}_{(1-d)(\varrho,\sigma)}$, and for integer $r > 0$ one has the estimates

$$\begin{aligned}
&\|\exp(\pm L_\chi)p - p\|_{(1-d)(\varrho,\sigma)} &&\leq d\varrho ,\\
&\|\exp(\pm L_\chi)q - q\|_{(1-d)(\varrho,\sigma)} &&\leq d\sigma ,\\
&\left\|\exp(\pm L_\chi)f - \textstyle\sum_{s=0}^{r} \frac{1}{s!} L_\chi^s f\right\|_{(1-d)(\varrho,\sigma)} &&\leq \frac{2}{e^2}\left(\frac{2e\|\chi\|_{(\varrho,\sigma)}}{d^2\varrho\sigma}\right)^{r+1} \|f\|_{(\varrho,\sigma)} .
\end{aligned} \tag{19}$$

Proof. By lemma 2 one has

$$\sum_{s\geq 0} \frac{1}{s!} \left\|L_\chi^s f\right\|_{(1-d)(\varrho,\sigma)} \leq \|f\|_{(\varrho,\sigma)} + \frac{1}{e^2} \sum_{s>0} \left(\frac{2e\|\chi\|_{(\varrho,\sigma)}}{d^2\varrho\sigma}\right)^s \|f\|_{(\varrho,\sigma)} ;$$

in view of condition 18 the series in the left hand side of the latter inequality is a convergent geometric series. Since the supremum norm is bounded by the weighted Fourier norm, this implies that the series $\exp(L_\chi)f$ is absolutely convergent on the domain $\mathcal{D}_{(1-d)(\varrho,\sigma)}$, as claimed. The third of 19 follows by simply computing the sum of the geometric series above starting from $s = r + 1$ and taking into account the condition 18. Concerning the estimate on the transformation of the coordinates p, q, a similar argument applies, with a minor difference: the first terms $L_\chi p$ and $L_\chi q$ must be estimated using 13 of lemma 1; then lemma 2 is used to estimate the remaining terms. Then the same convergence argument is used.

Proof of Proposition 1: By lemma 3, the series of analytic functions defining the canonical transformation are absolutely convergent in $\mathcal{D}_{(1-d)(\varrho,\sigma)}$. Thus, the same series are uniformly convergent on every compact subset of $\mathcal{D}_{(1-d)(\varrho,\sigma)}$. By Weierstrass theorem, this implies that the sums of the series are analytic functions on $\mathcal{D}_{(1-d)(\varrho,\sigma)}$, as claimed. The statement concerning the inclusions of the domains follows from the estimates 19.

A straightforward consequence of proposition 1 is the following: if $f(p,q)$ is analytic in $\mathcal{D}_{(\varrho,\sigma)}$, then the transformed function $f(\exp(L_\chi)p, \exp(L_\chi)q))$ is analytic in $\mathcal{D}_{(1-d)(\varrho,\sigma)}$, being a composition of analytic functions. However, as has been remarked in section 2, the transformed function is nothing but $\exp(L_\chi)f$. The last estimate 19 gives in fact a direct proof of the analyticity of $\exp(L_\chi)f$. Moreover, taking into account that an explicit computation must be truncated at some order r, it gives an estimate of the error.

A similar result holds for the Lie transform. I will give here only the statement, without entering the details of the proof. The technical tools of the proof are again the results at the end of section 2; however, some additional complication arises, due to the fact that the definition of the Lie transform is given in recursive form. The scheme of the proof can be found in (Giorgilli and Galgani, 1985), and requires only some trivial adaptation, due to the use of a different norm.

Proposition 2: Let f and the generating sequence $\chi = \{\chi_s\}_{s\geq 1}$ be analytic on the domain $\mathcal{D}_{(\varrho,\sigma)}$, and assume that $\|f\|_{(\varrho,\sigma)}$ is finite, and that

$$\|\chi_s\|_{(\varrho,\sigma)} \leq \frac{b^{s-1}}{s} G$$

for some $b \geq 0$ and $G > 0$. Then for every positive $d < 1/2$ the following statement holds true: if the condition

$$\frac{2eG}{d^2 \varrho\sigma} + b \leq \frac{1}{2} \tag{20}$$

is satisfied, then the operator T_χ and its inverse T_χ^{-1} define an analytic canonical transformation on the domain $\mathcal{D}_{(1-d)(\varrho,\sigma)}$ with the properties

$$\begin{aligned} \mathcal{D}_{(1-2d)(\varrho,\sigma)} &\subset T_\chi \mathcal{D}_{(1-d)(\varrho,\sigma)} \subset \mathcal{D}_{(\varrho,\sigma)} , \\ \mathcal{D}_{(1-2d)(\varrho,\sigma)} &\subset T_\chi^{-1} \mathcal{D}_{(1-d)(\varrho,\sigma)} \subset \mathcal{D}_{(\varrho,\sigma)} . \end{aligned}$$

Remark that the statement is essentially the same as for the Lie series. The difference is due to the use of a generating sequence instead of a function, which requires a condition on the behaviour of the sequence, and a corresponding minor modification in the convergence condition 20.

THE NEKHOROSHEV'S THEOREM

In this section I will give the lines of the proof of the theorem of Nekhoroshev. In order to have a well definite framework, consider the special nonautonomous canonical system with Hamiltonian

$$H(p,q,t) = \sum_{j=1}^{n} \frac{p_j^2}{2} + V(q,t) , \tag{21}$$

where $p \in \mathbf{R}^n$ and $q \in \mathbf{T}^n$ are action angle variables. The potential energy $V(q,t)$ is assumed to be real analytic and bounded in the complex strip $|\,Im\, x| < 2\sigma$ and $|\,Im\, t| < 2\sigma$ for some positive σ. The potential energy must be periodic in the angles, while no condition, apart from analyticity, is assumed for the dependence on time.

Such a formulation of the problem might look obscure in view of the absence of a small perturbation parameter. To clarify this point, one should observe that the potential energy in the Hamiltonian 21 is small in case the kinetic energy is large. The usual approach is recovered by rescaling the action variables and the time via the transformation

$$q = q' , \quad p = \frac{p'}{\varepsilon} , \quad t' = \varepsilon t ,$$

and considering the Hamiltonian

$$H'(q',p',t') = \sum_j \frac{p_j'^2}{2} + \varepsilon^2 V(q', t'/\varrho) .$$

This model has been studied in (Giorgilli and Zehnder, 1992), so that it is not necessary to give here a detailed exposition. Rather, I will try to bring into evidence some central ideas of the proof, without insisting on technical details.

Local Normal Form

Let me first recall in a few words the usual formulation, using the formalism of the Lie series. Starting with an Hamiltonian $H(p,q) = h(p) + \varepsilon f(p,q)$, one looks for a generating function $\chi(p,q)$ such that the transformed Hamiltonian

$$\exp(L_\chi)H = h_0 + \varepsilon\,(f + \{\chi, h_0\}) + \varepsilon^2 \ldots$$

has some desired property (e.g., it is integrable) up to terms of order ε^2. This means that one must solve for χ and an auxiliary function Z the equation

$$L_{h_0}\chi + Z = f \ ,$$

with some condition on Z (e.g., one requires $Z = Z(p)$). The latter equation is solved as follows. Since the q variables are angles, expand f in Fourier series as

$$f(p,q) = \sum_{k\in\mathbf{Z}^n} f_k(p)\exp(ik\cdot q) \ ,$$

with known coefficients $f_k(p)$, and consider the same expansion for χ and Z, with unknown coefficients $c_k(p)$ and $z_k(p)$ respectively. Then the equation above splits into the infinite system of equations for the coefficients of the Fourier expansion

$$ik\cdot\omega(p)c_k(p) + z_k(p) = f_k(p) \ , \quad \omega(p) = \frac{\partial h_0}{\partial p} \ .$$

where $\omega(p)$ are the frequencies of the unperturbed system. The key point here is that the coefficient $c_k(p)$ should be determined as

$$c_k(p) = -i\frac{f_k(p)}{k\cdot\omega(p)} \ ;$$

this is formally consistent only if the denominator $k\cdot\omega(p)$ does not vanish. Thus, the simplest rule seems to be the following: if $k\cdot\omega = 0$, then put $z_k = f_k$ and $c_k = 0$; else put $z_k = 0$ and determine c_k as above.

However, a careful consideration shows that such an approach is too naive. Indeed, the set $\omega \in \mathbf{R}^n$ such that $k\cdot\omega = 0$ for some nonzero $k \in \mathbf{Z}^n$ is clearly dense in $\mathbf{R}^n$; thus, unless some strong restriction is imposed on h_0 and/or on f, one has to expect that arbitrarily close to every value p of the actions there is some value, p', say, such that at least one denominator $k\cdot\omega(p')$ does vanish. This is indeed, in very rough terms, the basis of the Poincaré's theorem on the nonexistence of uniform integrals. The way out of these troubles consists in making some truncation of the Fourier expansion of the Hamiltonian (in fact, it is a common practice to consider only a *finite* set of Fourier components), and constructing a *local* theory (i.e., one restricts the action variables to some suitable domain).

A more precise formulation of these remarks is based on the following technical definitions.

1. A resonance module is defined as a subgroup $\mathcal{M} \in \mathbf{Z}^n$ satisfying $span(\mathcal{M})\cap\mathbf{Z}^n = \mathcal{M}$; here, both $\mathcal{M}$ and $\mathbf{Z}^n$ are considered as subset of $\mathbf{R}^n$, and $span(\mathcal{M})$ is the linear subspace in $\mathbf{R}^n$ spanned by $\mathcal{M}$.

2. A function $Z(p,q)$ is said to be in *normal form with respect to the resonance module* $\mathcal{M}$ in case its Fourier expansion has the form

   $$Z(p,q) = \sum_{k\in\mathcal{M}} z_k(p)\exp(ik\cdot q) \ ,$$

   namely, if it contains only harmonics belonging to $\mathcal{M}$ (resonant harmonics).

3. A set $\mathcal{G} \subset \mathbf{R}^n$ is said to be a nonresonance domain of type $(\mathcal{M}, \alpha, \varrho, N)$ in case one has

$$|k \cdot \omega| > \alpha \text{ for all } p \in \mathcal{G}_\varrho \, , \quad k \in \mathbf{Z}^n \setminus \mathcal{M} \text{ and } |k| \leq N \, .$$

Here, $|k| = |k_1| + \ldots + |k_n|$, α and ϱ are positive parameters, N is a positive integer, $\mathcal{M}$ is a resonance module, and $\mathcal{G}_\varrho$ is the complex extension of the real domain $\mathcal{G}$, as described in section 2.

The nonresonance domain appears as the subset of the action space where the Hamiltonian can be given a normal form with respect to the given resonance module. The only point that needs perhaps to be clarified is related to the use of a *finite* number harmonics in the Fourier expansion of the Hamiltonian. This can be justified by recalling that the coefficients of the Fourier expansion of an analytic function decay exponentially (this has been discussed in section 2). Thus, one can first consider only the harmonics k satisfying the condition $|k| \leq K$ for some K, and perform the first normalization step; then one can perform a second step considering harmonics up to order $2K$, and so on up to a *finite* number r of normalization steps, which takes into consideration harmonics up to order rK. On a nonresonance domain with $N = rK$ the procedure can be consistently performed. A formal statement concerning the Hamiltonian system 21 is given by the following

Proposition 3: Let r and K be positive integers, and $N = rK$. Moreover, let $\mathcal{G} \subset \mathbf{R}^n$ be a nonresonance domain of type $(\mathcal{M}, \alpha, \varrho, N)$. Then, there are positive constants A and B depending on $|V|_\sigma$, σ, ϱ and n such that if

$$\varepsilon := \left(\frac{rA}{\alpha} + 2e^{-K\sigma/2} \right) \leq \frac{1}{2} \tag{22}$$

then there exists a real analytic symplectic diffeomorphism φ such that φ and φ^{-1} are defined on the complex domain $\mathcal{G}_{(\varrho,\sigma)\frac{1}{2}}$ and satisfy

$$dist(p, \varphi(p)) \leq \frac{1}{4}\varrho \, , \quad dist(p, \varphi^{-1}(p)) \leq \frac{1}{4}\varrho \, . \tag{23}$$

The map φ transforms the Hamiltonian H into the following normal form on $\mathcal{G}_{\frac{1}{2}(\varrho,\sigma)}$:

$$H \circ \varphi = \frac{1}{2} \sum_j p_j^2 + Z_{\mathcal{M}} + \mathcal{R} \, , \tag{24}$$

where $Z_{\mathcal{M}}$ is in normal form with respect to the resonance module $\mathcal{M}$. The remainder $\mathcal{R}$ satisfies, on $\mathcal{G}_{\frac{1}{2}(\varrho,\sigma)}$, the estimate

$$|\mathcal{R}| \leq B\varepsilon^r \, . \tag{25}$$

Moreover:

$$\mathcal{G}_{(\varrho,\sigma)\frac{1}{4}} \subset \varphi(\mathcal{G}_{(\varrho,\sigma)\frac{1}{2}}) \subset \mathcal{G}_{(\varrho,\sigma)\frac{3}{4}} \, . \tag{26}$$

In addition, the same statement holds true for φ^{-1} instead of φ. The constants are given by

$$\begin{aligned} A &= \frac{2^5 en}{\varrho\sigma} \left[\left(\frac{1+e^{-\sigma/2}}{1-e^{-\sigma/2}} \right)^n |V|_{2\sigma} + \frac{e\varrho}{2} \right] \\ B &= 2 \left(\frac{1+e^{-\sigma/2}}{1-e^{-\sigma/2}} \right)^n |V|_{2\sigma} \, . \end{aligned}$$

The local normal form allows us to have some insight on the dynamics in the nonresonance domain. Indeed, forget for a moment the remainder $\mathcal{R}$, and write the Hamilton's equations for the normalized part of the Hamiltonian: it is immediately seen that $\dot{p}$ is expressed as a sum of vectors belonging to $span(\mathcal{M})$. The resulting picture is the following. Consider the nonresonance domain endowed with the new coordinates (i.e., the coordinates giving the Hamiltonian the normal form). Through any point p_0 belonging to the resonance domain draw the plane generated by $span(\mathcal{M})$: this will be called the *plane of fast drift.* Then the actions p evolve essentially along this plane, in the sense that the motion transversal to the plane is slow, being due only to the remainder $\mathcal{R}$. Try now to describe the motion in the original coordinates. The plane of fast drift is replaced by a surface which can be seen as generated by a *deformation* of a plane, due to the coordinate transformation; the size of such a deformation is estimated by 23. Thus, the orbit is contained in a kind of *cylinder* having the plane of fast drift as axis, and with radius estimated by 23; a small increase of the radius of the cylinder (e.g., multiply it by 2) takes into account the effect of the drift for a long time.

Thus, we know the behaviour of the orbit as far as it remains in the nonresonance domain. The natural question is: what happens if the motion along the plane of fast drift moves the orbit out of the nonresonance domain? The answer to this question is the achievement of the geometric part of the Nekhoroshev's theorem.

The Geometric Part

Let the domain $\mathcal{G}$ of the action variables be an open subset of $\mathbf{R}^n$, and assume that for some positive ϱ the set $\mathcal{G} - \varrho$ (i.e., the points $\in \mathcal{G}$ such that the ball of radius ϱ and center p is still contained in $\mathcal{G}$) is not empty. In the case of the Hamiltonian 21 the choice of $\mathcal{G}$ is quite arbitrary; it can be taken for instance to be a ball of radius larger than ϱ, or even the whole space $\mathbf{R}^n$.

Fix now a positive integer N, to be identified with the final truncation order rK of the analytic part. Let $\mathcal{M}$ be a resonance module with dimension s, and consider the set $\mathcal{M}_N = \{k \in \mathcal{M} \, : \, |k| \leq N\}$; we call $\mathcal{M}$ an N–module in case $\mathcal{M}_N$ contains s independent vectors k. In view of the results of the analytic part, we shall not need to consider resonance moduli which are not N–moduli. To a resonant N–module of dimension s we associate a real parameter β_s, with $\beta_0 < \ldots < \beta_n$, to be determined.

The aim of the geometric construction is to cover $\mathcal{G} - \varrho$ with nonresonance domains to which the results of the analytic part apply. This is the most difficult part. I give here the definitions for a generic unperturbed Hamiltonian $H_0(p)$; it is an useful exercise to consider the simple case of the Hamiltonian 21 for $n = 2$.

1. The *resonant zone* $\mathcal{Z}_\mathcal{M}$ associated to a N–module $\mathcal{M}$ of dimension s is defined as the set of points $p \in \mathcal{G} - \varrho$ satisfying $|k \cdot \omega(p)| < \beta_s$ for s independent vectors of $\mathcal{M}_N$, namely, the points close "within β_s" to resonance. The dimension s of $\mathcal{M}$ will be called the multiplicity of the resonance, or also, by abuse of language, the multiplicity of the zone. The union of all the resonant zones with the same multiplicity s will be denoted by $\mathcal{Z}^*_s$, and will be called the *resonant region of order s*. In rough terms, $\mathcal{Z}^*_s$ is the set of points which admit at least s resonances within β_s; further resonances are not excluded till now. In particular, resonant zones $\mathcal{Z}_\mathcal{M}$ corresponding to different $\mathcal{M}$'s can intersect, giving rise to regions characterized by resonances of higher multiplicity.

2. A resonant block $\mathcal{B}_\mathcal{M}$ associated to $\mathcal{M}$ of dimension s is defined as the set $\mathcal{B}_\mathcal{M} = \mathcal{Z}_\mathcal{M} \setminus \mathcal{Z}^*_{s+1}$. In view of this definition the block appears as the set of points admitting s resonances within β_s, but no further resonances; it is a good basis for the construction of nonresonant domains.

3. To every point p of a given resonant block $\mathcal{B}_{\mathcal{M}}$ we associate the plane in $\mathbf{R}^n$ passing through p and parallel to $span(\mathcal{M})$ (i.e., the plane of fast drift), and define the *extended plane* $P_{\mathcal{M},\delta_s}(y)$ as the set of points of $\mathbf{R}^n$ the distance of which from the plane above is less than $\delta_s = \beta_s/(2N)$. The *cylinder* $\mathcal{C}_{\mathcal{M},\delta_s}$ is defined as the intersection $\mathcal{P}_{\mathcal{M},\delta_s} \cap \mathcal{Z}_{\mathcal{M}}$ of the extended plane with the corresponding resonant zone, and the bases of the cylinder are defined as the intersection $\mathcal{P}_{\mathcal{M},\delta_s} \cap \partial\mathcal{Z}_{\mathcal{M}}$ of the extended plane with the border of the zone.

4. Finally, the extended resonant block $\mathcal{B}_{\mathcal{M},\delta_s}$ is defined as the union of the cylinders corresponding to all points of the block $\mathcal{B}_{\mathcal{M}}$.

Having completed this construction, one proves that with an hypothesis of convexity on the unperturbed Hamiltonian H_0 and with the constants $\beta_0, \ldots, \beta_n$ defined as

$$\begin{aligned} \beta_0 &= \left[2^{n+1}(n+1)!\, N^{(n^2+n-2)/2}\right]^{-1} \varrho \,, \\ \beta_s &= 2^s s!\, N^{[s(s-1)]/2} \beta_0 \quad for\ 1 \le s \le n \,, \end{aligned}$$

one has the following properties.

(i) The blocks are a covering of the domain $\mathcal{G} - \varrho$; that is, every point in $\mathcal{G} - \varrho$ belongs to at least one block.

(ii) The extended block $\mathcal{B}_{\mathcal{M},\delta_s}$ is a nonresonance domain of type $(\mathcal{M}, \beta_s/2, \delta_s, N)$ (remark that $\delta_s < \varrho$).

(iii) The complement of a resonant region $\mathcal{Z}_s^*$ of multiplicity s is covered by the union of all block of multiplicity less than s; that is, if a point does not admit a resonance of multiplicity s, then it belongs to at least one block of multiplicity less than s.

(iv) The closure of an extended block $\mathcal{B}_{\mathcal{M},\delta_s}$ and the resonant zone $\mathcal{Z}_{\mathcal{M}'}$ corresponding to a different N–module with the same dimension as $\mathcal{M}$ are disjoint; that is, the basis of a cylinder $\mathcal{C}_{\mathcal{M},\delta_s}(p)$ corresponding to a point $p \in \mathcal{B}_{\mathcal{M}}$ does not belong to any resonant zone of the same multiplicity $dim(\mathcal{M})$.

(v) The diameter of a cylinder does not exceed ϱ.

Property (ii) means that the extended block are the domains where the analytic theory of the first part can be consistently applied; in particular, setting $N = rK$ with a given K allows to perform r normalization steps in every block. Then one has the following picture of the motion: take an initial value p_0 for the actions; find a resonant block $\mathcal{B}_{\mathcal{M}}$ to which p_0 belongs (may be it is not unique, but this is not relevant), and consider the corresponding cylinder $\mathcal{C}_{\mathcal{M},\delta_s}(p_0)$. Then the orbit either is confined inside the cylinder up to a time of the order of ε^{-r}, or it leaves the cylinder through a basis. In the latter case the intersection of the orbit with the basis of the cylinder is a point not belonging to the resonant region of multiplicity s, by property (iv); moreover, by property (iii), this point belongs to a block of multiplicity less than s. It is now easy to conclude that there is a cylinder containing the orbit for a time interval of the order of ε^{-r}. For, assume that the orbit visits several blocks, and determine the block of minimal multiplicity; pick as initial point any point of the orbit belonging to that block, and consider the corresponding cylinder; this is the wanted cylinder, since the orbit could leave it only by entering a block of lower multiplicity, thus contradicting the assumption that the multiplicity of the block was minimal. Property (v) gives then a bound on the time evolution of the actions p, which can not exceed the diameter of the cylinder. The formal statement for the case of the Hamiltonian 21 is the following.

Proposition 4: Let $\mathcal{G}$ and ϱ be given, with $\mathcal{G} - 2\varrho$ nonempty; let r, K be positive integers, and $N = rK$; let moreover A and B be defined as in proposition 3, with ϱ replaced by δ_s, and let

$$\varepsilon_0 = \frac{2rA}{\beta_0} + 2e^{-K\sigma/2} \ .$$

Then for every solution with $p(0) \in \mathcal{G} - 2\varrho$ one has

$$dist(p(t), p(0)) < \varrho \quad if \quad |t| \le T_0^* = \frac{2\sigma}{B}\varepsilon_0^{-r} \ ,$$

provided $\varepsilon_0 \le 1/2$.

The Exponential Estimate

The proposition 4 gives a stability result depending on several parameters. The parameters σ, $|V|_{2\sigma}$ and ϱ depend on the analyticity properties of the Hamiltonian and on the choice of the domain of initial data, while the constants β_s and δ_s are given by the geometric part; these parameters can be considered as the natural ones of the system. The parameters K, namely the truncation of the Fourier expansion of the Hamiltonian, and r, the order of normalization, are instead arbitrary. So, it is natural to look for an optimal choice of these parameters, i.e., the choice giving the best estimate of the stability time. The key point is that the first term in the definition of ε_0, namely $2rA/\beta_0$, behaves roughly as $\varrho^{-1}r^a$, with some constant a (here the value of β_0 given by the geometric part must be used); thus, the estimate of the stability time has the form $r^{ar}\varrho^{-r}$. Following the argument given in the introduction, it is natural to set $N = (\varrho_0/\varrho)^{1/a}$ with a dimensional constant ϱ_0. Then, one determines K in such a way that the two contributions to ε_0 are of the same order. This immediately allows to realize that the stability time is indeed exponential in $(\varrho_0/\varrho)^{1/a}$. A more careful determination of the constants is just matter of æstetics, and leads to the final formulation of the stability theorem for the system 21.

Theorem 1: Let $\varphi^t(q(0), p(0)) = (q(t), p(t))$ be the flow of the Hamiltonian vector field 21 on $\mathbf{T}^n \times \mathbf{R}^n$. Assume the potential $V(q, t)$ is real analytic on $\mathbf{T}^n \times \mathbf{R}$ and has a bounded analytic extension to a complex strip $2\sigma > 0$. Then there are two positive constants ϱ_* and T_* depending on $|V|_{2\sigma}$, σ and the dimension n, such that for $\varrho \ge \varrho_*$ one has

$$dist(p(t), p(0)) < \varrho$$

for all t in

$$|t| \le T_* \exp\left(\frac{\varrho}{\varrho_*}\right)^{1/a} \quad with \quad a = \frac{n^2+n}{2} \ .$$

The constants are given by

$$\begin{aligned} T_* &= \frac{\sigma}{8e^2} \\ \varrho_* &= K_*^a \, \frac{2^{n+3}(n+1)!}{e} \left(\frac{1+e^{-\sigma/2}}{1-e^{-\sigma/2}}\right)^n |V|_{2\sigma} \end{aligned}$$

where K_* is the smallest positive integer satisfying

$$K_* \ge \frac{2^5 e^4 n}{\sigma} \left(\frac{1-e^{-\sigma/2}}{1+e^{-\sigma/2}}\right)^n \frac{1}{|V|_{2\sigma}} \quad and \quad K_* \ge \frac{5}{\sigma} \ .$$

REFERENCES

1. Deprit, A. (1969): "Canonical transformations depending on a small parameter", *Cel. Mech.* **1**, 12–30.

2. Gröbner, W. (1960): *Die Lie–Reihen und Ihre Anwendungen*, Springer Verlag, Berlin; it. transl.: *Le serie di Lie e le loro applicazioni*, Cremonese, Roma (1973).

3. Giorgilli, A. and Galgani, L. (1978): "Formal integrals for an autonomous Hamiltonian system near an equilibrium point", *Cel. Mech.* **17**, 267–180.

4. Giorgilli, A. and Galgani, L. (1985): "Rigorous estimates for the series expansions of Hamiltonian perturbation theory", *Cel. Mech.* **37**, 95–112.

5. Kolmogorov, A.N. (1954): *On the preservation of conditionally periodic motions, Doklady Akademia Nauk SSSR*, **96**, 527–530.

6. Hori, G. (1966):: "Theory of general perturbations with unspecified canonical variables", *Publ. Astron. Soc. Japan*, **18**, 287–296.

7. Littlewood, J.E. (1959): "On the equilateral configuration in the restricted problem of three bodies", *Proc. London Math. Soc.* **9**, 343–372;

8. Littlewood, J. E. (1959): "The Lagrange configuration in celestial mechanics", *Proc. London Math. Soc.* **9**, 525–543.

9. Moser, J., (1955): "Stabilitätsverhalten kanonisher differentialgleichungssysteme", *Nachr. Akad. Wiss. Göttingen, Math. Phys.* **K1 IIa**, nr.6, 87–120.

10. Nekhoroshev, N., N., (1977): "Exponential estimates of the stability time of near–integrable Hamiltonian systems.", *Russ. Math. Surveys*, **32**, 1-65.

11. Nekhoroshev, N., N., (1979): "Exponential estimates of the stability time of near–integrable Hamiltonian systems, 2.", *Trudy Sem. Petrovs.*, **5**, 5-50.

12. Poincaré, H. (1892): *Le méthodes nouvelles de la mćanique céléste*, Gauthier–Villard, Paris.

JACOBI GEOMETRY AND CHAOS IN N-BODY SYSTEMS

Piero Cipriani
Scuola di Dottorato di Ricerca in Fisica
and
International Center for Relativistic Astrophysics
University of Rome "La Sapienza", I-00185 Roma, Italy

Giuseppe Pucacco
II University of Rome "Tor Vergata", I-00133 Roma, Italy
and
International Center for Relativistic Astrophysics

INTRODUCTION

One of the most ambitious tasks of analytical dynamics is to assess the long-time behaviour of an Hamiltonian system and to try to make predictions about its stability. In the realm of Celestial Mechanics this is related to the fundamental problem of N bodies interacting via the gravitational force (leaving away, for simplicity, dissipative effects) as a model for the Solar System. To the celestial mechanician $N \simeq 10$ appears to be already a *big* number, at least since it is quite bigger than 3. The reasons for this attitude are justified broadly with theoretical and practical arguments, the most compelling being that, when N is really big, let us say $N \simeq 10^5$ (the order required to treat a globular cluster) or $N \simeq 10^{10}$ (that required to treat a galaxy), we are leaving the field of analytical mechanics and are entering the field of statistical mechanics. Easing the burden of such big systems is therefore accomplished by delivering the matter to another branch of physics. The problem is that the statistical mechanics of *non equilibrium* is still not a well grounded discipline and, worst of it, the statistical mechanics of large systems interacting via long-range forces cannot be carried out. So, even if many attempts were made to frame the matter into a stellar dynamical context, many points still remain obscure. Aim of this contribution is to signal the power of the technique based on the geometrization of dynamics to shed light on the behaviour of Hamiltonian systems with many degrees of freedom and to fill the gap of our understanding between them and small N systems.

RELAXATION TIME-SCALES

It is well known (see, e.g. [4]) that in the case of N body self-gravitating systems it is not possible to speak of *relaxation towards equilibrium*, since, in this context, the very meaning of equilibrium is an elusive concept. Therefore, we must content ourselves with the study of the processes through which the system gradually *forgets some* memory of its initial conditions. So, we cannot describe *"the"* relaxation to equilibrium, but, rather, to trace several histories, the *plots* of which have different players performing on different time-scales.

We surely know of two of these processes: the first one is the establishing of the *dynamical* equilibrium; the second corresponds, in a broad sense, to the approach to the *microcanonical* equilibrium of statistical mechanics, which, nonentheless, cannot yet be considered the final equilibrium. In order to reach the dynamical equilibrium (starting from an arbitrary initial bounded state) the system needs a time of the order of the crossing time ($\tau_C \sim R/\sigma$, where R is a characteristic lenght scale and σ is the RMS velocity)[19, 5], whereas, to reach equipartition of energy, it needs a time of the order of $N\tau_C/\ln N$ ($\sim \tau_B$, the binary relaxation time-scale)[4, 6]. It is a common opinion that, if a system tends to forget its initial conditions on a certain time-scale (say τ_S), it must also display a *chaotic behaviour*, namely its dynamics must be extremely sensitive to small changes in initial condition. A fundamental question is therefore the following: is τ_S related to τ_C or to τ_B or to some other (eventually intermediate) time-scale? There are two main lines of thinking on this topic:

a) $\tau_S \sim \tau_C$ (a result obtained by Kandrup[15, 16, 17] and by Goodman, Heggie and Hut[13], with different approaches and also with a different interpretation as far as the sources of instability);

b) $\tau_S \sim N^{1/3}\tau_C$ (Gurzadyan and Savvidy)[12].

The aims of the present contribution are to point out that the relationships between the *instability time-scale*, as measured from the *exponential divergence* of trajectories, the *mixing time, if it can be estimated in some "averaged" analytical way* and the *"relaxation" time* (**when** it is possible to speak about of *one* relaxation time) are, in general, highly non-trivial, but that, in a suitable framework, to many aspects of those relationships can be given a physical interpretation in specific classes of Hamiltonian systems[8, 22] and that, in this light, the question of the contradictions between analytical estimates and numerical simulations in gravitating systems can be addressed.

To test the sensitivity of an N-body system to changes in initial conditions the first simple approach is to compare the trajectories followed by the same system starting from a set of different, but very close, initial conditions: let us consider, for simplicity, two points in the phase space of the system and, choosing in some appropriate way a *"distance"* in the phase space, let us trace it in time as the two trajectories evolve from the initial points. If the system dynamics are sensitive to a set of initial conditions containing those two points, this distance grows very fast (indeed at an exponential rate). The problem with this simple approach is that the two trajectories compared explore different regions of phase space with a very different structure[22]: stated in another way the problem is that we have *two distinct* systems with no further knowledge of the fate of each other. A technique devised to overcome this difficulty is that of following the evolution of the perturbation to a *reference* system, namely to study the *tangent dynamics*: a version of this technique (firstly applied to gravitating systems by Miller[21]) is exploited by Goodman, Heggie and Hut[13]. Even with this approach an overfull problem is encountered (it can be partially overcome adopting the procedure scheme introduced by Benettin *et al.* [3].), since the solution of the variational equations attains quickly a non-linear regime. But, together with these troubles, there is another question, common to both the above approaches, concerning the reliability of extrapolations of numerical results: with $N \sim 10^3$ the mean and fluctuating fields are of comparable amplitude; so the techniques used, which are not able to disentangle the effects due to these distinct contributions, make predictions on the rate of instability that are dominated by the *more unstable* bunch of degrees of freedom, which are not necessarily representative of the average behaviour of the whole system. In addition, there is also the awkward problem raised by the interpretations of numerical N-body simulations performed until now: if these were intended to estimate something like the *Maximum Lyapounov exponent (MLe)* of the dynamical system used as the (model of a) stellar system, it is very embarassing to remember that, even for systems with few degrees of freedom, the time of integration needed to obtain a reliable estimate of MLe can be unimaginably long[9, 22], whereas *all* N-body simulations performed, stop after less than *ten* crossing times. Moreover, even when accurately determined, the reciprocal of the MLe has in general no direct and obvious relation with the *relaxation* time-scale.

On the other hand, the work of Gurzadyan and Savvidy[12] addressed from the very beginning to the role played by fluctuations in determining the instability, so that, in a sense, some of the claims argued there could be *recovered* if the meaning of the *"Collective relaxation time"* introduced in [12] is *reinterpreted* as a time scale for the *damping of fluctuations* in a system which *was already in equilibrium*; but, at the same time, they underestimated the effect of the mean bulk field[15, 16] and the peculiar structure of a realistic self-gravitating system, with its enormous density contrast and the strong anisotropy of the velocity distribution. On this basis, the averages performed in the configurations and velocity spaces are to be taken only as a first rough estimate of the instability growth rate.

In the light of the above considerations it appears clear that it is fully legitimate to ask whether does it make sense to look for a unique relaxation time for a system so far from homogeneity as a stellar system. To devise a reliable technique for the description of the instability we need therefore a tool suitable to describe the detailed evolution of the perturbation as a manifestation of a *global* instability. This tool is the *Jacobi Levi-Cìvita* equation of geodesic deviation on the configuration manifold in the Jacobi geometrical picture of dynamics.

THE JACOBI - LEVI-CIVITA EQUATION OF GEODESIC DEVIATION

The framework in which we settle for the study of the long term behaviour of N-body systems is that of exploiting the equivalence of the trajectories of a conservative Hamiltonian system and the geodesics of the Riemannian space with the corresponding Jacobi metric. Using the minimum action principle in the form given by Maupertuis and Jacobi, one can show[2, 14, 23] that the geodesics equation associated to the conformal-Euclidean metric

$$ds^2 = g_{ab} dq^a dq^b = [E - V(\mathbf{q})]\, \eta_{ab}\, dq^a dq^b \quad (a, b = 1, \ldots, 3N;\ \eta_{ab}\eta^{bc} \equiv \delta^c_a), \tag{1}$$

whose explicit form for the velocity vector $u^a = dq^a/ds$ is

$$\frac{Du^a}{ds} \equiv \frac{du^a}{ds} + \Gamma^a_{bc} u^b u^c = \frac{du^a}{ds} + \frac{1}{2(E-V)} \left(g^{ab} - 2u^a u^b\right) V_{,b} = 0, \tag{2}$$

coincides with the canonical equations with Hamiltonian function

$$H(\mathbf{p}, q) = \frac{1}{2}\eta^{ab} p_a p_b + V(\mathbf{q}) \equiv E. \tag{3}$$

With D/ds we indicate the covariant derivative of a vector field along the flow, E is the total energy of the system, $V(\mathbf{q})$ is the potential energy, the parameter along the geodesic is related to the physical time by $ds = \sqrt{2}(E-V)dt$ and the coordinates of the configuration manifold are related to the position of the $i-th$ particle $\vec{\mathbf{r}}_i$ by

$$q^a \equiv \sqrt{m^a}x^a, \quad m^{3i-2} = m^{3i-1} = m^{3i}, \quad \vec{\mathbf{r}}_i \equiv (x^{3i-2}, x^{3i-1}, x^{3i}), \quad i = 1, \ldots, N. \tag{4}$$

In this approach the stability properties of the dynamics are described by the Jacobi - Levi-Cìvita (JLC) equation of geodesic deviation

$$\frac{D^2\delta q^a}{ds^2} + \mathcal{R}^a{}_c\delta q^c = 0, \tag{5}$$

where δq^a is the perturbation vector in the positions and the *stability matrix* $\mathcal{R}^a{}_c$ is related to the Riemann curvature tensor associated to g_{ab} by

$$\mathcal{R}^a{}_c = R^a{}_{bcd}u^b u^d \;=\; \left[\frac{1}{4W^2}\left(\frac{dW}{ds}\right)^2 - \frac{1}{2W}\frac{d^2W}{ds^2}\right]\delta^a_c + \frac{(\overrightarrow{\nabla W})^2}{4W^3}u^a u_c + \tag{6}$$

$$+\frac{1}{2W}\left(W_{,bc}\,u^a u^b + g^{ab}W_{,bd}\,u^d u_c - g^{ab}W_{,bc}\right) + \tag{7}$$

$$-\frac{3}{4W^2}\left[\left(\frac{dW}{ds}\right)(u^a W_{,c} + g^{ab}W_{,b}\,u_c) - g^{ab}W_{,b}\,W_{,c}\right] \tag{8}$$

and $W(\mathbf{q}) = E - V$. The solutions of the JLC equation allow to predict the evolution of perturbations [22, 7] and to identify the geometric properties of the manifold related to the stochastic behaviour of the dynamics.

In almost all preceding studies[12, 15, 16, 13, 22] the analysis was limited to the study of the evolution equation for the (squared) *norm* of the perturbation

$$\parallel \delta q \parallel^2 \equiv g_{ab}\delta q^a \delta q^b \tag{9}$$

that is

$$\frac{d^2}{ds^2}\| n \|^2 - \frac{\| \nabla V \|^2}{2(E-V)^2}\| n \|^2 + \left(\frac{3}{2}\frac{V_{,a}V_{,b}}{(E-V)^2} + \frac{V_{,ab}}{E-V}\right)\left(u^a u^b \| n \|^2 + n^a n^b\right) = 2\left\|\frac{Dn}{Ds}\right\|^2. \tag{10}$$

where $n^a \equiv \delta q^a_\perp$ is the component of the perturbation normal to the velocity u^a, and we used the antisimmetry of Riemann tensor, which implies $\mathcal{R}^a_c l^c \equiv 0$ for every perturbation *along* a geodesic $l^a \equiv \delta q^a_\parallel$. Taking into account the positivity of term in the r.h.s., equation (10) can be rewritten as[12, 22]

$$\frac{d^2}{ds^2} \parallel n \parallel^2 + 2K(n,u) \parallel n \parallel^2 \geq 0, \tag{11}$$

so that, if the sectional curvature $K(n,u)$ is negative *everywhere and for every pair* (n,u), then $\|n\|$ grows exponentially (in terms of the "*proper time*" s) with a rate $\tau_s^{-1} \sim \sqrt{-K/2}$ (the systems where K is constant and negative are the classical Anosov[1]systems). However, it is worth stressing that the interaction potentials of physical importance have in general sectional curvatures of indefinite sign.

If instead we want to study the complete JLC equation (5), are of great importance the eigenvalues $\{\rho_{(k)}\}$ of the stability matrix which are defined by the equation:

$$\mathcal{R}^a{}_c v^c_{(k)} = \rho_{(k)} v^a_{(k)}, \quad k = 0, \ldots, 3N-1 \tag{12}$$

with $\rho_{(o)} = 0$, *and* $v^a_{(o)} = \delta q^a_\parallel$. Those eigenvalues, linked, trough the JLC equation, to the evolution of the covariant derivatives of the components of perturbations, and, as it is easy to see, to the *effective* growth of perturbations[7], give informations equivalent to that provided by other typical *stochasticity indicators* studied in dynamical systems theory, such as the *Lyapounov exponents*, but in a more manageable form.

APPLICATIONS OF THE JACOBI - LEVI-CIVITA EQUATION

As a preliminary test of this geometric picture we can compare its predictions with that obtained in the more usual frameworks. We have solved the *JLC* equation (5) in the classical case of the Henon-Heiles potential: here, being the system 2-dimensional, we have a *generalized Hill equation*, of the form

$$\frac{d^2 n}{ds^2} + \frac{1}{W^2}\left(2 + \| \nabla V \|^2\right) n = 0. \tag{13}$$

Adopting the renormalization algorithm[3] we compared the instability rate of the geodesic deviation with the estimates of the Lyapounov exponents resulting from the integration of the tangent dynamics equation. We have found[8] that in the regular regions both indicators tend to zero (respectively as s^{-1} and t^{-1}) and there is a very good agreement between the two estimates in the stochastic regions. A drawback of the use of the *JLC* equation in the case of systems with few degrees of freedom is the fact that the phase point explores frequently regions near the $V = E$ boundary: here, since W is small, by inspection of the equation (13) we can see that there is some computing trouble and that, to keep a good precision, the CPU time must be even longer of that already long of the ordinary procedure. This problem, however, lessens rapidly increasing the number of degrees of freedom.

Let us see therefore what can be said, with not too much effort in N-body systems. One of the interesting and immediate results of the geometric approach, when applied to the case of the dynamical (*"violent"*) relaxation of self-gravitating N-body systems, consists in the possibility of identifying, in a completely analytical way, a transition in the global stability properties of the dynamics, at the onset of the virial equilibrium, put forward by a damping of the fluctuations of the trace of the stability matrix (about a positive definite average value) which, through the solutions of *JLC* equation, indicates a transition towards a *less chaotic* behaviour, at least on a global scale. A procedure to simplify the analysis is to perform some average in the configuration or velocity space. This is precisely what has been done e.g. in [12, 15, 16] and has led to the prediction that the instability time is of the order of the crossing time, when the bulk mean forces constitutes the leading term. But it is easy to see that, under the assumption that the components of the perturbation are randomly chosen, namely if, for example, we assume that $n^a n^b \simeq \delta^{ab} \| n \|^2$, equation (10) reduces to

$$\frac{d^2}{ds^2}\| n \|^2 + \frac{1}{E-V}\left[\frac{d^2 V}{ds^2} + \frac{1}{2(E-V)}\left(\frac{dV}{ds}\right)^2 + \frac{3}{2}\| \nabla V \|^2 + \Delta V\right]\| n \|^2 \gtrsim 0. \tag{14}$$

where we introduced the derivative along the flow $dV/ds = V_{,a}u^a$ and posed $\Delta V = \delta^{ab} V_{,a} V_{,b}$.

Now, far from the virial equilibrium, the term with the flow derivative is the biggest and fast fluctuating one, so that it leads the growth rate of perturbations to be of the order of τ_C. But, after this stage is reached, this term has positive time average, and fluctuates with a magnitude comparable to that of the terms with the gradient and the Laplacian (and both are of indefinite sign), so that the evolution of the perturbations will proceed on a longer time-scale. It is therefore clear that there should exists a *transition*, when the process of *virialization* has been completed. This is well evident in most numerical simulations carried out for long enough time, for chains of coupled anharmonic oscillators, and it should be true also for generic many degrees of freedom Hamiltonians. As the fluctuations at equilibrium of the dominant term are of statistical nature, we would like also to stress that this gives an indication why there should exist an N-dependence on the evolutionary behaviour of fluctuations[7].

The results briefly sketched above seem, in our view, very promising. As told before in the discussion of the peculiar character of a gravitating system, the most direct way to gain some more accurate information about the *global* behaviour of perturbations, is to consider the complete equation for the components (5) with the diagonalized stability matrix (12). As shown elsewhere[8, 7], this equation shows how peculiar this dynamical system is, and how much significant is to speak about *multiple (instability) time-scales in stellar dynamics*: whether these time-scales can be also related to some kind of *'mixing'* or even *'relaxation'* processes is in our opinion a still unsolved question, which could be answered only looking at the behaviour of some meaningful quantities, univocally related to the approach to *some kind* of *meta-equilibrium*; see [20, 8, 10, 22]. In any case, the analysis of the eigenvalues of the stability matrix, shows the existence[8, 7, 22], for a system of N self-gravitating point masses, of a full hierarchy of time-scales, whereas, for the *usual* Hamiltonian systems with many degrees of freedom there exists a generic homogeneity of time-scales. As we have remarked above, the explicit expression of $\mathcal{R}^a{}_c$ (eq. 8) allows us to affirm that, also for the evolution of the *components of the perturbation*, the behaviour will change when the virialization is completed. In fact, the diagonal term in $\mathcal{R}$ is the biggest and fast fluctuating when the phase of *Violent Relaxation* is developping, and oscillates around a positive value, with magnitude of the order of the other terms in $\mathcal{R}$, when the *global dynamical equilibrium* has been attained. *But* we can say even more! It is straightforward to calculate the *trace* of $\mathcal{R}$, which is nothing else than the *Ricci curvature* in the direction of the flow and to see that, when the V.R. has been completed, at least in a time-averaged sense, $Ric(u) \equiv Tr\mathcal{R} \equiv \sum_i^{3N} \rho_{(i)} \geq 0$, that is the sum of the eigenvalues is positive; and this means, again at least in an "averaged" sense, that the dynamics become *less unstable*, and then, that the instability properties of the dynamics, undergo a transition after some dynamical (crossing) times.

It is also possible to show that the *spectrum* of eigenvalues of the stability matrix $\mathcal{R}$ describes in a coherent way the evolutionary properties of the system in the full phase space, giving a very effective tool for the determination of the stochastic character of the motion. Moreover, the study of the *statistical* behaviour of Hamiltonian systems with many ($N \gg 3$) degrees of freedom, passing by the study of the geometrical properties of the manifold on which motion takes place, leads to the assessment of possible transitions in the chaotic regime in dependence on the binding energy of the systems, a property which is *generic* to most of the well studied ones.

CONCLUSIONS

It should be clearly stated that, in general, there is no direct, or *a priori*, correlation between the time-scale(s) associated to the dynamical instability and any sort of *relaxation time*. The only case in which a trivial relation exists between the two time scales, corresponds to those abstract dynamical systems that can be described as flows on a manifold with a constant negative curvature, for which there exist only one *proper* instability time-scale. And for most *physical* system the two-dimensional gaussian curvatures are very far from being everywhere negative. When the curvature *is not constant*, as in our case (and in general), what should be estimated is not the mean value of the curvature, but, his minimum (negative, if it exists) absolute value; if the curvature is somewhere positive, then no rigorous result exists which allows to draw conclusions about *'mixing'* properties of the dynamics, even in the presence of a *sensitive dependence on small changes in initial conditions.* From this the importance of the *JLC* equation of geodesic deviation, which contains, when used in appropriate way, all informations needed in order to describe the evolution of the system in his 6N-dimensional phase space.

It should be interesting to verify the peculiarity of gravitational interaction with respect to the statistical mechanics, in connection with the existence (or not) of a Strong Stochasticity Threshold [22], as the (specific) energy of the system is increased. In any case, it must be stressed the idea that the approach to the equilibrium of a Hamiltonian system can't be associated only to the growth of a *bunch* of degrees of freedom, but should be related to the full stochastic behaviour of the stability tensor $\mathcal{R}$. There exist a twofold (at least) evolutionary behaviour of the instability properties of an N-body system (not only a self-gravitating one), which comes after the approach to the virial equilibrium. This is one of the reasons for which it is necessary to increase the number of particles used in numerical N-body simulations.

Two observations to conclude:

a) With the geometrical method it is possible to do an *importance sampling* (or something alike) of phase space of the system, without following the system actual evolution, overcoming the numerical growth of errors.

b) As shown by Pettini [22], the *geometrodynamical* properties of the configuration manifold of a Hamiltonian system with many degrees of freedom are very rapidly variable functions of the point (and then of time). This implies that the numerical simulations cannot have *any immunity* at all with respect to the *linearly-induced exponential growth* [8, 7] of perturbations due to the different values of curvature (for example) experimented by two different *realizations* of the same system. More; it is possible (with high probability) that the exponential growth of *errors* in N-body systems is related not (or not only) to the negativity of the curvature, but (also) to the parametric instability induced by the rapidly fluctuating geometrical properties: if this is the case, there is no rigorous result which connect the instability with "*mixing*".

REFERENCES

[1] Anosov, D.V. *Proc.Steklov Inst. Math.*, **90**, (1967).

[2] Arnold, V.I. *Mathematical Methods of Classical Mechanics*, (IInd Edition), (Springer-Verlag, 1989).

[3] Benettin,G. Galgani,L. Strelcyn,J.M. *Phys.Rev. A* **14 - n.6**, 2338, (1976).

[4] Binney,J. Tremaine,S. *Galactic Dynamics*, (Princeton Univ. Press, Princeton, N.J., 1987).

[5] Boccaletti,D. Pucacco,G. Ruffini,R. *Astron. Astroph.*, **244**, 48 (1991).

[6] Chandrasekhar, S. *Principles of Stellar Dynamics*, (Dover, 1960) (ristampa ediz. del 1943).

[7] Cipriani,P. Pucacco,G. *Nuovo Cimento B* (in press)

[8] Cipriani,P. - `Ph.D.Thesis` - *Univ. of Rome* (1993).

[9] Contopoulos,G. Barbanis,B. *Astron.Astroph.*, **222**, 329 (1989).

[10] Galgani,L. in *Nonlinear Evolution and Chaotic Phenomena.* Gallavotti, G. - P.F.Zweifel eds. - NATO-ASI Series:B 176 (Plenum 1988) pag. 147-159.

[11] Gaspard, P. *Preprint Universitè Libre de Bruxelles*, (1992).

[12] Gurzadyan, V.G. Savvidy, G.K. *Astron.Astroph.*, **160**, 203, (1986).

[13] Goodman, J. Heggie, D.C. Hut, P. *Ap.J.*, **415**, 715, (1993).

[14] Jacobi, C.G.J. 1884 *Vorlesungen über Dynamik*, IInd edition, (Reiner - Berlin - 1884, reprinted by Chelsea - New York - 1969)

[15] Kandrup, H.E. *Ap.J.*, **364**, 420, (1990).

[16] Kandrup, H.E. *Phys.Lett. A*, **140**, 97, (1989).

[17] Kandrup, H.E. Smith, H. *Ap.J.*, **374**, 255, (1991).

[18] Krylov, N.S. *Works on Foundations on Statistical Physics*, (Princeton Univ. Press, 1979).

[19] Lynden-Bell,D. *Mon. Not. R. A. S.*, **136**, 101 (1967).

[20] Ma, K.S. *Statistical Mechanics*, (World Scientific, 1988).

[21] Miller, R.H. *J. Comp. Phys.*, **8**, 449 (1971).

[22] Pettini, M. *Phys.Rev.* **E 47**, 828 (1993).

[23] Synge, J.L. *On the geometry of dynamics Handbuch der Physik*, **III-1**, (Springer - 1960)

PROPER ELEMENTS AND STABLE CHAOS

Andrea Milani
Space Mechanics Group
Department of Mathematics
University of Pisa, Italy
E-mail: milani@dm.unipi.it

ABSTRACT The long term evolution of the orbits of the asteroids is studied by means of proper elements, which are quasi–integrals of the motion. After a short review of the classical theories for secular perturbations, this paper presents the state of the art for the computation of proper elements. The recent theories have been extended to higher degree in the eccentricities and inclinations, and to the second order in the perturbing masses; they use new iterative algorithms to compute secular perturbations with fixed initial conditions but variable frequencies. This allows to compute proper elements stable over time spans of several million years, within a range of oscillations small enough to allow the identification of asteroid families; the same iterative algorithm can also be used to automatically detect secular resonances, that is to map the dynamical structure of the main asteroid belt. However the proper element theories approximate the true solution of the N-body problem with a conditionally periodic solution of a truncated problem, while the orbits of most asteroids are not conditionally periodic, but chaotic; positive Lyapounov exponents have been detected for a large number of real asteroids. The phenomenon of stable chaos occurs whenever the range of oscillations of the proper elements, as computed by state of the art theories, remains small for time spans of millions of years, while the Lyapounov time (in which the orbits diverge by a factor exp(1)) is much shorter, e.g. a few thousand years. This can be explained only by a theory which accounts correctly for the degeneracy of the unperturbed 2-body problem used as a first approximation. The two stages of computation of mean and proper elements are each subject to the phenomena of resonance and chaos; stable chaos occurs when a weak resonance affects the computation of mean elements, but the solution of the secular perturbation equations is regular.

INTRODUCTION

The development of the theory of proper elements is a success story for celestial mechanics. In the last few years we have progressed from every point of view: the new theories are pushed to a much higher order and degree, have proven their accuracy to a level fully compatible with the needs of the main applications in planetary science, use a sound mathematical formalism, and are fully documented in accessible publications. However, because of this dramatic progress, and also because of the enormous increase

From Newton to Chaos, Edited by A.E. Roy
and B.A. Steves, Plenum Press, New York, 1995

in the power of the digital computers, the computation of proper elements has reached the extreme limits of its domain of validity, both in terms of accuracy and of time span. In concluding the paper I presented in the Cortina meeting six years ago, I quoted Poincaré who said that *the epoch in which we will be forced to give up the old methods is without doubt very far away*, to argue that the epoch was now. I am now convinced that in the last few years we have already gone far beyond the use of the old methods; as a result, we are beginning to explore an uncharted territory, without the advantage we previously had, namely without a good mathematical theory capable of predicting what we might find.

One of the purposes of this paper is to convey to the readers the sense of this change, from old methods to new ones, from a celestial mechanics we can compute with our hands to one we can only tackle with the assistance of computers, from a situation in which we were just finding what was theoretically predicted to one in which we find new wonders at each new attempt. I shall begin with a short reminder of the classical theories of short and long periodic perturbations (on the orbits of asteroids, but it could be extended to other cases); this will be done in Sections 1 and 2. Then I will present the better understanding of the problems of secular perturbations and proper elements, which allowed us to produce the more advanced theories of the last years (Sections 3 and 4). Section 5 will be dedicated to an assessment of the current state of the art, namely to the quantitative accuracy and to the level of global understanding of the dynamics we can achieve with the current methods, which are already far beyond the traditional, formal perturbation theory. In Section 6 I will try to present the phenomenon of stable chaos, which is still poorly understood; actually the very name of stable chaos raised many controversies. Nevertheless I will show that the problem is there and cannot be ignored, if we want to discuss the stability of the proper elements, and even the stability of the asteroid belt, over a time span comparable to the age of the solar system. In Section 7 I will give the general outline of a theory capable of explaining the occurrence of stable chaos; although many details of such a theory have still to be worked out, I believe we are now beginning to understand the essential steps. The time when we will be able to use such a concept in a predictive, not only descriptive, way, is perhaps very close.

In the last section, I will present some arguments which I have not yet been able to put in a fully rigorous form. My opinion is that sometimes it is useful to present some heuristic arguments, rather than just confessing our ignorance on a problem we cannot sweep under the carpet and ignore. This presentation is more or less self contained, although I shall only mention some points which I have discussed somewhat at length in the papers I presented at the previous two Cortina meetings (Milani 1988, 1991). I shall try to use a rather informal language and to replace as much as possible formulae and brute force computations by arguments given in words, but of course some of these arguments can be fully appreciated only if the computations are redone in detail.

1. FIRST ORDER AVERAGING

Most of the classical results are based upon first order perturbation theory. We shall introduce the necessary formalism in 7 steps. As the first step, let us state the problem of the motion of an asteroid in the gravitational field of the Sun plus N planets in an abstract form, as a hamiltonian problem with Hamilton function:

$$H = H_0 - \mu R(E, E') \tag{1.1}$$

where E is some six dimensional vector of orbital elements of the asteroid which are canonical dynamical variables; E' is another vector of orbital elements of the perturbing planet (in practice we could well have more than one perturbing planet), which is assumed to be a known function of time: $E' = E'(t)$, e.g. it is in turn a solution of the corresponding problem for the orbits of the planets. μ, the perturbation parameter, is roughly speaking the ratio of the mass of the perturbing planet to the mass of the Sun (but more exactly it is one of the Roy–Walker small parameters, see Roy, 1982). H_0 is the hamiltonian of the unperturbed 2–body problem.

The second step is based upon the idea that half of the variables in E are angles: as an example, the angles can be the classical elements Ω, ω, ℓ (longitude of the node, argument of perihelion, mean anomaly); then the other three variables have to be (to be canonical, that is to preserve the form of the Hamilton equations) the Delaunay action variables $Z = k\sqrt{a(1-e^2)}\cos I, G = k\sqrt{a(1-e^2)}, L = k\sqrt{a}$ (where k is the Gauss constant, $k^2 = GM_\odot$ in some system of units). Then $H_0 = -k^4/(2L^2)$ is a function of one action only, and the perturbing function R can be expanded in a multiple Fourier series with arguments $\Omega, \omega, \ell, \Omega', \omega', \ell'$.

The third step arises from the recognition that not all the possible combinations – with integer coefficients– of the angular variables can actually have a non zero Fourier coefficient, and this for two reasons: first, the perturbing function R is invariant with respect to the isometries (rotations and mirror symmetries) of the configuration space; second, the arguments Ω, ω, ℓ can be singular for either $e = 0$ or $I = 0$, but the function R, when expressed in cartesian coordinates, is a perfectly regular function for both a circular orbit and for an orbit in the reference plane. From these two statements we can derive the following *D'Alembert rules*, in a form applicable to Delaunay variables:

I: A Fourier coefficient of R can be nonzero only if the sum of the coefficients of the longitudes of the nodes (both of the asteroid and of the planet(s)) is zero.

II: A Fourier coefficient of R can be nonzero only if the sum of the coefficients of the arguments of perihelion is even.

III: The Fourier coefficient with an argument containing $(p\ell + q\omega + r\Omega)$ is small with e, I, its Taylor series beginning with the monomial $e^{|p-q|}I^{|q-r|}$. The same statement applies to the arguments containing $(p'\ell' + q'\omega' + r'\Omega')$: their Fourier coefficients are of the order of $e'^{|p'-q'|}I'^{|q'-r'|}$. The following terms of the Taylor series have the degrees in either e, e' or I, I' increased by steps of 2.

IV: The Fourier series for R contains only the cosine terms.

Step number four would be the actual computation of these Fourier coefficients, as explicit Taylor series in the eccentricities and inclinations. I will not discuss this here; a very clear presentation can be found in Duriez (1990). I only would like to point out that the coefficients of this expansion can be explicitly computed by means of the so called *Laplace coefficients*, which are elliptic integrals and can be numerically computed, e.g. by a combination of hypergeometric series and recursion formulae. This means that each one of these coefficients can be estimated by means of an analytic computation which we can do by hand, but an accurate computation of a large set of these coefficients already requires a computer: even what we normally call an analytic theory is actually implemented as computer software.

Step number five is the *averaging principle*: to order one in μ, the average of the solution can be obtained by the solution of the averaged equations. Since H_0 contains only L, the variable ℓ is the only angle with a frequency n of order 0 in μ, that is with a period of a few years only. The angle ℓ' is also a function of time with a frequency n' of the same order. By averaging we mean integration with respect to the two fast variables ℓ, ℓ':

$$R = \overline{R} + \tilde{R} \quad ; \quad \overline{R} = \frac{1}{4\pi^2} \int_0^{2\pi} \int_0^{2\pi} R \, d\ell d\ell'. \tag{1.2}$$

Then the solution for every orbital element, e.g. G, can be expressed as the sum of a *long periodic* part $\overline{G}$ not containing fast frequencies (that is n, n') plus a *short periodic* part $\tilde{G}$ which does not evolve with time but only oscillates, with frequencies containing a combination of n, n'. Then the equations of motion for $\overline{G}$ are:

$$\frac{d\overline{G}}{dt} = \mu \frac{\partial \overline{R}}{\partial \omega}(E, E') + O(\mu^2). \tag{1.3}$$

Step number six is to study the new hamiltonian problem defined by the averaged Hamiltonian $H_0 - \mu\overline{R}$. This *secular hamiltonian* does not contain the variables ℓ, ℓ'; therefore the variable L is an integral, and the semimajor axis has no long periodic perturbations to first order in μ in the time derivative (Lagrange theorem). The dynamic variables are therefore reduced to 4: Ω, ω, Z, G, and they are all slow (with derivatives of order μ); the word *secular* has no deep meaning, but only hints at the long periods for these variables in the equations defined by the secular hamiltonian. The variable ℓ still has a secular equation, but it can be obtained by quadrature from a solution for the 4 variables quoted above.

The last step is the computation of the initial conditions in the space of *mean elements*, that is where the secular equations apply, as a function of the ordinary (*osculating*) elements. The main problem with (1.3) is not that it is complicated –it is actually much simpler than the original equations of motion for the osculating elements, since the large majority of the Fourier terms have been wiped away– but in its interpretation. Indeed, what exactly are $\overline{G}, \overline{Z}$ etc.? If the solution were representable as a multiple Fourier series with arguments having different frequencies, some containing n, n' and some with only slow frequencies, we could define $\overline{G}$ by removing the fast terms from the series. However, that the solution can be represented by a convergent Fourier series is a hypothesis we cannot prove, and which is in fact false for some orbits. Another way is to think to $\overline{G}$ as the result of the application to the function $G(t)$, known may be by numerical integration, of a low frequency pass digital filter. This has become a common way of thinking, since now computers are powerful enough to gives accurate solutions, for times of the order of the periods of the slow arguments, by brute force and in a few minutes; digital filtering is available even in public domain orbit computation software (such as my own *Orbit8v*). This can work even in the case of a chaotic orbit, since it does not depend upon the hypothesis of the existence of a Fourier series; however, there are cases in which even digital filters would not work, because every digital filter has a transition band in which the signal is distorted (Carpino et al., 1987). The classical solution to the problem of actually computing $\overline{G}$ etc. is either to ignore entirely the difference between G and $\overline{G}$, by arguing that in general it is small, or to compute $\tilde{G}$ by a rough analytic theory containing only the main terms, e.g. the ones with frequency $pn - (p+1)n'$. This is an approximation, nevertheless it works reasonably well in many cases.

2. CLASSICAL SECULAR PERTURBATION THEORY

Let us assume that, as a result of the computations discussed in the previous Section, the problem of the long term orbital evolution of some given asteroid has been reduced to a hamiltonian problem with only four dynamical variables:

$$K = K(\Omega_M, \omega_M, Z_M, G_M, t) = -\mu\overline{R} \tag{2.1}$$

where the subscript M indicates that the orbital elements have somehow been converted to the mean ones, with short periods already removed. The new hamiltonian K is the same as the long periodic part of the original one H, apart from the $H_0(L_M)$ portion which is irrelevant (with respect to the four remaining dynamical variables it is just a constant). The dependence upon time is not direct, but only through the (mean) elements of the perturbing planet(s) Ω', ω', Z', G'. To solve the problem, we need to set up a suitable perturbation scheme to solve the Hamilton equation. However, the secular hamiltonian K is not presented in (2.1) in a form: (hamiltonian of an integrable problem) + (small parameter)×(perturbing function). We need to select a portion of K as an integrable first approximation, and a new small parameter to replace μ (which is not small any more: the entire hamiltonian is of order 1 in μ).

It is important to realize that there is not a unique way to perform this separation into integrable part plus perturbation. The modern theories of secular perturbations use for this fundamental choice one of the two proposed by Kozai (1962) and Yuasa (1973) (or variations of these).

Kozai used as a first approximation a problem in which the eccentricities and the inclinations of the perturbing planets are exactly zero. Then the hamiltonian K can be expanded in a Taylor series with small parameters e', I':

$$K = K_0(\omega_M, G_M, Z_M) + e'K_{1e} + I'K_{1I} + O(I'^2, e'I', e'^2), \tag{2.2}$$

where K_{1e}, K_{1I} are functions of the four mean elements of both the asteroid and the planet(s). On the contrary it is possible to show, by using only the D'Alembert rules, that K_0 depends only upon one angle, namely ω_M. By rule **I**, a term with nonzero coefficient must have a combination of nodes with sum of the coefficients 0; on the other hand, there can be no ω' and no Ω' in the argument, unless there is some positive power of either e' or I', by rule **III**; but if the argument contains only the node of the asteroid, the only way to have a zero sum is to have all coefficients zero. Thus K_0 does not depend upon Ω, ω', Ω'; it does not contain e', I' either, thus it does not depend upon time in any way. By rules **II** and **IV**, only the terms with $\cos(2j\omega_M)$ can appear in K_0.

The Kozai choice of K_0 as first approximation leads to a consistent perturbation scheme because K_0 is integrable (as every hamiltonian depending upon only one coordinate ω_M and a number of momenta); Z is a first integral since $\partial K_0/\partial\Omega_M = 0$, and K_0 itself is an integral since $\partial K_0/\partial t = 0$. The portion of K of order 1 in the new small parameters e', I' is the perturbation, and to first order in e', I' it is possible to compute the solution by using the *Hori kernel* (Hori, 1966) consisting of the general solution to the integrable K_0 problem. A first order perturbation theory of this kind was computed by Williams (1969), although the formalism was not canonical; the results of the use of this method were the understanding of the behavior of the Earth–crossing asteroids, the computation of proper elements for main belt asteroids

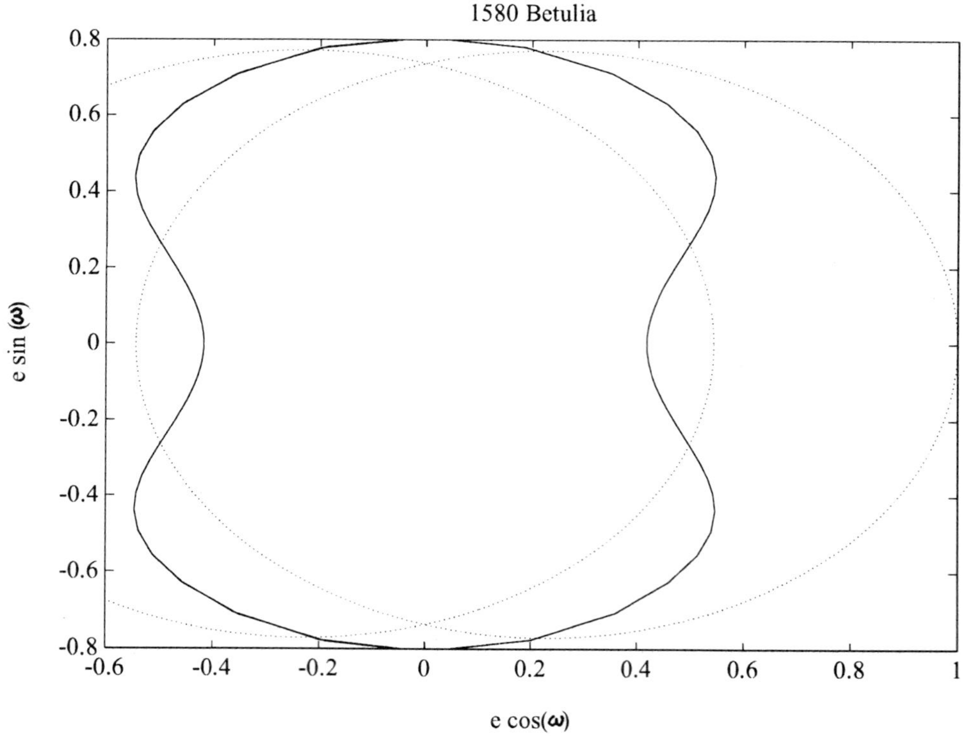

Figure 1. A solution curve of the Kozai approximation K_0 to the secular perturbation problem. In this plane, the radius from the origin is the eccentricity, and the polar angle is the argument of perihelion ω. The dotted lines indicate where orbital crossing with the Earth occurs; from this approximate computation, it is possible to conclude that *1580 Betulia* crosses the orbit of the Earth 8 times during each period of circulation of ω, a conclusion confirmed by numerical integrations of the full equations of motion (Milani et al., 1989).

and the determination of the location in phase space of the main (linear) secular resonances (Williams, 1973, 1979; Williams and Faulkner, 1981).

There are three main properties of the secular perturbation theories of the Kozai–Williams class. First, the geometry of the integrable first approximation K_0 is very different from both the 2–body approximation and from the more usual linear theory (discussed below); the main difference is that the eccentricity and inclination can undergo very large changes over one cycle of the argument 2ω. Figure 1 (from Baccili and Cattaneo 1993) shows a solution to K_0 for the initial conditions roughly corresponding to the asteroid *1580 Betúlia*, in the plane $(e\cos\omega, e\sin\omega)$: this solution curve is very different from the circle $e = const$, and also from the epicycles of Lagrange and Laplace (see below), and explains why *1580 Betulia*, an Amor asteroid with a present perihelion much above $1\,AU$, is an Earth crossing asteroid of the Geographos class (Milani et al., 1989).

Second, both a solution of the K_0 problem such as Figure 1, and the proper elements by Williams (1979, 1989), are obtained by a semianalytical method. The solution of the integrable K_0 problem is in principle an analytic function, but not one we know how to compute analytically. The solution can be obtained by quadrature, as stated by the Liouville–Arnold–Jost theorem (Arnold 1976), but in practice we need to perform the quadrature by numerical step-by-step integration. Henrard (1990) has shown that it is perfectly possible to perform all the operations of the canonical formalism, including a perturbation theory, by means of numerical quadratures along the solutions of the integrable Hori kernel. This procedure is of course somewhat more expensive, in terms of number of arithmetic operations, than an analytic one, but this is less and less relevant, giving the cost $\longrightarrow 0$ of computation.

Third, the choice of an integrable first approximation inside K is not a neutral one from the point of view of the accuracy. One approximation can be much better than another one in a given portion of the phase space; e.g. the Kozai type expansion results in the terms with $e'^2, e'I', I'^2$ being neglected as a "second order" perturbation, while there is no truncation whatsoever with respect to the eccentricity and inclination of the asteroid. Thus the Kozai–Williams approximation has to be adopted whenever e, I are large; on the contrary, when e, I are comparable in size to e', I', it is inconsistent to discard e'^2 and to retain e^2 in the same theory, and the scheme described below must be used. For other specific regions of the asteroid belt, even other perturbation schemes could be envisaged.

The other main perturbation scheme for (2.1) was first used, in the context of secular perturbations for asteroids, by Yuasa (1973); in the context of the secular perturbations for the major planets, the same scheme was used by Bretagnon (1974). The key idea is to look at K as a Taylor series in the variables e, I, e', I': in this way of thinking, what matters is the *degree* in the eccentricities and inclinations, does not matter if they belong to either the asteroid or the perturbing planet(s). Thus this approximation is consistent when the asteroid has a low to moderate e and I. By D'Alembert rules **II** and **III**, there are no odd degree terms, and by rule **III** the term of degree 0 does not contain any dynamical variable of the secular system, thus the lowest degree part of K is quadratic:

$$K = K_2 + K_4 + K_6 + \dots \tag{2.3}$$

where K_i is a homogeneous polynomial of degree i in the eccentricities and inclinations. The degree 2 part K_2 is integrable because a quadratic hamiltonian implies linear Hamilton equations. Actually the solution of K_2 alone is the classical secular perturbation theory of Lagrange and Laplace. The solution of the latter can be expressed as a sum of harmonics with frequencies g_i (one for each perturbing planet) and s_i (one less than the number of planets), g_0 (proper longitude of perihelion frequency) and s_0 (proper longitude of node frequency); each harmonic results from the solution of one of the decoupled linear oscillators appearing in a coordinate system in which the quadratic form K_2 is diagonalised. Thus the solution can be explicitly computed in terms of sine and cosine functions. Geometrically, the solution can be described by means of epicycles –the vectorial sum of one circular uniform motion per planet, plus one for the asteroid– in the planes with coordinates $(e \cos \varpi, e \sin \varpi)$ and $(\sin I \cos \Omega, \sin I \sin \Omega)$. The radii from the origin in these planes –the mean eccentricity and (sine of) inclination– can change significantly, as a result of secular perturbations, even in this linear theory: the changes have frequencies $g_0 - g_i$ for e and $s_0 - s_i$ for I, but the values remain small to moderate.

Yuasa (1973) first succeeded in computing the perturbations to this linear theory due to all the terms of degree 4, that is a "first order" perturbation theory for the perturbing function K_4; he also computed some contributions of order 2 in μ (see Section 3). However, his theory was never used to compute proper elements for a large catalogue of asteroids (a reduced version of the theory was used by Kozai, 1979). The formulae by Yuasa are very close to the limit of the computation which can be done by hand with a reasonable chance of being correct (actually, some mistakes and omissions were later corrected by Knežević, 1989, 1991). Once the formulae have been computed, and checked, they have been implemented as a computer code which can be tested in a very thorough way, as an example by comparison with numerical integrations (see Sections 4 and 5).

Before concluding the discussion of the "classical" results –by which I mean the ones already known 15–20 years ago– I would like to mention that Brouwer (1951) already used a theory containing some selected higher order and degree terms. When expanding K as in (2.3), the variables e', I' are assumed to be "of degree 1"; however, to obtain the solution they have to be replaced by some known function of time, and the latter is in turn the solution of the secular perturbation equations for the perturbing planets. If the secular perturbation theory for the planets is a linear one, then e', I' will be obtained as a sum of epicyclic motions, as in Laplace theory, with frequencies g_i, s_i. If, on the contrary, the theory for the planets contains some nonlinear terms, this can be represented as an additional Fourier term (epicycle) with a frequency combination of the fundamental ones. For the secular theory of the eccentricities of Jupiter and Saturn the most important nonlinear term is the one with frequency $2g_6 - g_5$; it is actually larger than some terms of the linear theory. If the variables e', ω' are replaced by their expression in a theory containing this term, then the solution of K_2 for e contains a forced Fourier term with frequency $g - 2g_6 + g_5$. In this way a solution to linear equations can contain a forced "nonlinear" term with more than two frequencies. It is typical of celestial mechanics, as opposed to some other branches of mathematics, to use theories which are not pushed to a given degree and order in a uniform way, but include selected terms of higher order and/or degree simply because they are large. The difficult job is to perform these choices aiming at the maximum accuracy, but in a consistent and fully documented way.

3. MEAN ELEMENTS AND SECULAR HAMILTONIAN

To present the contemporary theories of secular perturbations and proper elements we cannot avoid the use of some modern formalism and jargon. My favorite jargon is the theory of *Lie series*, introduced (at least for celestial mechanics problems) by Hori (1966). I shall not repeat the introduction of the method (see e.g. Milani 1988, 1990, Milani et al., 1987), but just take for granted that a canonical coordinate change can be defined by a *determining function* χ of the dynamical variables. In the new coordinate system the dynamics is still expressed by Hamilton equations, and the hamiltonian H in the old variables is replaced by the hamiltonian K in the new ones, which can be computed by the Lie series:

$$K = H + \{H, \chi\} + \frac{1}{2}\{\{H, \chi\}; \chi\} + \ldots, \qquad (3.1)$$

where $\{\ ,\ \}$ is the Poisson bracket.

If the hamiltonian is the sum of an integrable part plus a perturbation, as in (1.1), and also $\chi = \mu\chi_1 + \mu^2\chi_2 + \ldots$, then (3.1) can be rearranged as a power series in the small parameter μ:

$$
\begin{aligned}
K =& H_0 + \mu[-\overline{R} - \tilde{R} + \{H_0, \chi_1\}] + \\
& + \mu^2[-\{R, \chi_1\} + \frac{1}{2}\{\{H_0, \chi_1\}, \chi_1\} + \{H_0, \chi_2\}] + O(\mu^3).
\end{aligned}
\tag{3.2}
$$

Then the elimination of the short periodic terms to order one in μ can be performed by reducing what is in the first square bracket in (3.2) to the long periodic part $\overline{R}$, that is by solving the *homological equation*:

$$
\{H_0, \chi_1\} = \tilde{R}. \tag{3.3}
$$

To show that (3.3) can indeed be solved, we can expand both the known perturbing function R and the unknown first order determining function χ_1 in Fourier series, and then solve for the Fourier coefficients of χ_1 one by one (Milani, 1988,1990). This solution can be expressed in words as follows: the Fourier coefficient of χ_1 for a given argument, containing the combination $p\ell - q\ell'$, is the coefficient for the same argument in $-R$ divided by the *divisor* $\delta = pn - qn'$. This implies that the degree in e, I, e', I' of a term in χ_1 is the same of the corresponding term in R (that is, the D'Alembert rules **I, II, III** also apply to χ; however, **IV** is replaced by: the Fourier series only contains the sine terms). From the point of view of the celestial mechanician, who always looks at the order of magnitude of the terms, each term in χ_1 is of the same order of the corresponding term in R, unless the divisor δ is small, that is unless the ratio of the mean motions n/n' is close to the rational number q/p; for these *near resonant* terms, the factor $1/\delta$ can be more important than a factor e. Of course if some divisor is zero, that is if there is exact resonance, the removal of the corresponding term is not possible. The classical first order averaging of Section 1 is obtained by stopping here.

To push the approximation to second order in μ, we need to remove all the short periodic terms from the second square bracket of (3.2), by choosing a suitable χ_2. The essential point is just to show that this is possible; if we only want to compute to second order the secular hamiltonian K, an explicit computation of χ_2 is not needed. χ_2 would only be needed to explicitly compute the mean elements to second order, and this is not required by the levels of accuracy we are now aiming at (at least for the asteroids; see Knežević et al., 1988). To show that χ_2 exists we can write the homological equation of second order, which is of the same form as (3.3), namely $\{H_0, \chi_2\}$ =(known right hand side), then solve for the Fourier coefficients of χ_2 one by one, exactly in the same way used for χ_1; of course χ_2 can contain higher powers of the divisors $pn - qn'$. Therefore we can assume that the short periodic terms can be removed from K to order 2, and we need to understand which are the long periodic terms left, and whether they can be comparatively large even with the μ^2 factor.

Let us look at the first expression in the second order part of K: it is $-\{R, \chi_1\}$ (the other expression is very similar). The Poisson bracket of two terms, one with the cosine and the other with the sine, can result in a term with a long periodic argument only if the portion containing the fast variables of the two arguments is the same $p\ell - q\ell'$; this because it contains the difference of the arguments. Thus the Poisson bracket has as factors not only the powers of e, I, e', I' prescribed by the rule **III** for both functions R and χ_1, but also $1/\delta$; since $\delta = pn - qn'$ is a function of L, the Poisson bracket also has the derivative of $1/\delta$ with respect to L, and this contains $1/\delta^2$.

The above computation allows the following conclusion: there are long periodic terms in K with a factor μ^2/δ^2. Are these much smaller than the first order terms containing μ^1 ? The answer is that this is certainly not the case, if $\delta = O(\sqrt{\mu})$; that is, the procedure of elimination of the short periods not only fails for $\delta = 0$, but also for δ small enough, within a resonance band of width $O(\sqrt{\mu})$. What most concerns us is not what happens in the resonances, where the method has obviously to be changed, but what happens "near" the resonances, where the ratio μ/δ^2 is somewhat less than 1, but not very small. Then the elimination of the short periodic terms "order by order" is possible, but the contribution of the second order terms to the secular hamiltonian K can be substantial.

The explicit computation of these second order long periodic terms has been a challenge for a long time. Yuasa (1973) first begun the computation of second order terms, in particular the ones due to the $n - 2n'$ divisor and of degree 2 in e, I, e', I'. This computation was completed by Knežević (1989). Bretagnon (1974), Duriez (1979), Laskar (1986) and Milani et al. (1987) also succeeded in some second order computations of this class. However, the secular hamiltonian K including a consistent second order contribution was not used to compute asteroid proper elements until recently (Milani and Knežević, 1990, 1992, 1994; Lemaitre and Morbidelli, 1994). After this was done, it was possible to show that these second order terms were indeed necessary to obtain a secular perturbation theory applicable to the region of the asteroid main belt between $a \simeq 3\,AU$ and the $n - 2n'$ resonance with Jupiter ($a \simeq 3.28\,AU$).

The secular hamiltonian, when computed to second order, contains portions of different order (in μ) and degree (in e, I, e', I'; we are using here the Yuasa type expansion):

$$\begin{aligned} K =&\mu K_2^{(1)} + \mu K_4^{(1)} + \mu K_6^{(1)} + \ldots \\ &+ \mu^2 K_2^{(2)} + \mu^2 K_4^{(2)} + \mu^2 K_6^{(2)} + \ldots + O(\mu^3), \end{aligned} \tag{3.4}$$

where the subscripts indicate the degree, the superscripts the order in μ. The dots stand for higher degree terms; more of these can always be added, but it is important to realize that some truncation in powers of e, I, e', I' is necessary. Because of D'Alembert rule **III**, this truncation by degree ensures that the number of divisors appearing in K, and which can be small, is finite. Thus the secular hamiltonian we can define in a consistent way on some open set, and which we can practically compute, is defined by some truncation rule both in powers of the perturbing masses and in degree in the eccentricities and inclinations. Once this truncation is done, K needs to be rearranged into an integrable part and a perturbation, and this is done by degree, not by order, because after the removal of the short periods μ cannot be used as a perturbation parameter any more:

$$K = (\mu K_2^{(1)} + \mu^2 K_2^{(2)}) + (\mu K_4^{(1)} + \mu^2 K_4^{(2)}) + \ldots = K_2 + K_4 + \ldots . \tag{3.5}$$

After this reordering, the integrable hamiltonian K_2 contains both the first and the second order contributions. For the asteroids near (but by no means inside) the $n-2n'$ resonance with Jupiter, in the *Themis* region, the second order contribution to K_2 can be up to 50% of the first order one; thus the proper frequency g_0, computed with a *second order linear theory* such as the one by Milani and Knežević (1990), can be completely different from the one computed with the classical theories.

As for K_4, it also contains first and second order contributions. The second order contributions $K_4^{(2)}$ were first computed by Morbidelli and Henrard (1991), but only recently they have been incorporated into proper element computations (Milani and Knežević, 1994; Lemaitre and Morbidelli, 1994). Their effects can be important near the $n-3n'$ resonance with Jupiter ($a \simeq 2.50\,AU$). Computations equivalent to the second order contributions $K_4^{(2)}$ and $K_6^{(2)}$ where already done by Duriez (1979) and Laskar (1986) for the planetary case, but it is hard to compare the results because they did not use a canonical formalism; they found that these terms are very important in the computation of g_6, because of the $2n_5 - 5n_6$ near resonance.

The next step is to compute the perturbations resulting from K_4. This can be obtained by using again the Lie series formalism: we can look for a determining function $W = W_4 + W_6 + \ldots$ such that the transformed hamiltonian K^*, obtained by the Lie series:

$$K^* = K_2 + \left[K_4 + \{K_2, W_4\}\right] + \left[K_6 + \{K_4, W_4\} + \frac{1}{2}\{\{K_2, W_4\}, W_4\} + \{K_2, W_6\}\right] + \ldots \tag{3.6}$$

is in a simpler form. The problem is which simpler form should be aimed at in this context. Thinking in an abstract way, K_2 is integrable and we can use the angle–action variables of K_2, let us call them $\Omega_L, \omega_L, Z_L, G_L$; the subscript L and their technical name *proper linear elements* reminds us that the solution of the secular perturbation problem reduced to the linear portion of the equations is very simply expressed in these variables:

$$\begin{cases} Z_L = & const \\ G_L = & const \\ \Omega_L = & s_0 t + const \\ \omega_L = & (g_0 - s_0)t + const \end{cases} \tag{3.7}$$

where the frequencies g_0, s_0 are written in K_2 in this form:

$$K_2 = g_0 G_L + s_0(Z_L - G_L). \tag{3.8}$$

Then we can change coordinates to the proper linear elements, and define the simpler form to be the one without angles: as an example

$$K_4 = \overline{K_4}(Z_L, G_L) + \tilde{K_4}(\Omega_L, \omega_L, Z_L, G_L, t), \tag{3.9}$$

where the symbols overline and tilde indicate the integrable part (with the actions only) and the long periodic part (containing the slow angular variables); the meaning of these symbols is not the same we have used for the elimination of the short periods, here we are trying to eliminate all the angular variables. To understand why the $\tilde{K_4}$ part depends upon time we need to remember that the time dependent angles Ω', ω' also appear in K. A little more effort is needed to understand why $\partial\overline{K_4}/\partial t = 0$; this is because we assume that the elements of the perturbing planets are represented by some analytic theory, in turn the output of a secular perturbation theory (either analytic or numeric, this does not matter), in which there are as integration constants the *proper elements of the planets*. Only the proper eccentricities and inclinations of the planets, not the proper angles, occur in $\overline{K_4}$.

The above discussion is somewhat technical, but not difficult; however, when the change to proper linear elements has to be done, in a hamiltonian containing a large number of terms, it is in fact the most complicated step, because each term of the original hamiltonian generates by substitution a number of terms. This computation was first done by hand by Yuasa (1973); although it would be highly desirable to replace it by an entirely automatic symbolic manipulation, this has not been done yet. Once the secular hamiltonian K has been explicitly written in proper linear elements, the homological equation to remove all the angles from the degree 4 part of K is simply:

$$\{K_2, W_4\} = -\tilde{K_4}, \tag{3.10}$$

which can be considered as an equation between two Fourier series (in the slow angles only), and can be solved term by term exactly as we have done for (3.3); the Fourier coefficients of W_4 are obtained by dividing the corresponding coefficients of $-\tilde{K_4}$ by the *secular divisors*, combinations with integer coefficients of the slow frequencies g_0, g_i, s_0, s_i, and by replacing a cosine function with a sine. By the D'Alembert rules **II** and **III**, the allowed divisors are only the ones with either 2 or 4 frequencies: e.g. $g_0 - g_6, g_0 + s_0 - g_6 - s_6, g_0 - 2g_6 + g_5$, etc. Then, if we neglect the degree 6 terms (contained in the second square bracket in (3.6)) and the little dots standing for even higher degree terms, we can claim that the hamiltonian K^* is only a function of the action variables Z^*, G^* in the transformed coordinates, the *proper elements*, and the solution in this proper elements space is simply:

$$\begin{cases} Z^* = & const \\ G^* = & const \\ \Omega^* = & st + const \\ \omega^* = & (g-s)t + const \end{cases} \tag{3.11}$$

which looks deceivingly similar to (3.7), but is not the same. First, it is a solution to the hamiltonian problem K^*, which is, to degree 4 (and order 2 in μ), a solution to the original problem of the long term perturbations of an asteroid. Second, the frequencies g, s are:

$$s = \frac{\partial K^*}{\partial Z^*}(Z^*, G^*) \quad ; \quad g = \frac{\partial K^*}{\partial G^*}(Z^*, G^*) + s, \tag{3.12}$$

quite different from g_0, s_0 because they contain also the contribution from the degree four part $\overline{K_4}$. Anyway, the transformation from the osculating elements, to mean elements, to proper linear elements, to proper elements, can be explicitly computed (as a sequence of three transformations), and the final result includes the action like variables (L_M, G^*, Z^*), from which, by the usual formulae for Delaunay variables, we can derive the *proper semimajor axis, proper eccentricity and proper inclination*, which are approximate constants of the motion.

4. VARIABLE FREQUENCY PERTURBATION THEORY

The computation of proper elements, as discussed in the previous Section, is a computational *tour de force*, all the more remarkable in that it was completed before symbolic manipulation software powerful enough for these computations was available. When finally a computer program, written by Knežević, was available to perform the entire computation, it was tested by numerically integrating the orbits for a time span of the order of the long periods, and by checking the changes of the proper elements as a function of time. Since the main purpose of the whole exercise is to have quasi–integrals, the stability with time of the proper elements is the main criterion to assess their quality; actually we should never speak of accuracy of proper elements, since their values are arbitrarily defined, but only of stability. You can imagine the dismay, when the nonlinear proper elements computed with the full Yuasa theory, completed by Knežević, were found to be, in many cases, not significantly more stable than the classical proper linear elements.

The improvement of the proper elements by Milani and Knežević has been a long struggle, in which a large number of technical, and sometimes even exotic, corrections and additions have been applied; it is enough to say that the current version is numbered 6.8.5 (there have been more than 100 intermediate versions). Of course I am not going to discuss any of the technicalities, but I would only explain the main conceptual improvement which allowed to go from the very sophisticated, but practically not so successful, Yuasa type theory, to the new theory which has fully achieved, as we will see in Section 5, the design goals.

One possible way to explain the deceiving results of the Yuasa type theories is based on the divergence of the perturbative series. That such a divergence must necessarily occur was known to Poincarè (1892–1899); I have given a modernized version of the same argument in Milani (1988). Recently, we have been able to formally prove that divergence actually occurs in the very first step of the perturbative solution of the problem of secular perturbation of the asteroids (Milani and Làbianca, 1994). However, the habit of computing first and thinking to the convergence problems later is too strong for some people, therefore I have not been able to convince everybody. I would like to use here a completely different argument, which however leads to the same conclusion (Milani and Knežević, 1990).

The Yuasa theory provides a map from proper linear elements (which, in turn, can be computed from mean elements by a simple formula) to proper elements; it is defined by the determining function W_4 obtained as a solution of the homological equation (3.10). Let us look to this procedure as a way to compute the solution to the secular perturbation equations defined by the hamiltonian $K = K_2 + K_4$ (higher degree terms are neglected). Given the values of the proper elements Z_L, G_L, we solve the equations of motion with the hamiltonian $K^* = K_2 + \overline{K_4}$, in the space of the proper elements, by (3.11). Then we compute the map from proper to proper linear elements in such a way that the portion of the hamiltonian of degree 4 dependent upon the slow angles, namely $\tilde{K}_4$, is added back: that is, we need to use a determining function Y_4 such that:

$$K^* + \{K^*, Y_4\} + \ldots = K_2 + \overline{K_4} + \tilde{K}_4 + \ldots, \tag{4.1}$$

and this implies a new homological equation:

$$\{K^*, Y_4\} = \tilde{K}_4. \tag{4.2}$$

By comparing (3.10) and (4.2) we see that Y_4 is somewhat related to $-W_4$, but it is not the same. The main difference between the two Poisson brackets, with K_2 and with $K^* = K_2 + \overline{K_4}$, is that the former is the time derivative along the solution (3.7) of the linear secular equation, the latter is the time derivative along the solution (3.11) of the integrable dynamics of the space of proper elements. As a result, the Fourier coefficients of Y_4 contain as divisors the same combinations of either 2 or 4 frequencies, but with g, s instead of g_0, s_0. Thus the mapping from proper to mean elements defined by Y_4 is not the inverse of the mapping defined by W_4; the use of W_4 removes the long periodic perturbations, while the use of Y_4 adds them back, but the long periodic perturbations, as computed by the derivatives of the two determining functions, are not the same.

Since the higher degree terms are anyway truncated, the use of W_4 (that is, of the divisors containing g_0, s_0 as in Yuasa theory) and the use of Y_4 (that is, of the divisors with g, s) are both approximations. Now every approximation is a legitimate one, provided we know what is being approximated. To approximate something which does not exist is always a problem, while an approximation of a mathematically well defined entity is always possible, with more or less accuracy. Thus the crucial question is: which are the rigorously defined maps approximated, by truncation, by the Lie series defined by W_4 and, respectively, Y_4?

The map defined by Y_4 approximates the rigorously defined map whose existence is proven by the *Kolmogorov theorem* (Arnold, 1976, Appendix 8): if $K = K_2 + K_4 + \ldots$ is any hamiltonian (the higher degree terms do not matter, provided they form a convergent series), with K_2, the quadratic part, such that the equilibrium point $G_L = 0, Z_L = 0$ is stable in the linear approximation, then most (in the sense of measure theory) solutions (3.11) close enough to $G^* = 0, Z^* = 0$ are mapped by a smooth canonical map into solutions of K, which form *invariant tori* in the phase space. The values of G^*, Z^* which map into the invariant tori of K are the ones satisfying some *non resonance conditions* on the divisors formed with the frequencies g, s: the divisors not only cannot be zero, but must go to zero slowly enough as the degree grows. It can be shown, e.g. by using the proof of the Kolmogorov theorem given by Benettin et al. (1984), that Y_4 is the degree four portion of the determining function which realizes the map of the theorem. Thus the use of Y_4 only is a first approximation, which of course should be used only for low to moderate eccentricities and inclinations to be accurate.

On the contrary, W_4 is the first approximation to a map which does not exist. If there were a map, defined by a convergent series, obtained by the Lie series procedure whose first step gives W_4, it is easy to show that such a map would be a Taylor series in the eccentricities and inclinations. Then it should either converge on an open set, and the original dynamical system would be integrable, or nowhere; since the N–body problem is not integrable, the latter must be the case (Milani, 1988). The use of W_4 is a typical example of what Poincaré called *les méthodes anciens*: it can actually work in some very limited region (very low eccentricity and inclination), elsewhere it does not work any more, and we cannot understand why unless we take a new approach. Indeed the shape of the region where the old method is defined is a good indication that it is qualitatively wrong. The divisors appearing in W_4 contain g_0, s_0, which are functions of L_M, that is of the proper semimajor axes only; if a divisor such as $g_0 - g_6$ is zero for some a, W_4 is undefined for all values of eccentricities and inclinations. Thus *secular resonances* should occur along surfaces $a = const$, and this is well known not to be the case (Williams and Faulkner, 1981); on the contrary the divisors such as $g - g_6$ are zero along curved surfaces of the proper element space which project onto a wide range of values of the semimajor axis.

However, the map defined by Kolmogorov theorem goes the wrong way: given the values of g, s and of G^*, Z^* it is possible to compute Y_4, thus by the Lie series G_L, Z_L:

$$G_L = G^* + F_1(G^*, Z^*, \Omega^*, \omega^*, g, s), \tag{4.3}$$

and the same for the other elements. As a map from the proper elements to the proper linear (hence mean) ones, (4.3) is well defined wherever the divisors (computed with g, s) are nonzero, since g, s are functions of G^*, Z^* by (3.12). The inverse map is defined in an implicit way by (4.3) and the like, but no explicit analytic expression can be computed.

Milani and Knežević (1990) solved the problem of computing the inverse map of (4.3) by using a fixed point iteration (the people accustomed to teaching calculus, like myself, will remember that the fixed point iteration is the standard proof of the implicit function theorem, as it appears in most calculus textbooks). As a matter of principle, any other successive approximation method to obtain the inverse map could be used, e.g. Newton's method; the fixed point method is just the simplest. It is obtained by isolating G^* in (4.3):

$$G^* = G_L - F_1(G^*, Z^*, \Omega^*, \omega^*, g, s) \tag{4.4}$$

and the same for the other starred variables. We begin with some first approximation, e.g. $G^* = G_L, Z^* = Z_L$, then we compute a first approximation g, s from (3.12) and the long periodic perturbations F_1 from Y_4: this gives by (4.4) a second approximation for G^*, etc.; then g, s are recomputed from (3.12) with the new values of G^*, etc.; this gives a new computation of G^* by (4.4), and so on until either convergence or divergence occurs. If the iteration converges, it converges to the values of G^* etc. corresponding to the values of G_L etc. by the inverse map.

This formalism is called *variable frequency perturbation theory* because g, s change along the way, and are anyway changed from the very beginning with respect to the linear approximations g_0, s_0. On the contrary the initial conditions are fixed, that is we are computing the proper elements for a given asteroid, not for a variable point in the phase space (as it would be the case if the frequencies were held fixed). The difference between the Milani and Knežević (M&K) theory and the Yuasa theory is therefore conceptual, while the most cumbersome computations (such as (3.9)) are the same. Many formulae of M&K can be obtained from the Yuasa ones by replacing g_0, s_0 with g, s in the divisors, but they have to be read the other way round.

All the discussion from (3.4) onward was based upon the Yuasa type expansion (2.3) in powers of e, I, e', I'. However, the same ideas could be applied to the Kozai type expansion (2.2) in powers of e', I' only. Indeed Lemaitre and Morbidelli (1994) have applied the variable frequency formalism to a (second order) secular hamiltonian obtained by a semianalytic method, to compute a perturbation theory within the Kozai scheme. Although their theory has not yet been tested as thoroughly as the M&K one, the L&M proper elements have proven to be stable enough at very high eccentricities and inclinations to allow the identification of asteroid families, such as the one of Pallas. Thus the use of modern theories, extended to second order in the perturbing masses and using the variable frequency formalism, is not limited to the low/moderate eccentricities and inclinations. Once the overall setting of the problem has been understood, the computational effort of an advanced proper element theory is huge, but it can be done if needed.

5. LONG TERM STABILITY OF THE PROPER ELEMENTS

The reason why we rate the contemporary proper element theories a success is not just the improved understanding of the theoretical background, as discussed in the previous Section. In celestial mechanics, the success of a theory has to be measured by the accuracy and reliability with which the theory models the motion of real celestial

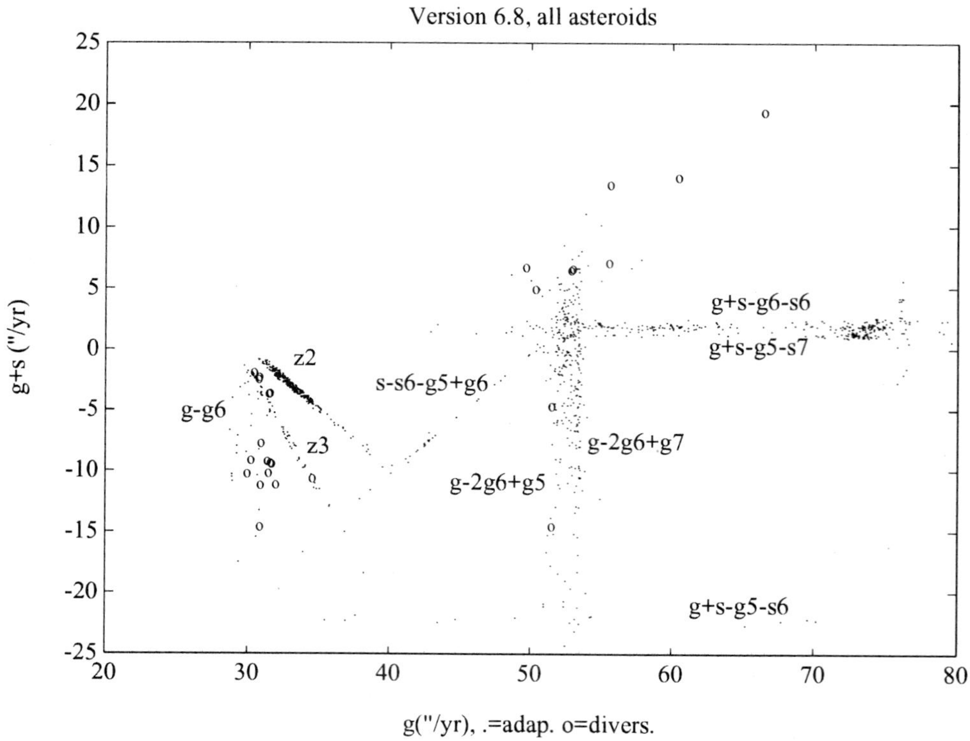

Figure 2. The asteroids for which the iterative procedure of the Milani and Knežević proper elements does not converge are plotted in the frequency plane g (proper perihelion frequency) vs. $g + s$ (s is the proper node frequency). The combination $g + s$ has been chosen because there are important secular resonances containing this combination; they appear as horizontal lines. The circles indicate the cases in which a nonlinear secular resonance responsible for the divergence has not been identified by the software; they are mostly due either to the $3 : 1$ mean motion resonance, or to the $g - g_6$ linear secular resonance.

bodies. Reliability does not necessarily mean that all the orbits can be handled with one and the same procedure; an important property of a theory is to be able to say no, this case is different, and cannot be computed like the others (at least, not with the same accuracy). The recent theories of proper elements do have this capability: proper elements are stable for several million years in most cases, and the cases in which the accuracy is degraded are automatically flagged, typically with a code indicating the resonance responsible for the problem.

To understand how this is done, we need to consider that, for proper elements computed by a variable frequency theory, the algorithm includes an iteration loop. The procedure can be terminated either when the variable frequencies g, s change, from one iteration to the next one, by less than a given, small amount (e.g. $0.001\ arcsec/yr$), or when the change in the last step is larger than in the previous one. In the latter case, it is useless to go on with the fixed point iteration, because the following steps would either experience an undamped oscillation or explode, diverging to absurd values. Lower accuracy proper elements can be obtained by the values at the iteration preceding the increase, but in such a case there is some more information to be extracted. The standard fixed point theorem states that convergence must occur, provided (a) the initial point is not too far, and (b) the functions like F_1 appearing in (4.4) are smooth. Inspection of the equations like (4.4) shows that the functions like F_1 are smooth everywhere in the proper element space with the only exception of the *secular resonance surfaces*, where some combination of the fundamental frequencies g, s, g_i, s_i is exactly zero: since they contain the divisors, they have polar singularities where a zero divide occurs. In a theory complete to degree 4, by D'Alembert rule **III** there are not only the secular resonances already occurring in the linear theory (with two frequencies, e.g. $g - g_6$), but also the *nonlinear* ones with four frequencies (e.g. $g + s - g_6 - s_6$). With initial conditions $G^* = G_L$ etc. the hypothesis (a) is satisfied (at least for low to moderate e, I), thus divergence indicates resonance.

Figure 2 has been prepared by using the output of the computation of proper elements for a large catalogue of asteroid orbits (more than 12,500) with the most recent version of the M&K algorithm (Milani and Knežević, 1994). All the points plotted correspond to one of the $\simeq 1,000$ cases in which the iterative procedure did not achieve a satisfactory convergence; in the frequency plane, these degraded accuracy cases are clearly aligned along lines of the form $hg + ks = const$ with h, k small integers, that is along the low degree secular resonance surfaces. The secular resonance $g - 2g_6 + g_5$ is especially prominent, also because of the strong "nonlinear forced term" discussed in Section 2. For all the cases marked with a point (97% of the divergent cases) the resonance responsible for the divergence, that is the divisor too close to zero, has been identified by a fully automated procedure; with human intelligence helping artificial stupidity, it is not difficult to explain the few cases not flagged properly by the software. The Lemaitre and Morbidelli proper elements have similar properties. Thus we can claim that the recent proper element theories are very reliable, the divergent cases being used to describe the so called "Arnold web" of resonances; for the "core belt" where most asteroids are located (at low to moderate e, I and between the $g - g_6$ and the 2 : 1 resonance), the nonlinear secular resonances which are "important" (that is, which can change the proper elements for more than 0.1% of the asteroids by a significant amount) are only 9, all visible in Figure 2:

$$\begin{aligned} g &= 2g_6 - g_5 & g &= 2g_6 - g_7 \\ g &= 3g_6 - 2g_5 & s &= s_6 + g_5 - g_6 \\ 2g + s &= 2g_6 + s_6 & 3g + s &= 3g_6 + s_6 \\ g + s &= g_6 + s_6 & g + s &= g_5 + s_7 \\ & & g + s = g_5 + s_6 & \end{aligned} \tag{5.1}$$

A better understanding of the geometry of resonances in the phase space can be obtained by directly plotting the surface $divisor = 0$ in the proper element space (Knežević et al., 1991); this is an explicit computation, without iterations, by (3.12). Figure 3 is a section of the 3–D space of proper elements (with constant proper e) showing the lines $divisor = \pm 0.5\, arcsec/yr$ for six nonlinear secular resonances; $g - g_6 = -2, 0, 2\, arcsec/yr$ are also plotted. To understand the relevance of these resonances for the distribution of asteroids, we have superimposed the points representing the proper elements of the known asteroids with proper e close to the value of the section. The clusters of asteroids in the proper element space are the *asteroid families*, which can be interpreted as the results of catastrophic disruptions of a parent asteroid by collisions (of course such an interpretation requires some care;

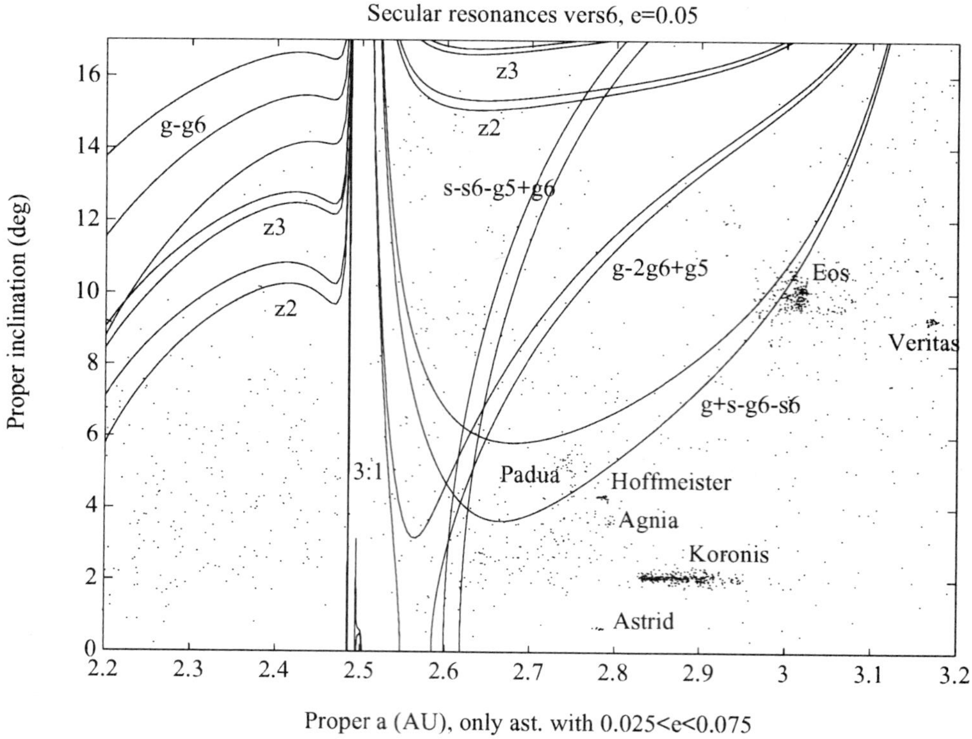

Figure 3. The secular resonances occurring on the section "proper $e = 0.1$" are superimposed to the proper elements a and $\sin I$ of the asteroids with proper e between 0.075 and 0.125. The names label asteroid families.

see Zappalà et al., 1990, 1994; Milani et al., 1992). The most visible families are labelled with their name (by convention, the name of the lowest numbered asteroid in the family). A conclusion we can draw from Figure 3 is that the nonlinear secular resonances do not result into large gaps in the distribution of the asteroids, as it is the case for both the main mean motion resonances and the $g - g_6$ linear secular one; on the contrary, there are families either crossed by (e.g. Eos) or entirely inside (e.g. Padua) the resonance slices. From this in turn we should deduce that the nonlinear secular resonances do not result in large scale instabilities. On the contrary very large instabilities are produced by the linear secular resonances and by the low degree mean motion resonances; they can even result in the transport of meteorites to the Earth (Farinella et al., 1993). However, some caution is required, since we do not know the age of most asteroid families (see Milani and Farinella, 1994).

In the cases in which the proper element computation is convergent, the stability of the proper elements has been extensively tested by piping the output of an accurate numerical integration into the computation of (mean, then) proper elements (Milani and Knežević, 1994; Knežević et al., 1994). These tests have been performed over time spans of $\simeq 5\,Myr$, and have shown that the changes in both the M&K proper elements $e, \sin I$ have typically a RMS ≤ 0.0015. Some increase in this instability only occurs near the boundaries of the "core belt", e.g. for a comparatively high inclination ($\simeq 15°$), for g close to g_6 ($g \leq 33\,arcsec/yr$), and near the $2:1$ mean motion resonance, and of course for the non convergent cases. This allows to identify in a reliable way asteroid families in almost the entire core belt, with the only problem area in the Flora region close to $g = g_6$, resulting from the large number of cases degraded by the $z_2 = 2(g - g_6) - (s - s_6)$ and $z_3 = 3(g - g_6) + (s - s_6)$ resonances (see the dense cluster of non convergent cases on the left in Figure 2). The z_2 and z_3 resonances do not appear in a Yuasa type theory; by rule **III** they occur only in the portion of degree 6 (respectively 8) of the hamiltonian (3.6); however their effect is amplified by the second order effect of the nearby $g - g_6$ resonance, and it has at least partially been accounted for in the latest version of the M&K proper elements. For higher inclinations, the L&M proper elements should be used, say for $I \geq 17°$; they cannot be as accurate as the M&K proper elements at low inclinations, but they can be consistently defined for arbitrarily high inclination, even in the region where ω librates. On the contrary, the difficulties in the computation of the proper elements near the 2 : 1 resonance are of a more fundamental nature, and cannot be solved by either adding some new terms or by changing the expansion procedure (see Sections 6 and 7).

The proper elements computed with the recent theories are all available on the `Internet` computer network, by `anonymous ftp` to `gauss.dm.unipi.it`, directory `pub/propel`. The software for the M&K theory is also public domain, together with software which can be used to perform numerical integration tests such as the ones quoted above (Milani et al., 1994a).

6. UNSTABLE CHAOS AND STABLE CHAOS

The successful computation of proper elements, by means of a variable frequency perturbation theory and for most asteroids, does not imply that the problem of the long term dynamical evolution of the asteroid orbits has been reduced to a trivial one. Proper elements are not integrals of the motion; they change very slowly, but over time scales much longer than those covered by the current numerical integration tests there is no guarantee that their values would not change significantly. Among the orbits which have been computed for the stability tests of the proper elements there are many examples of these long term instabilities; at least three very different phenomena can be observed.

The first kind of long term instability is the result of the presence of one of the secular resonances listed in (5.1). To be able to understand what happens, we use a theory "adapted" to one particular resonance, that is we remove from the theory -from the equations such as (4:3)- the near resonant term (the one with the small divisor); then the iterative procedure computes "semiproper" elements, belonging to a phase space in which the hamiltonian is $K^* + A\cos(\sigma) + \ldots$, with the critical argument σ of the resonance as the only angular variable. The behaviour of the orbit can still be ordered, e.g. by exhibiting a regular libration of σ; sometimes, however, the libration is not regular. In Figure 4 (from Milani and Knežević 1992) we show, among others, the orbit of the asteroid *221 Eos*, which is locked in the resonance with $g + s - g_6 - s_6 \simeq 0$. It is not possible, from this experiment over $\simeq 5\,Myr$, to decide if this orbit would remain locked in deep resonance (libration of σ) for a time comparable to the age of the solar system. However, the presence of a very large asteroid family around *221 Eos* suggests that this asteroid has been in the same region of the phase space for a very long time (at least several $10^8\,yr$).

The second kind of instability occurs when two resonances cross. The "adapted"

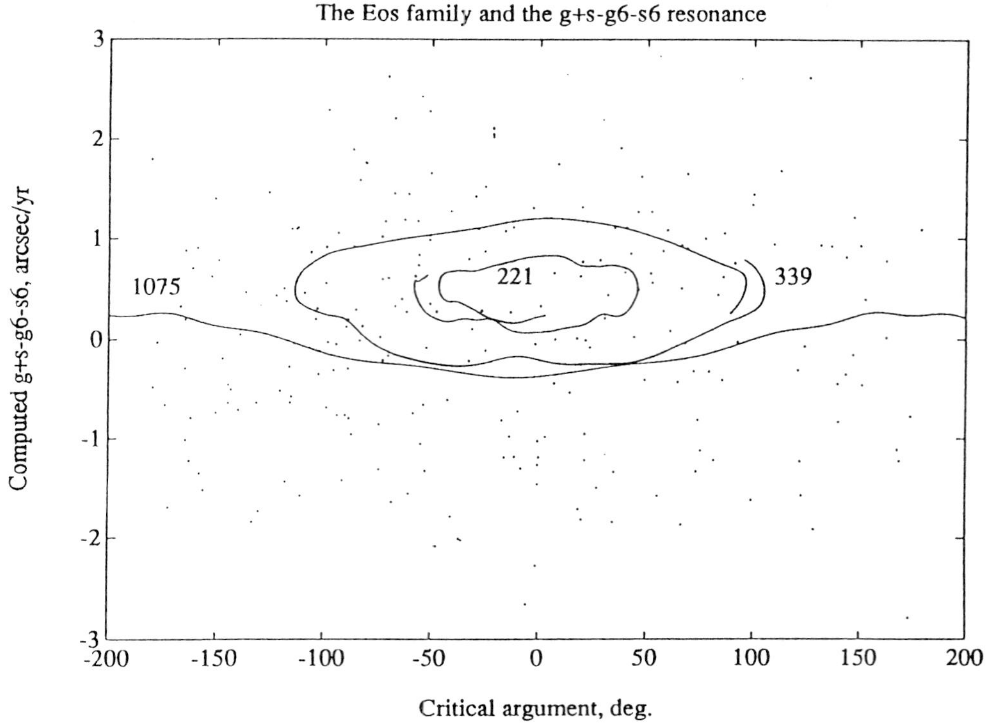

Figure 4. The orbits of the asteroids *221 Eos, 339 Dora* and *1075 Helina* for a time span of $\simeq 5\,Myr$ are plotted in the phase space of the $g+s-g_6-s_6$ resonance, that is critical argument σ vs. the frequency of σ as computed analytically. *221* and *339* are in libration, but the libration does not appear to be periodic. The dots represent the current values of the proper elements of other members of the Eos family.

theory in this case does not remove the terms with two critical angles; since the hamiltonian in the semiproper element space has, in this case, two angular variables, it is not integrable and very irregular behaviour can occur. As an example, the resonances $g-2g_6+g_5$ and $s-s_6-g_5+g_6$ cross in the very middle of the core asteroid belt (see Figure 3); the test case *214 Aschera*, discussed in Milani and Knežević (1994), has the two critical arguments coupled in such a way that one circulates when the other librates, and viceversa; at each exchange, the eccentricity and the inclination change in an unpredictable way (Figure 5). In such a case, we cannot exclude that the eccentricity could grow to planet crossing values, over a time scale of $10^8 \div 10^9\ yr$; on the other hand, we cannot state that such an increase must necessarily occur, because we do not know the size and shape of the chaotic region in which the orbit evolves. This example anyway shows that it is possible to have a main belt asteroid –by no means on a meteorite-like orbit– with a highly unstable orbit. We can only conjecture that such phenomena, being produced by the presence of two simultaneous resonances, can only occur in very limited portions of the phase space.

Having used the word, we need to give some definition of *chaos*. In the minds of many people, not only among the general public but also among qualified scientists, the notion of chaos is associated to the notion of instability. However, when requested for a rigorous mathematical definition of chaos, it is very difficult to give a definition

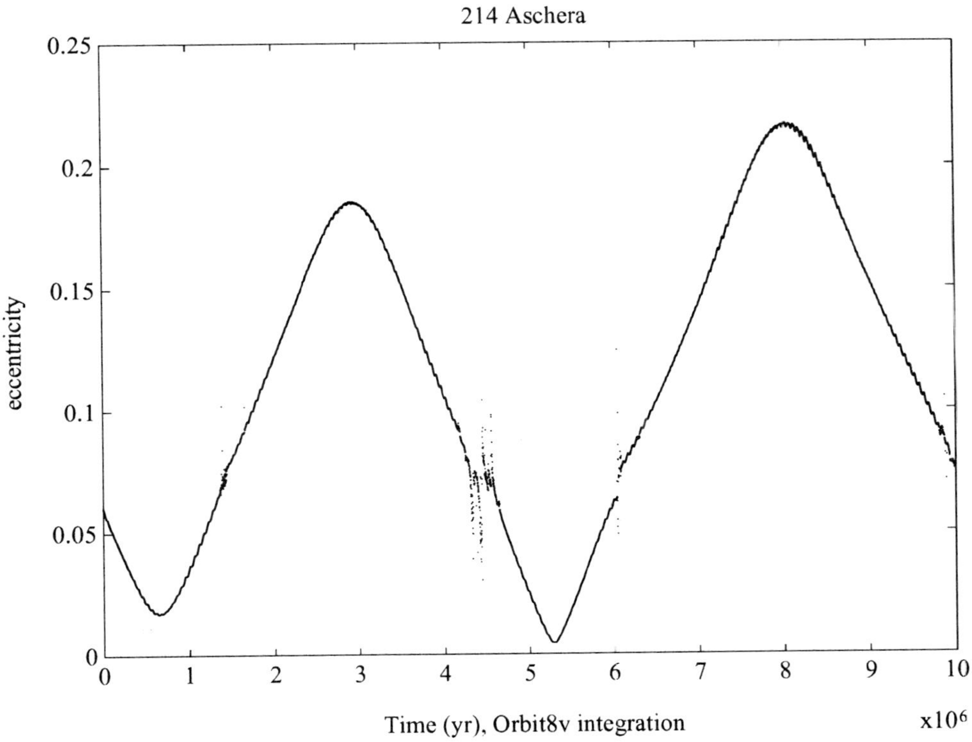

Figure 5. The semiproper eccentricity of *214 Aschera* over 10 Myr has been computed from the output of a numerical integration, by means of a theory "adapted" to both the $g - 2g_6 + g_5$ and the $s - s_6 - g_5 + g_6$ secular resonances. For most of the time the critical argument of the former resonance is in libration, and the eccentricity undergoes a large amplitude oscillation. Once in a while, at intervals of several Myr, the other resonance enters into a temporary libration, and the eccentricity undergoes irregular jumps; at these times, the inclination also changes in a nonperiodic way.

different from the one based upon the *Lyapounov characteristic exponents (LCE)*. It is possible to obtain an indication on the value of the maximum Lyapounov exponent by monitoring the function

$$\gamma(t) = \log_e \frac{|v(t)|}{|v_0|} \tag{6.1}$$

where $v(t)$ is the solution of the equations of relative motion, linearised around the solution under study, with initial conditions $v(0) = v_0$ chosen at random. Figure 6 shows such a function for the orbit of the asteroid *10 Hygiea*. The maximum Lyapounov exponent can be shown to be: $\lim_{t\to+\infty} \gamma(t)/t$ (with probability 1 over the sample space of the random initial displacements). Any calculation being finite, only an estimate can be obtained by means of a linear fit to $\gamma(t)$ for the available time span. For *10 Hygiea*, the best fit over a 5 Myr integration is with a slope of $6.3 \times 10^{-5}\, yr^{-1}$, i.e. the Lyapounov time T_L (the e-folding time for divergence of nearby orbits) is about $16,000\, yr$. Thus the orbit of an asteroid such as *10 Hygiea* cannot be predicted over a time span longer than $\simeq 20T_L$, because any small uncertainty in the initial conditions results into a very large error in the prediction, amplified by as much as a factor exp(20) (for a review on the Lyapounov exponents, see Froeschlé, 1984).

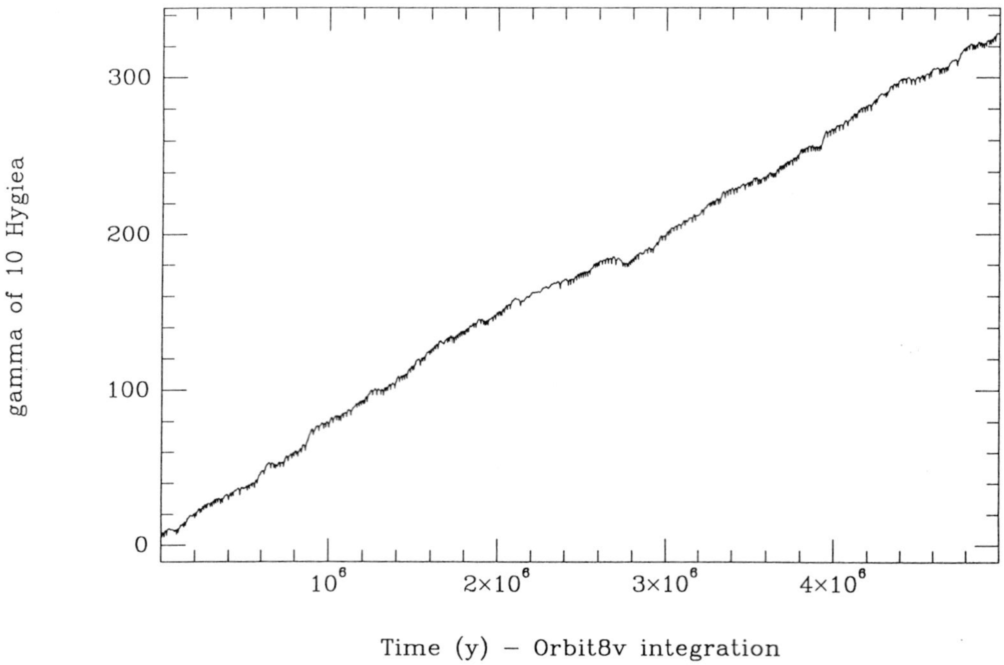

Figure 6. The divergence factor $\gamma(t)$ (see equation (6.1)) of *10 Hygiea* from the output of a numerical integration for $5\,Myr$. From this plot we can estimate that the maximum LCE is $\simeq 1/16,000\,yr$. The integrations used to generate the data of Figures 4–10 have been performed with the public domain software system *Orbit8v*.

Many non rigorous statements are often quoted about the properties of chaotic orbits, defined as the ones with maximum Lyapounov exponent larger than zero. However, the only rigorous result applicable to the N-body problem says that a conditionally periodic orbit, of the kind provided by the Kolmogorov theorem, has zero Lyapounov exponents. Thus we can assume that an orbit with a positive LCE is not conditionally periodic (although the contrary is not impossible, but only very unlikely). These orbits are unstable, but there is no rigorous result which allows to estimate the size of the instability; there is no rigorously proven relationship between the Lyapounov time T_L and the time in which an orbital change of a given size would occur. There is some empirical evidence that, in the same region of the phase space, an orbit with shorter T_L is more likely to undergo a large scale instability than one with longer T_L; but the T_L found in one region cannot be compared to the T_L found in another region, because different resonances contain chaotic orbits with different ranges of values for T_L. As an example, $T_L > 1\,Myr$ for *214 Aschera*, $T_L \simeq 70,000\,yr$ for *221 Eos*, and $T_L \simeq 16,000\,yr$ for *10 Hygiea*; the latter is the least unstable of the three (compare Figure 8 with Figure 5). The large scale instability of *214 Aschera* is the result of the action of two interacting secular resonances; the Lyapounov time is therefore of the order of the libration period of the critical arguments involved. For *221 Eos*, T_L is much less than the libration period of the critical argument of the $g + s - g_6 - s_6$ secular resonance, therefore the irregular oscillation of this crit-

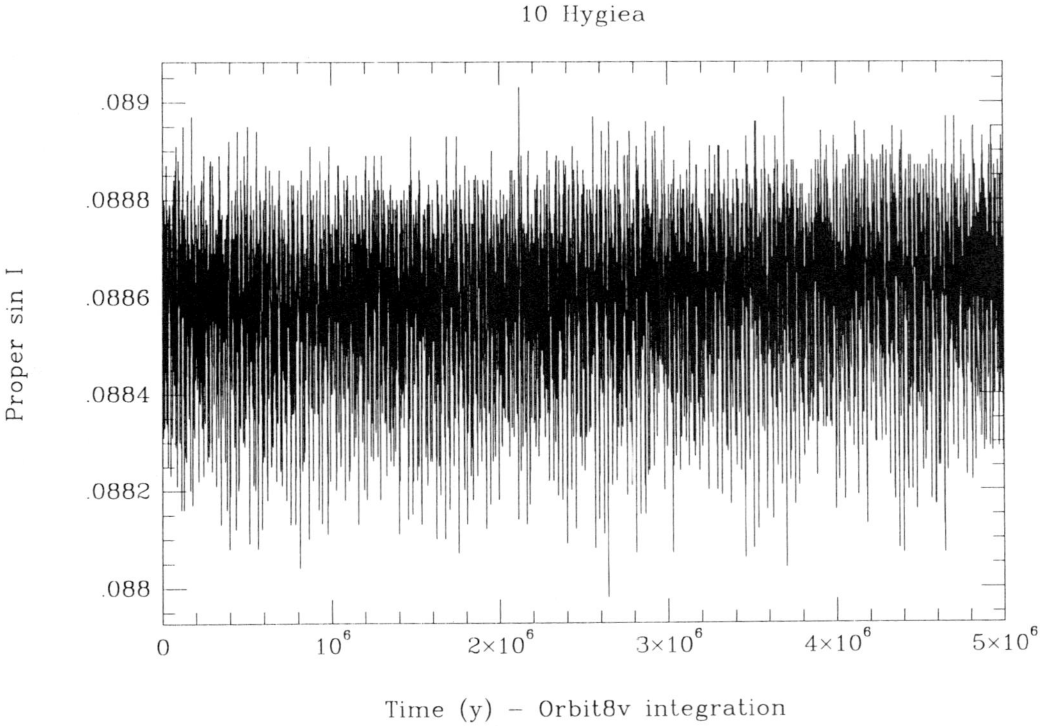

Figure 7. Proper $\sin I$ of *10 Hygiea*, computed (with the algorithm M&K vers. 6.8.4) from the output of a $5\,Myr$ numerical integration. The oscillations result from the truncations in the proper element theory, but the long term trend indicates chaotic diffusion.

ical argument is the effect, not the cause, of the instability occurring in some other argument.

I have chosen the case of *10 Hygiea*, among the ≥ 60 cases we know (Milani et al., 1994b), as a very striking example of the third kind of irregular behaviour, the so called *stable chaos* (Milani and Nobili, 1992). *10 Hygiea* is by no means a small object: with a diameter of more than $400\,Km$, it is the 4th largest asteroid in our solar system; this means it cannot be interpreted either as a comet, temporarily intruding in the main belt, or as a fragment injected in an unstable orbit by a collisional event. The very fact that *10 Hygiea* is in its present orbit proves that this orbit is stable over a time span longer than $4 \times 10^9\,yr$, at least in the weak sense that the orbit has always been a main belt asteroid orbit. Moreover, computations of the orbit for several millions of years show that the proper elements, as computed by the M&K algorithm, change only by small amounts, not significantly larger than those of ostensibly regular orbits. Over $5\,Myr$ the RMS of the changes is about $5 \times 10^{-4}\,AU$ for proper a, 10^{-3} for proper e, 2.4×10^{-4} for proper $\sin I$ (Figures 7, 8 and 9); even over a time span 15 times longer, the results do not change in a substantial way. Nevertheless, the long term stability (over a time span of the order of the age of the solar system) does depend upon the slow "diffusion" of the proper elements, which occurs over time scales enormously longer than T_L: as an example, by using a digital low frequency pass filter on the time series of the proper $\sin I$ (Figure 7), we can detect a long term trend with a slope of $\simeq 10^{-5}/Myr$.

It is common knowledge that chaos arises from resonances; therefore, when presented with an orbit with a large LCE such as this one, we would like to find one (or more) resonance "responsible" for the chaotic behaviour. It is clear that in a case such as *10 Hygiea* a secular resonance cannot be involved: first because the Lyapounov time is far too short with respect to the libration/circulation times of the critical secular arguments, second because when a secular resonance is involved, the instability of either proper e or proper $\sin I$ would be larger. Then we are looking for a mean motion resonance, that is a critical argument of the form $p\ell - q\ell' + \sigma$ (with σ a combination of slow angles allowed by D'Alembert rules) such that it can be locked in libration, at least part of the time. The classical technique for resonance hunting involves the expansion of the ratio n'/n in a continuous fraction; the truncations of the continuous fraction provide an approximation of n'/n by a sequence of rationals p/q rapidly convergent to n'/n. Let us try with *10 Hygiea*: the average value of $d\ell/dt$ over the $5\,Myr$ time span of the numerical integration used to prepare Figures 6–10 is $x = 0.17940\,rev/yr$ while the average rate of $\lambda - \lambda'$ is $y = 0.09520\,rev/yr$ (where $\lambda = \ell + \omega + \Omega$, and similarly for λ'; the use of this particular combinations of arguments ensures than any integer combination of x, y is the frequency of a D'Alembert argument).

$$[x/y] = 2\ ; \quad y_1 = 2y - x = .01100 \quad rate\ of\ \lambda - 2\lambda' + \varpi\ (period \simeq 90\,yr),$$

where $[w]$ is the integer nearest to w.

$$[y/y_1] = 9\ ; \quad y_2 = 9y_1 - y = .00380 \quad rate\ of\ 8\lambda - 17\lambda' + 9\varpi\ (period \simeq 260\,yr),$$

and we see that no low degree resonance is nearby; if we go on:

$$[y_1/y_2] = 3\ ; \quad y_3 = 3y_2 - y_1 = .00040 \quad rate\ of\ 23\lambda - 49\lambda' + 26\varpi\ (period \simeq 2,500\,yr).$$

The crucial point about this procedure is to decide when to stop. When a combination frequency is small to the point of becoming comparable to the secular frequencies, it is possible to use a different combination σ of the slow angles $\omega, \Omega, \omega', \Omega'$ together with the same combination of λ, λ' to obtain a smaller divisor: as an example, the average rate of $\varpi - \varpi'$ is $z = g - g_5 = 9.6 \times 10^{-5}\,rev/yr$:

$$[y_3/z] = 4\ ; \quad y_3 - 4z = 1.6 \times 10^{-5} \quad rate\ of\ 23\lambda - 49\lambda' + 22\varpi + 4\varpi'. \tag{6.2}$$

The choice done by stopping the continuous fraction expansion of the ratio x/y and using instead the secular frequencies needs to be justified. When there are more than two frequencies, the continuous fraction expansion has to be replaced by a more flexible procedure: at each step, the largest frequency has to be replaced by the remainder of a continuous fraction expansion of two of the frequencies, not necessarily the same at each step (this generalisation was discussed by Poincaré, 1884). At some stage in the continuous fraction expansion, it will always happen that a combination $px - qy$ will be smaller than the possible contribution of the secular frequencies. We can estimate the degree at which this happens, by taking into account that for every number x/y the continuous fraction expansion is convergent at least quadratically, that is:

$$|px - qy| < \frac{1}{q}|y| \tag{6.3}$$

The secular frequencies such as $z = g - g_5$ can be estimated by $g \simeq c\mu n$, where c is of the order of 1. Then a multiple rz can be used to generate a small divisor if

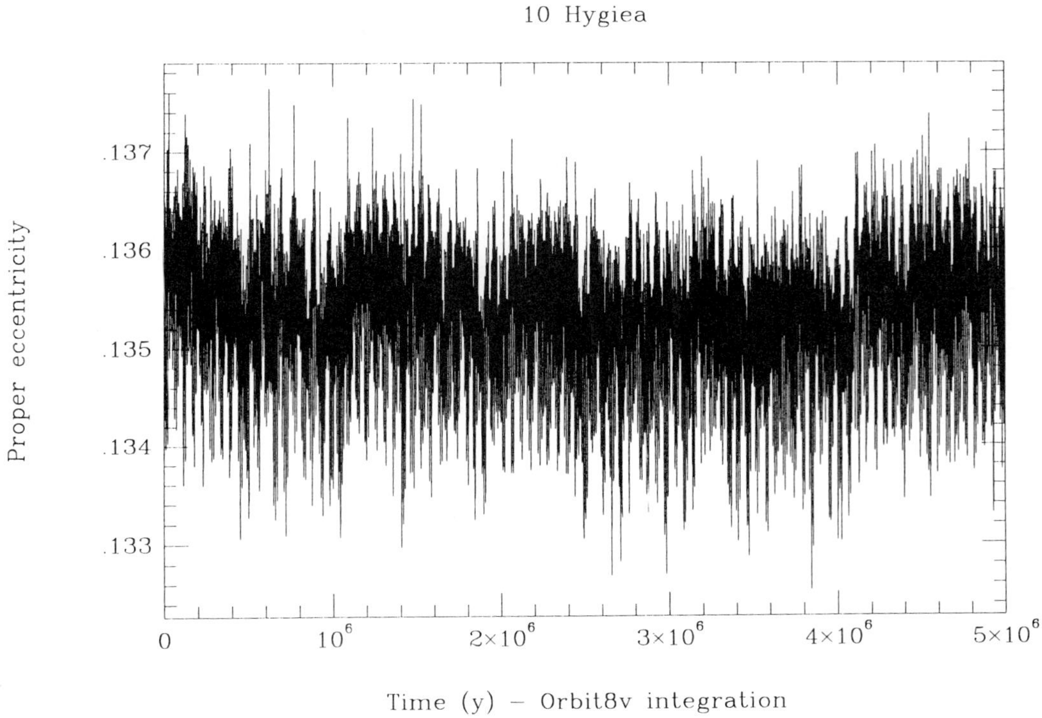

Figure 8. Proper eccentricity of *10 Hygiea* over $5\,Myr$. The oscillations due to the truncations of the theory and to the nonperiodic effects of chaotic random walk both contribute to the instability, which is anyway not much larger than in other ostensibly regular cases.

$r|z| \simeq |px - qy|$; in the region of *10 Hygiea*, where $n \simeq 2n'$, this is possible only if $qr > 1/2c\mu$; that is, if $r = p \simeq q/2$:

$$q \simeq \frac{1}{\sqrt{c\mu}} \tag{6.4}$$

With μ, the perturbation parameter (computed as in Roy, 1982) $\simeq 2.1 \times 10^{-4}$ in the *Hygiea* region, we obtain $q \simeq 70\sqrt{1/c}$. Since $c > 2$ in this region (also because of the second order effects discussed in Section 3), the small divisors like (6.2) are by no means exceptional, but their order is close to the typical value at which a resonance is generically found in this region.

The conclusion is that *10 Hygiea* is not locked into an especially remarkable resonance; it could be locked in a resonance of such a high degree that its occurrence is by no means a special property of this orbit, but rather a generic property of most orbits in this region of the phase space (that is, with the exception of a few orbits locked into resonances of a much lower degree).

What should we do with an example such as *10 Hygiea*? I think that different attitudes are acceptable, the only one which is antiscientific is to hide the head in the sand and pretend that such cases do not exist. It cannot even be claimed that such examples are rare; there are on the contrary some portions of the phase space in which most orbits have properties very similar to these: as an example, the region of the asteroid belt with $3 \le a \le 3.2\,AU$. If this is the case in which the presently available theories do not work, then this is the case to be studied. We would like to be able to *predict* for which orbits stable chaos will occur (instead of being hit by the result of each numerical test as an unmitigated surprise). We need to *explain* the dynamical mechanism allowing for a strongly chaotic, but localized behaviour, that is to isolate a subsystem, with a smaller number of degrees of freedom and a small number of terms in the hamiltonian, which exhibits the essential properties of such a behaviour. The planetary science applications (e.g., ages of the asteroid families) require that we be able to *model*, at least in a statistical sense, the "diffusion" process occurring over time scales much longer than T_L but still comparable to the asteroid family ages. All this will never happen, unless somebody tries.

7. TOWARDS A THEORY OF STABLE CHAOS

The goal of this section is to provide the framework for a theory of stable chaos. It is not possible to give details and to provide proofs of the statements of such a theory, and this not only for reasons of space, but also because not all this has been worked out yet. Nevertheless I would like to present here a first attempt of a new theory, which has the goal of explaining (possibly also predicting and modelling) stable chaos.

I shall begin with a definition: *stable chaos* is the occurrence of orbits such that (A) the maximum LCE is positive and large (that is, T_L is much less than the age of the solar system, say $T_L \le 100,000\,yr$); (B) the proper elements as computed by the appropriate algorithm (e.g. M&K for low inclination/eccentricity) are not significantly less stable than in the ostensibly regular cases, over a time span much longer than T_L (say RMS of changes over $5\,Myr$ should be $\le 3 \times 10^{-3}$ for proper $e, \sin I$); (C) all the features of chaotic motions are visible in the proper elements, such as intermittency (see Figures 8 and 9) and random walk (see Figures 7 and 10), but not necessarily in all the elements with the same time scales (as an example, Figure 9 displays little or no random walk behaviour); (D) no remarkable mean motion resonance is involved, that is, the only critical arguments which could be in libration are of very high degree, to the point of satisfying an equation like (6.4). Let me stress once more that in the real asteroid belt there is a very large number of examples of such orbits.

The analytical setting to explain stable chaos could be as follows. Proper elements are defined as constants of the motion for an integrable problem which is close to the real problem of asteroid dynamics. In both stages of the computation, osculating to mean elements and mean elements to proper elements, portions of both the transformation equations and the transformed hamiltonian are discarded, with the justification that they are of higher degree and/or order. We shall suppose that the hamiltonian H of the full problem can be described, in some system of elements (e.g. the mean elements) as a sum:

$$H = K - S, \tag{7.1}$$

where K is the truncated portion, whose integrals are the proper elements as computed, and S accounts for all the approximations. This is somewhat a simplification, since neglecting higher order terms in a Lie series results in a transformation which is not exactly canonical; a more comprehensive approach is possible, but let us keep things simple to start with. Then S contains the portion of the short periodic perturbations which are not removed during the computations of the mean elements (e.g. the Fourier terms with arguments containing $p\ell - p'\ell'$ with $|p - p'| \ge h$; by the

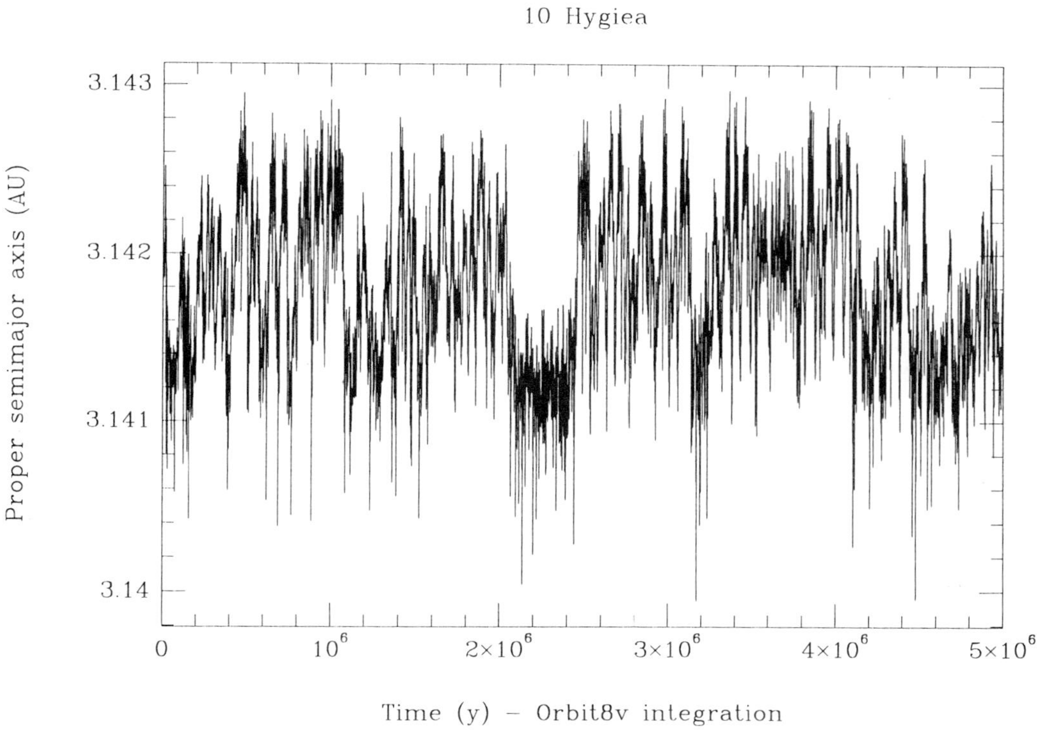

Figure 9. Proper semimajor axis of *10 Hygiea*. During a time span of 5 Myr the orbit visits in an irregular way several resonances belonging to the 23 : 49 multiplet, which occur at intervals in a of the order of 10^{-4} AU.

D'Alembert rules, they are at least of degree h in the eccentricities and inclinations); it also contains the portions of the secular hamiltonian of high degree, the portions of the Lie series (3.2) of order ≥ 3 in μ, and so on. The proper element algorithms provide a map from the mean elements to the proper elements G^*, Z^*, and transforms K into the integrable K^*; this map is defined not everywhere in the phase space of mean elements, but only "far enough" from the secular resonances. Indeed, even the truncated K is not integrable, but it has ordered and chaotic solutions. Typically, values of the proper elements close to one secular resonance surface result in proper elements which oscillate within a significant range (as in Figure 4); where two resonance surfaces cross, large scale instabilities can occur (as in Figure 5).

Let us assume that a particular set of values for L_M, G^*, Z^* corresponds to an ordered, quasiperiodic solution for K, with proper frequencies g, s far from any low degree resonance. Then the map to the proper linear elements is defined by the determining function Y, whose lowest degree part is Y_4, defined by the homological equation (4.2); the image of this map is a Kolmogorov invariant torus in the proper linear element space; by an explicitly known transformation, this torus can be given in the mean element space. Now comes the difficult step: can we compute the complete orbits, in the osculating elements space (hence in ordinary, cartesian coordinates), corresponding to this invariant torus? A map from mean elements to osculating elements is, in principle, the inverse of the computation of mean elements from the initial conditions discussed in Section 3. However, the computation of mean elements was only performed to first order in μ; to account for the higher order terms, we have to set up a rigorous procedure to define a map which adds back to the hamiltonian K the "perturbation" $-S$. To this purpose, we look for a determining function ρ such that:

$$K + \{K, \rho\} + \frac{1}{2}\{\{K, \rho\}, \rho\} + \ldots = H = K - S \qquad (7.2)$$

and assuming that $\rho = \rho_1 + \rho_2 + \ldots$ can be decomposed in parts of different "order" (the word has a different meaning in this context), the homological equation for ρ_1 is:

$$\{K, \rho_1\} = -S. \qquad (7.3)$$

The key to understand stable chaos in an N-body problem, and more generally the properties of the dynamics of a system which is a perturbation of a degenerate integrable one, is to realize how much different is equation (7.3) from equation (3.3) which was used for the first order computation of mean elements. Bv the same argument used in Section 4, ρ_1 is related to, but cannot be confused with $-\chi_1$. There are many technical differences, including the ones due to the fact that S is not just $\tilde{R}$, but also includes some secular terms resulting from the truncation of K to some degree; however, the main difference is in the structure of equation (7.3). Let us suppose we want to solve equation (7.3) by the usual method, by expanding in a Fourier series both the known S and the unknown ρ_1; let us concentrate on one particular Fourier term, with angular argument $\sigma = (p\ell - p'\ell' + q\omega - q'\omega' + r\Omega - r'\Omega')$, in S. Then the solution of (7.3) cannot be obtained by simply dividing by some divisor $pn - p'n'$. $\{K, \rho_1\}$ is the time derivative of ρ_1 along the solutions of K, that is on the invariant torus; on such a solution, the slow angles $\omega, \omega', \Omega, \Omega'$ are by no means constant, and the Fourier term with $\cos(\sigma)$ can be represented by a Fourier series with arguments linear functions of time, and with frequencies combinations not only of n, n' but also of g, s, g_i, s_i, the frequencies appearing in the "secular" solution of K. Thus all the combination frequencies allowed by D'Alembert rules can occur as divisors, and each mean motion resonance $pn - qn'$ is split into a multiplet of resonances involving different combinations of secular frequencies, as in (6.2). This is a very natural result, consistent with what is well known about the internal structure of the mean motion resonances.

However, the presence of a large number of resonances closely packed in the phase space does not necessarily imply chaos. Chaos occurs if these resonances overlap; in other words, the different resonances involving the same combination $p\lambda - q\lambda'$ need to be strong enough to interact. Since the resonances involved are generically of the very high degree prescribed by (6.4), the coefficient of $\cos(\sigma)$ contains a correspondingly high power of e, e', I, I' as prescribed by D'Alembert rule **III**, and we should expect that it is very small. Let us select the term in the original perturbing function with $p\ell - q\ell'$ and the eccentricity to power $r = |q - p|$; since the series in powers of the eccentricities has a radius of convergence e^*, the coefficient will be smaller than the 2-body hamiltonian by a factor of the order of $\mu(e/e^*)^r$. The standard theory of resonances (e.g. Henrard and Lemaitre 1983) shows that the corresponding change in mean motion, produced by the term under consideration, contains the square root of this small quantity:

$$\frac{\Delta n}{n} = \sqrt{\mu(\frac{e}{e^*})^r}. \qquad (7.4)$$

Overlap of two consecutive resonances in the same multiplet will occur if this change in n is larger than $g - g_5$, the change in average frequency obtained by replacing

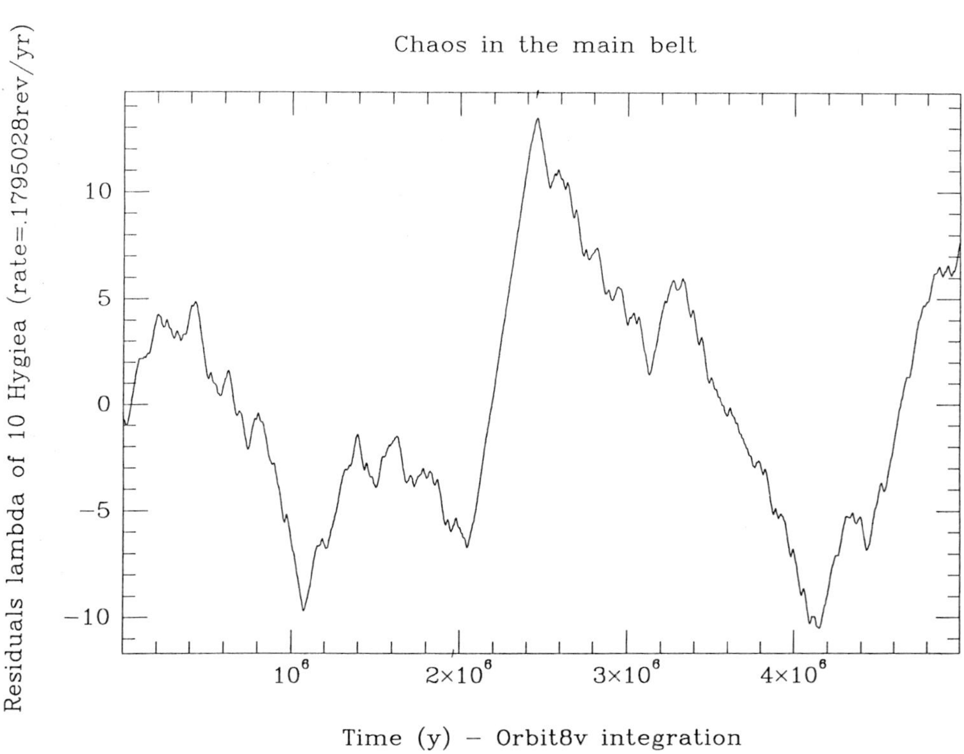

Figure 10. The residuals, in revolutions, in the mean longitude λ of *10 Hygiea*, after removal of the best fitting rate. Capture into different resonances (all belonging to the 23 : 49 multiplet) is apparent from the sharp changes in the slope. The size of the residuals is by no means small, that is the difference between a conditionally periodic solution and the actual solution is large; the chaotic behaviour results in a large instability, which is however confined to the along track direction.

one copy of ϖ with one copy of ϖ'. By using again $g \simeq c\mu n$ we obtain the overlap condition:

$$\sqrt{\mu\left(\frac{e}{e^*}\right)^r} \simeq c\mu \tag{7.5}$$

The next step is to replace r, the degree of the resonance, by the generic value $q/2$ with q given by (6.4) (at least for the *Hygiea* region): then we obtain an equation with only three variables, the perturbation parameter μ, the constant c, and the eccentricity e, which appears only through the ratio e/e^*; by solving for this ratio we obtain:

$$\frac{e}{e^*} = \left[c^2\mu\right]^{1/r}, \quad \text{with } r = \frac{1}{2\sqrt{c\mu}}. \tag{7.6}$$

For the values typical of the Hygiea region ($\mu \simeq 2.1 \times 10^{-4}$, $c \simeq 2.5$), we obtain $r \simeq 30$ and $e/e^* \simeq 0.8$. The conclusion is that all the orbits with eccentricity close to the radius of convergence of the expansion of the perturbing function in powers of e should be expected to be chaotic, although the resonances involved are of very high degree; this is a prediction of the generic occurrence of stable chaos in a comparatively wide region of the phase space.

The next step in the computation is obtained by finding the radius of convergence of the expansion of the perturbing function as a Taylor series in e. This is a classical problem, which has recently been revisited by Ferraz–Mello (the paper has also been presented at this meeting). The radius of convergence e^* can be considered, as a first approximation, a function of the semimajor axis only (in a second approximation, it also depends upon e', I, I'); its value decreases sharply across the range of values of a corresponding to the main asteroid belt, down to a value $\simeq 0.2$ at the $2:1$ resonance (see Figure 5 in Ferraz–Mello, 1994). Thus in the *Hygiea* region, just inside the $2:1$ resonance, the critical value of e for the generic occurrence of stable chaos is $\simeq 0.16$. The average value of the eccentricity is given by the proper eccentricity, and for *10 Hygiea* the proper $e = 0.135$. Given all the approximations used in this computation (as an example, the eccentricity can be larger than the proper eccentricity), I rate this as a very satisfactory result, which confirms that the behaviour of *10 Hygiea* is by no means exceptional. However, this is far from being a rigorous argument. The convergence of the series expansion of R as a Fourier series occurs for e larger than e^*, and for comparatively large e the value of a Fourier coefficient can be quite different from the monomial of lowest degree in e allowed by rule **III**. Moreover, asymptotic estimations cannot predict accurately the size of a specific coefficient, and the only way to rigorously explain the behaviour of cases such as *10 Hygiea* is to explicitly compute the strength of the resonances such as (6.2); this is possible, but not easy.

I would like to conclude by stating that what was presented in this Section is not a theory, but just the feasibility study for a theory of stable chaos. The result of the study is that such a theory is feasible, provided enough time and effort are applied to the task. Thus we do not need to face a phenomenon, occurring in a significant fraction of the cases, as a problem defying explanation, but just as work to be done in the near future.

Acknowledgements

I thank P. Farinella, Z. Knežević, J. Henrard and J. Laskar for comments and corrections which allowed me to improve this paper.

References

Arnold, V., 1976, "Méthodes Mathématiques de la Mécanique Classique", MIR, Moscou.

Baccili, S. and Cattaneo, L., 1993, Exam paper in "Advanced Mechanics", University of Pisa.

Benettin, G., Galgani, L. and Giorgilli, A., 1984, A proof of Kolmogorov theorem on invariant tori using canonical transformations without inversion, *Nuovo Cimento*, **79b**, 201–223.

Bretagnon, P., 1974, Termes à longues périodes dans le système solaire, *Astron. Astrophys.* **30**, 141–154.

Brouwer, D. 1951, Secular variations of the orbital elements of minor planets, *Astron. J.* **56**, 9–32.

Carpino, M., Milani, A. and Nobili, A.M., 1987, Long–term numerical integrations and synthetic theories for the motion of the outer planets, *Astron. Astrophys.*, **181**, 182–194.

Duriez, L., 1979, Approche d'une théorie générale planétaire en variables elliptiques héliocentriques, Ph.D. Thesis, Univ. Lille.

Duriez,L., 1990, Le dévelopment de la fonction perturbatrice, in *Modern methods in celestial mechanics*, Benest, D. and Froeschlé, C. eds., Editions Frontières, Gif-sur Yvette, pp. 35–62.

Farinella, P., Gonczi, R. and Froeschlé, Ch.,1993, Meteorites from the asteroid 6 Hebe, *Celest. Mech.*, **56**, 287–305.

Ferraz–Mello, S., 1994, The convergence domain of the laplacian expansion of the disturbing function, *Cel. Mech.*, **58**.

Froeschlé, C., 1984, The Lyapounov characteristic exponents and applications to the dimension of the invariant manifolds of a chaotic attractor, in *Stability of the solar system and its minor natural and artificial bodies*, Szebehely, V. ed., Reidel, Dordrecht, 265–282.

Henrard, J., 1990, A semi–numerical perturbation method for separable Hamiltonian systems, *Celest. Mech.* **49**, 43–67.

Henrard, J., and Lemaitre, A. 1983, A second fundamental model for resonance, *Celest. Mech.* **30**, 197–218.

Hori, G., 1966, Theory of general perturbations with unspecified canonical variables, *Publ. Astron. Soc. Japan* **18**, 287–296.

Knežević, Z., 1989, Asteroid long–periodic perturbations: the second order Hamiltonian, *Celes. Mech.* **46**, 147–158.

Knežević, Z., 1991, Asteroid long periodic perturbations; derivation of proper elements and assessment of their accuracy, *Astron. Astrophys.* **241**, 267–288.

Knežević, Z., Carpino, M., Farinella, P., Froeschlé, C., Froeschlé, Ch., Gonczi, R., Jovanović, B., Paolicchi, P. and Zappalà, V., 1988, Asteroid short–periodic perturbations and the accuracy of mean orbital elements, *Astron. Astrophys.* **192**, 360–369.

Knežević, Z., Milani, A., Farinella, P., Froeschlè, Ch. and Froeschlè, C., 1991, Secular resonances from 2 to 50 AU, *Icarus*, **93**, 316–330.

Knežević, Z., Froeschlé, Ch., Lemaître, A., Milani, A., and Morbidelli, A., 1994, Comparison of two theories for calculation of asteroid proper elements, in preparation.

Kozai, Y. 1962, Secular perturbations of asteroids with high inclinations and eccentricities, *Astron. J.* **67**, 591–598.

Kozai,Y., 1979, The dynamical evolution of the Hirayama families, in *Asteroids*, Gehrels, T. ed. (Univ. of Arizona Press), pp.334–358.

Laskar,J., 1986, Secular terms of classical planetary theories using the results of general theory, *Astron. Astrophys.*, **157**, 59–70.

Lemaitre, A., and A. Morbidelli, 1994, Calculation of proper elements for high inclined asteroidal orbits , *Celest. Mech.*, in press.

Milani, A., 1988, Secular perturbations of planetary orbits and their representation as series, in *Long Term Behaviour of Natural and Artificial N-Body Systems*, Roy, A.E., editor, Kluwer, Dordrecht, 73–108

Milani, A., 1990, Perturbation methods in Celestial Mechanics, in *Modern methods in celestial mechanics*, Benest, D. and Froeschlé, C. eds. , Editions Frontière, Gif-sur Yvette, pp. 109–150.

Milani, A., 1991, Chaos in the three body problem, in *Predictability, stability, and chaos in N-body dynamical systems*, Roy, A.E. ed., Plenum, New York, pp. 11–33.

Milani, A. and Farinella, P., 1994, Chaos as a clock: the age of the Veritas asteroid family, submitted for publication.

Milani, A., and Knežević, Z., 1990, Secular perturbation theory and computation of asteroid proper elements, *Celest. Mech.* **49**, 347–411.

Milani, A., and Knežević,Z., 1992, Asteroid proper elements and secular resonances, *Icarus* **98**, 211–232.

Milani, A., and Knežević, Z., 1994, Asteroid proper elements and the dynamical structure of the asteroid belt, *Icarus*, in press.

Milani, A. and Labianca, A., 1994, The radius of convergence of fixed frequency perturbation theories for the motion of the asteroids, in preparation.

Milani, A., and A. M. Nobili 1992. An example of stable chaos in the solar system. *Nature* **357**, 569–571.

Milani, A., Nobili, A.M. and Carpino, M. 1987, Secular variations of the semimajor axes: theory and experiments, *Astron. Astrophys.* **172**, 265–279.

Milani, A., Carpino, M., Hahn, G. and Nobili, A.M., 1989, Dynamics of planet–crossing asteroids: classes of orbital behaviour, Project SPACEGUARD, *Icarus* **78**, 212–269.

Milani, A., Farinella, P. and Knežević, Z., 1992, On the search for asteroid families, in *Interrelations between Physics and Dynamics for Minor Bodies in the Solar System*, D. Benest and C. Froeschlé eds., Editions Frontières, Gif–sur–Yvette, pp. 85–132.

Milani, A., Bowell, E., Knežević, Z., Lemaitre, A., Morbidelli, A. and Muinonen, K., 1994a, A composite catalogue of asteroid proper elements, in *Asteroids, Comets, Meteors 1993*, Milani, A., Di Martino, M. and Cellino, A. eds., Kluwer, Dordrecht, in press.

Milani, A., Nobili, A.M., and Knežević, Z., 1994b, Stable chaos in the asteroid belt, in preparation.

Morbidelli, A., and Henrard, J., 1991, Secular resonances in the asteroid belt: theoretical perturbation approach and the problem of their location, *Celestial Mechanics*, **51**, 131–168.

Poincaré, H., 1884, Sur une généralisation des fractions continues, *Comp. Rend. Acad. Sci. Paris* **99**, 1014–1016.

Poincaré H., 1892–1899, *Les Méthodes Nouvelles de la Méchanique Céleste*, Vol. I, 1892; Vol. II, 1893; Vol. III, 1899; Gauthier-Villars, Paris (reprinted by Blanchard, Paris, 1987).

Roy, A.E., 1982, The stability of N-body hierarchical dynamical systems, in *Applications of modern dynamics to celestial mechanics and astrodynamics*, Szebehely, V. ed., Kluwer, Dordrecht, pp. 103–130.

Williams, J.G., 1969, Secular perturbations in the Solar System, Ph.D. Thesis, Univ. California, Los Angeles.

Williams, J.G., 1973, Meteorites from the asteroid belt? *Eos*, **54**, 233.

Williams, J.G., 1979, Proper orbital elements and family memberships of the asteroids, in *Asteroids*, Gehrels,T. ed., Univ.Arizona Press, pp. 1040–1063.

Williams, J.G., 1989, Asteroid family identifications and proper elements, in *Asteroids II*, Binzel, R.P, Gehrels, T. and Matthews, M.S. eds., Univ.Arizona Press, Tucson, pp. 1034–1072.

Williams, J.G., and Faulkner, J., 1981, The position of secular resonance surfaces, *Icarus* **46**, 390–399.

Yuasa, M., 1973, Theory of secular perturbations of asteroids including terms of higher order and higher degree, *Publ. Astron. Soc. Japan* **25**, 399–445.

Zappalà, V., Cellino, A., Farinella, P. and Knežević, Z., 1990, Asteroid families I: identification by hierarchical clustering and reliability assessment, *Astron. J.* **100**, 2030–2046.

Zappalà, V., Cellino, A., Farinella, P. and Milani, A., 1994, Asteroid families: II. Extension to unnumbered multi-opposition asteroids, *Astron. J.* **107**, 772–801.

MECHANISMS OF GENERATION OF CHAOS IN THE SOLAR SYSTEM

John D. Hadjidemetriou

Department of Theoretical Mechanics, University of Thessaloniki
GR-540 06 Thessaloniki, Greece.
e-mail:Hadjidem@olymp.ccf.auth.gr

Abstract: The mechanism of generation of chaos in the Solar System is studied, by making use of a mapping model. This mapping is simple and we can study the mechanism of generation of chaos without beeing lost in unnecessary details. On the other hand, it is realistic as it contains all the main features of the original problem, i.e. the topology of its phase space is the same as that of the real system. The motion of asteroids near a resonance is used as the basic example. We start with a simple model with two degrees of freedom and we study how the appearance of chaos is affected as the model becomes more and more complex (and more realistic).

We start with the simplest nontrivial model, the restricted circular planar three body problem, with the Sun and Jupiter as primaries. A two dimensional mapping model is constructed, valid globally, and we study the topology of its phase space and the regions where chaotic motion is expected to appear. We focus our attention to the 2:1 and 3:1 resonances and their basic difference with respect to their chaotic properties is shown. Next, we proceed to the elliptic restricted planar three body problem , assuming that Jupiter is moving in a fixed elliptic orbit. The mapping now is four dimensional and we express it in the resonant action-angle variables S, σ, N, ν. It is explained how the coupling between the two degrees of freedom S, σ and N, ν generates chaos near a resonance. Inside the resonance we may also have ordered motion, depending on the initial phase (i.e. initial values of σ and ν). Finally, we include to our model the gravitational effect of Saturn on the orbit of Jupiter and show that the appearance of chaotic motion is now widespread inside the whole chaotic zone. It is also explained why the above nentiond chaotic motion appears only inside the resonance zone.

The effect of a nonrealistic model on the evolution of the system is studied. In particular, we study the consequences of not including in our model the high eccentricity resonances. As a byproduct of the present study, we show how the knowledge of the basic families of periodic orbits of the original system can be used to check the convergence of the perturbation series and the reality of a model.

From Newton to Chaos, Edited by A.E. Roy
and B.A. Steves, Plenum Press, New York, 1995

1 Introduction

The purpose of this presentation is to study the mechanism by which chaos is generated in a dynamical system that describes motion in the Solar System. The motion of the asteroids will be used as a basic example, but the same arguments apply to all other similar dynamical systems, for examle the motion of the satellites of the major planets.

The above systems are nonintegrable and consequently their analytic solution is not known. This means that we are oblidged to make approximations and use perturbation methods for the study of the long term evolution of the system. The model that we obtain by such a method is an essential part of the study. Different approaches will give different models, which do not have, in general, the same properties. Apart from direct numerical integrations, the averaging method and the construction of mapping models are the basic tools in the study of a nonintegrable dynamical system. In what follows, we start with the simplest nontrivial model for the motion of an asteroid and we increase the complexity of the model, to make it more realistic. In this process, we study how the basic properties of the model change and what are the basic factors that govern chaotic motion. The effect of a poor model on the long term evolution of the system is stressed and criteria are given to check how realistic is a model. We also see how useful is the knowledge of the basic families of periodic orbits of the real system in checking the validity of a model.

The basic subsystem in the Solar System is a *two body problem*, Sun - planet (or planet - satellite). This is an integrable system with two degrees of freedom and consequently, the motion is ordered everywhere. This however is a poor model for the study of the long term evolution. The real model is the general N-body problem, where the bodies are the Sun, all the planets and their satellites, with their real masses (not to mention tides, etc). This model is, evidently, very complicated and only recently it has been studied, by numerical integrations of the whole Solar System (Laskar, 1990, Nobili et al.,1989). But even in this case, we may be lost in the vast amount of output and miss the important factors that govern the evolution of the system and are responsible for the generation of chaos.

The problem we face now is how to chose our model between the above two extreme cases so that is is simple, but realistic. This means that our model must contain all the main features of the original system that it is supposed to represent. We start with the simplest nontrivial model, the *circular restricted planar three body problem*, with the Sun and Jupiter as primaries. This is a system with two degrees of freedom and we study the evolution of an asteroid in this framework. This is a *perturbed two body problem* Sun - asteroid, perturbed by Jupiter. Special emphasis is given to the difference between resonant and nonresonant motion. Two resonances are studied in detail, the 2:1 and the 3:1 mean motion resonance of the asteroid to Jupiter. The essential differences between them, as far as the appearance of chaos is concerned, are indicated.

The next approximation is to assume that Jupiter moves in an *elliptic* orbit. This is the *elliptic restricted three body problem.* This system is equivalent to a system with three degrees of freedom in many aspects. Finally, we assume that the orbit of Jupiter is not fixed, but varies due to the gravitational attraction from Saturn and we see how this affects the generation of chaos near and far from a resonance.

2 The Circular Restricted Three Body Problem

This is the simplest nontrivial model that has been used extensively in the study of the Solar System. The primaries are the Sun and Jupiter, that revolve around their common center of mass in circular orbits and a massless body (asteroid) moves in the orbital plane of the primaries under their gravitational attraction. For the three body problem see Szebehely, 1967 and Roy, 1982.

Since the mass of Jupiter is small compared to that of the Sun, the main attraction on the asteroid comes from the Sun and consequently the motion of the asteroid can be considered as a perturbed two body problem. This means that perturbation methods can be applied to study this problem. In partricular, the method of averaging has been widely used to study resonant motion of the asteroid.

The restricted circular planar three body problem is a dynamical system of two degrees of freedom. Such systems are now well understood and the use of this model to study asteroid motion will throw much light on the behaviour of the system, although, as we shall see in the following, this approximation is not satisfactory to study the long term evolution of the asteroids. But even so, this model is very useful because it provides the basic framework on top of which all other perturbations apply.

There are three different approaches to the study of the above mentioned system: (a) numerical integrations, (b) perturbation methods and in particular the method of averaging and (c) construction of mapping models. Usually, people who work on this problem restrict themselves to one such method only. However, we believe that each of the above methods has its merits and its limitations and a combination of all of them is very useful in the study of the problem. One of the purposes of this presentation is to show how the interaction between all three methods gives useful information not only for the particular problem but also on the validity of the perturbation method used and its range of applicability and also on the construction of simple but realistic models.

2.1 Numerical integrations

Poincaré mapping

For systems with two degrees of freedom a very useful numerical tool is the computation of the Poincaré mapping on a surface of section (for the details see Hénon, 1983, Lichtenberg and Liebermann, 1983). In the rotating xOy frame of reference, a convenient way to define the surface of section is to take $y = 0$ and $H = h_0$, where H is the energy (Jacobi) constant and h_0 the energy level at which we take the mapping. The Poincaré mapping is two dimensional, in the space $x, \dot{x}$. On this mapping, the consecutive points describe the evolution of the system and they define the topology of the phase space on the surface of section. These points may lie on a smooth curve, called *invariant curve*, if we are in an ordered region of phase space, or may be scattered if we are in a chaotic region. In many cases the distinction between these two extreme cases is sharp, but in some cases the difference may not be clear.

An important feature of the mapping is the existence of *fixed points*. These correspond to the *periodic orbits* of the system and their position and stability characteristics determine critically the topology of the phase space. This makes clear the importance of the knowledge of the families of periodic orbits in the study of the problem. We remark that their numerical computation is a relatively easy task and it involves a limited amount of computing time.

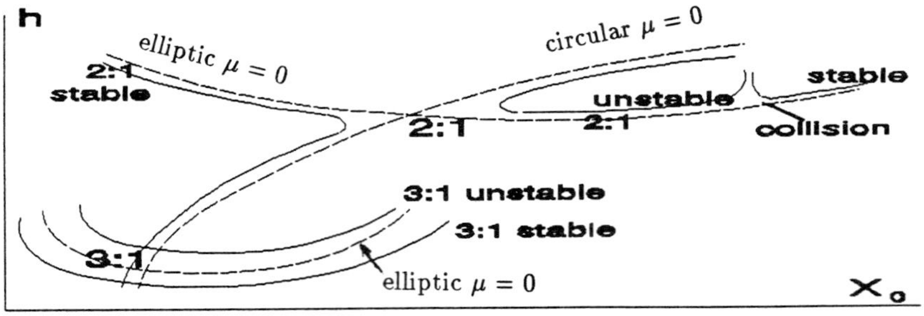

Figure 1. The characteristics of the families of periodic orbits of the first and second kind of the circular planar restricted three body problem, close to the 2:1 and 3:1 resonances (schematically). The dotted lines are the unperturbed families ($\mu = 0$).

Families of periodic orbits

We describe now briefly the basic families of periodic orbits of the circular restricted planar three body problem, with particular emphasis on the resonant periodic orbits at the 2:1 and 3:1 resonances. Their knowledge will help us form a clear picture of the topology of the phase space.

There are two types of families of periodic orbits in the restricted circular planar three body problem, with the Sun and Jupiter as primaries: The *periodic orbits of the first kind*, where the asteroid moves in a perturbed circular Keplerian orbit, and the *periodic orbits of the second kind*, where the asteroid moves in a perturbed elliptic Keplerian orbit. All these orbits are symmetric with respect to the rotating x-axis, i.e. their initial conditions are $x_0, y_0 = 0, \dot{x}_0 = 0, \dot{y}_0$. Consequently we can represent such a family of periodic orbits by a smooth curve in the space $x_0, \dot{y}_0$, or the space x_0, h, where h is the Jacobi constant, called *characteristic curve.* Along the family of periodic orbits of the first kind the ratio of the mean motion frequencies $n : n'$ of the asteroid and Jupiter, respectively, varies. At the points where this value is rational, we have a bifurcation of families of periodic orbits of the second kind. Along these latter families the ratio $n : n'$ is almost constant and the eccentricity of the asteroid increases, starting from zero values (for details see Hadjidemetriou, 1988, 1993b, Hadjidemetriou and Ichtiaroglou, 1984). Only the resonant families of low order are of practical importance. In Figure 1 we present the characteristics of the families of periodic orbits of the first and second kind of the restricted circular planar three body problem , close to the 2:1 and 3:1 resonances. These are two of the basic resonances in the asteroid problem that are of different nature and will receive special attention in the present study. The mass of Jupiter is taken equal to $\mu = 0.00095387535$ and the total mass of the Sun and Jupiter is equal to unity.

There is a basic difference in the continuation of the 2:1 and 3:1 resonant families of periodic orbits from $\mu = 0$ (dotted line in Figure 1) to $\mu > 0$: The unperturbed family of circular periodic orbits cannot be continued to $\mu > 0$ at the point $n : n' = 2 : 1$ and as a consequence the family of periodic orbits of the first kind has a discontinuity at that point, as is clearly seen in Figure 1. Two resonant 2:1 families of simple symmetric periodic orbits of the second kind appear, one stable and the other unstable. We remark also that on the unstable family there is a periodic orbit that corresponds to collision of the asteroid with Jupiter, at the energy level $h =\approx -1.37$, corresponding to $e =\approx 0.65$.

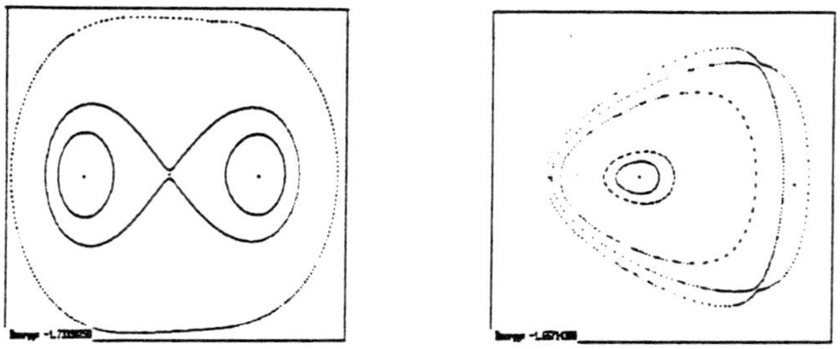

Figure 2. Poincaré surfaces of section at the energy levels $h = -1.733825$ and $h = -1.657430$, at the 3:1 resonance.

Beyond that point there is a complicated stucture of interchange between stable and unstable periodic orbits, confined to a very small region and after that all the orbits are stable. The collision orbit generates a discontinuity on the family that is shown in Figure 1.

At the 3:1 resonance of the circular family, the continuation to $\mu > 0$ is possible, but a small unstable region developes on the family of periodic orbits of the first kind. At the two ends of this region we have critical points as far as the stability is concerned, and from both these end points we have a bifurcation of a resonant 3:1 family of doubly symmetric periodic orbits of the second kind (they manifest themselves as double fixed points on the Poincaré surface of section). One of them is stable and the other is unstable (for more details see Hadjidemetriou, 1992, 1993b).

Poincaré mappings at the 3:1 and 2:1 resonance regions

The Poincaré surfaces of section close to the the 3:1 and the 2:1 resonances are shown in Figures 2 and 3, respectively, at different energy levels. The fixed points of the mapping correspond to the periodic orbits of the system. The importance of the position and stability character of the fixed points on the topology of the phase space is clear.

From the above numerical results we can draw the following conclusions:

- At low energy levels the system behaves, for all practical purposes as integrable, i.e. we have smooth invariant curves. In particular, the whole region at the 3:1 resonance appears as integrable, even for large values of the eccentricity of the asteroid.
- Chaotic motion appears at the 2:1 resonance, beyond an energy level, corresponding to values of the eccentricity larger than $e \approx 0.40$.
- The chaotic region starts at the unstable fixed points and spreads along the critical curves to cover a large part of phase space.
- Close to the smooth invariant curves corresponding to the 2:1 resonance, we have a chaotic region that originated from an unstable point corresponding to the 5:3 resonance. This is shown in Figure 3b.

2.2 Criteria for constructing a model

We have seen that the Poincaré mapping gives a clear picture of the topology of the phase space of the dynamical system. Any perturbation method, or any mapping model, should give the same topological structure of the phase space in order to be realistic. Thus, the Poincaré mappings given in Figures 2 and 3 can be used as a check for any model. This

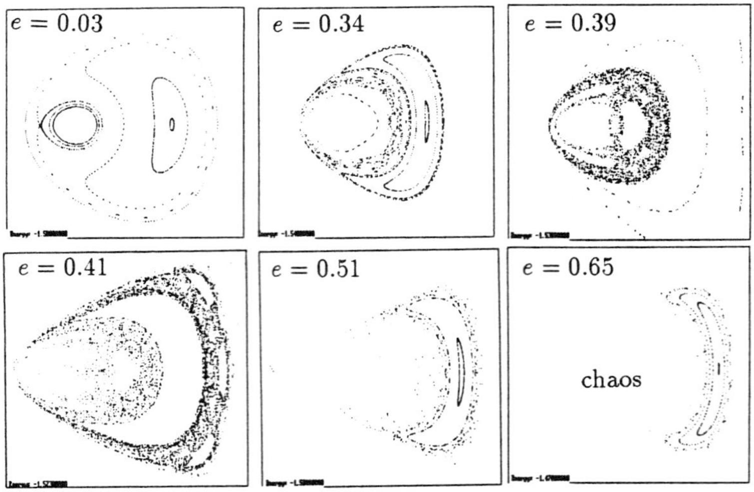

Figure 3. Poincaré surfaces of section at the energy levels h = -1.58, -1.54, -1.53, -1.523, -1.50, -1.47, at the 2:1 resonance. The eccentricity of the resonant fixed points is indicated.

means that we can state the following *necessary criteria* in order for a model to be realistic:

- Its fixed points must be the same as the fixed points/periodic orbits of the original system.
- The above fixed points must have the correct stability characteristics.
- It must have the same chaotic regions as the original system.

These will become clearer in the following.

2.3 The averaging method

The averaging method is a very powerful and useful method and has been used extensively in the study of resonant motion in a perturbed Keplerian motion (Ferraz-Mello, 1987, Henrard, 1988, Wisdom, 1985). For a description of this method see e.g. Hadjidemetriou, 1991. By this method the restricted circular planar three body problem reduces to an autonomous system with one degree of freedom only. Consequently, the averaged model that we obtain is integrable and no chaotic motion is expected. The averaged Hamiltonian depends on two variables only (for resonant motion these are the *action - angle* variables S, σ, that will be defined in the next section) and the level curves of the averaged Hamiltonian give in fact the motion. These level curves are easily obtained. If we know the averaged motion, we can then use the inverse transformation and go back and find the motion in the original variables. In particular, the fixed points of the averaged model correspond to the periodic orbits of the original system, or the fixed points of the Poincaré mapping.

The question now arises, how good is the averaged model. Since the perturbation method involves expansions in a small parameter, we have the problem of convergence (and the radius of convergence). A good check of how realistic is the averaged model and what is its range of validity is provided by comparing the level curves of the averaged Hamiltonian with the corresponding phase diagrammes of the Poincaré mapping. The necessary conditions are mentioned in section 2.2. We remark that the averaged system is integrable and consequently, no chaotic region can appear. Thus, *the averaged model can be a realistic model only in those regions of phase space where the system appears as integrable* (smooth invariant curves).

In what follows, we shall consider the averaging method and we shall comment on its validity in different cases, in view of the criteria mentioned above.

3:1 Resonance

The variables used in the study of resonant motion close to the 3:1 resonance are the resonant action-angle variables S, σ, N, ν defined by (Henrard and Caranicolas, 1990)

$$S = \sqrt{\mu_1 a}\left(1 - \sqrt{1-e^2}\right), \quad \sigma = \frac{1}{2}(3\lambda' - \lambda) - \omega \tag{1}$$

$$N = \sqrt{\mu_1 a}\left(3 - \sqrt{1-e^2}\right), \quad \nu = -\frac{1}{2}(3\lambda' - \lambda) + \omega' \tag{2}$$

where

$$\mu_1 = 1 - \mu, \quad e' = 0.048,$$

λ, ω, a are the mean longitude, the longitute of perihelion and the semimajor axis, respectively, of the asteroid and the corresponding primed quantities refer to Jupiter. The averaged Hamiltonian, truncated to first order in the eccentricity, is given by

$$H = H_0(S, N) + \mu H_1(\sigma, S, N) \tag{3}$$

where

$$H_0 = -\frac{2(1-\mu)^2}{(N-S)^2} - \frac{3}{2}(N - S), \tag{4}$$

$$H_1 = 2FS - b\frac{S}{N}\cos 2\sigma, \tag{5}$$

and F=-0.205070, b=2.392. We note that the angle ν is ignorable and consequently the corresponding action N is constant. Thus, the system is reduced to an autonomous system with only one degree of freedom and its motion is given by the level curves

$$H = \text{constant}. \tag{6}$$

We may thus present the motion on the S, σ plane, or, using the Poincaré variables

$$X = \sqrt{2S}\cos\sigma, \quad Y = \sqrt{2S}\sin\sigma, \tag{7}$$

in the XY plane. Either the action N or the energy constant can be used as the fixed parameter for such a phase diagramme. Typical phase diagrammes for different values of N are given by Henrard and Lemaitre (1983) and Lemaitre (1984). The different level curves in these diagrammes correspond to different values of the energy constant.

We ask now how realistic is the averaged model defined by the Hamiltonian (3). It can be proved (Hadjidemetriou 1992) that the families of fixed points of this model coincide, to a good accuracy, with the corresponding families of periodic orbits of the original system (or the fixed points of the Poncaré mapping), close to the 3:1 resonance. This is true not only for small values of the eccentricity of the asteroid, but for large values as well. For this comparison, we expressed the fixed points of the Poincaré mapping in the action angle variables S, σ defined by (1). Note that no chaotic regions exist in the original system (Figure 2) and this is, of course, true for the averaged model. Consequently, the Hamiltonian (3) is a realistic model for the study of the motion at the 3:1 resonance.

2:1 Resonance

The resonance variables at the 2:1 resonance are defined by (Henrard and Lemaitre, 1987)

$$\begin{aligned} S &= \sqrt{\mu_1 a}\left(1-\sqrt{1-e^2}\right), \quad \sigma = 2\lambda' - \lambda - \omega \\ N &= \sqrt{\mu_1 a}\left(2-\sqrt{1-e^2}\right), \quad \nu = -2\lambda' + \lambda + \omega' \end{aligned} \tag{8}$$

and the averaged Hamiltonian is

$$H = H_0(S,N) + \mu\left[H_1(\sigma,S,N) + H_2(\sigma,S,N) + ...\right] \tag{9}$$

The unperturbed part of the Hamiltonian is

$$H_0 = -\frac{(1-\mu)^2}{2(N-S)^2} - B(N-S), \tag{10}$$

and the functions H_1, H_2, ... are expansions in power series in the eccentricity, truncated to first, second, etc order, respectively. The first order term is

$$H_1 = A\sqrt{\frac{2S}{N}}\cos\sigma + \frac{2S}{N}D\cos 2\sigma + \frac{2S}{N}C \tag{11}$$

The values of the parameters in the above equations are: A=1.189, B=2.0008, C=-0.4273 and D=-1.169.

In order to ckeck the validity of this averaged model, we computed the families of fixed points of the Hamiltonian equations for the Hamiltonian (9), truncated, respectively to first, second, ... order in the eccentricity, up to the 8th order, using the expansions found by Henrard and Lemaitre (1987). These were compared with the corresponding families of fixed points of the Poincaré mapping of the original system (the restricted circular three body problem), obtained from the families of resonant periodic orbits at the 2:1 resonance that we computed (see Figure 1). We find that all models truncated to an even order are not realistic. Not only the position of the fixed points is not correct, but their stability is wrong. For the truncated models to odd order in the eccentricity, we find that the coincidence of their fixed points with those of the original model is good only for small values of the eccentricity, and that the addition to our model of higher approximations does not improve essentially the model. Note that, in addition to the above, chaotic motion appears in the original model, but this is not possible in the averaged model, which is integrable.

From all the above we see that the necessary criteria for a model to be realistic, stated in section 2.2, fail in the case of the 2:1 resonance for the averaged Hamiltonian (9). The chaotic motion that appears in the original system, at relatively low values of the eccentricity, is the main reason why the perturbation series in the averaging method do not converge.

2.4 Mapping Models

One of the main reasons why the averaged model fails, is that the averaging method reduces the number of degrees of freedom from two to one and consequently all chaotic

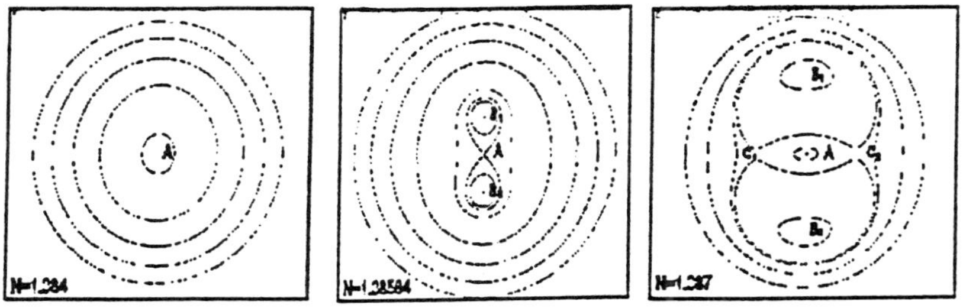

Figure 4. Phase diagrammes of the mapping (14).

motion disappears. This would not be important if the original model behaves as integrable (smooth invariant curves), as is the case with the 3:1 resonance. But if the original model has chaotic regions, the averaged model cannot be realistic. The question now arises, how to obtain in this latter case a realistic model? An obvious answer is to go back to a model with two degrees of freedom, provided of course that this latter model obeys the necessary criteria of section 2.2. This can be achieved by going from the two - dimensional level curves of the averaged model to a two - dimessional mapping model. This mapping is equivalent to the Poincaré mapping of the system and in this way we gain the degree of freedom that we lost in the averaging process. This idea has been used by Hadjidemetriou (1993a) to obtain a realistic model, valid globally, for the study of resonant motion. The averaged Hamiltonian is an esssential part of the method. We shall use here this mapping model to study the mechanism by which chaos is generated in the Solar System.

The mapping at the 3:1 resonance

This mapping is expressed in the variables S, σ, defined in (1) and is obtained from the generating function

$$W = \sigma_n S_{n+1} + TH(S_{n+1}, \sigma_n, N), \tag{12}$$

through the equations

$$\sigma_{n+1} = \partial W/\partial S_{n+1}, \quad S_n = \partial W/\partial \sigma_n. \tag{13}$$

The function H in Equation(12) is the averaged Hamiltonian (3) and $T = 2\pi$. The mapping equations are

$$\begin{aligned} S_{n+1} &= \frac{S_n}{1 + 2\mu T \frac{b}{N} \sin 2\sigma_n}, \\ \sigma_{n+1} &= \sigma_n - \frac{4T\mu_1^2}{(N - S_{n+1})^3} + \frac{3}{2} + 2\mu T F - \mu T \frac{b}{N} \cos 2\sigma_n. \end{aligned} \tag{14}$$

In these equations N is a fixed parameter. For each value of N we obtain a phase diagramme. There are three topologically different phase diagrammes, for different values of N, presented in Figure 4. Note that these phase diagrammes are topologically equivalent to the phase diagrammes of the Poincaré surface of section of Figure 2 and also one can find that the fixed points of both mappings coincide, to a good accuracy.

The mapping at the 2:1 resonance

The mapping model for the 2:1 resonance can be constructed by using the same method as for the 3:1 resonance. We must find a generating function, and then the

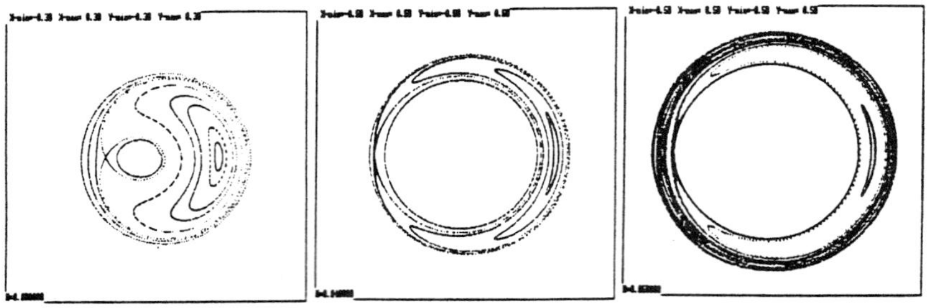

Figure 5. Phase diagrammes of the mapping model at the 2:1 resonance.

mapping equations can be obtained from the equations(13). In this case however, the averaged Hamiltonian (9) is not realistic and consequently, if we use this Hamiltonian in the generating function (12), we will not obtain a correct mapping model. We found that we can obtain a realistic mapping model valid near the 2:1 resonance if in (12) we use for H the function

$$H = H_0 + \mu H_1 + \mu H_7 + \mu H_c, \tag{15}$$

where $H_0 + \mu H_1$ is the averaged Hamiltonian, truncated to first order in the eccentricity, H_7 is the seventh order term in the expansion (9), given by (Henrard and Lemaitre, 1987)

$$H_7 = e^7 f \left[P \cos\sigma + Q\cos 3\sigma + R\cos 5\sigma + T\cos 7\sigma\right] \tag{16}$$

and H_c is a *correction term* given by

$$H_c = e^7 f T' \cos 8\sigma. \tag{17}$$

The values of the parameters are P=1.598, Q=6.964, R=-52.95, T=51.95 and T'=-20.1. Also, for the parameter D in H_1 we take the value D=-.30. It is found that this mapping model has the correct fixed points and the correct stability, up to the collision point with Jupiter, corresponding to $e \approx 0.60$. The additional parameter f is introduced to control the appearance of chaos and we found that we obtain the correct chaotic regions in the mapping if we take f=0.05. Some phase diagrammes are given in Figure 5. The above are preliminary results and work is still in progress.

In the above we have presented mapping models for the 2:1 and 3:1 resonance regions of the restricted circular three body problem. This model is not a good model to study the long term evolution in the Solar System. As we shall see in the next section, the introduction to our model of a nonzero eccentricity of Jupiter changes dramatically the properties of the system. However, a two dimensional mapping is still very useful, because it provides the basic framework on top of which all other perturbations apply. In going from the circular to the elliptic case we will be able to understand better the mechanism by which chaos is generated in the Solar System.

3 The next approximation: The elliptic restricted three-body problem

We shall introduce now to our model the nonzero eccentricity of Jupiter. This model is the *elliptic restricted three body problem* and this will be now our original model that we

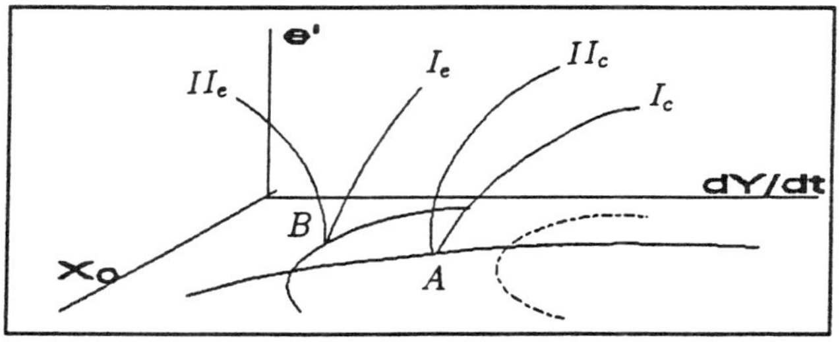

Figure 6. The families of resonante 3:1 families of periodic orbits of the elliptic restricted three body problem in the space $x\dot{y}e'$ (schematically).

wish to study. The 3:1 resonance will be used as the main example to study the effect of a nonzero eccentricity.

We assume that Jupiter moves in a *fixed* elliptic orbit, with eccentricity e'=0.048, the present value of Jupiter's eccentricity. We consider a rotating xOy plane whose origin O is at the center of mass of the Sun and Jupiter and the x-axis contains always these two bodies. Due to the elliptic motion, the Sun and Jupiter oscillate on the x-axis according to Keplerian motion, with period $T = 2\pi$ (in the usual normalized units). The Hamiltonian of the elliptic restricted problem depends explicitely on time and we can difine now a four dimensional mapping by considering the position and velocity coordinates x, y, $\dot{x}$, $\dot{y}$ of the asteroid in the rotating frame, at the times $t_n = nT$, where T is the period of the elliptic motion and n=1,2,3,...

We plan to study the motion of an asteroid close to the 3:1 resonance by constructing a four dimensional mapping model, using the same method as in the circular case. This mapping must obey the criteria stated in section 2.2 in order to be realistic. In particular, we expect this mapping to have the same fixed points, with the correct stability, as the original system. For this reason, we present briefly in the next section the resonant structure of the elliptic restricted three body problem at the 3:1 resonance. For more details see Hadjidemetriou (1992).

3.1 The resonant structure of the elliptic restricted three body problem at the 3:1 resonance

The resonant structure of the restricted problem is determined by the resonant families of periodic orbits. For a fixed value of the eccentricity e' of Jupiter the periodic orbits are isolated, but they can be considered to belong to families if we allow e' to vary. These are symmetric periodic orbits that start perpendicularly from the rotating x-axis at the time when Jupiter is at perihelion or aphelion (zero relative velocity on the x-axis) and consequently a periodic orbit is determined by its nonzero initial conditions x_0, $\dot{y}_0$, for each value of e'. These families can be represented by smooth curves, that we call *characteristic curves*, in the space $x_0, \dot{y}_0, e'$ and are shown in Figure 6. The families of periodic orbits of the circular problem appear in this diagramm as curves in the plane $x_0, \dot{y}_0$ and the families of the elliptic problem bifurcate from them at the points where the period of the periodic orbit of the circular problem is exactly equal to π or 2π. There are only two such bifurcation points, the point A in the unstable region of the family of periodic orbits of the first kind with period π and the point B in the stable family of resonant 3:1 periodic orbits of the second kind, with period 2π, at the eccentricity e=0.80. From the point A there bifurcate two families of the elliptic problem, I_c and II_c, that are both unstable and the eccentricity of the asteroid along them is close to zero. Two more families bifurcate from the point B, I_e and II_e, one stable and the other unstable. In both these families the eccentricity of the asteroid is large, close to 0.8. If we take the intersection of the families I_c, II_c, I_e, II_e with the plane $e' = 0.048$, we obtain four isolated points that represent the resonant 3:1 periodic orbits of the elliptic restricted three body problem, for the present value of the eccentricity of Jupiter.

3.2 The mapping model for the elliptic restricted three body problem

We shall construct now a four dimensional mapping model in the resonant action - angle variables S, σ, N, ν defined by the equations (1) and (2). This mapping will be considered as a model for the four dimensional Poincaré mapping mentioned in section 3 above. It is proved (Hadjidemetriou 1993a) that we obtain a realistic mapping if we use the generating function

$$W = \sigma_n S_{n+1} + \nu_n N_{n+1} + TH(S_{n+1}, \sigma_n, N_{n+1}, \nu_n), \tag{18}$$

through the equations

$$\sigma_{n+1} = \partial W/\partial S_{n+1},\ S_n = \partial W/\partial \sigma_n,\ \nu_{n+1} = \partial W/\partial N_{n+1},\ N_n = \partial W/\partial \nu_n, \tag{19}$$

where

$$H(S, \sigma, N, \nu) = H_0 + \mu H_1 + \mu e' H_2 + \mu H_c. \tag{20}$$

The function $H_0 + \mu H_1 + \mu e' H_2$ is the averaged Hamiltonian of the elliptic restricted three body problem at the 3:1 resonance, truncated to first order in the eccentricity of the asteroid and H_c is a *correction term*. The functions H_0 and H_1 are given by the equations (4) and (5) respectively, and

$$H_2 = \sqrt{2S}\,[G\cos(\sigma + \nu) + D\cos(\sigma - \nu)] + 2e' K \cos 2\nu, \tag{21}$$

where $G = 0.198705$, $D = 2.656407$, $K = -0.181477$, $e' = 0.048$ and

$$H_c = \mu F' \left(1 - F'' e' \cos(\sigma + \nu)\right) S^3, \tag{22}$$

where $F' = -3.4$, $F'' = 6.4$. The correction term H_c introduces to our model the high eccentricity resonances (i.e. the resonant periodic orbits I_e, II_e) that are missing from the averaged model $H_0 + \mu H_1 + \mu e' H_2$. This latter model is not valid for large values of the eccentricity and consequently it does not contain the high eccentricity resonances. We shall show in the following the consequences if this correction term is not included.

The four dimensional mapping is obtained from the equations (19) for the generating function (18) and is given by

$$\begin{aligned} S_n &= S_{n+1} + \mu T \frac{2b}{N_{n+1}} S_{n+1} \sin 2\sigma_n \\ &\quad - \mu e' T \sqrt{2S_{n+1}}\,(G\sin(\sigma_n + \nu_n) + D\sin(\sigma_n - \nu_n)) \\ &\quad + \mu T e' F' F'' S_{n+1}^3 \sin(\sigma_n + \nu_n), \end{aligned} \tag{23}$$

$$
\begin{aligned}
\sigma_{n+1} &= \sigma_n - \frac{4T\mu_1^2}{(N_{n+1}-S_{n+1})^3} + \frac{3}{2}T + 2\mu TF - \mu T\frac{b}{N_{n+1}}\cos 2\sigma_n \\
&+ \frac{\mu e' T}{\sqrt{2S_{n+1}}}\left(G\cos(\sigma_n+\nu_n) + D\cos(\sigma_n-\nu_n)\right) \\
&+ 3\mu TF'\left(1 - e'F''\cos(\sigma_n+\nu_n)\right)S_{n+1}^2,
\end{aligned} \tag{24}
$$

$$
\begin{aligned}
N_n &= N_{n+1} \\
&- \mu e' T\sqrt{2S_{n+1}}\left(G\sin(\sigma_n+\nu_n) - D\sin(\sigma_n-\nu_n)\right) \\
&+ \mu T e' F' F''\sin(\sigma_n+\nu_n)S_{n+1}^3 - 4T\mu e'^2 K\sin 2\nu_n,
\end{aligned} \tag{25}
$$

$$
\nu_{n+1} = \nu_n + \frac{4T\mu_1^2}{(N_{n+1}-S_{n+1})^3} - \frac{3}{2}T + \mu T\frac{bS_{n+1}}{N_{n+1}^2}\cos 2\sigma_n. \tag{26}
$$

Note that all equations are coupled to each other through e'. If e' were equal to zero, the first two equations would be uncoupled to the other two and also N would be constant. This coupling between the degrees of freedom (S, σ) and (N, ν) is the reason that chaotic motion is generated.

3.3 The evolution of an asteroid at the 3:1 resonance

High eccentricity resonances missing

The mapping (23)-(26) is four dimensional and cannot be easily presented geometrically. But in this case we can take advantage of the fact that we still have a fast and a slow angle so that we can consider a "mapping" of this mapping. The slow angle is $(\sigma + \nu) = -\omega$, as we find from equations (1) and (2), (taking $\omega' = 0$) and we can take the projection of the mapping (23)-(26) in the plane

$$
\xi = \sqrt{2S}\cos(\sigma+\nu), \quad \eta = \sqrt{2S}\sin(\sigma+\nu), \tag{27}
$$

at the points $\sigma = \pi/2$, $\Delta\sigma > 0$ (or $\Delta\sigma < 0$). (This is equivalent to a second averaging of the averaged Hamiltonian, considering $\sigma + \nu$ as the slow angle, that was made by Henrard and Lemaitre (1987) and Wisdom (1985)).

The evolution obtained in this way is shown if Figure 7. In order to show the effect on the model if the high eccentricity resonances are absent, we did not consider here the correction term H_c, i.e. we set $F' = F'' = 0$ in the mapping equations. We present four different cases, all inside the resonance zone, for the same initial values of S and N but different values of the angles σ, ν. Note that the resonance depends only on the actions S, N and not on the angles. We see that for some initial angles we have chaotic motion and for other angles we have ordered motion, which is an oscillation with an amplitude that may be in some cases large. We also show the time evolution of the eccentricity of the asteroid. We see that in the case of chaotic motion the eccentricity may stay small for a long time and suddenly may jump to high values at unpredictable instants. Outside the resonance zone the evolution is always like the one shown in Figure 7d.

High eccentricity resonances present

We include now to our model the high eccentricity resonances, by considering the correction term H_c. We find that as long as the value of the eccentricity remains small, the evolution is essentially the same with the evolution obtained by the model where the high eccentricity resonances are not present. If however the eccentricity becomes large

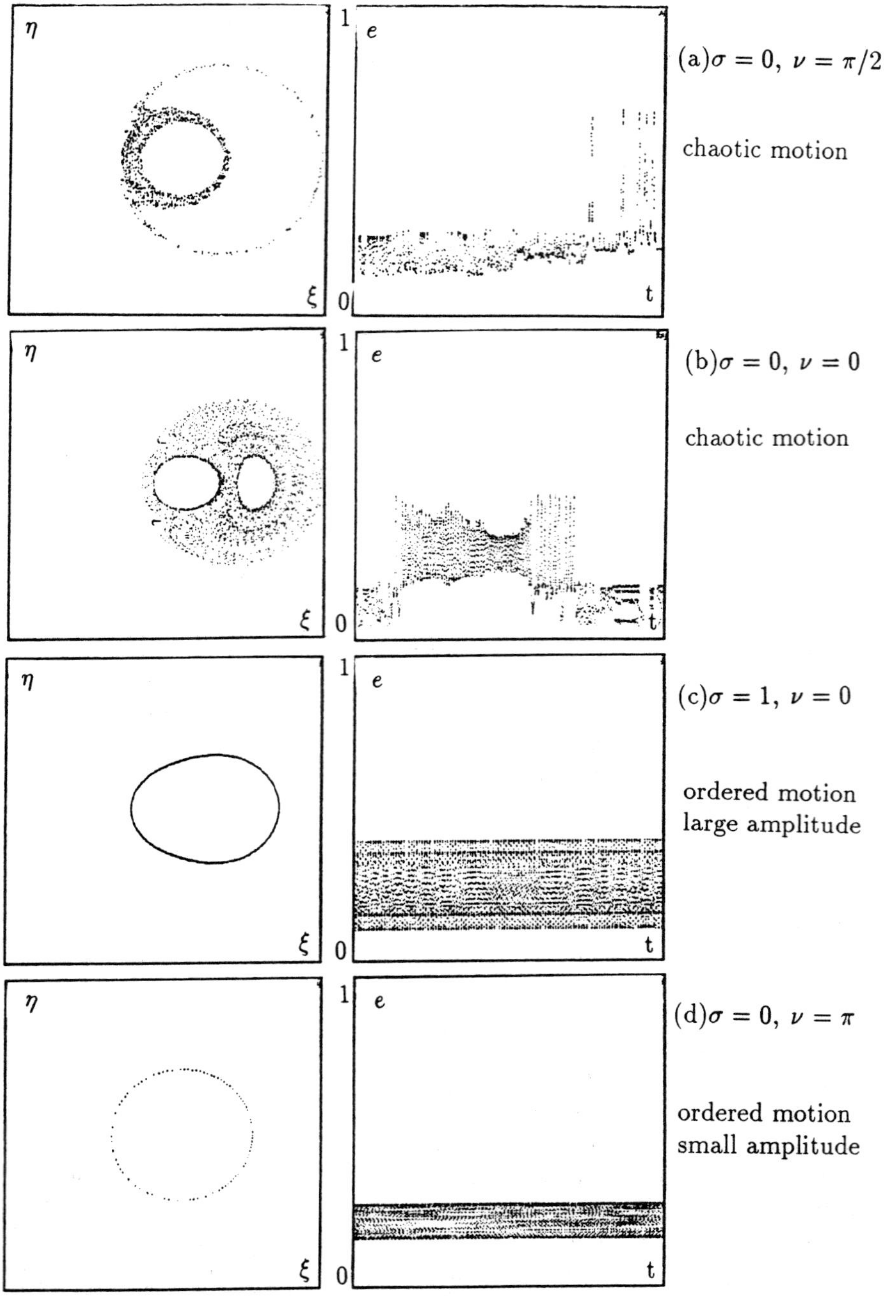

Figure 7. The evolution of an asteroid, in the $\xi\ \eta$ plane and the te plane for $S = 0.0116$, $N = 1.4$, and (a)$\sigma = 0$, $\nu = \pi/2$, (b)$\sigma = 0$, $\nu = 0$, (c)$\sigma = 1$, $\nu = 0$, (d)$\sigma = 0$, $\nu = \pi$. In all cases the evolution is for 3.6 million years. In the planes ξ, η the distance from the center of the figure is proportional to the eccentricity of the asteroid.

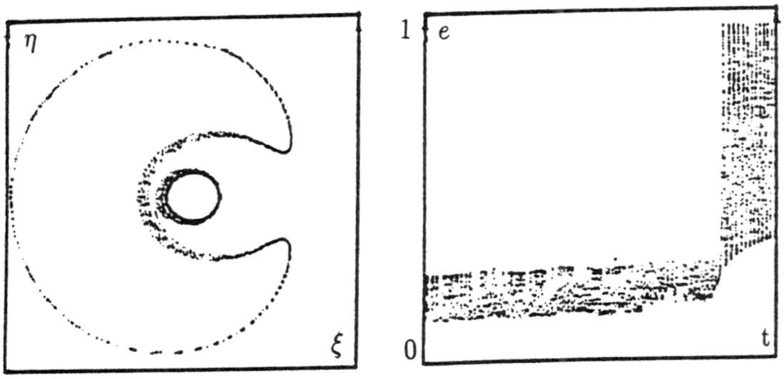

Figure 8. The evolution of an asteroid for the same initial conditions as in Figure 7a, with the high eccentricity resonances present. The scale here is larger.

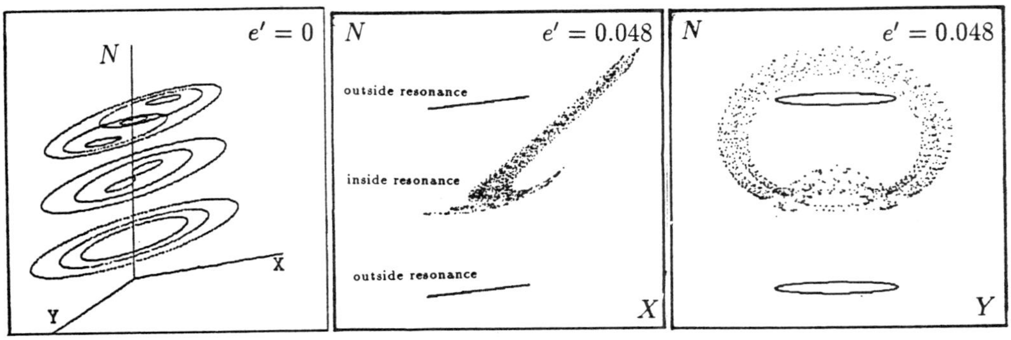

Figure 9. The evolution obtained by the mapping (23)-(26), in the space XYN, for $e' = 0$ and $e' \neq 0$.

enough, the effect of the high eccentricity resonances is important and the eccentricity is driven to values much higher than in the former case. This is clearly seen in Figure 8, for the initial conditions of Figure 7a. We can see that the inclusion of the high eccentricity resonances changes the topology of the phase space at large distances from the center of the $\xi\eta$ plane, that corresponds to large values of the eccentricity.

3.4 The mechanism of generation of chaos

As we mentioned before, the nonzero eccentricity of Jupiter is responsible for the coupling between the two degrees of freedom (S, σ) and (N, ν), so that N is no longer constant, as in the $e' = 0$ case. This slow variation of N is the basic reason that chaotic motion is generated near a resonance, if we are close to a critical curve. This can be visualized if we consider the mapping (23)-(26) in the three dimensional space $X\,Y\,N$, where X, Y are the Poincaré variables defined by (7). Consider first the case $e' = 0$. The first two equations (23), (24) are uncoupled from the other two and we have the two dimensional mapping (14). The third equation reduces in this case to $N_{n+1} = N_n$ =constant. The motion of the asteroid is restricted to a fixed plane N =constant (Figure 9a). The phase diagrammes on these planes are those shown in Figure 2. As soon however as e' is no longer zero, the equations in S, σ are coupled to the equations in N, ν. Consequently, N is no longer constant and the asteroid moves slowly from one N=constant plane to the next because, as we see from Equation (25), $(N_{n+1} - N_n)$ is proportional to $\mu e'$. The

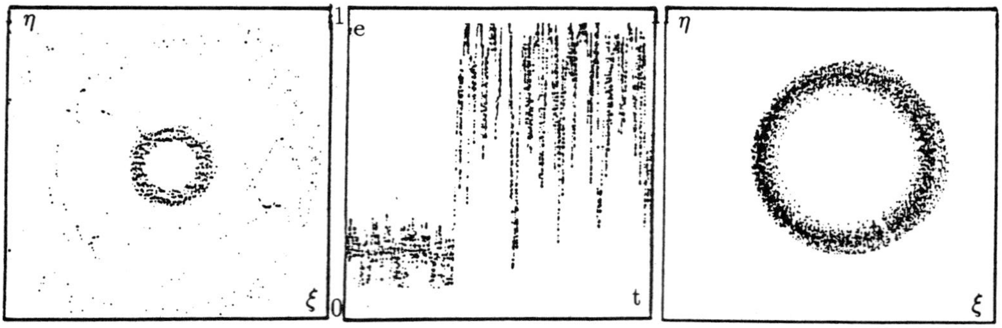

Figure 10. The effect of the gravitational attraction from Saturn.

variation of N can be quite large, as we see in Figures 9b,c, where we presented the evolution of the case of Figure 7b. Two projections are given, in the XN and the YN planes. This is a consequence of the fact that the topology of the phase space on the XY plane changes as N varies, due the existence of the resonance. We also show in these two figures the evolution of an asteroid outside the resonance zone, but close to it. This is shown by the smooth curves at the upper and lower parts of Figures (b) and (c). In this latter case the variation of N due to the coupling is very small. This is due to the fact that outside the resonance zone the topology of the phase space does not change as N varies. This shows clearly that the above mentioned mechanism of generation of chaos operates only inside the resonance zone.

4 The effect of the gravitational attraction from Saturn

In the mapping (23)-(26) the elements e' and ω' of Jupiter's orbit were considered as constant. This is not however realistic, because our computations extend to several million years, and during this period these elements change, due to the attraction from the other planets and mainly from Saturn. We shall include now into our model the variation of e' and ω' and we will study how this affects the evolution of the system, both for cases inside the resonance zone and outside. We use the variation of e' and ω' given by Laskar(1990). We studied again the evolution of the system by making use of the mapping (23)-(26), but now the variation of e' and ω' is also taken into account. The chaotic motion is, evidently, chaotic in this case also, but it is interesting to note that almost all cases of ordered motion *inside* the resonance zone become now chaotic. This is clearly seen in Figures 10a,b for the initial conditions of Figure 7d, where the motion is ordered and with small amplitude. The generation of chaos in this case can be interpreted by noting that the variation of ω' generates a *mixing of phases* and consequently the system is driven to that region of the resonance zone where chaotic motion appears. This mechanism does not work outside the resonance zone, as can be seen in Figure 10c, corresponding to initial conditions $S = 0.004$, $N = 1.386$, $\sigma = 0$ and $\nu = \pi/2$, which is just outside the resonance zone. For $e' = 0$ we have in this latter case a smooth invariant curve exactly like the one in Figure 7d of the resonant case.

We come thus to the conclusion that the effect of Saturn, i.e. the effect of the variation of a parameter, has as a consequence the spreading of the chaotic motion to the whole of the resonance zone, for all phases, but has no effect outside the resonance zone.

5 Discussion

We studied the generation of chaos in the motion of the asteroids in the Solar System, by making use of a mapping model. Several approximations were considered in constructing this model: first we assumed that the orbit of Jupiter is circular and studied the corresponding two dimensional mapping in the resonant action - angle variables S, σ. Next, we introduced a fixed nonzero eccentricity for Jupiter. This introduced to our model another degree of freedom and the mapping is now four dimensional, in the resonant action - angle variables S, σ, N, ν. The nonzero eccentricity of Jupiter introduced a coupling between the variables (S, σ) and (N, ν) and we showed how this coupling generated chaos inside the resonance zone, although in the $e' = 0$ case we had ordered motion. We explained why this mechanism operates only inside the resonance zone and has practically no effect outside. We also showed the consequences of using a poor model, i.e. one that does not include all the resonances of the real model, on the evolution of the system. Finally, we included in our model the variation of the elements of the orbit of Jupiter and showed that its effect is to spread the appearance of chaos to the whole of the resonance zone, but does not affect the properties of the system outside the resonance. Thus, in considering this hierarchy of perturbations, we were able to understand the basic mechanism of generation of chaos and the factors that are mainly responsible for it. The importance of the resonances in the evolution of a system was also explained.

We remark that the mapping model (23)-(26) that we used in this study is a realistic model for the actual system. The evolution obtained by this mapping is the same as that obtained by Wisdom (1983,1985), who used a different mapping, obtained by a different method. (We set $F' = F'' = 0$ in our model in the comparison with Wisdom's mapping). This is a clear indication that similar dynamical systems behave in the same way as far as the long term evolution is concerned and consequently one can use simple models to understand the properties of a dynamical system , provided they obey the criteria of section 2.2.

The above analysis helps us to develop criteria for the validity of a model, obtained by a perturbation method, compared to the real system. The asteroid motion was used as an example, but the basic conclusions apply also to all other similar dynamical systems.

References

[1] Ferraz-Mello, S. (1987): *Averaging the elliptic Asteroidal Problem near a First Order Resonance*, Astron. J. **94**, 208-212.

[2] Ferraz-Mello, S. (1988): *The High-Eccentricity Libration of the Hildas*, Astron.J.**96**, 400-408.

[3] Hadjidemetriou, J.D. (1988): *Periodic Orbits of the Planetary type and their stability*, Celest. Mec.**43**, 371-390.

[4] Hadjidemetriou, J.D. (1991): *Mapping Models for Hamiltonian Systems with application to Resonant Asteroid Motion*, in *Predictability, Stability and Chaos in N-Body Dynamical Systems*, A.E. Roy (ed), 157-175, Kluwer Publ.

[5] Hadjidemetriou, J.D. (1992): *The elliptic Restricted Problem at the 3:1 resonance*, Celest. Mech.**53**, 153-181.

[6] Hadjidemetriou, J.D. (1993a): *Asteroid Motion near the 3:1 Resonance*, Celest. Mech.**56**, 563-599.

[7] Hadjidemetriou, J.D. (1993b): *Resonant Motion in the Restricted Three Body Problem*, Celest. Mech.**56**, 201-219.

[8] Hadjidemetriou, J.D. and Ichtiaroglou, S. (1984): *A Qualitative Study of the Kirkwood Gaps in the Asteroids*, Astron. Astrophys. **131**, 20-32.

[9] Hénon, M. (1983):in G.Iooss, R.H.G.Helleman and R. Stora eds. *Chaotic Behaviour of Deterministic Systems*, North Holland Publ., 57-169.

[10] Henrard, J. (1988): *Resonances in the Planar Elliptic Restricted Problem*, in *Long Term Dynamical Behaviour of Natural and Artificial N-Body Systems*, A.E. Roy (ed), 405- 425, Kluwer Publ.

[11] Henrard, J. and Lemaitre, A. (1983): *A Mechanism of Formation of the Kirkwood Gaps*, Icarus **55**, 482-494.

[12] Henrard, J. and Lemaitre, A. (1987): *A Perturbative Treatment of the 2/1 Jovian Resonance*, Icarus **69**, 266- 279.

[13] Henrard, J. and Caranicolas, D. (1990): *Motion near the 3:1 Resonance of the Planar Elliptic Restricted Three-Body Problem*, Celest. Mech. **47**, 99-121.

[14] Laskar, J. (1990): *The chaotic motion of the Solar System*, Icarus **88**, 266-291.

[15] Lemaitre, A. (1984): *Higher Order Resonances in the Restricted Three-Body Problem*, Celest. Mech. **32**, 109- 126.

[16] Lemaitre, A. and Henrard, J (1990): *On the Origin of Chaotic Behaviour in the 2/1 Kirwood Gap*, Icarus **83**, 391-409.

[17] Lichtenberg, A.J. and Liebermann, M.A.: 1983, *Regular and Stochastic Motion*, Springer-Verlag.

[18] Nobili, A., Milani, A., Caprino, M. (1989): *Fundamental Frequencies and Small Divisors in the Orbits of The Outer Planets*, Astron. Astrophys. **210**, 313-336.

[19] Roy, A.E. (1982): *Orbital Motion*, Adam Hilger, (2nd ed.).

[20] Szebehely, V. (1967): *Theory of Orbits*, Academic Press.

[21] Wisdom, J. (1982): *The origin of the Kirkwood Gaps*, Astron. J. **87**, 577-593.

[22] Wisdom, J.(1983): *Chaotic Behaviour and the Origin of the 3/1 Kirkwood Gap*, Icarus **56**, 51-74.

[23] Wisdom, J. (1985): *A Perturbative Treatment of Motion Near the 3/1 Commensurability*, Icarus **63**, 272-289.

ON THE CONVERGENCE OF THE DISTURBING FUNCTION

S. Ferraz-Mello

[1]Universidade de São Paulo
Instituto Astronômico e Geofísico
Caixa Postal 9638
BR - 01065 - São Paulo, Brasil
e-mail Sylvio@iag.usp.ansp.br

New averaging techniques and new computational facilities for algebraic manipulation have brought new and promising possibilities for the analytical study of the motion of planets and asteroids. However, one difficulty to their general usage still remains. The usual series representations of the potential of the disturbing forces are not good enough to allow the study of large domains of the phase space. Indeed, Sundman [1] proved that the Laplacian expansion of the disturbing function converges only if

$$aF_1(e) < a'F_0(e')$$

where $a, a'e, e'$ are, respectively, semi-major axes ($a < a'$) and eccentricities.

$F_1(g)$ and $F_0(g)$ are the real functions

$$F_1(e) = \sqrt{1+e^2}\cosh w + e + \sinh w$$
$$F_0(e') = \sqrt{1+e'^2}\cosh w' - e' - \sinh w'$$

where $w = e\cosh w$ and $w' = e'\cosh w'$.

The consequences of this criterion, in the case of the asteroids perturbed by Jupiter, may be seen in Figure 1. One may note that many asteroids have eccentricities larger than the convergence radius and, therefore, the usual series expansion of the disturbing action of Jupiter on these asteroids is not valid. We have done a computer-assisted analysis of Sundman's determination of the the domain of absolute convergence of the Laplacian expansion of the disturbing function[2]. It is shown that the convergence condition is necessary and sufficient and holds even when the relative motion of the two perihelions is a libration, provided that the real variable $\varpi - \varpi'$ takes the values 0 or π.

From Newton to Chaos, Edited by A.E. Roy
and B.A. Steves, Plenum Press, New York, 1995

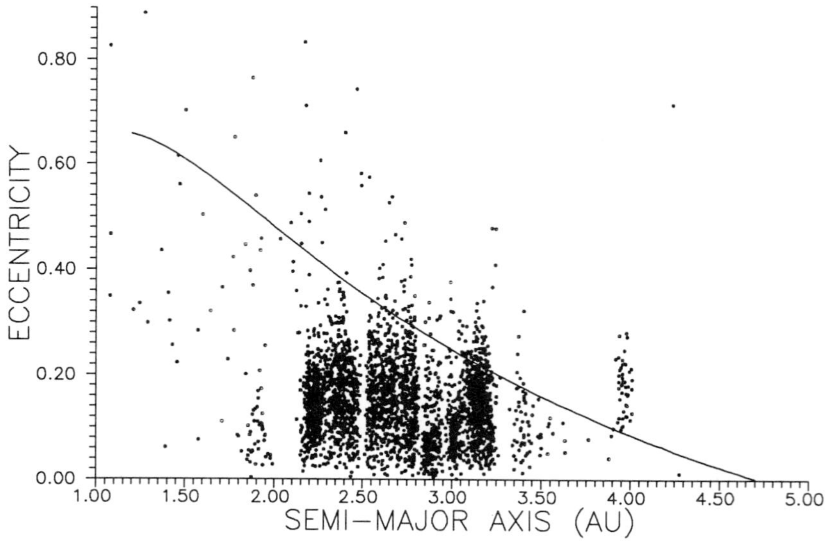

Figure 1. Limiting eccentricity for the convergence of the Laplacian expansion of the Jupiter-asteroid attraction. ($e' = 0.048$). Dots show current asteroid elements.

We have also shown that, in the case where the two orbits have a mutual inclination, it is also necessary that $|\sin\frac{I}{2}| \leq \min(\eta_I, \eta_{II}, \eta_{III}, \eta_{IV})$ where

$$\eta_I^2 = \frac{[\alpha(1+e) - (1-e')]^2}{4\alpha(1+e)(1-e')},$$

$$\eta_{II}^2 = \frac{[F_0(e') - \alpha F_1(e)][F_1(e') - \alpha F_0(e)]}{\alpha[F_0(e) + F_1(e)][F_0(e') + F_1(e')]},$$

$$\eta_{III}^2 = \frac{[\alpha(1+e) - \sqrt{1+e'^2}\cosh w']^2 - (\sinh w' + e')^2}{4\alpha(1+e)\sqrt{1+e'^2}\cosh w'}$$

and

$$\eta_{IV}^2 = \frac{[\alpha\sqrt{1+e^2}\cosh w - (1-e')]^2 - \alpha^2(\sinh w + e)^2}{4\alpha\sqrt{1+e^2}(1-e')\cosh w}.$$

($\alpha = a/a'$). We may conjecture that, in fact, this condition is necessary and sufficient since we have not found any evidence of a case where the convergence radius is less than that given by the above rule.

References

[1] K.F. Sundman, Sur les conditions nécessaires et suffisantes pour la convergence du développement de la fonction perturbatrice dans le mouvement plan, *Öfversigt Finska Vetenskaps-Soc. Förh.* **58 A** No.24 (1916).

[2] S. Ferraz-Mello, The convergence domain of the Laplacian expansion of the disturbing function, *Celest. Mech. Dyn. Astron.* **58** (1994).

IMPROVING THE CLASSICAL EXPANSION OF THE DISTURBING FUNCTION

Tadashi Yokoyama

Universidade Estadual Paulista IGCE DEMAC
Caixa Postal 178 CEP 13500-970 Rio Claro (SP)
Brasil

ABSTRACT

Following some ideas developed by Woltjer, Message and Yokoyama, an expansion of the disturbing function is given for high values of eccentricity and large amplitude of libration. The classical expansion is obtained as a particular case of the present model.

INTRODUCTION

Classical expansion of the disturbing function (R) of the restricted three body problem, when performed in the neighbourhood of the circular and planar case, may give very poor results even when only the eccentricity is not small (Henrard and Lemaître, 1987; Lemaître and Henrard, 1988). In order to circumvent this kind of trouble, many authors have tried to obtain a better representation of R (Ferraz-Mello, 1987, 1988; Ferraz-Mello and Sato, 1989; Duriez, 1988; Yokoyama, 1989; Message, 1989, 1993; Morbidelli and Giorgilli, 1990; etc.). The expansion proposed by Ferraz-Mello has been widely used in asteroidal problems by himself and collaborators. Recently, Morbidelli and Moons (1993) also used it in the study of secular resonances involving 2:1 and 3:2 comensurabilities. The development given here is a generalization of Ferraz-Mello's model, since it works even in the case of a large amplitude of libration. For the case of very slow convergence, Woltjer's transformation is also used to accelerate the fall-off of the coefficients in the series (Message, 1989; Yokoyama, 1989).

TAYLOR AND FOURIER EXPANSION

Here only a brief description will be given, since the full results should be reported elsewhere. In the disturbing function of the restricted three body problem (Sun-Jupiter-asteroid), let R be

From Newton to Chaos, Edited by A.E. Roy
and B.A. Steves, Plenum Press, New York, 1995

$$R = [r^2 + r\prime^2 - 2rr\prime\cos(v - v\prime)]^{-1/2} - rr\prime^{-2}\cos(v - v\prime)$$

where r is related to the heliocentric position vector of the asteroid, while its true longitude is indicated by v. The same variables with prime ($\prime$) are used for Jupiter's position.

The first step is to develop R in a Taylor's series around $e\prime = 0$ (Jupiter's eccentricity) and $e = e_0$ (asteroid's eccentricity). With the aid of some well known relations of the two bodies, all the partial derivatives of the Taylor series can be evaluated in closed form. Next, any term can be expanded in a double Fourier series (provided that some mathematical conditions are fulfilled) in the angles $\tilde{w} - \lambda\prime$ (longitude of the asteroid's argument of pericentre minus Jupiter's mean longitude) and l (mean anomaly of the asteroid). Since the derivatives of $e\prime$ are evaluated at $e\prime = 0$, no extra expansion is needed, and the resulting double- expanded series assumes a similar form to that of the classical case. On the other hand, the computation of Fourier coefficients may become very convenient if integration in l (mean anomaly) is switched to f (true anomaly). As a check, it is easy to see that the resulting expansion gives the classical one when $e_0 = 0$.

In the 3:1 resonance the ratio $a/a\prime$ is approximately 0.48, but in some resonances of the first order this ratio is not small and the fall-off of the terms in the Fourier series is very bad. In order to overcome this, it is possible in the case of libration to use Woltjer's transformation (Message, 1989; Yokoyama, 1989). Let us illustrate for the 2:1 resonance. Usually, since $a/a\prime$ is big, libration is a phenomenon of protection which avoids close approaches between Jupiter and an asteroid. Then if the end points of variation of the critical angle $2\lambda\prime - \lambda - \tilde{w}$ are $\pm\sigma$, let us define a new variable u:

$$\sigma \sin u = 2\lambda\prime - \lambda - \tilde{w}$$

Now it is easy to see that R can be expanded in u (2π - period) and in l (4π - period) through a Fourier series. Like before, except for the calculation of the coefficients, R can be analytically obtained up to any desired order. Some very nice combinations among the terms of the series will occur, so that this fact provides a very convenient way to write the equations of the motion.

To test the model, let us take some numbered asteroids, but fix their inclination at zero. Figure 1 shows the libration of the critical angle $\lambda - 3\lambda\prime + 2\tilde{w}$ for the asteroid Seneca ($e_0 = 0.585$). The dotted curve is obtained using the present model while the continuous one is obtained by direct integration of exact differential equations. When using the classical expansion, truncated at second or even at fourth order in e, Seneca's critical angle circulates (Klafke, 1988). For the 2:1 resonance, the best example is Zulu. Using the classical model of first order in e, its period of libration presents an error of about 40% and if second order terms are added, the centre of the libration is displaced from zero and topology is completely changed (Tsuchida, 1988).

Figure 2 shows the libration curves resulting from use of the present model with $\sigma = 1.5, e_0 = 0.478$. More detailed results and figures will be presented elsewhere.In this work only the planar problem is discussed, however the spatial case is very similar, but only requires much more calculation.

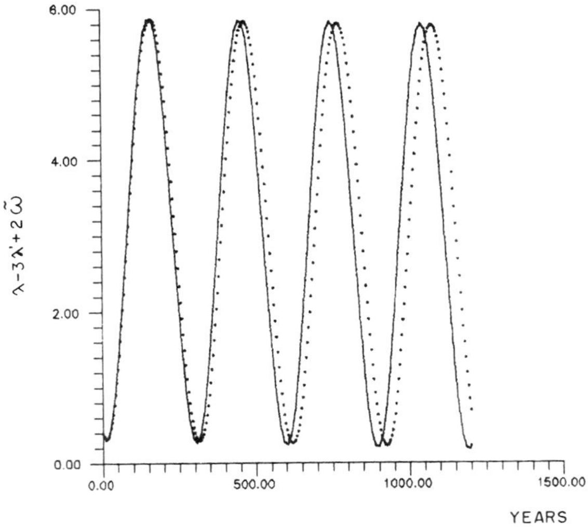

Figure 1. The asteroid Seneca. The dotted curve represents our model ($e_0 = 0.585$), while the continuous curve represents direct integration of the exact differential equations.

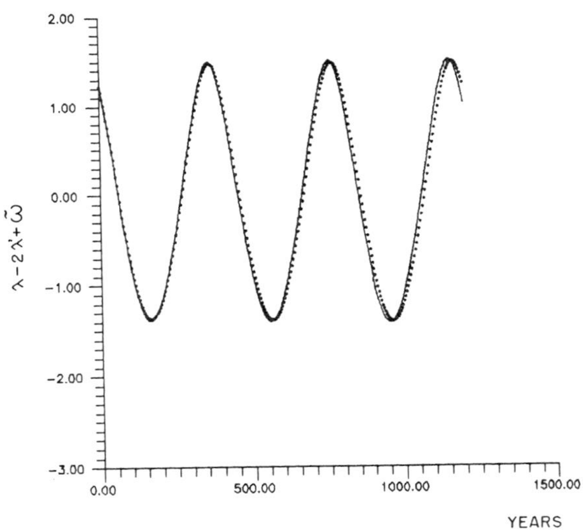

Figure 2. The asteroid Zulu. The dotted curve represents our model ($e_0 = 0.478, \sigma = 1.5$), while the continuous curve represents direct integration of the exact differential equations.

ACKNOWLEDGEMENTS

This work was supported by FAPESP through grants 91/4581-7 and 93/2240-3. The author thanks Drs S. Ferraz-Mello, P.J. Message and S.S. Fernandes for valuable discussions.

REFERENCES

1. Duriez, L., 1988, *Astron. Astrophys.* **194**:309

2. Ephemerides of Minor Planets, 1989, 1992, Institute of Theoretical Astronomy, Leningrad.

3. Ferraz-Mello, S., 1987, *Astron. Astrophys.* **183**:397

4. Ferraz-Mello, S., 1988, *Astron. Journal.* **96**:400

5. Ferraz-Mello, S. and Sato, M., 1989, *Astron. Astrophys.* **225**: 541

6. Klafke, J., 1988, Msc. Thesis, Universidade de São Paulo

7. Message, P.J., 1989, *Celest. Mech.* **45**:45

8. Message, P.J., 1993, *Celest. Mech. and Dyn. Astron.* **56**:277, 284.

9. Morbidelli, A. and Moons, M., 1993, "Secular Resonances in Mean Motion Commensurabilities: The 2/1 and 3/2 cases", pre-print.

10. Tsuchida, M., 1985, in "Resonances in the Motion of Planets, Satellites and Asteroids", S. Ferraz-Mello and W. Sessin, eds, Universidade de São Paulo, 149.

11. Yokoyama, T., 1989, in "Orbital Dynamics of Natural and Artificial Objects", R.V. Martins, D. Lazzaro and W. Sessin, eds., Observatório Nacional, Rio de Janeiro, 203.

12. Woltjer, J.J., 1928, *Ann. Sterr. Leiden,* **16**:3.

THE GREAT INEQUALITY IN A HAMILTONIAN PLANETARY THEORY

F. Varadi[1], M. Ghil[1,2] and W. M. Kaula[1,3]

[1]Institute of Geophysics and Planetary Physics
[2]Department of Atmospheric Sciences
[3]Department of Earth and Space Sciences
University of California, Los Angeles
Los Angeles, CA 90024

Abstract. The Jupiter–Saturn 2:5 near–commensurability is analyzed in a fully analytic Hamiltonian planetary theory. Computations for the Sun–Jupiter–Saturn system, extending to the third order of the masses and to the 8th degree in the eccentricities and inclinations, reveal an unexpectedly sensitive dependence of the solution on initial data and its likely nonconvergence. The source of the sensitivity and apparent lack of convergence is this near–commensurability, the so–called great inequality. This indicates that simple averaging, still common in current semi–analytic planetary theories, may not be an adequate technique to obtain information on the long-term dynamics of the Solar System. Preliminary results suggest that these difficulties can be overcome by using resonant normal forms.

INTRODUCTION

The long–term stability of the Solar System is one of the oldest unsolved problems in classical mechanics (Duncan and Quinn, 1993). Recent studies on the motion of the major planets use a variety of techniques. They range from the purely analytical (Duriez, 1979) through the semi–analytical (Laskar 1985, 1988; Wisdom and Holman, 1991) to the purely numerical (Carpino *et al.*, 1987; Applegate *et al.*, 1986; Quinn *et al.*, 1991). Our theory (Varadi and Ghil, 1993a,b) follows the ideas of Message (1982, 1988); it is analytic and fully Hamiltonian.

The dynamics of the Sun–Jupiter–Saturn system was recognized as problematic from the beginnings of perturbation theory. The problems are due to the so–called Great Inequality (GI) which is the Jupiter–Saturn 2:5 mean–motion near–commensurability. Brouwer and Van Woerkom (1950) (see also Knezevic, 1986), being aware of this, included some extra terms in their expansions, trying to account for the effects of the GI. We wanted to test our theory on this undoubtedly difficult case, and were interested in any signs of nonconvergence. Alternatively, in the case of apparent convergence, one would like to know the appropriate truncation of the expansions.

In some recent numerical integrations, evidence of chaos – defined as the presence of a positive Lyapunov exponent – has been found in the motion of the major planets

From Newton to Chaos, Edited by A.E. Roy
and B.A. Steves, Plenum Press, New York, 1995

(Laskar, 1990; Laskar *et al.*, 1992; Sussman and Wisdom, 1992). There is no consensus yet regarding the exact nature of this chaos or the unpredictability of planetary motion it might entail. It is natural to ask whether the GI is at least partly responsible for the apparently chaotic behavior of the Jovian planets.

Our computations extended to the 3rd order in the masses and to the 8th degree of eccentricities and inclinations. To our surprise we obtained vastly different values of the orbital parameters for small changes in the initial data. The expansions themselves are apparently nonconvergent. We repeated the computations without the GI terms. As a result, the sensitive dependence disappears and the expansions appear to converge. This is a positive result: perturbation theory does work for small perturbations. Averaging, however, as it is normally carried out using Lie–series or other methods for the necessary transformations in both Hamiltonian and non–Hamiltonian theories, is not adequate to deal with the GI.

One might infer from this evidence that the GI generates, through some yet unknown mechanism, chaotic motion. We do not think that this is the case since perturbation theory has its limitations. In order to illustrate these limitations, we present a simple example when a certain type of perturbation theory leads to the wrong conclusion.

We also have evidence that the GI can be dealt with existing techniques based on resonant normal forms. Preliminary computations, based on earlier results (Varadi, 1989), indicate that the nonconvergent behavior is not present when the appropriate normal form is used.

AN EXAMPLE OF AN INADEQUATE PERTURBATION THEORY

In order to put our GI results into the proper perspective, we present an example where classical time–dependent perturbation theory fails. This is intended to demonstrate that the failure of a particular perturbation theory does not necessarily mean that the problem cannot be solved by means of a better theory.

The issue of secular terms in the semi–major axes in non–Hamiltonian planetary theories occupied many researchers during the last century. We think that our deceptively simple example is a good demonstration of the basics of the phenomenon, without getting lost in the technical details that arise in actual planetary computations.

We applied the method of successive approximations (i.e., Picard iteration) used in classical perturbation theory (Brouwer and Clemence, 1961) to the case of a pendulum subject to a small gravitational force. The Hamiltonian is $p^2/2 + \epsilon \sin q$. The zeroth–order solution is $q = t, \quad p = 1$. The third–order solution is

$$q = t + \epsilon(-1 + \cos t) + O(\epsilon^4) \,, \tag{1a}$$

$$p = 1 - \epsilon \sin t + \epsilon^2 \left(-\frac{1}{2} + \cos t - \frac{\cos t^2}{2}\right) + \epsilon^3 \left(-\frac{t}{2} + \frac{7 \sin t}{8} - \frac{\sin 2t}{4} + \frac{\sin 3t}{24}\right) + O(\epsilon^4) \,. \tag{1b}$$

There is a secular term of type $\epsilon^3 t$ in the solution. This is clearly due to the perturbation method used, since p remains bounded in the actual system. The secular term appears in the third–order solution first, i.e., at the same order as in the classical perturbation theory of planetary motions.

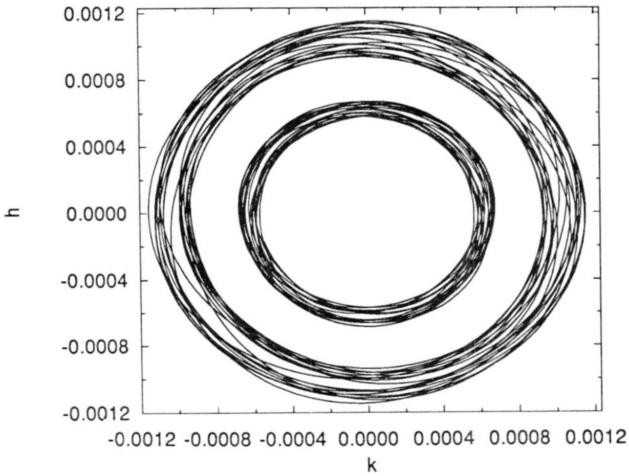

Figure 1. Jupiter's reduced (h, k) coordinates for runs A (inner curve) and Q (outer curve); see text for abbreviations.

THE EFFECTS OF THE GREAT INEQUALITY

The Poisson–series processor described in Varadi and Ghil (1993b) has been replaced by a more efficient one. In a typical run for the Sun–Jupiter–Saturn computation at hand, which takes about one day on a Sparc workstation, we are able to compute the Poisson bracket of two expansions having half a million terms each.

For the sake of brevity we use the letter O to refer to order in the masses, and the letter D to refer to degree in eccentricities and inclinations. The secular Hamiltonian includes all terms up to O1D8, O2D6 and O3D4. We call the result of the Birkhoff normalization of this secular system the full solution.

We used two sets of initial data. The first one, labeled A, is the initial data of the Applegate *et al.* (1986) integration, the second one, Q, is that of Quinn *et al.* (1991). These two are supposed to correspond to the same Solar System at slightly different

Table 1. Comparison accross runs.

	A	Q	AZ	QZ
Jupiter's semi–major axis	100.000	99.968	100.000	99.968
Saturn's semi–major axis	100.000	100.078	100.000	100.078
Period of GI	100.000	138.857	100.000	138.857
Frequency of f_5 at O1D2	100.000	99.775	100.000	99.775
Frequency of f_5 at O2D6	100.000	103.357	95.407	95.141
Frequency of f_6 at O2D6	100.000	112.950	81.520	81.195
Frequency of g_6 at O2D6	100.000	99.645	99.900	99.542
Amplitude of f_5 in J(h,k)	100.000	100.533	97.968	97.225
Amplitude of $-f_5 + 2f_6$ in J(h, k)	100.000	165.251	9.125	9.474
Amplitude of $-f_5 + 2f_6$ in S(h, k)	100.000	168.886	6.637	6.850
Amplitude of $-2f_5 + 3f_6$ in J(h, k)	100.000	211.582	0.381	0.411

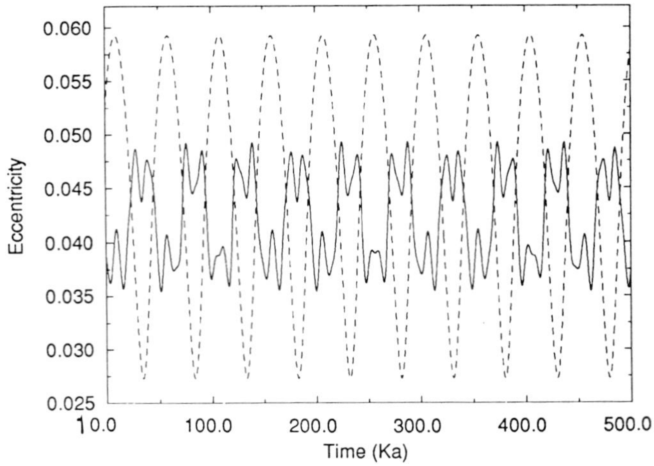

Figure 2. Jupiter's total (dashed line) and reduced (solid line) eccentricity for run Q.

epochs, i.e., most of the difference comes from different phases in the short–periodic perturbations. The transformation of initial data – necessitated by the use of Lie series (Henrard, 1970; Varadi, 1993) – was carried out only in the Birkhoff normalization of the secular system.

One advantage an analytic theory has over numerical integration is that one can eliminate certain terms from the expansions. We carried out computations where all terms related to the GI were eliminated. We refer to these calculations as runs AZ and QZ, respectively.

Solutions for the Jovian planets were computed at various levels of truncation. For initial data A we obtained the same results as in Varadi and Ghil (1993a,b), at the same level of truncation. In the light of the present results, it appears that we were lucky; the truncation happened to be such that the results were relatively close to the results of numerical integrations.

Our results are summarized in two tables and three figures. We use the classical notation of Brouwer and Clemence (1961), i.e., f refers to the frequency of the longitude of perihelion, while g refers to that of the ascending node.

Table 1 compares the solutions we obtained. The numbers are normalized with respect to the appropriate entries in the first column. The amplitudes are for the full

Table 2. Values of the secular Hamiltonian for runs Q and QZ.

	Degree 2		Degree 4		Degree 6		Degree 8	
	Q	QZ	Q	QZ	Q	QZ	Q	QZ
Order 1	100.000	100.000	0.5796	0.5796	0.3156e-2	0.3156e-2	1.0845e-4	1.0845e-4
Order 2	108.169	108.169	8.3584	0.8050	4.7429	0.3594e-2		
Order 3	108.172	108.306	4.2899	0.7936				

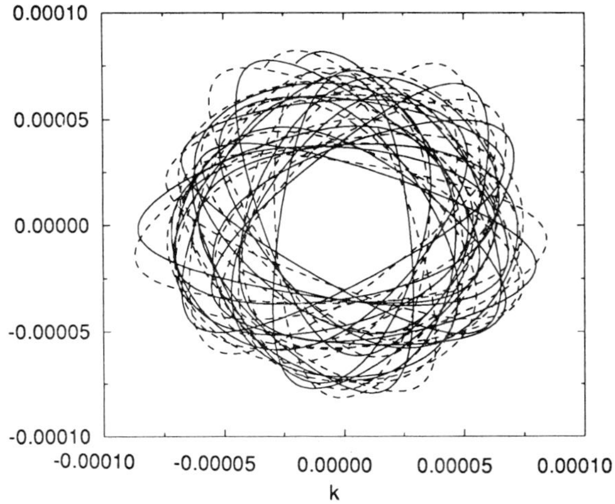

Figure 3. Jupiter's reduced (h, k) for runs AZ (solid line) and QZ (dashed line).

solution. A 0.19% change in initial, instantaneous semi-major axes can cause about 40% change in the period of the GI. For runs A and Q this leads to a 13% change in f_6. The amplitude of the $-f_5 + 2f_6$ term, which is the largest combination tone of the fundamental frequencies, changes by over 60% between A and Q. The smaller combination terms show even bigger differences, e.g., $-2f_5 + 3f_6$ differs by 111%. The cause of the large differences is obviously the GI, since the differences between AZ and QZ are much smaller and of the expected order for regular perturbations.

Table 2 lists the values of the secular Hamiltonian for Q and QZ. For each order we list the contribution of the various degrees; the numbers are normalized with respect to the value for O1D2, the classical Lagrange–Laplace secular system. The case of QZ shows what one would expect from a convergent expansion: rapidly decreasing values with respect to degree and small changes with respect to order. The case of Q shows none of these. The value of the secular Hamiltonian changes by 4% between O2 and O3 for the D4 terms and by another 4% between O2D4 and O2D6. The Lie-series generator for the Birkhoff normalization will obviously behave even worse. In fact, its value for the O2D6 terms is larger than for the O2D4 terms (not shown).

To visualize the nature of these differences, we plotted the solution for the variables $(h = e \sin \varpi, k = e \cos \varpi)$ over 500 thousand years in Figures 1–3, e being the eccentricity and ϖ the longitude of perihelion. When one plots the full solutions (not shown) the difference in the fundamental frequencies, i.e., f_5 and f_6, is clearly visible. We define as a reduced solution one omitting the f_5 and f_6 terms. The reduced solution curves are dominated by the $-f_5 + 2f_6$ term, so they exhibit the relatively narrow rings in Figure 1. The difference in the amplitudes of the two reduced solutions clearly generates two distinct rings. The picture is reminiscent of period-doubling bifurcation but we have not verified this to be the case.

Figure 2 shows how the fundamental terms and the reduced solution are related. We plotted the total eccentricity and the reduced solution's eccentricity for 500 thousand years; the latter is magnified by about a factor of 50 for better visualization. The reduced solution has a high and a low eccentricity state with regular and fast transitions between them. Its phase is opposite to that of the fundamental terms.

The behavior of the AZ and QZ solutions in Figure 3 shows only differences of phase and not of amplitudes. The scale of this plot is about one-tenth of that in Figure 1. Moreover, the solutions are not dominated by the $-f_5 + 2f_6$ term. Because of lack of space we are not able to show The eccentricity plot for AZ and QZ (not shown) does not exhibit the same bimodality as Figure 2, although the reduced solution still tends to counteract the fundamental terms.

Having shown that simple averaging does not work for the Sun-Jupiter-Saturn system, our next task is to carry out the details of a resonant normal form (Varadi, 1989) for the GI. Our preliminary results indicate that an apparently convergent theory can be obtained for all four Jovian planets. This brings us back to the quandary mentioned in the introduction: chaos in numerical integrations, on the one hand, and quasiperiodicity in an analytic theory, on the other, make for a very puzzling scenario, indeed.

Acknowledgments: The financial support of NSF Grant ATM90-13217 (M. Ghil and F. Varadi), NASA Grant NAGW-2269 (W. M. Kaula and F. Varadi) and a Guggenheim Fellowship (M. Ghil) is gratefully acknowledged.

REFERENCES

Applegate, J. H., Douglas, M. R., Gursel, Y., Sussman, G. J. and Wisdom, J., 1986, The outer Solar System for 200 million years, *Astron. J.* 92:176

Brouwer, D. and Clemence, G. M., 1961, "Methods of Celestial Mechanics", Academic Press, Orlando

Brouwer, D. and Van Woerkom, A. J. J., 1950, The secular variations of the orbital elements of the principal planets, *Astron. P. Amer. Eph.* 13(2):81

Carpino, M., Milani, A., and Nobili, A. M., 1987, Long-term numerical integrations and synthetic theories for the motion of the outer planets, *Astron. Astrophys.* 181:182

Duncan, M. J. and Quinn, T., 1993, The long-term dynamical evolution of the Solar System, *Annu. Rev. Astron. Astrophys.* 31:265

Duriez, L., 1979, Approche d'une théorie général planétaire en variables elliptiques héliocentriques, Thèse de doctorat, Lille

Henrard, J., 1970, On a perturbation theory using Lie transforms, *Celest. Mech.* 3:107

Knezevic, Z., 1986, Secular variations of the major planets' orbital elements, *Celest. Mech.* 38:123

Laskar, J., 1985, Accurate methods in general planetary theory, *Astron. Astrophys.* 144:133

Laskar, J., 1988, Secular evolution of the Solar System over 10 million years, *Astron. Astrophys.* 198:341

Laskar, J., 1990, The chaotic motion of the Solar System: a numerical estimate of the size of the chaotic zones, *Icarus* 88:266

Laskar, J., Quinn, T. and Tremaine, S., 1992, Confirmation of resonant structure in the Solar System, *Icarus* 95:148

Message, P. J., 1982, Asymptotic series for planetary motion in periodic terms in three dimensions, *Celest. Mech.* 26:25

Message, P. J., 1988, Planetary perturbation theory from Lie series, including resonance and critical arguments', *in* "Long-Term Dynamical Behaviour of Natural and Artificial N-body Systems", A. E. Roy, ed., Kluwer Academic Publ., Dordrecht

Quinn, T. Q., Tremaine, S. and Duncan, M., 1991, A three million year integration of the Earth's orbit, *Astron. J.*, 101:2287

Sussman, G. J. and Wisdom, J., 1992, Chaotic evolution of the Solar System, *Science*, 257:56

Varadi, F., 1989, Hamiltonian perturbation theory applied to planetary motions, Ph.D. Thesis, University of California, Los Angeles

Varadi, F., 1993, Branching solutions and Lie series, *Celest. Mech. Dyn. Astron.*, 57:517

Varadi, F. and Ghil, M., 1993a, Hamiltonian planetary theory. Part I: Theoretical preliminaries, submitted to *Celest. Mech. Dyn. Astron.*

Varadi, F. and Ghil, M., 1993b, Hamiltonian planetary theory. Part II: The secular system and numerical results, submitted to *Celest. Mech. Dyn. Astron.*

Wisdom, J. and Holman, M., 1991, Symplectic maps for the *n*-body problem, *Astron. J.*, 102:1528

STABILITY OVER EXPONENTIALLY LONG TIMES IN THE PLANETARY PROBLEM

Laurent Niederman
Astronomie et Systémes Dynamiques
3 rue Mazarine
75006 Paris
France
and
Université d'Evry Val d'Essonne
D.M.I.
Boulevard des Coquibus
91025 Evry Cedex
France

ABSTRACT

In this article, we apply to the planetary problem the theorems proved by Lochak about stability for near integrable Hamiltonian systems over timescales which are exponentially long with respect to the inverse of the size of the perturbation. The Lochak's theorems allow for important improvements of the original results obtained by Nekhorochev. Thus, we can give analytic results of stability in the problem with n planets (n arbitrary), whose eccentricities and inclinations are comparable with those met in our solar system. The threshold of application of these theorems on the masses of the planets are still low but can be improved and, perhaps, be pushed close to realistic values by making one (or more) numerical averaging on the system associated with the planetary problem before applying our theorems.

INTRODUCTION

By comparison with the solar system, we call the planetary problem the study of the dynamics of n bodies (which are still called planets) in gravitational interaction moving along nearly circular and coplanar trajectories in the same direction around a much heavier central body (which corresponds to the Sun).

If the interactions among the planets are neglected, the n+1 bodies system becomes uncoupled in n systems with 2 bodies - planet + sun - whose solutions are easily determined: two bodies in gravitational interaction describe conics around their centre

From Newton to Chaos, Edited by A.E. Roy
and B.A. Steves, Plenum Press, New York, 1995

of mass, and here the planets travel along ellipses in the unperturbed problem.

The stability of the planetary n-body-problem has been studied since the eighteenth century. Throughout the eighteenth and the nineteenth century, the analytic results of stability on this problem were obtained with classical perturbation methods. In the *Méthodes Nouvelles de la Mécanique Céleste*, Poincaré proved the divergence of the Lindstedt series which appears when we use classical perturbation methods to study a slightly perturbed integrable system. Thus, these methods can give a stability result only over finite times. More precisely, we obtain times of stability which are polynomial with respect to the inverse of the size of the perturbation when we study orbits whose initial conditions are located in an area without resonances of low order.

In 1963, Arnold [1] proved the existence of invariant tori in the planar planetary problem with two planets, this result has been extended recently to the spatial case, at least for inclinations comparable to those observed in the solar system, by Robutel [2]. This shows that many orbits in the planetary problem lie on an invariant torus and are quasi-periodic which ensures stability over infinite times, but there are still a lot of problems. At first, the proof of these theorems are valid only in an area of phase space composed of nearly circular trajectories. More, according to the estimates known at present, the masses of the planets must be extremely small compared to the mass of the central body. For example, these results cannot be applied to study the movement in the Sun-Jupiter-Saturn system, though the numerical explorations show a strong stability in the planetary three-body-problem with realistic parameters (see [2]). Finally, these methods cannot give effective results of stability in the planetary problem. Indeed, in a system with three or more degrees of freedom, an orbit not located on an invariant torus can drift in a large part of the phase space because of Arnold's diffusion or other mechanism (see Laskar [3]), and we cannot say if a particular orbit lies on an invariant torus or not, even if the frequency analysis method developed by Laskar in [3] should give important indications in this direction.

Giving up the search of some geometric sets invariant under the perturbed flow and going back to perturbation methods, N.N.Nekhorochev ([4]) proved a global stability result for near integrable Hamiltonian systems over timescales which are exponentially long with respect to the size of the perturbation. More precisely, he showed that action variables for the unperturbed system become quasi-integral for the perturbed system over a long time. This gives effective stability results over longer timescales than those obtained with classical perturbation methods. We obtain also lower bounds to the times over which the instabilities in systems with three or more degrees of freedom become important.

Nekhorochev also noticed that his theorem could be applied to the planetary n-body-problem.

In 1990, P.Lochak in [5] (see also [6], [7]) obtained the same kind of results with a proof which is completely different to the original one and gives important improvements of Nekhorochev's original results.

I have applied the Lochak's theorems to the planetary problem in an entirely constructive way. This work is presented here and has been developed in [8]. It will also be the subject of an article to come.

PLANETARY PROBLEM

A system of variables particularly fit for the planetary problem is given by the canonical heliocentric coordinates introduced by Poincaré. These coordinates have been entirely developed and implemented by Laskar in [9].

The masses of the planets (resp. to the Sun) are denoted as $(m_j)_{j=1,\ldots n}$, (resp. m_0).

The canonical heliocentric coordinates are denoted $(\mathbf{r}_j; \tilde{\mathbf{r}}_j)_{j=1,\ldots n}$, where we take the central body as the origin in the position space to define the position vector $\mathbf{r}_j$, and the conjugate quantites are $\tilde{\mathbf{r}}_j = m_j \dot{u}_j$ where $\dot{u}_j$ is the speed of the planet with respect to the barycentre of the system.

In these coordinates, the equations associated with the planetary problem are governed by the Hamiltonian $H = H_0 + H_1$ with :

$$H_0 = \sum_{j=1}^{n} H_{0j} \text{ where } H_{0j} = T_{0j} + U_{0j} = \frac{1}{2}||\tilde{\mathbf{r}}_j||^2(\frac{1}{m_0} + \frac{1}{m_j}) - G\frac{m_0 m_j}{||\mathbf{r}_j||}$$

and

$$H_1 = T_1 + U_1 = \sum_{0<j<k\leq n} \frac{\tilde{\mathbf{r}}_j \tilde{\mathbf{r}}_k}{m_0} - G \sum_{0<j<k\leq n} \frac{m_j m_k}{\Delta_{jk}} \text{ where } \Delta_{jk} = ||\mathbf{r}_j - \mathbf{r}_k||$$

The Hamiltonian H_{0j} (j = 1,...,n) corresponds to the study of the movement of a body of reduced mass $\beta_j = \frac{m_0 m_j}{m_0+m_j} = m_0\tilde{\beta}_j$ around a fixed center of mass $\mu_j = G(m_0 + m_j)$, this is the Kepler problem. We notice one advantage of the use of the heliocentric coordinates: it allows for an effective reduction of the mass centre and the Kepler problem appears directly instead of two-body-problems with one planet and the Sun.

H_1 is composed of a complementary part of kinetic energy and a term which concerns the mutual interactions among the planets.

The total Hamiltonian H is analytic on the phase space R^{6n} except on the manifold which corresponds to collisions between two bodies. In an area of the phase space where the bodies are not too close, the relative size of H_1 with respect to H_0 is of the order of $\beta = Sup(\tilde{\beta}_j)$; and the hypothesis $m_j << m_0$ for j=1,...,n becomes $\beta << 1$.

The solutions of the Kepler problem are known: the bodies of reduced mass $(\beta_j)_{j=1,\ldots,n}$ follow fixed ellipses in the space which is located with the classical elements:

a_j: semi-major axis; e_j: eccentricity;
i_j: inclination of the ellipse plane compared to the reference plane;
ω_j: longitude of the pericentre compared to the node axis;
Ω_j: longitude of the ascending node compared to the abcisses axis;
M_j: mean anomaly

The first five parameters remain constant, the angle M allows the location of the body on the ellipse formed and varies linearly with time.

Thus, the Hamiltonians H_{0k} are integrable and a set of action-angle variables for the Kepler is given by the Poincaré elliptic variables which are related to the classical elements of the osculating ellipses by the relations:

$$\Lambda_j = \beta_j\sqrt{\mu_j a_j}\ ;\ \lambda_j = M_j + \omega_j + \Omega_j;$$
$$\xi_j + \sqrt{-1}\eta_j = \sqrt{2\Lambda_j}\sqrt{1 - \sqrt{1-e_j^2}}\exp[-\sqrt{-1}(\omega_j + \Omega_j)];$$
$$p_j + \sqrt{-1}q_j = \sqrt{2\Lambda_j}\sqrt{\sqrt{1-e_j^2}(1-\cos(i_j))}\exp[-\sqrt{-1}\Omega_j];$$

The direct product of the previous transformation enables us to define a symplectic transformation on the phase space R^{6n}. The Poincaré elliptic variables are fit for the planetary problem because this transformation is also defined with zero eccentricities or inclinations and analytic when the eccentricities and inclinations are comparable to those observed in the solar system. In these variables, the planetary Hamiltonian becomes:

$$H(\Lambda, \lambda, p, q, \eta, \xi) = -\sum_{j=1}^{n} \frac{\mu_j^2 \beta_j^3}{2\Lambda_j^2} + \sum_{0<j<k\leq n} \frac{\tilde{\mathbf{r}}_j \tilde{\mathbf{r}}_k}{m_0} - G \sum_{0<j<k\leq n} \frac{m_j m_k}{\Delta_{jk}}$$

with the perturbation expressed in the Poincaré variables.

Finally, the problem is reduced to the study of a near integrable Hamiltonian. Of course, the size of the perturbation is given by the parameter β. In the real solar system, β is given by the reduced mass of Jupiter and is equal to about 10^{-3}.

STABILITY OVER EXPONENTIALLY LONG TIMES FOR NEAR INTEGRABLE HAMILTONIAN SYSTEMS

We look at systems governed by the Hamiltonian:

$$H(p,q) = h(p) + f(p,q) \text{ with } (p,q) \in G \times T^n, T = R/Z,$$

where G is a convex open set in R^n and (p,q) are action-angle variables of the Hamiltonian h.

Let ρ, σ be two positive constants and $|\Im(q)| = Sup_i(|\Im(q_i)|)$, we assume that H is *analytic* on the complex domain:

$$\mathcal{D} = \{(p,q) \in C^{2n}, dist(p,G) \leq \rho; \Re(q) \in T^n; |\Im(q)| \leq \sigma\}$$

which "thicken" the real domain of analyticity in C^{6n}; H is supposed to be continuous on the boundary of $\mathcal{D}$.

We also assume the h is a *convex* function ($\nabla^2 h(p)$ is a sign definite symmetric matrix). The perturbation f is of size ε.

Then for any initial conditions $(p(0), q(0))$ with $p(0) \in G$ not too close to the boundary, one has:

$$||p(t) - p(0)|| \leq R(\epsilon) \text{ for } |t| \leq \mathcal{T}(\epsilon), \tag{1}$$

provided the perturbation is small enough (i.e. $\varepsilon \leq \varepsilon_0$), with $R(\varepsilon)$ of order $\varepsilon^b, \mathcal{T}(\varepsilon)$ of order $exp(c/\varepsilon^a)$.

The positive numbers a and b will be called *stability exponents*, their value must be maximised to obtain the best result.

Such estimates as (1) were first obtained by N.N.Nekhorochev ([4]) in the case where h is assumed to be *steep*, a weaker requirement than convexity.

Recently, P.Lochak ([5]) reached the value $a = b = 1/2n$ in the convex case, these exponents improve Nekhorochev's ones. The value of this improvement lies mainly in that $1/2n$ is likely to be *optimal* in the sense that on longer timescales, Arnold's diffusion should switch on, leading to a drift of the action variables of order 1; this is predicted by heuristic reasonings and computations made by B.V.Chirikov and presented in [5].

But the main novelty introduced in his proof consists of focusing attention on periodic orbits, or rather periodic tori.

Let $\omega(p) = \nabla h(p) \in R^n$ be the frequency vector; if $\omega_0 = \omega(p_0)$ is rational (i.e. multiple of an integer one) then the torus $p = p_0$ is filled with closed orbits of the unperturbed system with a common period T such that $T\omega_0 \in Z^n$. With a time averaging procedure along these periodic orbits, we construct a resonant normal form with respect to ω_0 in the neighbourhood of the periodic torus which allows us to remove the non resonant harmonics up to an exponentially high order. Then, the canonicity of the equations prevents fast motions of the action variables in a direction normal to the unperturbed energy manifold. Finally, a simple geometric argument based on the convexity of the unperturbed Hamiltonian and the total energy conservation prevents also the motions of the actions tangent to the unperturbed energy surface. Thus, we obtain a local stability result in the vicinity of the periodic tori.

We must notice that the one phase averaging procedure used here is particularly simple and we have fine estimates on the analytic calculations which are made, so we will apply directly this local theorem in our applications.

The local theorem also allows us to deduce the global stability result announced. Given an arbitrary point in the phase space, or rather in the action space, it may be approximated by points corresponding to periodic tori and we can apply to the latter the local stability result. A simultaneous Diophantine approximation theorem of Dirichlet gives a minimal rate of approximation of an arbitrary vector by rational ones and the growth of the corresponding periods. This enables us to show that the open sets where the local result of stability over exponentially long times is valid recover the whole phase space.

We must notice that periodic tori correspond to resonances of maximal order (i.e. n-1) and in the vicinity of a manifold $\mathcal{R}$ associated with a resonance of order d, there is a relative abundance of periodic tori. Thus, close to the resonance surface $\mathcal{R}$, by modifying slightly the argument of simultaneous Diophantine approximation, we obtain a result of stability with the improved exponents

$$a = b = \frac{1}{2(n-d)}$$

The best times of stability appear when we study an orbit with a very resonant initial condition. This is very different from the usual proofs where resonances are viewed as a hindrance, whereas they become here an opportunity.

SPECIFIC DIFFICULTIES OF APPLICATION OF THESE THEOREMS TO THE PLANETARY PROBLEM

From now on, we state original results.

First, the Kepler problem is degenerate because the associated Hamiltonian does not depend on all the action variables but only on the semi-major-axes of the planets. Thus, we must extend the Lochak's theorems in the case of degenerate systems where the equations are governed by a Hamiltonian of the type:

$$H(p,q,x,y) = h(p) + \varepsilon f(p,q,x,y) \quad \begin{array}{l} p \in K \subset R^n \\ q \in T^n, T = R/Z \\ (x,y) \in L \subset R^{2m} \end{array}$$

with the conjugate variables (p,q) (called fast actions and fast angles) and (x,y) (degenerate variables).

Under the hypothesis:

- *h convex;*
- *H analytic* on the complex domain:

$$\mathcal{D} = \left\{ (p,q,x,y) \in c^{2(m+n)}; \begin{array}{l} -\Re eq \in T^n \\ -dist(p,K) < \rho_1 \\ -|\Im mq_j|_{j=1,\dots,n} < \sigma \\ -dist((x,y),L) < \rho_2 \end{array} \right\}$$

where ρ_1, ρ_2 and σ are positive constants;

- $x(t)$ and $y(t)$ *remain* in L during the time of stability;

then we can make the same reasoning as in the non-degenerate case. But here, the degenerate variables become parameters except in the averaging procedure, which imposes the condition of analyticity of the total Hamiltonian H with respect to the degenerate variables, the other parts of the reasoning concern only the dynamics of the fast action p.

So, we are reduced to study the dynamic associated with an n degrees of freedom system and we can give bounds to the variations of the fast actions with the exponents $a = b = 1/2n$. In the same way, the improved results of stability in the vicinity of the resonances are still valid, but they concern here only resonances between the fast frequencies associated with the actions p. On the other hand, all these theorems do not allow for any control on the degenerate variables.

Looking at the geometry involved, we must notice that the periodic tori considered here are n-dimensional (i.e. they are not of maximal dimension), and the local stability theorem is valid in the vicinity of the union of periodic tori contained in the analyticity domain of the total Hamiltonian and which have the same projection in the subspace of the fast action variables. This union forms a manifold of codimension n in the phase space $R^{2(n+m)}$.

As we have seen, the Kepler Hamiltonian in action angle variables is convex, thus the first hypothesis of the previous theorem is satisfied.

In the planetary problem, these theorems give bounds on the semi-major-axes of the planets which are the fast action variables, the degenerate variables correspond to the excentricites and the inclinations of the studied trajectories. Even if the semi-major-axes are very different, in case of large eccentricities, collisions or close encounters between the bodies are possible and the total Hamiltonian cannot be analytic or the size of the perturbation can be important. We use the conservation of the total angular momentum to avoid this problem.

The projection of the total angular momentum on the axis perpendicular to the reference plane (which is an integral of the system) is equal to:

$$\mathcal{N}(a_1,\dots,a_n,e_1,\dots,e_n,i_1,\dots,i_n) = \sum_{k=1}^{n} m_k\sqrt{a_k(1-e_k^2)}\cos i_k \le \mathcal{N}^+(a_1,\dots,a_n) = \sum_{k=1}^{n} m_k\sqrt{a_k}$$

and with fixed values of the semi-major-axes, the maximum of this quantity is obtained when the planets follow circular and coplanar trajectories, in the same direction. Thus, if this projection of the angular momentum for the studied trajectory is close enough to the maximum, one can expect that the eccentricities and the inclinations of the planets remain small. But we must notice that this method to control the eccentricities can be applied only if the masses of the planets are of the same order, otherwise the boundaries on the eccentricities of the heavy planets (resp. of the light planets) are too low (resp. too large) to be of practical use. Especially, this reasoning cannot be applied to the restricted three-body-problem.

Following the terminology of Nekhorochev, a domain of planetary motions will be a set in the phase space such that:

$$\mathcal{D} = \{\gamma_j \le a_j \le \delta_j \ j = 1, ..., n \text{ and } \mathcal{N}^- \le \mathcal{N}\}$$

with $\gamma_j, \delta_j (j = 1, ..., n)$ and $\mathcal{N}^-$ such that $0 < \gamma_1 \le \delta_1 < ... < \gamma_n \le \delta_n$ and $0 < \mathcal{N}^- < \mathcal{N}^+$ where $\mathcal{N}^+$ is the projection of the angular momentum in the circular planar problem with the radius δ_j. On $\mathcal{D}$, the semi-major-axes of the planets are different and if $\mathcal{N}^-$ is high enough, the eccentricities and the inclinations are small enough to prevent close encounter. Therefore, $\mathcal{D}$ is a real domain of analyticity of the total Hamiltonian where the perturbation is small.

Thus, we can apply our theorems of stability if we ensure that the trajectories remain in the domain of planetary motion during the time of stability. The invariance of the angular momentum warrants the condition $\mathcal{N}^- \le \mathcal{N}$ and the semi-major-axes stay in the disjoined intervals $]\gamma_j; \delta_j[$ if the confinement radii in our theorems are small enough.

We see that the eccentricities and the inclinations of the planets can take relatively high values (which are given by $\mathcal{N}^-$) and do not have to be considered as small parameters as in K.A.M.'s theory.

Now, we must determine a complex domain of analyticity of the planetary Hamiltonians which "thicken" the domain of planetary motion in C^{6n}. But the condition of analyticity and the size of the perturbation in the planetary Hamiltonian is given the mutual distances among the bodies which are easily expressed only in cartesian coordinates, whereas here we study the system in action-angle variables. Hence, we have to give lower bound to the distance between two points with complex Poincaré elliptic variables. The method developed here, which is presented in [8], allows us to give constructively precise estimates of the analyticity widths of the planetary Hamiltonian.

Beyond the determination of a domain of application for the theorem, this work gives a good analytical framework to apply perturbation methods to the planetary problem in a rigourous way. Indeed, we must express the perturbation function in the action-angle variables of the Kepler problem, this is only possible with the help of an analytic expansion. Of course, in the applications, we use truncate expansions of the perturbation and we must take accurately into account the influence of the remainder (which is not computed) on the dynamic. This is a problem, because the expansion of the perturbation is difficult to estimate and for the moment there is no majorizing series of this expansion. On the other hand, the use of Cauchy formulae on the complex domain determined here gives the desired estimates.

In that way, these methods to determine the analyticity widths and the deduced estimates have been used by Robutel in his application of the K.A.M. theory to the planetary problem ([2]).

RESULTS OBTAINED

Now, we have at our disposal all the parameters needed to apply the general theorems of stability to the planetary problem. Thus, we can give analytic results of stability over very long times in the problem with n planets (n arbitrary) whose eccentricities and inclinations are comparable with those met in our solar system.

We study the possibilities of application of these theorems in some particular cases.

First, we make an application of the local result of stability in the vicinity of a periodic torus in the problem with two planets. In the general case, the fast frequencies of the system on a periodic torus satisfy n-1 relations of commensurability. Equivalently, a periodic torus corresponds to a mean-motion resonance of order n-1. In the case of two planets, it is foliated in orbits of the unperturbed problem with fixed values of the semi-major-axes linked by a relation which corresponds to a single mean motion resonance. Here, the local theorem is valid in the vicinity of the union of periodic tori contained in the domain of planetary motion, associated to the same mean-motion resonance, and with a fixed value of the first semi-major-axis. We recall that this union forms a manifold of codimension n in the phase space.

The system composed of the Sun, Jupiter and Saturn in the present conditions is close to the 5/2 mean motion resonance, this incites us to apply the theorem of local stability in the neighbourhood of the periodic torus associated with this resonance with a first semi-major-axis equal to Jupiter's. In this case, the theorem ensures a time of stability of 6 billion years (which is comparable with the age of the solar system) if the ratio of the mass of the heavier planet over the Sun's is lower than 10^{-12}.

The theorem of global stability can be applied to the system: Sun, Jupiter and Saturn, with the time of stability of 6 billion years, if the mass ratio is lower than 10^{25}. This huge degradation of the threshold of application of the theorem is entirely due to the Diophantine approximation argument. We are confronted with an arithmetical problem, but the Dirichlet theorem we use is optimal if we do not assume any Diophantine conditions on the vector we approximate. In the applications, we had better determine a periodic torus in the vicinity of the initial condition of the studied trajectory and apply the local theorem around this torus. But, most of the time, we are obliged to use the local result in the neighbourhod of a torus foliated with closed orbits of long period and it follows a degradation of the threshold of application of the theorem (but less important than with the global theorem).

We can avoid this difficulty in the case of the problem with two planets or if we study solutions of the problem with n planets in the vicinity of a resonance of order n-2.

Let $\mathcal{F}_1$ be a family of periodic tori contained in the domain of planetary motion, associated to the same resonance and with a semi-major-axis of the first planet in an interval $I \subset R$. We notice that the tori in $\mathcal{F}_1$ are foliated in periodic orbits of same period. The union of tori which belong to $\mathcal{F}_1$ forms a manifold $\mathcal{M}$ of codimension n-1 in the phase space.

In the particular case of the problem with two planets, we obtain manifolds of codimension 1. By connexity, we can confine the trajectories in the portion of the total energy surface located between two manifolds $\mathcal{M}$ and $\mathcal{M}'$ which are associated to two close mean motion resonances if the masses of the planets are small enough to apply the local result of stability on the considered tori. The projection on the plane corresponding to the semi-major-axes of the area located between two manifolds $\mathcal{M}$ and $\mathcal{M}'$ provides a cone of stability; moreover, the unperturbed energy surface is a good approximation of the projection of the total energy surface. Thus, we give bounds to the variation of the semi-major-axes by estimating the intersection of a neighbourhood of the unperturbed energy surface and the cone of stability.

The same reasoning can be made if we study solutions of the problem with n planets in the vicinity of a resonance of order n-2 because of the abundance of periodic orbits in these areas.

These theorems allow us to bound the movements of a system whose initial conditions are those of the Sun, Jupiter and Saturn at present between tori associated to the 5/2 and 8/3 mean motion resonances if the mass ratio is lower than 10^{-13}. In the same way, we obtain results of stability in the problem with four planets close to two 2/1 mean motion resonances, this is very similar to the present conditions of the Galilean satellites of Jupiter, if the mass ratio is lower than 10^{-14}.

We must notice that the smallness of the thresholds of application of these theorems (even for the local results) comes essentially from the small analyticity widths of the planetary Hamiltonian even though they are carefully estimated in our study. Moreover, it must be difficult to improve the thresholds of application of the theorems of stability for a generic perturbation.

Finally, our results of stability obtained here with a direct application to the planetary Hamiltonian of the theorems proved for a generic perturbation are essentially optimal.

CONCLUSION

The theorems proved here can be compared only with the application of the K.A.M. theory to the same problem because it is the only previous analytic study of the planetary problem.

Unlike the present applications of the K.A.M. theory to the planetary problem, we have results of stability with n planets ($n \geq 2$); moreover, the eccentricities and the inclinations of the planet can be relatively high. Thus, we do not have theorems of existence of invariant tori in the two examples which have been studied here. Indeed, in the system (Sun, Jupiter, Saturn) the eccentricities of the planets are too important and in the system of the Galilean satellites we study the movement of four planets.

Finally, the thresholds of applications of our theorems are significantly greater than those which ensure the existence of invariant tori in the planetary problem (at least with the present estimates of these thresholds).

With the K.A.M. theory, it is possible to obtain, on certain problems, invariant tori with realistic parameters. To do this, it is necessary to make one, or several, preliminary numerical averaging, this allows us to exploit the specificities of the perturbation considered. But these methods have been applied on problems whose associated Hamiltonians are much simpler than the planetary Hamiltonian. Moreover, the thresholds needed for an application of the K.A.M. theory for a generic perturbation to the particular case of the planetary problem are much too small to succeed in increasing them up to realistic values with the previous methods.

On the other hand, the thresholds obtained here with stability theorems for a generic perturbation are perhaps high enough to be pushed close to a physically significant value with a preliminary numerical averaging.

REFERENCES

1. Arnol'd, V. I.: 1963, Small denominators and problems of stability of motion in classical and celestial mechanics, Russian Math. Surveys, **18**: 6, pp. 85-191.

2. Robutel, P.: 1993, Contribution à l'étude de la stabilité du problème planétaire à trois corps, Thèse de Doctorat de l'Observatoire de Paris.

3. Laskar, J.: 1993, Frequency analysis for multi-dimensional systems. Global dynamics and diffusion, Physica D **67**, pp. 257-281.

4. Nekhorochev, N. N.: 1977, An exponential estimate of the time of stability of nearly integrable Hamiltonian systems, Russian Math. Surveys, **32**, pp. 1-65.

5. Lochak, P.: 1992, Canonical Perturbation Theory via Simultaneous Approximation, Uspekhi. Math. Nauk,; English translation to be published in 1993 in Russian Math. Surveys.

6. Lochak, P., Neistadt, A.I.: 1992, Estimates of stability time for nearly integrable systems with a quasi-convex Hamiltonian, Chaos, **2**: 4, pp. 495-499.

7. Lochak, P., Neistadt, A.I., Neiderman, L.: 1992, Stability of nearly integrable convex Hamiltonian systems over exponentially long times, Proceedings of the 1991 Euler Institute Conference on Dynamical Systems, Birkhaüser.

8. Neiderman, L.: 1993, Résonances et stabilité dans le problème planétaire, Thèse de Doctorat de l'Université Paris 6.

9. Laskar, J.: 1991, Analytical framework in Poincaré variables for the motion of the solar system, Proceedings of the N.A.T.O. Advanced Study Institute, Predictability, Stability and Chaos in N-Body Dynamical Systems, Cortina, A.E.Roy (Ed), Plenum.

INTERLUDE

REMARKABLE AND SPECTACULAR COMETS

I.P.Williams

Astronomy Unit
Queen Mary & Westfield College
Mile End Rd
London E1 4NS

ABSTRACT

Comets have always been fascinating to the human race - even though most of us have never seen a truly spectacular comet. Yet, in the past there have been many that must have been a memorable sight, with some reported as being visible in daylight for a considerable period of time. This paper gives an account of a selection of past comets that the author regards as being spectacular or remarkable in some way.

INTRODUCTION

Of all the astronomical phenomena witnessed by the human race, none seems to fascinate it more than comets. Most other phenomena were recorded and explained in simple scientific terms early on, but comets were, and indeed still are in some circles, regarded as messengers of doom (or of good news, depending on your viewpoint, Harold's comet of doom was good news to William). On many occasions, this fascination with the predictive powers of the comet has got in the way of both accurate recording and meaningful descriptions of the actual appearance. Until fairly recently, the Chinese and Korean astronomers were the only people regularly recording appearances of comets (and other astronomical events) and this practice was open to ridicule by the west. For example, we have Frantz Kuhnert in 1888 (see Needham, 1959) writing

> "Probably another reason why many Europeans consider the Chinese such barbarians is on account of the support they give to their Astronomers - people regarded by our own cultivated Western mortals as **completely useless**"

In fact, for the purposes of this paper, the Chinese were probably too scientific. They recorded the time of appearances and locations on the sky but did not indulge in

From Newton to Chaos, Edited by A.E. Roy
and B.A. Steves, Plenum Press, New York, 1995

dramatic descriptions and it is the latter that is the main requirement here. However it would be unfair, in recognition of all their diligent hard work, not to start with the earliest record that I can find which perhaps refers to a comet of 1059 BC :- When King Wu-Wang waged a punitive war against King Chou, a broom star comet appeared with the handle of the broom star pointing East (See Ho Peng-PO, 1962).

The earliest record that gives an interesting description of what was seen that I have located is from Seneca, who described the comet of 147 BC in the following terms."It was as large as the Sun, reddish like fire and bright enough to dissipate the darkness". However, Seneca himself could not have seen this comet and neither was he around for the comet of 44BC, which was visible during the games that Octavian was holding in honour of the assassinated Julius Caesar and which is depicted on a Roman coin of that period.

In fact, Seneca made a very profound statement, and it is a great pity that it was not taken seriously or acted upon for 15 centuries. Seneca wrote (in Natural Questions)

> "It is essential that we have a record of all the appearances of Comets, for, on account of their infrequencies, their orbits cannot yet be discovered or examined in detail to see whether they observe periodic laws".

Seneca was also a very interesting person in his own right. He was born in Cordoba, Spain about 4BC but moved to Rome, the centre of the Empire. Being an intelligent and educated man, he soon had considerable influence on the Roman Senate, not always a good thing to have, especially when Caligula was Emperor. He presumably demonstrated his ability once too often and was condemned to death. At the time Seneca was sickly and not expected to live and fortunately Caligula was replaced before action was taken to actually kill him. Seneca's troubles were not over however; in AD 41 Messalina (third wife of Claudius) had Seneca exiled to Corsica because of alleged adultery with Julia Livilla (Caligula's sister). He spent eight years in exile before being recalled to Rome to tutor a 12 year old boy, called Nero, in Morals and literature, the former not very successfully. In due course, the boy got his revenge and in AD65 Seneca was ordered to commit suicide by Nero. This time there was no reprieve.

Seneca was born at roughly the same time as Jesus and a potentially interesting comet is that of 5BC, interpreted by some as being the star of Bethlehem (eg Humphreys 1991). The Chinese record of this comet is not in keeping with Christmas card images however, giving a bald "A broom star comet appeared near Capricornus for over 70 days". Giotto in his painting many centuries later of "The adoration of the Magi" placed a comet above the manger. Many think that this is because Giotto was inspired by the return of Halley's comet in 1301. Whether or not a comet is the explanation for the star of Bethlehem is open to debate, but we can be certain that the comet was not Halley's comet, for it appeared in 12BC, marking the death of the Roman General Agrippa, too early to be "The star" and again in AD66, well after the event

After the death of Seneca, we enter into the long period when essentially only the accurate but unromantic Chinese and Korean records exist and the few Roman records are equally brief. A slight exception is that of AD400 where a large comet was described as stretching from the sky to the ground. The Halley return of AD684 is claimed by some to be depicted in the Nuremburg Chronicles, though it is probable that this image is a stylized form for a *comet* used whenever a comet needed depicting in this era. In AD837, comet Halley made its closest known approach to Earth and the apparent rapid rotation of the tail caused by this is recorded, all four directions being covered in a few weeks. The Halley apparition of 1066 is also worthy of note, being described by a cleric in Italy in the following terms "it looked like an eclipsed Moon and its tail rose like smoke halfway up to the zenith" (Stein, 1910). This apparition is also famous because the event is depicted in the Bayeux tapestry and foretold the fall of Harold to William in the battle of Hastings. It is a much poorer scientific image of a comet than that of the Nuremburg Chronicles four centuries earlier and looks more like a jazzed up Saturn V rocket.

Some fifty years later in AD1106 another remarkable comet appeared and sightings of it were recorded in China, Japan, Korea and Europe. It was described as being the size of the mouth of a cup with its rays scattered in all directions as if broken into fragments, and is recorded as being very close to the Sun in early February. This comet may well be the progenitor of the Kreutz family of sun-grazing comets (Marsden, 1989).

Halley returned again in 1145 and this apparition, like that of 1066, lasted for an extraordinary length of time (78 days) after perihelion passage, the whole apparition occurring from April 15 to July 6. After this time, comet Halley of course appears regularly in the records and we have already commented on the 1301 apparition as the possible inspiration for Giotto. The next apparition worthy of note is that in 1531. The remarks on this may be coupled with those of a comet in 1532. Both comets were the subject of accurate observations, and indeed scientific measurements by Peter Apian at Ingolstadt. He measured accurate positions against background stars, gave tail lengths and directions and, more importantly, realized that the comet tail always pointed away from the Sun. The explanation for this latter phenomena had to wait until the discovery of the solar wind this century.

The comet of 1577 may be considered as the watershed in cometary science, certainly as far as Europe was concerned. It was discovered by (amongst others) Tycho Brahe, allegedly while fishing for his supper. In passing we should note that in common with many other cometary scientists, Tycho Brahe was a very interesting character and is usually remembered for the observational work he did which enabled Kepler to produce his three laws of planetary motion. He also wore an artificial nose, having lost his own in a sword duel with Parsbjerg over the question of which of the two of them was the best Mathematician. Perhaps in this example there lies a solution to the current age distribution of Astronomers in permanent jobs. The comet was reported as being as bright as the Moon and the vapour of its tail formed a white rainbow stretching from eastern Scorpius to western Aquarius, an arc of over sixty degrees. A number of prints from woodcuts of this comet survive. Mastalin, a pupil of Apian, was the first to point out that this comet had no apparent parallax and so could not be an atmospheric phenomenon, a result also confirmed by Tycho Brahe. Mastalin, Brahe and Horrocks tried to explain the motion of this comet in terms of closed orbits, but, rather surprisingly, none of them considered the possibility that the comet was periodic. In fact there seemed to be a marked reluctance amongst astronomers to consider the obvious. Kepler for example, having observed the comet of 1607 (Halley yet again), and produced laws about the motion of planets, nevertheless believed that comets somehow moved in straight lines through the solar system.

There were three observable comets in 1618, the first deserves mention for probably being the first comet to be observed through a telescope (by Kepler), while the third was a very impressive comet and is depicted over Heidelberg on a copper engraving, a reproduction of which can be seen for example on the cover of all three volumes of the ESA report SP 250, "The exploration of Halley's comet". For believers in the association of comets with doom and disaster, this comet marks the beginning of the "thirty years war", and before mentioning the next two spectacular comets to appear, it is interesting to digress slightly and ponder on the predictive powers of comets. John Gadbury, an astrologer in London, produced the following guide for what to expect if a comet first appeared in the relevant sign of the Zodiac. They were as follows.

Aries	Diseases affecting the head and eyes. Detriment to rich men. Sorrows and troubles to the vulgar.
Taurus	Sickness and great earthquakes. Death of a great man. Detriment to cattle, rotting of fruit.
Gemini	Diseases for children. Men commit fornication Many abortions. Prodigious wind.
Cancer	Famine, pestilence, wars. Abundance of locust, caterpillars and such worms that destroy fruit.
Leo	Vermin, rats. Detriment to great ladies. Dogs run mad in multitudes. Corn destroyed by worms.
Virgo	Detriment to merchants. Noble women subject to scandals, infamies and disgrace.
Libra	Portends thieves, housebreakers and highwaymen Extremes of heat and cold. Death of some King.
Scorpio	Wars and controversies among men. Rebellion. Scarcity of grain and fruit.
Sagittarius	Depression in noblemen.
Capricorn	Fornication and adulteries rife and common. Persecution of religious men.
Aquarius	Plague sweeps away a multitude. Terrible and durable wars. Death of a Prince or great female.
Pisces	Air repellent with prodigies. Destruction of fish.

As can be seen most are fairly nasty, though Capricorn and Virgo are not too bad, which is just as well for the great comets of 1664 and 1665 appeared in these two signs. Unfortunately, the comets themselves must have got it wrong, for London had the black death followed by the great fire, the former causing Newton to retreat to rural Cambridgeshire for safety, thus showing a certain lack of confidence in one of his creations. Newton had invented a concoction called Lucatello's Balsome, which consisted of Turpentine, Rosewater, beeswax, Olive Oil, Sack, Sandalwood and St John's Wort. This medication was to be applied externally for Colic, Burns, Bruises and Dog bites, but drank with broth for Measles, Smallpox and the Plague. It is as well to remember that as well as being a great mathematician, who ultimately gave us the laws to understand cometary motion, Newton spent a lot of time dabbling with other things, including Alchemy. A comet also gets a mention in Milton's Paradise Lost,

Satan stood
Unterrified and like a comet burn'd
That fires the length of Ophiuchus huge
In the artik sky,and from its horid hair
shakes pestilence and war.

Though comet Halley has already been referred to several times, chronologically, we are now at the time of Halley himself. We all know that Halley concluded, from the similarity of their orbits, that the comets of 1531, 1607 and 1682 were one and the same comet and that he predicted its return again for 1758. It was recovered by Messier on January 21, 1759, after a long search, but it transpired that Palitzsch had found the comet on December 27, 1758, though it is not clear that he was using any search pattern based on predictions to find the comet. In any event, the periodic nature of cometary orbits was finally established. Edmund Halley was the second Astronomer Royal, between Flamsteed and Bradley. His contributions to Science actually go far beyond cometary work, and indeed, he spent but a small part of his time working on comets. One could argue that his greatest contribution to the development of Science was to provide funds out of his own pocket to publish Newton's *Principia* when the Royal Society could not afford the publishing costs. He also designed an under-water diving bell and initiated work in life-assurance by publishing the first tables of life-expectancy.

We are now entering the age of the telescope and this clearly had a large effect on the detectability of comets, they no longer were naked eye objects that appeared without warning, but objects whose return were often expected or had been detected far away while still well below naked eye visibility. We have already mentioned that one of the comets in 1618 was probably the first to be observed through a telescope. The distinction of being the first comet to be detected by telescope belongs to the comet of 1680, when it was discovered by Kirch on November 14. One effect of being able to detect fainter comets was to give rise to what can best be described as "comet hunters". Most, like Messier and Pons did sterling work and early detection increased the time for which scientific observations could be carried out, so that the science of comets also progressed. However, one or two could not resist the temptation for instant fame, resulting in a few "discoveries" of bogus comets. One of the better known in the latter class is Chevalier D'Angos. In 1784, he announced the discovery of a comet and gave an orbit for it, but nobody else could find it. The mystery of the disappearing comet of 1784 was solved when Encke demonstrated that the 1784 orbit was in fact his (Encke) comet but with the geocentric distance multiplied by ten. In January 1798, D'Angos claimed to have observed a comet transiting the Sun. According to Lalande's standard text, the Sun was at the descending node of the orbit of the comet of 1672 at the time. Unfortunately for D'Angos, a misprint of 60° came to light in Lalande's book so that the Sun in reality was nowhere near the node of the cometary orbit!

This account is of spectacular and remarkable comets and so the vast amount of work on cometary orbits and comets in general that took place in the eighteenth century is excluded. Rather a large number of comets made a close approach to Earth during the eighteenth century, with 8 comets being closer than 2500 Earth radii. Despite this, not many were spectacular in the usual sense. The comet of 1744, Cheseaux's comet, was certainly an exception to this and is certainly worthy of mention in any century. It was brighter than Venus and visible to the naked eye only six minutes before sunrise, while it is reported as having six tails extending above the horizon.

If any century is to be regarded as exceptional in terms of the number of spectacular comets observed, then the nineteenth century must surely take the prize. The first true naked eye comet of the century was that of 1811, Flaugergue's comet. Had it not been for the succession of comets which followed, it would have ranked as a great comet, but judged by the nineteenth century it was pretty average as was Halley's return in 1835, compared to many of its previous returns, though we would have been happy with such a return in 1986. The comet of 1843 was spectacular in its own right, probably in the top ten of comets in the apparent brightness league. However, it also found fame through getting itself associated with another "end of the world" scenario. William Miller spent fourteen years studying the Bible, and at the end of this period, predicted in 1831, on the basis of a curious interpretation of the book of Daniel, that the Earth would come to an end by fire some time in 1843. By the beginning of 1843 there were over 50,000 true believers and many more who were perhaps apprehensive. The comet of 1843 first appeared in late February, visible in broad daylight, only a few degrees from the noonday Sun. By March, having moved away from the Sun, it was as bright as the brighter stars with a tail extending over an arc of 50° at least. Impressive engravings of the comet against Orion over both Paris and Blackheath are to be found. To the Millerites, this was indeed the end of the world. Fortunately for the rest of us, nothing untoward happened during 1843 and the Millerite movement gradually disappeared. This being the nineteenth century however, they did not have long to wait for the next comet, comet Biela in 1846. The Comet in itself was not spectacular, but its behaviour was certainly remarkable. As its name suggests, it was discovered (or more precisely recovered) by Biela (1826), an Austrian Army Captain, in 1826. Following observations and orbital calculations at this apparition, it was concluded that this was the same comet as those seen in 1772 and 1805 and that the orbital period was of order 6.7 years. The next return was calculated to be in September 1832, and it was duly recovered by John Herschel at Slough. The return of 1839 was very poorly placed in relation to the Sun and the comet was not seen, and so we come to the 1846 return. Santini (1844) predicted a return in 1846 which turned out to be within 10 hours of the correct time and the comet was recovered in late November 1845. By January 1846 it was clear that a fainter companion also existed 1 or 2 arcseconds North of the main object (Maury, 1846). The two comets travelled together, with tails parallel, for approximately 3 months, the maximum apparent separation being about 25000km. At the 1852 return, both comets were again present but fainter while the separation had increased to about 2 million kilometers. This was to be the last time that comet Biela or its companion were observed. Weiss (1868) calculated that the Earth would intersect the orbit of comet Biela close to the point where the comet would have been in 1872 and a strong meteor shower was observed at this time. Strong meteor showers were also observed in 1885, 1892 and 1899. The implication that comet Biela had first split into two and then totally disintegrated is obvious. An interesting twist to this story is that comet Biela probably passed through the densest part of the Leonid meteoroid stream around 1833 (Babadzhanov et al 1991). A second comet, not spectacular in its own right, but remarkable because of its associated visible effects is Comet Temple-Tuttle. This comet was discovered in 1866 and is associated with the Leonid meteor showers which produced such spectacular displays in 1799 and 1833.

Returning from this slight digression to the truly remarkable comets, the next such comet in the nineteenth century was 1858 VI, Donati's comet. This must be one of the most produced images of comets and reproductions may be found in virtually any book on comets. Three year later, comet (1861 II) Tebbutt was observed. According to Sir John Herschel,

> "it far exceeded the brightness of any comet that I have seen, those of 1811 (Flaugergue) and 1858 (Donati) not excepted. It certainly exceeded all the fixed stars or planets except perhaps Venus at maximum".

In just over a decade, comet Coggia came along in 1874, equal in brightness perhaps to Donati, but without the spectacular tail over Paris. In 1882, perhaps the brightest ever comet, Cruls came, seen in September in broad daylight only 12 degrees from the Sun. At perihelion it was claimed to be scarcely fainter than the Sun's limb and observers reported

> "a striking contrast between the comet's silvery radiance and the Sun's reddish yellow glare".

The twentieth century has had little to compare with this, though it started well with the great January comet of 1910, which many confuse with Halley also visible in that year but hardly more noteworthy than in 1986. Perhaps comet Bennett in 1970 and West in 1976 deserve to be included amongst the greats, but by this time civilization with its light pollution had ensured that these would only be visible in all their splendor to the adventurous few who sought out the darker corners of the Earth. By now we need a different definition of remarkable and comet Shoemaker-Levy 9 promises to produce this. It was discovered by Gene and Carolyn Shoemaker and David Levy on the night of 18 March 1993 with a Schmidt telescope at Mount Palomar. Subsequent images obtained by Scotti (1993) with the Spacewatch telescope at Kitt Peak showed that the nucleus was fragmented into numerous bits. It was soon realized that the reason for the fragmentation was a close passage to Jupiter in July 1992. As the number of observations increased, the orbital determination improved and it became apparent that the comet was orbiting Jupiter and further that it would collide with Jupiter in July 1994. Further observations have tied the orbit down further and it is confidently expected that the first fragment will collide with Jupiter on July 16 and the last on July 22nd (Sekanina, Chodas & Yeomans, 1994). The impact site is on the face of Jupiter that is away from the Earth, but near the limb, so that the rotation of Jupiter will bring the impact site into Earth visibility less than 20min after the impact. This will be the first observed collision of a comet with another solar system major body and so Comet Shoemaker-Levy 9 deserves to be included amongst our list of all time remarkable comets and is a fitting place to end this personal overview.

Any reader wishing for more detail on these and other comets are encouraged to read the excellent book by Yeomans (1991). Many descriptions are also to be found in Kronk (1984)

REFERENCES

1. Babadzhanov, P. D., Wu, Z., Williams, I.P. & Hughes, D.W., 1991, Mon. Not. R.ast. Soc., **253**, 69

2. Biela, W. von, 1826, Astronomische Nachrichten, **4**, 443.

3. Ho Peng-Yoke, 1962, Vistas in Astronomy, **5**, 127.

4. Humphreys, C. J., 1991, Q.Jl. R.astr. Soc.,**32**,389.

5. Kronk, G. W., 1984, Comets, a Descriptive Catalogue, Enslow Pub., New York.

6. Marsden, B. G., 1989, A. J., **98**, 2306.

7. Maury, M. F., 1846, Washington Observations, **2**, 422.

8. Needham, J., 1959, Science and Civilisation in China, Cambridge University Press, New York.

9. Santini, G., 1844, Astronomische Nachrichten, **21**, 171.

10. Sekanina, Z., Chodas,P.W.& Yeomans, D.K., 1994, A.J (in Press)

11. Scotti, J. V., 1993, IAU Circular 5725.

12. Stein, J., 1910, The Observatory, **33**, 234.

13. Weiss, E.,1868, Astronomische Nachrichten, **72**, 81.

14. Yeomans, D. K., 1991, Comets:a Chronological History of Observations, Science, Myth and Folklore, J Wiley Pub. Co, New York.

PART TWO

DYNAMICS OF ASTEROIDS, COMETS AND METEORS

FROM ORDER TO CHAOS : CLUSTERING AND DIFFUSION PROCESSES FOR SMALL BODIES IN SOLAR SYSTEM

C. Froeschlé

Observatoire de la Côte d'Azur - B.P. 229
06304 Nice Cedex 4 - France

ABSTRACT

We present some examples of dynamical problems in the solar system with strong physical implications and for which regular and chaotic behaviours are fundamental.

Due to their relatively small size, asteroids are mostly primordial (i.e. they have not been constrainted by geological processes) and therefore can give us essential information on the physical properties of the primitive matter in the solar system. Among asteroids, those belonging to the so called families (i.e. fragments of a parent body, with low enough ejection velocities to produce initially swarms of objects having similar orbital elements) give us a unique opportunity to get information about the interior of asteroidal bodies. The problem has two faces. First to obtain, under the assumption of quasi regularity, quasi first uniform integrals of the motion : the so called proper elements, and second, to use objective statistical methods of clusters analysis. Recently decisive improvements came from the application of high-order perturbation theories in computing the proper elements (Williams, Yuasa, Milani and Kneževič) and of the use of reliable cluster analysis methods (Zappalà et al. 1990, Bendjoya et al. 1991).

On the other hand, chaotic routes are essential to bring either asteroidal material or comets in the vicinity of Earth. To study the chaotic behaviour of asteroidal and cometary orbits, modelling becomes an aim and a tool. Difusion equations, Monte-Carlo mappings and Markov processes are the main tools which will be discussed in this paper as well as the strength and weakness of these methods versus the origin of chaos.

GENERAL INTRODUCTION

For centuries the aim of scientists working in celestial mechanics has been to solve Newton's equations which govern motions in the solar system. More precisely under the assumption that all systems are integrable by a sequence of canonical

From Newton to Chaos, Edited by A.E. Roy
and B.A. Steves, Plenum Press, New York, 1995

transformations the part of the Hamiltonian we are unable to handle analytically is pushed to higher and higher order. This approach culminated with the work of Poincaré who showed that the solutions of the n-body problem could be formally written as series of purely periodic terms thus solving at least apparently the old controversy about whether secular variations of the semi-major axis occur at any order of perturbation theory. But he also showed that this series were generally asymptotic series and were divergente due to the appearence of uncontrolled small divisors.

Actually until the 1960's Poincaré discoveries were not fully appreciated. The KAM theorem on the analytical front and the seminal numerical studies of Hénon and Heiles (1964) show the coexistence of regular quasiperiodic motions and wild chaotic zones. These "Janus like" two faces of Hamiltonian systems are at the origine of a new understanding and new interest for dynamical problems related to the physic evolution of the solar system. Indeed we are here at a third step of dynamical study of planetary science. The first one could be called the regular analytical approach. The second is connected with the new paradigm of dynamical systems and chaos. The third one being the interrelations between physics and dynamics especially for small bodies in the solar system. Actually asteroid families are good examples of such an interrelation between : a physical hypothesis on the origin of a family by catastrophic encounter, the search for quasi first integrals under the assumption of regularity and the use of reliable cluster analysis methods, (wavelet analysis, hierarchical clustering).

On the other hand, asteroidal material falling on the Earth and originating from such a catastrophic event is brought on Earth crossing orbits through a chaotic process. For carrying out such studies the ususal tools of celestial mechanics break down and methods are used coming from other fields like Monte-Carlo mappings and modelisations of Markov process.

1. ORDER AND MEMORY : AN ASTEROID FAMILY STORY

1.1. Introduction

The existence of the asteroid dynamical families has been known since 1918. They were discovered by Hirayama, who first noticed that some groupings are obvious in the space of the asteroid orbital elements a, e, i. He called these clusterings "*families*", and dedicated to this subject a series of papers from 1918 to 1933. The best known examples of families discovered by Hirayama are the famous Eos, Themis and Koronis families.

The interest about asteroid families is a consequence of the improvement of our knowledge of the overall process of collisional evolution of the asteroid belt. Indeed, the most natural explanation of the existence of the families is that they are the products of catastrophic impacts between pairs of asteroidal bodies. In other words, the present members of asteroids families should be fragments of single asteroids destroyed by impacts. They were ejected with comparatively low relative velocities, so that their orbital elements did not change too much with respect to each other and to the original parent body. This gave origin to the presently observed clusterings in the space of orbital elements. It is obvious that, in this scenario, asteroid families have a crucial importance: their number can be considered as an important constraint for our models of asteroid collisional evolution, while the overall physical properties of their members can give us important information on the processes of catastrophic breakups of asteroidal sized bodies, as well as on the internal composition of asteroids.

For these reasons, it is not surprising that various attempts to unambiguously identify families have been carried out for several years. As was already recognized by Hirayama, the gravitational perturbations due to the planets, mainly Jupiter, cause time variations of the orbital osculating elements of asteroids, hence these elements are not well suited for the purposes of families identification. However the motions look quasi periodic, reflecting "order" and therefore the search for quasi first integrals called proper elements (a', e', i') is natural. Obviously, these quantities which possibly change but extremely slowly with time (in the spirit of Nekhoroshev theory) provide a set of elements which, in principle, could be considered as constant and keep memory of initial conditions. For this reason proper elements have been computed and analyzed by the authors who have studied the problem of asteroid family identification during the last decades. Among them, and after the work of Hirayama, the most important studies have been published by Brouwer (1951), Arnold (1969), Lindblad and Southworth (1971), Carusi and Massaro (1978), Williams (1979), Kozai (1979).

In spite of the noticeable efforts produced by these authors, the situation concerning asteroid families has been far from satisfactory until very recently. This is due to the fact that a general agreement between the various proposed family lists has not been achieved neither for what concerns the number nor the memberships of these families. Only in a few cases, namely the three major families of Eos, Themis and Koronis a good agreement has been found, but this is not surprising, since these clusterings can be very easily identified in both the $a' - i'$ and $a' - e'$ planes.

The discrepancies existing among the quoted classifications are due to various reasons. In particular, they are a consequence of: (1) the different data bases considered by the different authors; (2) the different methods used to compute the orbital proper elements; (3) the different identification procedures used.

In this paper we do not deal in detail with issues like the theory of asteroid proper elements, the properties of the several clustering identification methods, or the precise implications of asteroids families studies to the problem of asteroid collisional evolution. Such arguments are treated in details by Milani et al. (1991), to which the reader will be often addressed to in the following. In the present chapter, after an overview of the different methods used to obtain proper elements, we mostly describe the two methods of asteroid families identification which have been most recently applied, namely the Hierarchical Clustering Method (HCM) and the Wavelet Analysis Method (WAM); we will show their respective results, in order to stress their general agreement as well as the problem which are presently open, and deserve further investigations. In particular, Section 1.3 will be devoted to a general description of the HCM, and its results, while Section 1.4 will present the WAM and the families identified in this way. Finally, Section 1.5 will be devoted to a comparison of the results obtained by means of these two methods, and to a general discussion of the open problems and future work.

1.2 The problematic of proper elements

Even looking on the distribution of osculating elements of asteroidal orbits there are some obvious groupings which could result from the fragmentation of a same parent body. To remove the change with time, due to the planets, of the osculating elements, i.e. to search invariants of the motion, called proper elements, becomes obviously one of the important parts of the dynamical analysis concerning the asteroidal families. Let us assume that we are able to calculate for each asteroid three invariants of the motion, depending only on the initial conditions, and staying constant for very long time. We can then imagine that planetesimals with very close proper elements could have been, in the past, physically very close to each other and perhaps fragments of the same important body

However, besides the calculation of the proper elements, remains the classification of the results (when are proper elements "close enough" to be candidates to a same parent ?) and the understanding of the fragmentation mechanism (are the fragments coming from a unique catastrophic event or are they fragments of fragments of successive impacts ?)

Using a linear perturbation theory Hirayama in 1918 has computed the first proper elements. Then Yuasa (1973) developed by hand a theory of secular motion up to degree 4 in eccentricity and inclination and order 2 in the mass of any perturbing planet. Later Knežević (1988, 1989) checked and completed (addition of the indirect part) the Yuasa's work. Then Milani and Knežević (1990) revisited the problem using a Lie formalism and adding an iterative procedure on the final calculation. On the other hand in a different spirit Williams (1969) developed an alternative theory of asteroid secular motion, and in 1979 he published a list of asteroid proper elements; these were computed without any expansion in eccentricity and inclination of the asteroid, and therefore should be particularly accurate for high inclined or eccentric asteroidal orbits.

In conclusion, nowadays two different main methods for the computation of proper elements exist : Milani and Knežević method and Williams' method. Actually a third one mainly in the spirit of Williams' method but more elaborated has been made by A. Lemaître and A. Morbidelli (1993).

In the last part of this paragraph we will closely follow the paper of Lemaître and Morbidelli (1993) to introduce the different methods for computing proper elements.

Let us analyze these methods to find out their limits and advantages.

First of all, we recall that the proper elements are the results of two averaging processes, the first one over the short periodic terms (to get the mean elements), the second one over the long periodic terms (to get the proper elements), on an initial three or many body problem. In between those two operations, before the elimination of the long periods, there is an important rearrangement of the terms. Theories may differ on either the adopted Hamiltonian for the model (i.e the number of terms taken into account) or on the perturbation method used to find out the approximate constants of motion. In other words, one has to decide, first, what must be considered as an "integrable approximation" and what must be considered as a "perturbation", and, later, which perturbation technique to adopt in order to take into account the effects produced by these "perturbations".

Milani and Knežević's method is a very accurate one for asteroids with small inclination and eccentricity. This is due to two different reasons. The first one is obvious and concerns the model : their program works on a Hamiltonian which is truncated at degree 4 in eccentricity, e, and inclination, i, which is a good approximation only providing that e and i stay small during all the secular evolution of the asteroid orbit. The second reason is that they consider e and i to be of the same order of magnitude as the eccentricities and inclinations of the main planets for what concerns the order of magnitude (which is perfectly compatible with the truncation hypothesis); so, what they choose as an integrable approximation is a quadratic form in e, i, e_p, i_p, the "$_p$" variables referring to the perturbing planet(s) (i.e. the Hamiltonian of the linear theory of Laplace). Therefore, the term that depends on the argument of perihelion, namely $e^2 i^2 \cos(2\omega)$ is considered as a "perturbation".

This choice is important because, as pointed out first by Kozai (1962), at high inclination, the dynamic associated with the motion of the argument of perihelion is the most relevant one, and a resonance, namely a region where ω librates, exists. By consequence, any method that pretends to consider the term $e^2 i^2 \cos(2\omega)$ as a perturbation, must fail at high inclination.

Let us see now the assumptions of Williams' theory, which are very different. First, no expansion in eccentricity or inclination of the asteroid is used; conversely, Williams uses a linear expansion in e_p and i_p. Therefore, one implicitly consider the asteroidal eccentricity or inclination to be much larger than those of the planets. Secondly, always following the same idea, the division between integrable part and perturbation is made in a completely different way : any term proportional to e_p or i_p is considered as a perturbation. The consequence of this choice is that the integrable part is a function of the argument of the perihelion and it is precisely the one studied by Kozai. Unfortunately, Williams did not include the second order in the perturbing masses in his work, and he calculated the proper elements directly on Lagrange equations, without the help of the Hamiltonian formalism and canonical transformations.

The method of Lemaître and Morbidelli for the computation of proper elements is very close to that of Williams. They make the same assumptions about the non expanded hamiltonian and they separate integrable and perturbation parts by the same parameter. There are differences on the following features:

- They have included the second order in the perturbing masses originated by the first averaging; the role of the quadratic terms (in mass) is relevant close to the main mean motion resonances; in particular it may significantly change the proper frequencies of the asteroid and, consequently, its relative position with respect to secular resonances. In the paper by Morbidelli and Henrard (1991), for example, the authors showed that the secular resonance of second order which affects the motion of the Eos family members cannot be found in the quadratic term if the masses are neglected.

- They have used more accurate values for the amplitudes, the frequencies and the phases of the secular planetary motion. Indeed Williams' work is based on the values given by Brouwer (1950), while their program uses the values computed numerically by Nobili et al. (1989) in the framework of the LONGSTOP project.

- Their perturbations method is written in a Hamiltonian formalism, using Arnold action angle canonical variables.

- They have averaged over the long periods by another semi numerical method.

- They have implemented, following Milani and Knežević an iterative procedure on the calculation of the proper elements. Actually the computation from mean elements to proper elements involves an approximation of the proper frequencies which are computed only at the end of the all computation. Hence the process is iterated like for the search of fixed point of the transformation.

So as Williams' one their method is particularly suited for high eccentric and, still more important, for high inclined orbits. Unfortunately the consideration of the quadratic terms in e_p, i_p is beyond the limits of their method.

It is obvious that the advantages of the method vanish for small inclinations and eccentricities. For these orbits Milani and Knežević's method turns out to be preferable for two reasons : it is much faster, and it takes into account the perturbations of 4 planets up to degree 2 in e_p, i_p. At this time (June 93) Lemaître and Morbidelli have not yet provided proper elements, therefore no statistical analysis has been performed.

A great deal has been achieved the last few years with the finding and the reliability of the proper elements. Let's now switch to questions beside the field of celestial mechanics : do the structures within the proper elements distribution reflects only a pure dynamical evolution or also some physical processes. To answer to this question one has to quantify the level of significance of the groupings versus chance fluctuations. Recently two methods of cluster analysis have given very similar results although their approach are different and complementary. The more classical one known as the single linkage hierarchical clustering method and the more recent one based on the wavelet transform are described in Zappala *et al.* 1990, Bendjoya *et al. 1991* and Bendjoya *et al.* 1993. We only recall in the following sections the main characteristics of these two methods.

1.3. The Hierarchical clustering method

The so-called Hierarchical Clustering method for asteroid families identification has been recently applied by Zappalà et al. (1990). This method is based on well known techniques of Multivariate data Analysis, which are explained in details, for example, in Murtagh and Heck (1987).

The basic idea of this method is to introduce a definition of distance in the space of orbital proper elements a', e', i'. When this is done, one can compute all the mutual distance $d(i, j)$ between the N objects of the considered data set. The two objects, i and j , having the smallest mutual distance are identified and an "agglomeration" is then performed, so that the two objects i and j are replaced by one new object $i + j$. All the mutual distances of the resulting $N - 1$ elements are computed again, with the assumption that the distance between the new (i, j) object and any other object, k, is taken as the minimum between $d(i, k)$ and $d(j, k)$; the couple of objects having the minimum mutual distance are again identified and agglomerated as in the previous step, and this process continues as long as only two objects remain, which are in fact two big agglomerations of the N elements of the original data set. In this

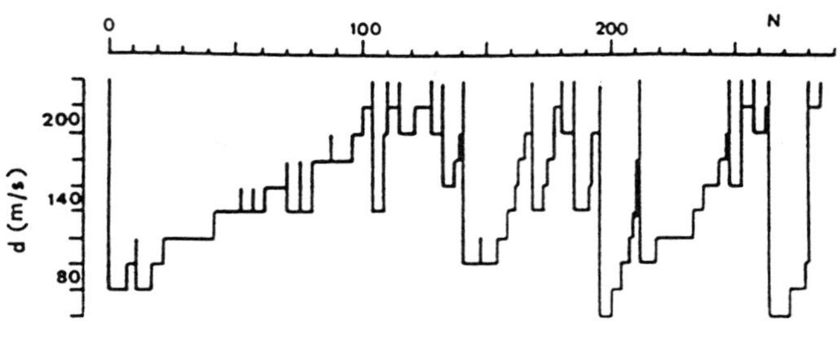

Figure 1. An example of stalactite diagram referring to a subsample of the asteroid data set analysed by Zappala et al. (1990).

this way, a *dendrogram*, as shown on Fig. 1, is created, which contains all necessary information for family identification purposes, since it gives immediately, for any given value d of the distance, all the existing agglomerations of objects, whose mutual distances are less than d. If one knows a threshold value of the distance, $\overline{d}$ for which no clustering can exist having a significant probability to be due to chance, then all the clusterings found at the $\overline{d}$ level can be reasonably called "families" . One must take into account that an eventual true collisional origin of these clusterings can be assessed only on the basis of further investigations concerning their overall physical properties (mass and velocity distributions, taxonomic types, etc ...). A typical shape of a dendrogram is a succesion of stalactites giving the number of groupings and their abundance as a function of the distance. Usually, at larger distances only a few big agglomerations exist, containing most of the objects of the sample. The deepest stalactites are, in general, families, since they correspond to intrinsically denser groupings or composed by objects mutually closer in the space of proper elements.

As can be seen, the procedure described above is in principle very easy, but contains two critical points: the definition of the metric in the space of proper elements, and the derivation of the threshold value $\bar{d}$ of the distance. For what concerns the first point, if we accept the idea of a collisional origin for the families, it is straightforward to introduce a definition of distance d in the space of the proper elements, in such a way that this distance can be put in relation with the ejection velocities of the fragments in an event of catastrophic disruption. In particular,indicating with δv_1, δv_2, δv_3 the three components of the velocity differences between two fragments of a collisional event along the directions which are, respectively, parallel, normal and binormal with respect to the direction of the orbital motion at the instant of breakup, the Gauss equations put in relation δv_1, δv_2, δv_3, with the differences in the orbital elements δa, δe, δi. Without entering into details which are extensively explained in both Milani et al.(1991), Zappalà et al. (1990) paper, a "natural" choice of the metric in the space of proper elements has been found to be given by

$$d = \sqrt{k_1(\frac{\delta a}{a})^2 + k_2(\delta e)^2 + k_3(\delta i)^2} \quad (1)$$

in which the three coefficients k_1, k_2, k_3, are given by 5/4, 2, 2, respectively. It has to be noticed that the distance defined in this way is dimensionally a velocity, and can be expressed in m/sec. The papers quoted above contain also exhaustive description of the procedure adopted to derive the value of the critical level of distance, $\bar{d}$, for which the existing clusterings can in general be assumed to be families. The idea is to create a fictitious population of objects randomly generated, but with the constraint to globally fit the a', e', i' distributions of the real population to be analyzed. The Hierarchical Clustering algorithm is then applied to this quasi-random population, and the critical level $\bar{d}$ is assumed to be the deepest value of d for which it is still possible to find one or more groupings of the fictitious population composed by at least 5 members. Having determined $\bar{d}$ in this way , the "families" are assumed to be all the clusterings of the real population having at least 5 members at distance levels $d \leq \bar{d}$, or having more than twice the number of members of the largest existing clustering of the fictitious population at the $\bar{d}$ level. In this way, not only the probability to find clusterings at low levels of distance, but also the probability to find noticeably large clusterings at the critical distance level are taken into account.

The method for families identification described above is fast and simple in principle, but in computer form it requires the computation of quite large matrices, whose elements are the mutual distances between each pair of objects belonging to the sample to be analyzed. For this reason, and also on the basis of physical considerations involving possible variations of the properties of different parts of the asteroid belt, due to both the presence of mean motion and secular resonances, and to different completeness of our asteroid sample at different heliocentric distances, the Hierarchical Clustering method has been applied separately to asteroids belonging to six different zones of the belt. These zones are delimited by the main mean motion resonances with Jupiter.

The sample of asteroids analyzed by Zappalà et al. (1990) has been the 4.2 version of the list of asteroid proper elements provided by Milani and Knežević (1990). Such a data set, including about 4100 asteroids, was the largest list ever used for families identification purposes; the same data set has been later used by Bendjoya et al. (1991), in their application of the Wavelet Analysis Method which is described in Section 1.4.

1.4. The wavelet analysis method

The wavelet cluster analysis is a new method able to extract from a set of points, structures with a degree of significance versus chance. It is a hierarchical method in the sense that the wavelet transform allows to point out the structures the size of which corresponds to the studied scale (a characteristic given size). By using a set of increasing scales, this analyzing method permits to obtain significant structures from the smallest ones to the biggest ones fitted into each other just as the well-known russian dolls. This method of analysis lies on the wavelet transform which is a mathematical tool which has already given very interesting results in different domains such as: the signal restauration and filtering, the fractal behavior of a signal, the compression of information and in the cluster analysis which is a peculiar application of the wavelet transform due to the peculiar studied signal since in this case the signal can be modelled by a set of points or more precisely a set of Dirac functions. The main advantage of the wavelet transform is to be able to give informations not only on the location of an event in a signal but also on the size of this event. The wavelet transform gives informations both in time and frequencies for a time-signal or both in space and size for a space-signal.

Let $s(x)$ represent a one-dimension space-signal in $L^2(\Re)$. The wavelet transform of this signal is a two dimension function $S(\sigma, b)$, where σ is a strictly positive scale-variable (size) and b a space variable (location). $S(\sigma, b)$ is the decomposition of $s(x)$ onto a base of functions obtained from the dilations and the translations of a unique function $\psi(x)$ of $L^2(\Re)$.

$$S(\sigma, b) = \sigma^{-\frac{1}{2}} \int_{-\infty}^{+\infty} s(x)\psi^*(\frac{x-b}{\sigma})dx \tag{2}$$

$\psi(x)$ is called the analyzing wavelet. $\sigma^{-\frac{1}{2}}$ is a $\mathrm{L}^2(\Re)$ normalization factor and $*$ denotes the complex conjugate.

The wavelet transform acts like a mathematical zoom and $S(\sigma, b)$ can be seen as the σ-sized details of $s(x)$ around the location b. By varying the scale variable σ, thinner and thinner details are extracted from the signal when σ decreases. In reality the wavelet transform allows to consider a signal from a new point of view consisting to consider the signal itself as the sum of all its own details at the different scales.

In order to be a suitable analyzing wavelet, the function $\psi(x)$ must satisfy the admissibility condition:

$$\text{i.e.} \quad \int_{-\infty}^{+\infty} \psi(x)dx = 0 \tag{3}$$

Also, $\psi(x)$ must be smooth enough and compact supported in order to give informations on the location.

These criteria make of the mexican hat function a good candidate for being an analyzing wavelet. This function comes from the second derivative function of a

Gaussian function:

$$\psi(x) = (n - \frac{r^2}{\sigma^2})\exp(-\frac{r^2}{\sigma^2}) \tag{4}$$

n is the dimension of the space in which the signal is defined and r is the distance defined from the metric chosen in the considered space.

This function is a zero integral one and, is considered equal to zero for $|r| > 5\sigma$, to express the compact support condition.
The wavelet analysis method has been applied to the issue of asteroids families identification by Bendjoya et al. (1991). In order to identify asteroids families, the wavelet analysis is carried out in the proper elements space, for reason explained in Section 1.2. Each asteroid is represented by a point in the (a', e', i') space (proper semi major axis, proper eccentricity, proper inclination). The signal to be studied is hence a set of Dirac functions $\delta(\vec{x} - \vec{\eta_j})$, $\vec{\eta_j}$ is the position of the jth point. In computer form, the wavelet transform is derived in a discrete way by computing the wavelet coefficients of the signal on each node of a frame superimposed to it.

Let us consider a plane of data points over which a network is superimposed. The wavelet coefficient for the node $\vec{b_i}$ is obtained by substituting in eq. 2 $s(x)$ by the set of Dirac functions $\delta(\vec{x} - \vec{\eta_j})$:

$$C(\sigma, \vec{b_i}) = \sum_{j=1}^{N} \psi(\frac{\vec{\eta_j} - \vec{b_i}}{\sigma}) \tag{5}$$

where N is the number of data points. $C(\sigma, \vec{b_i})$ is the wavelet coefficient at the node $\vec{b_i}$ for the scale σ. $C(\sigma, \vec{b_i})$ is in fact the sum of the weights w_j of the data points.

Considering the mexican hat function it results:

$$C(\sigma, \vec{b_i}) = \sum_{j=1}^{N} w_j$$

with:

$$w_j = (n - \frac{r_{ij}^2}{\sigma^2})\exp(-\frac{r_{ij}^2}{2\sigma^2}) \tag{6}$$

r_{ij} being the distance between the ith node $\vec{b_i}$ and the jth point $\vec{\eta_j}$. This distance is expressed on the basis of the metric fitted to the problem. Because of the admissibility condition eq.(4) it is easy to understand that a node centered on a uniformly populated area will have a near zero wavelet coefficient since the positive weights will be compensated by the negative ones, provided that the size of this area is big when compared to the studied scale. On the other hand a studied-scale-sized structure (resp. hole), will give wavelet coefficient stongly positive (resp. negative). A map of wavelet coefficients is so built in which all the positive coefficients are indicators of structures. Fig. 2 shows, but for one dimension, such an analysis.

The problem is to distinguish the physical structures from the structures due to chance. In order to do this, the idea is to build a "pseudo-random" distribution in the sense that the local density is conserved. Being given a certain number of boxes in the real distribution, the pseudo-random one is built by putting at random the same number of points in each corresponding boxes. Then the wavelet coefficient maps are computed for both distributions. The histogram of the wavelet coefficients of the pseudo-random distribution is derived in order to determine the value of the coefficient for which 99.5%, for example, of the other coefficients have a lower value than this threshold. It means that there are less than 5 chance over 1000 to find a higher value than this one in the pseudo-random wavelet coefficient map. By reporting this threshold in the real distribution wavelet coefficient map and by retaining all the coefficients higher than this value, it is easy to understand that the high value of these retained coefficients has less than 5 over 1000 risks to be due to chance. Consequently the structures associated to these kept coefficients have less than 5 over 1000 risks to be due to chance. This way of proceeding shows how it is possible to give a degree of significance versus chance to the detected structures. From the retained coefficients it is then possible to label all the asteroids whose weights have been taken into account in the computation of this coefficient and a significant structure can be so evidenced. By applying this procedure to a set of different scales with the suited networks, we get a hierarchy of structures with a degree of significance versus chance. Let us emphasize that such a procedure is part of the classical wavelet analysis for searching structures.

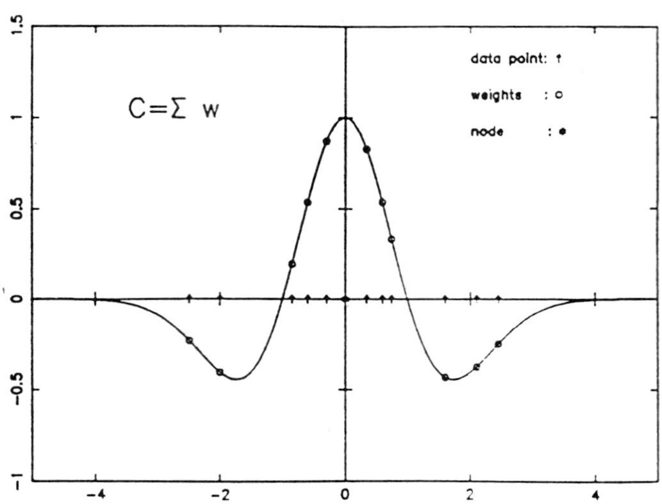

Figure 2. Wavelet coefficients computation

The two next steps are the choice of a metric well fitted to the considered problem and the way to cut the hierarchy in order to pass from structures to families.

The former problem is the same already encountered in the framework of the Hierarchical Clustering method; the solution adopted is the same, so the definition of distance in the space of orbital proper elements is still given by equation (2) introduced in Section 1.3. This metric has been adopted in the definition of the mexican hat function (eq. 5) and the analysis has been performed for the set of scale σ: 60, 80, 110, 160, 230, 320, 450, 640, 900 m/s. The ratio close to $\sqrt{2}$ between two successive scales is a condition imposed when using the mexican hat as analyzing wavelet, in order to loose the least of information between two successive scales (see Daubechi 1990). Significant structures are evidenced at each scale as shown on Fig. 3 and 4.

On the other hand, it is easy to understand that a criterion of cutting the hierarchy has to be chosen since different structures can merge into a unique one for a large enough scale. The chosen criterion for passing from the structure to families, lies on the philosophy of the wavelet transform that is the additional information got between two successive scales. The largest scale to consider in the studied area in order to distinguish families is the one for which the number of asteroids added at this scale to the already identified structures, is either zero or increases drastically. The first case indicates that the structures are well isolated and the second one indicates that the next scale is too large and bridges are made between structures which merge into a big cluster.

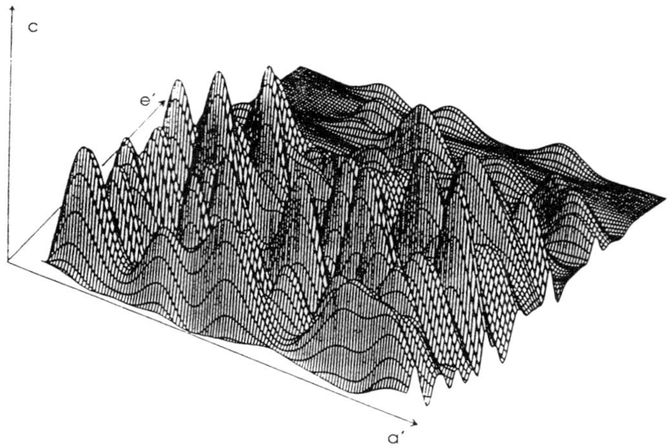

Figure 3. Wavelet coefficient maps in the (a', e') plane at the 160 m/s scale for the real distribution.

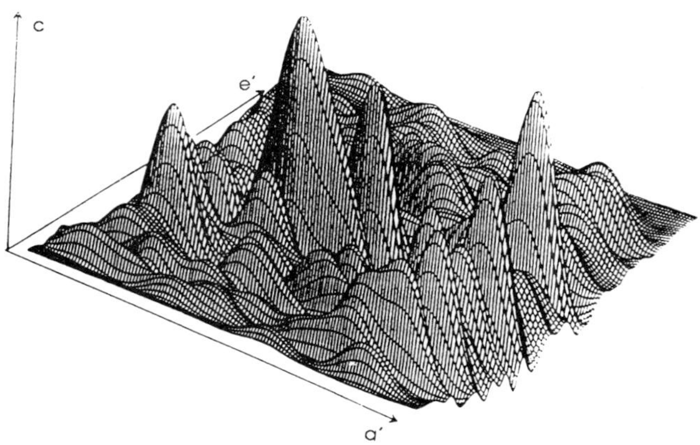

Figure 4. Wavelet coefficient maps in the (a', e') plane at the 160 m/s scale, for the pseudo random distribution.

1.5. Discussion

The two methods give results in a very good agreement for what concerns both the number of families found in the different zones of the belt and also for each family the percentage of common objects, i.e., the asteroids which have been found to belong to the same family on the basis of both WAM and HCM. Both methods exhibit, from a set of 4200 asteroids, about twenty families with a high quantified level of significance versus chance fluctuations. The agreement achieved by these two methods gives confidence on the reliability of the resulting asteroid families. Of course these families have been defined only in a dynamical and statistical sense, thus it is clear that a deeper study on the taxonomy (cosmochemical classification) of the members of these supposed families is obviously necessary. This will allow us to check whether these clusterings have cosmochemical compatible members , i.e., they are not in conflict with what we know about asteroid's composition and physics of catastrophic breakups.

The next step of this analysis has consisted in a check of the reliability of both the HCM and the WAM by means of suitable numerical simulations. The idea has been to create some "artificial families" located in different "artificial background". The aim was to analyze the efficiency of the two methods in various situations completely known *a priori* and reproducing some of the most common situations which can be encountered in the real belt (Bendjoya *et al.* 1993). The results seem to indicate that the methods described in the present paper can be considered sufficiently efficient for allowing us to believe that a significant improvement has been now achieved on the subject of asteroid families identification. Moreover, the fine tunings developped from these simulations have led to improved methods and a new classification, among about 6500 asteroids, has been performed by both methods (Zappala et al. 1993, Bendjoya 1993). The data rely on an improved proper element set (Milani and Knezevic 1992). Once again a good agreement between both methods is pointed out and an optimized list of families is actually to be proposed.

The overall physical properties of the families identified in the asteroid belt can now be analyzed and can give us important information on the collisional history of the belt itself and on the kinds of outcomes that are formed as a consequence of catastrophic breakups of asteroidal bodies.

If the search of order is at the basis of building proper elements and therefore, in finding families, the existence of chaos in some regions of the phase space is responsible of different phenomena. For example, chaos allows comets, coming either from the Oort cloud or from the Kuiper belt, to become observable. The same scenario occurs for meteorites which follow chaotic routes. Mappings are, so far, the main tools to study and model such processes. We will present in the following sections some basic recipes to "cook" such mappings.

2. STOCHASTIC MAPPINGS IN ASTRODYNAMICS

2.1. Introduction

Both the existence of Kirkwood gaps and the transfer of comets into observable orbits have been the main motivations for building mappings since they have many advantages over numerical integrations mainly with respect to computing time and to accuracy. However, even if the brutal force of super computer may in the near future solve these problems, building mappings will still be a challenge to researchers in Celestial Mechanics. Indeed to build mappings brings a deeper understanding of the general behaviour of non linear dynamical systems. In the particular case of mappings devised to study cometary motion , the origin of chaos is the key point.

2.2. Exogeneous stochastic mappings

2.2.1. A simple model of diffusion

2.2.1.1. Introduction. It is well known that the study of dynamical systems with n degrees of freedom can be reduced to the study of a $2n - 2$ measure preserving mapping, using the method of surface of section.

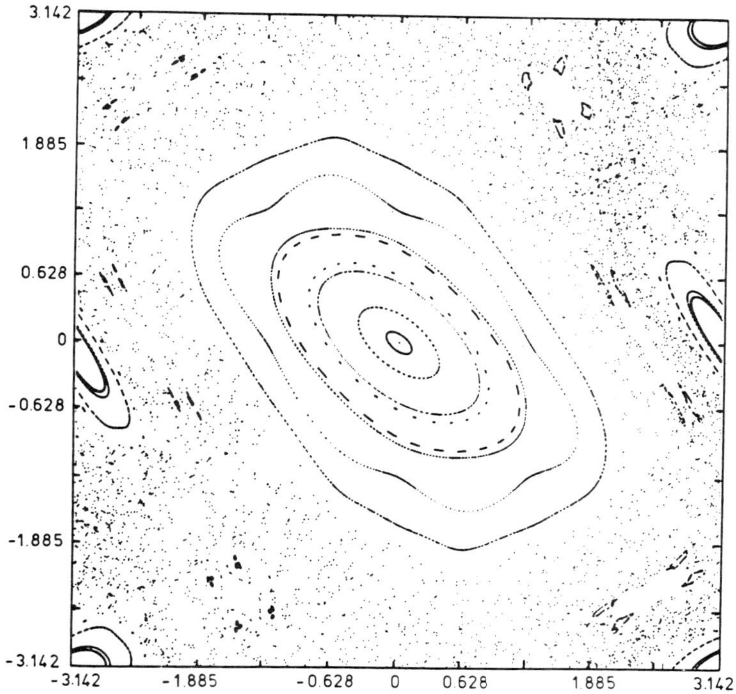

Figure 5. The area preserving mapping T_0 for $a_1 = -1.3$.

The present section deals with an area preserving mapping but in which a stochastic element has been introduced. (See Froeschlé 1975). Hence, we have researched how the classical situation of dynamical systems with two degrees of freedom was modified. In Section 2.2.1.2. we give some properties of the mapping T and its features In Section 2.2.1.3. we give graphical displays of the diffusion process. We define and estimate the diffusion time in Section 2.2.1.4. and in Section 2.2.1.5., its variations with the magnitude of the random term.

2.2.1.2. The mapping. We consider the mapping T_ε of the (x, y) plane over itself defined by

$$T_\varepsilon \begin{cases} x_1 = x_0 + a\sin(x_0 + y_0), \\ y_1 = x_0 + y_0 + \varepsilon. \end{cases} \quad (\mathrm{mod}\ 2\Pi) \tag{1}$$

where ε is a constant. If we make the change of variables

$$\begin{cases} X = x + \varepsilon\,, \\ Y = y - \varepsilon\,, \end{cases}$$

we get in the new axes XY the mapping T_0 hence the topology of T_ε is the same as that of T_0 within a translation of vector $(-\varepsilon, \varepsilon)$..

Figures 5 and 6 display typical sets of points for the mapping T_0. Figure 5 exhibits all the characteristics and well known features of problems with two degrees of freedom, i.e. invariant curves and islands which correspond to the existence of isolating integrals, and also wild zones, sometimes called ' chaotic' where the points seem to fill a broad region in the plane, and which correspond to the non-existence of isolating integrals. On the other hands, on Figure 6 isolating integrals seem to exist everywhere: this a case very close to an integrable case. All the points are either on *libration* curves (associated with the stable invariant point (0,0)), or on *circulation* curves (associated with the unstable invariant point $(\Pi, 0)$).

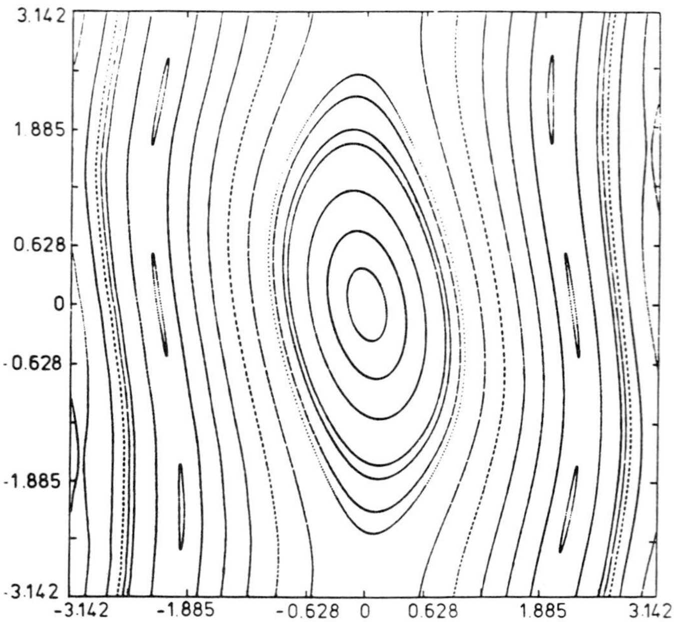

Figure 6. The area preserving mapping T_0 for $a_1 = -0.3$.

Now we study we study a random mapping F_ω for which $\omega = (\varepsilon_1 \varepsilon_2 \ldots)$ is an infinite sequence where each of the ε_k can take the value ε' or ε with equal probabilities. Let $T_{\varepsilon'}$ and T_ε be two area preserving mappings defined by the Equation (1). Let

$$F_\omega^n = T_{\varepsilon n} T_{\varepsilon n_1} \ldots T_{\varepsilon 1} \ , \ \varepsilon_n \quad \varepsilon_{n-1} \ldots T_{\varepsilon_1} \tag{2}$$

denote the random product of n mapping $T_{\varepsilon'}$ or T_ε.

2.2.1.3. Graphical displays of the diffusion process. In order to study the fluctuations of the invariant curves, the points $P_n = F_\omega^n(P_0)(n = 1 \text{ to } N)$ have been plotted for various values of N. Figure 7 shows the results for $a = -1.3$, $\varepsilon = 0$, $\varepsilon = 0.05$ and the initial conditions $x_0 = 0.4$, $y_0 = 0.4$, which correspond to a zone of invariant curves for both $T_{\varepsilon'}$ and T_ε. For N less than 2000, the points lie on strips including the invariant curves. The width of these strip increases with N. For N greater than 2000, the point are scattered and 'chaos' increases. Figure 8 shows the same experiments for $a = -0.3$. Again the points lie on strips including the invariant curves surrounding the origin, but for N greater than 2000 points still remain in a band along the y axis. The phenomenon of diffusion is much slower than the preceding case. This is due to the topology of the invariant curves for $a = -0.3$. Indeed as we no have longer a chaotic zone but some invariant curves nearly parallel to the y-axis, the effect of the perturbation which is also on y is strongly weakened.

2.2.1.4. Definition and estimation of the diffusion time. In order to have some quantitative information about this random walk, a measure D_j of the dimension of the invariant curve has been computed here, as given by

$$D_j = \sum_{m=j-99}^{m=j} (x_m^2 - a(y_m^2 + x_m y_m))/100 . \tag{3}$$

The quadratic term in this expression is constant, in the vicinity of the origin, in the linear approximation, in the case $\varepsilon' = \varepsilon = 0$.

Figure 9 shows the variations of D_j for the same case as for figure 7 (case 1) and 8 (case 2). The variations of D_j are quite large and in good agreement with the graphical results given by figure 7. Indeed we observe, for the values of j, corresponding in figures 7 and 8 to the scattering of the points, a sudden increase in the value of D_j. We call 'diffusion time' the number of iterations of the mapping F_ω which are necessary for the points $(x_n,\ y_n)$, starting in the integrable zone (or the libration zone) of T_0, to reach the wild (ergodic) zone (or the circulation zone). In other words, 'diffusion time' is the time necessary for disappearance of isolating integrals in the case 1. For estimating this time, we use the sudden change in the value of the measure D_j. As soon as D_j is greater or equal to 4, we say that diffusion has occured.

2.2.1.5. Variation of the diffusion time with the random term ε and with the distance of the initial point to the origin. We intend to estimate the diffusion time as a function of the magnitude of ε and as a function of the initial conditions.

We take the Gambler's ruin model (Feller, 1971) as an approximation of the diffusion process, since the problem is reduced to the study of the jumps of $(x_n,\ y_n)$ from one elliptic curve to another, up to the ergodic zone, which is considered as an absorbing barrier.

Indeed, the family of invariant curves surrounding the origin can be taken, in a first order approximation, as a continuous elliptic family (cf. Figures 5 and 6).

We consider the family of similar ellipses of equations

$$C = x^2 - a(y^2 + xy), \tag{4}$$

where C is a constant which can be considered as a generalized distance of the points $(x,\ y)$ to the origin. C takes all values from 0 to C_{max} corresponding to the largest ellipse, that is to say, the absorbing barrier. Let $N(C)$ be the expected number of iterations which are necessary for reaching the absorbing barrier when starting on the ellipse C, at the generalized distance C from the origin.

We have the fundamental equation

$$N(C) = 1 + \int N(C + \Delta C)\varrho(\Delta C)d(\Delta C), \tag{5}$$

where $\varrho(\Delta C)$ is the probability to jump from an ellipse C to an ellipse $C + \Delta C$.

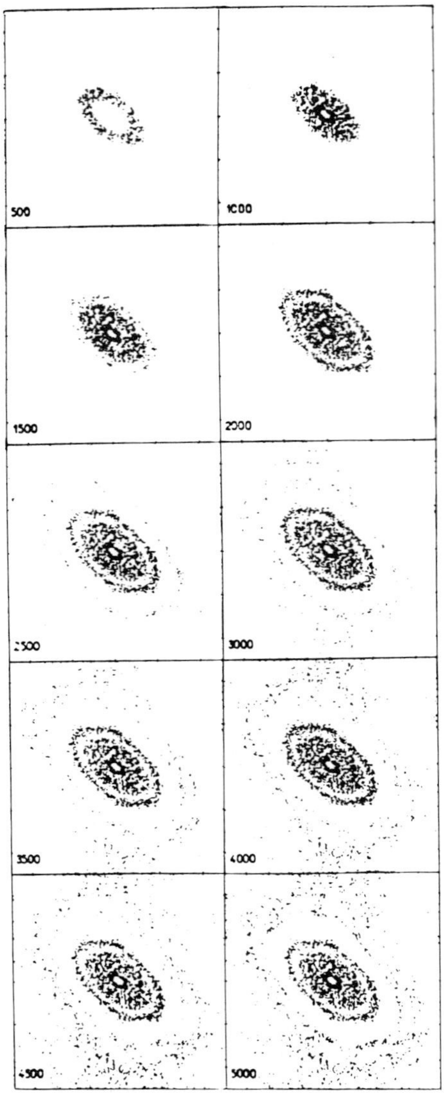

Figure 7. An orbit of the randon mapping F_ω where $a_1 = -1.3,\ x_0 = 0.4,\ y_0 = 0.4, \varepsilon' = 0,\ \varepsilon = 0.05$ for various values of N and a given realization of ω

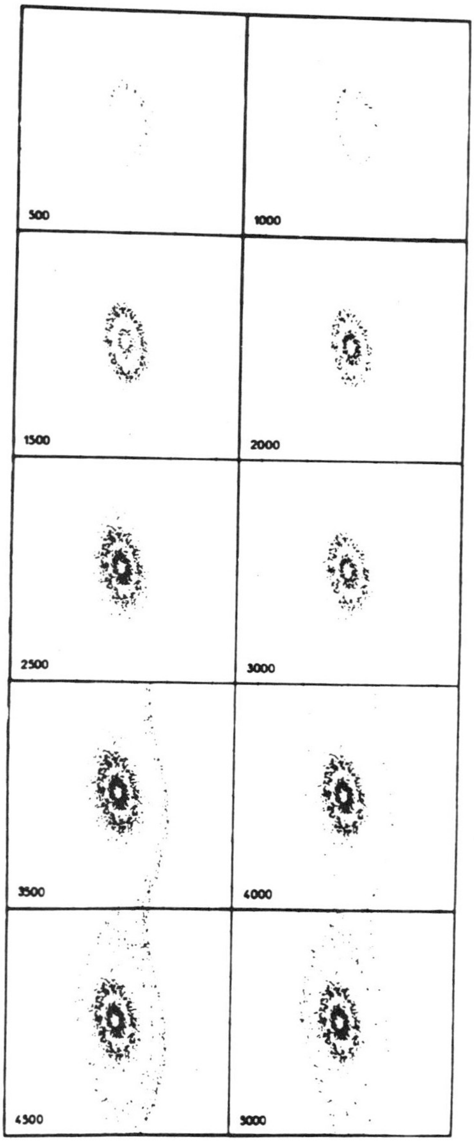

Figure 8. An orbit for the same initial conditions as for Figure 7 but for $a_1 = -0.3$.

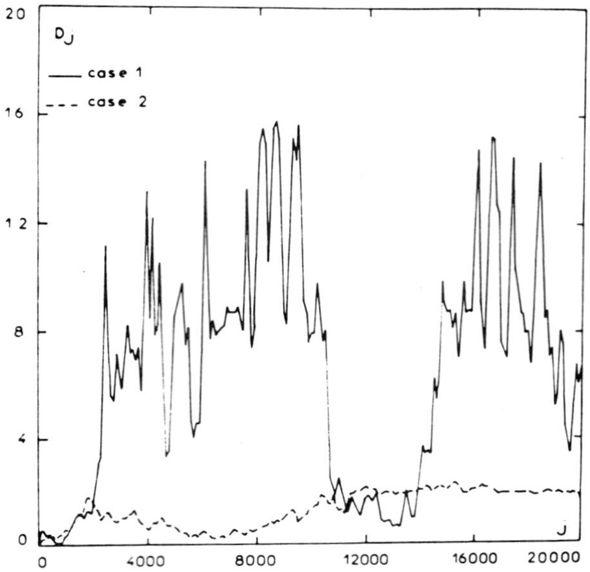

Figure 9. D_j, a measure of the dimension of the curves, against j.

Expanding to second order (see the Fokker-Planck method), we get

$$N(C) = 1 + N(C) + \frac{dN(C)}{dC} < \Delta C > + \frac{1}{2}\frac{d^2N(C)}{dC^2} < \Delta C^2 >, \tag{6}$$

we have

$$\Delta C = X^2 - a((y+V)^2 + x(y+V)) - x^2 + a(y^2+xy), \tag{7}$$

$$\Delta C = -a(x+2y)V - aV^2, \tag{8}$$

with V being a random variable taking the value 0 and ε with the probability 0.5. Furthermore the points $(x,\ y)$ are distributed on the ellipse C with a density proportional to $ds/\ \|\text{grad } H\ \|$, where

$$H(x,y) = x^2 - a(y^2 + xy). \tag{9}$$

In order to compute $< \Delta C >$ and $< (\Delta C)^2 >$ we take as new axes the principal axes of the ellipse C. After some algebra we obtain the equations

$$\begin{cases} x = X\cos\delta - Y\sin\delta, \\ y = X\sin\delta + Y\cos\delta, \end{cases} \tag{10}$$

where δ is given by the equations

$$\operatorname{tg}\delta = \frac{1 + a + \sqrt{\Delta}}{a}, \tag{11}$$

$$\Delta = 2a^2 + 2a + 1. \tag{12}$$

Then equations (4) become

$$\frac{X^2}{A^2} + \frac{Y^2}{B^2} = 1, \tag{13}$$

with

$$A^2 = \frac{2C}{1 - a - \sqrt{\Delta}}, \tag{14}$$

$$B^2 = \frac{2C}{1 - a - \sqrt{\Delta}}. \tag{15}$$

Let φ be the eccentric anomaly of the ellipse C, such that

$$X = A \cos \varphi, \quad Y = B \sin \varphi. \tag{16}$$

Then the density probability is uniform with respect to φ on $[0, 2\Pi]$ - i.e.,

$$P(\varphi) = \frac{1}{2\Pi} d\varphi. \tag{17}$$

From the Equation (8), (10), (17) we get immediately

$$< \Delta C >= -\frac{1}{2} a\varepsilon^2. \tag{18}$$

Neglecting terms of order greater than 2 in ε, we have

$$< \Delta C^2 >= \frac{1}{2} a^2 \varepsilon^2 < (x + 2y)^2 >. \tag{19}$$

From (4) and (17) we get

$$< (x + 2y)^2 >= -\frac{4C}{a} + \left(1 + \frac{4}{a}\right) < x^2 >; \tag{20}$$

and from (10), (14), (15) and (17)

$$< x^2 >= C\left(\frac{1 - a + \sqrt{\Delta} \cos 2\delta}{-a(a + 4)}\right). \tag{21}$$

Using (11) and (12) we obtain after some algebra

$$< x^2 >= \frac{2C}{a + 4}; \tag{22}$$

hence, from (20),

$$< (x + 2y)^2 >= -\frac{2C}{a}, \tag{23}$$

$$< \Delta C^2 >= -a\varepsilon^2 C. \tag{24}$$

The fundamental equation (6) becomes:

$$-\frac{a\varepsilon^2}{2} \frac{dN(C)}{dC} - \frac{a\varepsilon^2 C}{2} \frac{d^2 N(C)}{dC^2} = -1. \tag{25}$$

Hence, it follows by integration that:

$$\varepsilon^2 N(C) = -\frac{2}{a}(C_{\max} - C) + \varepsilon^2 K \log C\,, \tag{26}$$

where $C_{\max}$ and K are constants.

It is obvious that K is equal to zero to avoid the singularity at $C = 0$ which is without any physical meaning.

Finally for small ε we have:

$$\varepsilon^2 N(C) \cong -\frac{2}{a}(C_{\max} - C)\,. \tag{27}$$

2.2.1.6. Variations with the strength ε of random perturbation. Figure 10 shows the variation of $\varepsilon^2 N(C)$ plotted vs ε in the case $C = 0$. In order to eliminate the fluctuations due to the fact that $N(C)$ depends on the sequence of the ε, we take an average of 25 experiments for a given value for ε.

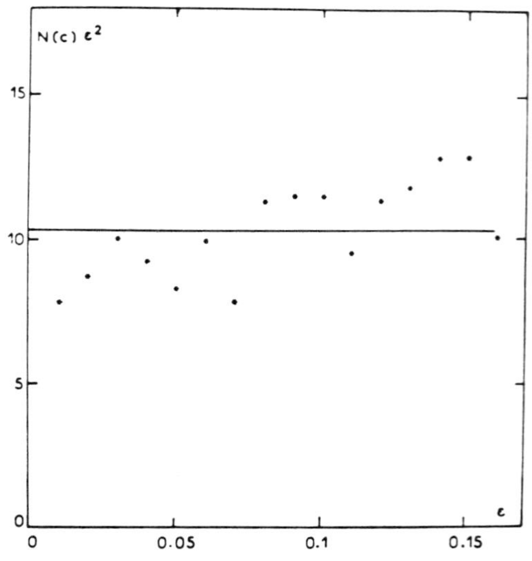

Figure 10. Verification of the diffusion law $N(C)\varepsilon^2 \cong$ constant for a given C.

Considering the crude approximations which have been made for obtaining e equation (1), the results are in rather good agreement with the Gambler's ruin model. From the numerical results, we get, taking the mean value of $\varepsilon^2 N(C)$ which is 10.37, the value 6.74 for $C_{\max}$. This value is in rather good agreement with the value we can estimate on Figure 5 taking into account the distortion along the Y axis of the invariant curves near the border of the ergodic zone.

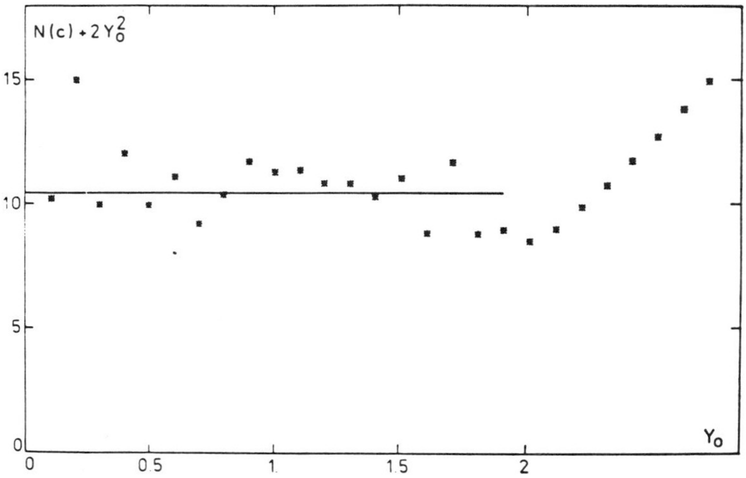

Figure 11. Verification of the diffusion law $N(C)\varepsilon^2 + 2y_0^2 \cong$ constant for $\varepsilon = 0.05$, $x_0 = 0$.

2.2.1.7. Variation with the initial conditions. We take a fixed value of the random term: $\varepsilon = 0.05$, and the initial condition $x_0 = 0$, y_0 being the parameter. From (27) we get

$$\varepsilon^2 N(C) + 2y_0^2 = -\frac{2}{a} C_{\max} . \tag{28}$$

Thus we plot the values of $\varepsilon^2 N(C) + 2y_0^2$ vs y_0 on Figure 11. Each value of $N(C)$ is computed in the same way as previously (average of 25 random sequences).

The straight line represents the average of the value of $\varepsilon^2 N(C) + 2y_0^2$ which is found to be equal to 10.44 and hence, to be very close to 10.37 given previously. This average has been taken for $y_0 \leq 2$ because for $y_0 > 2$ the orbit starts immediately in the chaotic zone of T_0 and $\varepsilon^2 N(C)$ is equal to zero.

Also in this case the Gambler's ruin model seems to be a good approximation of the diffusion problem.

2.2.2. Monte-Carlo mapping of long period comets dynamics. While it has recently been realized that an important part of the dynamics of Oort cloud comets arises from regular motion in the Galactic tidal field (Heisler and Tremaine 1986), a decisive role is nonetheless played by individual stellar encounters. Being given the physical parameters of such an encounter, its effect on the cometary motion is fully determined. However, the parameters of individual stellar encounters are unpredictable so the stellar perturbations impose a stochastic variation on the cometary orbits.

In Monte Carlo simulations of stellar perturbations (see, e.g., Weissman 1982; Remy and Mignard 1985) the dynamical evolution of a cloud of comets is studied as follows. At a given starting epoch, each comet is initialized by choosing a set of orbital elements. These are perturbed by the gravitational effect of passing stars. The geometrical parameters of the stellar encounters are chosen at random. During the passage of a star a comet receives a heliocentric impulse through the interaction of the star with the comet and the Sun. This induces a change in the cometary orbital elements, so these are updated and the comet moves along a new Keplerian ellipse until the next encounter with a random star. In Weissman's procedure the impulses are taken in an even more simplified way from a pre-determined distribution. In other words a stochastic mapping is iterated where the perturbations caused by random stars impose a stochastic process on the cometary orbital elements, which, therefore, undergo a random walk. It is obvious that all orbits are chaotic and correspondingly the largest LCE is strictly positive. The stochasticity is exogeneous since the stellar encounters occur at random.

2.3. Endogeneous stochastic mappings

2.3.1. Monte Carlo mapping of short-period comet dynamics. As comets are captured into short-period orbits, i.e. with orbital period of the same order as those of the perturbing planets, the situation changes in a fundamental way. We now have to deal with only the intrinsic stochasticity of a dynamical system which can usually be approximated by a three-body problem: Sun-planet-comet (Rickman and Froeschlé 1988). This stochasticity derives mainly from close encounters with the planet, and during the intervals between such encounters the cometary motion is quasi-regular and predictable. In more precise terms, this means that the motion is stochastic only over time scales longer than the typical interval between encounters with the perturbing planet. Thus the phase space domain of short-period comets presents chaotic and ordered regions in an intricate mixture.

Stochastic modelling of this motion is justified by the shadowing principle only if the time step is chosen long enough. Thus, we may consider the following procedure (Rickman and Vaghi 1976; Froeschlé and Rickman 1980): we pick a large number of initial cometary orbits at random into boxes of the (Q, q) plane (Q aphelion and q perihelion distances as shown on Fig. 12

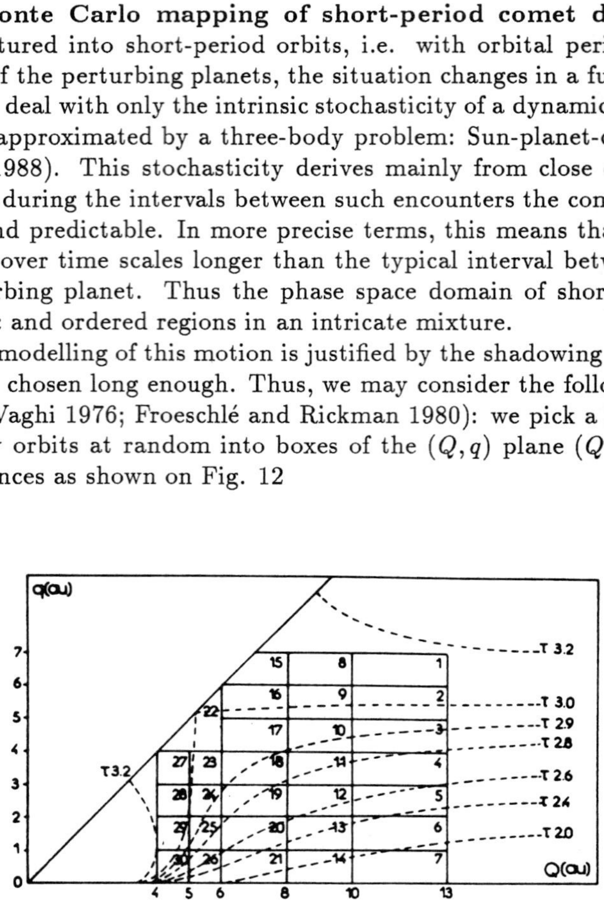

Figure 12. The 30 regions of the $(Q,\ q)$-plane. Evolutionary curves for seven different values of the Tisserand constant are shown (dashed curves). From Rickman and Froeschlé (1979).

and integrate these with Jupiter as the perturbing planet over a time Δt, which was chosen to be one unperturbed period. From the resulting set of perturbations, we construct stochastic orbital evolutions by picking independently up, a random perturbation, within a corresponding box, for each successive interval Δt. For these evolutions, in order to simulate real ones, we would have a constraint on the choice of Δt so that a minimum time step for the random walk could in principle be defined. However, there are practical problems of such a procedure which are not easy to solve.

These are connected with the question of the definition of the perturbation samples: how should one choose the "boxes" of phase space where the initial conditions are to be picked up ? As already mentioned, the dimension of the boxes can be reduced by symmetry arguments and in the present case a critical point is that short-period comet orbits are known to be generally of low inclination. One is therefore close to a planar problem with a four-dimensional phase space. Within the limits chosen for the inclination, the choice of inclination and orientation of the nodal line is statistically immaterial. Furthermore, since we are dealing both with Jupiter-crossing or Jupiter-tangent orbits and Jupiter's eccentricity is quite small, there is an approximately circular symmetry such that encounters with the planet can occur with equal probability, independently of the orientation of the apsidal line. The choice of the latter is hence also immaterial, and obviously the time-related parameter expressing Jupiter's position, at the time of the cometary perihelion passage, is the real stochastic variable of the problem which should be taken at random with its true probability distribution. We are left with two orbital parameters which can be taken as, e.g., semi-major axis and eccentricity (a, e), or aphelion and perihelion distances (Q, q).

As already noticed, boxes in the (Q, q)-plane were considered in the Monte Carlo simulations above mentioned. Let us note that a further reduction of the dimensions appears feasible. The Tisserand criterion for a co-planar cometary orbit :

$$T = \frac{2aJ}{Q+q} + 2\sqrt{\frac{2q\,q}{aJ(Q+q)}} = constant$$

might be used to restrict the random walk $(\Delta Q, \Delta q)$ to one-dimensional curves, and the Monte Carlo simulation would then consist of a set of independent, parallel simulations for the different values of T. In each of these one would consider a suitable orbital parameter varying along the curve, such as the inverse semi-major axis $z = 1/a$, and there would be a random walk with step size distribution $f_i(\Delta z)$ computed for interval $[z_{i-1}, z_i]$ along the z axis. However, although small (Froeschlé and Rickman 1981), the perturbations ΔT in the elliptic restricted problem are indeed important for the out-come of the low-velocity encounters with Jupiter occurring in low-eccentricity planet-tangent orbits, and those encounters appear essential for the dynamical transfer of comets.

Quite obviously the number of (Q, q)-boxes is limited in practice by the requirement to compute a sample of perturbations large enough to give a fair representation of the dynamics of the region in question. E.g., pronounced non-Gaussian tails are known to exist in the perturbation distributions and the sample must extent far enough into these tails to cover their significant parts. This means that the number of orbits to be integrated in each box may be very large [for a possible way to reduce this problem, see Rickman and Froeschlé (1983)]. At this time it was difficult to go much beyond the dynamical resolution of the 30 boxes. On the other hand, this already guarantees a certain "dynamical homogeneity" in the sense that the perturbation distribution over a short interval of time (one orbital period) should not vary too much from one side of a box to the opposite one.

However, within all the boxes we can expect to find smaller regions corresponding to "resonant strips" with $a \approx a_{res}$, where the comets are close to a simple mean motion resonance with Jupiter. Within such a region there may be slow circulation of the critical argument, in which case it is extremely difficult to define a mean interval between encounters and to find an integration interval Δt which is everywhere sufficient. But actually the situation is even worse, because there also exist integrable regions of phase space corresponding to stable librations of the critical argument, where encounters with Jupiter never occur.

The main result of a Monte Carlo simulation of cometary dynamics is a picture of the distribution of comets over the various orbits connected by the stochastic transfer process in question. If the short-period comets are viewed in their most general framework, the dynamics includes perturbations by all the planets as well as non gravitational perturbations, and it would then be interesting to estimate the number of comets trapped in such quasi-stable resonant regions. How this goal would be achieved is not yet clear: the problem is to find an appropriate definition of the perturbation sample boxes and corresponding integration intervals for such a detailed investigation. If on the other hand, one considers the dynamical transfer in the three-body problem (Sun-Jupiter-comet), the problem is instead to reach what may be called "topological homogeneity" (Froeschlé and Rickman 1988): comets should then follow only chaotic routes and the relevant boxes from which the initial conditions are to be picked are in fact the intersections of the usual $(Q,\ q)$-boxes with the chaotic part of phase space. If this restriction is not taken into account, the rate of the transfer is artificially slowed down by inclusion of irrelevant trappings or an overestimated probability of small perturbations.

2.3.2. Markov chain modelling.

a) Short-period comet dynamics. If indeed we consider the orbital distribution of comets as the principal result to be obtained, we can again argue that the random walks by individual sample comets considered in the Monte Carlo simulation contain too much information: the only interesting quantity is the number of comets in each $(Q,\ q)$-box either as a function of time or in a steady state. The same perturbation samples used for the Monte Carlo simulation can then be used to calculate "jump probabilities" between the various $(Q,\ q)$-boxes over a common time interval Δt. Defining the state vector $\mathbf{n}$ of the cometary population to be the set of numbers of comets in the different boxes, and calling P_{ij} the jump probability from box i to box j, n evolves according to a Markov chain:

$$\mathbf{n}(t + \Delta t) = \mathbf{n}(t).P$$

where $P = (P_{ij})$ is the transformation matrix. This method was first used in cometary dynamics by Rickman and Froeschlé (1979) for the same domain of the $(Q,\ q)$-plane as in the above-mentioned Monte Carlo simulation and later on for Oort cloud dynamics by Lago and Casenave (1983).

This Markov method has the advantage of extreme efficiency, in particular for finding steady-state solution where we have just to solve :

$$\mathbf{n}_{ss} = \mathbf{n}_{ss}.P$$

i.e., a system of linear equations. Its main disadvantage is that, just like for the Monte Carlo simulation, the number of $(Q,\ q)$-boxes is limited by the large number of integrated orbits required to obtain accurate estimates of the jump probabilities. If e.g., the 30 boxes of Rickman and Froeschlé are considered, the information obtained on the $(Q,\ q)$ distribution is strictly limited to the 30 sample points represented in the state vector.

b) The long-term dynamical behaviour of small bodies in the Kuiper belt. Again, like Rickman and Froeschlé (1979) but in different spirit, in order to study the slow diffusion of small bodies in the Kuiper belt, i.e. to determine the time scales of this process, Levinson (1991) has used Markov chain methods. Here we are not seeking any more steady states solutions but estimates of diffusion times which should be of the order of the age of the solar system.

The $(Q,\ q)$ plane is also divided into small bins for which transition probabilities are estimated through sample of 100 particles in each bin which are integrated for approximately 100 periods. In addition to the bins within the Kuiper belt, which is a transient region, two special absorbing bins act as border edges, an inner edge $q < 30$ AU where the objects become Neptune crossers and an outer edge left as a free parameter with corresponding probability $P_{ii} = 1$ and $P_{ij} = 0$ for $i \neq j$.

The probabilities matrix takes the form:

$$P = \begin{pmatrix} I & O \\ R & Q \end{pmatrix}$$

and the fundamental matrix

$$M = (M_{ij}) = (I - Q)^{-1}$$

gives the average number of time steps a particle spends in transient bin j before it is absorbed if it started in bin i. Then $t_i = \sum_j M_{ij}$ gives the average time a particle spends in all transient bins before absorption. Furthermore, the variance of the particle lifetime is $v = (2M - I)t - s$ where $s_i = t_i^2$. Finally $f = MR$ gives the probability that a particle starting in transient bin i enters absorbing bin j. Using these tools Levinson found that the Kuiper belt is a good candidate as a source of short-period comets. However, in addition to the already discussed possible draw backs of the method (size number of bins, time steps length) the stochasticity underlying the Markov process is indeed very small and may only be due to undetected long-period oscillations in the behaviour of q and Q.

In the previous study such a low stochasticity was indeed present as an artifact of the method but in the study concerning short-period comets dynamics a strong stochasticity due to strong interactions between the comet and Jupiter was mainly responsible for the diffusion, i.e. big-jumps occurred within Δt and therefore such an artifact could not invalidate the results.

2.4. An other approach the synthetic mapping

The validity of stochastic mappings is not yet unquestionable since it depends obviously of the intrinsic stochasticity of the region of the phase space under study. An other method also, purely numerical, has been used by Froeschlé and Petit (1990, paper I) who built a mapping valid everywhere in the phase space, following an idea already used by Varosi et al. (1987) but in the framework of non-Hamiltonian systems (i.e., systems where attractors do exist). The method consists of coarse-graining the phase space surface of section and then interpolating the value of the image of a point. Linear interpolation requires a rather fine graining of the phase space, hence it is necessary to compute a lot of points on the grid. However, Taylor expansions of order 3 and 5 can provide very good results as long as symmetrical interpolation formulae are applied, for which it is necessary to use an extended grid. Since there are cases where one cannot cross a given limit, asymmetrical interpolation formulae have been tested, but their accuracy was found to be inferior. Therefore Petit and Froeschlé (1993, paper II) have developed another type of interpolation, where the information, including that on the gradients, is stored to the same level of accuracy only for the nearest-neighbouring vertices. Thus, not only images of vertices are computed, but also tangential mappings at each vertex.

There are in any case two key parameters: the number of bins in each direction $N = (\text{total number of cells})^{1/D}$, where $D(= 2$ and 3 in papers I and II) is the dimension of the surface of section, and M the order of the Taylor expansion. In order to explore the validity of the synthetic approach, we have applied our method in two cases:

1) An algebraic area-preserving mapping for which the computation of orbits is very fast. This allows one to follow a large number of orbits and to carry out enough iterations for a meaningful comparison.

2) A special case of the restricted three-body problem, already studied by Duncan et al. (1989).

However contrary to stochastic mappings when the number of degrees of freedom D is greater than 3 the method becomes tedious. We are presently investigating mixed methods.

3. CONCLUSION

If the beautiful mathematical machinery of celestial mechanics is at the basis of the asteroid family story, the chaotic routes have however close connections with the finding of proper elements. Actually, the same machinery, but used in a complementary manner, gives the location and even the size of both mean motion and secular resonances and therefore the regions of the phase space where they overlap and generate chaos. Indeed slow chaotic diffusion, inducing increases of eccentricities result of such resonant interactions and are good candidates for bringing material to the inner solar system. This material must be understood as meteorites resulting of asteroid break up or as Kuiper belt comets. Close approaches with planets are at the origin of another source of chaos for which only crude models have been worked out so far. A better understanding of interactions between resonances and strong gravitational perturbations appears to be crucial for future models of chaotic routes through the solar system.

REFERENCES

Arnold, J.A., 1969, in "A. J". 38 : 1235

Bendjoya, Ph., Slézack, E., Froeschlé, Cl. : 1991, A&A 251, 312–330

Bendjoya Ph., Cellino A., Froeschlé Cl., Zappala V. : 1993 A&A 272, 651-670

Bendjoya Ph. : 1993 : A&A, in press.

Brouwer, D., 1951, in A. J. 56 : 9

Carusi, A., Massaro, E., 1978, in A&A Suppl." 34 : 81

Daubechie, I., 1990 I.E.E.E. Trans. on Information Theory

Duncan, M., Quinn, T., Tremaine, S. : 1989, Icarus 82, 402-418

Feller, W. 1971, An introduction to probability theory and its applications, Vol. 1, J. Wiley and Sons, New York.

Froeschlé, C., 1971, Astrophys. Space Sci. 14, 110.

Froeschlé, C., 1975, Astrophys. Space Sci. 37, 87

Froeschlé, C., Rickman, H. : 1980, A&A 82, 183

Froeschlé, C., Rickman, H. : 1981, Icarus 46, 400-414

Froeschlé, C., Rickman, H. : 1988, Celest. Mech. 43, 265-284

Froeschlé. C., Petit, J.M. : 1990, A&A 238, 413-423

Hadjidemetriou, J. : 1991, preprint

Heisler, J., Tremaine, S. : 1986, Icarus 65, 13-26

Hénon, M. and Heiles, C., 1964, astron. J. 69, 1, N° 1316

Hirayama, K., 1918 : A. J. 31 : 185
Knežević, Z. : 1988, Bull. Astron. Obs. Belgrade 139, 1-6
Knežević, Z. : 1989, Celest. Mech. 46, 147-158
Kozai, Y., 1979 in Asteroids, T. Gehrels. Ed., pp. 334-358, University of Arizona Press, Tucson
Lago, B., Cazenave, A. : 1983, Icarus 53, 68-83
Lemaître, A. Morbidelli, A. : 1993, submitted to Celest. Mech.
Levinson, H.F. : 1991, AJ., Vol 102, N° 2, pp 787-794
Lindblad. B.A., Southworth, R.B., 1971 in Physical Studies of Minor Planets, T. Gehrels, Ed. pp. 337-352, NASA SP-267
Meyer, Y., 1989, in Wavelets, time-frequency methods and phase space, J.M. Combes, A. Grossmann, Ph. Tchamitchian, Eds., Springer
Milani, A., Knežević, Z. : 1990, in Celest. Mech., 49 : 247
Milani, A. Farinella, P., Knežević, Z : 1991 in "Compte rendu de la quinzieme école de printemps d'astrophyique de Goutelas" Eds. Benest, D. and Froeschlé Cl. (Frontières).
Milani, A., Knežević, Z. : 1992 Icarus, 98 : 211-232
Morbidelli, A., Henrard, J. : 1991, Celest. Mech., 51, 131-167
Murtagh, F., Heck, A., 1987, Mutivariable data analysis, Reidel
Nobili, A. et al. : 1969, A&A, 210, 313-336
Petit, J.M., Froeschlé, C. : A&A, in press
Remy, F., Mignard, F. : 1985, Icarus 63, 1-19
Rickman, H., Froeschlé C. : 1979, AJ. 84, 1910-1917
Rickman, H., Froeschlé, C. : 1983, Moon and Planets 28, 69-86
Rickman, H., Froeschlé, C. : 1988, Celest. Mech. 43, 243-263
Rickman, H., Vaghi, S. : 1976, A&A 51, 327-342
Sinaï, G., 1970, Actes Congrès intern. Math., Tome 2, p. 929.
Varosi, F., Gebogi, V., Yorke. J.A. : 1987, Phys. Lett. A 124, 59-64
Weissman, P.R. : 1982, in Comets (ed. L.L. Wilkening), Univ. Arizona Press, Tucson, pp. 637-658
Williams, J.G. : 1969 "Secular perturbations in the solar system", Ph D thesis, Univ. California, los Angeles
Williams, J.G. : 1979, in Asteroids, T. Gerhels, Ed., pp.1040-1063 University of Arizona Press, Tucson
Yuasa, M. : 1973, Publ. Astr. Soc. Japan, 25, 399-445
Zappalà, V., Cellino, A, Farinella, P., Knežević, Z. : 1990, in A. J., 100, 2030
Zappalà V., Cellino A. Farinella P., Milani A.: 1993, A. J. in press

DEPLETION OF THE ASTEROID BELT AT RESONANCES

S. Ferraz-Mello[1], R. Dvorak[2] and T. A. Michtchenko[1]

[1]Universidade de São Paulo
Instituto Astronômico e Geofísico
Caixa Postal 9638
BR - 01065 - São Paulo, Brasil
e-mail Sylvio@iag.usp.ansp.br

[2]Institut für Astronomie
Universität Wien
Türkenschanzstrasse 17
A-1180 Wien, Austria
e-mail Dvorak@avia.una.ac.at

ABSTRACT. The existence of gaps and groups in the distribution of outer belt asteroid orbits, at resonances with Jupiter's orbit, is explained by different rates of destruction of the flow of regular motions by Saturn perturbations. This result completes the explanation given for the gaps of the inner belt where extended chaotic regions exist without the need of taking into account the action of Saturn. The 2/1 and 3/2 resonances are studied and it is shown how they are affected by Saturn. Lyapunov times are estimated from simulations over 10 Myr.

1. INTRODUCTION

One of the oldest problems in asteroids dynamics is the existence of gaps in the distribution of their average distances to the Sun in the values corresponding to orbits whose mean motions (mean angular velocities) are in the ratios 4/1, 3/1, 5/2, 7/3 and 2/1 to that of Jupiter [1–2].

Theories assuming that these gaps are caused by the gravitational evolution of asteroid orbits in the presence of the planets stand high in favor since Wisdom's work showing that the gap at the 3/1 resonance may be explained by the chaotic evolution of resonant orbits driving the asteroid toward close approaches to Mars in a very short time-scale ($10^5 - 10^6$ years) [3–5]. A real puzzle, however, is the fact that in the 3/2 resonance there is a group of asteroids instead of a gap: 56 of them are presently

From Newton to Chaos, Edited by A.E. Roy
and B.A. Steves, Plenum Press, New York, 1995

reckoned amongst the numbered asteroids [6]. Also, the only two known asteroids between this resonance and the Trojans are (279)Thule and (3551)1983 SA, both in the 4/3 resonance. A complete theory must therefore not only explain the existence of gaps at the 4/1, 3/1, 5/2, 7/3 and 2/1 resonances, but also the group at the 3/2 resonance.

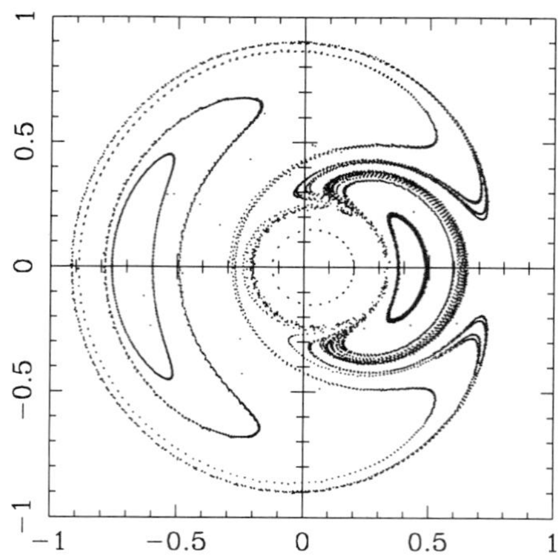

Figure 1. Poincaré map of the 3/1 resonance showing the heteroclinic bridge between chaotic regions of low and high eccentricity. The polar coordinates are the eccentricity and $(\varpi - \varpi_{Jup})$

The 3/1 gap was the first acceptably explained. The depletion mechanism was completely unraveled by Wisdom [3–5] who showed the existence of intermittencies appearing, in this situation, as sudden increases – jumps – of the eccentricities, clearly associated to chaos. During such jumps, the eccentricity reaches values larger than 0.35, the orbit crosses the orbit of Mars and the asteroid becomes liable to close approaches to the planet. The time scale associated to the intermittencies is small ($\sim 10^5$ years) as compared to the age of the Solar System. Later on, Ferraz-Mello and Klafke [7] mapped the structure of the phase space of the restricted elliptic model at this resonance, up to eccentricities as high as 0.9, and disclosed another chaotic region at high-eccentricities and, for some values of the energy, an heteroclinic bridge allowing the solutions to go from one chaotic region to another (Figure 1). The intermittencies associated with orbits going through this heteroclinic bridge can drive the asteroidal eccentricity to values as high as 0.9. The results on this resonance were recently confirmed by numerical simulations over 10^7 years by Saha [8].

Wisdom's mapping model and perturbative method were used also by several authors [9–13] to study the 5/2, 7/3, 2/1 and 3/2 resonances. However, many results were impaired by the fact that the second-order techniques devised by Wisdom to study the 3/1 resonance are based on an abridged Hamiltonian formed by the leading terms of the classical expansion of the disturbing function. The convergence radius of this expansion is $e \approx 0.35$ for the 3/1 resonance, but is smaller for the outermost resonances [14–15].

The only existing alternative for the outer asteroidal belt is to discard the use of classical expansions. One possibility is to model the long-term evolution of resonant motion using averaged potentials based on non-central expansions [7,16–19] or numerical simulations. The earlier simulations [20–21] could reach only some hundreds of centuries but the current investigations [8,22–24] are reaching 1 Myr or more and are leading to new views on the dynamics of the outermost resonances.

The topological dynamics modelling presented in this paper is a synthesis of the two approaches. It aims at putting together the high precision of numerical simulations and the knowledge of the topological dynamics provided by the semi-analytical modelling of the long-term evolution.

2. PRELIMINARY NUMERICAL EXPERIMENTS

We performed a series of numerical experiments with a high-order Lie integrator [25], including several planets. The initial conditions were taken so as to render the critical angle:

$$\sigma = (p+q)\lambda_{Jup} - p\lambda - q\varpi$$

close to its stable equilibrium value (λ and ϖ denote mean longitude and longitude of the perihelion, respectively). The initial semi-major axis and eccentricity correspond to periodic orbits of the restricted circular problem where the amplitude of the libration of the critical angle σ is equal to zero [26–27]. The initial conditions for the planets reproduce the current relative situation of their orbits. However, in almost all our experiments, a plane model was used in order to increase the chance for close approaches involving the asteroid. In view of the many parameters involved, this search was carried out quite unsystematically.

When Jupiter is the only planet, the resonance protects the asteroids against close approaches to Jupiter; as a consequence, they remain in the resonance during the whole integration time (usually 1 Myr). Eccentricities as large as $0.5 - 0.6$ were quasi-periodically attained by some orbits (typical periods: $40,000 - 60,000$ years). However, for initial eccentricities below 0.4, an eccentricity large enough to allow a collision with Mars to occur was never attained. This indicates that, as far as the 2/1 and 3/2 resonances are concerned, Mars is not able to scatter a librating asteroid, when Jupiter is kept in a fixed elliptic orbit. Saturn is needed to modify the orbit of Jupiter and, in this way, to introduce new modulations in the asteroid orbit. When Saturn is included in the calculations, the variations becomes less regular and the $\sigma-$libration amplitudes show large variations. Thus, when the inner planets are introduced, close approaches to them happen and many orbits escape the resonance (and become hyperbolic a short time thereafter). Two mechanisms were observed: (a) A very close approach to Mars ($\sim 10^{-5}$ AU) followed by hyperbolization; (b) A less close approach to Mars ($\sim 10^{-3}$ AU) and escape of the resonance; a subsequent approach to the Earth, Mars or Jupiter provokes the hyperbolization. In the time scale of our integrations (1 Myr), the observed hyperbolizations occurred for orbits starting close to the bifurcations shown in figures 3 and 6.

It is worth emphasizing that the results are highly sensitive to initial conditions and parameters when a close approach is involved. One curious example of this sensitivity was observed in integrations of (4179) Toutatis (in the 3/1 resonance). With a 7-planet model (all except Mercury and Pluto), the orbit went through the heteroclinic bridge found by Ferraz-Mello and Klafke [7] and escaped in less than 1 Myr; when Mercury was also taken into consideration, the same initial conditions led to an orbit that remained confined to the inner chaotic region and the asteroid did not leave the resonance for the whole integration time.

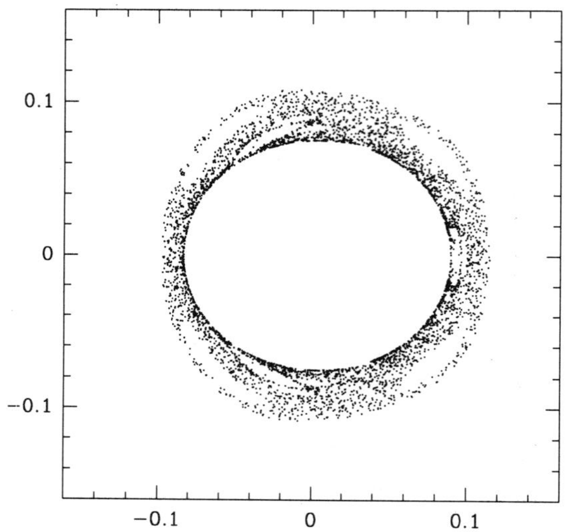

Figure 2. The low-eccentricity chaotic region of the 2/1 resonance. Polar coordinates as in Fig. 1

3. TOPOLOGICAL MODELLING

In the semi-analytical model of Ferraz-Mello and Klafke [7], the planar problem is reduced to a system with two degrees of freedom by means of an averaging over the mean synodic angle (difference between the mean longitudes of the asteroid and Jupiter). Brown and Shook [28–29] have shown that this averaged system has, in the neighbourhood of the $(p+1)/p$ resonance, an energy-like integral:

$$E = \frac{-\mu}{2a} - \frac{p+1}{p} n_1 n a^2 + R$$

(a and n are the asteroidal semi-major axis and mean motion, n_1 is Jupiter's mean motion, R is the potential of Jupiter's disturbing force on the asteroid and μ is the product of the gravitational constant and the mass of the Sun). The topological dynamics may then be shown on a surface of section.

The exact equations of the plane restricted elliptic problem were numerically integrated using Everhart's RA-15 integrator [30], and filtering out the short-period terms (periods less than 80 – 100 years) by using a digital low-pass filter [31–32]. The result is a numerical equivalent of the former procedure: the system is reduced to two degrees of freedom by means of an averaging done after the integration, instead of beforehand as usual. Brown and Shook's integral of the restricted elliptic problem averaged in the neighbourhood of resonances is used to allow us to draw the results on surfaces of section. The assumptions implicit in this technique were checked by constructing, in a regular region, several surfaces of section with small differences in the Brown and Shook averaged energy and making graphical composites. Small differences in the energy often lead to intersections of the invariant curves while no intersection is found for neighboring curves corresponding to the same energy. The level of discrimination is better than $0.0002\,(\mathrm{AU/yr})^2$ for averages done over at least 10^5 years. The dynamical topology of the 2/1 and 3/2 resonances could be systematically investigated in this way.

A similar procedure is adopted for including the effects of inclination and the action of Saturn. However, one must bear in mind that, now, the sections shown cannot be interpreted as Poincaré maps. The results correspond to systems with more than 2 degrees of freedom and the constant energy manifold has more than 3 dimensions. The possible existence of invariant (or quasi-invariant) curves can not be deduced from mere inspection of the sections. Other diagnostic procedures, such as the computation of the Lyapunov times and the analysis of the power spectrum of the solutions were generally used.

4. THE 2:1 GAP

The collision of a 2/1-resonant asteroid with Mars requires an eccentricity larger than 0.49. Figures 2 and 3 are surfaces of section constructed numerically as described in the preceding section, for $E = -18.0805\,(\mathrm{AU/yr})^2$. The region of low-eccentricity chaotic orbits found by Giffen [20] is shown in Figure 2, where we traced one orbit for 2 Myr. The confining invariant curves seen in Figure 3 correspond to 1-Myr integrations. In the frame of the plane restricted problem, the low-eccentricity chaos is not able to produce high-eccentricities. In the preliminary integrations, reported in section 2, orbits starting from smaller initial eccentricities and reaching 0.49 were found only when other planets were included. The situation shown in these sections repeats itself

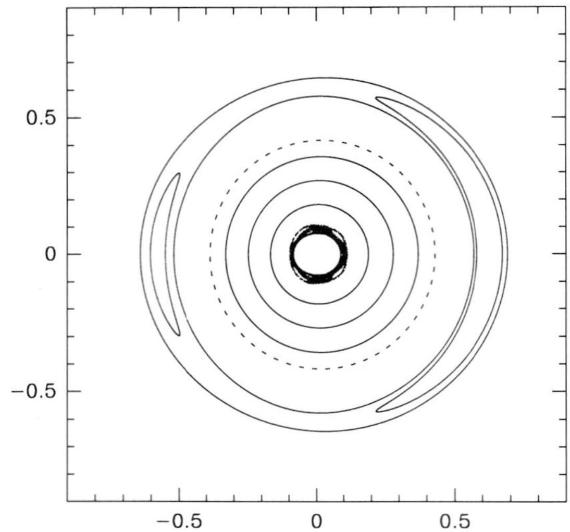

Figure 3. Poincaré map ($\sigma = 0, \dot{\sigma} > 0$) of the 2/1 resonance at $E = -18.0805\,(\mathrm{AU/yr})^2$. Polar coordinates as in Fig. 1

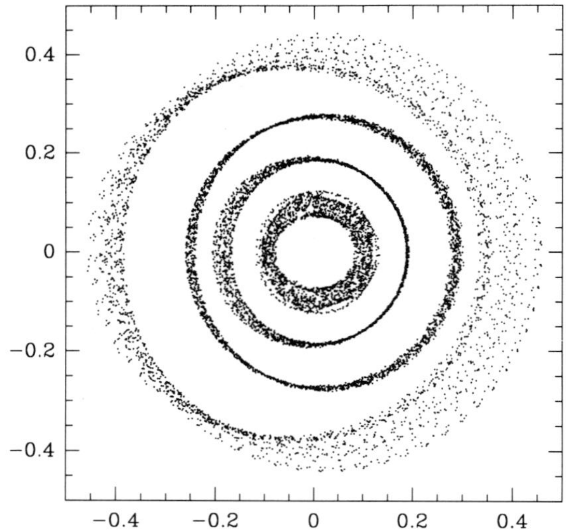

Figure 4. Evolution of some C-type orbits with an initial inclination of 5° in simulations including also Saturn. Polar coordinates as in Fig. 1

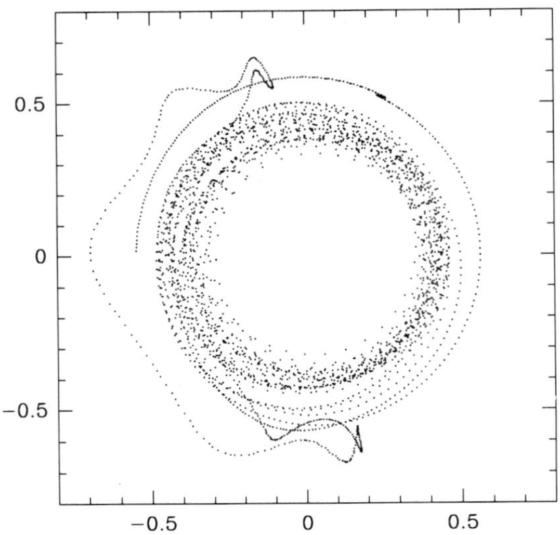

Figure 5. Evolution of the left L-type orbit with an initial inclination of 5° in simulations including also Saturn. Polar coordinates as in Fig. 1

with only minor variations for several values of E. We only note that the two lobes of perihelion libration (L-type solutions), shown in Figure 3, appear only in the interval $-18.0809 \le E \le -18.0801\,(\mathrm{AU/yr})^2$. For smaller values of E, only the right lobe appears, while for larger values of E, there is only the left one. The quasi-invariance of the curves may be verified by means of the computation of the Lyapunov times (inverse of the maximum Lyapunov characteristic numbers) with the rescaling procedure of Benettin et al. [33]:

$$T_{Lyap}(n\tau) = \left[\frac{1}{n\tau}\sum_{1}^{n} \ln \frac{d_i}{d_0}\right]^{-1}$$

(d_i/d_0 is the rate of divergence of the solutions in the ith integration interval of length τ). The Lyapunov time found for the inner chaotic orbit is 4,000 years, in agreement with the value computed by Froeschlé and Scholl [34]. The innermost invariant curves (solutions with circulating perihelion or C-type solutions) do not show any chaoticity during the integration interval (10 Myr); the values corresponding to the outermost curves are of the order of 1 Myr.

When the perturbations due to Saturn are included and the initial conditions assume an inclined orbit ($i = 5°$), all solutions become clearly chaotic. Lyapunov times were estimated from some fifty 5-Myr integrations for initial eccentricities in the range $0.1 < e_0 < 0.4$ and semi-major axes ranging from the middle to the border of the resonance; with only one exception, the results were less than 160,000 years (but generally less than 60,000 years). Parts of the 1-Myr surface of section for initial conditions corresponding to $E = -18.0805\,(\mathrm{AU/yr})^2$ are shown in Figures 4 and 5. However, even if this surface of section is computed in the same way as the previous ones, the introduction of Saturn and the initial inclined orbit mean that the system has, now, more than two degrees of freedom and the surface of section thus obtained is a projection onto a plane of a manifold of larger dimension. No escape from the resonance was noted in these 1-Myr integrations.

These results may be compared to those obtained with similar hypotheses (inclined orbits and Jupiter orbit affected by Saturn perturbations) by Morbidelli and Moons with an averaged model using non-central expansions. Their surfaces of section show a large region of regular motions, even for large libration amplitudes ($\Delta a \sim 0.15$ AU) and eccentricities as large as 0.5. The discrepancies with respect to our result may be due to the over-averaging necessary there to reduce the complete system to only two degrees of freedom.

5. THE GROUP AT THE 3/2 RESONANCE

The phenomena observed in the 2/1 resonance cannot repeat themselves in the 3/2 resonance. Otherwise, this resonance should be as depleted as the 2/1 resonance, at variance with some tens of asteroids observed there in libration about the exact resonance.

Figure 6 shows the 1-Myr surface of section corresponding to orbits which, after the digital filtering, have the same mean value of Brown and Shook averaged energy ($E = -14.9261\,(\mathrm{AU/yr})^2$). They differ from the surface of section obtained for the 2/1 resonance in several respects: (a) the bifurcation between the two modes of oscillation of the perihelion is the source of appreciable chaoticity, which spreads itself over the whole region of perihelion libration (L-type solutions) and also over the outer circulations; (b) the inner C-type solutions do not show chaotic activity; (c) there is one lobe of perihelion libration at the left-hand side (the inversion observed in the 2/1 resonance does not appear here). The study of the Lyapunov times in 10-Myr integrations shows that the

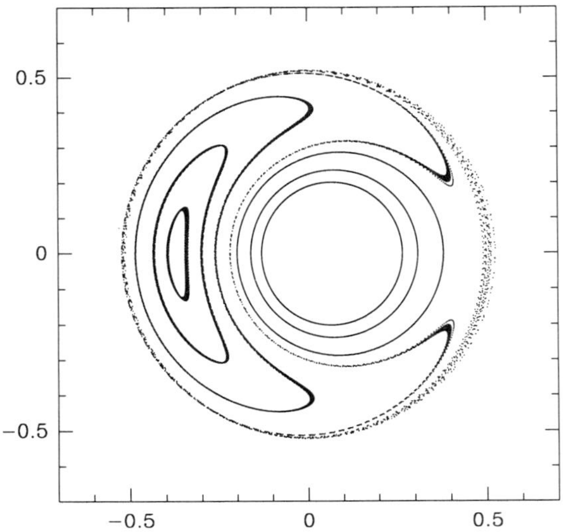

Figure 6. 1-Myr Poincaré map($\sigma = 0, \dot{\sigma} > 0$) of the 3/2 resonance at $E = -14.9261\,(\mathrm{AU/yr})^2$. Polar coordinates as in Fig. 1

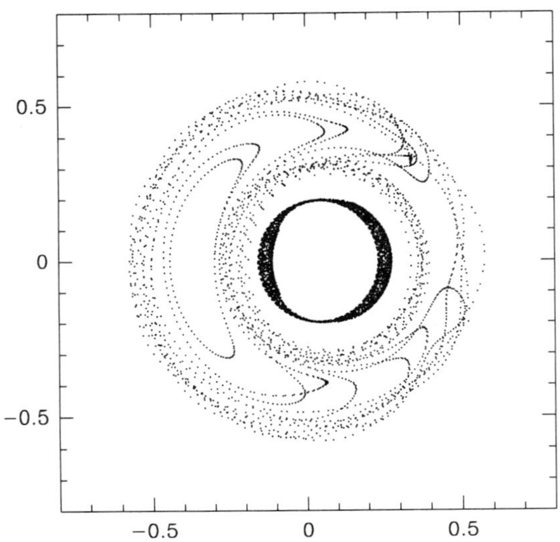

Figure 7. Evolution of one C-type and one L-type orbit with an initial inclination of 5° in simulations including also Saturn. Polar coordinates as in Fig. 1

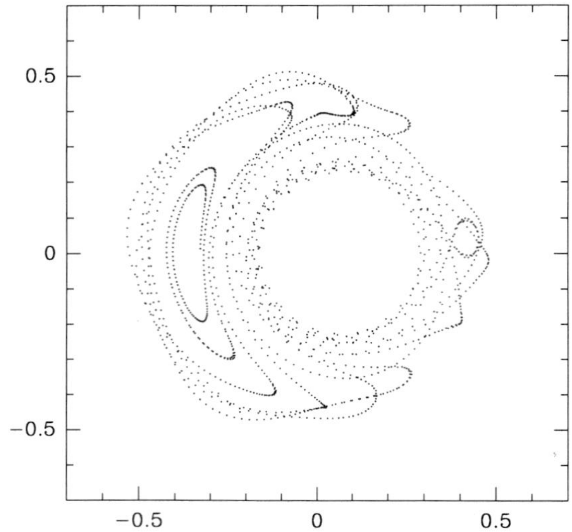

Figure 8. Evolution of one of the innermost L-type orbit with an initial inclination of 5° in simulations including also Saturn. Polar coordinates as in Fig. 1

inner C-type orbits are regular, in agreement with previous results [22]. However, a noticeable chaoticity (Lyapunov time of the order of 10,000 years) appear in the L-type orbits.

When we add Saturn and start the integration with the asteroid in an inclined orbit ($i = 5°$), the chaoticity of the L-type orbits is strongly increased. In fact, all of them escape the resonance in times of the order of half million years or less (see Figures 7 and 8). The computations of the Lyapunov times for the inner C-type orbits shows results in the range $10^{5.5} - 10^7$ years, after 10 Myr of integration, These results are slightly larger than those found by Franklin *et al.* [24]. They indicate that the time necessary for the occurrence of significant orbital transitions (the so-called crossing times) [35–37] is larger than the Solar System age. It is worth reminding that the width of the inner ring in Figure 7 cannot be interpreted as chaos since the surface of section, in this case, has more than two dimensions.

The large regular region obtained for this resonance by Morbidelli and Moons [19], with their averaged model, is in agreement with the Lyapunov times given above; however, our simulations including Saturn and inclined orbits did not show the regular orbits with librating perihelion found by them.

6. CONCLUSION

The results of the previous sections are strong evidences in favour of the theory of gravitational evolution and scattering by the inner planets for the origin of the Kirkwood gaps and of the importance of chaotic diffusion in this process. The results previously known were, in a certain sense, restricted to the understanding of the depletion process in the 3/1 and 5/2 resonances. Concerning the resonances in the outer part of the asteroidal belt, there were some results on the evolution of low-eccentricity orbits; but no general results about orbits whose eccentricities can reach values high enough to allow the asteroid to come close to Mars (0.49 and 0.58, respectively, in the 2/1 and 3/2 resonances).

The results reported here show that Saturn triggers the destruction of the tori seen in the 2/1 resonance in the frame of the restricted elliptic three-body problem The energy manifold becomes topologically and dynamically connected, even if KAM tori may exist, and an orbit can diffuse to high eccentricities and have a large probability of collision with the inner planets. The rate of this diffusion is very slow and, in some regions of the phase space an asteroid can remain in this resonance for a very long time before escaping. This result is consistent with the fact that 6 known asteroids are currently moving in the 2/1 resonance. It is not excluded that some of them be primordial. We may add that eccentricity jumps of the kind expected to deplete this resonance were, indeed, observed by Scholl (*personal communication*) in a number of simulations over 12 Myr including the perturbations of the four large planets.

In the 3/2 resonance, Saturn completely destroys the structure of the set of orbits with librating perihelion. Mars is not needed to explain the depletion of the orbits with mean eccentricities larger than ~ 0.3. However, the structure of the inner part of the phase space is not affected by Saturn, at least for times of the order of the Solar System age, and the asteroids situated there are protected against eccentricity jumps or close approaches to Jupiter. These results are consistent with the existence of 56 currently known asteroids moving in this resonance and with the limitations observed in their orbital eccentricities.

Acknowledgements: This investigation has been sponsored by the Research Foundation of the State of São Paulo (FAPESP). The authors are indebted to Drs. H.Eichhorn, M.Šidlichovský, J.Wisdom and T.Yokoyama for their comments on an early version of this paper.

References

[1] Cl. Froeschlé and R. Greenberg, Mean motion resonances, *in* "Asteroids", R.P.Binzel *et al.* eds., Univ. Arizona Press, Tucson, pp. 827-844 (1989).

[2] S. Ferraz-Mello, Kirkwood gaps and groups, *in* Proc. IAU Symposium 160 , A.Milani *et al.* eds., Kluwer, Dordrecht (in press).

[3] J. Wisdom, The origin of Kirkwood gaps: A mapping for asteroidal motion near the 3/1 commensurability, *Astron. J.* **85**, 1122-1133 (1982).

[4] J. Wisdom, Chaotic behaviour and the origin of the 3/1 Kirkwood gap, *Icarus* **56**, 51-74 (1983).

[5] J. Wisdom, A perturbative treatment of motion near the 3/1 commensurability, *Icarus* **63**, 272-289 (1985).

[6] "Efemeridi Malikh Planet na 1993 god", Inst. Teoret. Astron., St. Petersburg, Russia (1992).

[7] S. Ferraz-Mello and J.C. Klafke, A Model for the study of very-high-eccentricity asteroidal motion. The 3:1 resonance, *in* "Predictability, Stability and Chaos in N body Dynamical Systems", A.E.Roy, ed., Plenum Press, New York, pp. 177-184 (1991).

[8] Saha, P. "Simulating the 3:1 Kirkwood gap", *Icarus* **100**, 434-439 (1992).

[9] M.Šidlichovský and B. Melendo, Mapping for the 5/2 asteroidal commensurability, *Bull. Astron. Inst. Czechoslov.* **37**, 65-80 (1986).

[10] C.D. Murray, Structure of the 2:1 and 3:2 jovian resonances, *Icarus* **65**, 70-82 (1986).

[11] J. Henrard and A. Lemaître, A perturbative treatment of the 2/1 jovian resonance, *Icarus* **69**, 266-279 (1987).

[12] T. Yokoyama and J.M. Balthazar, Application of Wisdom's perturbative method for 5:2 and 7:3 resonances, *Icarus* **99**, 175-190 (1992).

[13] M. Šidlichovský, Chaotic behaviour of trajectories for the asteroidal resonances, *Celest. Mech. Dyn. Astron.* **56**, 143-152 (1993).

[14] K.F. Sundman, Sur les conditions nécessaires et suffisantes pour la convergence du développement de la fonction perturbatrice dans le mouvement plan, *Öfversigt Finska Vetenskaps-Soc. Förh.* **58 A** No. 24 (1916).

[15] S. Ferraz-Mello, The convergence domain of the Laplacian expansion of the disturbing function, *Celest. Mech. Dyn. Astron.* **58** (1994).

[16] S. Ferraz-Mello and M. Sato, A very-high-eccentricity asymmetric expansion of the disturbing function near resonances of any order, *Astron. Astrophys.* **225**, 541-547 (1989).

[17] J.C. Klafke, S. Ferraz-Mello and T. Michtchenko, Very-high-eccentricity librations at some higher-order resonances, *in* Proc. IAU Symposium 152, S.Ferraz-Mello, ed., Kluwer, Dordrecht, pp. 153-158 (1992).

[18] M. Moons and A. Morbidelli, The main mean-motion commensurabilities in the planar circular and elliptic problem, *Celest. Mech. Dyn. Astron.* **57**, 99-108 (1993).

[19] A. Morbidelli and M. Moons, Secular resonances in mean-motion commensurabilities. The 2/1 and 3/2 cases, *Icarus* **102**, 316-332 (1993).

[20] R. Giffen, A study of commensurable motion in the asteroidal belt, *Astron. Astrophys.* **23**, 387-403 (1973).

[21] Cl. Froeschlé and H. Scholl, On the dynamical topology of the Kirkwood gaps, *Astron. Astrophys.* **48**, 389-393 (1976).

[22] J. Wisdom, Chaotic dynamics in the Solar System, *Icarus* **72**, 241-275 (1987).

[23] S.P. Ipatov, Evolution of asteroidal orbits at the 5:2 resonance, *Icarus* **95**, 100-114 (1992).

[24] F. Franklin, M. Lecar and M. Murison, Chaotic orbits and long-term stability: an example from asteroids of the Hilda group, *Astron. J.* **105**, 2336-2343 (1993).

[25] A. Hanslemeier and R. Dvorak, Numerical integration with Lie series, *Astron. Astrophys.* **132**, 203-207 (1983).

[26] S. Ferraz-Mello, The high-eccentricity libration of the Hildas, *Astron. J.* **96**, 400-408 (1988).

[27] J. Hadjidemetriou, Resonant families of periodic orbits in the restricted 3-body problem at the 3:1 resonance and their relation to the averaged model. *Celest. Mech. Dyn. Astron.* **53**, 151-183 (1992).

[28] E.W. Brown and C.A. Shook, "Planetary Theory", (University Press, Cambridge), p.232 (1933).

[29] D. Brouwer, The problem of the Kirkwood gaps in the asteroidal belt, *Astron. J.* **68**, 152-159 (1963).

[30] E. Everhart, An efficient integrator that uses Gauss-Radau spacings, *in* "Dynamics of Comets, their origin and evolution", A. Carusi and G.B. Valsecchi, eds., Reidel, Dordrecht, 185-202 (1985).

[31] A. Milani, A. Nobilli and M. Carpino, Secular variations of the semimajor axes: theory and experiment, *Astron. Astrophys.* **172**, 265-279. (1987).

[32] T.A. Michtchenko and S. Ferraz-Mello, The high-eccentricity libration of the Hildas. II. Synthetic-theory approach, *Celest. Mech. Dynam. Astron.* **56**, 121-129 (1993).

[33] G. Benettin, L. Galgani and J.M. Strelcyn, Kolmogorov entropy and numerical experiments, *Phys Rev. A*, **14**, 2338-2345 (1976).

[34] Cl. Froeschlé and H. Scholl, The stochasticity of peculiar orbits in the 2/1 Kirkwood gap, *Astron. Astrophys.* **93**, 62-66 (1981).

[35] M. Lecar, F. Franklin and M. Murison, On predicting long-term orbital instability: A relation between the Lyapunov time and sudden orbital transitions, *Astron. J.* **104**, 1230-1236 (1992).

[36] H.A. Levison and M.J. Duncan, The gravitational sculpting of the Kuiper belt, *Astrophys. J. Lett.* **406**, L35-L38 (1993).

[37] M. Holman and J. Wisdom, Dynamical stability of the outer Solar System and the delivery of short-period comets, *Astron. J.* **105**, 1987-19990 (1993).

ON THE DEPLETION OF THE OUTER ASTEROID BELT

Judit Györgyey Ries

Department of Astronomy
The University of Texas at Austin
Austin, TX 78712

INTRODUCTION

In the last few years, reasonable success has been achieved in explaining the gaps in semimajor axis distribution of the asteroids. Wisdom (1982, 1983) introduced a mapping technique to study the behavior of bodies at the 3:1 mean motion resonance, which others modified and applied to other gaps (Murray, 1986). They and other researchers (Henrard et al, 1987, 1989; Lemaitre and Henrard, 1987) have also reached the conclusion that the formation of the gaps can be explained by removal of chaotic objects.

The distribution in the outer Asteroid Belt, outside of 3.4 AU, presents a different problem. Instead of gaps in the distribution of the semimajor axis, we have groups in the otherwise almost empty semimajor axis space. Several numerical works tested the hypothesis that Jupiter's direct perturbations ejected the majority of the primordial bodies from that part of the Solar System. (Lecar et al, 1973,1992; Franklin et al, 1989; Froeschlé and Scholl 1979; Milani and Nobili, 1985; Ries, 1992). They found that the region between the Hilda group (3:2 resonance) and Saturn depletes very quickly. Inside the Hilda group, the depletion was not sufficient to match the observed asteroid density. Also for the time intervals considered orbits survive at the 4:3 resonance (Yoshikawa, 1989, 1990; Ries, 1992).

This investigation's main goal was to qualitatively reproduce the observed distribution of the outer Asteroid Belt. Previous simulations used sample sizes of a few hundred; in this case, the behavior of about 3000 objects was considered to better approximate the conditions in the pre-planetary nebula. Instead of using long integrations, a numerical criterion was introduced in order to identify chaotic orbits. According to Wisdom (1982), a large change in the eccentricity of a chaotic orbit is likely to result in close encounters with other planets, which will probably remove the object from the belt. In addition, chaotic orbits will be more sensitive to other perturbations not taken into account in the three body problem. This could lead to the destruction of the protection mechanism of resonant outer belt asteroids against close encounters with Jupiter. Hence it was assumed that orbits which are not regular in the elliptic restricted three body problem would eventually be removed. Extensive reviews on the origin (Wetherill), mean motion resonances (Froeschlé and Greenberg) and dynamics of the outer belt (Nobili) can be found in Asteroids II (1989).

From Newton to Chaos, Edited by A.E. Roy
and B.A. Steves, Plenum Press, New York, 1995

THE DYNAMICAL MODEL

The planar elliptic restricted problem of three bodies was chosen as the model to approximate the dynamics of the asteroids, where a fixed ellipse with an eccentricity (e) of 0.48 was used to represent the orbit of Jupiter. The regularized equations of motion were derived using a Levi-Civita method (Szebehely, 1967) to remove the singularity at Jupiter, thus allowing the integration of asteroid orbits through close encounters with Jupiter.

The regularized coordinates u and v can be determined from the coordinates of the rotating-pulsating frame using

$$\xi = (u^2 - v^2) + (1-\mu) \tag{1a}$$

$$\eta = 2uv\,. \tag{1b}$$

The independent variable of the regularized equations, τ and the true anomaly of Jupiter, f are related according to the following equation:

$$\frac{df}{d\tau} = 4(u^2 + v^2). \tag{2}$$

Thus the regularized equations of motion are

$$\frac{d^2u}{d\tau^2} - 8(u^2 + v^2)\frac{dv}{d\tau} = 4\frac{\partial G}{\partial u} - 8u\mathrm{I} \tag{3a}$$

$$\frac{d^2v}{d\tau^2} + 8(u^2 + v^2)\frac{du}{d\tau} = 4\frac{\partial G}{\partial u} - 8v\mathrm{I}\,, \tag{3b}$$

where

$$G = \frac{1}{1 + e\cos f}\left\{\tfrac{1}{2}\left[(u^2 + v^2)^3 + 2(u^4 - v^4)(1-\mu) + (u^2 + v^2)(1-\mu)^2\right] + \frac{(u^2 + v^2)(1-\mu)}{\sqrt{(u^2 + v^2)^2 + 2(u^2 - v^2) + 1}} + \mu\right\} - \frac{(u^2 + v^2)}{2}C\,, \tag{4}$$

and

$$\mathrm{I} = e\int_{f_0}^{f_t} \frac{\sin f}{1+e\cos f}\,\Omega\; df. \tag{5}$$

The quantity I is integrated simultaneously with the equations of motion. C is a constant of motion determined from initial conditions, and Ω is the usual potential in the rotating-pulsating frame.

ORBIT CLASSIFICATION

Depletion studies have not yet achieved sufficient agreement with the observed semimajor axis distribution in the outer belt. Lecar et al. (1992) found an empirical relationship between the value of the largest LCN and the time required for an asteroid to cross the orbit of Jupiter. However, there is substantial scatter in their results, and the relationship is not valid at mean motion resonances where this study finds most of the survivors.

The working hypothesis for the purpose of this study was that asteroid trajectories which are chaotic in the elliptic restricted three body problem, would be removed from the corresponding orbits in the real Solar System. The properties of the Lyapunov characteristic numbers (exponents) were utilized for identification. Although chaos is usually associated with mean motion resonances, one could expect that very close to the center of the strong resonances we find regular, or at least weakly chaotic trajectories. For an example of a bounded, chaotic asteroid see Milani and Nobili's (1992) paper on 522 Helga.

Since only the outer Asteroid Belt's structure was investigated, a test particle was considered to be on an escape orbit if the object became a Saturn or a Mars crosser. When this happened, the integration was terminated, and the body was removed from the final distribution. In parallel with the integration of an orbit, a numerical estimate of the largest Lyapunov Characteristic Number (LCN) was calculated by integrating and periodically renormalizing a neighboring trajectory (Benettin and Galgani, 1976). The sum

$$\Gamma = \frac{1}{kT} \sum_{i=1}^{k} ln \frac{d_i}{d_0}; \tag{6}$$

should converge to the largest LCN as kT-> ∞. The choice of initial displacement (d_0) or how often one renormalizes the perturbed trajectory (T) did not seem to greatly influence the final LCN estimate. In some cases the numerical LCN estimates diverged. Some of those initial conditions were investigated again, and convergence to a positive LCN was obtained when the precision of the integration was increased. Hence, orbits with positive slopes were assumed to be chaotic.

Instead of using directly the value of the largest LCN estimates, the slope to the logarithm of those estimates were determined. For stable orbits in Hamiltonian systems, the largest, and thus all LCNs, are zero, and their logarithms are -∞; that is, their numerical estimates should always yield negative slopes on a log-log scale. When convergence to a non-zero value is reached, the slope is zero or numerically very small. Thus, only orbits with approximately constant negative slopes were classified as regular, since this suggests a largest LCN value of zero; the rest were considered chaotic. This method most likely underestimates the number of chaotic orbits. With further integration, the largest LCN estimates for orbits which had negative slopes still can level out, but they should be less than 0.0003. (Although the actual values of the largest LCN estimates were not used as the criterion for removal from the distribution, it was discovered that for the survivors there was a correlation with the semimajor axis. The closer a surviving body was to Jupiter, the smaller was the estimate for its largest LCN.)

NUMERICAL SIMULATION

The initial values of the semimajor axis, the argument of perihelion and the mean anomaly of the 2998 objects investigated were generated randomly from a uniform distribution, while the eccentricity distribution followed a Gaussian with its maximum at 0.15 and standard deviation 0.17. For the initial semimajor axes, values between 3.4 and 5.4 AU were chosen.

From all the objects considered, only 767 stayed within the specified region for the duration of the numerical integration. The resulting distribution of the osculating semimajor axes is compared to the observed distribution in Fig. 1. It is not surprising that this short integration does not reproduce the observed distribution, but it is interesting to note that all the excesses are at or near reasonably low order mean motion resonance with Jupiter. The observed asteroids are also in the neighborhood of resonances; 721 Tabora at the 7:4, 522 Helga at the 12:7, 1144 Oda at the 13:8, and 297 Thule at the 4:3 mean motion resonance.

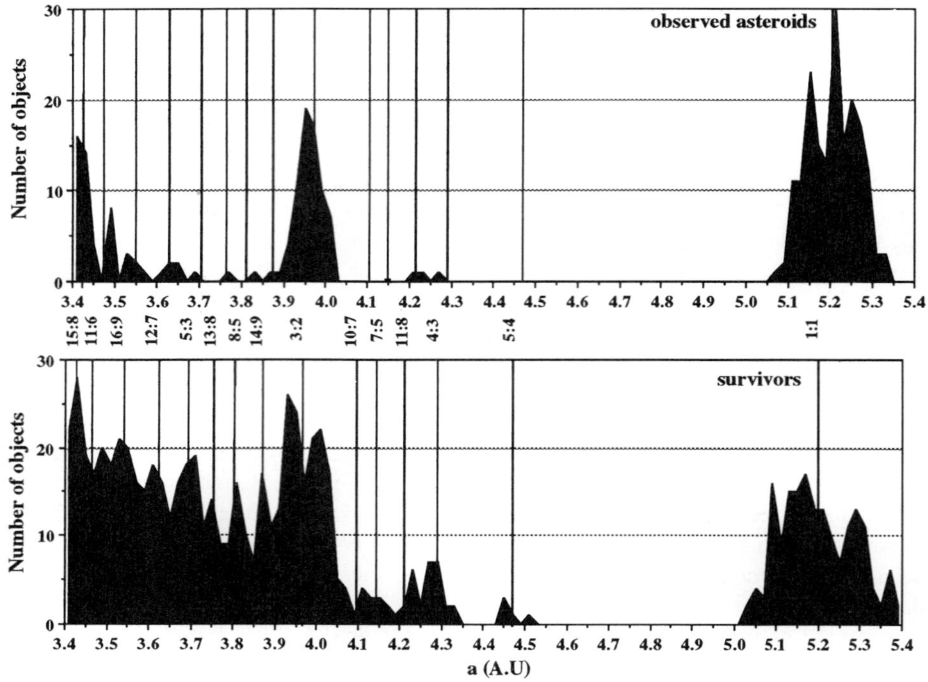

Figure 1. Distribution of osculating semimajor axis in the outer Asteroid Belt for the observed asteroids and the survivors of the simulation.

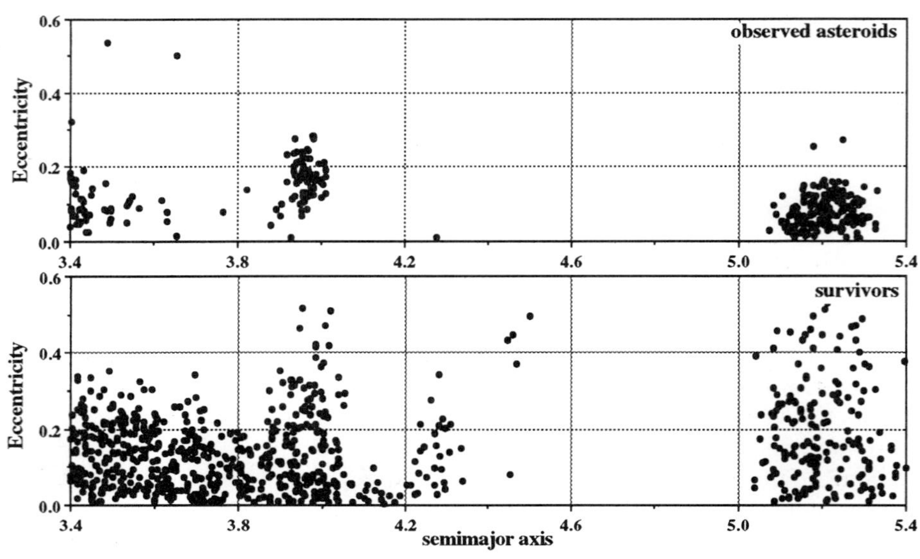

Figure 2. The eccentricity distribution of the observed outer Asteroid belt objects and of the survivors of the simulation as the function of the osculating semimajor axis.

Another way to look at the surviving orbits is shown on Fig. 2. The simulation shows an excess at high eccentricities at every semimajor axis, and fails to reproduce the gap between the Hildas and the main belt.

When the proposed stability criterion was applied only about a dozen Trojan type objects remained. Though it might be discouraging to find that most of the objects in the outer asteroid belt are chaotic, one should consider that the early Solar System was probably much more dense than what was investigated. This would probably have increased the number of objects classified as regular.

CONCLUSIONS

A numerical simulation which spans the age of the Solar System and approximates the early particle density is still not within reach. Since according to this study, survivors are associated with mean motion resonances of the Sun-Jupiter elliptic restricted problem, further studies should concentrate on secular resonances and how other planets might destroy weaker resonances where the objects, being chaotic, are more sensitive to additional perturbations.

REFERENCES

Benettin, G. Galgani, L., Strelcyn, J., 1976, Kolmogorov entropy and numerical experiments, *Phys. Rev. A*. 14, no. 6: 2338.

Franklin, F., Lecar, M., Soper, P., 1989, On the original distribution of the asteroids II, Do stable orbits exist between Jupiter and Saturn, *Icarus* 79:223.

Froeschlé, C., Scholl, H., 1979, New numerical experiments to deplete the outer part of the asteroidal belt, *Astron. Astrophys.* 72: 246.

Froeschlé, C., Greenberg, R., 1989, Mean motion resonances, *in:* "Asteroids II", R. Binzel, T. Gehrels. M.S. Matthews (eds.) The University of Arizona Press, Tucson.

Henrard, J.; Caranicolas, N.D., 1989, Motion near the 3/1 resonance of the planar elliptic restricted three body problem, *Celestial Mech. and Dyn. Astron.* 47:99.

Henrard, J., Lemaitre, A., 1987, A perturbative treatment for the 2/1 Jovian resonance, *Icarus* 69: 266.

Lecar, M., Franklin, F.A., 1973, On the original distribution of asteroids, *Icarus* 20:422.

Lecar, M., Franklin, F., Soper, P., 1992, On the original distribution of the asteroids IV, Numerical experiments in the outer Asteroid Belt, *Icarus* 96:234.

Lemaitre, A.; Henrard, J., 1987, The 3/2 resonance, *Celestial Mech.* 43, no.1-4:91.

Milani, A.; Nobili, A.M., 1985, The depletion of the outer asteroid belt, *Astron. Astrophys.* 144, no.2:261.

Milani, A.; Nobili, A.M., 1992, An example of stable chaos in the Solar System, *Nature* 357:569.

Nobili, A.M., 1989, Dynamics of the outer Asteroid Belt, *in:* "Asteroids II", R. Binzel, T. Gehrels. M.S. Matthews (eds.) The University of Arizona Press, Tucson.

Murray, C.D., 1986, Structure of the 2:1 and 3:2 Jovian resonances, *Icarus* 65, no.1:70.

Ries, J.G., 1992, On the Depletion in the Outer Asteroid Belt and the Structure of the 4:3 Mean Motion Resonance, Dissertation, The University of Texas at Austin.

Szebehely, V., 1967, Theory of Orbits, Academic Press, New York.

Wetherill,G.W., 1989, Origin of the Asteroid Belt, *in:* "Asteroids II", R. Binzel, T. Gehrels. M.S. Matthews (eds.) The University of Arizona Press, Tucson.

Wisdom, J., 1982, The origin of the Kirkwood gaps: a mapping for asteroidal motion near the 3/1 commensurability, *Astron. J.* 87:577.

Wisdom, J., 1983, Chaotic behavior and the origin of the 3/1 Kirkwood gap, *Icarus* 56, no.1: 51.

Yoshikawa, M., 1989, A survey of the motions of asteroids in the commensurabilities with Jupiter, *Astron. Astrophys.* 213, no.1-2,:436.

Yoshikawa, M., 1990, Motions of asteroids at the Kirkwood gaps. I. On the 3:1 resonance with Jupiter, *Icarus* 87:78.

THE CHAOTIC ZONE CONNECTED WITH THE 5/2 MEAN MOTION RESONANCE

David Nesvorný

Astronomical Institute
Boční II 1401, 141 31 Praha 4, Czech Republic

INTRODUCTION

The effective method - mapping - introduced by Wisdom (1982, 1983) allowed to study long-term behaviour in asteroidal resonances. It was applied to the 3/1 resonance and phase space of the problem was investigated there. The existence of the large chaotic zone and the regular islands was then explained by perturbative treatment (Wisdom, 1985). Murray (1986) derived the mapping for first-order resonances 2/1 and 3/2, Šidlichovský at al. (1986) derived the mapping for 5/2 and later Šidlichovský (1992) presented the generalized method for the second, third and fourth-order resonances. The mapping is derived for the Hamiltonian of the restricted three-body problem in the planar-elliptic case, and the truncation to some degree in eccentricities must be used. This gives a good representation for low eccentricities but the approximation is not acceptable for high eccentricities.

This difficulty was overcome by Klafke et al. (1992) where local expansions of disturbing function were employed and averaged equations were then integrated numerically. The method was applied to the 3/1, 4/1 and 5/2 resonances and the high-eccentricity chaotic lobe was found. The analogical approach for high-eccentricity area description was presented in Šidlichovský's (1993) application of symplectic mapping (Yoshida, 1990) to the 5/2 resonance.

In this article we propose the method based on the symmetric multistep integrator (SMI) which was introduced by Quinlan et al. (1990). We will apply this method to the 5/2 resonance and compare results with Wisdom's mapping (WM) and the symplectic mapping (SM).

THE METHOD

We choose the system of units $n_1 = 1$, $a_1 = 1$ and $\mu_0 + \mu_1 = 1$, where n_1 and a_1 are the mean motion and the semi-major axis of Jupiter, μ_0 and μ_1 are the masses of the Sun and Jupiter. The heliocentric radius vector of an asteroid is $\vec{r}$ and corresponding variables of Jupiter will be distinguished by subscript 1, so that $\vec{r}_1$ is the heliocentric radius vector of Jupiter. If we adopt the restricted three-body model we can write the

From Newton to Chaos, Edited by A.E. Roy
and B.A. Steves, Plenum Press, New York, 1995

equation of motion in heliocentric coordinates.

$$\ddot{\vec{r}} = -\frac{\mu_0}{r^3}\vec{r} + \nabla R, \tag{1}$$

where

$$R = \frac{1}{|\vec{r} - \vec{r_1}|} - \frac{\vec{r_1} \cdot \vec{r}}{r_1^3}. \tag{2}$$

This equation can be clearly written in the Hamiltonian form. Then the Laplacian expansion of disturbing function R can be used, and following Wisdom's (1982) calculation for the planar and elliptic problem the new set of canonical variables can be introduced. We define (Jupiter's pericenter is fixed as the reference line; $\tilde{\omega}_1 = 0$)

$$\begin{aligned}
\Phi &= \sqrt{\mu_0 a}/p, \\
\phi &= p\lambda - (p+q)\lambda_1, \\
x &= \sqrt{2\sqrt{\mu_0 a}(1 - \sqrt{1-e^2})}\cos\tilde{\omega}, \\
y &= -\sqrt{2\sqrt{\mu_0 a}(1 - \sqrt{1-e^2})}\sin\tilde{\omega}
\end{aligned} \tag{3}$$

for resonance $(p+q)/p$, where e is eccentricity, λ mean longitude and $\tilde{\omega}$ longitude of pericenter. Than we perform the averaging over short-periodic terms by cancelling the non-resonant short-periodic part of R and obtain Hamiltonian H' composed from three parts

$$H' = H_0(\Phi) - H_0(\Phi_r) + H_s(x,y) + A(x,y)\cos(\phi - \phi_0(x,y)). \tag{4}$$

$H_0(\Phi)$ refers to the unperturbed problem ($H_0(\Phi_r)$ is the constant term; H_0 evaluated at resonant value of Φ), $H_s(x,y)$ is the secular part of perturbing function R and the last term is the resonant part of R. The evolution is then described by the Hamilton canonical equations

$$\dot{\Phi} = -\frac{\partial H'}{\partial \phi},\ \dot{\phi} = \frac{\partial H'}{\partial \Phi},\ \dot{x} = -\frac{\partial H'}{\partial y},\ \dot{y} = \frac{\partial H'}{\partial x} \tag{5}$$

with integral of motion H'. One can employ numerical integration of these equations, or equivalent but much faster mapping, in order to investigate the topology of trajectories in four-dimensional phase space and obtain the general view of orbit evolution. But these methods are based on the expansion of the perturbing function and thus, they can be used only for low-eccentricity area. It can be proved that the Laplacian expansion is absolutely convergent only under the certain eccentricity limit. In the case of resonance 5/2 with Jupiter ($e_1 = 0.048$) the limiting value is ~ 0.283. See discussion in Ferraz-Mello (1993).

Another approach should be used if we wish to describe the high-eccentricity area. We propose to employ symmetric multistep method (Quinlan at al., 1990) and integrate directly Equation (1). This numerical method is specially convenient for near-Keplerian system investigations. It is an ordinary multistep integrator for the second-order differential equations, but the coefficients are devised to suppress the cumulation of the local truncation error to linear dependence on time. We developed the programming package in C language for the quite general constellation of bodies. This package also includes the method of numerical filtration (Carpino at al., 1987) which is needed for averaging. The whole applied procedure was composed from the numerical integration using the symmetric integrator of the twelfth order, numerical averaging and transformation to the resonant variables. This allowed us to describe the whole phase space of the problem.

THE 5/2 TOPOLOGY

Using the method explained in the previous section we performed several numerical integrations for the 5/2 resonance. From Wisdom's (1985) argumentation we can expect ϕ either librate about $\phi_0(x, y)$ or circulate in the time scale which is much shorter then the time scale of x and y evolution. It is convenient to choose $\phi = \phi_0(x, y)$ for surfaces of section. The set of trajectories with certain value of H' will be featured in (x, y)-plane recording a point when the trajectory crosses $\phi = \phi_0(x, y)$. In other words, we will gain the representation of the long-term evolution in $\tilde{\omega}$ and e.

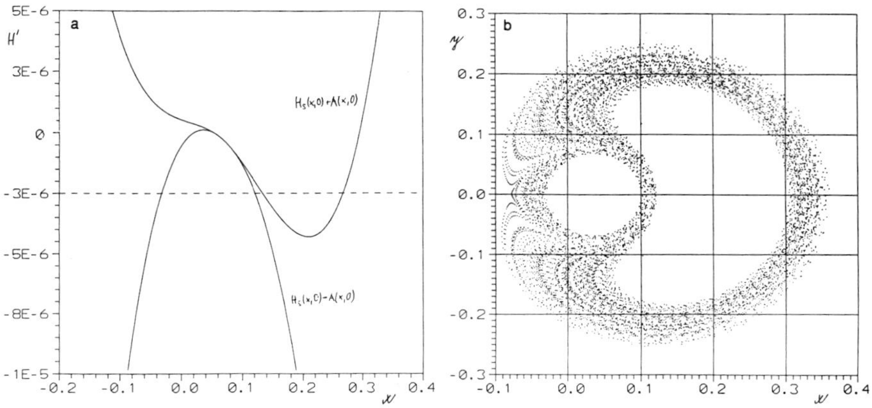

Figure 1. The 5/2 energy diagram and the chaotic zone location on $H' = -3.10^{-6}$ (WM used).

The energy diagram (Šidlichovský, 1992), the dependence of $H_s(x,0) + A(x,0)$ and $H_s(x,0) - A(x,0)$ on x, is presented in Figure 1a. The region between curves refers to the ϕ-libration area, for trajectories below both curves ϕ circulates and the value x on the x-axis is inaccessible for values H' above both curves. This diagram uses the truncated Laplacian series approximation to the third degree.

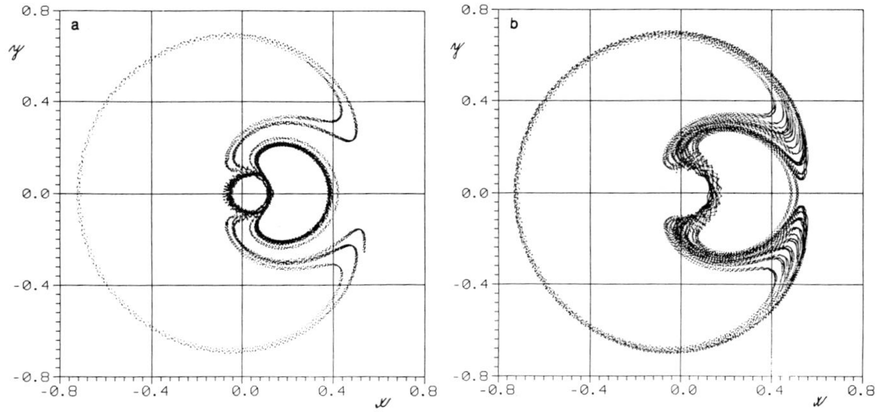

Figure 2. The real chaotic zone computed by SMI (2a) compared to SM integration (2b). Both pictures are projections on $H = -3.10^{-6}$ surface.

The corresponding surface of section is obtained by WM for $H' = -3.10^{-6}$. We present there only the trajectory which is apparently chaotic. It divides three regions with regular behaviour; interior $\tilde{\omega}$-circulation area, banana $\tilde{\omega}$-libration orbits and exterior $\tilde{\omega}$-circulation area. The relation between the shape of this chaotic zone and the location of boundary between libration and circulation of ϕ (the lower curve in the energy diagram) can be explained on the basis of adiabatic semi-analytical theory (Wisdom, 1985). The thing which should be underlined here is, that when drawing this pictures one uses the expansion in e and the topology can be severely destroyed for $e > 0.283$ (or even for lower eccentricities due to truncation).

Now, let us compare Figure 1b with Figure 2a where the same plane and the same trajectory is presented but the SMI described in Section 2 is used. The length of integration is 10^6 years. As we did not mention any substantial difference between surfaces of section and direct projections on (x, y)-plane we present projections here and on all subsequent pictures. There is another high-eccentricity lobe present which forms the new regular island with $\tilde{\omega}$-libration about $\tilde{\omega} = \pi$. An asteroid located inside the chaotic zone can reach the eccentricity above 0.7 and becomes the Earth-crossing body. On the other hand, it also closely approaches Jupiter's orbit during the high-eccentricity passage. It is exactly what was observed in both Klafke at al. (1992) and Šidlichovský (1993).

Figure 2b presents the fourth-order symplectic mapping (Kinoshita et al., 1991) calculation for the same initial conditions and the same time span as Figure 2a using one hundred steps per Jupiter's period. Again, the high-eccentricity lobe appears and the trajectory seems to fill the similar zone. The difference in trajectories follows from the error estimation of SM integration in the the chaotic region with maximum Lyapunov characteristic exponent $\sim 10^{-4}$ y^{-1} (Šidlichovský, 1992). The order of error is much higher than units after 10^{-6} years and the trajectory substantially differs from the exact one. On the other hand, the precise numerical integration using SMI (4000 steps per Jupiter's period) was payed with the computer time needed. When trying to increase the step of SMI we meet the boundary where SMI looses its stability and we did not succeed to accelerate our integration.

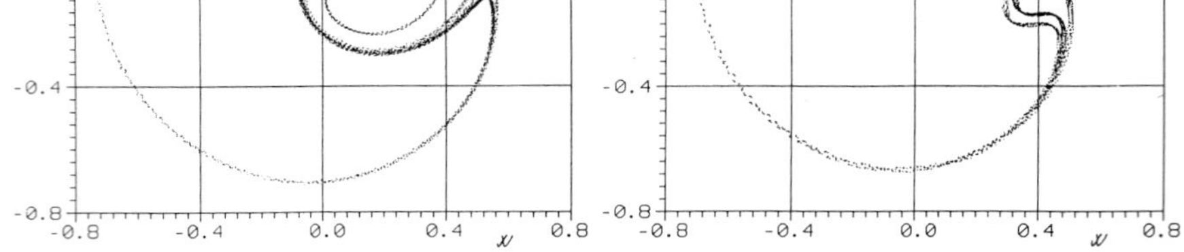

Figure 3. Projections for $H' = -1.10^{-6}$ (3a) and $H' = -1.5\,10^{-5}$ (3b).

We have also calculated several other surfaces of section for different values of H'. At first, we increased H' to -1.10^{-6} and found the chaotic trajectory. This trajectory was then integrated for 5.10^5 years (Figure 3a). It can be seen that the typical shape of chaotic zone is roughly unchanged but the interior $\tilde{\omega}$-libration area (not completely confined) is getting smaller.

It is known from Klafke at al. (1993) that the center surrounded with banana orbits comes to higher eccentricities when the value of H' decreases and this area gradually diminishes. Figure 3b refers to this case. The shape of chaotic zone is much different to what was observed before and banana orbits are now located in the small exterior fold around $x = 0.45$ and $y = 0$. The time of integration 3.10^{-6} years seems to be not sufficient here to describe the complete pattern of the chaotic zone.

CONCLUSION

Presented analysis of the 5/2 mean motion resonance with Jupiter confirmed the existence of the high-eccentricity lobe with e going up to ~ 0.8. An asteroid located inside the chaotic zone is closely reaching Jupiter's orbit.

The symmetric multistep integrator is very precise but time-consuming tool for resonance analysis. It cannot be used for qualitative and fast examination due to the instability when longer step is used.

REFERENCES

Carpino, M., Milani, A., and Nobili, A.M., 1987, Long-term numerical integrations and synthetic theories for the motion of the outer planets, *Astron. Astrophys.* 181:182.

Ferraz-Mello, S., 1993, The convergence domain of the Laplacian expansion of the disturbing function, submitted to *Celest. Mech.*

Kinoshita, H., Yoshida, H., and Nakai, H., 1991, Symplectic integrators and their application to dynamical astronomy, *Celest. Mech.* 50:59.

Klafke, J.C., Ferraz-Mello, S., and Mitchtchenko, T., 1992, Very-high-eccentricity librations at some higher order resonances, *in:* "Chaos, Resonance and Collective Dynamical Phenomena in the Solar System (IAU Symposium 152)," S. Ferraz-Mello, ed., Kluwer, Dordrecht.

Murray, C.D., 1986, Structure of the 2:1 and 3:1 Jovian resonances, *Icarus* 65:70.

Quinlan, G.D., and Tremaine, S., 1990, Symmetric multistep methods for the numerical integration of planetary orbits, *Astron. J.* 100:1694.

Šidlichovský, M., and Melendo, B., 1986, Mapping for 5/2 asteroidal commensurability, *Bull. Astron. Inst. Czechosl.* 37:65.

Šidlichovský, 1992, Mapping for the asteroidal resonances, *Astron. Astrophys.* 259:341.

Šidlichovský, 1993, Chaotic behaviour of trajectories for the asteroidal resonances, *Celest. Mech.* 56:143.

Wisdom, J., 1982, The origin of the Kirkwood gaps: A mapping for asteroidal motion near the 3/1 commensurability, *Astron. J.* 87:577.

Wisdom, J., 1983, Chaotic behavior and the origin of the 3/1 Kirkwood gap, *Icarus* 56:51.

Wisdom, J., 1985, A perturbative treatment of motion near the 3/1 commensurability, *Icarus* 63:272.

Yoshida, H., 1990, Construction of higher order symplectic integrators, *Phys. Let. A* 150:262.

THE INFLUENCE OF CHAOS ON THE EVOLUTION OF THE MINOR BODIES OF THE SOLAR SYSTEM

R.O. Vicente

Faculty of Sciences
Lisbon University, Portugal

ABSTRACT

Some logical evolutionary models for the unviverse are described, starting from the cosmological chaos. The evolution of the main components of the solar system is described with special relevance for the minor bodies (asteroids, comets and meteor streams). The theory that some of these bodies are not permanent members, is described and supported by relevant facts; as a consequence, the origin and evolution of comets is made easier. The comets are considered important for the origin of certain types of asteroids and meteor streams, and these bodies will disappear with the extinction of comets.

LOGICAL EVOLUTIONARY MODELS

Let us start from the beginning, that is, from chaos (the corresponding Greek word is $XAO\Sigma$) and see its evolution, following some logical principles which have been laid down by Greek philosophy.

At the moment, we are concerned about the origin and evolution of the whole universe, that is, we are thinking in cosmological terms. In agreement with our present ideas, we consider the beginning of the solar system to have happened about 5×10^9 years ago.

There are different definitions and concepts of chaos available in the literature on the subject. The definition of chaos we are thinking about is not amenable to any dynamical or statistical formulation, because neither the initial stage nor the final stage of the universe can be dealt with by any mathematical formulation, due to the fact that they are two unique and really chaotic events. We call such a chaos the cosmological chaos.

There are two possible cases:

α) chaos remains chaos forever

β) chaos evolves into some order, which has two logical possibilities:

$\beta 1$) order remains order forever; $\beta 2$) order changes into chaos.

From Newton to Chaos, Edited by A.E. Roy
and B.A. Steves, Plenum Press, New York, 1995

In agreement with our scientific principles, any theory should be checked by the available observations. Our observations show that the second logical conclusion corresponds to the actual state of the universe. It is obvious that we cannot, from a purely logical point of view, select between possibilities $\beta 1$ or $\beta 2$. But, I think, we can make valid scientific inference, based on the available evidence, and infer that order will change into chaos, that is, the final stage of the universe.

We are going to consider now a small portion of the universe, and restrict the scientific inquiries to our galaxy and, in particular, to the solar system.

Once chaos has evolved into some sort of order, our observations show that the Galaxy and the solar system adopted a configuration corresponding to a very flattened ellipsoid.

Considering the solar system to be constituted by planets, satellites, asteroids, comets and meteor streams, there are the following logical steps:

1) all these bodies were created simultaneously
2) some of these bodies were created at a later stage.

The answer to these two logical steps depends on the cosmological concepts about the evolution of the solar system.

After they have been created, we can consider the following cases:
a) all these bodies will exist until the final stages of the evolution of the solar system
b) some of these bodies will disappear before the disintegration of the solar system: b1) forever; b2) but might be created later on.

These are all the main possible logical steps that can be considered.

It is interesting to notice that the prevalent astronomical ideas, during the last 200 years, consider the great majority of the above-mentioned bodies to be created practically at the same time, and they will last until the end of the solar system, that is, options 1) and a).

It is convenient for our purposes to consider the following classification of time intervals:

$\leq 10^3$ years	VERY SHORT
$\leq 10^5$ years	SHORT
$\leq 10^7$ years	LONG
$\leq 10^9$ years	VERY LONG

Considering their evolution and possible chaotic behaviour, we can separate the main components of the solar system into category P (planets, satellites) and category M (asteroids, comets and meteors).

In agreement with current thinking, bodies of category P are considered to follow the logical steps above designated by 1) and a).

There are a number of definitions of chaos, for instance, linked to nonlinear dynamics. Chaotic behaviour might be called "ergodic behaviour" in a conservative system or "turbulent behaviour" in a dissipative system.

From our point of view, we are interested in the chaotic behaviour of solar system orbits. For bodies of category P (planets and satellites) there are not many possibilities of studying the implications of their chaotic behaviour because they are so few.

We are considering the main stage of the evolution of the solar system which will correspond to the greatest time span for the existence of the system, therefore, the chaotic behaviour of planets has been considered for future time intervals of 10^8 years, that is, very long time intervals.

Table 1. Some features of minor bodies

Minor bodies	Total number forecasts	Orbital Elements a (A.U.)	i (degrees)	Known Orbits (approx.)
ASTEROIDS	30,000			5,500
A1. Trans-jovian		> 5	6-42	30
A2. Main asteroidal belt				
Several families		2.16-3.07	0.5-27	1,000
A3. Crossing Earth's orbit	1,300		≤ 64	100
Apollo				50
COMETS	10^{13}			1,300
C1. Short period			3-162	
C2. Long period			≤ 178	
METEOR STREAMS	150	1-59	≤ 173	60

We can reverse this time interval and consider what has happened in the last 10^8 years. The conclusion will be that any chaotic behaviour of the planets and their satellites resulted in the present stable configuration of the system.

What is more interesting is to try to discover bodies of the solar system corresponding to cases 2) and b), that is, bodies that have been created at later stages of the evolution and might disappear during the main stage, corresponding to life times from 10^5 up to 10^8 years. The only possible candidates are the minor bodies.

STATISTICS OF MINOR BODIES

Table 1 gives us some idea about the numbers, composition and orbital elements of these bodies, so that we can infer some conclusions about their evolution, and possible relationships among them.

A quick inspection of these data leads us already to some useful conclusions. First of all, we see that the number of known orbits constitutes a very small sample in comparison with the forecasts of the total number of asteroids and comets, and, therefore, we have to be cautious about inferring conclusions from such a small sample. The case of meteor streams, in the vicinity of the Earth, is still more complicated because we have really few in comparison with the numbers of asteroids and comets.

The forecasts, about the total number of minor bodies, depend very much on the cosmological concepts dealing with the formation and evolution of these bodies.

Looking at the inclinations of the orbits, we see that they show a wide range of values for all the minor bodies in comparison with the planets and the same conclusion applies to the eccentricities.

The values of the semi-major axis also show a wide range starting from the comets, very far away from the planets, down to values well inside the Earth's orbit for some asteroids and meteor streams.

This simple table shows one of the main features of the minor bodies, that is, the researches referring to their motions should be done in 3 dimensions, whenever possible, in order to arrive at more valid conclusions. The studies about the chaotic behaviour of their orbits is more interesting and varied than for other bodies of the solar system.

ASTEROIDS

The first ones to be observed were evidently the bigger ones, with greater masses, and they were called minor planets. There are about 100 with diameters bigger than 130 km, that is, a very small number in comparison with the total number of these bodies which are evaluated, by some statistical estimates, at about 30,000 accessible to our present day instruments. A catalogue of orbits of unnumbered minor planets mentions about 14,000 entries, showing the difficulties of inferring quick conclusions about their distribution.

We have already a selection effect because the great majority of these bodies have small dimensions as in the case of the Aten group.

We can still consider the broad band of debris, which probably resulted from chaotic behaviour leading to collisions, having still smaller masses, and orbits around the Sun. Can we call them minor planets? The name asteroids is more appropriate, considering the overall features of these small bodies.

Let us try to deduce some relationships between the morphological aspects of the orbits and the evolution of the asteroids, considering the possible contribution of chaotic behaviour.

We adopted the following classification (Table 1):

- A1 - asteroids beyond Jupiter: there is a wide range of inclinations, but there are not many known so far because of observational difficulties.

- A2 - main asteroidal belt: the elements of their orbits are more similar among them than classes A1 and A3 but, even so, they show a three dimensional distribution.

- A3 - Earth-crossing orbits: the total population is estimated at about 1300 up to absolute visual magnitude 18; they comprise the Apollo group (about 50 known orbits) and the Aten group. They are very difficult to observe and, therefore, we have an important selection effect.

It should be mentioned that the observations of these bodies have been neglected in the past, because they were not considered worthwhile observing. Lately, there has been renewed interest in their observations because they are so numerous and, therefore, very interesting from the points of view of evolution and chaotic behaviour.

This simple grouping of the asteroids in 3 classes raises already several questions: Why is class A2 far more numerous than A1 and A3?

Why do classes A1 and A3 present greater variations in their inclinations?

We can consider several logical answers: a) all the classes were formed at the same time, having two possiblilities: a1) same origin; a2) different origins
b) some classes were formed at different times.

The possibility, that asteroids A1 and A3 originated at different and later stages of the evolution of the solar system, is not considered often in the available literature which generally emphasizes the permanent duration of these bodies. We might attribute this classical way of thinking to the lack of knowledge about the chaotic behaviour of nonlinear dynamics.

We can envisage different types of evolution during the main stage of the solar system:

1. if they were formed at the same time and with the same origin, classes A1 and A3 can only be explained by some evolution from the main asteroidal belt (A2), and

this will imply some chaotic behaviour. This chaotic behaviour has to be such that either the initial conditions vary to produce orbits of classes A1 and A3 or collisions provoked them. The several resonances existing in the main asteroidal belt might be sensitive regions for producing chaotic behaviour. Only future researches on this subject might support such guesses.

2. if classes A2 and A3 are formed at different times, we have to look for an evolution depending on other types of bodies, distinct from the main asteroidal belt.

Another question relating to classes A1 and A3 refers to their duration. Their orbits show signs of instability (Vincente, 1986), and, therefore, we can infer there are fewer members of these classes because they do not last long. In this case, we shall have an example of bodies that do not last until the final stages of the solar system, and they might not even last longer than the main stage of the evolution.

Those who like to consider that all the components of the system will last until the end will have a difficult time trying to discover a source of material in such a way that will keep forming these type of asteroids.

The evolutionary question is further complicated by the existence of the so-called asteroidal debris. The range of masses of the asteroids is so great that it is difficult to define what are the dimensions separating asteroids from asteroidal debris which, of course, are very difficult to observe. Recently, a tiny asteroid (1993 KA_2) has been discovered with a probable diameter of about 10 metres.

COMETS

Comets correspond to a type of minor bodies with unique features. First of all, the same problem is presented of a very small sample of known comets in comparison with the statistical guesses about their total numbers (Table 1).

It should be pointed out that the total number indicated (10^{13}) is based on the theory that they are permanent members of the system, and associated with the hypothesis of the existence of the so-called Oort's cloud, which will be discussed later on.

So far about 1300 comets have been observed, and most of them correspond to long-period orbits with periods exceeding 200 years. The better studied ones are the short period comets but we only know of a very small number of them.

The remarks we made about biased samples of the asteroid population apply even more to the cometary population. The observations are difficult to make and they exhibit greater variations in their appearances and orbits. The available sample of orbits is seriously biased by observational selection, mainly on account of the considerable brightness changes of comets with heliocentric distances.

There is abundant documentaries about the appearances of comets throughout human history (Williams, 1994), but they have been considered to play a minor role in the evolution of the solar system. Nonetheless, the discussions about their type of orbits (Roy, 1988) were lively, specially the question if they really belonged to the solar system.

We know that comets lose mass whenever they approach the Sun, and that is one of the facts researchers on the subject are in total agreement about.

The existence of short and long period comets can be explained by two possibilities:
1) both have the same origin
2) the origin of the short period comets is different from the long period comets.

There are arguments in favour of any of these possibilities, and the main reason for that is the lack of sufficent observations (Bailey et al, 1986).

Considering that the lifetimes of comets are limited, for instance, to the main stage of the evolution of the system, then we have to discuss the behaviour of their evolution until either their final disappearance or transformation into other minor bodies. There are several interesting examples:
α)asteroids showing cometary features, like Chiron
β)asteroids with orbital elements similar to comets (Hidalgo)
γ)asteroids related to meteor showers, for instance Adonis.

We can consider two cases:
a) they disappear because they lose their masses completely
b) the end products of comets might be b1) certain types of asteroids (classes A1 and A3); or b2) meteor streams.

RELATIONSHIPS AMONG MINOR BODIES

To examine the possible relations among the minor bodies, it is interesting to consider a diagram in which is plotted the semi-major axis versus eccentricity of the minor bodies (Vicente, 1988). We can consider the diagram divided into several regions but there are no well defined boundaries and, as the number of observations of these celestial bodies increase, we shall probably see a blending of the regions. We consider the following regions: A) Apollo objects; B) asteroids; C) comets mainly influenced by Jupiter (Jupiter family); D) minor bodies in Jupiter's weak domain; E) trans-jovian minor bodies.

We can infer from such a diagram that there is a relationship between the stability and the inclination of their orbits (Szebehely et al, 1983). We also see the need to consider 3 dimensional solutions because regions A, C, D and E correspond to minor bodies with higher inclinations.

There might be a connection between asteroids in regions A and E with short period comets, but it is difficult to draw conclusions on this subject with such a small sample of data.

Let us consider, in detail, the more interesting case about the transformation of the end products of comets, because it will have more consequences for the evolution of some other minor bodies. If they evolve into asteroids, we have to think about the dynamical initial conditions necessary for producing asteroidal orbits beyond Jupiter (class A1) and crossing the Earth's orbit (class A3). These initial conditions are more likely to be satisfied for class A3, because that corresponds to the region of perihelion passages of comets, and, therefore, of greater perturbation of the physical and dynamical parameters of the comets, originating asteroids of the Apollo and Aten groups. But, of course, we can also envisage cases where the perturbations are such that the orbits of these end products of comets become trans-jovian.

Researches on the chaotic behaviour of cometary orbits, near perihelion passages, will be useful either to prove or disprove the possibility of the appearance of trans-jovian and Earth's crossing asteroids.

We see that comets are important components of the solar system because they give rise to other types of minor bodies, and they can be considered as sources of distribution of mass in different regions of the solar system during the main stage of its evolution.

This is quite unique for any component of the system due to their orbital instability and physical structure. In spite of the very short span of reliable astronomical observations, we have seen:

- the splitting of comets, like Biela, Brooks 2, and very recently, the peculiar comet Shoemaker-Levy 1993e now composed of a string of nuclear fragments spread out along the orbit.

- comet dust trails, formed by particles up to the centimeter size, and they are spread out along the orbits.

EVOLUTION OF COMETS

Considering that quite a number of cometary orbits are unstable, their origin and evolution can be explained by the following possible cases:

1. they are permanent members of the solar system, originating during the initial or main stages of its evolution.

2. they are not permanent members of the solar system, appearing or disappearing during the initial, main or final stages of its evolution.

Case I

In this case we have to set up a theory implying their formation from the initial stage of the solar system, and that there is enough mass to keep them forever, because we know that individual comets have a limited duration.

From a very limited sample of orbits, that seemed to originate from a very localized region of the celestial sphere, Oort (1950) considered that they form a cloud of great dimensions and enough mass, in such a way that it corresponds to a huge reservoir of comets producing them until the final stages of the solar system. This cloud is called Oort's cloud.

We can assume, from the cosmological point of view, that this cloud originated at the initial stage or, more likely, during the main stage of the evolution. In any case, we have to postulate a huge amount of comets in such a cloud, because the current idea is that they will last forever, and that is the reason for the large number of 10^{13} comets supposed to exist.

More elaborate theories, dealing with the difficulties of the Oort cloud, consider an inner core of the classical Oort cloud as a possible source of replenishment. The origin of the Jupiter family of comets is suggested either from a flux of classical Oort cloud comets or from the inner core of the Oort cloud and the trans-neptunian Kuiper disk (Rickman, 1992).

One of the difficulties about the Oort cloud refers to the initial conditions necessary to get the comets moving all the time from such a huge reservoir. There are elaborate theories on the subject, for instance, the perturbations caused in the cloud by passing stars and molecular clouds, with adequate trajectories, and the existence of galactic tides (Fernandez, 1992).

Since the time of Oort's paper, our knowledge about the existence and density of interstellar matter has increased in a very significant way, and some authors have pointed out that, due to the dimensions of such a cloud, it is very likely that at least the outer shell of the cloud has disappeared in the encounters with such interstellar clouds. This is a great difficulty for any theory that considers comets as permanent members.

Case II

This is an interesting hypothesis that, unfortunately, has not been considered very often. In this case, the Oort cloud does not need to have huge dimensions, containing an enormous amount of comets, because comets are not permanent members and, therefore, they will finish when the appropriate conditions for producing comets within the finite mass of the cloud are exhausted.

The encounter of the Oort cloud with any interstellar clouds does not cause any problems, and it only means that the duration of comets will be abbreviated. Of course, after they are exhausted, there will not be any more comets to be observed, and they will disappear forever.

Another possibility is that the formation of comets is the result of the interaction of molecular clouds with the solar system in such dynamical conditions that they will produce, at least, some types of present-day comets.

As we can see, Case II is more interesting and offers an easier solution; it does not need such stringent conditions to be fulfilled as in Case I.

METEOR STREAMS

There seems to be a consensus that these minor bodies are not very numerous and that at least some of them have a close connection with certain types of comets.

We are very much hampered by selection offsets in their observations made from the Earth. There are two difficulties with an Earth observatory: 1) one can only have one sample per year, when the Earth orbit intersects the meteor stream; 2) a narrow meteor stream may not be well observed, because of the rotation of the Earth, if the observer happens to be on the wrong side of the Earth as it passes through the stream.

If we adopt the theory that all meteor streams originate from comets, then we can state that they are not original components of the solar system, because the first meteor streams could only be formed after comets evolved their orbits around the Sun in such a way that they became unstable. We could start observing meteor streams only when this orbital instability appeared and, therefore, after the initial stage of the planetary system had entered the main stage.

Another statement we can make about them refers to the fact that they are really the only minor bodies for which we have observed the variation of their orbital elements during the very short interval of accurate astronomical observations.

We can also say that they are not very numerous because, after their separation from the parent comets, their orbits are so unstable that they have a very short lifetime in comparison with the main stages of the evolution of the solar system, and therefore there will always be a very limited number of meteor streams around the Earth.

Considering the hypothesis that comets are not permanent members of the solar system, the associated meteor streams will disappear as components of the solar system shortly after the vanishing of the last group of comets.

CONCLUSIONS

The hypothesis that all the main components of the solar system have to be permanent members appears in most papers dealing with the cosmology of the solar system. In fact, we can compare this hypothesis of permanence in the solar system until the final stages of its evolution with the hypothesis of ether, permeating the universe, that was accepted in physics at the end of last century.

Table 2 gives a summary of the possible time scales for the origin of the minor bodies within the framework of the different stages of the evolution of the solar system.

Table 2. Evolution of minor bodies

	Origin and lifetimes
PERMANENT MEMBERS	
A2. Main asteroidal belt	initial, main and final stages
NON-PERMANENT MEMBERS	
A1. Trans-jovian	main stage
A3. Crossing Earth's orbit	main stage
Comets	main stage
Meteor streams	main stage (short lifetimes)

We can consider the possible evolutionary diagram:

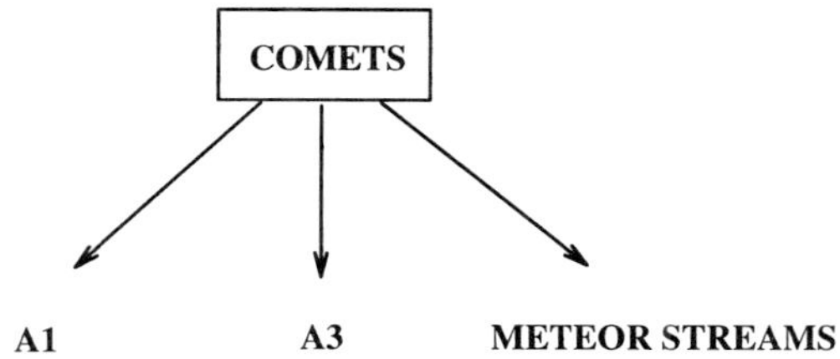

The disappearance of the comets will have, as a consequence, the impossibility of originating more of these minor bodies and therefore in due time, they will also disappear from the solar system.

REFERENCES

1. Bailey, M.E., Clube, S.V.M. and Napier, W.M., 1986: The Origin of Comets, Vistas in Astron. 29:53

2. Fernández, J.A., 1992: Comet Showers in : "Chaos, Resonance and Collective Dynamical Phenomena in the Solar System", S. Ferraz-Mello, ed., IAU, Holland

3. Oort, J.H., 1950: The structure of the cloud of comets surrounding the solar system and a hypothesis concerning its origin, Bull. Astron. Inst. Neth. 11:91.

4. Rickman, H., 1992, Physico-Dynamical evolution of aging comets, in: "Interrelations Between Physics and Dynamics for Minor Bodies in the Solar System", D. Benest and C. Froeschle, ed., Ed. Frontières, France.

5. Roy, A.E., 1988, "Orbital Motion", IOP Publishing, England.

6. Szebehely, V., Vicente, R.O., and Lundberg, J., 1983, Regions of stability of asteroids, in: "Dynamical Trapping and Evolution in the Solar System", V.V. Markellos and Y. Kozai, eds., D. Reidel Publ. Co., Holland.

7. Vicente, R.O., 1986, Stability and evolution of the minor bodies of the solar system, in: "Space Dynamics and Celestial Mechanics", K.B. Bhatnager, ed., D. Reidel Publ. Co., Holland.

8. Vicente, R.O., 1988, Orbital stability and origin of the minor bodies of the solar system, Atti. Accad. Sci. Torino, Suppl. 122:55

9. Williams, I.P., 1994, Remarkable and Spectacular Comets, in: "From Newton to Chaos: Modern Techniques for Understanding and Coping with Chaos in N-Body Dynamical Systems", A.E. Roy and B. Steves, eds., Plenum, London.

PROJECT CRISS-CROSS: A PRELIMINARY ANALYSIS

Othon C. Winter[1,2] and Carl D. Murray[1]

[1] Astronomy Unit, Queen Mary and Westfield College,
Mile End Road, London E1 4NS, United Kingdom
[2] Departamento de Matemática, Campus de Guaratinguetá, UNESP,
CP 205, CEP 12500-000, Guaratinguetá, São Paulo, Brazil

Abstract. A brief description of a project to determine the chaotic regions of the inner and outer solar system (Project CRISS-CROSS) is presented. Taking the planar, circular restricted three-body problem, for a mass ratio $\mu = 10^{-3}$ we have determined the regular and chaotic regions of the phase space for a variety of starting conditions using the techniques of Poincaré Surface of Section (PSS) and the calculation of the Liapunov Characteristic Exponent (LCE). A preliminary comparison of these two techniques shows that there is some advantage in using the PSS method and that it can help to understand the LCE results.

INTRODUCTION

The purpose of Project CRISS-CROSS (Chaotic Regions of the Inner Solar System – Chaotic Regions of the Outer Solar System) is to determine the regular and chaotic regions of the solar system using numerical and analytical methods. The project was undertaken in order to understand some of the dynamical problems related to the motion of solar system bodies. These include (i) the location of possible reservoirs of objects with applications to the Kuiper belt, (ii) the diffusion of objects throughout the solar system and (iii) the structure of the asteroid belt.

The first step in this project is a study of the planar, circular restricted three-body problem (CRTBP). Although this is a relatively simple dynamical system which is only a crude approximation to the real motion of solar system bodies, it has to be well understood before proceeding to a more realistic model; we believe that the structure of its phase space has to be mapped out and then understood via an analytical theory.

In order to determine the structure of the phase space of the CRTBP in terms of the location and size of the regular and chaotic regions, we are using the techniques of Poincaré Surface of Section (PSS) and Liapunov Characteristic Exponent (LCE). We are in the process of mapping out the phase space for a specific mass ratio, $\mu = 10^{-3}$, similar to that of the Jupiter–Sun system. Once such structure is known we intend

From Newton to Chaos, Edited by A.E. Roy
and B.A. Steves, Plenum Press, New York, 1995

to understand its general characteristics and then make predictions for any general mass ratio. This part of the work is in development and will be published elsewhere.

The present paper is mainly concerned with one aspect of the project which relates to a comparative study of the two numerical techniques. We present some preliminary analysis where features that appear in the LCE plots are clarified by looking at the time evolution of the equivalent PSS.

THE TECHNIQUES

Consider the planar, circular restricted three-body problem consisting of two bodies of masses m_1 and m_2 and a test particle. The equations of motion of the particle in a uniformly rotating frame are given by

$$\begin{aligned} \ddot{x} - 2\dot{y} - x &= -\mu_1 \frac{x+\mu_2}{r_1^3} - \mu_2 \frac{x-\mu_1}{r_2^3} \\ \ddot{y} + 2\dot{x} - y &= -\left(\frac{\mu_1}{r_1^3} + \frac{\mu_2}{r_2^3}\right) y \end{aligned} \tag{1}$$

where $\mu_1 = m_1/(m_1+m_2)$, $\mu_2 = m_2/(m_1+m_2) = 10^{-3}$,

$$r_1^2 = (x+\mu_2)^2 + y^2 \qquad r_2^2 = (x-\mu_1)^2 + y^2 \tag{2}$$

and we have chosen units such that the mean motion of the mass m_2 about m_1 is unity. In this system 2π time units correspond to one orbital period of m_2. This dynamical system has a constant of the motion called the Jacobi Constant, C_j, given by

$$C_j = x^2 + y^2 + 2\left(\frac{\mu_1}{r_1} + \frac{\mu_2}{r_2}\right) - \dot{x}^2 - \dot{y}^2. \tag{3}$$

In order to determine the orbital elements of a particle at any instant it is necessary to know its position (x, y) and velocity $(\dot{x}, \dot{y})$. For a fixed value of the Jacobi constant only three of this four quantities are needed. In our computations the starting conditions were chosen such that for each value of C_j the values of x were selected by taking $y = \dot{x} = 0$ with $\dot{y} > 0$. Therefore the integration always started at either the pericentre or apocentre of the particle's orbit.

The computation of the LCE was based on the definition of its maximum value, γ, given by Benettin *et al.* (1980) as

$$\gamma = \lim_{t\to\infty} \frac{\ln(d/d_0)}{t - t_0} \tag{4}$$

where d_0 is the initial separation of the two nearby trajectories at time t_0 and d is the separation at time t. The value of γ is used as an indicator of chaotic or regular behaviour. The maximum LCE gives the rate of divergence of two nearby trajectories and it has to be computed by the numerical integration of equations (1), ideally for infinite time, with its 'final' value determining the nature of the trajectory. It is customary to plot $\log\gamma$ versus $\log t$ for a long but finite time and monitor the 'final' slope. This should be approximately -1 for a regular orbit and ~ 0 for a chaotic one. In our computations the value of γ was computed every 10 orbital periods of m_2 after which the separation vector of the nearby trajectory was rescaled in order to maintain measurement of the local divergence of the two orbits.

During the numerical integration the values of x and $\dot{x}$ were computed whenever the trajectory crossed the plane $y = 0$ with $\dot{y} > 0$. The PSS are plots of $\dot{x}$ versus x. This technique is good at determining the regular or chaotic nature of the trajectory. If a PSS shows smooth, well defined islands, the trajectory is regular and the islands

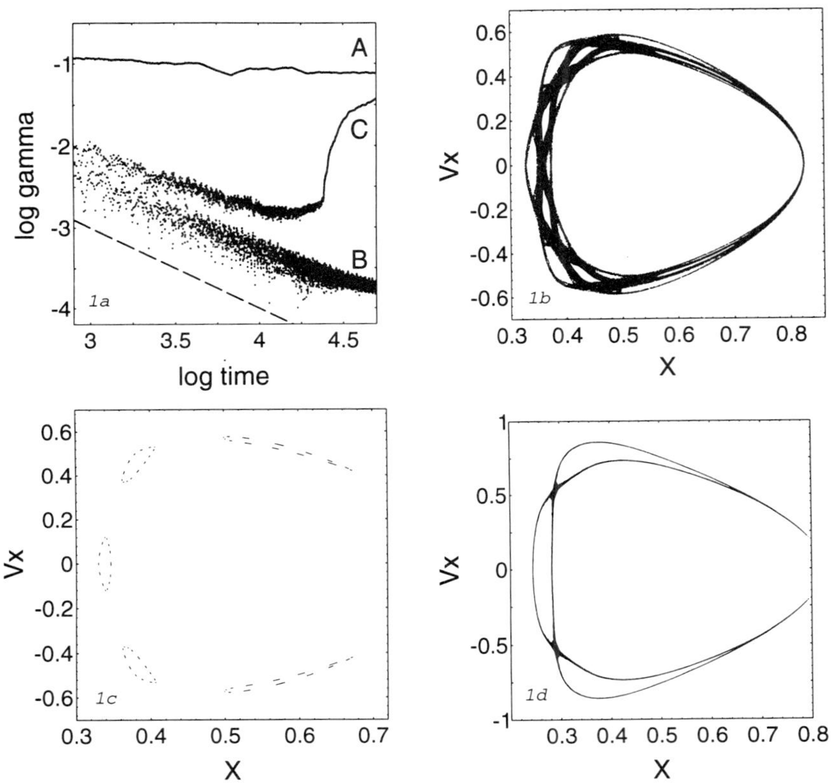

Figure 1. (a) Plot of LCE for three different starting conditions A ($x_0 = 0.350$), B ($x_0 = 0.330$) and C ($x_0 = 0.282$) with the same Jacobi constant, $C_j = 3.110$. (b) PSS for trajectory A. (c) PSS for trajectory B. (d) PSS for trajectory C. The x component of the velocity vector is denoted by Vx.

correspond to libration around an exact resonance between the mean motions of the perturber, m_2, and the particle; the number of islands corresponds to the order of the resonance. Any 'fuzzy' distribution of points in the PSS implies that the trajectory is chaotic. The PSS is also useful in demonstrating the excursions of the particle throughout the phase space.

COMPARISON OF THE TWO TECHNIQUES

One crucial parameter in the computation of the LCE is the length of the integration. How long should the system be integrated in order to determine the character of the trajectory using the LCE method? If, after some integration time, the LCE plot shows a slope far from -1 indicative of chaotic behaviour, then the trajectory

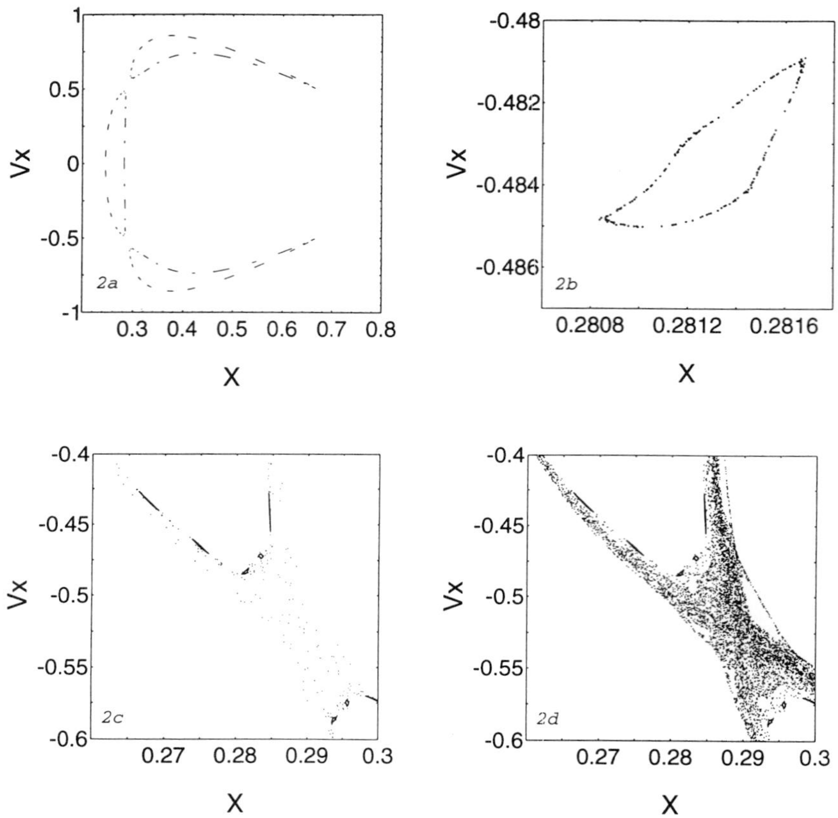

Figure 2: (a) PSS for trajectory C (Fig.1(a)) using points up to $t = 10^4$. (b) Enlargement of a small portion of Fig.2(a). (c) Part of the PSS for trajectory C using points up to $t = 10^{4.3}$. (d) Same as Fig.2(c) but for $t = 10^{4.7}$. The x component of the velocity vector is denoted Vx.

is chaotic. However, if the plot gives the appearance of regular behaviour, with a slope close to -1, then the trajectory could be either chaotic or regular, since an extended integration may reveal a chaotic nature. Many chaotic trajectories in the case of the CRTBP appear regular for extended time intervals, so it is important to have a qualitative understanding of this phenomenon.

The LCE plots for three different starting conditions but the same Jacobi constant ($C_j = 3.110$) are shown in Fig.1(a). The dashed line has a slope of -1. Trajectory A is a clear example of a chaotic orbit with a final value $\log\gamma \approx -1.1$. The surface of section for the same trajectory is shown in Fig.1(b) and this confirms the chaotic nature of the orbit. The trajectory is in a chaotic sea around islands of stability; the latter are due to the 9:4, 11:5 and 13:6 resonances. Trajectory B, with a final value of

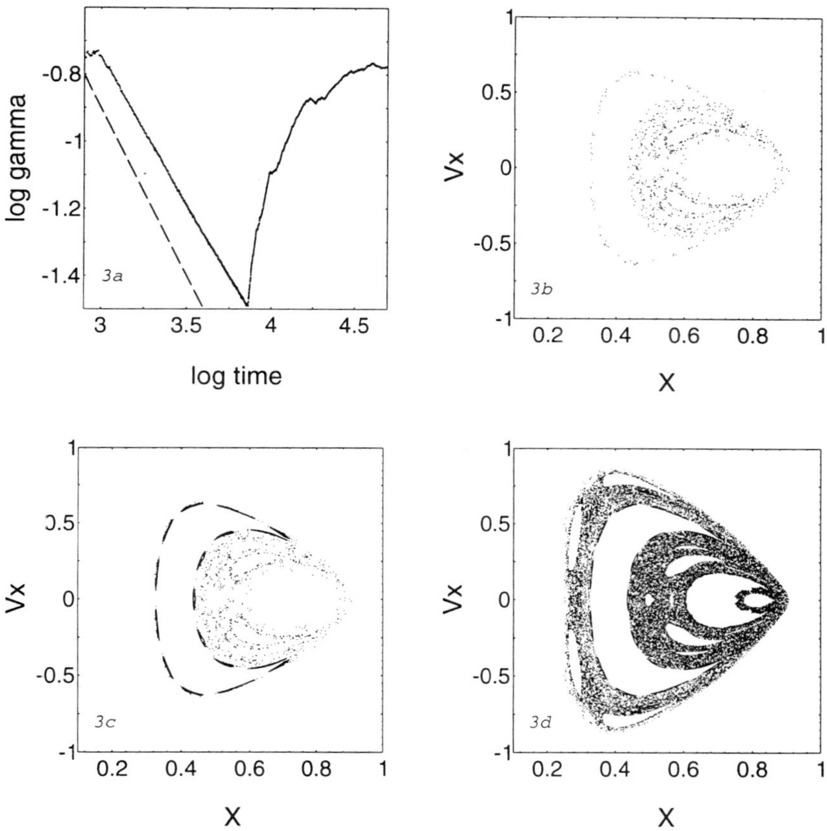

Figure 3. (a) Plot of LCE for the starting condition $x_0 = 0.898$, $C_j = 3.045$. (b) PSS for the same starting condition of Fig.3(a) with points up to $t = 10^3$. (c) Same as fig 3b, but for $t = 7 \times 10^3$. (d) Same as Fig.3(b), but for $t = 10^{4.7}$. The x component of the velocity vector is denoted Vx.

$\log\gamma \approx -3.8$, has a LCE plot with slope close to -1. Therefore, given the appearance of regular behaviour confirmed by its PSS in Fig.1(c), we can see that the trajectory is regular, librating in a high order secondary resonance around the 9:4 resonance. Trajectory C is more unusual, with a final value of $\log\gamma \approx -1.4$ and a LCE plot that gives the appearance of regular behaviour (similar to trajectory B) up to time $t = t_1 \approx 10^4$. However, after time $t = t_2 \approx 10^{4.3}$ the value of γ increases sharply and gives clear evidence of a chaotic trajectory. The PSS for this trajectory is shown in Fig.1(d), where it is obvious that the trajectory is close to the separatrix of the 5:2 resonance.

In order to understand the origin of the unusual LCE plot of trajectory C we have analysed the appearance of its PSS as a function of time. Fig.2(a) shows the PSS for the time interval $0 < t < t_1$. The particle is clearly librating close to a high order, secondary resonance around the main 5:2 resonance. However, a closer examination (Fig.2(b)) reveals that there is no well defined, smooth island. Instead the PSS displays a 'fuzzy' distribution of points close to the largest islands of this secondary resonance. In the PSS for $0 < t < t_2$ (Fig.2(c)) the dots are spread out over a larger region and the trajectory is not so close to the islands; this continues until the end of the integration (Fig.2(d)). This phenomenon of PSS points 'sticking' to the edges of islands for some time has been observed in a variety of dynamical systems and is quite well understood (see, for example, Mackay *et al.*, 1984); this is the phenomenon that is responsible for the temporary, regular appearance of the LCE plot.

In the case of trajectory C, the starting condition is such that the trajectory begins close to a secondary island and a region where the 'sticking' phenomenon occurs. Here it is the *initial* trajectory that gives the appearance of regular behaviour in the LCE plot. However, this feature could also appear at any time in the integration. This is illustrated in Fig.3 for $x_0 = 0.898$, $C_j = 3.045$. The LCE plot (Fig.3(a)) shows that in the time interval $10^3 < t < 10^{3.85}$ the trajectory seems to have a regular behaviour with a slope close to -1. An analysis of the PSS plots makes the behaviour more understandable. For the interval $0 < t < 10^3$ (Fig.3(b)) the PSS shows that the trajectory is clearly chaotic with the dots spread over a large area. In the interval $10^3 < t < 10^{3.85}$ the trajectory enters a region where the 'sticking' phenomenon occurs around a high order secondary resonance near the 2:1 resonance (Fig.3(c)). Finally, the trajectory leaves this region and the chaotic trajectory wanders over a large region of the phase space (Fig.3(d)).

It is important to note that in both the examples given above we have a much earlier definition of the character of the trajectory through the use of PSS compared with the LCE. This is a clear advantage of the PSS method. However, the two dimensional representations of the phase space used in the PSS method are not applicable to dynamical systems with more degrees of freedom.

ACKNOWLEDGEMENTS

Othon Winter thanks CAPES for a doctoral scholarship (Proc. 445/90-5). Carl Murray is grateful to the SERC for the award of an Advanced Fellowship and the allocation of supercomputer time on the ULCC Convex machines.

REFERENCES

Benettin,G., Galgani, L., Giorgilli, A., and Strelcyn, J.-M, 1980. Lyapunov characteristic exponents for smooth dynamical systems and for Hamiltonian systems; a method for computing all of them. Part 1: Theory. *Meccanica* **15**, 9–20.

Mackay,R.S., Meiss, J.D., and Percival, I.C., 1984. Transport in Hamiltonian systems. *Physica* **13D**, 55–81.

METEOROID STREAM DYNAMICS

I.P.Williams

Astronomy Unit
Queen Mary & Westfield College
Mile End Rd
London E1 4NS

SUMMARY

The formation of meteoroid streams, the forces affecting their subsequent motion and the techniques used in order to understand their dynamical evolution are discussed. One example of Chaotic motion is given in the final section.

INTRODUCTION

Rather surprisingly, until the 18th century little attention was paid to meteors and meteorites. Indeed, there was a belief, based mainly on religious grounds, that the universe had to be perfect and so such debris could not exist. At the end of the 18th century, Benzenberg and Brandes (1800) observed 22 meteors simultaneously from locations a few kilometers apart and deduced that their mean height was 89 kilometers, a value that is remarkably close to the accepted value. While this result does not prove the extra-terrestrial origin of meteors, it places them well above normal atmospheric phenomena.

The study of meteor astronomy was probably initiated by the very spectacular Leonid display witnessed in North America on 1833 November 13. From observations of this display, Olmstead (1834) and Twining (1834) established that the meteors appeared to emerge from a point in the constellation of Leo, and that the meteors were moving on nearly parallel lines when they entered the Earth's atmosphere. Locke (1834) pointed out that the Perseid meteors appeared to emerge from a fixed point lying near the star Algol.

The notion emerged that there was a relationship between comets and meteor showers. Le Verrier (1867) assumed a period of 33.25 years for the orbit of the Leonid meteoroids, based on the average period between spectacular displays and noted the similarity of this orbit with that of the newly discovered comet 1866I, Tempel-Tuttle.

From Newton to Chaos, Edited by A.E. Roy
and B.A. Steves, Plenum Press, New York, 1995

However, the first actual identification of a comet-meteoroid stream pair was by Schiaparelli (1867), who showed that the Perseid stream had a similar orbit to that of comet 1862 III, Swift-Tuttle, a comet recovered again in October 1992 after 130 years.

Thus, before the end of the 19th century, the cause of meteors showers had been identified as families of small particles burning in the upper atmosphere of the Earth at times when the Earth passed through this family on its heliocentric orbit. It is important to make the point that we shall be dealing with meteor showers and the associated meteoroid streams, and not with the general population of small interplanetary particles which can be detected as the cause of the zodiacal light or as sporadic meteors. Meteorites and any association of meteorites are also excluded While it is clear that meteoroid streams must contribute particles, primarily through collisions, to the general dust population, they are clearly not the only source feeding particles into this general background. Care must be taken in extending conclusions reached about the general background to meteoroid streams and vice-versa. On many considerations, it is clear that the survival time of meteoroids in such a stream is far less than the age of the Solar System and so such streams must form and evolve within the system essentially as it is today. There are two almost distinct problems which need to be addressed in a general discussion; the formation problem, that is the determination of the orbital elements of meteoroids soon after ejection from the parent and the subsequent evolution of this swarm of meteoroids. The second problem is of more relevance to a dynamical meeting concerned with chaos and most of the remaining discussion will be devoted to this. However, it is of some value to obtain in a very general way the properties of the initial orbit.

THE INITIAL ORBITS

On any formation model, it is difficult to imagine a situation where the speed of a meteoroid relative to its parent can exceed about a kilometer per second. Indeed most models would produce speeds smaller than this by a factor of up to ten (For a discussion of ejection mechanisms, see Williams 1992, Steel 1994). The ejection speed is thus at the most a few percent of the orbital speed of the parent and, in most realistic cases, considerably smaller than this. If the speed relative to the nucleus is denoted by v and the ejection direction making an angle q with the direction of the orbital velocity of the nucleus, denoted by V, then the kinetic energy per unit mass of the meteoroid will differ from that of the parent by $\frac{1}{2}(v^2 + 2vV\cos\theta)$.

If the meteoroid is small, then solar radiation pressure may be important and this will cause a difference also in the potential energy per unit mass of $\frac{\beta G M_0}{r}$, where β is the ratio of radiation pressure to gravity and r the heliocentric distance of the point of ejection. For a meteoroid of radius s and density σ, β=5.74x10-5/ss, in cgs units. Hence,the total energy per unit mass of a meteoroid, by E, is given by

$$2E = 2E_0 + v^2 + 2vV\cos\theta + \frac{2GM_0\beta}{r} \tag{1}$$

where E_o is the energy per unit mass of the parent. If the parent was moving on an elliptical orbit of semi-major axis a_o, then

$$2Eo = -\frac{GM_0}{a_o}$$

It can be seen that the energy of the meteoroid becomes positive if $\beta > \frac{r}{2a_0}$ even if the ejection velocity v is zero, a result first pointed out by Kresak (1974). In general, ejection will have occurred close to perihelion so that we can write $a_0(1-e_0)$ for r and meteoroids will then escape if $2\beta > (1-e_0)$. Inserting relevant numerical values shows that meteoroids of radius 5×10^{-3}cm will escape provided $e_0 > 0.9$. For streams with a higher eccentricity, even larger grains will escape. Meteoroid streams do not therefor contain grains as small as interstellar grains or typical i nter-planetary grains and in consequence, the value for β, the ratio of radiation pressure to gravity is small.

The energy difference between the grain and the comet is obtained by rearranging equation 1 to give

$$2\Delta E = 2E - 2E_0 = v^2 + 2vV\cos\theta + \frac{2GM_0\beta}{r} \tag{2}$$

If we confine ourselves to the larger meteoroids so that β can be ignored together with the term in v^2, we also obtain from standard equations

$$\frac{\Delta E}{E} = -\frac{\Delta a}{a} = -\frac{2\Delta P}{3P} \tag{3}$$

For a standard elliptical orbit, the orbital speed at heliocentric distance r is

$$V^2 = GM_0\left(\frac{2}{r} - \frac{1}{a}\right) \tag{4}$$

From these equations and the simplified equation 2, we obtain

$$\frac{1}{a} - \frac{1}{a_0} = 2E_0 - 2E = -2vV\cos\theta$$

We can also obtain for the first part of equation 3

$$\frac{\Delta a}{a_0} = -\frac{2vV\cos\theta}{E_0} = \frac{2(1+e)v\cos\theta}{V_q(1-e)} \tag{5}$$

if we assume that the ejection occurs at perihelion so that

$$V_q^2 = \frac{GM_0(1+e_0)}{a_0(1-e_0)}$$

Also, considering angular momentum per unit mass,for such an ejection,

$$\frac{\Delta h}{h_0} = \frac{v\cos\theta}{V} = \frac{\Delta a}{2a_0} - \frac{e_0\Delta e}{(1-e_0^2)} \tag{6}$$

since $h^2 = GM_0a(1-e^2)$. Combining these two equations 5 and 6 gives

$$\frac{e_0\Delta e}{(1-e_0^2)} = \frac{2e_0v\cos\theta}{V_q(1-e_o)}$$

which simplifies to

$$\Delta e = \frac{2v(1+e_0)\cos\theta}{V_q} = \frac{2vV_q\cos\theta a_0(1-e_0)}{GM_0} \tag{7}$$

There are some general conclusions that can be drawn from equations 4 and 6. First, under our assumption that $\frac{v\cos\theta}{V}$ is at most of order 0.02, then the difference in e is at most of the order of 0.08. This could lead to meteoroids escaping if the eccentricity of the parent orbit was close to unity, but not otherwise, If, as is likely to be the case in reality, the ejection is primarily towards the sun, then $\cos\theta$ will also be small, which leads to a smaller difference in the eccentricity of the meteoroid compared to that of the parent comet. Turning to the semi-major axis, again if the eccentricity is not close to unity, then the percentage change in $\frac{\Delta a}{a_0}$ will be of order $\frac{v\cos\theta}{V}$, in general no more than a few percent. When e_0 approaches unity, then the change can be arbitrarily large, with escape onto hyperbolic orbits again possible.

If we consider real meteoroid streams, of the six best known, the Perseids, η Aquarid and Orionids have a value of e_0 sufficiently close to unity for escape of many meteoroids to be probable. For the Leonids and Geminids, escape is possible but not probable and for the Quadrantids, very unlikely. This raises a very interesting point as far as meteor astronomy is concerned, namely whether we are interested in the evolution of the whole stream and the relevance this has, or might have, on interplanetary space and collision there with other objects, or with interactions with Earth, leading to a visible meteor shower. In the former category are investigations of whether a passage of comet Biela through the Leonid stream in 1832 was responsible for the breakup of the comet (Babadzhanov et al 1991) or whether the comet Halley persistently suffers high speed collisions with meteoroids from its own streams (Williams et al 1992). Most other investigations fall into the other category, where the prime consideration is explaining observed events on Earth, for example predicting the observed flux on Earth due to the Perseid stream (Wu & Williams 1993). In either event, from Kepler's law we have

$$\frac{\Delta a}{a_0} = \frac{2\Delta P}{3P}$$

where P is the orbital period. It is obvious that different major axis implies different periods, but more important differences in period are additive orbit after orbit, so that in a relatively short period of time, meteoroids can arrive at the ascending node (or any other specified point of the orbit at any time, giving rise to a regular shower visible from Earth at the same time each year. From the point of view of orbital dynamics, which we consider next, this is important, for it means that every meteoroid has a slightly different interaction geometry with Jupiter and all the other perturbers. In particular, using a mean orbit to determine the evolution of a stream by integrating over one period will miss many interesting point. It is also possible for some meteoroids to be in orbital resonance with Jupiter but protected from close encounters, while others on the same orbit will experience a close encounter every orbit.

ORBIT DYNAMICS - PHYSICAL CONSIDERATIONS

In principle, there are a number of forces that can operate on a meteoroid once it has been released from its parent and starts moving on its own independent orbit. The most dominant is solar gravitation, and to a very good approximation, meteoroids move on the same initial elliptic orbit about the Sun. Since the meteoroids are small, radiation pressure has some effect on them, and this was mentioned in the section on the formation process. Following Kresak(1974) it is easy to obtain

$$a' = a_0 r(1-\beta)(r-2a_0\beta)^{-1} \tag{8}$$

$$e' = 1-(1-e^2)(r-2a_0\beta)(1-\beta)^{-2}r^{-1} \tag{9}$$

as the new semi-major axis and eccentricity of the orbit when radiation pressure is included.

Radiation also produces a continuous "drag", known as the Poynting-Robertson effect (Robertson 1937). Mathematical expressions for the rate of change of the orbital parameters have been given by Wyatt & Whipple (1950), Williams (1983). The most relevant are

$$a\frac{da}{dt} = -\gamma(2 + 3e^2)(1 - e^2)^{\frac{3}{2}} \tag{10}$$

$$\frac{de}{dt} = -2.5\gamma(1 - e^2)^{-\frac{1}{2}}a^{-2} \tag{11}$$

where $\gamma = \frac{GM_0\beta}{c}$ and c is the speed of light. In cgs units, γ has a value of $1.3\times10^{11}/s\sigma$ and so significant changes in both a and e take place on a timescale of $a^2s\sigma\times10^{-11}$ sec. For a typical meteor stream a is of order 5AU . Hence, for visible and radar meteors, the timescale is generally larger than the age of the stream. It is therefor general practice to ignore the Poynting-Robertson effect in orbital calculations. However care should be taken with this argument, if the definition of a significant change is modified to mean a change of $a/100$ say, and a itself is only of the order of 1AU, then the timescale gets reduced by a factor of 2500. It may well now be comparable to the age of the stream.

The gravitational perturbations of the planets. have also to be considered. In principle, these are easy to understand in a physical sense, the force from each planet is simply the inverse square law.Application is however harder and we will leave discussion of this until the next section when techniques for solution are discussed.

The final effect that may be important is inter-meteoroid collisions. It is undoubtedly true that meteoroids will collide, both with other meteoroids in the stream and with interplanetary dust. Fortunately, for the dynamics of the stream, this may not be important for the following reason. Collisions will significantly alter both the energy and angular momentum of the meteoroid concerned and hence significantly alter their orbits so that they will not be recognized as stream meteoroids any more in general.Those meteoroids that we recognize as stream will not have suffered a major collision. Of course, for stream evolution in general, and the mass distribution in particular, collisions will be important, but here we are only concerned with the dynamical evolution.

These are the major effects that have to be included, as indicated in a discussion of meteoroid stream evolution. We now discuss briefly some methods that have been used to study the dynamical evolution before concluding with a few remarks on a particular meteoroid stream, the Quadrantids, which shows many indications of chaotic behaviour.

ORBITAL DYNAMICS - TECHNIQUES USED

At a very early stage, the effects of planetary perturbations had been included in discussion. For example, in order to determine the rate of advancement of the nodes of the Leonid stream, Newton (1863), included some planetary perturbations. In early days, calculation were extremely tedious, with only mechanical calculators as an aid. The simplest model to use, is the one particle model. A single stream meteoroid is taken to represent the stream, and placed at a known point on the orbit. Values for the gravitational forces are then calculated by hand and the resulting change in the motion of the particle computed. The process is then repeated as often as is necessary. An advancement on this idea is to average the perturbations over a complete orbit, so that only secular variations remain. Brouwer (1947) generated such a method and this was used by Whipple & Hamid (1950) to show that 4700 years ago the orbits of the Taurid stream and that of comet Encke were very similar.

Using secular methods, Plavec (1950) to show that the orbit of the Geminid stream is evolving very rapidly, a result confirmed by many other integrations since (for example Hunt et al, 1985, Jones & Hawkes, 1986, Gustafson, 1989). The most popular of the secular perturbation methods is Gauss-Halphen-Goryachev method (see for example Hagihara, 1972). Because the averaging reduces considerably the amount of computation that has to be undertaken, these methods still represent the most efficient way of investigating the long term behavior of the mean meteor stream. Their disadvantage is that through the averaging, peculiar short term behavior of individual particles can be missed.

With the arrival of the electronic computer, it became possible to represent a meteor stream in terms of a number of individual test particles and to numerically integrate the equations of motion of each particle and to include radiation pressure as well as gravitational effects. A brief description of some of the numerical methods is given later. One of the first to use these methods were Hamid & Youssef (1963) to investigate the evolution of six actual meteors belonging to the Quadrantid stream.They concluded that their orbit must have changed drastically over the last millennium. A numerical integration based on Cowell's method which was used by Levin et al (1972) to show that perturbations by Jupiter on meteor streams produces a general increase in their widths. This highlights the difference between secular perturbation methods which deals with the evolution of a single orbit into an other single orbit and the direct approach where particles initially on the same orbit maybe perturbed onto a set of different orbits. The first significant increase in the number of particles used was perhaps by Hughes et al (1981) when 210 test particles were used to investigate the nodal retrogression rate of the Quadrantid stream. Though the specific term was not used, this investigation also threw up a hint that part of the Quadrantid stream could be chaotic, of which more will be said later. By 1982, Fox et al (1982) had increased the number of particles to 1000 in an attempt to obtain the cross-section of the Geminid meteor stream. The integration interval was only 150 years however. A further dramatic increase in computational effort was obtained by Fox et al (1983), when 500 000 particles were integrated over 500 years. This resulted in a moderately realistic stream representation being produced By now, the use of direct numerical integration is very widespread, being used for example by Jones & McIntosh (1986), Gustafson (1989) and Williams & Wu (1993).

Thus, the last 20 years has been one of steady progress in terms of understanding the evolution of meteor streams under the action of gravitational perturbations and it can rightly be claimed that the main effects are now well understood. One development, only recently beginning to influence research, has been the realization, mainly through the work of Lindblad (1989) that many of the mean orbits given in the working list of Cook (1973) are based on very few actual meteor orbits, the extreme example perhaps being the η Aquarid stream where the orbit in Cook, and used by many other workers since, is based on only one meteor orbit. A new orbit for this stream is given by Lindblad (1989). Lindblad has also collected together at Lund the catalogue of meteor orbits which may be used to determine new orbits for any stream, or indeed, to be able to use actual meteors in sufficient numbers, rather than using hypothetical test particles, for the modelling of stream evolution (see for example Wu & Williams 1992)

The integration methods that are currently in common use can be divided into three broad categories, though the which side of the division line some of the more sophisticated methods actually lies may be a matter of personal preference rather than an absolute definition.The actual name used to describe the category may also vary from person to person These three broad categories are

(i) Taylor series methods
(ii) Single step methods
(iii) Predictor-corrector methods.

All the methods reduce the equations of motion to be integrated into a set of first order differential equations in the standard way and solve these equations in sequence. The Taylor series method, as its name implies, assumes that the right hand side of all the differential equations can be differentiated, either analytically or numerically, as often as is required. Taylor series expansions for the variables are then generated so that the numerical value of each variable can then be calculated at a point other than the starting point. The process is repeated as often as is necessary. In principle, this is a very efficient method, now that computer algebra is widely accessible to allow the machine to produce the complex differentiation necessary.

The Runge-Kutta method is the best known of the single step methods. The fourth order method is described in most numerical analysis text books and is generally available as a sub-routine in most computer libraries. For problems associated with the evolution of meteor streams a variant such as the Runge-Kutta-Nystrom is generally used. In such a modification, two orders of the expansion, are generally produced and the difference between them used to estimate the error involved in the lower order. This allows the optimum step length to be determined and automatically adjusted for each step. This is of considerable advantage when dealing with a small number of bodies on a highly eccentric orbits as the rate of change of the parameters close to perihelion is fast but very slow near aphelion. This advantage is lost if there are many meteoroids in the model stream since at least one is always close to perihelion.A good high order Runge-Kutta-Nystrom method has been published by Dormand et al (1987).

The third group of methods are those of the predictor-corrector type and they require values of the dependent variables being available at a number of different time points. The available data is used in order to predict the next value of the dependent variables in the sequence. With this value a second equation is used in an iterative way to "correct", that is improve, the accuracy of the predicted value. A typical example of this group is the Adams-Bashforth method. Many variants of this method exist with names hyphenated with Gauss. The main disadvantage of such methods for the solution of problems associated with meteor streams is that a single step method has to be used in order to determine a set of starting values and the accuracy of the final solution is then governed in part by the accuracy of the initial single step method.

All such methods are capable of integrating the equations of motion of a large set of meteoroids over several thousand years.

THE QUADRANTIDS-AN EXAMPLE OF CHAOS?

The Quadrantid meteor shower has occurred around January 3 - 4 every year for the last century or so with refreshing regularity. This regularity would at first sight mitigate against the existence of Chaos, but in reality shows that apparent regularity and chaotic motion are not mutually exclusive. The shower radiant is located in a region of the sky which used to belong to a constellation called Quadrans Muralis, hence the shower name. However in an exercises to rationalize constellation names, this constellation was deleted by the I.A.U ant the region of sky containing the Quadrantid radiant became part of the constellation of Bootes, and lies at its northeastern corner. The shower should therefore be called the Bootids, but fortunately common sense prevailed and the much more distinctive Quadrantid has been retained. Bad visibility in January often hampers observations, but on clear nights free from moonlight, around 30 meteors per hour may actually be seen. (Wood,1989) This translates to a zenith hourly rate of upward of 110, so that the Quadrantid shower is a truly major shower with an activity level comparable with the Perseids and the Geminids.

Many of the meteor showers that we see today are old, that is ,there exists records of their visibility extending back over considerable periods. The Perseids for example were recorded at least a dozen times between AD 36 and AD 1451.(see Lovell, 1954). Even the Leonid stream, regarded by many as being a dynamically young stream, has observations dating back at least to the eighth century (Yeomans, 1981). However, Quetelet (1839) appears to contain the first mention of the Quadrantid shower, while Wartmann (1841) also comments that the first few nights of January were noteworthy for their abundance of meteors. Since these dates however, observations have been plentiful and records of these may be found in Fisher (1930) and Prentice (1940). This observational fact was explained by Murray et al (1980) in terms of the evolution of the orbit of the Quadrantids as a consequence of gravitational perturbations by Jupiter. Hughes et al (1979) investigated the evolution of the stream since 1830 and concluded that the gravitational perturbations of Jupiter in causing a nodal retrogression, matched very well the nodal retrogression rate observed for visible meteors over the same period. They also pointed out that neither the visible observations nor the theory matched very well with the longitude of the ascending node found from the radio meteor data by Poole et al (1972). but offered no explanation for this occurrence.

Hughes et al (1981) reviewed the available data from both visible observations and radio reflections and re-iterated the conclusions of Hughes et al (1979) that a significant difference existed between the rates of retrogression of the nodes of the two types of meteors. However, in this paper possible explanations for the phenomenon were explored and it was suggested that as the aphelion of the stream is very close to the orbit of Jupiter, there existed a variation in the effect of perturbations as the closeness of the encounter with Jupiter varied They also suggested that radiation pressure may have an effect in causing a slight change in aphelion distance between large (visible) meteors and small (radio) meteors. Whether or not the suggestion regarding different aphelia for different sized particles is valid remains an open question, but they did show conclusively that the rate of nodal retrogression is very sensitive to the shortest distance between the meteoroid orbit and the Jovian orbit. Investigation of the evolution by Froeschle & Scholl (1981) also show some evidence for chaotic behaviour amongst the hypothetical meteoroids of their models. Potentially peculiar behaviour in the Quadrantid stream, this time due to the closeness of the 2:1 resonance with Jupiter, was also pointed out by Froeschle & Scholl (1986). In fact, this is not the only orbital resonance that lies within the region occupied by the Quadrantid With this possibility both of having close encounters with Jupiter and having commensurabilities with Jupiter, it may be expected that parts of the Quadrantid stream could be chaotic.

Wu & Williams (1992) searched the records of meteor orbits in the catalogue of the IAU Meteor Data Center, at Lund (see Lindblad, 1987)and found 122 possible members of the Quadrantid stream. A long term integration on each orbit was performed, starting from the observed date and integrating backwards to BC5000. Fifteen of these 122 orbits showed definite chaotic behaviour, which suggests that about 10% of the Quadrantid meteoroids could be moving chaotically.

REFERENCES

1. Babadzhanov, P. D., Wu, Z., Williams, I.P. & Hughes, D.W., 1991, Mon. Not. R. ast. Soc., 253, 69.

2. Benzenberg, J. F. & Brandes, H. W., 1800, Annalen der Phys., 6, 224.

3. Brouwer, D., 1947, Astr. Jl., 52, 190.

4. Cook, A. F., 1973, Evolutionary and Physical Properties of Meteoroids, 283, NASA SP-319.

5. Dormand, J.R., El-Mikkai, M.E.A. & Prince, P.J., 1987, J. Numer. Anal., 7, 423.

6. Fisher, W.J., 1930.Harvard College Observatory, Circular No.346, 1.

7. Fox, K.,Williams, I.P. & Hughes, D.W., 1982, Mon. Not. R. astr. Soc., 200, 313.

8. Fox, K.,Williams, I.P. & Hughes, D.W., 1983, Mon. Not. R. astr. Soc., 205, 1155.

9. Froeschle, Cl & Scholl, H., 1981, Astron. Astroph., 93, 62.

10. Froeschle, Cl. & Scholl, H., 1986, Asteroids Comets Meteors II, Ed Lagerkvist, C.-I., Lindblad, B. A.,Lundstedt, H. & Rickman, H. , Uppsala Univ. Reprocentralen.

11. Gustafson, B.A.S., 1989, Astron. Astrophys., 225, 533.

12. Hamid, S. E.&Youssef, M. N., 1963, Smithson. Cont. Astrophys., 7, 309.

13. Hagihara, Y., 1972, Celestial Mechanics vol 2. MIT, London.

14. Hughes, D. W., Williams, I.P. & Murray, C. D., 1979, Mon. Not. R. astr. Soc., 189, 493.

15. Hughes, D. W., Williams, I.P. & Fox, K., 1981, Mon. Not. R. astr. Soc., 195, 625.

16. Hunt, J., Williams, I.P. & Fox, K., 1985, Mon. Not. R. astr. Soc., 217, 533.

17. Jones, J. & Hawkes, R., 1986, Mon. Not. R. astr. Soc., 223, 479.

18. Jones, J. & McIntosh, B. A., 1986, Exploration of Comet Halley, 233, ESA SP-250.

19. Kresak, L., 1974, Bull. astr. Inst. Csl., 13, 176.

20. Le Verrier, U. J. J., 1867, Comptes Rendus, 64, 94.

21. Levin, B. Y., Simonenko, A. N. & Sherbaum, L. M., 1972, in The Motion, Evolution of Orbits and Origin of Comets, ed Chebotarev, G. A. et al., D. Reidel Pub. Co.

22. Lindblad, B. A., 1989, Asteroids Comets Meteors III, Ed Lagerkvist, C.-I., H.Rickman, Lindblad, B. A. and Lindgren M. , 551 Uppsala Univ. Reprocentralen.

23. Lovell, A. C. B., 1954, Meteor Astronomy, Oxford University Press.

24. Locke, J., 1834, Cincinati Daily Gazette, Aug 11 and 12.

25. Murray, C.D., Hughes, D.W. & Williams, I.P., 1980.Mon.Not.R.astr.Soc., 190, 733.

26. Newton, H. A., 1863, American Jl. of Sci. and Arts, 36, 145.

27. Olmstead, D., 1834, American Jl. of Science, 29, 376.

28. Plavec, M., 1950, Nature, 165, 362.

29. Poole, L.M.G., Hughes, D.W. & Kaiser, T.R., 1972, Mon.Not.R.astr.Soc., 156, 223.

30. Prentice, J.P.M., 1940. J.Brit. Astr. Ass., 51, 19.

31. Quetelet, A., 1839, Catalogue des principales apparitiones d'etoiles filantes, Bruxelles.

32. Robertson, H.P.,1937, Mon. Not. R. astr.Soc., 97, 423.

33. Schiaparelli, G. V., 1867, Astronomische Nachrichten, 68, 331.

34. Steel, D.,1994, Asteroid Comets Meteors 93 (in Press).

35. Twining, A. C., 1834, American Jl. Sciences, 26, 320.

36. Wartmann, M., 1841, Bull. Ac. R. Brux., 8, 226.

37. Whipple, F. L. & Hamid, S. E. D., 1950, Sky and Telescope, 9, 248.

38. Williams, I.P., 1983, in Dynamical Trapping and evolution in the Solar System Eds Markellos,V.V. & Kozai,Y., D Reidel Pub Co., 83.

39. Williams, I.P., 1992, in Meteoroids and their parent bodies, Eds Stohl, J. &Williams, I.P.,Slovak Acad. Sci., Bratislava, 31.

40. Williams, I. P., Murray, C. D. & Hughes, D. W., 1979, Mon. Not. R. astr. Soc., 189, 483.

41. Williams, I.P., Hughes, D.W., McBride, N., & Wu, Z.,1993, Mon. Not. R. astr. Soc., 260, 43.

42. Williams, I.P. & Wu, Z., 1993, Mon. Not. R. astr.Soc., 262, 231.

43. Wood, J., 1989, WGN, 17, 210.

44. Wu, Z. & Williams, I.P., 1992, Mon. Not. R. astr. Soc., 259, 617.

45. Wu, Z. & Williams I.P., 1993, Mon. Not. R. astr. Soc., 264, 980.

46. Wyatt, S.P. & Whipple, F.L., 1950, Ap.J., 111 134.

47. Yeomans, D. K., 1981, Icarus, 47, 492.

INTERLUDE

FROM NEWTON'S ADJUSTABLE CLOCK TO POINCARÉ'S CHAOS

Victor Szebehely

The Department of Aerospace Engineering
and Engineering Mechanics
The University of Texas at Austin
Austin, Texas 78712-1085

INTRODUCTION

More that one hundred years ago Poincaré (1907) pointed out the limited predictability in dynamics. This fundamentally important new aspect of dynamics was ignored at that time for several reasons. One reason was its negative aspect since in classical dynamics predictability was considered one of the major accomplishments. To destroy the stronghold of dynamics, (i.e. predictability) was not a popular subject and Poincaré's paper was ignored, probably for this psychological reason. Other reasons for postponing actions concerning limited predictability in dynamics were the new fields of mechanics which gained suddenly enormous popularity and attention; quantum mechanics (Planck, 1900) and relativity (Einstein, 1914). Instead of "destroying" the beautiful classical methods and results of dynamics, scientists preferred to turn to the above-mentioned two new fields.

Even today we hear and read remarks like "Newton has solved all problems of the dynamics of the solar system." I do not wish to mention references regarding such remarks since the recent results and acceptance of limited predictability in dynamics make such statements meaningless.

On the other side, an interesting formal relation might be mentioned between Heisenberg's (1927) uncertainty principle and the librational motion in the restricted problem of three bodies. It is well known that the triangular equilibrium points are stable if the value of the mass parameter $\left(\mu = m_2 / (m_1 + m_2)\right)$ satisfies the inequality, $\mu \leq \left(\frac{1}{2}\right)\left(1 - \sqrt{69/9}\right)$, where $m_1 \geq m_2$ are the masses of the primaries.

In a synodic coordinate system, rotating with the masses m_1 and m_2, the initial conditions for the third body when placed at a triangular equilibrium point L_4 or L_5 are $x_0 = \mu - 1/2, y_0 = \pm\sqrt{3/2}, \quad \dot{x}_0 = \dot{y}_0 = 0$. If these conditions are not satisfied the resulting motion is called libration and it takes place in the vicinity of the equilibrium point, provided the deviations satisfy the inequality $\Delta r \, \Delta v \leqq a$, where a is a numerically established constant, depending on μ and $(\Delta r, \Delta v)$ are the initial coordinate and velocity deviations.

Such numerically established results of course depend on the length of time and reliability of integrations, on the definition of librational motion, etc. This leads us to chaos of our non-integrable dynamical system (the restricted three-body problem) since the long-time predictability of motions with initial conditions outside the above mentioned $\Delta r \, \Delta v \leq a$ region is highly questionable. This relation shows formal similarity to Heisenberg's relation but of course there is no physical relation between the two inequalities. One reason this result is mentioned is that the evaluation of the consequences of errors

From Newton to Chaos, Edited by A.E. Roy
and B.A. Steves, Plenum Press, New York, 1995

in the initial conditions is an example of new dynamics where it is realized that the initial conditions are never known exactly in practically important problems and that the effect of such deviations should be evaluated. Another reason to mention this example of limited predictability in classical dynamics is to relate, at least formally, quantum mechanics and the motion of Trojan asteroids.

To relate the concept of uncertainty to dynamics on a more global and cosmological level Prigogine (1984) should be quoted referring to a new uncertainty principle: "Systems lose their initial conditions and cannot be recovered or reversed, the future cannot be determined. We can say what may happen but not what will happen."

NEWTON'S VIEW OF PREDICTABILITY IN DYNAMICS

There seem to be several misunderstandings concerning Newton's ideas of dynamics. He was the first to formulate the laws of dynamics in mathematical forms and he gets unquestionably credits for this. At the same time his ideas concerning the limitations of the science established by him are often forgotten. The dynamics of the Solar System is a good example to demonstrate Newton's concept of predictability. Here we need his laws of dynamics combined with his law of gravity, resulting in an extremely complicated "clock work". Not knowing exactly the positions of the planets at a given time, their exact positions at a later time cannot be established and the clockwork needs adjustments according to Newton. Leibnitz' reaction to this was that he referred to Newton's God as a "second-rate watch maker" who had to keep cleaning and rewinding the mechanism.

Today we realize that precise and detailed planetary predictions require observations which improve the published almanacs by using slightly modified and continuously updated initial conditions. This might be considered as adjusting the clock.

In connection with Newton's clockwork it should also be mentioned that such a device, originated in 70 BC was found in the ocean near the Greek island of Antikythera around 1900. The clock-work consisted of 32 gears and it was made operational by Derek de Solla in 1972. This orrery was apparently used to predict the motion of the Sun-Earth-Moon system (Zeeman, 1986 and Stewart, 1990).

Returning to the predictability of the Solar System, Newton's ideas of limitations are valid even today when numerical integrations with modern, sophisticated super-computers are accumulating errors at every integration step. Poincaré (1892), of course was on Newton's side when he established the non-integrability and the divergence of series solutions of gravitational n-body system for $n \geq 3$.

An interesting similarity between Newton's and Poincaré's reactions to non-integrability and to chaos might be mentioned. Note that Newton was not aware of non-integrability and Poincaré's knowledge of the numerical details of chaoticity was rather limited. Nevertheless, Newton's complaints, headaches and ailments were related to his attempts to "solve" the Sun-Earth-Moon three-body problem. Poincaré's comment on chaos was that "these things are so bizarre that I cannot bear to contemplate them".

After the agreement between Newton and Poincaré is established we can look at the development and history of dynamics and celestial mechanics during the years of 1700 and 1912, as well as the eighty years following Poincaré and leading to today's chaos.

CLASSICAL AND MODERN DYNAMICS

Leibnitz, Laplace and Hamilton

Leibnitz's reaction, making fun of Newton's God as a clock-maker is no surprise and it fits well into the general Leibnitz-Newton controversy.

Laplace claims predictability without giving precise laws and initial conditions and, at the same time announcing that his demon can predict the behavior of any dynamical system for any length of time provided the laws and the initial conditions are given. He also attacks Newton by stating that God would start things but he would not intervene once the Universe was started (Lighthill, 1986). His theorem (Laplace, 1773) concerning the stability of the Solar System (no secular terms in the formulas for the semi-major axis) is of fundamental importance and received recent verification by long-time numerical integrations

(Milani and Nobili, 1992). Euler's lunar theory resulted in the first lunar ephemeris (1797) with generally accepted time limitations regarding its validity.

Hamiltonian dynamics (1834) is still one of the basic tools of analytical work in celestial mechanics, in spite of its limited applicability to non-integrable dynamical systems, resulting in divergent series solutions. Real physical systems do not satisfy the reversibility built in Hamiltonian systems. Furthermore Siegel's (1971) theorem states clearly that almost all analytic Hamiltonian systems are non-integrable. Another way to put this statement is to recall Poincaré's warning that Hamiltonian systems have in general no well-behaved constraints other than the constancy of the Hamiltonian function. Combining this with Nekhorshev's theorem, according to which all non-integrable systems contain chaotic regions in the phase space, we arrive at the generally chaotic behavior of Hamiltonian systems. For an excellent view of the intrinsic existence of chaos and the non-existence of global solutions of non-linear dynamical systems, see Ge (1992).

Periodic phenomena

When we wish to compare classical and modern dynamics, several fundamental concepts enter the picture. One is the change from infinite to finite predictability. By this we mean that the length of time and the accuracy of our predictions are limited. One step further and we encounter chaos when we not only have limited accuracy but we loose the relation between future state and present state and the concept of single trajectories is replaced by a bundle of trajectories. The two dynamics (classical and modern) might be compared using the concept of periodic orbits. If such orbits can be established for our non-linear dynamical system, predictability becomes, once again unlimited, at least for those initial conditions which give periodic orbits. This idea of periodic (and predictable behavior) versus chaos need clarification. The existence of densely distributed periodic orbits excludes chaoticity in the region of the phase-space where periodic orbits prevail. The establishment of families of periodic orbits for the restricted problem at three bodies goes back to Strömgren's Copenhagen school (1913-1939). Once again, attention must be directed to the practical aspects of periodicity as the exorciser of chaos. The members of the families of periodic orbits show considerable sensitivity to the parameters of the model used. The restricted problem of three bodies is an idealized model requiring several restrictions which are never exactly satisfied by real dynamical systems. Nevertheless, Lorenz's statement restricting infinite predictability only to periodic phenomena is theoretically, completely acceptable (Stewart, 1990).

Another aspect of classical versus modern dynamics might be found concerning the methods and tools being used. Neither Newton nor Poincaré used computers but relied heavily on analysis. Not only numerical integrations, but actual establishments of series solutions are presently done by using computers. Classical analysis is often replaced by topological approaches. Neither the high accuracy of powerful computers nor sophisticated mathematics can change the non-integrability of important dynamical systems, therefore, limited predictability - even on a qualitative level - is here to stay.

Direct and inverse dynamics

It is difficult to decide if the direct or the inverse problem of dynamics belongs to the classical approach. Newton's law of gravity is probably the most celebrated solution of an inverse problem. His work on the dynamics of the Sun-Earth-Moon system, on the other hand is an example of the direct problem. There seems to be little change in this double approach since today we still emphasize both problems. Finding the Earth's gravitational field and its higher order gravitational coefficients from satellite observations is probably even more popular than finding the dynamical behavior of the Earth's artificial satellites. Following Laplace's approach and representing the field by Legendre polynomials, often leading to divergent series, strongly suggests the use of a more general inverse approach.

At this point we might realize that there is a fundamental problem concerning the inverse approach associated with chaotic orbits of modern dynamics. If the system behaves chaotically, individual trajectories loose their meaning and the field must be determined from bundles of trajectories. Chaoticity, consequently leads to force fields which cannot be determined by the inverse method.

STABILITY AND CHAOTICITY

Stability investigations were always important in dynamics and both theoretical and numerical approaches go back to Lagrange and even prior to him. Today we prefer speaking about regular and chaotic regions of the phase space rather than stable or unstable regions. Furthermore, we can distinguish between chaos due to slight changes in the initial conditions (corresponding to standard stability) and chaos due to slight changes in the parameters describing the system (corresponding to structural stability). In this way we might define standard and structural chaos. The very important concept of stable chaos should be mentioned, which according to Murray (1992) corresponds to chaotic behavior confined to a particular region. The precise predictability is lost, but the system's deviation is limited. An excellent example is the behavior of the Asteroid 522 Helga as discussed by Milani and Nobili (1992).

The chaoticity in the attitude of Hyperion was studied by Wisdom (1095) who also found that its orbit was not chaotic. Today the chaotic orbital behavior of Pluto is generally accepted. The behavior of the members of the Solar System is studied by Laskar (1989, 1990) and stable chaoticity, especially for the inner planets seems to be a rule. I wish to add that chaotic regions exist in the restricted problem of three bodies. If the corresponding Jacobian constant is such that the motion takes place inside a zero velocity curve, we speak, once again about stable chaos.

The large number of definitions for chaotic motion will not be included in this paper but a few more recent one should be mentioned. The one which probably describes the idea of chaos best is contributed by Percival (1987), "local exponential divergence of trajectories accompanied by global confinement in the phase space". The two other versions which are very close to Percival's definition are "local exponential instability of non-integrable systems" and "stochastic behavior occurring in a deterministic system." Ford's (1983) definition brings in some rather practical aspects: "chaotic orbits are random and incalculable; from the past the future cannot be predicted". A greatly simplified definition, very popular in the literature is "sensitive dependence on the initial conditions". Finally another simple version, recommended by this writer is "exponential (regular or structural) instability in a bounded region".

CONCLUDING REMARKS; DETERMINISM, REVERSIBILITY AND LINEARIZATION

Prior to Poincaré roulette, dice and coin toss were considered completely random, unpredictable and non-deterministic. Today, we use different terminology but the unpredictability remains. It is clear that for precise initial conditions the result of a coin toss or dice are easily predictable, the dynamical systems being deterministic and simple. The problem of predictability in roulette might be more difficult because of the somewhat more complex nature of the dynamical system, but it is still deterministic and for given initial conditions, predictable (Poincaré, 1907).

It might be proper at this point to mention Marchal's (1991) and Ge's (1992) definition of a deterministic dynamical system. Using the mathematical approach a dynamical system is deterministic if the conditions of existence and uniqueness of the solutions are satisfied.

Another aspect of classical dynamics is the reversibility of conservative systems. Today it is realized that reversibility is a purely theoretical aspect since the property of non-integrability does not allow the establishment of exact trajectories. Consequently, when numerical integration is performed from t_0 to t, the reverse integration form t to t_0 does not result in the original initial conditions.

The dangers of linearization have been recognized in modern dynamics, the reason being that the linearized results might be erroneous, even from a qualitative point of view. A delightful statement by Holmes (1990) says that "the more we know about nonlinear analysis, the less we are tempted to linearize at the outset".

The emphasis on quantitative analysis has been replaced by qualitative approaches, in spite of the well-known remark that "qualitative is poor quantitative". Poincaré's (1890) emphasis on qualitative methods compensates for the above belittling remark. A good example is the use of Hill's zero-velocity curves in the restricted problem of three bodies.

This qualitative approach in a non-integrable dynamical system eliminates the need for numerical integration and allows qualitative predictions for arbitrary time duration.

To summarize the classical versus the modern approaches in dynamics we should realize that the uncertainty cannot be eliminated by uncritical devotion (or fanaticism) and that the creative approach must accept uncertainty and limited predictability. In fact extreme sensitivity as the reason for chaos can also be used to describe a creative person with one of the major distinguishing characteristics of extreme sensitivity to certain nuances of feeling, perception and thought (Briggs and Peat, 1989).

ACKNOWLEDGMENTS

The author wishes to express his appreciation to Professor Archie E. Roy for his kind invitation to prepare and deliver this lecture. The research effort was sponsored by the author's R. B. Curran Centennial Chair at the University of Texas. Sponsorship of the US Air Force Space Command is gratefully acknowledged. The author's graduate students and colleagues should get credit for the many useful scientific conferences and discussions.

REFERENCES

Briggs, J. and Peat, F. D., "Turbulent Mirror", Harper and Row, 1980.

Ford, J., "How random is a coin toss?", Physics Today, April, 40-47, 1983.

Ge, Y. C. , "From Newton to Chaos and Modern Physics", Vistas in Astronomy, 35, 281-314, 1992.

Hamilton, W. R., "On a General Method in Dynamics", Phil. Trans. Roy. Soc. London, Pt. 2, 247, 1834.

Heisenberg, W. "Über den anschaulichen Inhalt der quantentheoretischen Kinematik und Mechanik", Z. Phys. 43, 172-196, 1927.

Holmes, P., "Poincaré, Celestial Mechanics, Dynamical-Systems Theory and Chaos", Phys. Reports, 193, 137-163, 1990.

Laplace, P. S. "Traité de Mécanique Céleste", Duprat, Courrier and Bachelier, Paris, 1799-1825.

Laskar, J., "A Numerical Experiment on the Chaotic Behavior of the Solar System", Nature, 338, 237-238, 1989. Also Icarus, 88, 266-291, 1990.

Lighthill, J. "The Recently Recognized Failure of Predictability in Newtonian Dynamics", Proc. Roy. Svc. of London, A407, 35-50, 1986.

Milani, A. and Nobili A. M., "An Example of Stable Chaos in the Solar System", Nature, 357, 569-571, 1992.

Murray, D. D., "Wandering on a Leash", Nature, 375, 542-543, 1992.

Percival, I. C., "Chaos in Hamiltonian Systems", Proc. Roy. Soc. London, A413, 131-144, 1987.

Poincaré, H. "Le Hasard", La Revue de Mois, III, 257-276, 1907.

Poincaré, H. "Les Méthodes Nouvelles de la Mécanique Céleste", Gauthier-Villars, Paris, 1892-1899.

Poincaré, H. "Sur les équations de la dynamique et le problème des trois corps", Acta Math., 13, 1-270, 1890.

Prigogine, I and Stengers, I., "Order out of Chaos", Bantam Books, 1984.

Siegel, C. L. and Moser, J. K., "Lectures on Celestial Mechanics", Springer-Verlag, 1971.

Stewart, I. "Does God Play Dice? The Mathematics of Chaos", Penguin Books, London, 1990.

Strömgren, E., " Connaisance actuelle des orbits dans le problème des trois corps", Bull. Astron. 9, 87, 1935.

Szebehely, V. "Theory of Orbits", Academic Press, NY, 1967.

Wisdom, J., Nature, 315, 731-733, 1985.

Zeeman C. "Gears from the Greeks", Proc. Royal Inst. 58, 1986.

PART THREE

DYNAMICS OF NATURAL AND ARTIFICIAL SATELLITES

CHAOTIC MOTION IN HILL'S LUNAR PROBLEM

Jörg Waldvogel and Franz Spirig

Swiss Federal Institute of Technology ETH
CH-8092 Zürich, Switzerland

Abstract

Hill's lunar problem is a model for the motion of the moon around the earth under the additional influence of the sun. However, it also models the relative motion of a pair of co-orbital satellites near their close encounters. In spite of the simplicity of the differential equations their solutions show a remarkable degree of complexity. In this paper we will discuss the asymptotic behavior of the solutions and outline adequate methods for their numerical integration. Then, based on the notion of the Poincaré map, some particular periodic solutions will be considered. Finally, for a family of homoclinic solutions the intersection angle α in the range $|\alpha| \in (10^{-8}, 10^{-2})$ between invariant manifolds is numerically calculated.

1. INTRODUCTION

Among the numerous particular cases of the celebrated problem of three bodies Hill's lunar problem seems to be the simplest one. It is described by a remarkably simple Hamiltonian that may even be transformed to a polynomial of degree no more than 6. Yet its solutions display all the properties of complex dynamics, although non-integrability has not been rigorously proven.

It is suggested that Hill's lunar problem is used as a benchmark problem for testing and refining various methods or objects such as asymptotic expansions, numerical integration, iteration of the Poincaré map, periodic orbits and tori, homoclinic points etc.

In this paper we will first discuss the equations of motion together with the asymptotic behavior of the solutions. Then, some aspects of numerical integration will be considered. Finally, in a particular family of homoclinic orbits the intersection angle between the invariant manifolds will be calculated numerically.

2. EQUATIONS OF MOTION

Hill's lunar equations describe the relative motion of two particles m_1, m_2 on close near-circular orbits in the gravitational field of a large central body m_0. In order to obtain autonomous differential equations a rotating coordinate system has to be

From Newton to Chaos, Edited by A.E. Roy
and B.A. Steves, Plenum Press, New York, 1995

adopted, where x or y denote the radial or the tangential coordinates, respectively. The equations of motion in the limit

(1) $$\epsilon := (m_1 + m_2)/m_0 \to 0$$

are found to be [11]

(2) $$\begin{aligned} \ddot{x} - 2\dot{y} - 3x + xr^{-3} &= 0, \quad r = \sqrt{x^2 + y^2} \\ \ddot{y} + 2\dot{x} + yr^{-3} &= 0, \end{aligned}$$

where dots denote derivatives with respect to time t. Equally true, x and y may be interpreted as the scaled coordinates of the moon m_2 with respect to the earth m_1 in a rotating geocentric coordinate system whose x-axis always points to the sun m_0. This is the original situation considered by Hill [ ], whereas the more modern view mentioned earlier is inspired by the pairs of Saturnian co-orbital satellites, see Figure 1.

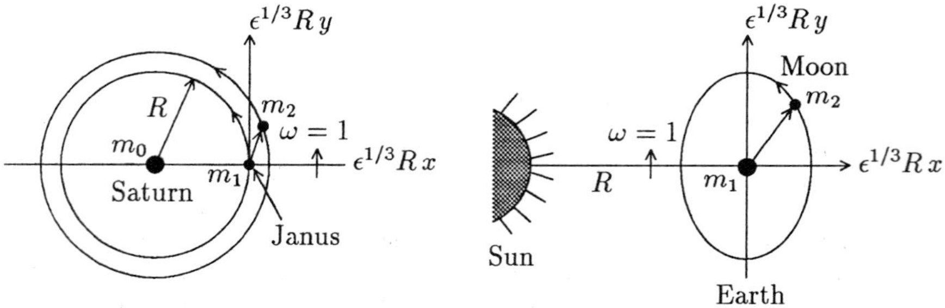

Figure 1. Co-orbital satellites and Hill's lunar problem

Just as the restricted problem of three bodies Hill's lunar problem has an energy integral, referred to as the Jacobi integral. Multiplying the equations (2) by $\dot{x}$ or $\dot{y}$, respectively, adding the products and integrating yields

(3) $$\tfrac{1}{2}\,(\dot{x}^2 + \dot{y}^2) - \tfrac{3}{2}\,x^2 - \tfrac{1}{r} = h = \text{const},$$

where h is the Jacobi constant. It turns out to be useful to consider families of orbits corresponding to a fixed value of h.

In the remaining part of this section we discuss the behavior of the solutions of Hill's problem (2) at infinity.

First, Equ. (2) are rewritten as a system of 4 differential equations of first order

(4) $$\dot{\mathbf{x}} = A\mathbf{x} + \mathbf{f}(\mathbf{x})\,,$$

where $\mathbf{x} = (x, y, \dot{x}, \dot{y})^T$, $\mathbf{f}(\mathbf{x}) = (0, 0, -xr^{-3}, -yr^{-3})^T$ and A is the matrix

(5) $$A = \begin{bmatrix} 0 & 0 & 1 & 0 \\ 0 & 0 & 0 & 1 \\ 3 & 0 & 0 & 2 \\ 0 & 0 & -2 & 0 \end{bmatrix},$$

a non-diagonalizable matrix with the characteristic polynomial [ ]

(6) $$p(\lambda) = \lambda^2(\lambda^2 + 1)\,.$$

The second term on the right-hand side contains all the nonlinearities, and it satisfies $\mathbf{f}(\mathbf{x}) = O(r^{-2})$ as $r \to \infty$. Therefore the solution $\mathbf{x}_L(t)$ of the linear part

$$\dot{\mathbf{x}}_L = A\mathbf{x}_L$$

provides an approximation of the solution of Hill's problem at infinity. We obtain

$$\mathbf{x}_L(t) = e^{At}\,\mathbf{x}_0 \tag{7}$$

where $\mathbf{x}_0 = (x_0, y_0, \dot{x}_0, \dot{y}_0)$ is a constant vector containing four orbital elements uniquely characterizing a solution of Hill's problem. The Taylor series for the matrix exponential together with the relation $p(A) = A^4 + A^2 = 0$ yields

$$e^{At} = I + At + A^2(1 - \cos t) + A^3(t - \sin t) \tag{8}$$

with

$$A^2 = \begin{bmatrix} 3 & 0 & 0 & 2 \\ 0 & 0 & -2 & 0 \\ 0 & 0 & -1 & 0 \\ -6 & 0 & 0 & -4 \end{bmatrix}, \quad A^3 = \begin{bmatrix} 0 & 0 & -1 & 0 \\ -6 & 0 & 0 & -4 \\ -3 & 0 & 0 & -2 \\ 0 & 0 & 2 & 0 \end{bmatrix}.$$

Therefore Equ. (7) implies

$$\begin{aligned} x_L(t) &= 2(2x_0 + \dot{y}_0) - (3x_0 + 2\dot{y}_0)\ \cos t + \dot{x}_0\ \sin t \\ y_L(t) &= y_0 - 2\dot{x}_0 - 3(2x_0 + \dot{y}_0)\,t + 2[\dot{x}_0\ \cos t + (3x_0 + 2\dot{y}_0)\ \sin t]\,, \end{aligned} \tag{9}$$

and we introduce the energy h_L of the orbit $\mathbf{x}_L$ as

$$h_L := \tfrac{1}{2}\,(\dot{x}_L^2 + \dot{y}_L^2) - \tfrac{3}{2}\,x_L^2 = \tfrac{1}{2}\,(\dot{x}_0^2 + \dot{y}_0^2) - \tfrac{3}{2}\,x_0^2\,. \tag{10}$$

The orbit (9) is a distorted cycloid with $x_0, \dot{x}_0, \dot{y}_0$ as essential parameters, whereas the parameter y_0 may be normalized to $y_0 = 0$ with no loss of generality since it may be absorbed into a time shift.

An important subfamily of solutions is obtained when the cycloid degenerates to a straight line, i.e. from $\dot{x}_0 = 0$, $3x_0 + 2\dot{y}_0 = 0$. The family contains one essential parameter, e.g. $x_0 =: c$; with $\dot{x}_0 = y_0 = 0$, $\dot{y}_0 = -\frac{3}{2}\,c$ we obtain from (9),(10):

$$x_L(t) = c, \quad y_L(t) = -\tfrac{3}{2}\,ct, \quad h_L = -\tfrac{3}{8}\,c^2\,. \tag{11}$$

Orbits asymptotic to (11) as $t \to -\infty$ are referred to as nonoscillating orbits, and c is called the *impact parameter* of the orbit.

The full asymptotic expansion for nonoscillating orbits as $t \to -\infty$ was derived in [14], and it will be given here for convenience. In order to avoid logarithmic terms an appropriate parameter s is used instead of time t as the independent variable. Also, for convenience the notation

$$b := \tfrac{3}{2}\,c, \quad B := b^6, \quad h = -\frac{b^2}{6} \tag{12}$$

is used; then s is defined such that

$$t = b^{-3}(s - 3\,\ell n\,|\,b^{-2}(s + 3)|)\,. \tag{13}$$

The coordinates are expressed in terms of the auxiliary quantities z_1, z_2, z_3, z_4 as

$$\begin{aligned} x &= \tfrac{2}{3}\,z_2 - z_4, & y &= z_1 + 2z_3 \\ \dot{x} &= z_3, & \dot{y} &= -z_2 + 2z_4\,. \end{aligned} \tag{14}$$

Expansions as $s \to -\infty$ up to order 5 for the auxiliary quantities are

$$
\begin{aligned}
z_1 &= b^{-2}[-s - \tfrac{9}{2}\, s^{-1} + (\tfrac{27}{4} + \tfrac{B}{3})\, s^{-2} - (\tfrac{81}{8} + \tfrac{5}{3}\, B)\, s^{-3} \\
&\quad + (\tfrac{729}{32} + \tfrac{13}{4}\, B - \tfrac{B^2}{6})\, s^{-4} + (-\tfrac{2187}{32} + \tfrac{3}{4}\, B + \tfrac{83}{90}\, B^2)\, s^{-5} + O(s^{-6})] \\
z_2 &= b\Big[1 + 3s^{-1} - \tfrac{9}{2}\, s^{-2} - \tfrac{2}{3}\, Bs^{-3} + \big(\tfrac{81}{8} - 7B\big)\, s^{-4} \\
&\quad + (22B + \tfrac{2}{9}\, B^2)\, s^{-5} + O(s^{-6})\Big] \\
z_3 &= b^{-2}\, B[-2s^{-2} + (18 + \tfrac{34}{3}\, B)\, s^{-4} + (-27 + \tfrac{278}{3}\, B)\, s^{-5} + O(s^{-6})] \\
z_4 &= b\, B[\tfrac{10}{3}\, s^{-3} + 10s^{-4} - (60 + \tfrac{404}{9}\, B)\, s^{-5} + O(s^{-6})]
\end{aligned}
\tag{15}
$$

These expansions may be used to generate accurate initial conditions (e.g. with $s = -200$) for numerically integrating nonoscillating orbits.

3. SYMPLECTIC INTEGRATORS FOR HILL's EQUATIONS

The goal of this section is to carefully exploit the structure of Hill's lunar problem in order to design special numerical integrators with good conservation properties.

To this end the system (2) is first written in Hamiltonian form. By using the coordinates and conjugated momenta

$$
q_1 = x, \quad q_2 = y, \quad p_1 = \dot{q}_1 - q_2, \quad p_2 = \dot{q}_2 + q_1 \tag{16}
$$

Hill's lunar problem is represented by the Hamiltonian

$$
H(q,p) = \tfrac{1}{2}\,(p_1^2 + p_2^2) + p_1\, q_2 - p_2\, q_1 - q_1^2 + \tfrac{1}{2}\, q_2^2 - \frac{1}{r}, \quad r = \sqrt{q_1^2 + q_2^2}\,. \tag{17}
$$

Notice that the slightly simpler Hamiltonian $H_1 = H + q_1^2 - \frac{1}{2}\, q_2^2$ is integrable and corresponds to the Kepler motion in rotating coordinates.

For any large-scale numerical exploration of Hill's lunar problem regularization of the singularity at the origin $q_1 = q_2 = 0$ is essential since collision orbits play a crucial role. Levi-Civita's regularization [5] has the additional property of producing a polynomial Hamiltonian [12]. This is seen as follows. Levi-Civita's regularization consists of introducing a new independent variable s (instead of time t) and a new Hamiltonian K according to

$$
dt = r\, ds, \quad K = r(H - h)\,,
$$

where h is the fixed energy $H(q,p) = h$ of the orbit under consideration. Furthermore, new coordinates u_1, u_2 and new momenta v_1, v_2 are introduced according to the canonical transformation

$$
q_1 + i\, q_2 = (u_1 + i\, u_2)^2, \quad p_1 + i\, p_2 = \frac{v_1 + i\, v_2}{2(u_1 - i\, u_2)}\,, \tag{18}
$$

where complex notation is used. In terms of the new coordinates and momenta the new Hamiltonian becomes

$$
K(u,v) = \tfrac{1}{8}\,(v_1^2 + v_2^2) - (u_1^2 + u_2^2)\left[\tfrac{1}{2}\,(u_1 v_2 - u_2 v_1) + h + u_1^4 - 4u_1^2\, u_2^2 + u_2^4\right] - 1\,, \tag{19}
$$

and the equations of motion (in vector notation) are

$$\frac{du}{ds} = \frac{\partial K}{\partial v}, \quad \frac{dv}{ds} = -\frac{\partial K}{\partial u}, \quad \frac{dt}{ds} = u_1^2 + u_2^2 \tag{20}$$

The Jacobi constant h must be chosen such that $K = 0$ for the initial data; then K vanishes on the entire orbit.

Next, we will show that the so-called composition methods discussed by Trotter [13], Ruth [8], and McLachlan and Atela [6] can be used for the numerical integration of Hill's problem in the form (20). Since composition methods preserve the symplectic structure they are expected to be more accurate in long-term integrations. They also preserve the structure of the solutions in a better way than the traditionally used explicit Runge-Kutta methods [4, 15].

Composition methods for integrating the system of differential equations

$$\frac{dx}{ds} = X(x), \quad x \in \mathbb{R}^n \tag{21}$$

are based on an additive splitting of the vector field, $X(x) = A(x) + B(x)$, such that the differential equations with the vector fields $A(x)$, $B(x)$, respectively can be solved explicitly:

$$\dot{x} = A(x), \;\; x(0) = x_0 \Longrightarrow x(s) =: \exp(sA)(x_0) \tag{22}$$

$$\dot{y} = B(x), \;\; y(0) = y_0 \Longrightarrow y(s) =: \exp(sB)(y_0)\,. \tag{23}$$

The notation introduced above is inspired by the linear case $\dot{x} = Ax$ with a constant matrix A. Then $\exp(sA)$ is the matrix exponential; in general $\exp(sA)$ denotes the flow induced by the differential equation. One integration step of a composition method of order m for the system (21) consists of a finite interlacing composition of the maps (22) and (23) with different values of s:

$$x(s) = \exp(a_1\, sA)\, \exp(b_1\, sB)\, \exp(a_2\, sA) \ldots (x_0) + O(s^{m+1})\,. \tag{24}$$

A popular method of order $m = 2$ is the leapfrog method

$$x(s) = \exp(\tfrac{s}{2}\, A)\, \exp(sB)\, \exp(\tfrac{s}{2}\, A) + O(s^3)\,. \tag{25}$$

A simple symmetric Runge-Kutta-Nyström method [7] of order $m = 4$ that may be applied to Hills' problem with the splitting to be discussed below is

$$\begin{aligned} x(s) &= \exp(a_1\, sA)\, \exp(sB)\, \exp(a_2\, sA)\, \exp(-\tfrac{s}{2}\, B)\, \exp(\tfrac{2}{3}\, sA) \\ &\quad \exp(-\tfrac{s}{2}\, B)\, \exp(a_2\, sA)\, \exp(sB)\, \exp(a_1\, sA)(x_0) + O(s^5) \end{aligned} \tag{26}$$

with

$$a_1 = \tfrac{1}{2} - \sqrt{\tfrac{7}{72}}, \quad a_2 = \sqrt{\tfrac{7}{72}} - \tfrac{1}{3}\,.$$

Composition methods have been explicitly constructed up to the order $m = 8$ [7, 9].

The Hamiltonian (19) of the regularized Hill problem suggests the obvious splitting

$$K(u,v) = K_1(u,v) - K_2(u), \;\; K_2(u) = (u_1^2 + u_2^2)(u_1^4 - 4u_1^2\, u_2^2 + u_2^4) \tag{27}$$

where $K_1(u,v)$ contains all the terms of degree ≤ 4 in u and is only quadratic in v. The Hamiltonian system with K_2 is trivially integrable; with the initial conditions

$$u(0) = u_0,\ v(0) = v_0,\ t(0) = t_0 \tag{28}$$

we obtain

$$\begin{aligned} u(s) &= u_0 \\ v(s) &= v_0 + s\cdot\frac{\partial K_2}{\partial u}(u_0), \quad \frac{\partial K_2}{\partial u} = \begin{pmatrix} 6u_1(u_1^4 - 2u_1^2u_2^2 - u_2^4) \\ 6u_2(u_2^4 - 2u_1^2u_2^2 - u_1^4) \end{pmatrix}. \\ t(s) &= t_0 \end{aligned} \tag{29}$$

The Hamiltonian system with the Hamiltonian $K_1(u,v)$ describes the regularized Kepler motion of given total energy h in rotating coordinates and is therefore integrable too. Using complex rotation $u = u_1 + i\,u_2,\ v = v_1 + i\,v_2$ this system of differential equations reads as

$$\begin{aligned} \frac{du}{ds} &= \tfrac{v}{4} - \frac{u\,\overline{u}}{2}\,iu \\ \frac{dv}{ds} &= u(2h + \mathrm{Im}\ \overline{u}v) - \frac{u\,\overline{u}}{2}\,iv \\ \frac{dt}{ds} &= u\,\overline{u}\,, \end{aligned} \tag{30}$$

and it is to be solved again with the initial conditions (28). Instead of backtracking over the numerous coordinate transformations (which is prone to mistakes) we temporarily introduce the new coordinates

$$U = e^{it/2}\,u, \quad V = e^{it/2}\,v\,, \tag{31}$$

hence

$$U(0) = e^{it_0/2}\,u_0, \quad V(0) = e^{it_0/2}\,v_0\,. \tag{32}$$

The system (30) is transformed into

$$\frac{dU}{ds} = \frac{V}{4}, \quad \frac{dV}{ds} = (2h + U_1V_2 - U_2V_1)\,U, \quad \frac{dt}{ds} = U\,\overline{U}\,. \tag{33}$$

These equations imply $\dfrac{d}{ds}(U_1V_2 - U_2V_1) = 0$, therefore

$$U_1V_2 - U_2V_1 = \mathrm{Im}\,(\overline{U}V) = \mathrm{Im}\,(\overline{u}_0\,v_0) = \text{const} \tag{34}$$

is an integral of motion (the angular momentum). With

$$\omega := \sqrt{-2h - \mathrm{Im}\,(\overline{u}_0\,v_0)} \tag{35}$$

the solution of (33) with initial conditions (32) becomes

$$\begin{aligned} U(s) &= e^{i\frac{t_0}{2}}\,[u_0\cos(\tfrac{\omega}{2}s) + \frac{v_0}{2\omega}\sin(\tfrac{\omega}{2}s)] \\ V(s) &= e^{i\frac{t_0}{2}}\,[v_0\cos(\tfrac{\omega}{2}s) - 2\omega\,u_0\sin(\tfrac{\omega}{2}s)]\,. \end{aligned}$$

Finally, integration of the third equation in (33) and transforming back to the original coordinates yields

$$
(36)\quad
\begin{aligned}
t(s) &= t_0 + \tfrac{s}{2}\left(u_0\,\overline{u}_0 + \frac{v_0\,\overline{v}_0}{4\omega^2}\right) + \tfrac{1}{2}\left(u_0\overline{u}_0 - \frac{v_0\,\overline{v}_0}{4\omega^2}\right)\frac{\sin(\omega s)}{\omega} \\
&\quad + \frac{u_0\,\overline{v}_0 + \overline{u}_0\,v_0}{4}\;\frac{1-\cos(\omega s)}{\omega^2} \\
u(s) &= e^{-i\,\frac{t(s)-t_0}{2}}\left[u_0\,\cos(\tfrac{\omega}{2}\,s) + \frac{v_0}{2\omega}\,\sin(\tfrac{\omega}{2}\,s)\right] \\
v(s) &= e^{-i\,\frac{t(s)-t_0}{2}}\left[v_0\,\cos(\tfrac{\omega}{2}\,s) - 2\omega\,u_0\,\sin(\tfrac{\omega}{2}\,s)\right].
\end{aligned}
$$

This is the explicit solution of the Hamiltonian system with Hamiltonian $K_1(u,v)$. In the case of an imaginary value $\omega = i\Omega$, $\Omega \in \mathbb{R}$ the relations

$$
\frac{\sin(\omega s)}{\omega} = \frac{\sinh(\Omega s)}{\Omega}, \qquad \frac{1-\cos(ws)}{\omega^2} = \frac{\cosh(\Omega s)-1}{\Omega^2}
$$

may be used, and in the limiting case $\omega \to 0$ we have

$$
\frac{\sin(\omega s)}{\omega} \to s, \qquad \frac{1-\cos(ws)}{\omega^2} \to \tfrac{s^2}{2} \quad \text{as } \omega \to 0\,.
$$

Comparisons of symplectic integrators based on these ideas with other techniques such as Dormand-Prince, Taylor series etc. will be described in a later paper.

4. PERIODIC ORBITS

As in other dynamical systems the structure of phase space in Hill's problem is intimately connected with periodic orbits. These, in turn, are best seen as fixed points of the Poincaré map.

As surface of section S in the phase space with points $\mathbf{x} = (x, y, \dot{x}, \dot{y})$ we choose the non-negative x-axis $x > 0$, $y = 0$, $\dot{y} < 0$. In regularized coordinates this corresponds to the hyperplane $u_2 = 0$. In the following we will consider the family of orbits C_h of the fixed energy h, i.e. orbits on the energy surface E_h. The intersection $P_h = S \cap E_h$ is a 2-dimensional manifold; we will use $x, \dot{x}$ as coordinates on P_h. Let $\mathbf{X}_0 = (x_0, \dot{x}_0) \in P_h$ be a point in P_h, consider the (unique) orbit $C_h \subset E_h$ through $\mathbf{X}_0$ and find its next intersection $\mathbf{X}_1 \in P_h$ with P_h in forward time direction. The Poincaré map $\mathbf{f}_h$ is defined as

$$
(37)\qquad \mathbf{f}_h : \mathbf{X}_0 \in P_h \longrightarrow \mathbf{X}_1 \in P_h, \quad \mathbf{X}_1 = \mathbf{f}_h(\mathbf{X}_0)\,.
$$

As a consequence of the Hamiltonian nature of the system $\mathbf{f}_h$ is *area-preserving*.

The numerical construction of the Poincaré map involves integrating the orbit through the initial point

$$
(38)\qquad \mathrm{x} = (x_0, 0, \dot{x}_0, \dot{y}_0), \quad \dot{y}_0 = -\sqrt{2h + 3x_0^2 + \frac{2}{|x_0|} - \dot{x}_0^2}
$$

and searching for its next intersection with P_h, i.e. the next point with $u_2 = 0$ in regularized coordinates. The Jacobi matrix M of $\mathbf{f}_h$, i.e. the linearization

$$
(39)\qquad \mathbf{f}_h(\mathbf{X} + \boldsymbol{\Delta}) = \mathbf{f}_h(\mathbf{X}) + M \cdot \boldsymbol{\Delta} + O(|\boldsymbol{\Delta}|^2)\,,
$$

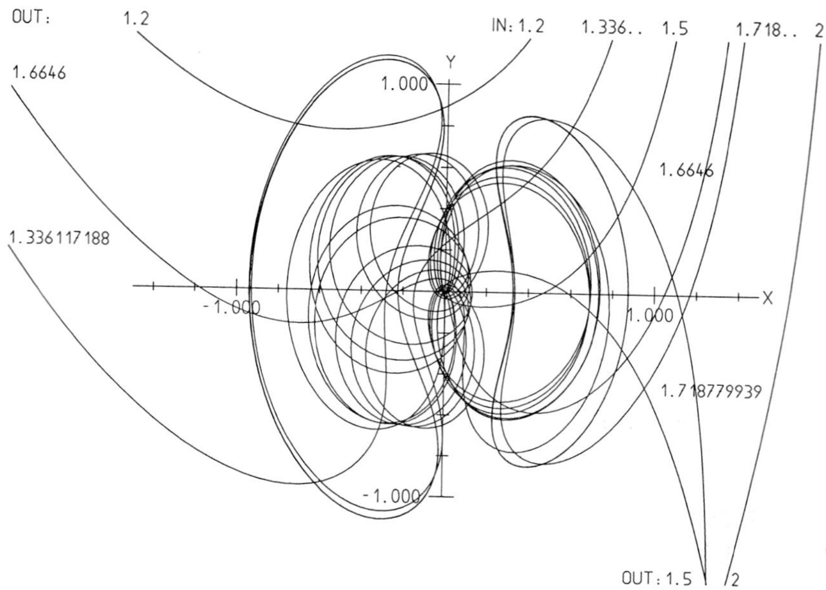

Figure 2. Six nonoscillating orbits in Hill's lunar problem superimposed. The values of the impact parameter c are 1.2, 1.33611 71883, 1.5, 1.6646, 1.71877 99380, 2 .

may be computed by integrating together with Hill's problem also its variational equation.

Clearly, a periodic orbit of energy h corresponds to a fixed point of $\mathbf{f}_h$ or of the n-th iterate $\mathbf{f}_h^n$, $n \in \mathbb{N}$, i.e. the point $\mathbf{X}_0 \in P_h$ leads to a periodic if and only if

$$\mathbf{f}_h^n(\mathbf{X}_0) = \mathbf{X}_0 \tag{40}$$

for some natural number $n \in \mathbb{N}$.

Periodic orbits in Hill's lunar problem have been extensively studied in the literature. Hill [3] was mainly interested in periodic small ovals about the origin that can serve as models of the motion of the moon. Hill showed that for sufficiently large negative energies $h < 0$, $|h| >> 1$ both posigrade and retrograde periodic orbits exist; these can be explicitly represented by convergent Fourier series. A concise account of this topic is found in Siegel-Moser [10].

The first major numerical exploration of families of periodic orbits in Hill's problem has been done by Hénon [1]. Clearly, an infinite number of families must be expected due to the infinite number of choices for n in (40). Here two families are of particular interest.

First, we mention the kidney or bean shaped orbits that were discussed in [14]. These orbits are asymptotically approached by the nonoscillating orbits that limit the transition zones mentioned in [2, 14]. In Figure 2, where six typical nonoscillating orbits are superimposed, two of these periodic orbits can be seen as limiting orbits, corresponding to the energy $h = -0.375\,c^2$ (see Equ. (11)), where the impact parameter is

$$c = 1.33611\,71883 \quad \text{or} \quad c = 1.71877\,99380\ .$$

Since the fixed points corresponding to the periodic orbits are hyperbolic points the actual orbits, as seen in Figure 2, diverge away from the periodic orbits after a few revolutions. The situation in the plane P_h near a fixed point is illustrated in Figure 3.

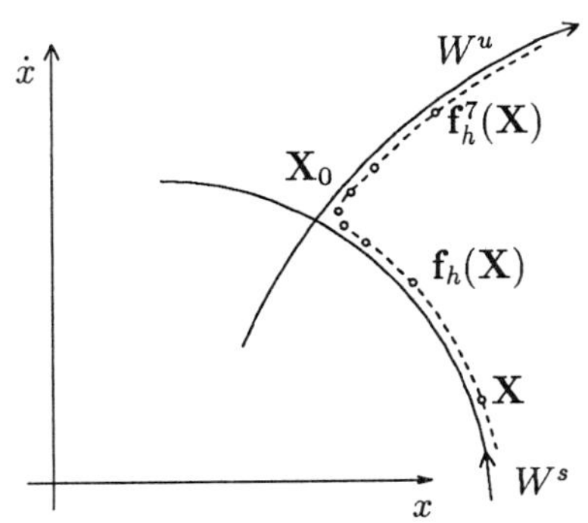

Figure 3. A hyperbolic fixed point $\mathbf{X}_0 = \mathbf{f}_h(\mathbf{X}_0)$ with its stable and unstable manifolds W^s and W^u. 7 iterates of a point $\mathbf{X}$ near W^s under the Poincaré map $\mathbf{f}_h$ are sketched.

These periodic orbits are members of two families originating from the equilibrium solutions corresponding to the well known L_4 and L_5 equilibria of the restricted problem of three bodies. In Hill's problem these equilibria $x =$ const, $y = \dot{x} = \dot{y} = 0$, $r = |x|$ are given by

$$r = 3^{-1/3} = .69336\,12744, \quad h = -\frac{1.5}{r} = -2.16337\,43555\ . \tag{41}$$

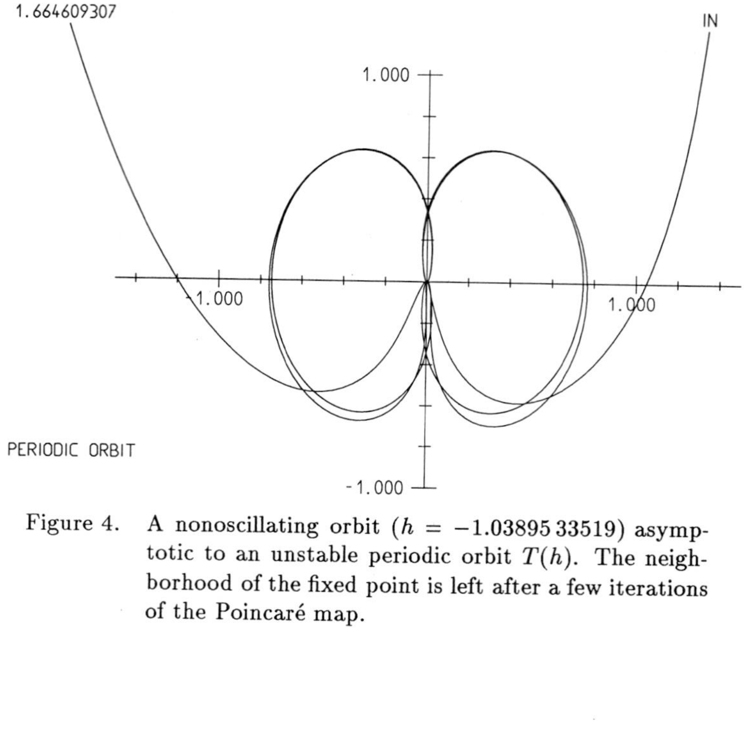

Figure 4. A nonoscillating orbit ($h = -1.03895\,33519$) asymptotic to an unstable periodic orbit $T(h)$. The neighborhood of the fixed point is left after a few iterations of the Poincaré map.

Furthermore, the family contains two collision orbits that are approached as one of the points of intersection with the x-axis approaches the origin.

The second family considered here bifurcates from Hill's (retrograde) ovals with the three-fold period.

A member of the family can be seen in Figure 4, again asymptotically approached by a nonoscillating orbit - and left again due to the hyperbolicity of the corresponding fixed point. In this case, however, it is a fixed point of the third iterate $\mathbf{f}_h^3(\mathbf{X})$ of the Poincaré map. This family of periodic orbits is the basis of the homoclinic orbits to be discussed in the next section.

5. HOMOCLINIC ORBITS

The family of periodic orbits $T(h)$ discussed at the end of Section 4 gives rise to homoclinic orbits in Hill's lunar problem. A rigorous proof of this statement would imply non-integrability of Hill's problem. Here we will pursue the less ambitious goal of numerically calculating the small homoclinic intersection angle $\alpha(h)$ - as a function of the energy parameter h - between two invariant manifolds related to the periodic orbit $T(h)$.

Due to the frequent close encounters with the origin the use of Levi-Civita's regularized variables u, v according to (18) is essential. If u, u' are used as coordinates in the Poincaré sections P_h the fixed points $\mathbf{U}_j = (u_j, u'_j)$ corresponding to the orbit $T(h)$ are of order 6, i.e. $n = 6$ is the smallest exponent with

$$\mathbf{f}_h^n(\mathbf{U}_j) = \mathbf{U}_j, \quad j = 0, 1, ..., 5 .$$

The picture in the section P_h is symmetric with respect to both coordinate axes, as is seen in Figure 7. In Figure 5 the 6 fixed points $\mathbf{U}_j = (u_j, u'_j)$ are sketched, together with the stable manifolds W_j^s of $\mathbf{U}_j$ and the unstable manifolds W_j^u of $\mathbf{U}_j$.

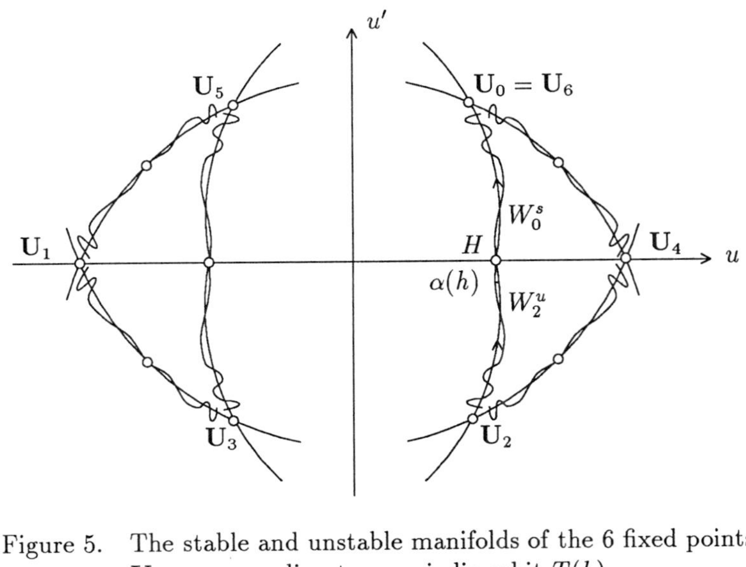

Figure 5. The stable and unstable manifolds of the 6 fixed points $\mathbf{U}_j$ corresponding to a periodic orbit $T(h)$.

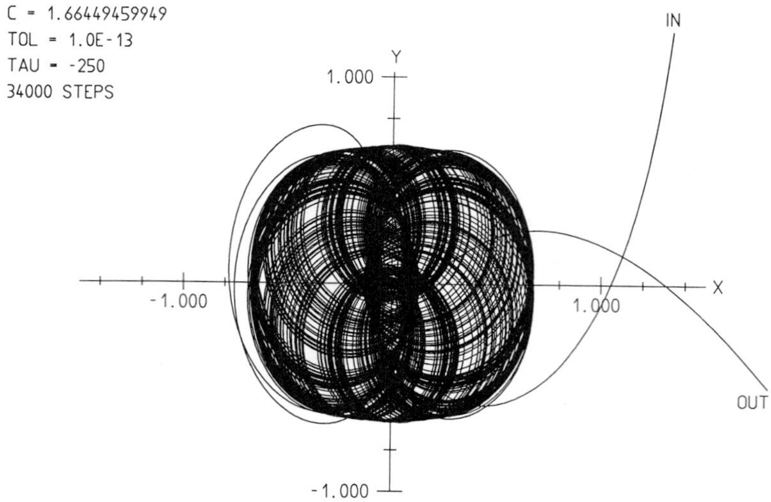

Figure 6. Numerical approximation of a homoclinic orbit with energy $h = -\frac{3}{8}c^2 = -1.03895\,33519$. The approximation remains close to the homoclinic orbit for 34000 integration steps of Runge-Kutta-Fehlberg - 7/8 at a tolerance of 10^{-13}.

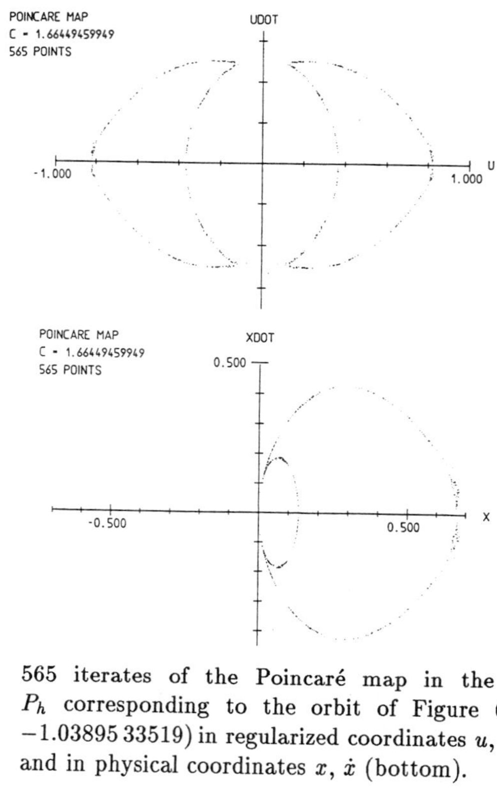

Figure 7. 565 iterates of the Poincaré map in the section P_h corresponding to the orbit of Figure 6 ($h = -1.03895\,33519$) in regularized coordinates u, u' (top) and in physical coordinates x, $\dot{x}$ (bottom).

The true situation is difficult to visualize since the expansion factors of $\mathbf{f}_h^6$ along the unstable manifolds are in the range of 100 000.

We now describe the numerical procedure that allowed us to accurately compute one of the homoclinic intersection angles, $\alpha(h)$, between W_2^u and W_0^s. These two invariant manifolds were chosen since due to the symmetry one of their points of intersection, H, lies on the u-axis. We begin by fixing the energy in the range $h \in (-\frac{9}{8}, -\frac{1}{4})$. Then, the corresponding periodic orbit $T(h)$ is computed by means of Newton's method using the Jacobi matrix M of the Poincaré map. Due to the symmetry it suffices to solve the equation $\mathbf{f}^3(\mathbf{U}) + \mathbf{U} = 0$ by means of the Jacobi matrix $M^{(3)}(\mathbf{U})$ of $\mathbf{f}_h^3$. Using the larger eigenvalue λ_2 and the corresponding unit eigenvector $\mathbf{W}_2^u$ of $M^{(3)}(\mathbf{U}_2)$ a linearized approximation

$$\mathbf{U} = \mathbf{U}_2 + \mu\,\mathbf{W}_2^u, \quad \mu > 0, \quad \mu << 1$$

of the unstable manifold W_2^u near $\mathbf{U}_2$ is constructed. Next, we choose a *geometric* sequence of N points on W_2^u close to $\mathbf{U}_2$, e.g.

(42) $$\mathbf{U}^{(k)} = \mathbf{U}_2 + \mu_0\,\rho^k\,\mathbf{W}_2^u, \; k = 0, 1, ..., N-1, \; \rho = |\lambda_2|^{1/N} .$$

This choice ensures a smooth transition in passing to the next section on W_2^u when $\mathbf{f}_h^3$ is iterated. To obtain sufficient accuracy numerical integration was done with a tolerance of 10^{-13}, whereas μ_0 was chosen in the range of 10^{-8}. λ_2 is in the order of 10^2, therefore the choice of $N = 30$ yields $\rho \approx 1.2$ which allows for a fairly detailed description of W_2^u.

The final step is to map the sequence of points (42) repeatedly until W_2^u reaches the u-axis. Since the intersection at H is nearly perpendicular the variable u' is a good local parameter on W_2^u near H. By interpolation we obtain the polynomial

(43) $$u(u', h) = c_0(h) + c_1(h)\,u' + c_2(h)(u')^2 + O((u')^3)$$

approximating W_2^u for sufficiently small values of u'. The results for seven different values of h are given below. All digits are believed to be correct.

h	u_2	u_2'	$-\lambda_2$	$c_0(h)$	$c_1(h)$	$c_2(h)$
-.25	.66199 58535	-.18764 51222	5.21313	.68367 58231	$-3.89 \cdot 10^{-8}$	-.59295
-.375	.57404 94861	-.27675 79914	12.03554	.62486 98736	$-2.739 \cdot 10^{-5}$	-.60623
-.5	.47964 46464	-.35217 95345	26.39627	.56997 05201	$-4.4831 \cdot 10^{-4}$	-.61908
- 2/3	.34786 60382	-.42710 60477	69.15090	.50151 46703	$-2.7436 \cdot 10^{-3}$	-.64350
-.75	.28136 71085	-.45355 59774	108.33455	.46883 01567	$-4.7356 \cdot 10^{-3}$	-.66059
- 1	.08765 35061	-.49569 13828	376.06211	.37472 07698	$-1.3276 \cdot 10^{-2}$	-.73149
- 10/9	.00591 90439	-.49998 03974	630.67853	.33390 56724	$-1.8145 \cdot 10^{-2}$	-.77272
- 1.125	-.00410 52364	-.49999 05662	672.07601			

Notice that the coefficient $c_0(h)$ defines the position of the homoclinic point H, whereas $c_1(h) = \tan(\frac{1}{2}\alpha(h))$ is roughly one half of the homoclinic intersection angle.

In Figures 6 and 7 we finally show a nonoscillating orbit that was actually computed by an explicit Runge-Kutta-Fehlberg-7/8 integrator with a local error tolerance of 10^{-13}. The orbit follows (shadows!) a homoclinic orbit for more than 500 iterations of the Poincaré map. Therefore, the homoclinic orbit itself, although of chaotic appearance, must inherently be quite insensitive to perturbations.

To turn these computations into a rigorous proof of the existence of homoclinic orbits seems difficult. More promising is the search for other regions of phase space with possibly simpler types of chaotic motion.

References

[1] Hénon, M., 1969: Numerical exploration of the restricted problem V. Hill's case: periodic orbits and their stability. Astron. and Astrophys. **1**, 223-238.

[2] Hénon, M. and Petit, J.M., 1986: Series expansion for encounter-type solutions of Hill's problem. Celest. Mech. **38**, 67-100.

[3] Hill, G.W., 1886: On the part of the motion of the lunar perigee which is a fraction of the mean motions of the sun and moon. Acta Math. **8**, 1; also: Collected Mathematical works of G.W. Hill. Vol. 1. 1, p. 243. Cargenie Inst. of Wash., Washington, D.C., 1905.

[4] Lasagni, F., 1988: Canonical Runge-Kutta methods. ZAMP **39**, 952-953.

[5] Levi-Civita, T., 1920: Sur la régularisation du problème des trois corps. Acta Math. **42**, 99-144.

[6] McLachlan, R.I. and Atela, P., 1992: The accuracy of symplectic integrators. Nonlinearity **5**, 541-562.

[7] McLachlan, R.I., 1994: On the numerical integration of ordinary differential equations by symmetric composition methods. To appear in SIAM J. Sci. Comp.

[8] Ruth, R.D., 1983: A canonical integration technique. IEEE Trans. Nucl. Sci. NS-**30**, 2669-2671.

[9] Sanz-Serna, J.M., 1988: Runge-Kutta schemes for Hamiltonian systems. BIT **28**, 877-883.

[10] Siegel, C.L. and Moser, J.K., 1971: Lectures on Celestial Mechanics. Springer-Verlag, Berlin.

[11] Spirig, F. and Waldvogel, J., 1985: The three-body problem with two small masses: A singular-perturbation approach to the problem of Saturn's co-orbiting satellites. In: V. Szebehely (ed.), Stability of the solar system and its minor natural and artificial bodies. Reidel, 53-63.

[12] Spirig, F. and Waldvogel, J., 1991: Chaos in co-orbital motion. In: A.E. Roy (ed.), Predictability, stability, and chaos in N-body dynamical systems.

[13] Trotter, H.F., 1959: On the product of semi-group of operators. Proc. Am. Math. Soc. **10**, 545-551.

[14] Waldvogel, J. and Spirig, F., 1988: Co-orbital satellites and Hill's lunar problem. In: A.E. Roy (ed.), Long-term dynamical behavior of natural and artificial N-body systems. Kluwer, 223-243.

[15] Yoshida, H., 1993: Recent progress in the theory and application of symplectic integrators. Celest. Mech. **56**, 27-43.

HUNTING FOR PERIODIC ORBITS CLOSE TO THAT OF THE MOON IN THE RESTRICTED CIRCULAR THREE-BODY PROBLEM

G.B. Valsecchi[1], E. Perozzi[2], A.E. Roy[3], and B.A. Steves[4]

[1]I.A.S. – Planetologia, viale dell'Università 11, 00185 Roma, Italy
[2]Telespazio s.p.a., via Tiburtina 965, I-00156 Roma, Italy
[3]Department of Physics and Astronomy, University of Glasgow
Glasgow G12 8QQ, U.K.
[4]Department of Mathematics, Glasgow Caledonian University
Glasgow, U.K.

INTRODUCTION

The role of high-integer near commensurabilities among lunar months – like the long known Saros cycle – in the dynamics of the Moon has been examined in previous papers (Perozzi et al., 1991; Roy et al., 1991; Steves et al., 1993). A by-product of this study has been the discovery that the lunar orbit is very close to a set of 8 long-period periodic orbits of the restricted circular 3-dimensional Sun-Earth-Moon problem in which also the secular motion of the argument of perigee ω is involved (Valsecchi et al., 1993a). In each of these periodic orbits 223 synodic months are equal to 239 anomalistic and 242 nodical ones, a relationship that approximately holds in the case of the observed Saros cycle, and the various orbits differ from each other for the initial phases. Note that these integer ratios imply that, in one cycle of the periodic orbit, the argument of perigee ω makes exactly 3 revolutions, i.e. the difference between the 242 nodical and the 239 anomalistic months (these two months differ from each other just for the prograde rotation of ω).

The periodic orbits associated with the Saros cycle, even if of long duration when compared to those usually found in literature, are by no means the longest ones that can possibly be found close to that of the Moon: according to Poincaré's conjecture there should be infinitely many, of longer and longer period. We have used Delaunay's expressions for the motion of the lunar perigee and node (Delaunay, 1872) to compute the possible values of $\bar{e}$ and $\bar{i}$ that result in an integer number of anomalistic months and of revolutions of ω within an integer number of lunar synodic months. In this way we have found how these orbits are arranged in the space of the geocentric mean orbital elements $\bar{e}$ and $\bar{i}$ (Valsecchi et al., 1993b).

Here we show that this knowledge is actually a useful guide to perform a search for

From Newton to Chaos, Edited by A.E. Roy
and B.A. Steves, Plenum Press, New York, 1995

Table 1. For the sets of periodic orbits with the shown integer coefficients are here given the values of $\bar{e}$ and $\bar{i}$ found by using Delaunay's theory, and those of a, e and i, found numerically, for the periodic orbit, in each set, passing through $\Delta\lambda = 0$, $M = 0$ and $\omega = 0$. For the Moon, $\bar{e} = 0.0549$ and $\bar{i} = 5.^\circ 13$.

$N_{\Delta\lambda}$	N_M	N_ω	$\bar{e}$	$\bar{i}$	a	e	i
223	239	3	0.0623	5°.57	0.00258099716	0.083356670	5°.6908146
516	553	7	0.0911	2°.97	0.00257824274	0.116279731	2°.9861113
528	566	7	0.0894	8°.70	0.00257870093	0.113013393	8°.9049840
599	642	8	0.0091	7°.07	0.00258570991	0.021657128	7°.2204390
681	730	9	0.0611	9°.07	0.00258131449	0.081122309	9°.2671774
739	792	10	0.0835	3°.95	0.00257899210	0.107515485	4°.0074516
751	805	10	0.0823	7°.90	0.00257930293	0.105317505	8°.0785855
7571	8114	102	0.0548	5°.13	0.00258164637	0.074986882	5°.2443334

the initial conditions of longer duration periodic orbits close to that of the Moon. We start by finding some of the periodic orbits of shortest duration, and then go on to look for a longer one whose $\bar{e}$ and $\bar{i}$ are much closer to those of the lunar orbit than the $\bar{e}$ and $\bar{i}$ of the periodic orbit associated to the Saros. The time behaviour of the osculating elements of these two periodic orbits are then compared to that of the Moon.

PERIODIC ORBITS OF LONGER DURATION

As discussed in Roy et al. (1991) and Valsecchi et al. (1993a), in the restricted circular 3-dimensional 3-body problem there are 16 combinations of ω, M and $\Delta\lambda$ (i.e. the argument of pericentre, the mean anomaly of the small body, and the angle between the mean longitude of the latter and the line joining the two primaries) that result in a mirror configuration. Since in a symmetric periodic orbit the system passes through two mirror configurations, there is a total of 8 possible individual periodic orbits, differing from each other for the values of the above mentioned angles, for every set of lunar frequencies (synodic, anomalistic and nodical) that are commensurable. Of these 8 orbits, 4 are really independent, since the other 4 can be obtained by exchanging the ascending and descending nodes of the lunar orbit (the plane of motion of the primaries is a simmetry plane for all these periodic orbits).

Valsecchi et al. (1993b) have used Delaunay's expressions for the motion of the lunar perigee and node (Delaunay, 1872) to find, for the present synodic lunar frequency, the values of $\bar{e}$ and $\bar{i}$ at which the anomalistic and nodical frequencies of the Moon are commensurable with each other and with the synodic one; in this way, they found about 10 000 sets of mean elements corresponding to commensurable frequencies, in a box extending from 0 to 0.1 in $\bar{e}$ and from 0° to 10° in $\bar{i}$, for durations of the associated periodic orbits up to 115 revolutions of ω, about 700 yr. In rows 1-7 of Table 1 are given the sets of $\bar{e}$ and $\bar{i}$ that correspond to the seven shortest sets of possible periodic orbits found. The absolutely shortest one is that associated with the Saros cycle.

In the Table are also reported the starting conditions – osculating a, e and i – of the periodic orbits that belong to the sets with the given frequency ratios and pass through the mirror configuration in which $\Delta\lambda = 0$, $M = 0$ and $\omega = 0$ (or $\omega = \pi$ that, as we have seen, makes no difference). Note that the osculating values of e and i are arranged in the same way as the mean ones, so that the latter can be used – and in

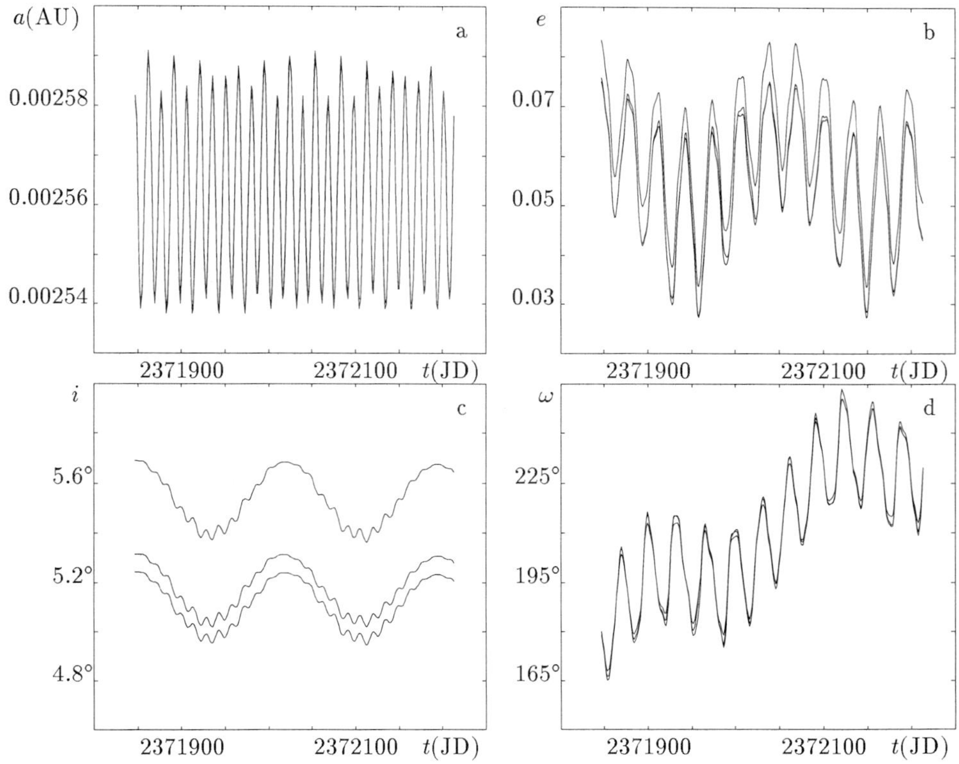

Figure 1. **a**) Time evolution of the osculating semimajor axis a of the Moon, as given by the JPL Ephemeris DE 102 and as given by two periodic orbits (for details, see text), over about one year. Although practically indistinguishable, there are actually three curves in the figure. **b**) Same as a) for the osculating eccentricity e: the upper curve refers to the periodic orbit associated with the Saros, the lower ones to the "long" periodic orbit and to the DE 102 Moon. **c**) Same as a) for the osculating inclination i: the upper curve refers to the periodic orbit associated with the Saros, the central one to the DE 102 orbit, and the lowest to the "long" periodic orbit. **d**) Same as a) for the osculating argument of perigee ω: the individual curves here are difficult to distinguish.

fact *have been used* – to guess suitable tentative values of osculating e and i to start the periodic orbit finding procedure described in Valsecchi et al. (1993a).

We then go on to look for a periodic orbit whose $\overline{e}$ and $\overline{i}$ are very close to those of the Moon. As shown by Valsecchi et al. (1993a), the time behaviours of the osculating elements of the periodic orbit associated with the Saros are indeed very close to those of the Moon, but one wonders whether something better is achievable. If the motion of the Moon could be accurately modelled by the restricted circular 3-dimensional 3-body problem, then Poincaré's conjecture would suggest to try longer periodic orbits, since among these there should be some that are closer to the motion under study. In practice a realistic model of the lunar motion is much more complex, as it must take into account the mass and figure of the Moon, the figure of the Earth, planetary perturbations and so on. Therefore, beyond a certain duration, the advantages obtained going to longer orbits would be negated by the effects of the perturbations neglected in the circular restricted problem used here.

We have thus searched among the orbits of Valsecchi et al. (1993b) the one with $\overline{e}$

and $\bar{i}$ closest to those of the real Moon (that are $\bar{e} = 0.0549$ and $\bar{i} = 5^\circ\!.13$) and, as for the others, we have computed the corresponding osculating a, e, and i; the results are given in the last line of Table 1.

DISCUSSION

How much better is the orbit of the last line of Table 1 in approximating the real motion of the Moon than the one associated with the Saros? To address this question we have plotted in Fig. 1 the time evolution of a, e, i and ω for both these periodic orbits and the orbit of the Moon taken from JPL DE 102 (Newhall et al., 1983) for about one year starting from JD 2 371 846.872 (we use this date because the Sun-Earth-Moon system was then very close to a mirror configuration, see Valsecchi et al., 1993a).

As it is possible to see, the longer periodic orbit is closer to that of the real Moon in both eccentricity and inclination; in semimajor axis and argument of perigee the improvement is less noticeable, due to the scale of the Figure, but still present. However, it may not be worth the effort to try to go to a much longer orbit, since a preliminary analysis of the residuals shows that there is, already on this short time-span of one year, a significant trend that should be tied to having neglected the eccentricity of the Earth, and thus that, to improve the approximation, it is really time to go to a more realistic model like, for instance, the elliptic restricted 3-dimensional 3-body problem.

REFERENCES

Delaunay, Ch., 1872, Note sur les mouvements du périgée et du noeud de la Lune, *Compt. rend. hebdom. Acad. Sci.* 74:17.

Newhall, XX, Standish, E.M., and Williams, J.G., 1983, DE 102: a numerically integrated ephemeris of the Moon and planets spanning forty-four centuries, *Astron. Astrophys.* 125:150.

Perozzi, E., Roy, A.E., Steves, B.A., and Valsecchi, G.B., 1991, Significant high number commensurabilities in the main lunar problem. I: the Saros as a near periodicity of the Moon's orbit, *Celest. Mech. & Dynam. Astron.* 52:241.

Roy, A.E., Steves, B.A., Valsecchi, G.B., and Perozzi, E., 1991, Significant high number commensurabilities in the Main Lunar Problem: a postscript to a discovery of the ancient Chaldeans, *in* "Predictability, Stability and Chaos in N-Body Dynamical Systems", A.E. Roy ed., Plenum, London.

Steves, B.A., Valsecchi, G.B., Perozzi, E., and Roy, A.E., 1993, Significant high number commensurabilities in the Main Lunar Problem. II: The occurrence of Saros-like near-periodicities, *Celest. Mech. & Dynam. Astron.* 57:341.

Valsecchi, G.B., Perozzi, E., Roy, A.E., and Steves, B.A., 1993a, Periodic orbits close to that of the Moon, *Astron. Astrophys.* 271:308.

Valsecchi, G.B., Perozzi, E., Roy, A.E., and Steves, B.A., 1993b, The arrangement in mean elements' space of the periodic orbits close to that of the Moon, *Celest. Mech. & Dynam. Astron.* 56:373.

ORBITAL ELEMENTS OF A SATELLITE MOVING IN THE POTENTIAL OF AN OBLATE SPHEROID

B. Zafiropoulos and Ch. Stavliotis

Department of Physics
University of Patras
26110 Patras
Greece

ABSTRACT

Various forms of potentials produced by rings, disks, oblate or prolate spheroids, electric or magnetic dipoles etc., can be expressed by means of Legendre polynomials. We choose an oblate spheroid to demonstrate a general method for obtaining the perturbed orbital elements of a satellite moving in any of the above potentials. We substitute the disturbing accelerations due to an oblate speroid into the Gauss-Lagrange form of the planetary equations to yield both secular and periodic terms for the orbital elements.

INTRODUCTION

Potentials produced by rings, disks, oblate or prolate spheroids, electric or magnetic dipoles etc. are of fundamental importance to Astronomy. For the potentials given below, r represents the distance from the centre of the body and ϑ stands for the co-latitude.

(I) Ring Potential

For a circular ring of radius A, the potential is (Taff, 1985)

$$U = -\frac{\mu}{r} \, , \Big[1 + \sum_{\nu=1}^{\infty} \, , \frac{(-1)^{\nu}(2\nu - 1)!!}{(2\nu)!!}\Big(\frac{A}{r}\Big)^{2\nu} P_{2\nu}(\cos\vartheta)\Big], \quad r > A, \tag{1}$$

where $P_{2\nu}$ denote the Legendre polynomials, μ is the product of the ring's mass times the gravitational constant, and

$$(2\nu - 1)!! = (2\nu - 1)(2\nu - 3)(2\nu - 5)\ldots 1. \tag{2}$$

(II) Disk Potential

The gravitational potential of a circular disk of radius A is (Ramsey, 1981)

From Newton to Chaos, Edited by A.E. Roy
and B.A. Steves, Plenum Press, New York, 1995

$$U = -\frac{2\mu}{r}\Big[\frac{1}{2} + \sum_{\nu=1}^{\infty} \frac{(-1)^{\nu}(2\nu-1)!!}{(2\nu)!!}\Big(\frac{A}{r}\Big)^{2\nu} P_{2\nu}(\cos\vartheta)\Big], \quad r > A, \tag{3}$$

(III) Oblate Spheroid

For a homogeneous oblate spheroid, the potential at a point with distance r greater than the semi-major axes A is (MacMillan, 1958)

$$U = -\frac{3\mu}{r}\sum_{\nu=0}^{\infty} \frac{(-1)^{\nu}}{(2\nu+1)(2\nu+3)}\Big(\frac{A\varepsilon}{r}\Big)^{2\nu} P_{2\nu}(\cos\vartheta), \tag{4}$$

where ε is the eccentricity of a meridian section.

(IV) Potential of a Magnet

The relative expression is (MacRobert, 1967)

$$U = \frac{M}{r^2}\sum_{\nu=1}^{\infty}\Big(\frac{\alpha}{r}\Big)^{2\nu-2} P_{2\nu-1}(\cos\vartheta), \quad r > \alpha, \tag{5}$$

where $M = 2\alpha m$ represents the magnetic moment of the dipole of size 2α and magnetic quantities $(+m, -m)$.

The potentials of all the above bodies possess an axis of symmetry. Similar expressions can be obtained for a homogeneous hemisphere, the potential of a magnetic shell, etc.

GAUSS-LAGRANGE PLANETARY EQUATIONS

We shall assume that a satellite of negligible mass moves around an oblate spheroid; however, the method is general and can be applied to any of the potentials given above.

In order to describe the motion of the satellite, we use six independent parameters, or elements. These elements will be the semi-major axis α, the eccentricity e, the longitude of the ascending node Ω, the inclination ι of the orbit to equatorial plane of the spheroid, the argument of periapse ω and χ the mean anomaly at time $t = 0$.

We employ the osculating ellipse parameters in order to describe the motion of the satellite. The Gauss-Lagrange planetary equations using the above elements are (cf., e.g., Smart 1953)

$$\frac{d\alpha}{dt} = \frac{2\alpha^2}{rH}\{e\,r\sin v\,R + pS\}, \tag{6}$$

$$\frac{de}{dt} = \frac{1}{H}\{p\sin v\,R + (p+r)\cos v\,S + e\,r\,S\}, \tag{7}$$

$$\frac{d\Omega}{dt} = \frac{r\sin u}{H\sin\iota}W, \tag{8}$$

$$\frac{d\iota}{dt} = \frac{r\cos u}{H}W, \tag{9}$$

$$\frac{d\omega}{dt} = \frac{1}{eH}\{-p\cos v\,R + (p+r)\sin v\,S\} - \cos\iota\frac{d\Omega}{dt}, \tag{10}$$

$$\frac{d\chi}{dt} = \frac{1}{e\sqrt{\mu\alpha}}\{(p\cos v - 2er)R - (p+r)\sin v\,S\}, \tag{11}$$

where R, S and W are the components of the disturbing accelerations, u and v denote the true anomaly measured from the ascending node and the periastron, respectively, H

is the angular momentum of the satellite about the centre of force given by $H = \sqrt{\mu p}$, and r stands for the magnitude of the position vector, defined by

$$r = \frac{\alpha(1 - e^2)}{1 + e\cos v} = \frac{p}{1 + e\cos v}. \tag{12}$$

THE DISTURBING ACCELERATIONS R, S AND W

By means of Equation (4) we consider the disturbing potential due to the asphericity of the oblate spheroid to be

$$V = U - U_0 = U - \frac{\mu}{r} = -\frac{3\mu}{r}\sum_{\nu=1}^{\infty}\frac{(-1)^{\nu}}{(2\nu+1)(2\nu+3)}\left(\frac{A\varepsilon}{r}\right)^{2\nu}P_{2\nu}(\cos\vartheta). \tag{13}$$

Through simple spherical trigonometry, we obtain for the components R, S and W of the disturbing accelerations

$$R = \frac{\partial V}{\partial r}, \tag{14}$$

$$S = -\frac{\cos u \sin\iota}{r\sin\vartheta}\frac{\partial V}{\partial\vartheta}, \tag{15}$$

$$W = -\frac{\cos\iota}{r\sin\vartheta}\frac{\partial V}{\partial\vartheta}. \tag{16}$$

where $\cos\vartheta = \sin\iota\sin u$.

Performing the differentiations with respect to r in (14), we obtain for the component R

$$R = \frac{\mu}{r^2}\sum_{\nu=1}^{\infty}\left(\frac{A\varepsilon}{r}\right)^{2\nu}\sum_{j=0}^{\nu}\rho_{2\nu,j}\cos(2ju). \tag{17}$$

where $f = \sin^2\iota$ and

$$\rho_{2,0} = \frac{3}{10}(1 - \frac{3}{2}f), \quad \rho_{2,1} = \frac{9f}{20}, \tag{18}$$

$$\rho_{4,0} = \frac{15}{8}(\frac{3}{35} - \frac{3f}{7} + \frac{3f^2}{8}), \quad \rho_{4,1} = \frac{15f}{56}(3 - \frac{7f}{2}), \quad \rho_{4,2} = \frac{15f^2}{64}, \quad etc. \tag{19}$$

Similarly the component S of the disturbing acceleration becomes

$$S = -\frac{\mu f\cos u}{r^2}\sum_{\nu=1}^{\infty}\left(\frac{A\varepsilon}{r}\right)^{2\nu}\sum_{j=0}^{\nu-1}\sigma_{2\nu,j}\sin[(2j+1)u]. \tag{20}$$

where we have put

$$\sigma_{2,0} = -\frac{3}{5}, \quad \sigma_{4,0} = -\frac{9}{2}\left(\frac{1}{7} - \frac{f}{4}\right), \quad \sigma_{4,1} = -\frac{3f}{8}, \quad etc. \tag{21}$$

Finally, the component W, in the compact form of summation is

$$W = -\frac{\mu}{2r^2}\sin 2\iota\sum_{\nu=1}^{\infty}\left(\frac{A\varepsilon}{r}\right)^{2\nu}\sum_{j=0}^{\nu-1}\sigma_{2\nu,j}\sin[(2j+1)u]. \tag{22}$$

THE PERTURBED ELEMENTS OF THE ORBIT

In order to obtain the orbital elements we change the independent variable from time t to true anomaly u, by means of the relation

$$\frac{dt}{du} = \frac{r^2}{H}. \tag{23}$$

We present them in the compact form of summations by means of Hansen coefficients of zero order $X_0^{n,m}(e)$ (Zafiropoulos and Kopal, 1982, Zafiropoulos and Zafiropoulos, 1982). The relation which permits us to introduce the zero-order Hansen coefficients into the expressions for the perturbed elements of the orbit has the form

$$\left(\frac{\alpha}{r}\right)^n = \sqrt{1-e^2}\sum_{k=0}^{n}(2-\delta_{0,k})\,X_0^{-(n+2),k}\cos k\,v, \tag{24}$$

where the symbol $\delta_{0,k}$ denotes the Kronecker's delta.

Substituting into the Gauss form of Lagrange planetary equations for the components R, S and W, by means of Equations (17), (20) and (22) respectively, and (α/r) through (24), the following results have been obtained:

$$\begin{aligned}\delta\alpha &= \alpha-\alpha_0 = \frac{\alpha\sqrt{1-e^2}}{2}\sum_{\nu=1}^{\infty}\left(\frac{A\varepsilon}{\alpha}\right)^{2\nu}\times\\ &\times\Big\{f\sum_{j=0}^{\nu-1}\sigma_{2\nu,j}\sum_{k=0}^{2\nu+1}(2-\delta_{0,k})\,X_0^{-(2\nu+3),k}\,C^1_{jk}(u,\omega)-\\ &-\frac{e\alpha}{p}\sum_{j=0}^{\nu}\rho_{2\nu,j}\sum_{k=0}^{2\nu}(2-\delta_{0,k})\,X_0^{-(2\nu+2),k}\,C^2_{jk}(u,\omega)\Big\}\Big|_{u=0}^{u=u},\end{aligned} \tag{25}$$

where

$$\begin{aligned}C^1_{jk}(u,\omega) &= \frac{\cos[(2j+k)u-k\omega]}{(2j+k)}+\frac{\cos[(2j-k)u+k\omega]}{(2j-k)}+\\ &+\frac{\cos[(2j+2+k)u-k\omega]}{(2j+2+k)}+\frac{\cos[(2j+2-k)u+k\omega]}{(2j+2-k)},\end{aligned}$$

$$\begin{aligned}C^2_{jk}(u,\omega) &= \frac{\cos[(2j+1+k)u-(1+k)\omega]}{(2j+1+k)}+\frac{\cos[(2j+1-k)u-(1-k)\omega]}{(2j+1-k)}-\\ &-\frac{\cos[(2j-1+k)u+(1-k)\omega]}{(2j-1+k)}-\frac{\cos[(2j-1-k)u+(1+k)\omega]}{(2j-1-k)}.\end{aligned}$$

$$\begin{aligned}\delta e &= e-e_0 = \frac{\sqrt{1-e^2}}{4}\sum_{\nu=1}^{\infty}\left(\frac{A\varepsilon}{\alpha}\right)^{2\nu}\Big\{\frac{\alpha f}{p}\sum_{j=0}^{\nu-1}\sigma_{2\nu,j}\sum_{k=0}^{2\nu-1}(2-\delta_{0,k})\times\\ &\times X_0^{-(2\nu+1),k}\Big[\frac{3e}{2}C^1_{jk}(u,\omega)+C^3_{jk}(u,\omega)+\frac{e}{4}C^4_{jk}(u,\omega)\Big]-\\ &-\sum_{j=0}^{\nu}\rho_{2\nu,j}\sum_{k=0}^{2\nu}(2-\delta_{0,k})\,X_0^{-(2\nu+2),k}\,C^2_{jk}(u,\omega)\Big\}\Big|_{u=0}^{u=u},\end{aligned} \tag{26}$$

where

$$C^3_{jk}(u,\omega) = 2\cos\omega\Big\{\frac{\cos[(2j+1+k)u-k\omega]}{(2j+1+k)}+\frac{\cos[(2j+1-k)u+k\omega]}{(2j+1-k)}\Big\}+$$
$$+\frac{\cos[(2j-1+k)u+(1-k)\omega]}{(2j-1+k)}+\frac{\cos[(2j-1-k)u+(1+k)\omega]}{(2j-1-k)}+$$
$$+\frac{\cos[(2j+3+k)u-(1+k)\omega]}{(2j+3+k)}+\frac{\cos[(2j+3-k)u-(1-k)\omega]}{(2j+3-k)},$$

$$C^4_{jk}(u,\omega) = \frac{\cos[(2j+k)u+(2-k)\omega]}{(2j+k)}+\frac{\cos[(2j-k)u+(2+k)\omega]}{(2j-k)}+$$
$$+\frac{\cos[(2j+2+k)u-(2+k)\omega]}{(2j+2+k)}+\frac{\cos[(2j+2-k)u-(2-k)\omega]}{(2j+2-k)}+$$
$$+\frac{\cos[(2j-2+k)u+(2-k)\omega]}{(2j-2+k)}+\frac{\cos[(2j-2-k)u+(2+k)\omega]}{(2j-2-k)}+$$
$$+\frac{\cos[(2j+4+k)u-(2+k)\omega]}{(2j+4+k)}+\frac{\cos[(2j+4-k)u-(2-k)\omega]}{(2j+4-k)}.$$

$$\delta\Omega = \Omega-\Omega_0 = -\frac{\alpha\cos\iota\sqrt{1-e^2}}{4p}\sum_{\nu=1}^{\infty}\Big(\frac{A\varepsilon}{\alpha}\Big)^{2\nu}\sum_{j=0}^{\nu-1}\sigma_{2\nu,j}\times$$
$$\times\sum_{k=0}^{2\nu-1}(2-\delta_{0,k})X_0^{-(2\nu+1),k}S^1_{jk}(u,\omega)\Big|_{u=0}^{u=u}, \qquad (27)$$

where

$$S^1_{jk}(u,\omega) = \frac{\sin[(2j+k)u-k\omega]}{(2j+k)}+\frac{\sin[(2j-k)u+k\omega]}{(2j-k)}-$$
$$-\frac{\sin[(2j+2+k)u-k\omega]}{(2j+2+k)}-\frac{\sin[(2j+2-k)u+k\omega]}{(2j+2-k)}.$$

$$\delta\iota = \iota-\iota_0 = \frac{\sin 2\iota}{8\sqrt{1-e^2}}\sum_{\nu=1}^{\infty}\Big(\frac{A\varepsilon}{\alpha}\Big)^{2\nu}\sum_{j=0}^{\nu-1}\sigma_{2\nu,j}\times$$
$$\times\sum_{k=0}^{2\nu-1}(2-\delta_{0,k})X_0^{-(2\nu+1),k}C^1_{jk}(u,\omega)\Big|_{u=0}^{u=u}. \qquad (28)$$

$$\delta\omega = \omega-\omega_0 = -\frac{\sqrt{1-e^2}}{4e}\sum_{\nu=1}^{\infty}\Big(\frac{A\varepsilon}{\alpha}\Big)^{2\nu}\Big\{\sum_{j=0}^{\nu}\rho_{2\nu,j}\sum_{k=0}^{2\nu}(2-\delta_{0,k})X_0^{-(2\nu+2),k}\times$$
$$\times S^2_{jk}(u,\omega)+\frac{f}{2}\Big[\sum_{k=0}^{2\nu}(2-\delta_{0,k})X_0^{-(2\nu+2),k}+\frac{\alpha}{p}\sum_{k=0}^{2\nu-1}(2-\delta_{0,k})\times$$
$$\times X_0^{-(2\nu+1),k}\Big]\sum_{j=0}^{\nu-1}\sigma_{2\nu,j}S^3_{jk}(u,\omega)\Big\}\Big|_{u=0}^{u=u}-\cos\iota\,\delta\Omega, \qquad (29)$$

where

$$S_{jk}^2(u,\omega) = \frac{sin[(2j+1+k)u-(1+k)\omega]}{(2j+1+k)} + \frac{sin[(2j+1-k)u-(1-k)\omega]}{(2j+1-k)} + \\ + \frac{sin[(2j-1+k)u+(1-k)\omega]}{(2j-1+k)} + \frac{sin[(2j-1-k)u+(1+k)\omega]}{(2j-1-k)},$$

$$S_{jk}^3(u,\omega) = 2sin\,\omega\left\{\frac{cos[(2j+1+k)u-k\omega]}{(2j+1+k)} + \frac{cos[(2j+1-k)u+k\omega]}{(2j+1-k)}\right\} + \\ + \frac{sin[(2j-1+k)u+(1-k)\omega]}{(2j-1+k)} + \frac{sin[(2j-1-k)u+(1+k)\omega]}{(2j-1-k)} - \\ - \frac{sin[(2j+3+k)u-(1+k)\omega]}{(2j+3+k)} - \frac{sin[(2j+3-k)u-(1-k)\omega]}{(2j+3-k)}.$$

$$\delta\chi = \chi - \chi_0 = \sum_{\nu=1}^{\infty}\left(\frac{A\varepsilon}{\alpha}\right)^{2\nu}\left\{\frac{(1-e^2)}{4e}\Big[\sum_{j=0}^{\nu}\rho_{2\nu,j}\,S_{jk}^2(u,\omega) + \right. \\ + \frac{f}{2}\sum_{j=0}^{\nu-1}\sigma_{2\nu,j}\,S_{jk}^3(u,\omega)\Big]\sum_{k=0}^{2\nu}(2-\delta_{0,k})\,X_0^{-(2\nu+2),k} + \Big[\frac{f}{8e}\sum_{j=0}^{\nu-1}\sigma_{2\nu,j}\,S_{jk}^3(u,\omega) - \\ \left. - \sum_{j=0}^{\nu}\rho_{2\nu,j}\,S_{jk}^4(u,\omega)\Big]\sum_{k=0}^{2\nu-1}(2-\delta_{0,k})\,X_0^{-(2\nu+1),k}\right\}\Big|_{u=0}^{u=u}, \tag{30}$$

where

$$S_{jk}^4(u,\omega) = \frac{sin[(2j+k)u-k\omega]}{(2j+k)} + \frac{sin[(2j-k)u+k\omega]}{(2j-k)}.$$

Equations (25) to (30) represent the perturbed elements of a satellite moving in the potential field of an oblate spheroid. They have been expressed by means of Hansen coefficients in the form of summations and include both secular and periodic terms. The procedure applied gives the elements in their most general form. The perturbations due to any order spherical harmonic can be obtained by means of the above relations.

Tests have been applied that prove the correctness of the calculated elements. It is possible to separate secular from periodic terms. The procedure can be easily applied to similar problems of satellite motion.

REFERENCES

1. MacMillan, W.D., 1958, *The Theory of the Potential*, Dover Publ. Inc., N.Y.
2. MacRobert, T.M., 1967, *Spherical Harmonics*, Pergamon Press Ltd, 3rd Ed.
3. Ramsey, A.S., 1981, *Newtonian Attraction*, Cambridge Univ. Press, Cambridge.
4. Smart, W.M., 1953, *Celestial Mechanics*, Longmans, London.
5. Taff, L.G.,1985, *Celestial Mechanics*, John Wiley and Sons, N.Y.
6. Zafiropoulos, B. and Kopal, Z., 1982, Astrophys. Space Sci. **88**, 355.
7. Zafiropoulos, B. and Zafiropoulos, F., 1982, Astrophys. Space Sci. **88**, 401.

INTEGRABILITY AND CHAOS IN THE CLASSICAL 2-CENTRE FIELD AND ITS J_2-APPROXIMATION

Yan Chao Ge

Dept. of Mathematics, University of Southampton
Southampton, S09 5NH, U.K.

INTRODUCTION

The Henon-Heiles problem (Henon and Heiles, 1964) is a celebrated example of Hamiltonian systems which exhibit chaos. What makes it more interesting is that the Henon-Heiles Hamiltonian is a Taylor approximation of an integrable Hamiltonian, namely, the Toda lattice (Toda, 1970). The integrability of the Toda Hamiltonian is due to the existence of an additional hidden symmetry, which has no geometrical meaning in the configuration space. More recently, it has been proved that at any finite order of the Taylor expansion, the Toda Hamiltonian is not integrable (Yoshida, 1988). This example raises the following questions concerning integrability, chaos and approximation. Is the loss of integrability due to approximation a common phenomenon? What kind of symmetries may be lost (preserved) in the course of approximation? Is it possible to find an approximation method which retains all symmetries of the original system?

To investigate these problems, we need a proper understanding of the general relation between integrals and symmetries. A consistent discussion on symmetries of phase space (which may or may not be symmetries of configuration space) requires the coordinate independent language of modern Hamiltonian system, which goes beyond the conventional canonical formalism (Marsden, 1992). From this analysis, we come to the tentative conclusion that *integrals nonlinear in momenta (respectively, coordinates) are lost if the Hamiltonian function is approximated with respect to coordinates (respectively, momenta) only*. Although no rigorous proof is given here, supporting examples are abundant. Two of them are the classical 2-centre problem and the geodesics in Kerr space-time mentioned in Ge (1992). Here we present numerical results on the former example. The Poincare surface of section is used to demonstrate the invariant curves of the 2-centre problem and chaotic structure of its J_2-truncation.

From Newton to Chaos, Edited by A.E. Roy
and B.A. Steves, Plenum Press, New York, 1995

SYMMETRY, FIRST INTEGRAL AND HAMILTONIAN SYSTEM ON MANIFOLD

Special coordinate systems such as Cartesian coordinates are usually used in physics. Even in terms of vectors and curvilinear coordinates, the background space is often limited to Euclidean space, which is a linear space with many implicit structures. It was relativity that finally liberated physics to the more natural framework of manifold. Surprisingly, it is a rather recent event that Hamiltonian dynamics has started evolving beyond canonical coordinates, which are but a special class of coordinates in the same way that Cartesian's in Riemannian geometry. Differential manifold offers a natural approach to coordinate independent Hamiltonian dynamics.

By a manifold we emphasise that it is a smooth space covered by *local* coordinate charts $\{x^i\}$, irrespective of how complicated the global structure may be. The natural objects one can define on manifolds are *curves* $x^i(\lambda)$ and (scalar) *functions* $f(x^i)$. It is useful to keep in mind that both curves and functions are objects independent of the choice of coordinates, and the coordinate variables x^i can be regarded as either curve-parameters or functions. Then the derivative $df/d\lambda$, which is also coordinate independent, may be introduced as in calculus. The critical step is to regard $d/d\lambda=(dx^i/d\lambda)(\partial/\partial x^i)$ as a linear operator on functions. This operator is the *vector* tangent to a curve; and at a point, the tangent vectors to all curves form a linear space. There is a natural dual space to any linear space. The dual of the tangent space is the cotangent space consisting of *1-forms*, thereby vectors and 1-forms are linear operators upon each other. A 1-form may be associated with any function via. $<d/d\lambda,\mathbf{df}>=df/d\lambda$, thus $\mathbf{df}=(\partial f/\partial x^i)\mathbf{dx}^i$. Tensors may then be constructed accordingly. One advantage of using manifold language is that all quantities are presented in local linear space so that the higher order infinitesimals occur in conventional approach disappear automatically.

Conventionally, a Hamiltonian system is defined by the Hamiltonian function, canonical equations and phase space constructed from a configuration space. Although coordinates and momenta are given equal weight in transformation theory, the transformations are still limited to those among canonical coordinates having special fundamental Poisson brackets. In fact, any independent vector fields having commutative Lie brackets can be chosen simultaneously as the coordinates of phase space. In arbitrary coordinates, the equations of motion is not necessarily canonical, but a deeper insight may be gained from the modern formulation. In this new language, a Hamiltonian system is defined by a Hamiltonian function on a hierarchy of spaces, namely, {Poisson manifold} $\supset$ {Symplectic manifold} $\supset$ {Cotangent bundle T^*Q}, where Q is the configuration space and T^*Q the phase space. On *Poisson manifolds*, everything starts from functions and a *Poisson bracket* is defined formally for any two functions. Since functions define 1-forms, the Poisson bracket naturally defines an anti-symmetric tensor $\mathbf{\Omega}$ via. $\{f,g\}=\mathbf{\Omega}(\mathbf{df},\mathbf{dg}) =\Omega^{ij}f_{,i}g_{,j}=\mathbf{\Omega}(\mathbf{dx}^i,\mathbf{dx}^j)f_{,i}g_{,j}=\{x^i,x^j\}f_{,i}g_{,j}$. Similar to the metric tensor, the 2-vector $\mathbf{\Omega}$ maps a 1-form to a vector. For example, the associated vector of the 1-form $\mathbf{dg}$, or the *Hamiltonian vector field* of the function g, is $\mathbf{Vg}=\mathbf{\Omega}(\ ,\mathbf{dg})=\{\ ,g\}=-\mathbf{\Omega}(\mathbf{dg},\ )=-\{g,\ \}$, hence $\{f,g\} = \mathbf{\Omega}(\mathbf{df},\mathbf{dg}) = <\mathbf{df},\mathbf{Vg}>$

$=-<\mathbf{d}g, \mathbf{V}f>$. We also have the relation between Poisson bracket and Lie bracket, namely, $[\mathbf{V}f, \mathbf{V}g] = \mathbf{\Omega}(\,, \mathbf{d}\{f,g\}) = \mathbf{V}\{f,g\}$, thus the Hamiltonian vector fields form a Lie algebra. However, the inverse mapping from vector to 1-form requires $\mathbf{\Omega}$ to be non-degenerate. A *symplectic manifold* is a Poisson manifold having a closed non-degenerate 2-form ω ($\mathbf{\Omega}=\omega^{-1}$, thus $\{f,g\}=\omega(\mathbf{V}f,\mathbf{V}g)$). Since coordinates $\{q^i\}$ of Q induce coordinates $\{q^i,p_i\}$ of T*Q, the *cotangent bundle* has a natural symplectic structure $\omega=\mathbf{d}q^i \wedge \mathbf{d}p_i$, called the canonical 2-form (so we have $\mathbf{V}q=-\partial/\partial p$, $\mathbf{V}p=\partial/\partial q$). This tensor formulation of Hamiltonian structure may be used to generalise Poisson bracket to arbitrary coordinates of phase space. It can be shown that the rank of $\mathbf{\Omega}$ is an even number, and every symplectic manifold locally looks like T*Q.

This modern formulation does not only go beyond canonical coordinates, but also applies to where canonical theory is no longer available, eg. Hamiltonian structures in field theory. We see that the *Hamiltonian structure*, ie. the background manifold with a Poisson bracket, is independent of the Hamiltonian function. A *Hamiltonian dynamical system* is the Hamiltonian vector field of the Hamiltonian function H, namely, $d/dt=\mathbf{V}H=\mathbf{\Omega}(\,,\mathbf{d}H)=\{ \,,H\}$ (ie. the Liouville operator of ensemble theory). The component form of it is the *equations of motion* $dx^i/dt=\mathbf{\Omega}(\mathbf{d}x^i,\mathbf{d}H)=\{x^i,H\}$. Here care must be taken to see that the conventional definition of Poisson bracket can only be used in canonical coordinates. The tensor notion is easier to use in non-canonical coordinates. In the modern language, any (generating) function F generates an infinitesimal transformation, $d/d\epsilon = \{ \,,F\}$, which is just the Hamiltonian vector of F. Then we have the identity $dF/dt=\{F,H\}=-dH/d\epsilon$.

The last relation makes the link between invariant transformation and first integral (ie. $dH/d\epsilon=0 \Leftrightarrow dF/dt=0$). According to manifold theory, any vector field may be selected as a coordinate basis. Therefore the Hamiltonian vector of an integral may be adopted into a coordinate system. It follows that in phase space there is an ignorable coordinate for any integral. However, such coordinate cannot always be found by transformations of the configuration space (ie. extended point transformation of the type $Q=Q(q)$). Only integrals linear in momenta can be found by extended point transformation and has a meaning in configuration space; the invariant transformations of integrals nonlinear in momenta involve both p and q and have no obvious meaning in configuration space. Approximation is usually a coordinate dependent operation which only respects the symmetries expressed as ignorable variables. Taylor expansion of H(q,p) with respect to q does not respect symmetries which are not explicitly expressed as ignorable coordinates. Therefore nonlinear integrals in p are very likely to be lost in the course of expansion and truncation. Similarly, integrals nonlinear in q are lost if expansion is made with respect to p. In principle, it must be possible to preserve symmetries if approximation is made with respect to p and q simultaneously. It remains to investigate the effect of approximation on linear integrals whose Poisson brackets do not vanish.

It is a common practice to expand the potential field in Legendre polynomials and truncate at some finite order. However, such approximation does not necessarily simplify the original problem, but may lead to the loss of certain symmetries and integrals. The point may be demonstrated in axially symmetric fields whose integrability is due to integrals with no obvious meaning in configuration space. An example exists in relativity: the geodesics of Kerr space-time. The integrability of this problem is due to an integral quadratic in p, which was shown to be lost in the q-approximation of the Kerr metric (Ge, 1993). In a sense, this example is simpler than the Toda lattice and the 2-centre problem because the integral still has a geometrical meaning in the q-space, namely, irreducible Killing tensor.

NUMERICAL INVESTIGATION ON THE TRUNCATED 2-CENTRE PROBLEM

It is well-known that the 2-centre problem is separable in elliptical coordinates; and the additional integral is linked to the Laplace vector and degeneracy of the Kepler problem (Ge, 1992). However, there is no ignorable coordinate in configuration space to explain this nonlinear integral. Therefore the integrability may be lost in the approximated 2-centre field problem. In spherical coordinates $\{r,\theta,\phi\}$ and cylindrical coordinates $\{\rho,\phi,z\}$, the Hamiltonian $H=T+U$ of an axially symmetric field are respectively

$$H=\frac{1}{2m}\left[p_r^2+\frac{1}{r^2}\left(p_\theta^2+\frac{p_\phi^2}{\sin^2\theta}\right)\right]+U(r,\theta)=\frac{1}{2m}(p_\rho^2+p_z^2)+\frac{1}{2m}\frac{p_\phi^2}{\rho^2}+U(\rho,z)$$

If the two central masses $M_\pm$ are located at $\{\rho,\phi,z\}=\{0,0,\pm c\}$, then $-U=mM_+/d_+ + mM_-/d_-$. For $c/r<1$, the potential energy may be expanded as follows

$$\frac{1}{d_\pm}=\frac{1}{\sqrt{r^2+c^2\mp 2rc\cos\theta}}=\sum_{n=0}^{+\infty}\frac{c^n}{r^{n+1}}P_n(\pm\cos\theta)$$

which is extended to cover $c/r\geq 1$ for our interests here. Recall the first few Legendre polynomials $P_0(x)=1$, $P_1(x)=x$ and $P_2(x)=0.5(3x^2-1)$, then we have the approximated potential energy $-U = -U_0-U_1-U_2-\ldots$, where

$$-U_0=\frac{m(M_+ + M_-)}{r},\quad -U_1=\frac{m(M_+ - M_-)c}{r^2}\cos\theta,\quad -U_2=\frac{m(M_+ + M_-)c^2}{2r^3}(3\cos^2\theta-1)$$

Obviously, $H=T+U_0$ is the Kepler problem, and $H=T+U_0+U_1$ possesses the first integral $p_\theta^2+(p_\phi^2/\sin^2\theta)-2m^2c(M_+-M_-)\cos\theta$. However, $H=T+U_0+U_1+U_2$ is no longer integrable because of the P_2 term.

Without loss of generality, we numerically integrate the case with $m=M_+=M_-=1$, $c=22.2$, $P_\phi=3.3$, $H=-0.01$ and present the results on the surface of section. The orbits are computed in the cylindrical coordinates using a first order explicit symplectic integrator. Because of the numerical method used, chaos is suppressed. However, the phase space structure (eg. fixed points and their types) are well captured (Yoshida, 1993). The invariant curves of the 2-centre problem and heteroclinic chaos (circle) around the period two hyperbolic points of the truncated problem are evident in Figure 1 a and b. For comparison purpose, the J_2 problem of the Earth's satellite motion is also included (Figure 1c). The potential energy is

$$-U \doteq -U_0 - U_2 = \frac{mM}{r} - \frac{m(C-A)}{2r^3}(3\cos^2\theta - 1)$$

where M is the mass of the Earth, and A, B, C (with C>A=B) are the principal inertia of the Earth. If we look at the case M=2 and C-A=$2c^2$, then the J_2 terms of the 2-equal-mass field and the Earth's field are only different by sign. The period three fixed points and islands (circle) are shown in Figure 1c (where P_ϕ=11.1, H=-0.008).

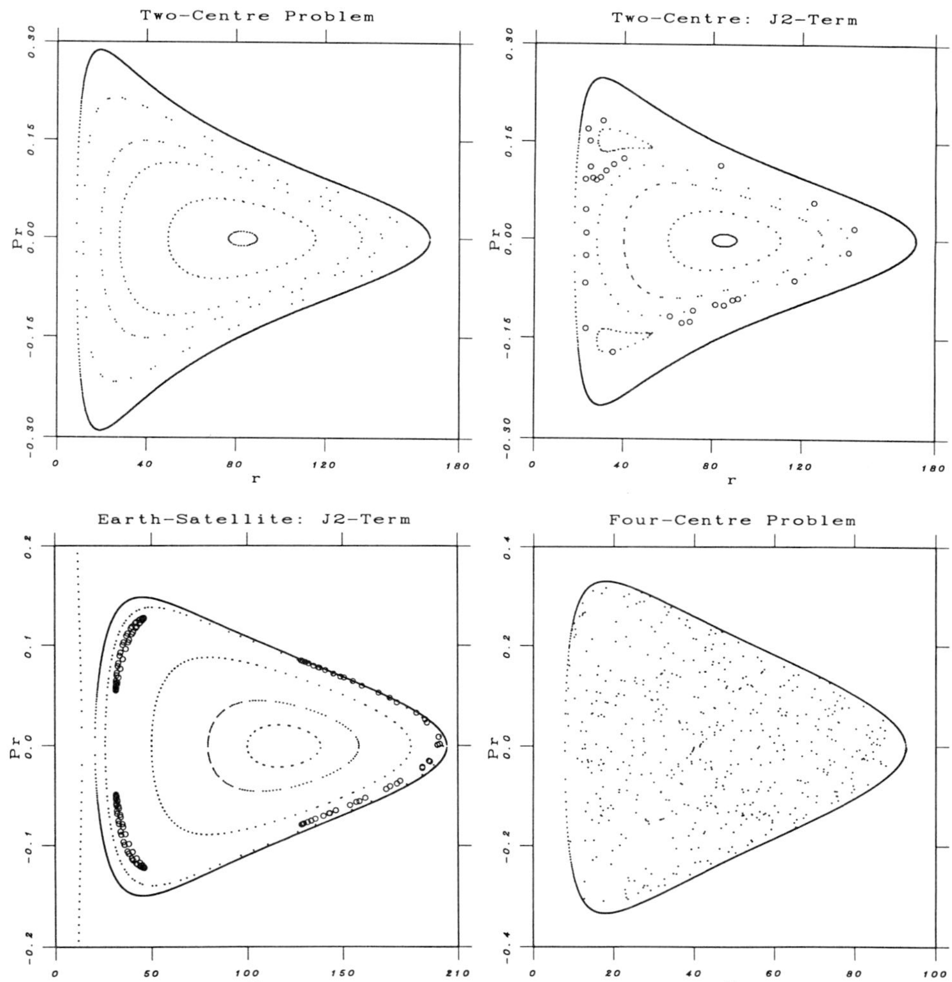

Figure 1. The $\{\rho, P_\rho\}$ surface of section at z=0, P_z>0. The outermost curves are the zero-velocity curves.

The J_2 problem of the Earth's satellite has been numerically investigated in detail by Broucke (1993) and proved to be non-integrable by Irigoyen and Simo (1993) using Ziglin's method. It is straightforward to verify that the proof of Irigoyen and Simo does not change for a sign difference of the J_2 term. We may further *conjecture that the J_n problem ($n>2$) are also chaotic*. In Figure 1d, an aligned 4-centre problem is also included to show how the number of centres change the integrability of the problem. The fixed central masses are $M_1=M_2=1$ located at {0, 0, ±22.2} and $M_3=M_4=1$ located at {0, 0, ±44.4}, and m=1, $P_\phi=3.3$, H=-0.04

CONCLUSION AND DISCUSSION

In addition to the classical Toda lattice and Henon-Heiles problem, we numerically investigated the 2-centre problem and its J_2-approximation, with the latter found to be chaotic. We conjecture that the J_n problem ($n>2$) are also chaotic. Compared with the Earth's potential, the 2-centre field is a good example to study how chaos disappears as $n\rightarrow\infty$. As symmetry preserving numerical integrators (eg. symplectic integration) are under active approach recently, we hope to stimulate interests in a symmetry preserving approximation approach.

ACKNOWLEDGEMENT

My sincere thanks go to Professor A.E. Roy for his encouragement and support.

REFERENCES

Broucke, R.A., 1993, Numerical Integration of Periodic Orbits in the Main Problem of Artificial Satellite Theory, *Celest. Mech.* in press.

Ge, Y.C., 1992, From Newton to Chaos and Modern Physics - I, *Vistas in Astron.* **35**:281-314.

Ge, Y.C., 1993, Integrability and Approximation in Kerr Space-Time, *Phys. Let. A*, in press.

Henon, M. and Heiles, C., 1964, The Applicability of the Third Integral of Motion, *Astron. J.* **69**:73-79.

Irigoyen, M. and Simo, C., 1993, Non-Integrability of the J_2 Problem, *Celest. Mech.* **55**:281-287.

Marsden, J.E., 1992, Lectures on Mechanics, *London Mathematical Society Lecture Note Series*, **174**, Cambridge University Press.

Toda, M., 1970, Waves in Nonlinear Lattice, *Prog. Theor. Phys. Suppl.* **45**:174-200.

Yoshida, H., 1988, Non-Integrability of the Truncated Toda Lattice Hamiltonian at Any Order, *Commun. Math. Phys.* **116**:529-538.

Yoshida, H., 1993, Recent Progress in the Theory and Application of Symplectic Integrators, *Celest. Mech.* **56**:7-43.

ANALYTIC NON-INTEGRABILITY AND THE J_2-PROBLEM

Maylis Irigoyen

Université Paris 2
92, rue d'Assas
F-75006 Paris

INTRODUCTION

We consider[1] the family of Hamiltonians with three degrees of freedom, defined by:

$$H = \frac{1}{2}\left(p_x^2 + p_y^2 + p_z^2\right) - \frac{\mu}{r} + \frac{1}{r^3}\left(\alpha\left(\frac{z}{r}\right)^2 + \beta\right)$$

where $r^2 = x^2 + y^2 + z^2$ and α and β are parameters proportional to the physical adimensional coefficient J_2. These parameters are such that: $\alpha/\beta = -3$. This Hamiltonian system corresponds to the motion of an artificial satellite around an oblate planet, when one keeps only the second order term in the expansion of the potential in spherical harmonics.

Numerical experiments were performed by C. Simó[2], and revealed the non-integrability of this problem, with the existence of stochastic zones. The size of these zones is so small, for the realistic orbits, that the lack of integrability can be neglected for all practical applications.

We shall prove the analytical non-integrability of this problem, i.e. the non-existence of a meromorphic first integral independent of the energy and angular momentum. This proof uses Ziglin's theorem of 1982-83[3], which gives very restrictive necessary conditions to be satisfied by a Hamiltonian system which is integrable. We first give a formulation of this theorem, which is adapted to the Hamiltonian systems with two degrees of freedom.

ZIGLIN'S THEOREM[3] (1982-83)

Assume that a Hamiltonian system has a family of particular solutions Γ_h, parametrized by elliptic functions of the complex time, and depending analytically on a real parameter $h \in (h_1, h_2)$. Let G be the monodromy group of the normal variational equation associated to the solution Γ_h, (Definition: We say that $g \in G$ is non resonant if no one of its eigenvalues is a root of unity). If the Hamiltonian system has a meromorphic integral F, functionally independent of H, in a neighbourhood of Γ_h, and if G contains a non resonant element g_1, then, for any $g_2 \in G$, the commutator $g^* = g_2^{-1}.g_1^{-1}.g_2.g_1$ satisfies: either $g^* = Id$, or $g^* = (g_1)^2$.

From Newton to Chaos, Edited by A.E. Roy
and B.A. Steves, Plenum Press, New York, 1995

It is sufficient that these necessary conditions not be satisfied by a Hamiltonian system, to ensure that it is not analytically integrable. This happens, in particular, if one can find two different non resonant monodromy matrices g_1 and g_2, such that they do not commute. It is the case when (for instance) the following conditions (C) (sufficient conditions of non-integrability) are satisfied:

$$(C)\ \begin{cases} \exists\, g_1 \in G,\ \exists\, g_2 \in G \text{ such that:} \\ g_1.g_2 \neq g_2.g_1, \text{ and } \operatorname{tr} g_1 > 2, \text{ and } \operatorname{tr} g_2 > 2. \end{cases}$$

THE SPATIAL PROBLEM WITH A FIXED ANGULAR MOMENTUM

As the J_2-problem admits a revolution symmetry, we can write the Hamiltonian in cylindrical coordinates:

$$H = \frac{1}{2}\left(p_\rho^2 + \frac{p_\theta^2}{\rho^2} + p_z^2\right) - \frac{\mu}{r} + \frac{1}{r^3}\left(\alpha\left(\frac{z}{r}\right)^2 + \beta\right) \tag{1}$$

where: $r^2 = \rho^2 + z^2$.

The first integral of angular momentum: $p_\theta = c$ allows us to split the couple (θ, p_θ), and to reduce (1) to the following Hamiltonian with two degrees of freedom:

$$\tilde{H}(\rho,z,p_\rho,p_z) = \frac{1}{2}\left(p_\rho^2 + p_z^2\right) - \frac{\mu}{r} + \frac{c^2}{2\rho^2} + \frac{1}{r^3}\left(\alpha\left(\frac{z}{r}\right)^2 + \beta\right) \tag{2}$$

In the equatorial plane of the configuration space (z = 0), the reduced system is a Hamiltonian system with one degree of freedom, and is integrable. The first integral: $\tilde{H} = h$ defines the solutions Γ_h by:

$$p_\rho^2 = 2h + \frac{2\mu}{\rho} - \frac{c^2}{\rho^2} - \frac{2\beta}{\rho^3}, \quad \text{where } \frac{d\rho}{dt} = p_\rho \tag{3}$$

The change of time: $dt = \rho^2.ds$ shows that [4,5] the solutions Γ_h are defined by elliptic functions of the complex time s, because (3) can be written as:

$$s = \int \frac{d\rho}{\sqrt{2h\rho^4 + 2\mu\rho^3 - c^2\rho^2 - 2\beta\rho}} \tag{4}$$

These Jacobi elliptic functions are doubly periodic, with a real period ω_1 and an imaginary one ω_2, and two simple poles in each parallelogram of periods. These solutions define a Riemann surface Γ_h[6], and they are also particular solutions of the spatial problem defined by the Hamiltonian (1).

We shall apply Ziglin's theory to these solutions Γ_h and we shall study the variational equations along Γ_h. Then we shall prove that the reduced Hamiltonian $\tilde{H}$, with two degrees of freedom, does not admit a first integral which is functionally independent of the energy, in a neighbourhood of the particular solutions Γ_h.

For this proof, we need to write the normal variational equation along Γ_h, and to study its monodromy group. But this group cannot be known explicitly. So, we first study the limit problem when the energy $h \to -\infty$ and we shall characterize the monodromy group in this case. Later, we shall extend the results concerning the limit problem (i.e. its non-integrability) to the problem with finite values of h.

We first perform a change of scale of the variables, defined by:

$$\begin{cases} \rho = h^{-1/3}.\,\beta^{1/3}.\,\varphi \\ z = h^{-1/3}.\,\beta^{1/3}.\,\psi \\ t = 3^{-1/2}.(-h)^{-5/6}.\,\beta^{1/3}.\tau \end{cases}$$

always assuming: $h < 0$.
We shall put: $\varepsilon = h^{-1/3}$.

The family of particular solutions Γ_h is then defined by:

$$-\frac{3}{2}\left(\frac{d\varphi}{d\tau}\right)^2 = 1 - \frac{1}{\varphi^3} - \frac{c^2}{2\beta^{2/3}.\varphi^2}\,\varepsilon + \frac{\mu}{\beta^{1/3}.\varphi}\,\varepsilon^2 \qquad (5)$$

Along these solutions, the normal variational equation can be written as:

$$\frac{d^2\eta}{d\tau^2} + ((1 - \frac{2\alpha}{3\beta})\frac{1}{\varphi^5} - \frac{\mu}{3\beta^{1/3}.\varphi^3}\,\varepsilon^2)\,\eta = 0 \qquad (6)$$

where $\varphi(\tau)$ is defined by the equation (5).

NON-INTEGRABILITY OF THE LIMIT-PROBLEM: $h \to -\infty$ $(\varepsilon \to 0)$

For this limit problem, the particular solutions $\Gamma_{-\infty}$ are also defined by Jacobi elliptic functions, and along $\Gamma_{-\infty}$ the normal variational equation is:

$$\frac{d^2\eta}{d\tau^2} + (1 - \frac{2\alpha}{3\beta})\,\frac{\eta}{\varphi^5} = 0.$$

This is exactly the problem, studied by Yoshida[7] (1987) defined by a Hamiltonian system with a homogeneous potential of degree $k = -3$. Yoshida proved that such a linear equation can be transformed into the Gauss hypergeometric equation, whose monodromy group is known explicitly. Two generators of this group are the symplectic matrices associated to particular loops on the Riemann surface $\Gamma_{-\infty}$. These matrices are the transfert matrices associated to the periods ω_1 and ω_2 of the Jacobi function, and they depend only on the homotopy class of the loops.

Yoshida studied the sufficient conditions (C) of non-integrability based on a particular choice of the generators g_1 and g_2, and he gave a criterion of non-integrability which can be applied to our limit problem:

Proposition (Yoshida[7]) :

If the coefficient $\left(1-\frac{2\alpha}{3\beta}\right) \in S_{-3}$, with: $S_{-3} = \ldots\cup(-6,-4)\cup(-1,0)\cup(1,\infty)$, then there exist, in the monodromy group G, two elements $g_1(\varepsilon = 0)$ and $g_2(0)$ which satisfy the sufficient conditions (C) of non-integrability, i.e. :

$g_1(0).g_2(0) \neq g_2(0).g_1(0)$ and $\mathrm{tr}\, g_1(0) > 2$ and $\mathrm{tr}\, g_2(0) > 2$.

In our limit problem, we have: $1-\frac{2\alpha}{3\beta} = 3 \in S_{-3}$, and the J_2-problem is analytically non-integrable if $h \to -\infty$ (i.e. $\varepsilon \to 0$).

NON-INTEGRABILITY FOR FINITE VALUES OF h

If $\varepsilon \neq 0$, it is impossible to know explicitly the monodromy group of the normal variational equation. However, we shall show that the previous proposition is still true for finite values of h:

Theorem :

There exists a value of the energy, $h_o < 0$, such that for $h \in (-\infty, h_o)$ the Hamiltonian system (1) has no first integral independent of H and p_θ, and analytic, provided $\left(1-\frac{2\alpha}{3\beta}\right) \in S_{-3}$.

Proof:

When we pass from $\varepsilon = 0$ to $\varepsilon \neq 0$, the family of particular solutions Γ_h is still described by Jacobi elliptic functions, because in both cases, the polynomial in the elliptic integral (4) is of the forth degree[5].

As the second member of (5) is analytic in ε, the two simple poles of the elliptic function $\varphi(s)$ are as close as desired to the poles of the limit problem, provided $|h|$ is large enough. Then, the fundamental group of the Riemann surface Γ_h is the same as the one of $\Gamma_{-\infty}$.

So, we can define, for $\varepsilon \neq 0$ small enough, the generators $g_1(\varepsilon)$ and $g_2(\varepsilon)$ of the monodromy group of the normal variational equation (6) by using the same loops as in the case $\varepsilon = 0$. Because in the equation (6) the coefficient of η is analytic in ε along the loops, the matrices $g_1(\varepsilon)$ and $g_2(\varepsilon)$ are continuous with respect to ε. Hence, if $\left(1-\frac{2\alpha}{3\beta}\right) \in S_{-3}$, there exists $\varepsilon_o < 0$ such that the conditions (C) (which are open conditions): $g_1(\varepsilon).g_2(\varepsilon) \neq g_2(\varepsilon).g_1(\varepsilon)$, $\mathrm{tr}\, g_1(\varepsilon) > 2$, $\mathrm{tr}\, g_2(\varepsilon) > 2$ hold if $\varepsilon \in (\varepsilon_o, 0)$. We can define h_o by: $h_o = (\varepsilon_o)^{-3}$, and then, the Hamiltonian system has no first integral independent of H and p_θ, provided $h \in (-\infty, h_o)$.

From this theorem, we can deduce that the J_2-problem has no global analytic integral, independent of the energy and the angular momentum.

REFERENCES

1. M. Irigoyen and C. Simó: Cel. Mech. and Dyn. Astr. **55**: 281-287, 1993.
2. C. Simó: Predictability, Stability and Chaos in N-Body Dynamical Systems, Plenum Press, 305-309, 1991.
3. S.L. Ziglin: Funct. Anal. Appl. **16**, 181-189, and **17**, 6-17, 1983.
4. E. Hille: Ordinary Differential Equations, Dover Publ. New York, 1976.
5. E.T. Whittaker and G.N. Watson: A Course of Modern Analysis, Cambridge Univ. Press, 1927.
6. J. Plemelj: Problems in the Sense of Riemann and Klein, Wiley Interscience, New York, 1964.
7. H. Yoshida: Physica 29D, 128-142, 1987.

GUIDELINES FOR A GENERAL TREATMENT OF THE J_2 PROBLEM IN DS VARIABLES WITH TRUE–ANOMALY TYPE FICTITIOUS TIME

Luis Floría

Dep. de Matemática Aplicada a la Ingeniería
E. T. S. de Ingenieros Industriales
Universidad de Valladolid
E–47011 Valladolid, Spain

ABSTRACT

A generic canonical set of unspecified Delaunay–Similar (DS) elements using an independent variable proportional to the Keplerian true anomaly is applied to formulate the Main Problem in Satellite Theory. This way of proceeding renders a compact treatment of this problem in the said DS variables feasible. Preliminary steps are taken towards the construction of an approximate analytical solution within this approach.

UNSPECIFIED DS VARIABLES AND TIME TRANSFORMATION

In Floría (1993), one has shown how a modification of the *Delaunay–Levi-Cività transformation* (Deprit (1981a, 1981b), Deprit and Rom (1970); Levi–Cività (1913)) provides a unifying pattern for systematically deriving the Delaunay–Similar (DS) canonical sets introduced by Scheifele and his collaborators (Scheifele (1970a, §2.2; 1970b), Scheifele and Graf (1974), Scheifele and Stiefel (1972)). Starting from polar nodal variables in the extended phase space, the new elements are obtained with the help of a generating function S:

$$\text{EXTENDED POLAR NODAL VARIABLES} \qquad\qquad \text{SCHEIFELE DS SETS}$$

$$(r, \theta, \nu, t; p_r, p_\theta, p_\nu, p_0) \qquad \xrightarrow{S} \quad (\varphi, l, g, h; \Phi, L, G, H)$$

$$S \equiv S_{DS}(r, \theta, \nu, t; \Phi, L, G, H) = \theta G + \nu H + t L + \int_{r_0}^{r} \sqrt{Q}\, dr ,$$

$$Q \equiv Q_{DS} = \frac{2\mu}{r} - 2L - \frac{\gamma^2}{r^2} \implies \text{DS variables} ,$$

where $\gamma \equiv \gamma(\Phi, L, G, H)$ is a generic function of the new canonical momenta, and r_0 is the *lowest positive root* of the r–equation $Q(r; \Phi, L, G, H) = 0$.

From Newton to Chaos, Edited by A.E. Roy
and B.A. Steves, Plenum Press, New York, 1995

As a consequence, the equations of the transformation generated by S are

$$p_r = \sqrt{Q}\,, \qquad \varphi = -\frac{\partial\gamma}{\partial\Phi} f\,,$$
$$p_\theta = G\,, \qquad g = \theta - \frac{\partial\gamma}{\partial G} f\,,$$
$$p_\nu = H\,, \qquad h = \nu - \frac{\partial\gamma}{\partial H} f\,,$$
$$p_0 == L\,, \qquad l = t - \frac{\mu}{(2L)^{3/2}}(E - e\sin E) - \frac{\partial\gamma}{\partial L} f\,,$$

where the quantities $a \equiv a(L)$, $n \equiv n(L)$, $p \equiv p(\gamma)$ and $e \equiv e(L, \gamma)$ depending on the new momenta, and the auxiliary variables E and f are used:

$$a = \frac{\mu}{2L}\,, \qquad n = \sqrt{\frac{\mu}{a^3}} = \frac{(2L)^{3/2}}{\mu}\,, \qquad \gamma^2 = \mu a(1 - e^2) = \mu p\,,$$
$$p = \frac{\gamma^2}{\mu} = a(1 - e^2)\,, \qquad e^2 = 1 - \frac{2L}{\mu} p = 1 - \frac{2L\gamma^2}{\mu^2}\,,$$
$$r = a(1 - e\cos E)\,, \qquad r = \frac{p}{1 + e\cos f}\,.$$

The application of these variables to the homogeneous Keplerian Hamiltonian

$$\mathcal{H}_{0,h} \equiv \mathcal{H}_h = \frac{1}{2}\left(p_r^2 + \frac{p_\theta^2}{r^2}\right) - \frac{\mu}{r} + p_0$$

converts it into the function

$$\widetilde{\mathcal{H}}_{0,h} \equiv \widetilde{\mathcal{H}}_h = \frac{1}{2r^2}\left(G^2 - \gamma^2\right).$$

A *new independent variable* different from the physical time t is introduced. The time transformation $t \longrightarrow \tau$ defined by the differential relation $dt = \tilde{f}\,d\tau$, with $\tilde{f} = r^2/\mathcal{G}$, where $\mathcal{G}$ is a function of the new variables, is performed. Then the equations of the motion parametrized by τ are governed by the transformed Hamiltonian

$$\mathcal{K}_{0,h} \equiv \mathcal{K}_0 = \widetilde{\mathcal{H}}_h \tilde{f} = \frac{1}{2\mathcal{G}}\left(G^2 - \gamma^2\right),$$

whose integration is simplified if $\mathcal{G}$ reduces to a function of the new momenta. Accordingly, the *frequencies* are

$$C_{(\Phi)} \equiv \frac{\partial\mathcal{K}_0}{\partial\Phi} = -\frac{1}{\mathcal{G}}\left[\gamma\frac{\partial\gamma}{\partial\Phi} + \mathcal{K}_0\frac{\partial\mathcal{G}}{\partial\Phi}\right],$$
$$C_{(L)} \equiv \frac{\partial\mathcal{K}_0}{\partial L} = -\frac{1}{\mathcal{G}}\left[\gamma\frac{\partial\gamma}{\partial L} + \mathcal{K}_0\frac{\partial\mathcal{G}}{\partial L}\right],$$
$$C_{(G)} \equiv \frac{\partial\mathcal{K}_0}{\partial G} = -\frac{1}{\mathcal{G}}\left[\gamma\frac{\partial\gamma}{\partial G} + \mathcal{K}_0\frac{\partial\mathcal{G}}{\partial G} - G\right],$$
$$C_{(H)} \equiv \frac{\partial\mathcal{K}_0}{\partial H} = -\frac{1}{\mathcal{G}}\left[\gamma\frac{\partial\gamma}{\partial H} + \mathcal{K}_0\frac{\partial\mathcal{G}}{\partial H}\right],$$

and the *solution* to the Kepler problem posed by $\mathcal{K}_0$ is

$$\begin{aligned} \varphi &= C_{(\Phi)}\,\tau + \varphi_0\,, & \Phi &= \Phi_0\,, \\ l &= C_{(L)}\,\tau + l_0\,, & L &= L_0\,, \\ g &= C_{(G)}\,\tau + g_0\,, & G &= G_0\,, \\ h &= C_{(H)}\,\tau + h_0\,, & H &= H_0\,. \end{aligned}$$

The nature of this solution reveals that one has obtained, *in a systematic way and with a great freedom of choice*, a general and unified derivation of the *canonical sets of* DS *elements* (e. g., in the sense of Stiefel and Scheifele (1971), §18) for pure Kepler motion. The pseudo–time τ, introduced as the independent variable by the above process, will be *proportional to the Keplerian true anomaly.*

Under appropriate selections of the reparametrizing function $\tilde{f}$ (and, consequently, of $\mathcal{G}$) the Hamiltonian $\mathcal{K}_0$ can be brought into special expressions exhibiting a simple functional form, as shown in Floría (1993), §1.3.

THE MAIN PROBLEM IN UNSPECIFIED DS–TYPE VARIABLES

In polar nodal variables, the J_2 problem of the Satellite Theory is given by

$$\begin{aligned} \mathcal{M} &= \mathcal{H}_0(r, -, -; p_r, p_\theta, -) + \varepsilon\,\mathcal{M}_1(r, \theta, -; p_r, p_\theta, p_\nu) \\ &= \frac{1}{2}\left[p_r^2 + \frac{p_\theta^2}{r^2}\right] - \frac{\mu}{r} + \varepsilon\,\frac{\mu R_e^2}{4 r^3}\{(3c^2 - 1) + 3s^2\cos 2\theta\}, \end{aligned}$$

with the customary convention for the functions of the inclination $I \equiv I(p_\theta, p_\nu)$, $c \equiv \cos I = p_\nu / p_\theta$, $s \equiv \sin I$. The corresponding homogeneous Hamiltonian in the extended polar nodal set is $\mathcal{M}_h = \mathcal{M} + p_0$.

The preceding transformation to unspecified DS elements maps $\mathcal{M}_h$ onto the function

$$\begin{aligned} \widetilde{\mathcal{M}}_h &= \widetilde{\mathcal{H}}_0(-, -, -, -; \Phi, L, G, H) + \varepsilon\widetilde{\mathcal{M}}_1(\varphi, -, g, -; \Phi, L, G, H) \\ &= \frac{1}{2r^2}\left(G^2 - \gamma^2\right) + \varepsilon\,\frac{\mu R_e^2}{4r^3}\left\{(3c^2 - 1) + 3s^2\cos 2\theta\right\}, \end{aligned}$$

where r and f must be regarded as functions of the canonical DS variables through the Keplerian relations defining E and f, bearing in mind that, here, E acts as a parameter. In fact, E plays the role of an intermediate variable connecting the remaining anomalies with each other by means of the *Kepler equation* and the so–called *Gauss equation.*

By introducing the new time parameter τ one gets the general expression

$$\mathcal{M}_h^* = \frac{1}{2\mathcal{G}}\left(G^2 - \gamma^2\right) + \varepsilon\,\frac{\mu R_e^2}{4\,\mathcal{G}\,r}\left\{(3c^2 - 1) + 3s^2\cos 2\theta\right\};$$

once again, use must be made of the Keplerian relations so as to have all the quantities occurring in this formula expressed in terms of the generic DS variables. The result is

$$\mathcal{M}_h^* = \mathcal{K}_0 + \varepsilon\frac{\mu^2 R_e^2}{2}\,\mathcal{M}_1^* = \left(\mathcal{K}_0 + \varepsilon\frac{\mu^2 R_e^2}{2}\,\mathcal{S}_1\right) + \varepsilon\,\frac{\mu^2 R_e^2}{2}\,\mathcal{P}_1\,,$$

$$\mathcal{M}_h^* = \frac{1}{2\mathcal{G}} (G^2 - \gamma^2) + \varepsilon \frac{\mu^2 R_e^2}{2} \frac{(3c^2 - 1)}{2\gamma^2 \mathcal{G}} +$$

$$+ \varepsilon \frac{\mu^2 R_e^2}{4\gamma^2 \mathcal{G}} \left[\frac{3es^2}{2} \cos(2\theta + f) + 3s^2 \cos 2\theta + \right.$$

$$\left. + \frac{3es^2}{2} \cos(2\theta - f) + (3c^2 - 1) e \cos f \right],$$

where, for the sake of conciseness, one has written

$$\theta = g + f\gamma_G, \quad \varphi = -f\gamma_\Phi, \quad \gamma_G \equiv \frac{\partial\gamma}{\partial G}, \quad \gamma_\Phi \equiv \frac{\partial\gamma}{\partial\Phi},$$

$$\mathcal{P}_1 = \frac{1}{2\gamma^2\mathcal{G}} \Lambda(\varphi, -, g, -; \Phi, L, G, H),$$

$$\Lambda = (3c^2 - 1) e \cos f + 3s^2 \cos 2\theta + \frac{3es^2}{2} [\cos(2\theta + f) + \cos(2\theta - f)].$$

The calligraphic letters $\mathcal{S}$ and $\mathcal{P}$ denote *secular* and *periodic* part, respectively.

To arrive at a *general and systematic treatment* of the Main Satellite Problem *in unspecified DS variables*, some considerations concerning its integration are presented. In this study, interest will be restricted to the first order. To this end, the problem given by $\mathcal{M}_h^*$ will be approached by means of a canonical transformation that generates ignorable coordinates. The resulting secular Hamiltonian will be integrated in the new variables.

THE AVERAGED HAMILTONIAN AND THE SECULAR MOTION

As a first step towards an approximate analytical integration of this Hamiltonian, the solution to its secular part will be developed; at a later stage, the first-order periodic perturbations undergone by the secular motion will be evaluated.

The averaged Hamiltonian is the result of the elimination of the periodic terms depending on the angles φ and g. To this end, one defines a *near-identity canonical transformation*, $(\varphi, l, g, h; \Phi, L, G, H) \longrightarrow (\varphi', l', g', h'; \Phi', L', G', H')$, governed by a generating function $W = \sum_{j\geq 1} \varepsilon^j W_j$ that depends only on the new variables. Up to a constant factor of the form $\mu^2 R_e^2/2$, the first-order part of this function is determined as

$$W_1 = \int_{UP} \mathcal{P}_1 \, d\tau = \int_{UP} \frac{1}{2\gamma^2\mathcal{G}} \Lambda(\varphi, -, g, -; \Phi, L, G, H) \, d\tau,$$

the integral being calculated along the solution to the unperturbed problem (UP), which for the present purposes is the *Keplerian solution* generated by $\mathcal{K}_0$, as shown in Section 2. The *generating function* leading to the elimination of the first-order periodic terms is

$$W_1 = \frac{1}{2\gamma^2\mathcal{G}} \left\{ -\frac{\gamma_\Phi}{C_{(\Phi)}} (3c^2 - 1) e \sin f + \frac{3s^2}{2} \frac{\sin 2\theta}{\left[C_{(G)} - \frac{\gamma_G}{\gamma_\Phi} C_{(\Phi)} \right]} + \right.$$

$$\left. + \frac{3es^2}{4} \left(\frac{\sin(2\theta + f)}{\left[C_{(G)} - \frac{1}{\gamma_\Phi} \left(\gamma_G + \frac{1}{2} \right) C_{(\Phi)} \right]} + \frac{\sin(2\theta - f)}{\left[C_{(G)} - \frac{1}{\gamma_\Phi} \left(\gamma_G - \frac{1}{2} \right) C_{(\Phi)} \right]} \right) \right\}.$$

Calculating the first-order periodic perturbations involves the partial derivatives of the first-order term in the generating function W (essentially, those of W_1). For convenience, the constant factor will not be written during the intermediate reckoning work.

Under this transformation, the Hamiltonian $\mathcal{M}_h^*$ becomes

$$\begin{aligned}\mathcal{M}'(-,-,-,-;\Phi',L',G',H') &\equiv \mathcal{S}'(-,-,-,-;\Phi',L',G',H') \\ &= \mathcal{K}_0' + \varepsilon \frac{\mu^2 R_e{}^2}{2} \mathcal{S}_1' .\end{aligned}$$

The quantities occurring in it depend on the new, *primed* variables. When there is no danger of confusion, for the sake of brevity, and following the usual practice, the *prime* symbol will be omitted in these new variables. The integration of this Hamiltonian is now straightforward: the coordinates are linear functions of the independent variable, and the momenta are constants of the motion.

To determine the partial derivatives of $\mathcal{S}'$ with respect to the momenta, the derivatives of $\mathcal{K}_0'$ exhibit the same formal expression as in the case of the corresponding unprimed expressions (see Section 1); thus, it suffices to build the derivatives of the first-order term. Consequently, in terms of the new variables the *solution to the averaged Hamiltonian* is given by the formulae

$$\begin{aligned}
\varphi^{(s)} &= \varphi^{(0)} - \left\{ \frac{1}{\mathcal{G}} \left[\gamma \frac{\partial \gamma}{\partial \Phi} + \mathcal{K}_0 \frac{\partial \mathcal{G}}{\partial \Phi} \right] + \right. \\
&\quad \left. + \varepsilon \frac{\mu^2 R_e{}^2}{2} \left[\frac{2}{\gamma} \frac{\partial \gamma}{\partial \Phi} + \frac{1}{\mathcal{G}} \frac{\partial \mathcal{G}}{\partial \Phi} \right] \mathcal{S}_1 \right\} \tau , \\
l^{(s)} &= l^{(0)} - \left\{ \frac{1}{\mathcal{G}} \left[\gamma \frac{\partial \gamma}{\partial L} + \mathcal{K}_0 \frac{\partial \mathcal{G}}{\partial L} \right] + \right. \\
&\quad \left. + \varepsilon \frac{\mu^2 R_e{}^2}{2} \left[\frac{2}{\gamma} \frac{\partial \gamma}{\partial L} + \frac{1}{\mathcal{G}} \frac{\partial \mathcal{G}}{\partial L} \right] \mathcal{S}_1 \right\} \tau , \\
g^{(s)} &= g^{(0)} - \left\{ \frac{1}{\mathcal{G}} \left[\gamma \frac{\partial \gamma}{\partial G} + \mathcal{K}_0 \frac{\partial \mathcal{G}}{\partial G} - G \right] + \right. \\
&\quad \left. + \varepsilon \frac{\mu^2 R_e{}^2}{2} \left(\left[\frac{2}{\gamma} \frac{\partial \gamma}{\partial G} + \frac{1}{\mathcal{G}} \frac{\partial \mathcal{G}}{\partial G} \right] \mathcal{S}_1 + \frac{3 c^2}{G \mathcal{G} \gamma^2} \right) \right\} \tau , \\
h^{(s)} &= h^{(0)} - \left\{ \frac{1}{\mathcal{G}} \left[\gamma \frac{\partial \gamma}{\partial H} + \mathcal{K}_0 \frac{\partial \mathcal{G}}{\partial H} \right] + \right. \\
&\quad \left. + \varepsilon \frac{\mu^2 R_e{}^2}{2} \left(\left[\frac{2}{\gamma} \frac{\partial \gamma}{\partial H} + \frac{1}{\mathcal{G}} \frac{\partial \mathcal{G}}{\partial H} \right] \mathcal{S}_1 - \frac{3 c^2}{H \mathcal{G} \gamma^2} \right) \right\} \tau ,
\end{aligned}$$

together with the constant momenta

$$\Phi^{(s)} = \Phi^{(0)}, \quad L^{(s)} = L^{(0)}, \quad G^{(s)} = G^{(0)}, \quad H^{(s)} = H^{(0)} .$$

The *solution to the averaged part* $\mathcal{S}'$ undergoes the effect of the periodic perturbations of the first order, calculated with the help of the partial derivatives of the function

W_1 and imposed on the preceding secular solution. Such effects are accounted for according to the general formulae (Cary (1981), Deprit (1969), Hori (1966))

$$q = q' + \varepsilon \frac{\mu^2 R_e{}^2}{2} \frac{\partial W_1}{\partial p'}, \quad p = p' - \varepsilon \frac{\mu^2 R_e{}^2}{2} \frac{\partial W_1}{\partial q'},$$

where the notation (q, p) represents a pair of canonically conjugate variables of the original set, and (q', p') has the analogous meaning in the new set of variables obtained after application of the transformation generated by W_1.

A detailed account of the use of the preceding, generic DS elements will be given elsewhere.

ACKNOWLEDGEMENTS

This paper is based on a part of the Ph. D. Thesis (Floría (1993)) at the University of Valladolid. The author thanks his supervisor Prof. José Manuel Ferrándiz for introducing him to this field of work, and for advice and helpful discussion. Much of this version was written while the author enjoyed the hospitality of the Mission Analysis Section at ESOC in Darmstadt (Germany), on leave of absence from the University of Valladolid thanks to partial financial support from this University (Plan de Movilidad del Personal Investigador, Convocatoria de 1992). Thanks also are due to ESOC for the facilities made available during the author's visit.

REFERENCES

Cary, J. R., 1981, Lie Transform Perturbation Theory for Hamiltonian Systems, *Physics Reports (Review Section of Physics Letters)*, 79:129.

Deprit, A., 1969, Canonical Transformations Depending on a Small Parameter, *Celest. Mech.*, 1:12.

Deprit, A., 1981a, A Note Concerning the TR–Transformation, *Celest. Mech.*, 23: 299.

Deprit, A., 1981b, The Elimination of the Parallax in Satellite Theory, *Celest. Mech.*, 24:111.

Deprit, A. and Rom, A., 1970, The Main Problem of Artificial Satellite Theory for Small and Moderate Eccentricities, *Celest. Mech.*, 2:166.

Floría, L., 1993, "Intermediarios radiales y generalizaciones de las variables de tipo Delaunay–Scheifele. Aplicación al movimiento orbital de satélites artificiales"(Doctoral Dissertation). University of Valladolid (Spain).

Hori, G.–i., 1966, Theory of General Perturbations with Unspecified Canonical Variables, *Publ. Astron. Soc. Japan*, 18:287.

Levi–Civitá, T., 1913, Nuovo sistema canonico di elementi ellittici, *Annali di Matematica Pura ed Applicata*, Serie III, 20:153.

Scheifele, G., 1970a, On Nonclassical Canonical Systems, *Celest. Mech.*, 2:296.

Scheifele, G., 1970b, Généralisation des éléments de Delaunay en Mécanique Céleste. Application au mouvement d'un satellite artificiel, *Comptes Rendus de l'Académie des Sciences de Paris*, Série A, 271:729.

Scheifele, G. and Graf, O., 1974, Analytical satellite theories based on a new set of canonical elements, *AIAA Paper 74–838*, 19pp.

Scheifele, G. and Stiefel, E., 1972, "Canonical Satellite Theory Based on Independent Variables Different from Time", Report to ESRO under ESOC–contract No. 219/70/AR, ETH, Zürich.

Stiefel, E. L. and Scheifele G., 1971, "Linear and Regular Celestial Mechanics", Springer–Verlag, Berlin.

SECOND ORDER PERTURBATIONS DUE TO THE GRAVITY POTENTIAL OF A PLANET

Edwin Wnuk

Astronomical Observatory
of the A.Mickiewicz University, Poznań, Poland

ABSTRACT

A theory of perturbations due to the gravitational potential of a planet is derived applying the Hori-Lie algorithm and planet's gravitational potential expressed using generalized lumped coefficients. The first and the second order secular, short- and long-period, and partly third order short-period perturbations are calculated including all zonal and tesseral coefficients. Formulas for perturbations include the eccentricity function and the inclination function in their general form and therefore can be applied for orbits with arbitrary eccentricity and arbitrary inclination, except the critical inclination.

1. INTRODUCTION

The gravitational potential of a planet expressed in terms of spherical coordinates r, φ, λ is usually described by the following formula:

$$V = \frac{\mu}{r}\left[1 + \sum_{n=2}^{\infty} \sum_{m=0}^{n} \left(\frac{r_p}{r}\right)^n \overline{P}_{nm}(\sin\varphi)\left(\overline{C}_{nm}\cos m\lambda + \overline{S}_{nm}\sin m\lambda\right)\right] \tag{1}$$

where $\mu = GM_p$ is the product of the gravitational constant G and the mass of a planet M_p, r_p is the planet's mean equatorial radius, $\overline{P}_{nm}$ are the normalized Legendre associated functions of degree n and order m, r is the distance from the centre of mass, φ is the geocentric latitude, λ is the longitude, and $\overline{C}_{nm}$ and $\overline{S}_{nm}$ are the normalized spherical harmonic coefficients. The planet's gravity model is defined by the values of GM_p and r_p constants, and by $\overline{C}_{nm}$ and $\overline{S}_{nm}$ coefficients. Contemporary Earth's gravity models contain the coefficients up to 70 degree and order (for "satellite only" models) and up to 360 degree and order (for "combined" models). New Venus and Mars models contain coefficients up to 50 degree and order. For other planets only first zonal harmonic coefficients are determined.

From Newton to Chaos, Edited by A.E. Roy
and B.A. Steves, Plenum Press, New York, 1995

High precision determination of artificial satellite position requires a theory of motion which includes the influence of all gravity model coefficients and which is of appropriately high order for zonal and tesseral coefficients as well as for products of zonals and tesserals. There are many different order theories of Earth's artificial satellite motion which include the influence of particular parts of geopotential (e.g. Brouwer,1959, Kozai,1959,1962, Kaula, 1966, Aksnes,1970, Deprit and Rom,1970, Coffey and Deprit,1982, Kinoshita,1977, Emelyanov,1992, Metris et al.,1993), but it does not exist a general theory which includes all effects on an accuracy level sufficient for contemporary applications.

In this paper, the form of gravitational potential expressed using generalized lumped coefficients, has been applied in deriving the first, the second and some of the third order perturbations for arbitrary degree and order coefficients. Particularly, perturbations due to products of the second zonal harmonic and tesserals are considered.

2. THE MODEL

If the influence of all spherical harmonic coefficients of a planet's potential is considered, and if the second zonal harmonic coefficient $J_2 = -C_{2,0}$ plays the role of a small parameter ε, then the Hamiltonian $\mathcal{H}$ is the sum of three parts:

$$\mathcal{H} = \mathcal{H}_0 + \mathcal{H}_1 + \mathcal{H}_2, \tag{2}$$

where

$$\mathcal{H}_0 = \frac{\mu^2}{2L^2} - \omega_p H, \tag{3}$$

$$\mathcal{H}_1 = \mathcal{H}_1(L, G, H, \ell, g, -; C_{2,0}), \tag{4}$$

$$\mathcal{H}_2 = \mathcal{H}_2(L, G, H, \ell, g, h; C_{nm}, S_{nm}) \tag{5}$$

for $n = 2,3,\ldots,N$, $m = 0,1,\ldots,n$, with $n > 2$ for $m = 0$. N is the maximum order and degree of $\bar{C}_{nm}$ and $\bar{S}_{nm}$ coefficients taken into account. The Hamiltonian $\mathcal{H}_1$ is of the order of ε, and the Hamiltonian $\mathcal{H}_2$ is of the order of $\varepsilon^2 = J_2^2$.

Using the form of gravity potential expressed in terms of orbital elements (Wnuk,1988) $\mathcal{H}_1$ and $\mathcal{H}_2$ can be written in the form:

$$\mathcal{H}_1 = \frac{\mu^2}{L^2} \sum_{k=-2}^{2}{}^{*} \sum_{q=-Q}^{Q} \gamma_{k,0} \bar{\boldsymbol{C}}_0^{k,q} \cos \psi_{0,k,q}, \tag{6}$$

$$\mathcal{H}_2 = \mathcal{H}_2^Z(L, G, H, \ell, g, -) + \mathcal{H}_2^T(L, G, H, \ell, g, h), \tag{7}$$

where

$$\mathcal{H}_2^Z = \frac{\mu^2}{L^2} \sum_{k=-N}^{N} \sum_{q=-Q}^{Q} \gamma_{k,0} \bar{\boldsymbol{C}}_0^{k,q} \cos \psi_{0,k,q}, \tag{8}$$

$$\mathcal{H}_2^T = \frac{\mu^2}{L^2} \sum_{m=1}^{N} \sum_{k=-N}^{N} \sum_{q=-Q}^{Q} \gamma_{k,m} \left[\bar{\boldsymbol{C}}_m^{k,q} \cos \psi_{m,k,q} + \bar{\boldsymbol{S}}_m^{k,q} \sin \psi_{m,k,q} \right], \tag{9}$$

and where

$$\bar{\boldsymbol{C}}_m^{k,q} = \sum_{j=j_1}^{N}{}^{*} \bar{Q}_{j,m}^{k,q} \bar{C}_{j,m}, \quad \bar{\boldsymbol{S}}_m^{k,q} = \sum_{j=j_1}^{N}{}^{*} \bar{Q}_{j,m}^{k,q} \bar{S}_{j,m}, \tag{10}$$

$$\overline{Q}^{k,q}_{j,m} = \left(\frac{r_p \mu}{L^2}\right)^j \overline{A}^k_{j,m} G_{j,(j-k)/2,q} , \tag{11}$$

$$\psi_{m,k,q} = kg + (k+q)\ell + mh + (k-m)\frac{\pi}{2} , \tag{12}$$

$$\gamma_{k,m} = (-1)^{E\left(\frac{k-m+1}{2}\right)} ,$$

$$j_1 = \max\left[k_1, k_1 + 2E\left(\frac{m-k_1+1}{2}\right)\right], \quad k_1 = |k| + 2\left(\delta_{0,|k|} + \delta_{1,|k|}\right),$$

$\overline{A}^k_{l,m}(I) = \overline{F}_{l,m,(l-k)/2}(I)$ is the normalized inclination function, $G_{l,(l-k)/2,q}(e)$ is the eccentricity function, Q is the maximum limit of the index q, $E(x)$ is the Entier function and $\delta_{l,k}$ is the Kronecker delta function. The symbol $\sum^*$ stands for the summation with a step 2. Coefficients $\overline{C}^{k,q}_m$ and $\overline{S}^{k,q}_m$ given by formulas (10) group all amplitudes of the terms with the same frequency and are a generalization of lumped coefficients known in the theory of resonance perturbations in the sense that they include the eccentricity function $G_{l,p,q}(e)$ in its general form. The generalized lumped coefficients (10) are not expanded in powers of eccentricity or inclination, permitting , for example, a very large value of the eccentricity.

The set of Delaunay variables L, G, H, ℓ, g, h with respect to a frame rotating with the planet has been adopted as the set of canonical variables. The relations between these variables and the Keplerian elements $a, e, I, \omega, \Omega, M$ are as follows: $\ell = M$, $g = \omega$, $h = \Omega - \omega_p t$, $L = \sqrt{\mu a}$, $G = L\sqrt{1-e^2}$, $H = G\cos I$, where ω_p is the planet's mean angular velocity. Because of their complex form the eccentricity function and the inclination function are not expressed here in terms of Delaunay variables and since that the Hamiltonian $\mathcal{H}$ depends not only on the Delaunay variables, but on the inclination I and eccentricity e as well. The partial derivatives with respect to L, G, and H are then calculated using the following formulas:

$$\begin{aligned} \frac{\partial}{\partial L} &= \left(\frac{\partial}{\partial L}\right) + \frac{1-e^2}{eL}\frac{\partial}{\partial e}, \\ \frac{\partial}{\partial G} &= -\frac{\sqrt{1-e^2}}{eL}\frac{\partial}{\partial e} + \frac{1}{L\sqrt{1-e^2}}\frac{\cos I}{\sin I}\frac{\partial}{\partial I}, \\ \frac{\partial}{\partial H} &= -\frac{1}{L\sqrt{1-e^2}\sin I}\frac{\partial}{\partial I}. \end{aligned} \tag{13}$$

3. CANONICAL TRANSFORMATIONS

The problem of a planet's satellite motion described by the following canonical equations:

$$\frac{d}{dt}(L,G,H) = \frac{\partial \mathcal{H}}{\partial(\ell,g,h)}, \qquad \frac{d}{dt}(\ell,g,h) = -\frac{\partial \mathcal{H}}{\partial(L,G,H)}, \tag{14}$$

where the Hamiltonian $\mathcal{H}$ is given by Eq. (2) can be solved by applying two canonical transformations. The first transformation

$$(L,G,H,\ell,g,h) \rightarrow (L',G',H',\ell',g',h') \tag{15}$$

transforms Eqs. (14) into

$$\frac{d}{dt}(L',G',H') = \frac{\partial \mathcal{H}^*}{\partial(\ell',g',h')}, \qquad \frac{d}{dt}(\ell',g',h') = -\frac{\partial \mathcal{H}^*}{\partial(L',G',H')}, \tag{16}$$

where the new Hamiltonian $\mathcal{H}^* = \mathcal{H}^*(L',G',H',-,g',-)$, averaged over the fast variables ℓ' and h', does not contain short-period terms and is of the following form:

$$\mathcal{H}^* = \mathcal{H}_0^* + \mathcal{H}_1^* + \mathcal{H}_2^* + \ldots, \qquad \mathcal{H}_n^* \approx \varepsilon^n, \tag{17}$$

and the generating function $\mathcal{S}^*(L',G',H',\ell',g',h')$ of this transformation is of the form:

$$\mathcal{S}^* = \mathcal{S}_1^* + \mathcal{S}_2^* + \mathcal{S}_3^* + \ldots, \qquad \mathcal{S}_n^* \approx \varepsilon^n. \tag{18}$$

The second transformation

$$(L',G',H',\ell',g',h') \rightarrow (L'',G'',H'',\ell'',g'',h'') \tag{19}$$

transforms Eqs. (16) into

$$\frac{d}{dt}(L'',G'',H'') = \frac{\partial \mathcal{H}^{**}}{\partial(\ell'',g'',h'')}, \qquad \frac{d}{dt}(\ell'',g'',h'') = -\frac{\partial \mathcal{H}^{**}}{\partial(L'',G'',H'')}, \tag{20}$$

where the new Hamiltonian $\mathcal{H}^{**} = \mathcal{H}^{**}(L'',G'',H'',-,-,-)$, averaged over g', does not contain both short- and long-period terms and is of the form:

$$\mathcal{H}^{**} = \mathcal{H}_0^{**} + \mathcal{H}_1^{**} + \mathcal{H}_2^{**} + \ldots, \qquad \mathcal{H}_n^{**} \approx \varepsilon^n, \tag{21}$$

and the generating function $\mathcal{S}^{**}(L'',G'',H'',\ell'',g'',h'')$ of this transformation is of the form:

$$\mathcal{S}^{**} = \mathcal{S}_1^{**} + \mathcal{S}_2^{**} + \mathcal{S}_3^{**} + \ldots, \qquad \mathcal{S}_n^{**} \approx \varepsilon^n. \tag{22}$$

The above canonical transformations can be realised applying either Hori (1966) or Deprit (1969) version of the Lie transformation. In this paper we calculate particular terms of the Hamiltonians $\mathcal{H}^*$ and $\mathcal{H}^{**}$ as well as of the generating functions $\mathcal{S}^*$ and $\mathcal{S}^{**}$ using the Hori-Lie algorithm. We use $\langle f \rangle_x$ symbol for averaged over the variable x function f (do not contain terms depending on x), $(f)_x$ symbol for the rest $f - \langle f \rangle_x$ (the part of f containing terms with x), and $\{f,g\}$ symbol for Poisson bracket. Formulas for the zero, the first and the second order Hamiltonians $\mathcal{H}^*$ and $\mathcal{H}^{**}$, the generating functions $\mathcal{S}_i^*$, $i = 1,2,3$, up to the third order, and the generating functions $\mathcal{S}_i^{**}$, $i = 1,2$ up to the second order have the following form:

The first transformation

Order 0: $\mathcal{H}_0^* = \mathcal{H}_0.$ (23)

Order 1: $\mathcal{H}_1^* = \mathcal{H}_1^*(C_{2,0}) = \langle \mathcal{H}_1 \rangle_{\ell'}$ (24)

$$\mathcal{S}_1^* = \frac{L'^3}{\mu^2} \int (\mathcal{H}_1)_{\ell'} d\ell'. \tag{25}$$

Order 2: $\mathcal{H}_2^* = \mathcal{H}_2^{*1}\left((C_{2,0})^2\right) + \mathcal{H}_2^{*2}(C_{n,0})_{n>2},$ (26)

where $\quad \mathcal{H}_2^{*1} = \frac{1}{2}\left\langle \left\{ \mathcal{H}_1 + \mathcal{H}_1^*, \mathcal{S}_1^* \right\} \right\rangle_{\ell'}, \quad \mathcal{H}_2^{*2} = \left\langle \mathcal{H}_2^Z \right\rangle_{\ell'},$

and $\quad \mathcal{S}_2^* = \mathcal{S}_2^{*1}\left((C_{2,0})^2\right) + \mathcal{S}_2^{*2}(C_{n,0})_{n>2} + \mathcal{S}_2^{*3}(C_{n,m}, S_{n,m})_{m>0},$ (27)

where

$$\mathcal{S}_2^{*1} = \frac{1}{2}\frac{L'^3}{\mu^2} \int \left(\left\{ \mathcal{H}_1 + \mathcal{H}_1^*, \mathcal{S}_1^* \right\}\right)_{\ell'} d\ell', \quad \mathcal{S}_2^{*2} = \frac{L'^3}{\mu^2} \int \left(\mathcal{H}_2^Z\right)_{\ell'} d\ell', \quad \mathcal{S}_2^{*3} = \frac{L'^3}{\mu^2} \int \left(\mathcal{H}_2^T\right)_{\ell'} d\ell'.$$

Order 3: $\mathcal{S}_3^* = \mathcal{S}_3^{*1}\left((C_{2,0})^3\right) + \mathcal{S}_3^{*2}(C_{2,0} \times C_{n,0})_{n>2} + \mathcal{S}_3^{*3}\left(C_{2,0} \times (C_{n,m}, S_{n,m})_{m>0}\right),$ (28)

where

$$\begin{aligned}
\mathcal{S}_3^{*1} &= \frac{1}{2}\frac{L'^3}{\mu^2} \int \left(\left\{ \mathcal{H}_1 + \mathcal{H}_1^*, \mathcal{S}_2^{*1} \right\}\right)_{\ell'} d\ell' + \frac{1}{2}\frac{L'^3}{\mu^2} \int \left(\left\{ \mathcal{H}_2^{*1}, \mathcal{S}_1^* \right\}\right)_{\ell'} d\ell' \\
&+ \frac{1}{12}\frac{L'^3}{\mu^2} \int \left(\left\{\left\{ \mathcal{H}_1 - \mathcal{H}_1^*, \mathcal{S}_1^* \right\}, \mathcal{S}_1^* \right\}\right)_{\ell'} d\ell',
\end{aligned}$$

$$\begin{aligned}
\mathcal{S}_3^{*2} &= \frac{1}{2}\frac{L'^3}{\mu^2} \int \left(\left\{ \mathcal{H}_1 + \mathcal{H}_1^*, \mathcal{S}_2^{*2} \right\}\right)_{\ell'} d\ell' + \frac{1}{2}\frac{L'^3}{\mu^2} \int \left(\left\{ \mathcal{H}_2^Z, \mathcal{S}_1^* \right\}\right)_{\ell'} d\ell' \\
&+ \frac{1}{2}\frac{L'^3}{\mu^2} \int \left(\left\{ \mathcal{H}_2^{*2}, \mathcal{S}_1^* \right\}\right)_{\ell'} d\ell',
\end{aligned}$$

$$\mathcal{S}_3^{*3} = \frac{1}{2}\frac{L'^3}{\mu^2} \int \left(\left\{ \mathcal{H}_1 + \mathcal{H}_1^*, \mathcal{S}_2^{*3} \right\}\right)_{\ell'} d\ell' + \frac{1}{2}\frac{L'^3}{\mu^2} \int \left(\left\{ \mathcal{H}_2^T, \mathcal{S}_1^* \right\}\right)_{\ell'} d\ell'.$$

The second transformation

Order 0: $\mathcal{H}_0^{**} = \mathcal{H}_0^*$ (29)

Order 1: $\mathcal{H}_1^{**} = \mathcal{H}_1^*$ (30)

Order 2: $\mathcal{H}_2^{**} = \mathcal{H}_2^{**1}\left(\left(C_{2,0}\right)^2\right) + \mathcal{H}_2^{**2}\left(C_{n,0}\right)_{n>2}$, (31)

where $\mathcal{H}_2^{**1} = \left\langle \mathcal{H}_2^{*1} \right\rangle_{g''}$, $\mathcal{H}_2^{**2} = \left\langle \mathcal{H}_2^{*2} \right\rangle_{g''}$,

and
$$\mathcal{S}_1^{**} = \mathcal{S}_1^{**1}\left(C_{2,0}\right) + \mathcal{S}_1^{**2}\left(\frac{C_{n,0}}{C_{2,0}}\right)_{n>2} \tag{32}$$

where $\mathcal{S}_1^{**1} = -\left(\frac{\partial \mathcal{H}_1^*}{\partial G''}\right)^{-1} \int \left(\mathcal{H}_2^{*1}\right)_{g''} dg''$, $\mathcal{S}_1^{**2} = -\left(\frac{\partial \mathcal{H}_1^*}{\partial G''}\right)^{-1} \int \left(\mathcal{H}_2^{*2}\right)_{g''} dg''$.

The higher order Hamiltonians and generating functions can be derived in same way using next steps of the Hori-Lie algorithm.

4. PERTURBATIONS

The secular perturbations

By integration of the Eqs.(20) we have:

$$\begin{aligned} &L'' = const. = L_0'', && G'' = const = G_0'', && H'' = const = H_0'', \\ &\ell'' = -\frac{\partial \mathcal{H}^{**}}{\partial L''}(t - t_0) + \ell_0'', && g'' = -\frac{\partial \mathcal{H}^{**}}{\partial G''}(t - t_0) + g_0'', && h'' = -\frac{\partial \mathcal{H}^{**}}{\partial H''}(t - t_0) + h_0''. \end{aligned} \tag{33}$$

where $L_0'' = L''(t_0)$, $G_0'' = G''(t_0)$, $H_0'' = H''(t_0)$, $\ell_0'' = \ell''(t_0)$, $g_0'' = g''(t_0)$, $h_0'' = h''(t_0)$.

The periodic perturbarions

If α is an arbitrary function of canonical variables (e.g. an orbital element or a component of the radius vector) then short- and long-period perturbations are calculated using appropriate generating function.

The long-period perturbations

$$\alpha' - \alpha'' = \Delta\alpha'' = \left\{\alpha'', \mathcal{S}^{**}\right\} + \frac{1}{2}\left\{\left\{\alpha'', \mathcal{S}^{**}\right\}, \mathcal{S}^{**}\right\} + \ldots = {}^{I}\Delta\alpha'' + {}^{II}\Delta\alpha'' + \ldots \,. \tag{34}$$

The first order:

$${}^{I}\Delta\alpha'' = {}^{I}\Delta\alpha''^{1}\left(C_{2,0}\right) + {}^{I}\Delta\alpha''^{2}\left(\frac{C_{n,0}}{C_{2,0}}\right)_{n>2}, \tag{35}$$

where ${}^{I}\Delta\alpha''^{1} = \left\{\alpha'', \mathcal{S}_1^{**1}\right\}$, ${}^{I}\Delta\alpha''^{2} = \left\{\alpha'', \mathcal{S}_1^{**2}\right\}$.

The second order:

$$ {}^{II}\Delta\alpha'' = {}^{II}\Delta\alpha''^{1}\left(\left(C_{2,0}\right)^{2}\right) + {}^{II}\Delta\alpha''^{2}\left(C_{n,0}\right)_{n>2} + {}^{II}\Delta\alpha''^{3}\left(\frac{C_{n,0}\times C_{n',0}}{\left(C_{2,0}\right)^{2}}\right)_{n,n'>2}, \tag{36} $$

where $$ {}^{II}\Delta\alpha''^{1} = \left\{\alpha'', S_2^{**1}\right\} + \frac{1}{2}\left\{\left\{\alpha'', S_1^{**1}\right\}, S_1^{**1}\right\}, $$

$$ {}^{II}\Delta\alpha''^{2} = \left\{\alpha'', S_2^{**2}\right\} + \frac{1}{2}\left\{\left\{\alpha'', S_1^{**1}\right\}, S_1^{**2}\right\} + \frac{1}{2}\left\{\left\{\alpha'', S_2^{**2}\right\}, S_1^{**1}\right\}, $$

$$ {}^{II}\Delta\alpha''^{3} = \left\{\alpha'', S_2^{**3}\right\} + \frac{1}{2}\left\{\left\{\alpha'', S_2^{**2}\right\}, S_2^{**2}\right\}. $$

The short period perturbations

$$ \alpha - \alpha' = \Delta\alpha' = \left\{\alpha', S^{*}\right\} + \frac{1}{2}\left\{\left\{\alpha', S^{*}\right\}, S^{*}\right\} + \ldots = {}^{I}\Delta\alpha' + {}^{II}\Delta\alpha' + \ldots \tag{37} $$

The first order:

$$ {}^{I}\Delta\alpha' = {}^{I}\Delta\alpha'\left(C_{2,0}\right) = \left\{\alpha', S_1^{*}\right\}. \tag{38} $$

The second order:

$$ {}^{II}\Delta\alpha' = {}^{II}\Delta\alpha'^{1}\left(\left(C_{2,0}\right)^{2}\right) + {}^{II}\Delta\alpha'^{2}\left(C_{n,0}\right)_{n>2} + {}^{II}\Delta\alpha'^{3}\left(C_{n,m}, S_{n,m}\right)_{m>0}, \tag{39} $$

where

$$ {}^{II}\Delta\alpha'^{1} = \left\{\alpha', S_2^{*1}\right\} + \frac{1}{2}\left\{\left\{\alpha', S_1^{*}\right\}, S_1^{*}\right\}, $$

$$ {}^{II}\Delta\alpha'^{2} = \left\{\alpha', S_2^{*2}\right\}, \qquad {}^{II}\Delta\alpha'^{3} = \left\{\alpha', S_2^{*3}\right\}. $$

The third order (only perturbations due to products $C_{2,0} \times \left(C_{l,m}, S_{l,m}\right)$):

$$ {}^{III}\Delta\alpha'^{3} = {}^{III}\Delta\alpha'^{3}\left(C_{2,0} \times \left(C_{n,m}, S_{n,m}\right)_{m>0}\right) = \left\{\alpha', S_3^{*3}\right\} + \frac{1}{2}\left\{\left\{\alpha', S_1^{*}\right\}, S_2^{*3}\right\} + \frac{1}{2}\left\{\left\{\alpha', S_2^{*3}\right\}, S_1^{*}\right\}. \tag{40} $$

Substituting appropriate expressions for Hamiltonians H_0, H_1 and H_2 we derive explicit expressions for successive parts of Hamiltonians, generating functions and finally for perturbations. For example the general form of the first order short-period perturbations ${}^{I}\Delta\alpha'$ is:

$$ {}^{I}\Delta\alpha' = \sum_{k=-2}^{2}{}^{*}\sum_{q=-Q}^{Q} A_{kq}^{I\alpha}\begin{bmatrix}\cos\\ \sin\end{bmatrix}\psi_{0,k,q}, \tag{41} $$

where the amplitudes $A^{I}_{\alpha_{k,q}}$ are functions of the generalized lumped coefficients $\overline{C}_0^{k,q}$ and their derivatives. For the second order short-period perturbations due to tesseral harmonics we have the following formula:

$$ {}^{II}\Delta\alpha'^{3} = \sum_{m=1}^{N} \sum_{k=-N}^{N} \sum_{q=-Q}^{Q} \left(A_{mkq}^{II\alpha} \cos \psi_{mkq} + B_{mkq}^{II\alpha} \sin \psi_{mkq} \right), \qquad (42) $$

where the amplitudes $A_{mkq}^{II\alpha}$ and $B_{mkq}^{II\alpha}$ are functions of generalized lumped coefficients $\overline{C}_m^{kq}$ and $\overline{S}_m^{kq}$ and their derivatives. The third order short-period perturbations due to products of the second zonal and tesseral harmonics are given by the following formulas:

$$ {}^{III}\Delta\alpha'^{3} = \sum_{m=1}^{N} \sum_{k_1=-N}^{N} \sum_{q_1=-Q_1}^{Q_1} \sum_{k_2=-2}^{2} \sum_{q_2=-Q_2}^{Q_2} \left(A_{mk_1q_1k_2q_2}^{III\alpha} \cos \psi_{mk_1q_1k_2q_2} + B_{mk_1q_1k_2q_2}^{III\alpha} \sin \psi_{mk_1q_1k_2q_2} \right), \qquad (43) $$

where amplitudes $A_{mk_1q_1k_2q_2}^{III\alpha}$ and $B_{mk_1q_1k_2q_2}^{III\alpha}$ are functions of the products of $\overline{C}_0^{k_2q_2} \times \overline{C}_m^{k_1q_1}$ and $\overline{C}_0^{k_2q_2} \times \overline{S}_m^{k_1q_1}$ and derivatives of these products over L', G' and H'. The summation in Eq. (44) can be changed into summation which groups terms with the same frequency:

$$ {}^{III}\Delta\alpha'^{3} = \sum_{m=1}^{N} \sum_{k=k_{min}}^{k_{max}} \sum_{q=q_{min}}^{q_{max}} \left(A_{mkq}^{III\alpha} \cos \psi_{mkq} + B_{mkq}^{III\alpha} \sin \psi_{mkq} \right), \qquad (44) $$

where $k = k_1 + k_2$, $q = q_1 + q_2$ and

$$ A_{mkq}^{III\alpha} = \sum_{k_1=-N}^{N} \sum_{q_1=-Q_1}^{Q_1} A_{m,k_1,q_1,(k-k_1),(q-q_1)}, \quad B_{mkq}^{III\alpha} = \sum_{k_1=-N}^{N} \sum_{q_1=-Q_1}^{Q_1} B_{m,k_1,q_1,(k-k_1),(q-q_1)}. $$

Similar formulas we can obtain for other types of perturbations.

5. CONCLUSIONS

Formulas for perturbations (e.g. Eqs. (41), (42), (43)), which are the result of the application of the Hori-Lie algorithm and the gravitational potential expressed through generalized lumped coefficients, include the influence of all planet's gravitational potential coefficients. Since the eccentricity function and the inclination function are included into theory in their general forms, these formulas can be used for an arbitrary eccentricity and an arbitrary inclination except the critical inclination. Perturbations can be described in different sets of orbital elements (e.g. in the Keplerian elements or in non-singular elements) or in quantities that are functions of orbital elements, (e.g. in components of the radius vector). Since particular terms in formulas for perturbations represent unique frequency terms, the spectral decomposition of the perturbations is very simple. The theory, which here is described up to the second order and partly to the third order only can be extended to the higher orders.

Formulas (41) - (44) were applied in calculations of perturbations in orbits of different Earth's artificial satellites and for orbits of Mars' satellites. The examples of the perturbation calculations for very low Earth's artificial satellite using 360 degree and order are given by Wnuk and Jopek (1993).

ACKNOWLEDGEMENTS

This work has been partly supported by Polish Committee for Scientific Research under the grant No 2 1239 91 01.

REFERENCES

Aksnes K., 1970, "A second-order artificial satellite theory based on an intermediate orbit" *Astron. Journ.*, vol. 75, pp. 1066-1076.

Brouwer, D., 1959, "Solution of the problem of artificial satellite theory without drag", *Astron. Journ.*, vol.64, p.378-397.

Coffey S. and A. Deprit, 1982, " Third-order solution to the main problem in satellite theory", *Journal of Guidance, Control, and Dynamics*, Vol. 5, pp. 366-371.

Deprit A., and A. Rom, 1970, "The main problem of artificial satellite theory for small and moderate eccentricities", *Celest. Mech.*, vol. 2, pp. 166-206.

Emelyanov N.V., 1992, "The acting analytical theory of artificial earth satellites", *Astronomical and Astrophysical Transactions*, vol. 1, pp. 119-127.

Kaula, W.M., 1966, "Theory of Satellite Geodesy", Blaisdell Publ.Co., Walthman, Mass., 1966.

Kinoshita H., 1977, "Third order solution of an artificial satellite theory", *SAO Special Report*, No. 379.

Kozai, Y., 1959, "The motion of a close Earth satellite", *Astron. Journ.*, vol. 64, pp. 367-377.

Kozai, Y., 1962, "Second-order solution of artificial satellite theory without air drag", *Astron. Journ.*, vol.67, pp. 446-461.

Metris G., P. Exertier, Y.Boudon, and F.Barlier, 1993, " Long period variations of the mo tion of a satellite due to non-resonant tesseral harmonics of a gravity potential", *Celest. Mech.* Vol. 57, pp.175-188.

Wnuk E., 1988, "Tesseral harmonic perturbations for high order and degree harmonics", *Celest. Mech.*, vol. 44, pp. 179-191.

Wnuk,E., Jopek,T., 1993, "Satellite orbit calculations using geopotential coefficients up to high degree and order", *Advances in Space Research* (in press).

PERTURBATION EXPANSIONS AROUND ELLIPTIC FIXED POINTS IN THE SPIN–ORBIT PROBLEM

Alessandra Celletti

Dipt. di Matematica Pura e Applicata
Universitá dell'Aquila
Coppito (L'Aquila)-Italy
celletti@40282.decnet.cern.ch

INTRODUCTION

The stability of nearly–integrable systems can be studied using perturbation theory techniques. One of the most powerful results in this field is provided by the Kolmogorov–Arnold–Moser theory[1,5,6]. Under very general assumptions, KAM theory ensures the persistence under perturbation of an invariant surface (with preassigned rotation number). In this paper we want to develop perturbation techniques in order to establish the stability of planet–satellite systems trapped in a so–called "spin–orbit resonance".

In order to introduce the problem in a more general framework, we consider the following model–problem, described by the Hamiltonian function (in action–angle variables)

$$H(y,x,t) \;=\; h(y)+\varepsilon\, f(x,t) \;=\; \frac{y^2}{2}+\varepsilon\, f(x,t), \qquad y\in \mathbf{R},\;\; (x,t)\in(\mathbf{R}/2\pi\mathbf{Z})^2\equiv T^2\ , \tag{1}$$

with one degree of freedom and an explicit time dependence. We assume that the *perturbing function* $f(x,t)$ is 2π–periodic in the angle variables x, t. Here ε is a small parameter which measures the strength of the perturbation. In the integrable case (i.e., setting $\varepsilon=0$), Hamilton's equations

$$\begin{cases} \dot y = 0 \\ \dot x = \frac{dh(y)}{dy} = y \end{cases}$$

are trivially integrated as

$$\begin{cases} y = y_0 \\ x = x_0 + y_0 t\ , \end{cases}$$

where we shall refer to the quantity $\omega \equiv \frac{dh(y_0)}{dy} = \omega(y_0)$ as the *rotation number*. The

From Newton to Chaos, Edited by A.E. Roy
and B.A. Steves, Plenum Press, New York, 1995

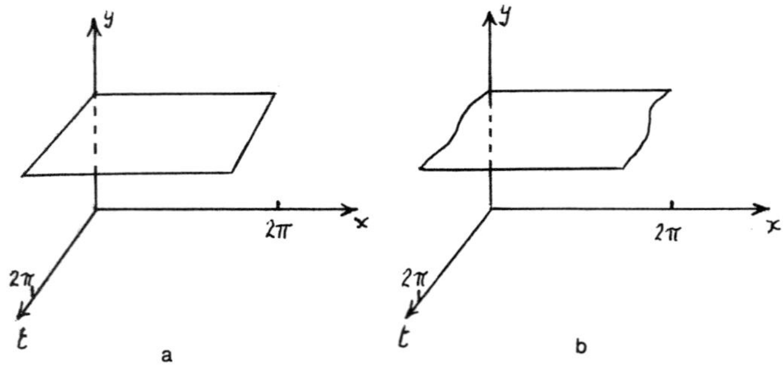

Figure 1. a) An invariant plane in the unperturbed case $\varepsilon = 0$; b) schematic representation of the invariant surface in the perturbed situation, i.e. $\varepsilon \neq 0$.

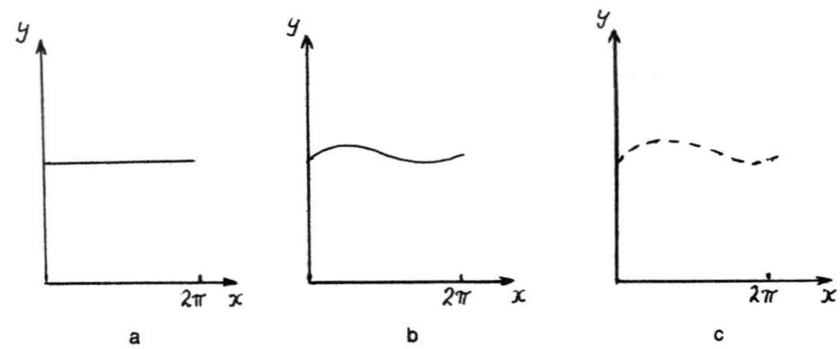

Figure 2. Poincaré map of Hamilton's equations associated to equation 1: a) $\varepsilon = 0$; b) $\varepsilon < \varepsilon_c(\omega)$; c) $\varepsilon = \varepsilon_c(\omega)$.

invariant surfaces of the unperturbed system, i.e. $\mathcal{T}_0(\omega) \equiv \{y_0\} \times T^2$, are described by the planes $y = y_0$ on which periodic or quasi–periodic motions take place.

When the perturbing parameter ε is different from zero, KAM theorem states (under suitable assumptions) the existence of an invariant surface with rotation number ω, say $\mathcal{T}_\varepsilon(\omega)$, for the perturbed system. This surface will be more deformed and displaced as ε grows. When ε reaches a critical value, say $\varepsilon_c = \varepsilon_c(\omega)$, the invariant surface $\mathcal{T}_\varepsilon(\omega)$ breaks–down and gives way to a so–called *Mather set* (which is a closed invariant set, graph of a Cantor set).

For Hamiltonian systems of the type described by equation 1, one has an important stability property, since the corresponding phase space

$$\mathcal{S} = \{(y, x, t)/\ y \in \mathbf{R},\ \ (x, t) \in T^2\}$$

has dimension 3, while the invariant surfaces $\mathcal{T}_\varepsilon(\omega)$ have dimension 2. Therefore if there exists an invariant torus $\mathcal{T}_\varepsilon(\omega)$, it will divide $\mathcal{S}$ in two separate regions with the property that any motion starting in one of these regions will remain forever in it. When $\varepsilon = \varepsilon_c(\omega)$, then $\mathcal{T}_\varepsilon(\omega)$ breaks down and the orbits can diffuse through the gaps of the Cantor set.

We remark that the confinement property holds for systems with at most $l \leq 2$ degrees of freedom. In fact, the phase space has dimension $2l$, the constant energy hyperspace has dimension $2l - 1$, while the invariant surfaces have dimension l and they do not separate the phase space when $l > 2$.

In order to determine the critical break–down threshold $\varepsilon_c(\omega)$, one can apply numerical or rigorous methods. Among the numerical techniques the most reliable is provided by Greene's method, which is based on the conjecture that the break–down of an invariant torus is related to a sudden change from stability to instability of those periodic orbits whose frequency is given by the rational approximants to the irrational rotation number ω.

Concerning the rigorous techniques, the KAM theory provides an explicit constructive algorithm to give an estimate on the perturbing parameter, say $\varepsilon_r = \varepsilon_r(\omega)$, which allows one to state the existence of $\mathcal{T}_\varepsilon(\omega)$ for any $\varepsilon \leq \varepsilon_r(\omega)$. The KAM theorem can be applied provided the following conditions are satisfied:

i) the unperturbed Hamiltonian $h(y)$ is not–degenerate, i.e.

$$\frac{d^2 h(y)}{dy^2} \neq 0 \qquad \forall y \in \mathbf{R}$$

(notice that, for equation 1, $\frac{d^2 h(y)}{dy^2} \equiv 1$);

ii) the rotation number ω satisfies a strong non–resonance condition, i.e. the diophantine condition

$$|\omega - \frac{p}{q}|^{-1} \leq Cq^2 \qquad \forall p, q \in \mathbf{Z},\ q \neq 0\ ,$$

for some positive constant C.

We intend to apply the KAM theorem to a physical model in Celestial Mechanics. Our aim is to obtain the persistence of some invariant surfaces for the realistic (i.e., the astronomical) values of the parameters. However, the application of standard versions of the KAM theorem usually provides results which are not useful in concrete applications. Significative improvements can be obtained refining the estimates involved in the proof of the theorem or using a direct parametrization of the invariant surface, together with a computer–assisted approach.

SPIN–ORBIT RESONANCES

We consider a two–body system P_1, P_2 and denote by T_{rev} and T_{rot} the period of revolution of P_2 around P_1 and the period of rotation of P_2 around its spin–axis.

Definition: One has a spin–orbit resonance of the *first kind*, if there exist $p, q \in \mathbf{Z}$ ($(p, q) \neq (0, 0)$), such that

$$qT_{rev} - pT_{rot} = 0\ ,$$

namely

$$\frac{T_{rev}}{T_{rot}} = \frac{p}{q}\ .$$

In this case one speaks of a spin–orbit resonance of type $p : q$.

The Earth–Moon system is the most familiar example of a 1:1 spin–orbit resonance, since the periods of rotation and revolution of the Moon coincide, in such a way that the Moon always points the same face to the Earth. As it is well known, all the tidally evolved planets or satellites of the solar system are trapped in a 1:1 resonance. The only exception to this rule is provided by the Mercury–Sun system, since the ratio between the period of revolution of Mercury around the Sun and the period of rotation about its spin–axis amounts to $\frac{3}{2}(1 \pm 10^{-4})$.

Another kind of resonance involves the coupling between the spin of a planet with the orbit of another planet. More precisely, let P_1 and P_2 be two bodies revolving around a central body S with revolutional periods T^i_{rev}, $i = 1, 2$ (and frequencies $\omega^i_{rev} = \frac{2\pi}{T^i_{rev}}$) and rotational periods T^i_{rot} (and frequencies $\omega^i_{rot} = \frac{2\pi}{T^i_{rot}}$).

Definition: A spin–orbit resonance of the *second kind* is a commensurability among the rotational and revolutional frequencies of a planet with the orbital frequency of another planet, i.e.

$$p\omega^1_{rev} + q\omega^1_{rot} + r\omega^2_{rev} = 0 , \qquad p, q, r \in \mathbf{Z}/\{0\} .$$

The Sun–Venus–Earth system is nearly trapped in a spin–orbit resonance of the second kind, since, up to a high degree of approximation, the following relation holds

$$\omega^V_{rot} = 4\omega^V_{rev} - 5\omega^E_{rev} .$$

This condition implies that anytime Earth and Venus are in an inferior conjunction with the Sun, Venus shows the same face to the Earth. In the following chapters we will introduce the main equations describing the above types of spin–orbit resonances.

SPIN–ORBIT RESONANCES OF THE FIRST KIND

We introduce a mathematical model describing an approximation of the spin–orbit problem. We consider a triaxial ellipsoidal satellite with principal moments of inertia $A < B < C$. We assume that

1. the satellite moves in a Keplerian orbit around a primary body;
2. the spin–axis coincides with the shortest physical axis (i.e., the axis whose moment of inertia is largest);
3. the spin–axis is perpendicular to the orbit plane;
4. dissipative forces as well as perturbations due to other bodies are neglected.

Under these assumptions the equation of motion can be derived from Euler's equations for a rigid body. Normalizing the mean motion to one, one has

$$\ddot{x} + \frac{3}{2}\frac{B-A}{C}\left(\frac{a}{r}\right)^3 \sin(2x - 2f) = 0 , \tag{2}$$

where, referring to figure 3, x is the angle between the longest axis of the ellipsoid and the periapsis line, r is the instantaneous orbital radius and f is the true anomaly.

Notice that a rotation of the satellite by 180° (i.e., $x' = x + \pi$) gives an equivalent configuration and that the above equation is trivially integrable in the case of equatorial symmetry (i.e., $A = B$) as well as for circular orbits.

Since the orbit of the satellite is Keplerian, both r and f are known 2π–periodic functions of the time. Therefore one can expand equation 2 as

$$\ddot{x} + \varepsilon \sum_{m \neq 0, m=-\infty}^{\infty} W\left(\frac{m}{2}, e\right) \sin(2x - mt) = 0 , \tag{3}$$

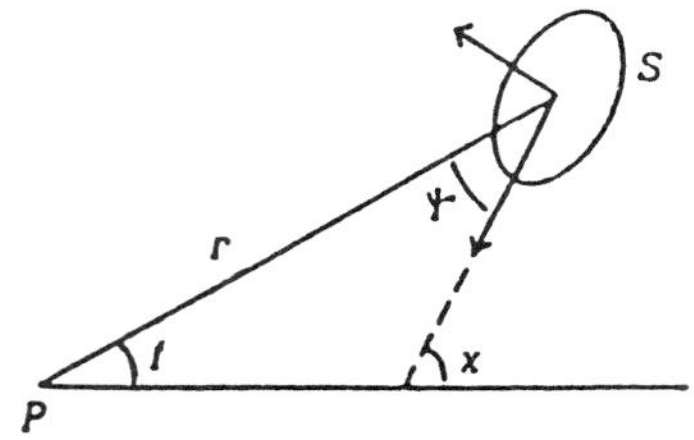

Figure 3. The spin–orbit geometry.

where we introduced the parameter $\varepsilon = \frac{3}{2}\frac{B-A}{C}$, which is proportional to the so–called *ellipticity* $\frac{B-A}{C}$ and where the coefficients $W(\frac{m}{2}, e)$ decay as powers of the eccentricity e. For example, the first few coefficients are given by

$$\begin{aligned} W(\tfrac{1}{2}, e) &= -\tfrac{e}{2} + \tfrac{e^3}{16} - \tfrac{5}{384}e^5 - \tfrac{143}{18432}e^7 + O(e^9) \\ W(1, e) &= 1 - \tfrac{5}{2}e^2 + \tfrac{13}{16}e^4 - \tfrac{35}{288}e^6 + O(e^8) \\ W(\tfrac{3}{2}, e) &= \tfrac{7}{2}e - \tfrac{123}{16}e^3 + \tfrac{489}{128}e^5 - \tfrac{1763}{2048}e^7 + O(e^9) \\ W(2, e) &= \tfrac{17}{2}e^2 - \tfrac{115}{6}e^4 + \tfrac{601}{48}e^6 + O(e^8) \ . \end{aligned}$$

To simplify further the model, we truncate equation 3, taking into account that according to assumption 4) we have neglected all the dissipative forces among which the *tidal torque*, due to the internal non–rigidity of the satellite, provides the strongest contribution. However this force is small compared to the gravitational term and we decide to neglect in the series expansion of equation 3 those terms whose size is comparable to the average effect of the tidal torque. Therefore we obtain an equation of the form

$$\ddot{x} + \varepsilon \sum_{m \neq 0, m=N_1}^{N_2} W(\frac{m}{2}, e) \sin(2x - mt) = 0 \ , \tag{4}$$

where N_1 and N_2 are some integers, depending on the physical and orbital elements of the satellite.

We remark that the Poincaré map associated to equation 4 has a pendulum–like structure, in which the periodic orbit is surrounded by *librational* curves. A chaotic separatrix divides the librational regime from the region in which *rotational* invariant curves take place.

Due to its conservative character, equation 4 can be derived from Hamilton's equations associated to the Hamiltonian function

$$H(y, x, t) = \frac{y^2}{2} - \frac{\varepsilon}{2} \sum_{m \neq 0, m=N_1}^{N_2} W(\frac{m}{2}, e) \cos(2x - mt) \ , \tag{5}$$

where $y \in \mathbf{R}$, $(x, t) \in T^2$. Notice that the above nearly–integrable Hamiltonian is of the same type as equation 1. Due to the confinement property illustrated in §1, in

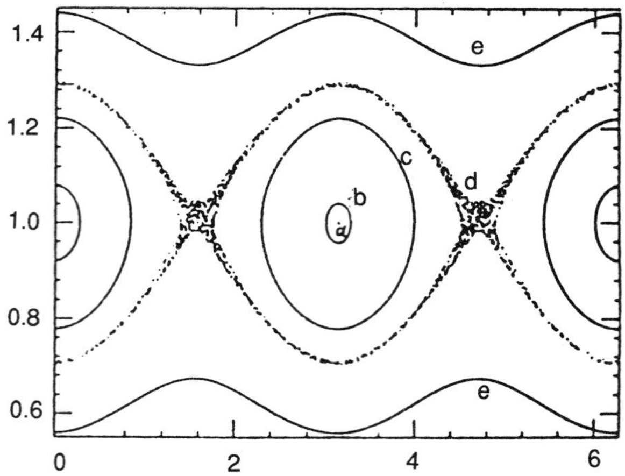

Figure 4. The Poincaré map around the synchronous periodic orbit associated to equation 4, with $N_1 = 1$ and $N_2 = 5$.

order to obtain the confinement (and therefore the stability) of a periodic orbit with frequency $\frac{p}{q}$, one can try to establish the existence of two invariant surfaces $\mathcal{T}_\varepsilon(\omega_1)$, $\mathcal{T}_\varepsilon(\omega_2)$ with $\omega_1 < \frac{p}{q} < \omega_2$, which trap the periodic orbit from above and below. A more powerful confinement in phase space can be obtained by establishing the existence of a librational invariant curve, surrounding the resonance.

The first statement has been investigated by the author[2]; here the existence of trapping rotational surfaces has been proved in the cases of the Moon, Mercury and some satellites of Saturn. By KAM theory one can prove the persistence under perturbation of such invariant surfaces for the realistic value of the perturbing parameter. We refer the reader to [2] for further details.

On the other hand, the existence of librational trapping surfaces can be proved by applying a Birkhoff–KAM scheme along the following lines (see [3] for technical details). We start by performing some symplectic changes of variables, in order to put the Hamiltonian equation 5 to a form which is suitable for the description of librational motions.

To this end we rewrite the Hamiltonian equation 5 putting into evidence the term corresponding to the synchronous resonance, i.e.

$$H(y,x,t) \;=\; \frac{y^2}{2} \;-\; \varepsilon a\;\cos(2x-2t) \;-\; \frac{\varepsilon}{2}\sum_{m\neq 0,2} W(\frac{m}{2},e)\;\cos(2x-mt)\;,$$

where $a \equiv \frac{1}{2}W(1,e)$. We next center the resonance on the point (0,0) by means of the coordinate change $x' = 2x - 2t$, $y' = \frac{1}{2}y - 1$. Next we expand in Taylor series around the origin and diagonalize the time independent quadratic terms by means of the symplectic transformation

$$\begin{cases} p = \alpha\, y' \\ q = \beta\, x'\;, \end{cases}$$

with $\alpha = \frac{\sqrt{2}}{(\varepsilon a)^{1/4}}$, $\beta = \frac{(\varepsilon a)^{1/4}}{\sqrt{2}}$, so that the new Hamiltonian is given by

$$\begin{aligned} H(p,q,t) \;&=\; \tfrac{\omega}{2}(p^2+q^2) \;-\varepsilon a\;(\tfrac{q^4}{4!\beta^4} - \tfrac{q^6}{6!\beta^6} + \ldots) \\ &\quad -\tfrac{\mu}{2}\; \textstyle\sum_{m\neq 0,-2} \tilde{W}(\tfrac{m+2}{2},e)\;[\cos(mt)\;(1 - \tfrac{q^2}{2\beta^2} + \tfrac{q^4}{4!\beta^4} + \ldots) \\ &\quad +\; \sin(mt)\;(\tfrac{q}{\beta} - \tfrac{q^3}{3!\beta^3} + \tfrac{q^5}{5!\beta^5} + \ldots)]\;, \end{aligned}$$

where $\omega = 2\sqrt{\varepsilon a}$ is the frequency of the harmonic oscillation, while the coefficients W have been rescaled as $\tilde{W}(\frac{m+2}{2}, e) = \frac{1}{e} W(\frac{m+2}{2}, e)$. Finally we transform to action–angle variables (I, ϕ) by means of the change of variables

$$\begin{cases} p = \sqrt{2I}\ \cos\phi \\ q = \sqrt{2I}\ \sin\phi\ . \end{cases}$$

Then the final Hamiltonian is given by

$$\begin{aligned} H\ \ (I,\phi,t)\ &=\ \omega I\ -\varepsilon a\ (\tfrac{I^2}{16\beta^4} - \tfrac{5I^3}{2\cdot 6!\beta^6} + \ldots) \\ &-\varepsilon a\ [-\tfrac{I^2}{12\beta^4}\cos 2\phi + \tfrac{I^2}{48\beta^4}\cos 4\phi + \tfrac{I^3}{4\cdot 6!\beta^6}\cdot(15\cos 2\phi - 6\cos 4\phi + \cos 6\phi) + \ldots] \\ &-\tfrac{\varepsilon e}{2}\ \textstyle\sum_{m\neq 0,-2}\tilde{W}(\tfrac{m+2}{2}, e)\ \{\cos(mt)\ [1 - \tfrac{I}{2!\beta^2}(1-\cos 2\phi) + \tfrac{I^2}{8\cdot 3!\beta^4}\cdot \\ &\cdot(3 - 4\cos 2\phi + \cos 4\phi) - \tfrac{I^3}{4\cdot 6!\beta^6}\ (10 - 15\cos 2\phi + 6\cos 4\phi - \cos 6\phi) + \ldots] \\ &+\ \sin(mt)\ [\tfrac{\sqrt{2I}}{\beta}\sin\phi - \tfrac{\sqrt{2}\ I^{3/2}}{12\beta^3}(3\sin\phi - \sin 3\phi) \\ &+\tfrac{\sqrt{2}I^{5/2}}{4\cdot 5!\beta^5}\ (10\sin\phi - 5\sin 3\phi + \sin 5\phi) + \ldots]\}\ . \end{aligned}$$

We can rewrite the above Hamiltonian in a compact form as

$$H(I,\phi,t)\ =\ \omega I + \varepsilon\bar{h}(I) + \varepsilon h(I,\phi) + \varepsilon e f(I,\phi,t)\ .$$

We apply a Birkhoff normal form so to reduce the size of the perturbation $g(I,\phi,t) \equiv h(I,\phi) + ef(I,\phi,t)$. This can be attained by using a close–to–identity transformation with generating function $I'\phi + \Phi(I',\phi,t)$, where one can expand the unknown function $\Phi(I',\phi,t)$ as $\Phi(I',\phi,t) = \sum_{j=1}^{k} \Phi_j(I',\phi,t)$, for some integer k. With this normalization, the transformed Hamiltonian takes the form

$$H_k(I',\phi',t)\ =\ h_k(I';\varepsilon) + \varepsilon^{k+1} R_k(I',\phi',t;\varepsilon)\ ,$$

where the functions h_k and R_k can be explicitly determined. At this point one can apply the KAM theorem in order to establish the existence of a librational torus with preassigned rotation number. In particular we used[3] the refined estimates developed in [4]. We remark that due to the lengthy computations needed to perform the normal form transformation and to check the KAM estimates, the proof of the existence of a librational torus is computer–assisted. However in order to control the numerical errors we implemented the so–called *interval arithmetic* technique.

The results concern two examples (i.e., the Moon–Earth and the Rhea–Saturn systems), both observed in a synchronous resonance.

Moon–Earth: We consider the system described by the Hamiltonian equation 5 with $N_1 = -1$, $N_2 = 5$ and $e = 0.0549$. Let $\varepsilon_{obs} = 3.45 \cdot 10^{-4}$ (i.e., the physical value of the ellipticity of the Moon); then there exists an invariant torus corresponding to a libration of $8^o.79$ for any $\varepsilon \leq \varepsilon_{obs}/5.26$.

Rhea–Saturn: Consider the Hamiltonian function equation 5 with $N_1 = 1$, $N_2 = 5$ and $e = 0.00098$. Let $\varepsilon_{obs} = 3.45 \cdot 10^{-3}$ (i.e., the observed value); then there exists an invariant torus corresponding to a libration of $1^o.95$ for any $\varepsilon \leq \varepsilon_{obs}$.

SPIN–ORBIT RESONANCES OF THE SECOND KIND

The only known example of a spin–orbit resonance of the second kind is provided by the Venus–Earth–Sun system. In fact, up to a good degree of approximation, the following relation holds:

$$\omega^V_{rot}\ =\ 4\omega^V_{rev} - 5\omega^E_{rev}\ .$$

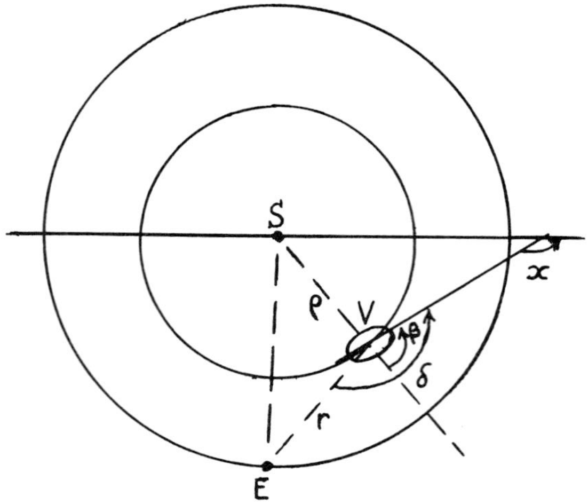

Figure 5. The geometry of spin–orbit resonances of the second kind.

This equality implies that Venus always points the same face to the Earth at any inferior conjunction.

We can assume, like in the study of spin–orbit resonances of the first kind, that the spin–axis of Venus is perpendicular to the orbit plane. However we cannot assume that the orbit of Venus is Keplerian, since the gravitational interactions due to the Earth on the orbital elements of Venus are of the same order of magnitude of the force exerted by the Earth on the figure of Venus. Assuming in a first approximation, that the orbits of Venus and the Earth are Keplerian circles around the Sun, the equation of motion is given by

$$\ddot{x} + \frac{3}{2}\frac{B-A}{C}\frac{GM_E}{r^3}\sin(2\delta) + \frac{3}{2}\frac{B-A}{C}\frac{GM_S}{\rho^3}\sin(2\beta) \;=\; 0\ ,$$

where G is the gravitational constant, M_E and M_S denote the masses of Earth and Sun respectively, r and ρ are the distances of Venus from the Earth and the Sun, while δ and β are depicted in figure 5.

In order to obtain a more realistic model we let Venus and the Earth move on Keplerian ellipses and then we include the perturbation of the Earth on the orbital parameters of Venus by adding the integration of Hamilton's equations of the corresponding restricted three body problem.

The results of a numerical integration show that the resonance in which this system is nearly trapped, is stable as far as $\frac{B-A}{C} < 0.005$, which is compatible with the actual value of the ellipticity of Venus.

REFERENCES

1. V.I. Arnold, *Proof of a Theorem by A.N. Kolmogorov on the invariance of quasi-periodic motions under small perturbations of the Hamiltonian, Russ. Math. Surveys **18**, 9 (1963)*

2. A. Celletti, *Analysis of resonances in the spin–orbit problem in Celestial Mechanics: The synchronous resonance (Part I), J. Appl. Math. and Phys. (ZAMP) **41**, 174 (1990)*

3. A. Celletti, *Construction of librational invariant tori in the spin–orbit problem, to appear in J. of Appl. Math. and Phys. (ZAMP)*

4. A. Celletti, L. Chierchia, *Rigorous estimates for a computer-assisted KAM theory, J. Math. Phys.* ***28****, 2078 (1987)*

5. A.N. Kolmogorov, *On the conservation of conditionally periodic motions under small perturbation of the Hamiltonian, Dokl. Akad. Nauk. SSR* ***98****, 469 (1954)*

6. J. Moser, *On invariant curves of area-preserving mappings of an annulus, Nach. Akad. Wiss. Göttingen, Math. Phys. Kl. II* ***1****, 1 (1962)*

MOMENT FORMALISM FOR THE RADIATIVE FORCE EVALUATION

David Vokrouhlický

Observatoire de la Côte d'Azur, Dept. CERGA
Av N Copernic, F-06130 Grasse, France [1]

ABSTRACT: We overview a theoretical scheme enabling evaluation of the radiative force acting on different celestial bodies (artificial satellites, interplanetary dust grains) based on the moment formalism introduced in the theory of radiative transfer. This approach allows an elegant and powerful algorithm for the radiative force investigation in complex situations (e.g. the Earth shadow penumbra transitions). It accounts for the influence of general radiative field generated by an extended source with arbitrary surface emissivity. A number of tutorial computations of the simplified configurations already met in the literature are presented.

Recently, we have applied the moment formalism to the shadow penumbra theory in context of the artificial satellite dynamics. Here we follow several ideas generalizing previous model in different routes: towards the (i) first order terms in presumably small parameter v/c (v stands for the object velocity, while c for the velocity of light) generally known as the Poynting-Robertson terms, (ii) arbitrary distances from the planet.

1. INTRODUCTION AND MOTIVATION

Celestial mechanics is generally thought to deal primarily with the dynamics of an isolated system of gravitationally interacting bodies. This scenario was predetermined by Newton's discovery of dynamics in one line with gravitational law. Even now, celestial mechanics considered from this viewpoint remains a crucial problem of any theory of gravity (Damour et al., 1991). Nevertheless, the idea of the non-gravitational forces influence appeared progressively in context of the celestial bodies motion. Cometary motion and secular acceleration of the Moon are archetypes of those phenomena. More recently, dynamics of interplanetary dust particles, planetary rings as well as the motion of artificial satellites involve non-gravitational forces as an important part.

Among all of them, radiative forces – those consisting in momentum exchange of a radiative field and considered body – play a special role. They are often the most prominent (e.g. direct radiative pressure in context of motion of solar system dust particles, middle and high orbit artificial satellites) or they give rise tiny secular phenomena (e.g. the Poynting–Robertson effect for the dust particles, thermal thrust in case of artificial satellites).

In this paper, we would like to avoid making an overview of different situations where radiative force enters as an important ingredient. Instead of that, we want to concentrate on exposing an approach well suited for evaluation of the radiative force even in case of complex configurations. We think that such tool has been missing in celestial mechanics up to now. Most of the previous contributions involving radiative forces were based on simple treatment of the radiative fields while concentrating on the dynamical aspects

[1] On leave from Astronomical Institute, Charles University Prague, Švédská 8, CZ-15000 Prague 5, Czech Republic.

From Newton to Chaos, Edited by A.E. Roy
and B.A. Steves, Plenum Press, New York, 1995

(with only a few exceptions, e.g. Guess, 1962). As a result a detailed theory of the planetary shadow penumbra effects (for example) has not been elaborated up to recently.

In Sec. 2 we overview our approach to the radiative force evaluation. In the last section we briefly overview the main results connected with the penumbra theory for the artificial satellites achieved by our complex approach to the radiative force evaluation (Vokrouhlický et al., 1993b, 1994a) and mention our plans to expand them for the dynamics of the solar system dust grains (Vokrouhlický et al., 1994b).

2. RADIATIVE FORCE – METHOD OF EVALUATION

In what follows we shall generally speak about a particle, having in mind arbitrary body influenced by the radiative force (i.e. dust grain, artificial satellite). However, our approach being macroscopic excludes systems which would need quantum mechanical treatment (e.g. molecules, atoms).

As the considered problem involves photons it is most appropriate to use relativistic approach. On the other hand it is well known that up to the first order in a small parameter v/c, where v is the particle velocity and c light velocity, the problem is essentially 'classic' (Burns et al., 1979). However from classical or semiclassical arguments lot of misinterpretation has followed (see e.g. Lyttleton, 1976 and references therein). As a result we choose to solve the problem in rather formal and strictly relativistic manner. Interestingly, it is very similar to that used by Robertson (1937) apart of generalizations mentioned later on.

The total radiative force essentially results from global momentum budget of the ingoing and outgoing radiation in the particle's local frame. One usually distinguish absorption, characterized by the fact that all radiation removed from the incident beam – absorbed – is reemited isotropically in the local frame, and scattering, characterized by instantaneous anisotropic emission of the removed radiation. We shall follow this scheme mainly from pedagogical standpoint. Moreover, one can meet more complicated situations when true absorption is followed with anisotropic reemission in the local frame (e.g. Lyttleton, 1976). The other example are Yarkovsky type effects (Burns et al., 1979; Rubincam, 1988).

Our formalism, allowing to treat the situations with arbitrary incident radiative field, is based on moment decomposition of the specific intensity (e.g. Mihalas, 1970; Thorne, 1981). This tool is widely used in the theory of radiative transfer. Here, we shall expose only such material which is necessary for understanding our approach.

In the following series of subsections we shall gradually develop a general scheme for the force evaluation.

I. Reference frames introduced. We start with definition of the reference frames to be used later on. First, we introduce global coordinate system x^α ($\alpha = 1, 2, 3$) which serves for global description of the particle motion (e.g. heliocentric, geocentric). We construct a reference frames induced by the coordinate system x^α in a usual manner – each of the frame base vectors is tangent to one of the coordinates x^α – at each point of the space. We shall denote such frames generally by $\mathcal{S}$.

It should be noted that this introduction of the basic reference frames is the most elementary. One often meets more general situation, when the frames $\mathcal{S}$ are transported in the space by preferred class of 'observers' (e.g. expanding or collapsing; see Thorne, 1981), but for the purposes of applications in celestial mechanics we shall restrict to 'static' frames $\mathcal{S}$ introduced earlier.

Together with frames $\mathcal{S}$ we define a system of frames $\mathcal{S}'$ each of them being instantaneously comoving with the particle. It should be understood that we do not have in mind absolutely comoving frame with the particle, as this one will not be inertial, while $\mathcal{S}'$ is supposed to be inertial (at least locally when general relativistic point of view is applied). One concludes from the preceding that the transformation between $\mathcal{S}$ and $\mathcal{S}'$ frames is simply given by a Lorentz boost with instantaneous coordinate velocity $\vec{v}$ of the particle. In what follows we shall systematically denote physical quantities measured in the comoving frames $\mathcal{S}'$ with prime.

II. Radiative field characterized. Essential quantity characterizing properties of the radiative field is specific intensity (or just intensity) $I_\nu(\vec{n})$. To be exact it represents a scalar field in the phase space of photon states if polarization is disregarded (Mihalas, 1970; Misner et al., 1973). Translated into more common language $I_\nu(\vec{n}) \times (\vec{n}.\vec{n}_s)$ gives radiative energy passing (i) through unit surface area with normal $\vec{n}_s$ (ii) per unit of time (iii) in unit frequency interval centered around ν (iv) into unit solid angle with the axe given by $\vec{n}$. Specific intensity depends on (i) position x^α of the local frame $\mathcal{S}$ origin, (ii) local direction n^α and (iii) frequency ν, so that extend notation will read $I(x^\alpha, n^\alpha, \nu)$. [2] In agreement with widely used notation we shall suppress position vector x^α and write the ν parameter as an index.

[2] Note, that arguments (n^α, ν) can be replaced with photon momentum p^α as frequency ν is directly related to the total momentum of considered photon.

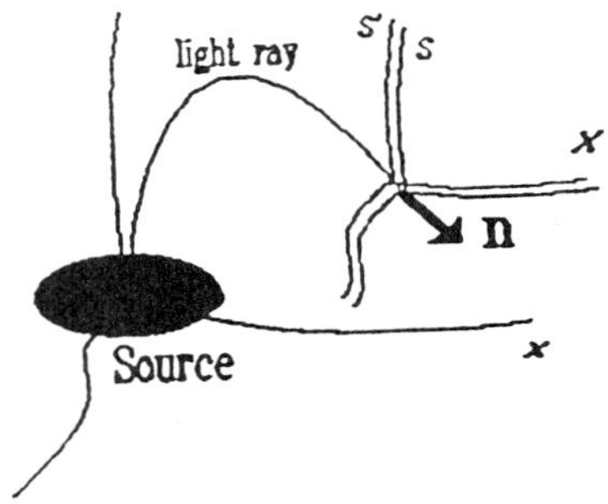

Figure 1. Geometrical objects introduced in the text: global coordinate system, induced local (static) frame $\mathcal{S}$, local (instantaneously) comoving frame $\mathcal{S}'$, light ray emitted by the radiative source characterized by local directional vector $\mathbf{n}$.

Before we start effectively the computation of the radiation force on a particle, we have first to go through two preliminary steps, not addressed in full detail here. They can be expressed by the following problems:

- *How the radiative field is constructed provided one knows emissivity of the radiative sources?* In order to solve it one introduces the concept of *light rays*, which form important guidelines along which intensity is transported. Light rays correspond to the photon trajectories in considered medium as far as methods of geometrical optics are physically accessible. In celestial mechanics one often disregards refractive properties of the background medium and considers light rays as straight lines. Sometimes however, one faces to the problem of solving nontrivial light rays (e.g. the case of the planetary shadow penumbra effect, Vokrouhlický et al., 1993b, 1994a). One usually employs Fermat principle (e.g. Landau and Lifschitz, 1962) or sometimes solves directly the dynamical equations of the photon motion (e.g. Abramowicz et al., 1991; Vokrouhlický and Karas, 1991). It is worth noting that solving for the light rays is essentially 'initial problem' and *not* 'boundary problem'. A direction $\mathbf{n}$ in the local system $\mathcal{S}$ uniquely determines a light ray. On contrary, there may exist several light rays connecting the same element of the radiative source and origin of the chosen local frame $\mathcal{S}$. Such effects result in interesting focusing effects due to the planetary atmospheres leading to the radiative flux increase (to be discussed in detail in Vokrouhlický et al., 1994b).

- *How is the specific intensity transported along the light rays?* This is the central problem of the radiative transfer theory (e.g. Chandrasekhar, 1950; Mihalas, 1970). We shall not present here a complete formulation of corresponding equations. It is sufficient to know that there exist an unambiguous 'shifting law' which determines the value of the specific intensity transported along given light ray provided its boundary value on the radiative source is given (in most solar system cases this means in the solar photosphere or on the planetary surface). In celestial mechanics, one usually disregards extinction by the medium in between the radiative source and considered particle resulting in conservation of the specific intensity along straight light rays. The other often met example is simple exponential damping of the specific intensity, which correspond to the formal solution of the radiative transfer theory in case of absorbing media with constant absorption coefficient (see Mihalas, 1970; Vokrouhlický et al., 1993b).

III. Multipole moment formalism. Due to general theorems in analysis one can develop the angular dependence of the specific intensity into multipole series. Corresponding coefficients – multipole moments – then contain full information on the radiative field. Moreover, low order moments have straightforward physical meaning (e.g. Mihalas, 1970). This leads us to use extensively moment formalism as basic mathematical tool.

Moments of the radiative intensity are defined according to the following scheme (see Thorne, 1981)

$$M_{\alpha_1\alpha_2\ldots\alpha_k} = \int_{(\theta,\phi)} d(\cos\theta)d\phi \; n_{\alpha_1} n_{\alpha_2} \ldots n_{\alpha_k} \, I(\theta,\phi) \; , \tag{1}$$

where α_i are indexes spanning the values $(1,2,3)$ referring to the axes of the reference frame[3], and $n_{\alpha_1}(\theta,\phi)$ are the components of the unit vector in the direction (θ,ϕ) [i.e., we have $\vec{n}^T(\theta,\phi) = (\sin\theta\cos\phi, \sin\theta\sin\phi, \cos\theta)$]. It is supposed that spatial events in the frame $\mathcal{S}$ are referred to spherical coordinates (θ,ϕ) introduced in usual manner. Note, that we define moments of the frequency integrated intensity $I = \int d\nu\, I_\nu$ and not of the quantity I_ν itself (compare with Thorne, 1981). Moreover, it can be shown (Thorne, 1981), that full information about the radiative field is contained already in symmetric and trace-free part of tensors (1). We shall extensively use only moments up to order two ($k = 2$). At this level the general result mentioned is equivalent to trivial observation that $M_{\alpha\beta} = M^{\rm STF}_{\alpha\beta} + \frac{1}{3}M\delta_{\alpha\beta}$, where $M^{\rm STF}_{\alpha\beta}$ symmetric trace-free part of tensor $M_{\alpha\beta}$. Practically it means, that instead the set $(M, M_\alpha, M_{\alpha\beta})$ necessary for evaluation of the radiative force [see Eqs. (12), (21) below], one needs only the reduced set $(M, M_\alpha, M^{\rm STF}_{\alpha\beta})$ having one component less.

IV. Radiative force evaluation. The essential scheme to evaluate the radiative force acting on the dust particle consists in deriving this force in the instantaneously comoving frame $\mathcal{S}'$, which is the frame where unambiguous results are available and more meaningful (e.g. Mie scattering theory in case of the microscopic dust grains, specular reflection or Lambert diffusion in case of artificial satellites). Then, we have to transform the radiative force to the static local frames $\mathcal{S}$, which serve as 'a skeleton' for the description of global motion of considered particle.

We found tutorial to treat separately cases of ideally absorbing particle on one hand and purely scattering particle on the other hand. However, it should be noted that the scheme is the same and we shall finally put both results together to express the radiative force acting on partially absorbing and partially scattering particle.

V. Ideally absorbing particle. We start by focusing on the description of the situation on the local comoving frame $\mathcal{S}'$. Here, we shall assume that the particle is spherical resulting in isotropy of its optical properties and we shall denote the effective cross-section for removal of the incident energy due to the true absorption by $C_{\rm ab} = AQ_{\rm ab}$, where $Q_{\rm ab}$ is the efficiency factor and A geometrical cross-section of the particle. Both quantities $C_{\rm ab}$ and $Q_{\rm ab}$ could generally show frequency dependence. In case of microscopic systems such as interplanetary dust grains the efficiency factors are known from the Mie theory (e.g. van de Hulst, 1957; Wickramasinghe, 1973), while in case of macroscopic bodies such as artificial satellites one usually accepts phenomenological approach built on presumably unknown absorption coefficients (e.g. Milani et al., 1987, Klinkrad et al., 1990). We shall also assume that the accumulated thermal energy is reradiated isotropically in this system, so that final momentum budget is only due to the absorbed incident radiation. Resulting formula for the radiative force reads

$$f'^{(\rm R)}_\alpha = \frac{A}{c}\int d\omega' \int d\nu'\, Q_{\rm ab}(\nu')\, I_{\nu'}(\vec{n}')\, n'_\alpha \; , \tag{2}$$

where the two integrals are performed over: (i) all spatial directions of the rays constituting the local radiative field ($d\omega'$ stands for element of the solid angle), (ii) all frequencies.

One should understand clearly at this point that although the theory provides the value of $Q_{\rm ab}(\nu')$ we do not know a priori the radiative field $I_{\nu'}(\vec{n}')$, as this is 'distorted' by aberration and the Doppler effect when comparing to the radiative field $I_\nu(\vec{n})$ in the static frame $\mathcal{S}$ constructed by the procedure of the previous sections. In order to transform parameters of integration in (2) for 'the unprimed' one employs relations (see Thomas, 1930; Mihalas, 1970)

$$I_{\nu'}(\vec{n}')\, d\omega'\, d\nu' = \gamma^2 \left(1 - \frac{n_\beta v_\beta}{c}\right)^2 I_\nu(\vec{n})\, d\omega\, d\nu \; , \tag{3}$$

$$n'_\alpha = \gamma^{-1}\left(1 - \frac{n_\beta v_\beta}{c}\right)^{-1}\left[n_\alpha + (\gamma - 1)\frac{n_\beta v_\beta}{v^2} v_\alpha - \gamma\frac{v_\alpha}{c}\right] \; , \tag{4}$$

where $\gamma = \left[1 - (v/c)^2\right]^{-1/2}$ (summation of quantities with common index is assumed implicitly). Truncating previous formulas at the first order of the (presumably) small parameter (v/c) and employing Doppler shift law $\nu' \to \nu\left(1 - \frac{n_\beta v_\beta}{c}\right) + \mathcal{O}(2)$ [here $\mathcal{O}(n) = \mathcal{O}(c^{-n})$] we arrive at the following formula for the force (2) on a spherical particle

$$f'^{(\rm R)}_\alpha = A\frac{Q_{\rm ab}}{c}\left[M_\alpha - M\frac{v_\alpha}{c} - M_{\alpha\beta}\frac{v_\beta}{c}\right] - A\frac{Q^{(1)}_{\rm ab}}{c} M_{\alpha\beta}\frac{v_\beta}{c} + \mathcal{O}(2) \; , \tag{5}$$

[3] We shall generally use greek letters for spatial indexes, but keeping letter ν for frequency and γ for the obvious relativistic factor (see bellow).

where we introduced 'frequency averaged' factors

$$Q_{\rm ab} \equiv \frac{\int d\nu\, Q_{\rm ab}(\nu) I_\nu}{\int d\nu\, I_\nu} \,, \tag{6}$$

$$Q^{(1)}_{\rm ab} \equiv \frac{\int d\nu\, \nu\, (\partial Q_{\rm ab}(\nu)/\partial \nu)\, I_\nu}{\int d\nu\, I_\nu} \,. \tag{7}$$

It should be understood that in successive steps from formula (2) to (5) we just changed parameters of integration in (2); a priori 'unknown' quantities $I_{\nu'}(\mathbf{n}')$ and n'_α has been changed to construct radiative field linked to the quantities $I_\nu(\mathbf{n})$ and n_α.

In the last step we have to transform the force components (5) from the comoving frame $\mathcal{S}'$ to the static one $\mathcal{S}$. We learn from the classical relativistic dynamics (e.g. Landau and Lifschitz, 1962) that the following simple rule holds

$$f^{\rm (R)}_\alpha = f'^{\rm (R)}_\alpha + \mathcal{O}(2) \,. \tag{8}$$

We thus conclude that Eq. (5) gives already the final expression of the radiative force acting on spherical particle up to specified order.

It may be interesting to discuss briefly features of the right hand sides of formula (5). At first, we note that truncation to the zeroth order in quoted parameter (v/c) leads to

$$f^{\rm (R)}_\alpha = A \frac{Q_{\rm ab}}{c} M_\alpha + \mathcal{O}(1) \,. \tag{9}$$

Formula (9) represents the radiative pressure exerted on a spherical particle as is well known from the literature (see e.g. Vokrouhlický et al., 1993a,b). We have used it extensively for the description of the shadow penumbra effect in context of the artificial satellite motion.

The last term in the right hand side of Eq. (5) is usually omitted in the theory as the frequency dependence of the absorption efficiency factor $Q_{\rm ab}(\nu)$ is either not great or shows oscillations around the mean value, which effectively average out its contribution. Second approximation to be reminded here is incident the plane wave with direction $\vec{\mathbf{n}}^{(0)}$. Under such conditions we may write $I_\nu(\vec{\mathbf{n}}) = \Phi_\nu \delta(\vec{\mathbf{n}} - \vec{\mathbf{n}}^{(0)})$ for the incident radiative field with δ standing for Dirac's delta-function. Realizing that

$$M_{\alpha_1 \alpha_2 \ldots \alpha_k} = \Phi\, n^{(0)}_{\alpha_1} n^{(0)}_{\alpha_2} \ldots n^{(0)}_{\alpha_k} \,, \tag{10}$$

for all $k = 0, 1, \ldots$ and $\Phi = \int d\nu\, \Phi_\nu$, we get the final form of radiative force [the second term on right hand side of (5) omitted]

$$f^{\rm (R)}_\alpha = A \frac{Q_{\rm ab}}{c} \Phi \left[n^{(0)}_\alpha \left(1 - \frac{n^{(0)}_\beta v_\beta}{c} \right) - \frac{v_\alpha}{c} \right] + \mathcal{O}(2) \,, \tag{11}$$

well known for the expression of the Poynting–Robertson effect (e.g. Robertson, 1937; Burns et al., 1979).

VI. Scattering particle. In this section we turn to the case of purely scattering particles. The treatment is essentially similar to that in previous section, but here the net momentum budget of the outgoing scattered radiation is generally nonzero and should be accounted for. The effective cross-section for removal of the incident radiation by scattering will be denoted by $C_{\rm sc} = A Q_{\rm sc}$, where $Q_{\rm sc}$ is the corresponding efficiency factor and A is geometrical cross-section of the particle. Next, we denote the scattering phase function by $S_\nu(\Theta)$ which gives the probability distribution of the scattering in the direction deflected by angle Θ from the incident plane wave mode (e.g. van de Hulst, 1980).

As for the momentum budget of the removed radiation from the incident beam we can apply the same procedure as in previous section devoted to the absorbing particle provided factors $Q_{\rm ab}$ and $Q^{(1)}_{\rm ab}$ are replaced with integral scattering efficiency factors

$$Q_{\rm sc} \equiv \frac{\int d\nu\, Q_{\rm sc}(\nu) I_\nu}{\int d\nu\, I_\nu} \,, \tag{12}$$

$$Q^{(1)}_{\rm sc} \equiv \frac{\int d\nu\, \nu\, (\partial Q_{\rm sc}(\nu)/\partial \nu)\, I_\nu}{\int d\nu\, I_\nu} \,. \tag{13}$$

Let us now turn to consider the scattered radiative field. As noted previously, each of the "infinitesimal plane modes" $I_{\nu'}(\vec{\mathbf{n}}')$ with given $\vec{\mathbf{n}}'$, which constitute the incident radiative field, give rise to new radiative field $I^{\rm sc}_{\nu'}(\vec{\mathbf{n}}'') \equiv S_{\nu'}(\Theta) I_{\nu'}(\vec{\mathbf{n}}')$, with $(\vec{\mathbf{n}}''.\vec{\mathbf{n}}' = \cos\Theta)$, which is clearly axially symmetric around the incident direction $\vec{n}'$. Notice, that the scattering redistribution in effectively only spatial (Θ) and not in frequencies (ν' is the same on both sides of previous equation), which is well satisfied for scattering on dust grains, but might become erroneous on molecular level (see discussion in Mihalas, 1970). Next step consists in integrating the total scattering field over all incident infinitesimal plane modes (one implicitly uses the assumption of spherical shape of the particle) and substituting it into expression similar to Eq. (2). Then changing integration symbols one arrives at the following expression for the radiative force due to the scattered radiative field

$$f_\alpha'^{(\mathrm{R})} = \frac{A}{c}\int d\omega' \int d\nu'\, Q_{\mathrm{sc}}(\nu')\, g(\nu')\, I_{\nu'}(\vec{n}')\, n'_\alpha \; . \tag{14}$$

where we define the phase parameter $g(\nu)$ by

$$g(\nu) = \int_{-1}^{1} d\,(\cos\Theta)\cos\Theta\; S_\nu\,(\Theta) \; . \tag{15}$$

Note the sign minus (coming from derivative of cosine in (15)) which expresses the fact that scattered photons take away momentum from the system.

In the last step the algebra is the same as in the case of the absorbing particle to transform the force components (14) into the static frame $\mathcal{S}$. Final expression reads

$$f_\alpha^{(\mathrm{R})} = A\frac{Q_{\mathrm{sc}}}{c} g\left[M_\alpha - M\frac{v_\alpha}{c} - M_{\alpha\beta}\frac{v_\beta}{c}\right] - A\frac{Q_{\mathrm{sc}}^{(1)}}{c} g^{(1)} M_{\alpha\beta}\frac{v_\beta}{c} + \mathcal{O}(2) \; , \tag{16}$$

where we introduce frequency integrated versions of the phase parameter

$$g\;(= <\cos\theta>) \equiv \frac{\int d\nu\; Q_{\mathrm{sc}}(\nu) g(\nu) I_\nu}{\int d\nu\; Q_{\mathrm{sc}}(\nu) I_\nu} \; , \tag{17}$$

$$g^{(1)} \equiv \frac{\int d\nu\;\nu\,[(\partial Q_{\mathrm{sc}}(\nu)/\partial\nu)\, g(\nu) + (\partial g(\nu)/\partial\nu)\, Q_{\mathrm{sc}}(\nu)]\, I_\nu}{\int d\nu\; (\partial Q_{\mathrm{sc}}(\nu)/\partial\nu)\, I_\nu} \; . \tag{18}$$

Putting together results (8) and (15) we finally obtain expression for the total radiative force acting on the particle partially absorbing (with efficiency Q_{ab}) and partially scattering (with efficiency Q_{sc} and phase function $S_\nu(\Theta)$) the incident radiation accurate to the first order in small parameter (v/c) in form

$$f_\alpha^{(\mathrm{R})} = A\frac{Q_{\mathrm{ab}} + Q_{\mathrm{sc}}\,(1+g)}{c}\left[M_\alpha - M\frac{v_\alpha}{c} - M_{\alpha\beta}\frac{v_\beta}{c}\right] - A\frac{Q_{\mathrm{ab}}^{(1)} + Q_{\mathrm{sc}}^{(1)}\,(1+g^{(1)})}{c} M_{\alpha\beta}\frac{v_\beta}{c} + \mathcal{O}(2) \; . \tag{19}$$

VII. 'Cooking book' for the radiative force evaluation. We thus suggest the following algorithm for the evaluation of the radiative force on celestial bodies sufficiently general even for complex situations:

- Choice of the global coordinate system, which (i) serves for the description of the global motion of the particle under consideration, (ii) induced local frames $\mathcal{S}$ (e.g. heliocentric cartesian system);
- choice of the radiative sources together with description of their emissivity (e.g. solar photosphere models);
- construction of the radiative field in each of the local systems $\mathcal{S}$ using the concept of the light rays (Fermat principle) and theory of the specific intensity transport along given ray (radiative transfer equations);
- computation of the radiative intensity moments (1) up to order 2 in local systems $\mathcal{S}$;
- computation of the radiative force using one of the formulas (5), (9), (16), (19) or (20) (depending on the considered problem).

3. SOME PREVIOUS RESULTS

In this section, we shall mention a few applications of the previous theory in celestial mechanics. Let us remark that according to our opinion moment formalism allows to treat these complex problems with maximal possible clarity.

Formula (19) is valid for arbitrary extended source of radiation and partially absorbing and scattering particle.

It is worth noting that up to zeroth order in the parameter (v/c) we again get simple expression

$$f_\alpha^{(\mathrm{R})} = A \frac{Q_{\mathrm{ab}} + Q_{\mathrm{sc}}\,(1+g)}{c} M_\alpha + \mathcal{O}(1) \,, \tag{20}$$

for the usual radiation pressure.

Let us demonstrate, its validity on the known case of polished sphere, reflecting all the radiation specularly, met in the theory of the artificial satellites. In that case $Q_{\mathrm{ab}} = 0$ and $Q_{\mathrm{sc}} = 1$. Moreover, one can easily show that $S_\nu(\Theta) \approx 1$ (see e.g. Landau and Lifschitz, 1960) and thus $g = 0$. As a result $f_\alpha^{(\mathrm{R})} = \frac{A}{c} M_\alpha$ which corresponds to earlier results (e.g. Vokrouhlický et al., 1993a).

In Eq. (19) the bracket in the first term and part $M_{\alpha\beta}\frac{v_\beta}{c}$ in the second term depend on the local radiative field (i.e. expressed in $\mathcal{S}$ frame) and particle's velocity with respect to this frame. Hence they are independent on the optical properties of the particle itself. On contrary, terms involving efficiency parameters "Q" depend on the frequency distribution of the incident radiative field and the optical features of the particle. It can be also said, that they represent "inertiality parameters" characterizing resistance of the particle to the dynamical changes under the influence of the radiative field similarly to the inertial mass in the theory of gravitation. We would like to stress, that they might not be perfectly constant due to their dependence on the frequency spectrum of the incident radiation. One generally takes into account solar spectrum out of the atmosphere to evaluate integral efficiency factors of absorption Q_{ab} and scattering Q_{sc} (e.g. Lamy, 1974; Burns et al., 1979). This is perfectly justified in most part of the particle trajectories around planets. But it may not be exactly met in the period of planetary shadow penumbra penetration, when the solar spectrum is significantly altered due to the passage through the planetary atmosphere.

Recently, we have elaborated detailed theory of the Earth shadow penumbra theory aimed for the artificial satellite dynamics based on previous formalism (Vokrouhlický et al., 1993b). Importance of such theory is given by the fact that solar radiation pressure is the dominant non-gravitational effect perturbing satellite orbits at altitudes higher than about 400 km. Uncertainties of modeling of this effect crucially affect the orbit predictions (e.g. Fliegel et al., 1992; Ries et al., 1992).

It is interesting to note that eclipsing of the solar disk by the Earth body plays only minor role on forming the penumbra decrease of the radiative perturbations. As a result previous studies based mainly on this 'geometrical' approach could not reach high precision. The atmospheric effects – refraction and extinction – dominantly influence the penumbra effect. We have identified several tiny effects of interesting physical origin which constitute decrease of the radiative flux when passing through the shadow penumbra. Let us note for example refractive attenuation of the solar incident flux of pure geometrical origin, which was also discussed as early as in the beginning of the century in context of the theory of lunar eclipses (e.g. Link, 1933).

Probably the most interesting result obtained so far concerns long periodic perturbations of the LAGEOS semimajor axis. We shall not remind here extensive literature on this appealing puzzle of the satellite orbital motion (for introduction to the problem see e.g. Smith and Dunn, 1980; Barlier et al., 1986). We have found that additional perturbations due to exact treatment of the shadow penumbra phenomena could account under plausible assumptions on the north/south asymmetry of the meteorological conditions for about 10 – 20 % of the observed residuals (see Vokrouhlický et al., 1994a). This is rather unexpected result regarding to the fact that the penumbra transitions cover maximally 5 % of the orbital period (about 3.75 hours).

We have also investigated the radiative perturbations for the dust grains captured to the Earth vicinity (Vokrouhlický et al., 1994b). Here arbitrary distances from the Earth have to be considered. This leads to much richer set phenomena, such as occurrence of higher order solar images, 'atmospheric flux lensing' etc. The possible dynamical consequences of those effects remain to be established.

Acknowledgements: The author thanks F. Mignard for careful reading of the manuscript and valuable comments. The staff of CERGA (Grasse, France) is acknowledged for its kind hospitality. This work has been supported in part by EC grant 2836.

REFERENCES

Abramowicz, M. A., Ellis, G. F. R., and Lanza, A, 1991, Relativistic effects in superluminal jets and neutron star winds, *Astroph. J.* 361:470.

Barlier, F., Carpino, M., Farinella, P., Mignard, F., Milani, A., and Nobili, A. M, 1986, Non-gravitational perturbations on the semimajor axis of LAGEOS, *Ann. Geophys.* 4:193.

Burns, J. A., Lamy, P. L., and Soter, S, 1979, Radiation forces on small particles in the solar system, *Icarus* 40:1.

Chandrasekhar, S, 1950, "Radiative Transfer", Oxford Univ. Press, Oxford.

Damour, T., Soffel, M., and Xu, Ch, 1991, General-relativistic celestial mechanics. I. Method and definition of reference systems, *Phys. Rev.* D43:3273.

Fliegel, H. F., Gallini, T. E., and Swift, E. R., 1992, Global positioning system radiation force model for geodetic applications, *J. Geoph. Res.* 97:559.

Guess, A, 1962, Poynting-Robertson effect for a spherical source of radiation, *Astroph. J.* 135:855.

Klinkrad, H., Koeck, Ch., and Renard, P, 1990, Precise satellite skin-force modelling by means of Monte-Carlo ray tracing, *ESA Journal* 14:409.

Lamy, P. L, 1974, Interaction of interplanetary dust grains with the solar radiation field, *A&A* 35:197.

Landau, L. D., and Lifschitz E. M, 1960, "Mechanics", Addison-Wesley, Reading, Mass.

Landau, L. D., and Lifschitz E. M, 1962, "The classical theory of fields", Addison-Wesley, Reading, Mass.

Link, F, 1933, Théorie photométrique des éclipses de Lune, *Bull. astron.* 8:77.

Lyttleton, R. A, 1976, Effects of solar radiation on the orbits of small particles, *Astrophys. Space Sci.* 44:119.

Mihalas, D, 1970, "Stellar atmospheres", W.H. Freeman and Co., San Francisco.

Misner, Ch., Thorne, K. S., and Wheeler, J. A, 1973, "Gravitation", W. H. Freeman & Co., San Francisco.

Milani, A., Nobili, A. M., and Farinella, P, 1987, " Non-gravitational perturbations and satellite geodesy", A. Hilger, Bristol.

Ries, J. C., Shum, C. K., and Tapley, B. D, 1992, Surface modeling for precision orbit determination, paper submitted to the IUGG XX General Assembly.

Robertson, H. P, 1937, Dynamical effects of radiation in the solar system, *Mon. Not. R. astr. Soc.* 97:423.

Rubincam, D. P, 1988, Yarkovsky thermal drag on LAGEOS, *J. Geoph. Res.* 93:13805.

Smith, D. E., and Dunn, P. J, 1980, Long term evolution of the LAGEOS orbit, *Geophys. Res. Lett.* 7:437.

Thomas, L. H, 1930, The radiation field in a fluid in motion, *Quart. J. Math. (Oxford)* 1:239.

Thorne, K. S, 1981, Relativistic radiative transfer: moment formalisms, *Mon. Not. R. astr. Soc.* 194:439.

Van de Hulst, H. C, 1957, "Light scattering by small particles", J. Willey & Sons, New York.

Van de Hulst, H. C, 1980, "Multiple Light Scattering (Tables, Formulas, and Applications)", Academic Press, New York.

Vokrouhlický, D., and Karas, V, 1991, General relativistic effects in astrophysical jets, *A&A* 252:835.

Vokrouhlický, D., Farinella, P., and Lucchesi, D, 1993a, Albedo perturbation models: General formalism and applications to LAGEOS, *Celest. Mech.* 57:225.

Vokrouhlický, D., Farinella, P., and Mignard, F, 1993b, Solar radiation pressure perturbations for Earth satellites: I. A complete theory including penumbra transitions, *A&A* 280:295.

Vokrouhlický, D., Farinella, P., and Mignard, F, 1994a, Solar radiation pressure perturbations for Earth satellites: II. An approximative approach to the Earth shadow penumbra theory, *A&A* 285:333.

Vokrouhlický, D., Farinella, P., and Mignard, F, 1994b, Radiative forces on a dust particle: A complex model for the planetary shadow penumbra phase, in preparation.

Wickramasinghe, N. C, 1973, "Light scattering functions for small particles with applications in astronomy", A. Hilger, London.

OPTIMIZATION OF SPACECRAFT TRAJECTORIES USING NONLINEAR PROGRAMMING

Bruce A. Conway

Dept. of Aeronautical & Astronautical Engineering
306 Talbot Laboratory
University of Illinois
Urbana, IL 61801

INTRODUCTION

This paper describes a recently developed method for the determination of optimal, e.g. minimum-fuel or minimum-time, trajectories for spacecraft. The method will be introduced after a brief discussion of methods previously used by researchers for trajectory optimization. However, the objective here is not to discuss all such methods or provide a comprehensive description of them but rather to provide sufficient information regarding extant methods that what is novel and beneficial about the solution method we have developed may be recognized. Several example optimal trajectories for spacecraft using low-thrust propulsion will be presented.

The subject has a surprisingly long history considering that it is only in the last 30 years that there has been any need for such trajectories. In 1925 Hohmann[1] described the minimum fuel impulsive thrust transfer between coplanar circular orbits which now bears his name. The general space trajectory optimization problem is easily formulated but, except in special circumstances, such as when the Hohmann transfer is minimizing, difficult to solve. The equations of motion may be written as, at a minimum, a system of six first-order, coupled, nonlinear differential equations. Three of the six system variables represent position and three represent velocity. When the variation of spacecraft mass with propellant expulsion is considered another variable is added to the system.

METHOD

The fundamental approach to solving the problem is to apply the calculus of variations. The types of boundary conditions which are usually applied to the problem, e.g. fixed final time, fixed final position, fixed final velocity and combinations thereof, cause the necessary conditions for an optimal trajectory to be the Euler-Lagrange equations[2]. In this formulation of the optimization problem the system differential equations are adjoined to the system Hamiltonian through the "adjoint variables" (also known as the costate variables or Lagrange multipliers) as will be described later in this paper. Among the Euler-Lagrange necessary conditions are then differential equations governing the history of these adjoint variables and (terminal) boundary conditions on these adjoint variables. The optimization problem is thus a two-point-boundary-value problem (TPBVP). Such TPBVP's are generally very difficult to solve. Solution methods are broadly categorized as being of two types, "indirect" and "direct"[3].

From Newton to Chaos, Edited by A.E. Roy
and B.A. Steves, Plenum Press, New York, 1995

The earliest and most obvious of the indirect methods is to guess the unspecified boundary conditions on the state and adjoint variables (at either end) and numerically integrate the Euler-Lagrange equations (to the other end). The approximate initial conditions are then iteratively adjusted to improve the satisfaction of the terminal conditions. In the context of the optimal control problem these are called "extremal field" methods because if the integration is done backwards (in time), with the terminal constraints of the Euler-Lagrange equations satisfied by the "guessed" values of states and Lagrange multipliers, then the integrated trajectory satisfies the necessary conditions for an optimal trajectory for the resulting initial conditions, which will not in general be the desired initial state conditions. A common difficulty with these so-called "shooting" methods is that if the initial guesses for the missing boundary conditions are not close enough to the unknown correct values the integrated trajectories can be dramatically different from the desired trajectory. The Euler-Lagrange equations are particularly sensitive as TPBVP's go, so that very accurate initial guesses are required. There are examples, for specific problems, of successful algorithms for approximating the unknown terminal values of the state and adjoint variables, e.g. a low-thrust transfer to geosynchronous orbit from LEO found by Redding and Breakwell[4].

A qualitatively different method, which can yet be referred to as "indirect", employs "primer vector" theory. This theory, developed in the 1950's by Lawden[5], enables the optimal control strategy, for some vehicles and some cost functions (fortunately including the most reasonable and practical), to be determined simply in terms of the magnitude and direction of the "primer vector", which is the negative of a vector formed from the three adjoint variables which are conjugate to the three velocity variables. As an example, for the impulsive thrust (instantaneous Δv) case (in an inverse-square gravitational field) minimizing fuel consumption, three of the necessary conditions for an optimal trajectory are that: (i) the primer vector be continuous and have continuous first derivatives during the time of the flight, (ii) the magnitude of the primer vector be less than or equal to one, with impulses applied when the primer vector magnitude equals one, and (iii) that the direction of the thrust at the time(s) of the impulse be the direction of the primer vector. (The latter two necessary conditions arise from applying the Pontryagin minimum principle to the system[2].)

If the primer vector history is then obtained for a candidate control history, e.g. by integrating the Euler-Lagrange equations backward in time using the candidate control history, these necessary conditions on the primer vector allow it to be determined if the trajectory is optimal. The use of the primer vector history along a resulting non-optimal trajectory to improve the trajectory was investigated by Lion and Handelsman[6] and Jezewski and Rozendaal[7]. As an example; if the primer vector magnitude exceeds unity on the candidate trajectory, this is taken as an indication that an additional impulse would reduce the total cost. The precise time of the midcourse impulse, its position, and its magnitude can be determined iteratively, and the process of improving the trajectory continued until all of the necessary conditions are satisfied, using methods described by Lion and Handelsman; a starting approximation which is frequently successful is to apply the midcourse impulse where the primer vector had its largest amplitude.

The resulting optimal trajectories are frequently non-intuitive. As an example, the optimal, impulsive, minimum-fuel transfer between coplanar circular orbits is the Hohmann transfer only while the ratio of the circular orbit radii is less than 11.94 [8]. For larger ratios of the outer to inner orbit radius a three impulse (bi-elliptic) transfer is minimizing; as could easily be determined from the use of the primer vector theory. This is a case where the optimal control employs more impulses than are required to satisfy the boundary conditions. Another example is a minimum-time, low thrust escape from LEO[9]. The problem is equivalent to maximizing mechanical energy in a given time so it is surprising (at first glance) that the optimal thrust direction is not parallel to the velocity vector. In fact, the optimal thrust pointing direction oscillates about the velocity vector and for brief periods has a radially inward component!

So-called "direct" methods of solution for the optimal trajectory ignore the calculus of variations first-order necessary conditions and approximate the original optimal control problem with a mathematical programming problem. There are many such methods[10,11,12]. Perhaps the simplest method, though the least economical of computer time, is to assume some simple form for the time history of the optimal control, e.g. polynomials parametrized in time. The system equations may then be integrated forward in time from known initial conditions. This forward integration is accomplished within a nonlinear programming problem solver, which adjusts the problem parameters (typically the coefficients of the control polynomials and perhaps some initial states) so that the terminal constraints are satisfied while simultaneously minimizing the cost function. Unfortunately the explicit numerical integration of the entire trajectory which must occur at each evaluation of the constraints and cost function can result in sensitivity problems, and may consume large amounts of computer time.

An apparently new direct method was recently introduced by Hargraves and Paris[13], though it is qualitatively similar to earlier so-called "collocation" methods. The state and control histories are represented by discrete values on a mesh. The state history between mesh points (for each state variable) is represented using Hermite cubic polynomials. The control history may be represented linearly or with a cubic polynomial (in time). The equations of motion are enforced at the mesh points, i.e., the slopes of the Hermite cubic polynomials representing the states are in agreement with the system differential equations evaluated at the mesh points. The equations of motion are further enforced using a collocation scheme; a collocation point is chosen at the center of each mesh segment, i.e., the Hermite cubic state polynomial is constrained to satisfy the system differential equations between mesh points as well as at mesh points. This yields nonlinear constraint equations involving the state and control variables. Additional constraints, linear and nonlinear, equality and inequality, may be present involving (typically) initial and terminal values of the states and magnitudes of the controls. This method is illustrated in Figures 1 and 2. It has proven to be efficient and robust in the determination of a variety of optimal space trajectories.

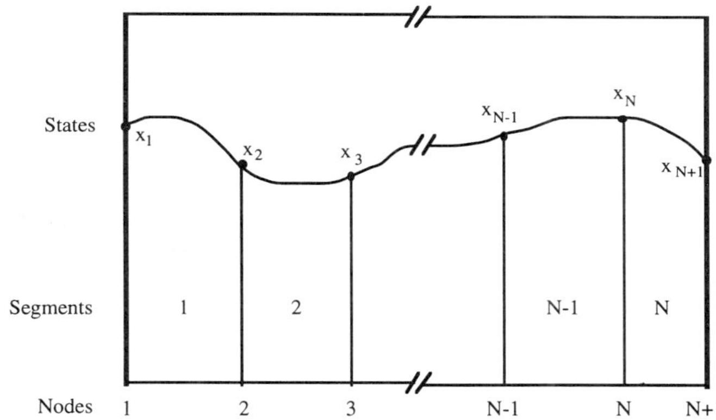

Figure 1. Illustrating the mesh (for one variable) on which the continuous problem is discretized.

IMPROVEMENTS TO THE METHOD

Improvements have been made, by the author and his graduate research assistants, to the collocation method of Hargraves and Paris[13]. This work is documented in papers and theses[3,9,14-20]. The bulk of this research deals with trajectories in which the space vehicle employs a continuous low-thrust propulsion. Flight times are correspondingly large so that for an accurate discretization of the continuous problem a very large mesh is required. Application of the conventional (Hargraves and Paris) discretization yielded a nonlinear programming problem with so many free state and control parameters that it was much too large to be solved with the software tools available.

Enright[3] and Enright and Conway[19,20] describe several techniques which may be used to reduce the size of the NLP problem while retaining the same accuracy for the solution. One technique is helpful in the case where the thrust is either impulsive or continuous (when on) but with periods of no thrust (termed "coast arcs"). For the two-body case, when there is no thrust the equations of motion are integrable, thus it is a waste of resources to discretize the trajectory over the coast arcs since an exact solution for the state variables is known. The details of the method may be found in the references cited; essentially, nonlinear constraints are formed using constants of the motion on the coast arc which must be satisfied by the state and control variables at the respective ends of the arc and the coasting time. A very large number of variables and nonlinear constraints may thus be saved as compared to a conventional discretization.

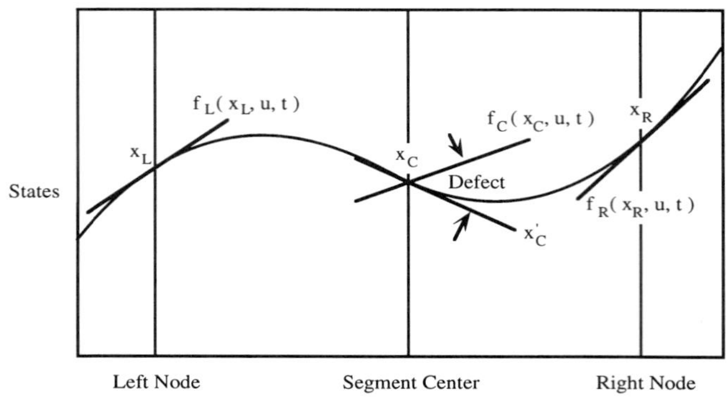

Figure 2. Illustrating the collocation method described by Hargraves and Paris. If the cubic polynomial representing the state within a segment of the mesh satisfies the system differential equation at the center, which has a value f_c whose magnitude is represented by the slope of the line, the slope of the cubic at this point should be the same. The nonlinear programming problem solver adjusts the values of the states at the boundaries and the controls at all three points to accomplish this.

A second technique which may also substantially decrease the size of the NLP problem is to completely replace the implicit integration of the Hargraves and Paris method (which is the Simpson method) with a different integrator, and use fewer nonlinear constraints while retaining the same size mesh for accuracy[3]. This is illustrated in Figure 3.

A third technique developed to economize on the number of NLP problem variables required to achieve a given accuracy is automatic mesh refinement[9,18]. A mesh with equal spacing of nodes (in time), which is conventionally used, is sub-optimal. In this method, qualitatively similar to that used in computational fluid dynamics or in structural mechanics, segments in the nominal mesh are subdivided, if necessary, based on an error test. The process is continued in a loop, with subdivided segments perhaps further subdivided, until the desired level of integration error is achieved. The resulting mesh with nodes unequally spaced in time can require fewer nodes altogether, yielding a smaller NLP problem for a given solution accuracy, than a mesh with equally spaced nodes.

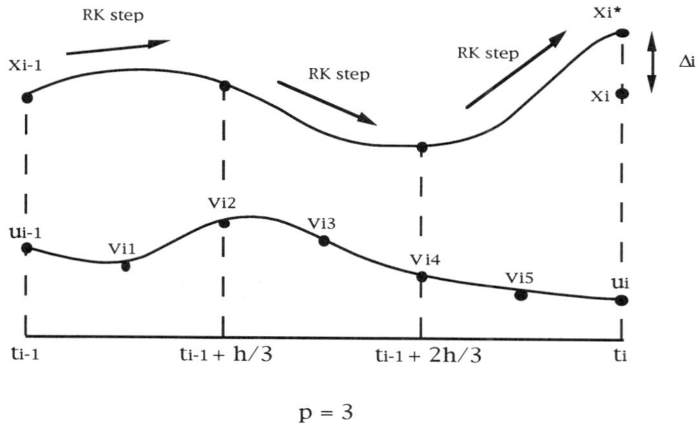

Figure 3. To economize on the number of nonlinear "defect" constraints and discrete state variables, the same mesh size as in a Hargraves and Paris type solution may be used but a different integrator employed. Instead of implicit Simpson rule integration an explicit Runge-Kutta integrator is used. Rather than form a nonlinear constraint in each segment it is done only after advancing forward several (here 3) intervals. The constraint becomes the difference between the current value of the state variable known to the NLP problem solver (at the ith time), x_i, and the result from the forward stepping assuming discrete controls (the u's and v's) which is x_i*.

EXAMPLE OPTIMAL TRAJECTORIES

All three of the techniques mentioned above have the same goal, reducing the number of free parameters in the discretization of the continuous optimal control problem. Since the execution time of the NLP problem solver increases exponentially with the number of free parameters this method can yield substantial savings. In addition, the smaller problems tend to be more robust, so that decreasing the size of the problem with one (or more) of these methods may mean the difference between obtaining a solution and not obtaining one. (The practical upper limit of the size of the NLP problem which may be solved with existing software is approximately 700-900 free variables with an equal number of nonlinear constraint equations.)

As mentioned previously, this upper limit of problem size was being exceeded by a class of problems we attempted, low-thrust orbit transfer (or interplanetary transfer) with necessarily long times of flight. These developments we have made, while interesting in themselves, have been required for the solution of such problems. Two examples will be presented here; others are available in recent literature.

The first example is very-low-thrust orbit raising, from LEO to GEO[14,15]. The initial low Earth circular orbit has an altitude of 270 km. The initial thrust acceleration is 325 micro-g. All motion was assumed to be in the equatorial plane and hence equinoctial variables were used to avoid singularities. Five state variables are required altogether; the semimajor axis, the true longitude, the equinoctial variables P_1 and P_2, and the thrust acceleration, which increases from the initial value as propellant is consumed. There is only one control variable, thrust pointing angle. The performance index is the final time, which is to be minimized. Since the engine operates continuously, minimizing final time is equivalent to minimizing fuel consumed.

The Runge-Kutta (R-K) transcription illustrated in Figure 3 is employed to reduce the size of the problem, which is nonetheless substantial. 60 segments with 6 R-K steps per segment are used, thus there are 61 x 5 = 305 state variables and 60 x 13 = 780 discrete control variables (spaced approx. 50 deg. apart in longitude in the resulting solution) for a total of 1086 free NLP problem parameters, when the unknown final time is added as a single parameter. This is a very large NLP problem by present standards. Without the R-K transcription, e.g. if the Hargraves and Paris transcription illustrated in Figure 2 had been used on the same mesh, 360 segments would have been required, yielding 1805 state variables alone and an insoluble problem!

The optimal trajectory, illustrated in part in Figures 4 and 5, requires 100.6 revolutions of the Earth and 14.75 days. The solution required 18.1 minutes of CPU time on a Cray Y-MP. The form of the optimal control (Fig. 4) and of one of the state variable histories, the eccentricity, shown in Fig. 5, clearly indicate that a discretization with a coarser mesh could not capture correctly the time scale of the solution. There are significant variations in both the control and the eccentricity during just one orbit, so that spacing nodes more widely than the approximately 50 degrees in true longitude used here would not allow the character of the solution to be revealed.

One might question how it is that we can be confident that we have the true optimal control history using the present discretization. The optimality of the solution may be checked in several ways, but the most straightforward is by a backward numerical integration of the necessary conditions, which are the Euler-Lagrange equations. As described in the previous section, the "direct" solution methods, of which this example is representative, discretize the continuous problem and convert the determination of the optimal control into a nonlinear programming problem; Lagrange multipliers do not explicitly appear. The Lagrange multipliers are a part of the Euler-Lagrange equations however. So, how are their time histories numerically integrated when they do not explicitly appear in the solution?

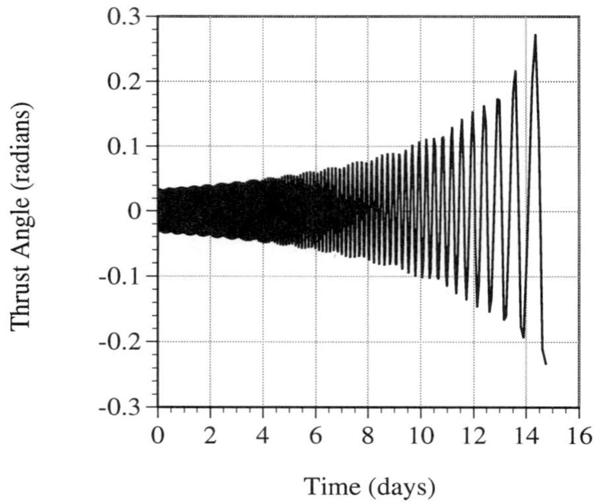

Figure 4. The control history, thrust angle vs. time, for the low-thrust LEO to GEO transfer found using the method of direct collocation with nonlinear programming.

The program which solves our NLP problem, NPSOL[21], does calculate Lagrange multipliers internally and prints them out with the solution. It does this because the Lagrange multipliers have a very useful interpretation as the sensitivity of the cost function to changes in the constraints. Enright[3] showed that these Lagrange multipliers are equivalent to the Lagrange multipliers of the continuous problem. Thus, the final values of the Lagrange multipliers, as well as final values of the state variables, which are the "initial conditions" for the backward integration are available in the solution. If the solution satisfies the necessary conditions for (local) optimality the backward integration should recover all of the known initial conditions for the state variables very accurately. We routinely perform this kind of check on our supposed optimal control histories.

Another example optimal trajectory also uses the Runge-Kutta transcription we have developed. This is a low-thrust interplanetary transfer with given launch date, minimizing flight time and hence fuel required[16,17]. A patched-conic approximation is made, i.e., the spacecraft is assumed to be within the gravitational influence of only one body at a time. To solve the problem, local coordinates, with origin at the center of the attracting body, must be used at all times. For scaling purposes canonical units are also used, i.e., the gravitational parameter of each body is equal to unity in local coordinate units. A coordinate transformation must thus be performed, for example, as the spacecraft leaves the Earth's sphere of influence and enters the Sun's. This is accomplished in a method qualitatively similar to that described in the previous section for the case in which integrable arcs were a feature of the trajectory. Nonlinear equations which govern the transformation from, for example, geocentric position and velocity to heliocentric position and velocity are included in the NLP problem as constraints.

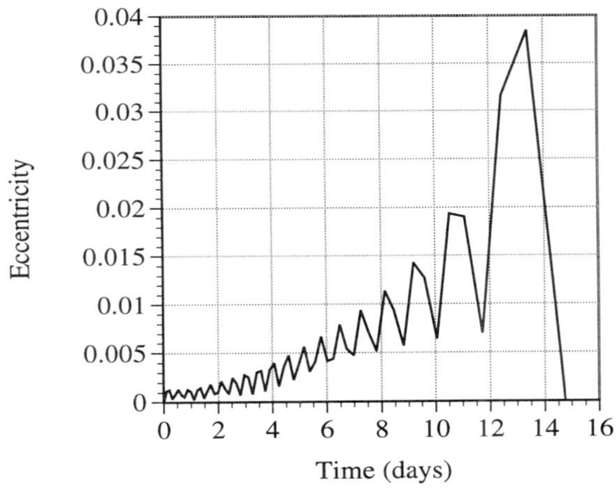

Figure 5. Time history of the eccentricity of the transfer orbit for the low-thrust LEO to GEO transfer.

Figures 6 through 8 show the trajectories and control histories for an Earth-Mars transfer beginning in GEO and ending in a Martian synchronous orbit. The thrust acceleration is one milli-g and total flight time is 222.14 days. The discretization used 918 state and control variables and included 728 (mostly nonlinear) constraints. It required 5 hours of CPU time on a Convex C240 computer. This is another example of a problem which simply could not have been solved with a conventional transcription, e.g. that of Hargraves and Paris, which would have yielded an NLP problem with several thousand parameters; beyond the capability of present algorithms to solve.

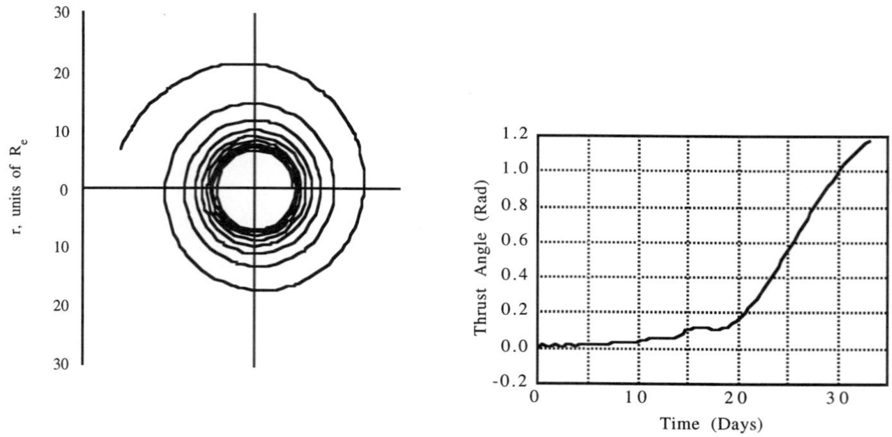

Figure 6. Trajectory and control (thrust pointing angle) for the Earth departure phase of an optimal Earth to Mars transfer. The time of flight is 33.08 days.

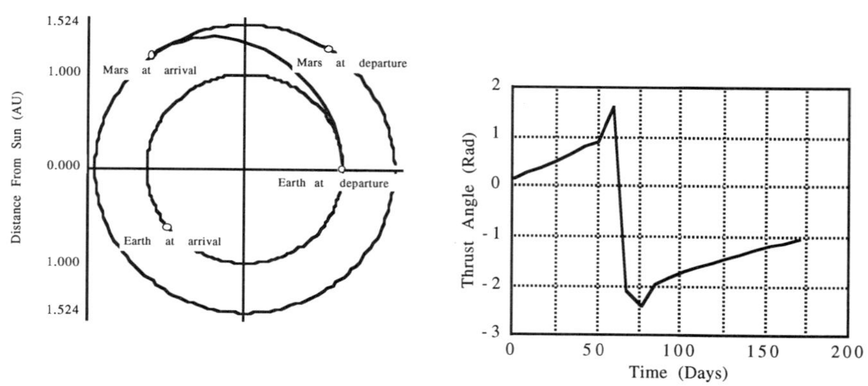

Figure 7. Trajectory and control (thrust pointing angle) for the heliocentric flight phase of an optimal Earth to Mars transfer. The time of flight is 169.78 days.

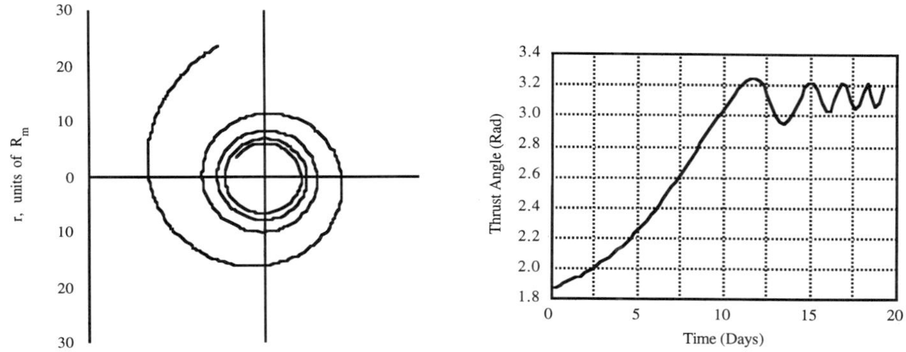

Figure 8. Trajectory and control (thrust pointing angle) for the Mars arrival phase of an optimal Earth to Mars transfer. The time of flight is 19.28 days.

SUMMARY

A very brief introduction to the theory of trajectory optimization has been presented. The application of the "indirect" solution method, i.e. the calculus of variations, yields problems which are either very difficult to solve or are insoluble, owing to the sensitivity of the Lagrange multipliers of the problem to changes in initial conditions. A recently developed method for the discretization of such problems, which we term direct collocation with nonlinear programming, has been successfully applied to a number of optimal orbit transfer problems, particularly problems using low-thrust propulsion which have continuously varying controls and long flight times.

Several techniques for reducing the size of the resulting NLP problem have been described. This is important for three reasons; (i) the execution time of the program which solves the NLP problem increases exponentially with problem size, (ii) there is a practical upper limit of 700-900 variables if a solution is to be found using existing programs, and (iii) smaller problems are generally more robust, that is, more likely to converge to a solution as problem parameters such as initial conditions and thrust magnitude are varied.

Two examples are shown, both of which are ambitious problems representing the "state of the art" in trajectory optimization. One problem is a low-thrust LEO to GEO transfer requiring a flight time of 15 days including 101 orbits of the Earth. The second is an interplanetary transfer including an optimal departure trajectory from Earth orbit and arrival into Martian synchronous orbit.

REFERENCES

1. W. F. Hohmann. "Die Erreichbarkeit der Himmelskorper," Oldenbourg, Munich (1925).
2. A. E. Bryson and Y. C. Ho. "Applied Optimal Control," Hemisphere Publ., Washington DC (1975).
3. P. J. Enright. "Optimal Finite-Thrust Spacecraft Trajectories Using Direct Transcription and Nonlinear Programming," Ph. D. Thesis, University of Illinois at Urbana-Champaign (1991).
4. D. C. Redding and J. V. Breakwell, Optimal low-thrust transfers to synchronous orbit, AIAA J. of Guidance and Control, Vol. 7, 148-155 (1984).
5. D. F. Lawden. "Optimal Trajectories for Space Navigation," Butterworths, London (1963).

6. P. M. Lion and M. Handelsman, The primer vector on fixed-time impulsive trajectories, *AIAA Journal*, Vol. 6, 127-132 (1968)
7. D. J. Jezewski and H. L. Rosendaal, An efficient method for calculating optimal free-space N-impulsive trajectories, *AIAA Journal*, Vol. 6, 2160-2165 (1968).
8. J. E. Prussing and B. A. Conway. "Orbital Mechanics," Oxford University Press, New York (1993).
9. A. L. Herman and B. A. Conway, An automatic node placement strategy for optimal control problems discretized using third-degree Hermite polynomials, Paper AIAA 92-4511, AIAA/AAS Astrodynamics Conf., Hilton Head, SC (1992).
10. M. D. Canon, C. D. Cullum and E. Polak. "Theory of Optimal Control and Mathematical Programming," McGraw-Hill, New York (1970).
11. F. T. Johnson, Approximate finite-thrust trajectory optimization, *AIAA Journal*, Vol. 7, 993-997 (1969).
12. C. R. Hargraves, F. T. Johnson, S. W. Paris, and I. Rettie, Numerical computation of optimal atmospheric trajectories, *AIAA J. of Guidance and Control*, Vol. 4, 406-414 (1981).
13. C. R. Hargraves and S. W. Paris, Direct trajectory optimization using nonlinear programming and collocation, *AIAA J. of Guidance and Control*, Vol. 10, 338-342 (1987).
14. W. A. Scheel. "Optimization of Very-Low-Thrust, Many-Revolution Spacecraft Trajectories Using Nonlinear Programming," M. S. Thesis, University of Illinois at Urbana-Champaign (1993).
15. W. A. Scheel and B. A. Conway, Optimization of very-low-thrust, many revolution spacecraft trajectories, to appear in *AIAA J. of Guidance, Control and Dynamics* (1994).
16. S. Tang. "Optimization of Interplanetary Trajectories Using Direct Collocation and Nonlinear Programming," M. S. Thesis, University of Illinois at Urbana-Champaign (1993).
17. S. Tang and B. A. Conway, Optimization of low-thrust interplanetary trajectories using collocation and nonlinear programming, to appear in *AIAA J. of Guidance, Control and Dynamics* (1994).
18. A. L. Herman and B. A. Conway, Direct solutions of optimal orbit transfers using collocation based on Jacobi polynomials, to be presented at the 1994 AAS/AIAA Space Flight Mechanics Conference, Cocoa Beach, FL, February (1994)
19. P. J. Enright and B. A. Conway, Optimal finite-thrust spacecraft trajectories using collocation and nonlinear programming, *AIAA J. of Guidance, Control and Dynamics*, Vol. 14, 981-985 (1991).
20. P. J. Enright and B. A. Conway, Discrete approximations to optimal trajectories using direct transcription and nonlinear programming, *AIAA J. of Guidance, Control and Dynamics*, Vol. 15, 994-1002 (1992).
21. P. E. Gill, W. Murray, M. A. Saunders, and M. H. Wright. "User's Guide for NPSOL (4.0), Technical Report SOL 86-2, Stanford University (1986).

PART FOUR

THE THREE BODY PROBLEM

ORDER OUT OF CHAOS IN THE THREE-BODY PROBLEM: REGIONS OF ESCAPE

K. Zare and V. Szebehely

The Department of Aerospace Engineering
And Engineering Mechanics
The University of Texas at Austin
Austin, Texas 78712

ABSTRACT

The dynamical behavior of the general three-body problem with equal masses is studied in the regions where the motion leads to triple close approaches (e.g., near equilateral triangular position with zero initial velocities). Due to the strong third body effect, the classical perturbation approach is no longer applicable in these regions. This paper consists of two parts. In the first part a qualitative analysis of a simpler case, namely the planar isosceles three-body problem with negative energy is offered. Various aspects of this problem have been investigated by many authors in the recent years. In this paper, it is shown that a necessary and sufficient condition for escape is a syzygy crossing immediately after a binary collision. Furthermore, we have shown that the initial conditions leading to escape form open sets (i.e., connected regions) in the phase space. On each set the number of syzygy crossings as well as the number of binary collisions between consecutive syzygy crossings are invariant. In the second part the results are confirmed by finding some of these regions, using numerical integration. In each region the parameters of escape (i.e., orbital elements of the binary, velocity of the escaping body, etc.) are continuous functions of the initial conditions, and the boundary provides the measure for dynamical sensitivity of initial conditions. A comparison of the measure for dynamical sensitivity with a measure for numerical sensitivity (i.e. time reversal test) enable us to give a precise definition for computability. According to our numerical results, the higher number of syzygy crossings leads to higher dynamical sensitivity (i.e., smaller regions). This suggests the non-computability of the escape solutions with very high number of syzygy crossings, and the difficulty of detecting the regions of escape in these cases. Finally, using numerical integration, the results are extended to a more general case where the isosceles symmetry is broken and a special set of initial conditions for the planar three-body problem is considered.

INTRODUCTION

Poincaré's (1892) classical approach to the behavior and to the non-integrability of the general problem of three bodies is based on perturbations and periodic motions. He used the classical Hamiltonian perturbation analysis which is applicable to special initial conditions, such as planetary systems. It took about one hundred years of research to find that the behavior of three bodies is basically controlled by their triple close approaches and escapes, rather than their periodic motions, once these special restrictions are removed.

From Newton to Chaos, Edited by A.E. Roy
and B.A. Steves, Plenum Press, New York, 1995

The existence of initial conditions with negative total energy resulting in an escape was shown by Birkhoff (1927). Since then many authors have improved Birkhoff's escape conditions; Merman (1958), Tevzadze (1962), Standish (1971), Marchal (1974), Zare (1981), Laskar (1984) and Marchal (1990), just to mention a few. Birkhoff (1927) was also the first to mention the importance of the triple close approach as a necessary mechanism for escape, by showing that the moment of inertia increases indefinitely in both senses of time if its minimum becomes less than a critical value. As pointed out by Zare (1981), Birkhoff's example is restricted to the solutions with no interplay between the participating bodies, and it is essentially a perturbed elliptic-hyperbolic solution before, at and after the triple close approach. However, due to the strong third-body effect, in general there are interplays and to obtain the solution requires numerical integration.

The numerical integration has been made possible by regularization of binary collisions (e.g., Aarseth (1971), Bettis and Szebehely (1971), Waldvogal (1972), Aarseth and Zare (1974), Zare (1974), Heggie (1974), Susin (1989), Mikkola and Aarseth (1990), etc.). These regularizations not only remove the singularities due to the binary collisions, but they also improve the numerical treatment of triple close approaches. For more detail and a new formulation see the Appendix of this paper.

Burrau's problem (also referred to as the Pythagorean problem) was the first general demonstration of a triple close approach followed by escape. Its "complete solution" was given by Szebehely and Peters (1967). The solution of Burrau's problem leads to several interplays between participating bodies as well as triple close approaches, before reaching its outcome. Since then many other investigations (e.g., Agekian and Anosova (1967), Agekian and Martinova (1973), Szebehely (1974), Aarseth and Heggie (1976), etc.) support the opinion that arbitrary initial conditions usually lead to triple close approaches resulting in the escape of one of the participating bodies and the binary formation of the remaining bodies. There might be several triple close approaches followed by ejection of one of the bodies without escape but eventually a final triple close approach will occur followed by an escape. It should be emphasized that all of the above mentioned numerical investigations and results are based on statistical approaches of choosing random initial conditions and computing the probability of binary formation, and escape.

A more comprehensive approach followed in this paper is the study of the geometric structure of the initial conditions in the phase space. Do the initial conditions leading to escape form open sets (i.e., connected regions) or discrete sets (i.e., isolated points) in the phase space? Is there any rule for their distribution in the phase space? Can we introduce a measure for the sensitivity with respect to the initial conditions? The answers to these questions enable us to determine whether a given initial condition is in an ordered or in a chaotic region, and to decide on the computability of the solution. This paper offers the initial steps toward this goal, and it consists of two parts.

In the first part, the planar isosceles three-body problem with negative energy is investigated analytically. Many authors have contributed in understanding various aspects of this problem, for example see Broucke (1979) for a numerical investigation of periodic orbits, Devaney (1980) and Simo (1981) for the analysis of triple collision, Simo and Martinez (1988) for a qualitative study on a fixed negative energy manifold using symbolic dynamics, and Diacu (1991) for a discussion of chaotic behavior also using symbolic dynamics.

In the second part, the results of numerical integration are presented. The numerical results are obtained for the planar isosceles case, as well as, for a more general case of the planar three-body problem.

DESCRIPTION OF THE DYNAMICAL SYSTEM

The equations of motion of the problem of three bodies in Jacobian variables become

$$\ddot{\mathbf{r}} + \frac{G\mu}{|\mathbf{r}|^3}\mathbf{r} = G\frac{m_3\mu}{m_1 m_2}\,\frac{\partial F}{\partial \mathbf{r}}, \tag{1}$$

$$\ddot{\rho} = G\frac{M}{\mu}\,\frac{\partial F}{\partial \rho}. \tag{2}$$

Here $F = \frac{m_1}{\left|\rho + \frac{m_2}{\mu} \boldsymbol{r}\right|} + \frac{m_2}{\left|\rho - \frac{m_1}{\mu} \boldsymbol{r}\right|}$, $\mu = m_1 + m_2$, and $M = \mu + m_3$.

The Jacobian vectors are $\boldsymbol{r}$ and ρ, the first connecting m_1 and m_2 and the second the center of mass of m_1 and m_2 with m_3. The integrals of the system are the conservation of the energy

$$h = \frac{1}{2}\left(\frac{m_1 m_2}{\mu}\dot{\boldsymbol{r}}^2 + \frac{m_3\mu}{M}\dot{\rho}^2\right) - \frac{m_1 m_2}{|\boldsymbol{r}|} - m_3 F \tag{3}$$

and the conservation of the angular momentum

$$c = \frac{m_1 m_2}{\mu}\boldsymbol{r} \times \dot{\boldsymbol{r}} + \frac{m_3 \mu}{M}\rho \times \dot{\rho}. \tag{4}$$

The initial conditions for our planar problem are represented by the points in the 8-dimensional phase space. For the geometric visualization of the results, in this study, we only search a 2-dimensional subspace (X_3, Y_3), namely the initial positions of m_3, by imposing the following restrictions on the initial conditions: (i) the initial velocities are zero, and (ii) the initial positions of m_1 and m_2 are fixed $\left(X_1 = -\frac{1}{2},\ Y_1 = 0, X_2 = \frac{1}{2},\ Y_2 = 0\right)$. Our study is also limited to the system of unit masses $(m_1 = m_2 = m_3 = 1)$, where there is no dominant mass, and where we may take the advantage of symmetry.

ANALYSIS

To simplify the analysis in this section a further restriction (iii) of the initial conditions is introduced by constraining the initial position of m_3 to the y-axis. This will reduce our search to the one-dimensional subspace. From conditions (i), (ii), and (iii), together with the equality of m_1 and m_2, it follows that the subsequent positions of m_3 are on the y-axis. Therefore, the problem is essentially reduced to the so-called planar isosceles three-body problem, with specified initial conditions. The equations of motion for this case, using $G = m_1 = m_2 = m_3 = 1$, and the symmetry, simplify to

$$\ddot{x} + \frac{1}{4x^2} = \frac{\partial F}{\partial x}, \tag{5}$$

$$\ddot{y} = 3\frac{\partial F}{\partial y}, \tag{6}$$

$$F(x, y) = \frac{1}{\sqrt{x^2 + y^2}}.$$

The variables x and y are the distance of the center of mass of m_1 and m_2 to m_2 and m_3 respectively.

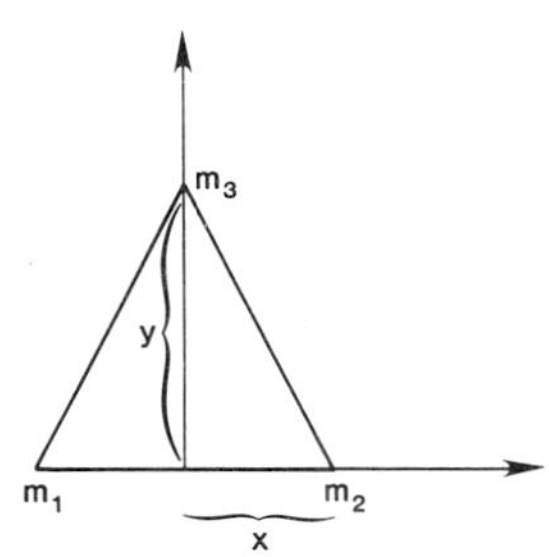

Figure 1. Geometry of planar isosceles three-body problem.

Using the energy integral, we may show that

$$x \leq \frac{3}{|h|} \quad , \quad |h| = 1 + \frac{2}{\sqrt{\frac{1}{4} + y_0^2}}. \tag{7}$$

where h is the constant of energy, and y_0 is the initial position of m_3. For $x > 0$, from Eq. (5) we have $\ddot{x} < 0$. Therefore the curve $x(t)$ is concave, and it must have a maximum, since x is bounded. This implies that the forward and the backward motion of m_1 and m_2 leads to binary collisions (i.e., there exist a $t_1 < 0$, and a $t_2 > 0$, such that $x(t_1) = 0$, and $x(t_2) = 0$). Since binary collisions are not essential singularities, we may continue the solution after the collision. In fact using the time transformation $\frac{dt}{d\tau} = x$, we obtain

$$x'' = \frac{1}{4} + \left(h - \frac{2}{3}Z\right) x - \frac{x^3}{\left(x^2 + y^2\right)^{3/2}}, \tag{8}$$

$$x'^2 = \frac{1}{2}x + \left(h - \frac{2}{3}Z\right)x^2, \tag{9}$$

where $\quad Z = \frac{1}{2}\left(\frac{y'}{x}\right)^2 - \frac{3}{\left(x^2+y^2\right)^{1/2}}, \quad$ and $\quad y' = x\,\dot{y}\,. \quad$ (10)

Here Z represents the energy function for m_3 . At the time of a binary collision τ_c we have $x(\tau_c) = 0$, $y(\tau_c) \neq 0$, $x'(\tau_c) = 0$, and $x''(\tau_c) = \frac{1}{4}$. Therefore, x has a minimum at τ_c. This implies that $x > x(\tau_c) = 0$, so after the collision $\ddot{x}$ is negative, and m_1, and m_2 experience repeated binary collisions as time increases. As follows from Eq. (9), the maximum value of x between two consecutive collisions (i.e., the amplitude of oscillation) is

$$x_m = \frac{1}{2\left(\frac{2}{3}Z - h\right)}. \tag{11}$$

This value depends on the energy function Z and consequently on the motion of m_3. The time between two consecutive collisions T (i.e., the period of oscillation) is bounded according to

$$\frac{2\pi\, x_m^{3/2}}{\sqrt{10}} \leq T \leq \frac{2\pi\, x_m^{3/2}}{\sqrt{2}}. \tag{12}$$

These bounds follow from $\ddot{x}_a < \ddot{x} < \ddot{x}_b$, where $\ddot{x}_a = -\frac{5}{4x_a^2}$, $\ddot{x}_b = -\frac{1}{4x_b^2}$, and they are the periods for these limiting two-body cases. The lower bound is the period for the case in which m_3 is at rest and it is located at the center of mass of m_1 and m_2 (i.e., $y = 0$), and the upper bound is the period for the case in which m_3 is removed from the system (i.e., $y \to \infty$).

Now we proceed to describe the motion of m_3 on the y-axis. From Eq.(6), we have that if $y > 0$, $\ddot{y} < 0$, and if $y < 0$, $\ddot{y} > 0$. Therefore the curve $y(t)$ is concave downward or upward according to $y > 0$ or $y < 0$. If no escape occurs along the positive or negative y-axis, then at some time $\dot{y} = 0$, and the curve $y(t)$ has a maximum or a minimum respectively. This implies that the motion of m_3 leads either to a finite or to an infinite number of syzygy crossings (i.e., $y = 0$). The first case represents motion that leads to an escape (or sometimes to a triple collision) after a finite number of oscillations, and the second case represents a bounded motion. From Eq.(6), it also follows that

$$\frac{d}{dt}\dot{y}^2 = \frac{-6y\dot{y}}{\left(x^2+y^2\right)^{3/2}}. \tag{13}$$

Therefore, the speed of m_3 (i.e., $|\dot{y}|$) increases if m_3 is approaching the syzygy crossing $(y\dot{y} < 0)$, and it decreases if m_3 is receding from the syzygy crossing $(y\dot{y} > 0)$. To establish the escape of m_3 at some time T, it is sufficient to show that $|\dot{y}| > 0$ for all $t > T$, or equivalently to show that $Z > 0$ for all $t > T$.

Differentiation of Eq.(10) leads to

$$\frac{dZ}{dt} = \frac{3x\dot{x}}{\left(x^2+y^2\right)^{3/2}}. \tag{14}$$

Therefore, m_3 gains or looses energy when m_1 and m_2 are receding from each other (i.e., $\dot{x} > 0$) or approaching each other (i.e., $\dot{x} < 0$) respectively.

In fact the energy gain $(\Delta Z)_k$ in a cycle between k and $k+1$ binary collision is given by

$$(\Delta Z)_k = \int_0^{x_k} \frac{3x dx}{\left(x^2+y^2\right)^{3/2}} + \int_{x_k}^{0} \frac{3x dx}{\left(x^2+y^2\right)^{3/2}}, \tag{15}$$

where x_k is the maximum value of x in that cycle. Assuming that $a < |y| < b$ in the first interval, and $c < |y| < d$ in the second interval, we obtain

$$\left[\frac{3}{\sqrt{x^2+c^2}} - \frac{3}{\sqrt{x^2+b^2}}\right]_0^{x_k} \le (\Delta Z)_k \le \left[\frac{3}{\sqrt{x^2+d^2}} - \frac{3}{\sqrt{x^2+a^2}}\right]_0^{x_k}. \tag{16}$$

If $|y|$ is increasing during the cycle then $b = c$ and $(\Delta Z)_k \ge 0$, and if $|y|$ is decreasing during the cycle then a = d and $(\Delta Z)_k \le 0$.

This implies that m_3 respectively gains or looses energy during a cycle, if it is receding from or approaching to the syzygy crossing. Using the mean value theorem, we may also show that

$$|(\Delta Z)_k| < \int_0^{x_k} \frac{3x dx}{\left(x^2+y^2\right)^{3/2}} = \frac{3\bar{x}_k}{\left(\bar{x}_k^2+\bar{y}_k^2\right)^{3/2}} x_k < \frac{3x_k^2}{|\bar{y}_k|^3}. \tag{17}$$

where $0 < \bar{x}_k < x_k$, $|y_k| < |\bar{y}_k| < |y_{k+1}|$ if $|y|$ is increasing, and $|y_{k+1}| < |\bar{y}_k| < |y_k|$ if $|y|$ is decreasing. Note that y_k and y_{k+1} are the values of y at the beginning of the k and $k+1$ cycles. The energy gain during the cycle which contains the syzygy crossing plays a critical role. If this energy gain is sufficiently large to make $Z > 0$, then Z remains positive for all time, and consequently m_3 escapes (i.e., $|y| \to \infty$ as $t \to \infty$).

Now let us assume that, the syzygy crossing occurs immediately after k number of binary collision, $x = x^*$ at the time of syzygy crossing, and $y = \lambda x^*$ at the time of binary collision. Under these conditions, the energy gain during this cycle is

$$(\Delta Z)_k = \int_0^{x^*} \frac{3xdx}{(x^2+y^2)^{3/2}} + \int_{x^*}^{x_k} \frac{3xdx}{(x^2+y^2)^{3/2}} + \int_{x_k}^{0} \frac{3xdx}{(x^2+y^2)^{3/2}}, \tag{18}$$

and the following lower bound estimate may be obtained

$$(\Delta Z)_k > \frac{3}{2(1+\lambda^2)^{3/2}}\left(\frac{1}{x^*}\right) - \left[\frac{3}{\sqrt{x^2+b^2}}\right]_{x^*}^{x_k} + \left[\frac{3}{\sqrt{x^2+c^2}}\right]_0^{x_k}. \tag{19}$$

It follows that as $x^* \to 0$, $(\Delta Z)_k \to \infty$, and a sufficiently small x^* (i.e., $x^* < \varepsilon$) will result in the escape of m_3. Notice that the total loss of energy in the preceding cycles remains bounded as $x^* \to 0$, since $\bar{y}_{k-1} \neq 0$. Similarly if the syzygy crossing occurs immediately before $k+1$ binary collision, then

$$(\Delta Z)_k < -\left[\frac{3}{\sqrt{x^2+a^2}}\right]_0^{x_k} + \left[\frac{3}{\sqrt{x^2+d^2}}\right]_{x^*}^{x_k} - \frac{3}{2(1+\lambda^2)^{3/2}}\left(\frac{1}{x^*}\right). \tag{20}$$

In this case as $x^* \to 0$, $(\Delta Z)_k \to -\infty$, and for a sufficiently small x^*, Z will decrease to an arbitrarily high negative value. Assuming m_3 will escape, then Z has to become positive at some time later. The total energy gain in the succeeding cycles is

$$\Delta Z = \sum_{i=k+1}^{\infty} (\Delta Z)_i < \sum_{i=k+1}^{\infty} \frac{3x_i^2}{|\bar{y}_i|^3} < 3\alpha^2 \sum_{i=k+1}^{\infty} \frac{1}{|\bar{y}_i|^3} < 3\alpha^2\left(\frac{1}{|\bar{y}_{k+1}|^3} + \frac{1}{2\delta|\bar{y}_{k+1}|^2}\right), \tag{21}$$

where $\alpha = \sup\{x_i\}$, $\delta = \inf\{|y_{i+1} - y_i|\}$. Since $\bar{y}_{k+1} \neq 0$, ΔZ is bounded and for a sufficiently small x^*, Z will remain negative. This implies that m_3 does not escape in this case and it will return at some time after the syzygy crossing. Therefore, the escape of m_3 does not depend <u>only</u> on how close the particles are at the triple close approach (i.e., on the value of x^*), but also on whether the syzygy crossing occurs before or after the binary collision (i.e. on the sign of $\dot{x}^*$).

To investigate the relationship between the initial conditions and the escape of m_3, let $x = f(y_0, t)$ and $y = g(y_0, t)$ be the solutions of Eq. (5) and Eq. (6), with y_0 representing the initial condition. Let t^* be the time of a syzygy crossing for a particular solution given by the initial condition y_0^* (i.e., $g(y_0^*, t^*) = 0$). Let us assume that $x^* = f(y_0^*, t^*) < \varepsilon$, and $\dot{x}^* = \frac{\partial f}{\partial t}(y_0^*, t^*) > 0$, so that this solution leads to the escape of m_3. The velocity of m_3 at the syzygy crossing is not zero, i.e.

$$\dot{y} = \frac{\partial g}{\partial t}(y_0^*, t^*) \neq 0. \tag{22}$$

Then by the implicit function theorem, there exist an open interval $U(y_0^* \in U)$, an open interval $V(t^* \in V)$, and a continuously differentiable function $t = h(y_0)$ for all $y_0 \in U$ such that $g(y_0, h(y_0)) = 0$. Therefore, x and $\dot{x}$ at the syzygy crossing are given by

$$x = f(y_0, h(y_0)) = f_1(y_0),\ \dot{x} = \frac{\partial f}{\partial t}(y_0, h(y_0)) = f_2(y_0), \tag{23}$$

which are continuous functions for all $y_0 \in U$. This continuous dependence on y_0 implies the existence of an open interval U^* $(y_0^* \in U^* \subset U)$ such that $f_1 < \varepsilon$, and $f_2 > 0$ for all $y_0 \in U^*$. This proves that the set of initial conditions leading to the escape of m_3 is the union of open intervals on the y-axis. On each interval the number of syzygy crossings (n) as well as, the number of binary collisions between consecutive syzygy crossings $\{k_i, i = 1, n\}$ are invariant. The length of the interval and its relation to these invariants will be investigated numerically in the next section. Since this set is an open set, its complement (i.e. the initial conditions representing bounded motion and in particular, periodic motion) can not be densely distributed over the entire y-axis.

Now, let us denote by E, B, and P the set of initial positions of m_3 which lead to escape, bounded motion, and periodic motion respectively. Clearly B is the complement of E, and P is a subset of B. We have already shown that E is an open set. If in addition we show that E is dense, then escape is a generic property. E is dense if and only if B is discrete. While it is difficult to show that B is an open set or a discrete set, it is easy to do so for P.

Let y_0^* represent a periodic solution with period T^*, then we have

$$f(y_0^*, T^*) = \frac{1}{2},\ \ \frac{\partial f}{\partial t}(y_0^*, T^*) = 0. \tag{24}$$

For an initial condition y_0 and a time T sufficiently close to y_0^* and T^*, we may write

$$f(y_0, T) = \frac{1}{2} + \frac{\partial f}{\partial y_0}(y_0^*, T^*)\,(y_0 - y_0^*). \tag{25}$$

The solution is periodic if $f(y_0, T) = \frac{1}{2}$. This implies that either $y_0 = y_0^*$ or (y_0^*, T^*) is a critical point of f (i.e. $\frac{\partial f}{\partial y_0}(y_0^*, T^*) = \dot{0}, \frac{\partial f}{\partial t}(y_0^*, T^*) = 0$). In either case, we conclude that y_0^* is an isolated point and the points in its immediate neighborhood do not represent a continuous family of periodic orbits.

RESULTS OF NUMERICAL INTEGRATIONS

A. Isosceles Case

The initial position of m_3 is varied on the y-axis, and the trajectories of the particles are obtained by numerical integration for each case. The integration is terminated as soon as a test of escape is satisfied or a pre specified time limit is reached. When escape occurs, the parameters of escape (i.e., orbital elements of the remaining binary, velocity of the escaping body, number of syzygy crossing, time of the last syzygy crossing, etc.) are evaluated. As a measure of numerical sensitivity, time reversal integration are performed for all cases.

First we examine in detail the neighborhood of the equilateral triangular position (i.e., $y_0 = \frac{\sqrt{3}}{2}$). This position leads to a triple collision at $t_c = \frac{\pi}{\sqrt{24}}$. However its neighborhood consists of two separate intervals leading to the escape of m_3. The first interval is above the equilateral position (i.e., $\frac{\sqrt{3}}{2} < y_0 < .994$). For an initial condition in this interval the first syzygy crossing occurs immediately after the binary collision between m_1 and m_2, and consequently m_3 escapes after the first syzygy crossing $(n=1)$ along the negative y-axis. We refer to this type of escape as type 1. The second interval is below the equilateral position (i.e., $.851 < y < \frac{\sqrt{3}}{2}$) and it is smaller than the first by a factor of ten. For an initial condition in this interval the first syzygy crossing occurs immediately before the binary collision between m_1 and m_2 and therefore m_3 will quickly return to the second syzygy crossing after the binary collisions between m_1 and m_2, and consequently m_3 escapes after the second syzygy crossing (n=2) along the positive y-axis. We refer to this type of escape as type 2. The variations of the semi-major axis of the remaining binary (a), the velocity of the escaping body (v), and the time of the last syzygy crossing (t_c) in these intervals are shown in Figures 2 a,b,c. Notice that the continuous dependence on the initial conditions does not hold at $y = \frac{\sqrt{3}}{2}$.

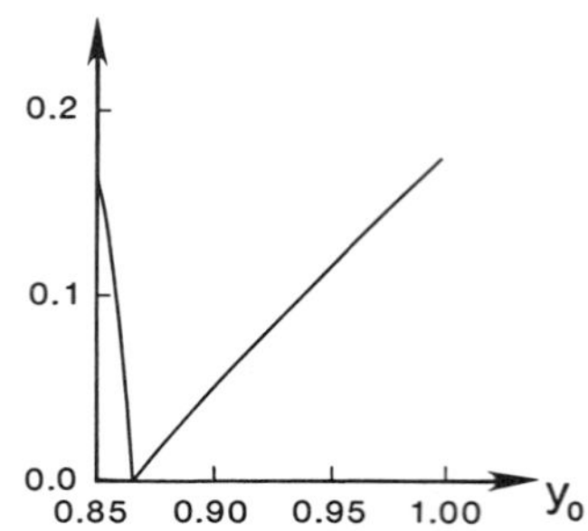

Figure 2a The semi-major axis of the binary as a function of the initial position of m_3.

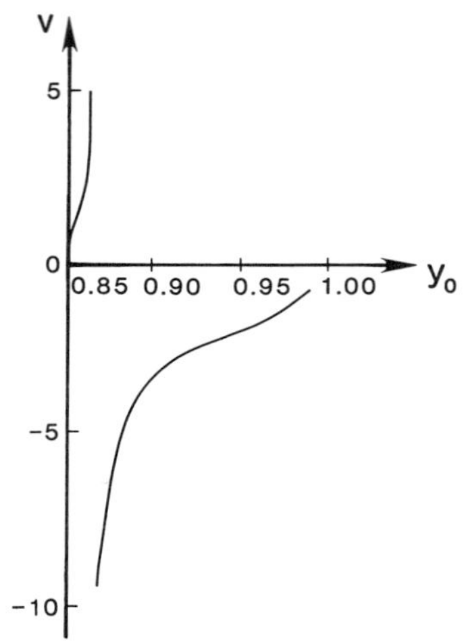

Figure 2b The velocity of escape as a function of the initial position of m_3.

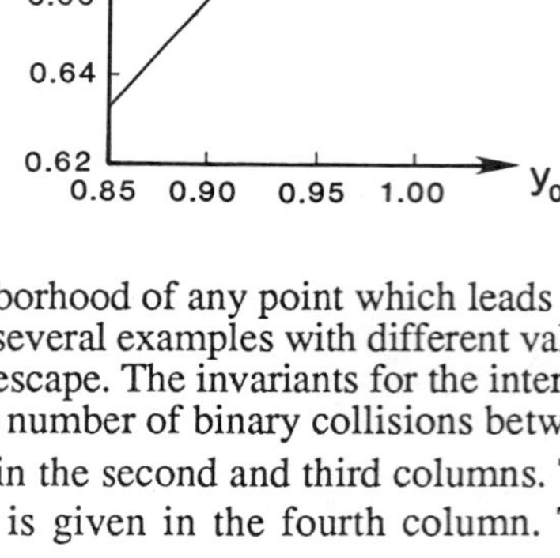

Figure 2c The time of syzygy crossing as a function of the initial position of m_3.

The above scenario repeats itself in the neighborhood of any point which leads to a triple collision at n syzygy crossing. Table I contains several examples with different values of n. The first column in Table I gives the interval of escape. The invariants for the interval, namely, the number of syzygy crossings (n), and the number of binary collisions between consecutive syzygy crossings $\{k_i, i=1, n\}$ are given in the second and third columns. The half length of the interval is denoted by L, and it is given in the fourth column. The accuracy of time reversal test for an initial position of m_3 at the midpoint of the interval is denoted by L', and it is listed in the fifth column. Ten cases are presented in Table I, the first seven cases are type 1 escape, and the last three are type 2.

Table I. Intervals of escape with different number of syzygy crossings

	interval	n	$\{K_i, i=1,n\}$	L	L'
Type 1	$\frac{\sqrt{3}}{2}-.994$	1	{1}	$5\text{x}10^{-2}$	$1\text{x}10^{-12}$
	$2.23-2.33$	1	{2}	$5\text{x}10^{-2}$	$2\text{x}10^{-12}$
	$1.097-1.09905$	2	{1,4}	$1\text{x}10^{-3}$	$6\text{x}10^{-11}$
	$1.081-1.082$	2	{1,5}	$5\text{x}10^{-4}$	$1\text{x}10^{-10}$
	$1.11971-1.11990$	3	{1,3,2}	$1\text{x}10^{-4}$	$5\text{x}10^{-9}$
	$1.1006-1.1007$	7	{1,3,1,1,1,1,2}	$5\text{x}10^{-5}$	$9\text{x}10^{-9}$
	$1.10049-1.10050$	9	{1,3,1,1,1,1,1,1,3}	$5\text{x}10^{-6}$	$9\text{x}10^{-8}$
Type 2	$0.851-\frac{\sqrt{3}}{2}$	2	{0,1}	$5\text{x}10^{-3}$	$3\text{x}10^{-8}$
	$1.09905-1.09913$	3	{1,3,1}	$4\text{x}10^{-5}$	$1\text{x}10^{-5}$
	$1.11970-1.11971$	4	{1,3,1,1}	$5\text{x}10^{-6}$	$5\text{x}10^{-5}$

We define L as a measure of dynamical sensitivity, and L' as a measure of numerical sensitivity. Furthermore we call the solution 'computable' if the numerical sensitivity is smaller than the dynamical sensitivity (i.e., L' < L). Clearly with this definition, the qualitative outcome of the system is preserved, despite numerical uncertainties. The following remarks are based on the results presented in Table I.

1. The higher number of syzygy crossings (n) leads to higher dynamical sensitivity (i.e., smaller L), as well as higher numerical sensitivity (i.e., larger L'). This suggests the existence of a large number N, such that L' > L for all n > N. In other words, the escape orbits with high number of syzygy crossings are not computable, and their corresponding intervals cannot be detected by numerical integration.
2. The dynamical sensitivity does not seem to have strong dependency on the number of binary collision. This may be seen by comparing the values of L in case 1 and case 2, or in case 3 and case 4.
3. In general type 2 escape is numerically more sensitive than type 1. This may be seen by comparing the values of L' in case 8 and in case 3. See also case 10, for an example of non-computability (i.e., L' > L).

B. Planar Case

Next the initial position of m_3 is varied in the two-dimensional region around the equilateral triangular position, and the trajectories of the particles are obtained by numerical integration for each case. The results are shown in Figures 3 a,b,c. The outcome of the motion in the immediate vicinity of $(0,\frac{\sqrt{3}}{2})$ is shown in Figure 3a. The regions marked by I, II, and III show the initial positions of m_3 resulting in type 1 escape of m_1, m_2 and m_3 after the first triple close approach (n=1). The narrow region between I and II shows the initial positions of m_3 resulting in the quick return of m_3 after the first triple close approach and type 2 escape of m_3 after the second triple close approach (n=2). Similarly the narrow regions between II and III, and between III and I are the initial positions of m_3 resulting respectively in type 2 escape of m_1 and m_2 after the second triple close approach.

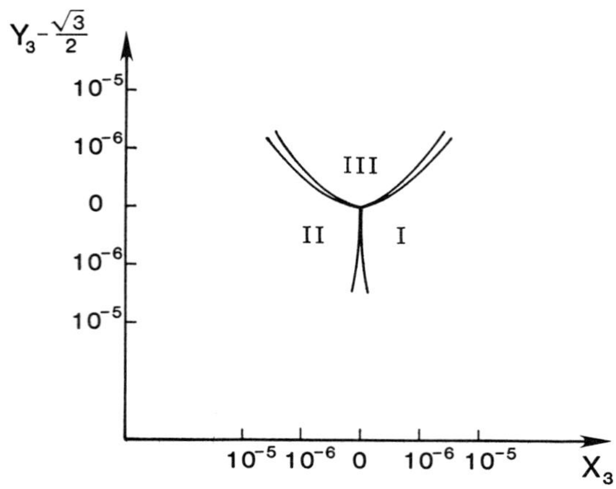

Figure 3a The regions of escape in the immediate vicinity of the equilateral triangular position.

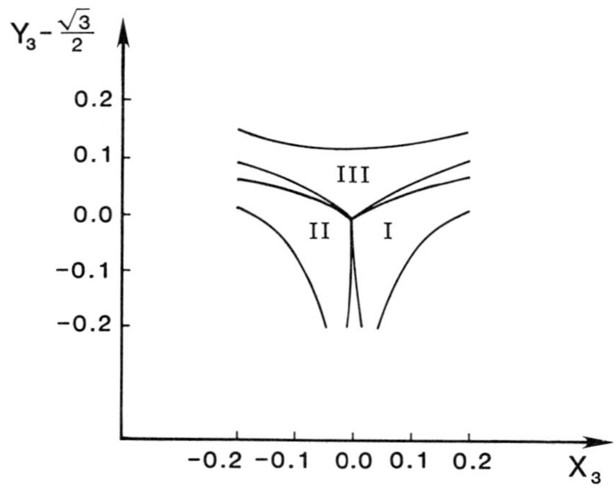

Figure 3b The more extended regions of escape near the equilateral triangular position.

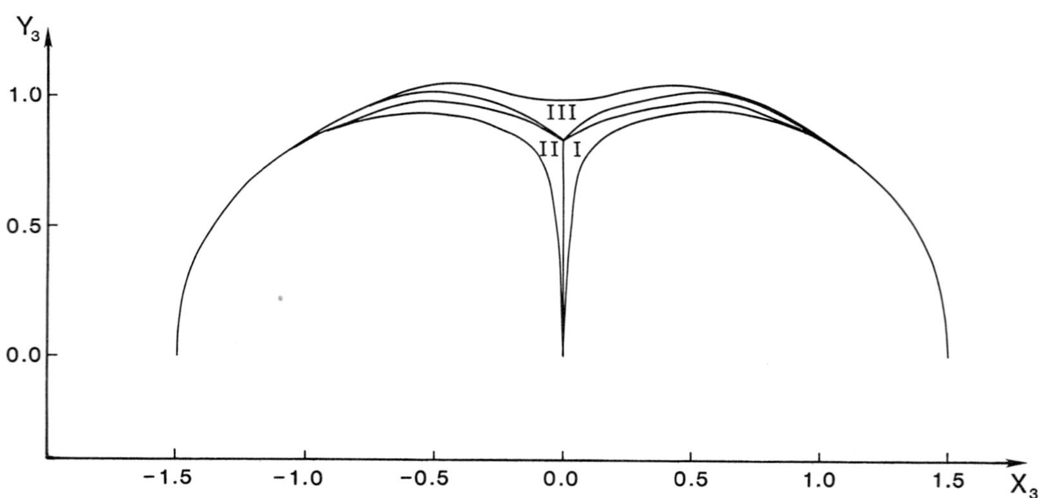

Figure 3c The extension far away from the equilateral triangular position.

Figure 3b shows the more extended regions where the outer boundaries start to appear. Here the narrow regions separating I,II, and III represent the initial positions of m_3 which lead to the variety of more complex motion (e.g. escape after several triple close approaches) than the motion corresponding to I,II, and III.

Figure 3c shows that I,II, and III extend far away from the Lagrangian triangular position, and their continuation terminate with narrow tongues at the Eulerian rectilinear positions. As in the isosceles case, the parameters of escape in each region depend continuously on the initial position of m_3.

The regions of escape corresponding to n triple close approaches (n >1) have smaller areas than those for n=1 but they have similar structure. They form complicated networks of areas connected at the nodes similar to Figure 3c. The initial position of m_3 at any of these nodes leads to a triple collision at n triple close approach. The existence of these nodes is due to our special initial conditions with zero angular momentum. In the case of non-zero angular momentum the regions of escape still exist, but the nodes will disappear. As in the isosceles case the regions of escape will shrink as n increases and they will be detectable only for small n.

Numerical Methods Used

Most of the orbits were established using three different computers, IBM-PC, VAX, and CRAY. The purpose was not to compare the efficiency of the various computers, but rather to obtain consistent and reliable numerical results.

The numerical integration was performed using a Runge-Kutta-Fehlberg 7(8) integrator with a new set of globally regularized equations. Although this global regularization is similar to those of Heggie (1974) and Susin (1989), it is different in some aspects and it is discussed in the Appendix.

All escape were verified by Standish's (1971) escape criteria. However, a combination of Sundman's function and Standish's criteria leads to a new test of escape based on the value of Sundman's function. This test provides an earlier detection of escape and it was used when the semi-major axis of the binary was very small. Such a utilization of Sundman's function with Birkhoff's (1927) escape criteria is shown by Zare (1981).

ACKNOWLEGMENT

The authors wish to thank Prof. A. E. Roy for the invitation to deliver this paper at NATO ASI. Support of the R. B. Curran Centennial Chair in Engineering of The University of Texas is gratefully acknowleged.

REFERENCES

Aarseth, S. J., 1971, *Astrophysics and Space Science*, **14**:118.
Aarseth, S. J. and Heggie, D. C., 1976, *Astron. Astrophysics*, **53**:259.
Aarseth, S. J. and Zare, K., 1974, *Celest. Mech*, **10**: 217.
Agekian, T. A. and Anasova, Zh. P. 1967, *Astron. Zh*, **44**:1261.
Agekian, T. A. and Martinova, A. I, 1973. University of Leningrad Publ., **1**:122
Bettis, D. G. and Szebehely, V., 1971, *Astrophysics and Space Science*, **14**:133.
Birkhoff, G. D., 1927, Dynamical Systems, Am. Math. Soc. Publ., Providence RI.
Broucke, R., 1979 *Astron. and Astrophysics*, **73**:303.
Devaney, R. 1980, *Invent. Math.*, **60**:249.
Diacu, F. N., 1991, *Celest. Mech. and Dyn. Astron.*, **50**:313.
Heggie, D. C., 1974, *Celest. Mech.*, **10**:217
Laskar, J. and Marchal, C., 1984, *Celest., Mech.*, **32**:15.
Marchal, C., 1974, *Celest. Mech.*, **10**:278.
Marchal, C., 1990, "The Three-body Problem", Elsevier, Amesterdan.
McGehee, R., 1974, *Invent. Math.*, **27**:191.
Merman, G. A., 1958, Byull. Inst. Teor. Astron., **6**:687.
Mikkola, S. and Aarseth, S. J., 1990, *Cel Mech. and Dyn. Astron.*, **47**:375.
Poincaré, H., 1892, "Les Methods Nouvelles de la Méchanique Céleste", Volumes **1-3**, Gauthier-Villars, Paris.

Simo, C., 1981, in "Classical Mechanics and Dynamical Systems", Marcel Dekker, 203.
Simo, C. and Martinez, R., 1988, *Celest. Mech.*, **41**:179.
Standish, M., 1971, *Celest. Mech.* **4**:44.
Stiefel, E. and Scheifele, G., 1971, "Regular and Linear Celestial Mechanics", Springer-Verlag Berlin.
Susin, A., 1989, Congress of Applied Mathematics, University of Malaga.
Szebehely, V. and Peters, F., 1967, *Astron. J.*, **72**:876.
Szebehely, V., 1974, *Astron. J.*, **79**:981, and 1449.
Tevzadze, F. A., 1962, IZV. and Akad. Nauk Armyan, SSR **15**:67.
Waldvogel, J., 1972, *Celest. Mech.*, 6:221.
Waldvogel, J., 1979, Instabilities in Dynamical Systems, Szebehely, V., ed., 263.
Zare, K., 1974, *Celest. Mech.*, **10**:207.
Zare, K., 1981, *Celest. Mech.*, **24**:345.

APPENDIX

Regularization

Let q_i' , p_i' represent the coordinates and the momenta of m_i (i = 1,2,3). The Hamiltonian may be written as

$$H = \sum_{i=1}^{3} \frac{p_i'^2}{2m_i} - \sum_{1 \le i < j \le 3} \frac{m_i m_j}{|q_i' - q_j'|}. \tag{A-1}$$

The Hamiltonian may be expressed in terms of the relative coordinates using the following extended point transformation

$$q_i = \frac{\partial W_1}{\partial p_i} \quad (i=0,1,2,3) \ , \qquad p_i' = \frac{\partial W_1}{\partial q_i'} \quad (i=1,2,3\ ), \tag{A-2}$$

where the generating function W_1 is given by

$$W_1 = p_1^T(q_3' - q_2') + p_2^T(q_1' - q_3') + p_3^T(q_2' - q_1')$$
$$+ p_0^T\,(m_1 q_1' + m_2 q_2' + m_3 q_3')\,/\,M,\ \ M = m_1 + m_2 + m_3.$$

Here q_i are the relative coordinates, and p_i are the new momenta. Furthermore the additional variables q_0 and p_0 are the coordinates and momenta of the center of mass.

Substituting (A-2) into (A-1) leads to the new Hamiltonian

$$H = \frac{1}{2M} p_0^2 + \sum_{\bullet} \left(\frac{p_i^2}{2\mu_{jk}} - \frac{1}{m_i} p_j^T p_k - \frac{m_j m_k}{|q_i|} \right). \tag{A-3}$$

Here the summation $\sum_{\bullet}$ stands for the cyclic interchange of the subscripts, and $\mu_{ij} = \frac{m_i m_j}{(m_i + m_j)}$. Without loss of generality we may adopt $q_0 = 0$, $p_0 = 0$, which implies that the coordinates and momenta are expressed with respect to the center of mass.

Next we introduce the Kustaanheimo and Stiefel (KS) transformation defined by

$$q_i = \frac{\partial W_2}{\partial P_i}, \quad P_i = \frac{\partial W_2}{\partial Q_i} \quad (i = 1,2,3), \tag{A-4}$$

where the generating function W_2 is given by

$$W_2 = \sum_{i=1}^{3} P_i^T\, L\,(Q_i)\, Q_i\,,$$

and

$$L\,(Q_i) = \begin{bmatrix} Q_{i1} & -Q_{i2} & -Q_{i3} & Q_{i4} \\ Q_{i2} & Q_{i1} & -Q_{i4} & -Q_{i3} \\ Q_{i3} & Q_{i4} & Q_{i1} & Q_{i2} \\ Q_{i4} & -Q_{i3} & Q_{i2} & -Q_{i1} \end{bmatrix}$$

is the generalized Levi-civita matrix , see Stiefel and Scheifele (1971).

Substituting (A-4) into (A-3), we obtain a new Hamiltonian

$$H = \sum_{*} \frac{P_i^2}{8\mu_{jk} Q_i^2} - \frac{P_j^T L^T(Q_j) L(Q_k) P_k^T}{4 m_i Q_j^2 Q_k^2} - \frac{m_j m_k}{Q_i^2}. \tag{A-5}$$

The introduction of the independent variable transformation

$$\frac{dt}{d\tau} = Q_1^2 Q_2^2 Q_3^2 \tag{A-6}$$

leads to the following Hamiltonian system of equations

$$\frac{dQ_i}{d\tau} = \frac{\partial \Gamma^*}{\partial P_i}, \quad \frac{dP_i}{d\tau} = -\frac{\partial \Gamma^*}{\partial Q_i} \quad (i = 1,2,3) \tag{A-7}$$

where

$$\Gamma^* = Q_1^2 Q_2^2 Q_3^2 \,(-h + H) =$$

$$-h Q_1^2 Q_2^2 Q_3^2 + \sum_{*} \frac{P_i^2 Q_j^2 Q_k^2}{8\mu_{jk}} - \frac{Q_i^2 \left(P_j^T L^T(Q_j) L(Q_k) P_k\right)}{4\mathrm{m_i}} - m_j m_k Q_j^2 Q_k^2.$$

The equations of motion (A-7) are regular for all binary collisions. This is the global regularization given by Heggie (1974). However, the new independent variable in the above formulation increases indefinitely ($\tau \to \infty$) as a triple collision is approached.

We introduce the scaled variables

$$u_i = \frac{Q_i}{r}, \quad v_i = P_i, \quad dw = r^3 d\tau \tag{A-8}$$

where $r^2 = Q_1^2 + Q_2^2 + Q_3^2$ is the perimeter of the triangle formed by the three bodies. This change of variables with the moment of inertia instead of the perimeter as the scaling factor was first introduced by McGehee (1974) for the collinear problem and it was followed by Waldvogel (1979) for the planar problem and by Susin (1982) for the spatial problem. With these variables, the equations of motion become

$$\frac{dr}{dw} = r\,f,$$

$$\frac{du_i}{dw} = f_i - f\,u_i, \quad \frac{dv_i}{dw} = g_i, \quad (i = 1,2,3) \tag{A-9}$$

$$\frac{dt}{dw} = r^3\, u_1^2\, u_2^2\, u_3^2,$$

where $f = \left(u_1^T f_1 + u_2^T f_2 + u_3^T f_3\right)$, and

$$f_i = \left(\frac{u_j^2 u_k^2}{4\mu_{jk}}\right) v_i - \left(\frac{u_j^2}{4 m_j}\right) L^T(u_j)\, L(u_k)\, v_k - \left(\frac{u_k^2}{4 m_k}\right) L^T(u_i)\, L(u_j)\, v_j,$$

$$g_i = \left(-2h\, r^2 u_j^2 u_k^2 - \frac{v_j^2 u_k^2}{4\mu_{ki}} - \frac{v_k^2 u_j^2}{4\mu_{ij}} + \frac{v_j^T L^T(u_j) L(u_k) v_k}{2m_i} + 2m_k m_i u_k^2 + 2m_i m_j u_j^2 \right) u_i$$

$$+ Z(v_i) \left[\left(\frac{u_j^2}{4m_j} \right) L(u_k) v_k + \left(\frac{u_k^2}{4m_k} \right) L(u_j) v_j \right],$$

$$Z(v_i) = \begin{bmatrix} v_{i1} & v_{i2} & v_{i3} & -v_{i4} \\ -v_{i2} & v_{i1} & v_{i4} & v_{i3} \\ -v_{i3} & -v_{i4} & v_{i1} & -v_{i2} \\ v_{i4} & -v_{i3} & v_{i2} & v_{i1} \end{bmatrix}$$

Here f_i and $g_i (i = 1,2,3)$ are obtained by cyclic interchange of the subscripts. The equations of motion (A-9) are regular for all binary collisions. The invariance of triple collision manifold (i.e.*r=0*), as well as, the equivalence of the equations of motion on the triple collision manifold (*r=0*) and on the zero energy manifold (*h=0*) follows easly from equations (A-9). However, the new independent variable in this formulation also increases indefinitely ($w \to \infty$) as a triple collision is approached ($r \to 0$).

If we introduce different scaled variables given by

$$x_i = r\, u_i = Q_i, \quad y_i = r\, v_i = r\, P_i, \quad ds = r\, dw = r^4 d\tau \ , \tag{A-10}$$

the equations of motion become

$$\frac{dx_i}{ds} = f_i, \quad \frac{dy_i}{ds} = g_i + f\, v_i, \quad (i = 1,2,3)$$

$$\frac{dt}{ds} = r^2 u_1^2 u_2^2 u_3^2, \tag{A-11}$$

where $r = \left(x_1^2 + x_2^2 + x_3^2 \right)^{1/2}$.

The equations of motion (A-11) are regular for all binary collisions. Moreover the new independent variables in this formulation does not increase indefinitely as $r \to 0$, and therefore triple collisions as well as near triple collisions can be reached in finite values of s. The vector field defined by equations (A-11) is continuous (but not continuously differentiable) at $r = 0$. This condition garantees the existence but not the uniqueness of a triple collision solution. For a similar discussion on the non-uniqueness of a triple collision solution see Diacu (1991). All numerical results presented in this paper are obtained by numerical integration of equations (A-11).

CHAOS IN THE HYPERBOLIC RESTRICTED 3-BODY PROBLEM

Josep M. Cors[1] and Jaume Llibre[2]

[1]Departament de Matemàtica Aplicada III, Escola Universitària Politècnica de Manresa, Universiat Politècnica de Catalunya 08240 Manresa, Barcelona, Spain
[2]Departament de Matemàtiques, Universitat Autònoma de Barcelona 08193 Bellaterra, Barcelona, Spain

Let $m_1 = m_2 = 1$ be two mass points moving under Newton's law of attraction in hyperbolic orbits in the xy-plane while their center of mass is fixed at the origin of coordinates. As usual these two masses are called *primaries*. We suppose that the hyperbolic orbit is a non-collision orbit; that is, the motion of m_1 and m_2 is not on a straight line, or equivalently, their angular momentum is different from zero. We consider a third mass point with infinitely small mass moving on the z-axis (see Figure 1). Since $m_3 = 0$ the motion of the first two mass points is not affected by the third and from the symmetry of the motion it is clear that the third mass point will remain on the z-axis. The problem is to describe the motion of the infinitely small mass, and then we have a restricted three-body problem that we call *the hyperbolic restricted problem*. The equation of motion of this problem in the phase space $(z, \dot{z}, t)$ is

$$\ddot{z} = -\frac{16z}{(x(t)^2 + y(t)^2 + 4z^2)^{\frac{3}{2}}}, \tag{1}$$

where $(x(t), y(t))$ is a non-rectilinear hyperbolic orbit solution of the 2-body problem.

This note is a summary of the paper Cors and Llibre[1]. In Cors and Llibre[1] we showed that the flow of the hyperbolic restricted problem is complete, that is each solution of the hyperbolic restricted problem is defined for all time t when $-\infty < t < +\infty$.

From Newton to Chaos, Edited by A.E. Roy
and B.A. Steves, Plenum Press, New York, 1995

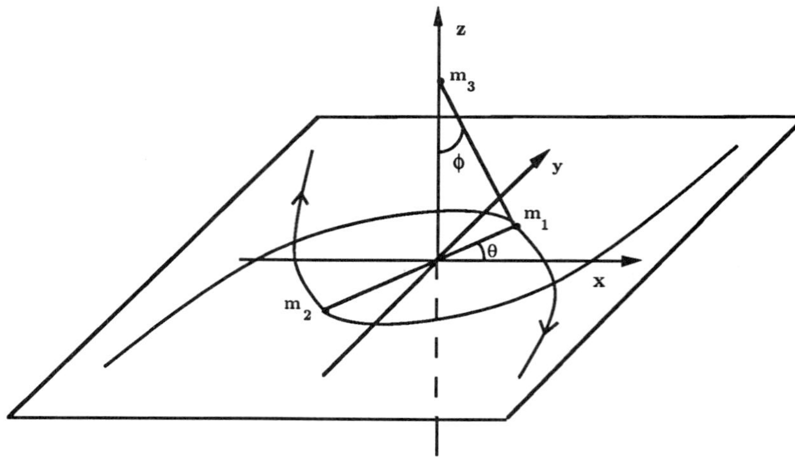

Figure 1. The hyperbolic restricted problem.

In order to extend the flow of the hyperbolic restricted problem to the infinite manifold associated to the primaries we use a new transformation introduced by Wang[2], see also Meyer and Wang[3]. The difference between this transformation and the one due to McGehee[4] is that in the last one the scale factor is given by the momentum of inertia while in the first one it is essentially twice the potential function, i.e.

$$u(t)^{-1} = \frac{4}{(x(t)^2 + y(t)^2)^{\frac{1}{2}}} + 2h,$$

where h is the energy of the primaries. We consider the new variables

$$\begin{aligned} \phi &= \cot^{-1} \frac{2z}{(x(t)^2 + y(t)^2)^{\frac{1}{2}}}, \\ G &= u(t)^{\frac{1}{2}} \dot{z}, \end{aligned}$$

where $0 < \phi < \pi$ and $G \in \mathbf{R}$. We define the functions $G_1(t) = u(t)^{\frac{1}{2}} \dot{x}(t)$, $G_2(t) = u(t)^{\frac{1}{2}} \dot{y}(t)$, $\theta(t) = \tan^{-1} \frac{y(t)}{x(t)}$ and $\xi(t) = 1 - 2u(t)h$. Then in the new variables the equation of motion (1) becomes

$$(2) \quad \begin{aligned} \frac{d\phi}{d\tau} &= \sin\phi \left(\frac{1}{2} [G_1(t) \cos\theta(t) + G_2(t) \sin\theta(t)] \cos\phi - G \sin\phi \right), \\ \frac{dG}{d\tau} &= \frac{\xi(t)}{4} \left([G_1(t) \cos\theta(t) + G_2(t) \sin\theta(t)] G - 4 \sin^2\phi \cos\phi \right), \end{aligned}$$

where $\xi(t), \theta(t), G_1(t), G_2(t)$ are evaluated on the hyperbolic orbit solution of the 2-body problem, and the new time τ is defined by $dt = \frac{2u(t)^{3/2}}{\xi(t)} d\tau$. The phase space of system (2) is $D \times \mathbf{R} = (0, \pi) \times \mathbf{R} \times \mathbf{R}$ associated to the variables (ϕ, G, t). The flow of (2) can be extended to the phase space $[0, \pi] \times \mathbf{R}^2$.

The following proposition shows that the infinitesimal body when $t \to \pm\infty$, either escapes to infinity (*hyperbolic motion*) or tends to a finite position (*parabolic motion*).

PROPOSITION 1. *Let $z(t)$ be a solution of (1). Then, either $z(t) = At + B \ln t + C + O(t^{-1} \ln t)$, or $z(t) = C + O(t^{-1} \ln t)$ when $t \to \pm\infty$, where A, B, C are real constants with $A \cdot B < 0$.*

It is an immediate consequence of Proposition 1 that the hyperbolic restricted 3-body problem has no periodic orbits.

We remark that a parabolic orbit $(\phi(t), G(t))$ for $t \to +\infty$ (respectively $t \to -\infty$) satisfies that $(\phi(t), G(t)) \to (\pi/2, 0)$ when $t \to +\infty$ (respectively $t \to -\infty$). In fact, the point $K = (\pi/2, 0)$ is an equilibrium point of system (2).

To study the flow near the equilibrium point K we introduce another transformation

$$\begin{aligned} \eta &= \cos\phi/\xi, \\ \lambda &= G/\xi. \end{aligned}$$

Equations of motion (2) in the new variables η and λ become

$$\begin{aligned} \frac{d\eta}{d\tau} &= \frac{1}{2}\eta\xi(t)[\eta^2\xi(t) - 1][G_1(t)\cos\theta(t) + G_2(t)\sin\theta(t)] + \lambda[1 - \eta^2\xi^2(t)]^{3/2}, \\ \frac{d\lambda}{d\tau} &= \frac{1}{2}\lambda[1 - \frac{1}{2}\xi(t)][G_1(t)\cos\theta(t) + G_2(t)\sin\theta(t)] - \eta\xi(t)(1 - \eta^2\xi^2(t)). \end{aligned} \tag{3}$$

Now a parabolic orbit $(\eta(t), \lambda(t))$ for $t \to +\infty$ (respectively $t \to -\infty$) satisfies that $(\eta(t), \lambda(t))$ tends to a point of the straight line of equilibrium points $\lambda = 0$ of system (3) (remark that $\xi(t) \to 0$ when $t \to \pm\infty$), and this condition characterizes the set of parabolic orbits. More precisely, for any equilibrium point $(\eta_0, 0)$ there is one and only one parabolic orbit $(\eta(t), \lambda(t))$ such that $\eta(t) \to \eta_0$ when $t \to +\infty$. A similar result occurs for the parabolic orbits when $t \to -\infty$. Furthermore, by using the normally hyperbolic theory (see for instance Hirsch, Pugh and Shub[5]) we can show doing another change of variables that the set of parabolic orbits is a 2-dimensional submanifold of the phase space $(z, \dot{z}, t)$.

We say that an orbit of the infinitesimal mass has type $\mathcal{P}^- n \mathcal{H}^+$ if the orbit is parabolic when $t \to -\infty$, hyperbolic when $t \to +\infty$, and intersects exactly n times the plane which contains the motion of the two primaries (i.e. the plane $z = 0$) when t goes from $-\infty$ to $+\infty$. In a similar way we define the types $\mathcal{P}^- n \mathcal{P}^+$, $\mathcal{H}^- n \mathcal{P}^+$ and $\mathcal{H}^- n \mathcal{H}^+$. Our main result is the following theorem which characterizes the global flow of the hyperbolic restricted 3-boby problem.

THEOREM. *For all $n \in \mathbb{N}$ the following statements hold.*

1. *There is at least one orbit of type $\mathcal{P}^- n \mathcal{P}^+$ with n odd.*
2. *There are at least an 1-dimensional continuum of orbits of type $\mathcal{P}^- n \mathcal{H}^+$ and $\mathcal{H}^- n \mathcal{P}^+$.*
3. *There are at least a 2-dimensional continuum of orbits of type $\mathcal{H}^- n \mathcal{H}^+$.*

This theorem is proved by using the Poincaré map associated to the surface $z = 0$.

There are two problems nearby to our problem that have been studied: the *elliptic Sitnikov problem* and the *hyperbolic collision restricted problem*. More precisely, in the first, we have two mass points of equal mass $m_1 = m_2 = 1$ moving under Newton's law of attraction in elliptic orbits while their center of mass is at the origen. We consider a third mass point of mass $m_3 = 0$ moving on the straight line perpendicular to the plane of motion of the first two mass points and going through the center of mass. The Sitnikov problem is to describe the motion of the third mass point which is periodically excited by the first two mass points, see Moser[6]. The hyperbolic collision restricted problem was studied by Meyer and Wang[3]. Here we put two gravitational particles with masses $m_1 = m_2 = 1$ in the x-axis moving apart from each other with positive energy and their center of mass is fixed at the origin. The problem is to study the motion of an infinitely small mass point moving on the y-axis under the gravitacional field created by m_1 and m_2. In Meyer and Wang[3] the authors studied the global flow of the hyperbolic collision restricted problem in forward time.

REFERENCES

1. J. M. Cors and J. Llibre, *The global flow of the hyperbolic restricted three-body problem*, preprint 1993.

2. Q. Wang, *Qualitative study of n-body problem: Unitized momentum transformation and its application restricted isosceles three-body problem with positive energy*, "Space Dynamics and Celestial Mechanics," K.B. Bhatnagar, 1986, pp. 61-69.

3. R. K. Meyer and Q. Wang, *Global phase structure of the restricted isosceles three-body problem with positive energy*, to appear in Trans. Amer. Math. Soc.

4. R. McGehee, *Triple collision in the collinear three-body problem*, Invent. Math. **27** (1974), 191-227.

5. M. W. Hirsch, C. C. Pugh and M. Shub, "Invariant Manifolds," Springer-Verlag, 1977.

6. J. Moser, "Stable and random motions in dynamical systems with special emphasis on Celestial Mechanics," Annals of Mathematics Studies Number 77, Princeton University Press, New Jersey, 1973.

TYPES OF MOTION AND STABILITY OF HIERARCHICAL TRIPLE STAR SYSTEMS

L.G. Kiseleva[1], V.V. Orlov[2] and P.P. Eggleton[1]

[1]Institute of Astronomy, Madingley Road, Cambridge CB3 OHA, UK
[2]Astronomical Institute, St. Petersburg State University, Bibliotechnaya pl. 2
198904 St. Petersburg Peterhof, Russia

ABSTRACT

Hierarchical triple star systems with planar prograde motions are considered. Initially circular orbits are constructed for inner and outer binaries. The mass ratios of both inner and outer binaries are chosen between 1:1 and 100:1. The regions of dynamical stability and instability are investigated. For the Hill-type stability, a comparison between the empirical and analytical criteria is made. The types of motion in different regions are described. Some escapes without any close triple approach are discovered.

INTRODUCTION

The gravitational three-body problem is still an attractive and exciting puzzle (see, e.g. Marchal, 1990). This problem is very simple in formulation; however it contains a large number of difficulties and diversities in its realization. Many advances in science have been extracted from this problem. One of the fundamental problems is the transition from stability to instability of the motion. In this case, the three-body problem is a good template for study. There are some numerical and analytical criteria for stability of hierarchical stellar and planetary triple systems (see, e.g. Golubev 1967, 1968; Bozis 1976; Marchal and Bozis 1982; Golubev and Grebenikov 1985; Marchal 1990; Ge 1991; Kiseleva, Eggleton and Anosova 1993; and references therein).

We may represent a hierarchical triple system as a superposition of two subsystems: binary and distant body. For each subsystem, there is a region of possible motions, similar to Roche lobes in the Hill problem. In principle, four states are possible:

1. the lobes are closed and non-intersecting;

2. the lobes are open to infinity and non-intersecting;

3. the lobes are closed but intersecting;

From Newton to Chaos, Edited by A.E. Roy
and B.A. Steves, Plenum Press, New York, 1995

4. the lobes are open and intersecting.

The first and second cases correspond to so-called Hill-type stability in the three-body problem, when an exchange of components is impossible. However, in the second case, the escape of the distant body is possible without exchange. The third case gives quasiperiodical motions when exchanges of the bodies take place but escape is impossible. Finally, in the last case, both exchange and escape may be realised. The above stability criteria serve to separate the states described. Note that by the method of Hill stability one always finds open lobes (i.e. there is always a possibility of escape). The Sundman inequality plays an important role (see, e.g. Marchal, 1990). Application of this inequality allows us to separate the two first and two last cases.

The analytical criterion for the Hill-type stability (with respect to exchange) in the general three-body problem was independently derived by Golubev (1967, 1968; Golubev and Grebenikov, 1985), Bozis (1976), Szebehely and Zare (1977). This criterion is based on the quantity

$$s = -c^2 H/(G^2 \overline{m}^5) \tag{1}$$

where c is the angular momentum of the triple system, H its total energy, G the constant of gravity, and $\overline{m}$ the mean mass of the components. If the actual value of s exceeds a critical one, s_{cr}, then the triple system is stable with respect to exchange. The task is to separate the first and second states, as well as the third and fourth ones. A convenient example for such studies is the general three-body problem with initially circular orbits.

In the previous work (Kiseleva et al. 1993, thereafter KEA) we defined numerically the regions of hierarchical stability for planar triple star systems with prograde initially circular motions and out of phase initial configuration of inner and outer binaries.

Two parameters for the mass ratios were employed,

$$\alpha = log_{10}(m_1/m_2), \quad \beta = log_{10}((m_1 + m_2)/m_3) \tag{2}$$

Here m_1, m_2 (with $m_1 \geq m_2$) are the masses of the inner binary components and m_3 is the mass of the distant body. A third parameter X_0, the initial period ratio (outer/inner), was taken as in a sequence of values, from large values, for which the system appeared stable, to successively smaller values until the system broke up. The decrease ΔX in X_0 between successive calculations was usually taken to be 0.01. The values $\alpha = 0.0, 0.2, \cdots 2.0$ and $\beta = -2.0, -1.8, \cdots 2.0$ were considered.

Table 1 summarises the results we obtained over the whole range of α, β that we investigated in KEA. The Table 1 gives $\overline{X}_{min}$, the lowest value of the mean period ratio (outer/inner) $\overline{X}$ for which the system was apparently stable (the ratio $\overline{X}$ was calculated as an average during the evolution for 100 outer initial orbital periods and in a number of cases for 1000 or 10000 P_{out}). This was not necessarily the value of $\overline{X}$ at the lowest stable value (say, X_0^{min}) of X_0, since $\overline{X}(X_0)$ might reach a local minimum shortly before the first clearly disruptive case. The choice of $\overline{X}$ instead of the ratio of the semi-major axes used by other authors was connected with the fact that the observational determination of periods (and their ratio) is more reliable than the semi-major axes in real systems. X_0 cannot be measured in observed systems at all. X_0 and $\overline{X}$ do not usually differ by more than about 5% for stable cases. Note that as a rule X_0 is a little bit greater than $\overline{X}$ for all values of X_0 which are not close to critical values for the out-of-phase initial configurations which we considered. In situations with a 'disruptive resonance' described below, we give two values for $\overline{X}_{min}$, one being the least value found in the region above the disruptive resonance, and the other in the region below.

Table 1. Lowest values of the average period ratio $\overline{X}$ for stable triple systems

α	0.0	0.2	0.4	0.6	0.8	1.0	1.2	1.4	1.6	1.8	2.0
m_2/m_{12}	*.50*	*.39*	*.28*	*.20*	*.14*	*.09*	*.06*	*.04*	*.025*	*.016*	*.01*
β (m_3/m_{12})											
-2.0 (*100*)	5.90	5.90	6.03	6.10	6.15	6.17	6.19	6.21	6.21	6.23	6.20
-1.8 (*63*)	5.90	5.95	6.07	6.14	6.18	6.12	6.18	6.24	6.24	6.24	6.23
-1.6 (*40*)	5.90	5.95	6.07	6.16	6.20	6.26	6.22	6.29	6.29	6.28	6.30
-1.4 (*25*)	5.90	6.00	6.08	6.18	6.23	6.26	6.28	6.29	6.29	6.32	6.32
-1.2 (*16*)	5.89	5.95	6.09	6.15	6.23	6.27	6.25	6.30	6.33	6.32	6.32
-1.0 (*10*)	5.75	5.90	5.98	6.14	6.18	6.23	6.29	6.31	6.31	6.31	6.32
-0.8 (*6.3*)	5.67	5.67	5.91	6.03	6.11	6.17	6.21	6.24	6.22	6.21	6.22
-0.6 (*4.0*)	5.45	5.60	5.74	5.88	5.99	6.05	6.09	6.10	6.09	6.10	6.10
-0.4 (*2.5*)	5.20	5.31	5.47	5.62	5.73	5.79	5.84	5.87	5.87	5.88	5.88
-0.2 (*1.6*)	4.81	4.93	5.07	5.20	5.32	5.38	5.42	5.46	5.47	5.47	5.47
0.0 (*1.0*)	4.37	4.40	4.50	4.68	4.77	4.81	4.87	4.90	4.91	4.92	4.92
0.2 (*.63*)	4.27	4.25	4.21	4.40	4.11	4.11	4.17	4.21	4.24	4.24	4.25
				4.15							
0.4 (*.40*)	**4.31**	**4.31**	4.25	4.20	4.14	4.06	4.00	3.95	3.61	3.50	3.52
										3.44	*3.46*
0.6 (*.25*)	**4.32**	**4.32**	**4.30**	**4.22**	**4.18**	4.11	3.94	3.91	3.89	3.87	3.85
	3.67	**3.71**	*3.68*	*3.65*	*3.62*					*3.73*	*3.73*
0.8 (*.16*)	**4.32**	**4.33**	**4.29**	**4.25**	**4.20**	3.95	3.92	4.00	3.89	3.90	3.88
	3.63	**3.65**	**3.62**	**3.55**	*3.53*	*3.45*	*3.42*	*3.35*	*3.28*	*3.30*	*3.22*
1.0 (*.10*)	**4.28**	**4.30**	**4.29**	**4.23**	**3.45**	3.37	3.30	3.26	3.21	3.19	3.17
	3.58	**3.58**	**3.54**	**3.50**							
1.2 (*.063*)	**4.29**	**4.29**	**4.25**	3.45	**3.39**	**3.32**	3.25	3.21	3.17	3.14	3.10
	3.54	**3.52**	**3.50**								
1.4 (*.040*)	**4.24**	3.48	**3.46**	**3.40**	3.35	3.25	3.21	3.16	3.07	3.05	3.04
	3.51										
1.6 (*.025*)	**3.47**	**3.47**	**3.43**	**3.36**	**3.31**	**3.24**	**3.17**	**3.10**	**3.03**	2.59	2.58
									2.64		
1.8 (*.016*)	**3.44**	**3.43**	**3.39**	**3.34**	**3.28**	**3.21**	**3.12**	**3.06**	**3.04**	2.57	2.43
								2.61	**2.58**	2.42	
2.0 (*.010*)	**3.41**	**3.41**	**3.37**	**3.31**	**3.25**	**3.18**	**3.10**	**3.02**	2.59	2.37	2.39
								2.59			

It is not easy to assess the accuracy of these entries, most of which are based on computations lasting 100 outer orbits. In several of the cases which looked most interesting, we explored the boundary between stability and instability using longer integrations, up to 10000 outer orbits. This usually moved the boundary of stability by 0.02 - 0.04 in X_0, but although we think this is likely to be representative of the Table 1 as a whole we cannot claim this to be established. The main trend in Table 1 is from larger values $\overline{X}_{min} \sim 6$ at negative β to smaller values $\overline{X}_{min} < 3$ at positive β, with a minor dependence on α.

Table 2. Values X_0^{min} for last stable configurations in KEA simulations

α		0.00	0.20	0.40	0.60	0.80	1.00	1.20	1.40	1.60	1.80	2.00
m_1/μ		*0.50*	*0.39*	*0.28*	*0.20*	*0.14*	*0.09*	*0.06*	*0.04*	*0.025*	*0.016*	*0.01*
β	m_3/μ											
-2.00	*100*	6.09	6.14	6.28	6.31	6.37	6.40	6.45	6.46	6.45	6.47	6.49
-1.80	*63*	6.09	6.19	6.30	6.38	6.42	6.45	6.48	6.50	6.53	6.53	6.54
-1.60	*40*	6.09	6.22	6.30	6.41	6.50	6.54	6.53	6.59	6.60	6.60	6.59
-1.40	*25*	6.07	6.20	6.36	6.46	6.51	6.58	6.60	6.60	6.62	6.64	6.67
-1.20	*16*	6.11	6.20	6.37	6.42	6.52	6.57	6.58	6.63	6.66	6.66	6.67
-1.00	*10*	5.99	6.10	6.26	6.39	6.47	6.52	6.59	6.60	6.64	6.65	6.66
-0.80	*6.3*	5.86	6.00	6.15	6.29	6.37	6.44	6.50	6.54	6.54	6.55	6.57
-0.60	*4.0*	5.60	5.80	5.95	6.10	6.20	6.29	6.34	6.37	6.40	6.42	6.43
-0.40	*2.5*	5.29	5.47	5.64	5.80	5.92	6.01	6.07	6.11	6.13	6.15	6.14
-0.20	*1.6*	4.88	5.04	5.19	5.35	5.48	5.56	5.62	5.66	5.69	5.70	5.71
0.0	*1.0*	4.37	4.56	4.57	4.82	4.90	4.97	5.03	5.07	5.09	5.11	5.12
0.20	*0.63*	4.29	4.31	4.28	4.53	4.15	4.18	4.26	4.32	4.36	4.38	4.39
0.40	*0.40*	4.37	4.38	4.33	4.25	4.26	4.12	4.06	4.01	3.64	3.51	3.55
0.60	*0.25*	4.39	4.39	4.34	4.29	4.22	4.18	4.04	4.00	3.96	3.96	3.94
0.80	*0.16*	4.37	4.38	4.32	4.27	4.18	4.05	4.00	3.96	3.90	3.96	3.88
1.00	*0.10*	4.32	4.33	4.28	4.22	3.45	3.39	3.31	3.24	3.16	3.18	3.13
1.20	*0.063*	4.31	4.29	4.26	3.47	3.38	3.30	3.21	3.12	3.09	3.06	2.99
1.40	*0.040*	4.27	3.54	3.48	3.40	3.31	3.20	3.15	3.08	2.93	2.88	2.83
1.60	*0.025*	3.50	3.50	3.42	3.33	3.26	3.18	3.05	2.98	2.90	2.47	2.43
1.80	*0.016*	3.48	3.48	3.39	3.29	3.22	3.12	3.03	2.98	2.87	2.46	2.06
2.00	*0.010*	3.44	3.42	3.38	3.26	3.17	3.11	3.03	2.91	2.49	2.07	2.01

Note that only for sets of α and β in the area shown in boldface in Table 1 do we have an escape of the distant component without exchange near X_0^{min}.

For all other mass ratios we considered we observe different types of exchange which lead to the formation of a new close binary in the triple system. This new binary can have short or long life-time, but in any case the initial hierarchical configuration is broken in these cases. But there is a possibility that a very narrow range of X_0 just below X_0^{min}, where an escape of distant body without exchange takes place, exists for systems with $\beta \leq 0.4$ (i.e., when $m_3 < m_1 + m_2$); a step of $\Delta X < 0.01$ must be used in order to discover this.

KEA found a region of 'disruptive resonances' on their (α, β) plan. In this region the system disrupts in a narrow range of initial period ratio X_0, before settling down to the same steady progression of stable orbits as X_0 decreases further. Only at some smaller value still for X_0 does the system disrupt again, apparently for good.

In the present work, we compare analytical (Golubev 1967, 1968; Golubev and Grebenikov 1985; Szebehely and Zare 1977) and numerical (KEA) conditions for dynamical stability of hierarchical triple systems. We also study the dynamical behaviour of unstable triple systems near the limit of stability. In our simulations, the TRIPLE code of Aarseth (see Aarseth and Zare 1974) is used. The initial conditions are chosen as in KEA.

RESULTS AND DISCUSSION

First, let us check the c^2H (Szebehely and Zare 1977) stability criterion for the critical period ratios found in KEA (Table 2). The results are given in Table 3.

Table 3. Differences $X_0^{cr} - X_0^{min}$

α / β	0.00	0.20	0.40	0.60	0.80	1.00	1.20	1.40	1.60	1.80	2.00
-2.00	0.00	0.05	0.00	0.04	0.02	0.03	0.00	0.01	0.03	0.01	0.00
-1.80	0.01	0.03	0.02	0.01	0.03	0.04	0.03	0.03	0.01	0.02	0.01
-1.60	0.02	0.03	0.06	0.04	0.01	0.01	0.05	0.01	0.01	0.02	0.03
-1.40	0.04	0.07	0.04	0.04	0.06	0.04	0.05	0.07	0.06	0.05	0.02
-1.20	0.00	0.09	0.07	0.13	0.11	0.11	0.13	0.10	0.08	0.09	0.09
-1.00	0.10	0.20	0.21	0.21	0.21	0.21	0.17	0.18	0.15	0.15	0.14
-0.80	0.18	0.29	0.33	0.33	0.33	0.31	0.28	0.26	0.27	0.26	0.25
-0.60	0.35	0.43	0.50	0.50	0.48	0.44	0.41	0.39	0.37	0.35	0.34
-0.40	0.52	0.65	0.72	0.71	0.68	0.62	0.57	0.53	0.51	0.49	0.50
-0.20	0.72	0.91	1.01	1.01	0.95	0.88	0.82	0.77	0.73	0.71	0.69
0.00	0.96	1.14	1.40	1.30	1.27	1.19	1.11	1.04	0.99	0.95	0.93
0.20	0.73	1.09	1.40	1.28	1.69	1.62	1.49	1.37	1.29	1.24	1.21
0.40	0.33	0.71	1.02	1.22	1.21	1.28	1.25	1.22	1.52	1.61	1.54
0.60	0.01	0.39	0.70	0.84	0.87	0.81	0.83	0.76	0.71	0.65	0.62
0.80	**-0.22**	0.14	0.44	0.55	0.58	0.56	0.46	0.35	0.30	0.16	0.18
1.00	**-0.36**	**-0.01**	0.25	0.35	1.02	0.91	0.79	0.69	0.63	0.51	0.49
1.20	**-0.49**	**-0.13**	0.10	0.91	0.87	0.74	0.61	0.49	0.36	0.27	0.25
1.40	**-0.55**	0.52	0.76	0.84	0.78	0.65	0.45	0.28	0.24	0.15	0.10
1.60	0.15	0.48	0.74	0.81	0.71	0.53	0.39	0.20	0.07	0.34	0.25
1.80	0.13	0.45	0.71	0.78	0.67	0.50	0.29	0.07	**-0.06**	0.17	0.43
2.00	0.14	0.48	0.68	0.77	0.67	0.45	0.22	0.05	0.21	0.43	0.34

Table 2 contains the critical quantities X_0^{min} for each pair (α, β) considered in KEA. Here we indicate the ratios of the masses m_1/μ and m_3/μ where $\mu = m_1 + m_2$. In Table 3 are the differences between the values X_0^{cr} and the initial ratios X_0^{min} of the periods for the last stable cases found in the experiments described in KEA. The values X_0^{cr} are calculated according to the c^2H -criterion using the initial coordinates and velocities of the bodies. Let us note that the two-body approximations used by Szebehely and Zare (1977) give smaller ratios X_0^{cr} than an exact formula derived in the present work using (1) and the third Keplerian law (see Table 4). This exact formula connects the values s and X for the case when we start with the two circular orbits out of phase with each other by 90° (see KEA).

The approximate formula connecting s_{cr} and X_0^{cr} does not reveal a number of cases unstable with respect to exchanges. The corresponding differences between X_0^{cr} calculated with exact and approximate formulae are shown in italics in Table 3. In these

Table 4. Differences between X_0^{cr} calculated by exact and approximate formulae

α / β	0.00	0.20	0.40	0.60	0.80	1.00	1.20	1.40	1.60	1.80	2.00
-2.00	*0.29*	*0.28*	*0.28*	*0.28*	*0.27*	*0.27*	*0.27*	*0.27*	*0.27*	*0.27*	*0.27*
-1.80	*0.29*	*0.29*	*0.28*	*0.27*	*0.27*	*0.27*	*0.26*	*0.27*	*0.26*	*0.27*	*0.26*
-1.60	*0.29*	*0.29*	*0.27*	*0.27*	*0.27*	*0.26*	*0.26*	*0.26*	*0.26*	*0.26*	*0.26*
-1.40	*0.28*	*0.28*	*0.27*	*0.26*	*0.26*	*0.26*	*0.26*	*0.26*	*0.25*	*0.26*	*0.25*
-1.20	*0.28*	*0.27*	*0.26*	*0.25*	*0.25*	*0.25*	*0.24*	*0.24*	*0.24*	*0.24*	*0.25*
-1.00	*0.27*	*0.26*	*0.25*	*0.25*	*0.24*	*0.24*	*0.23*	*0.23*	*0.23*	*0.23*	*0.23*
-0.80	*0.26*	0.26	0.24	0.24	0.23	0.22	0.22	0.22	0.22	0.22	0.22
-0.60	0.25	0.24	0.23	0.22	0.21	0.21	0.21	0.21	0.21	0.21	0.20
-0.40	0.24	0.22	0.21	0.20	0.20	0.19	0.19	0.18	0.18	0.19	0.19
-0.20	0.22	0.21	0.19	0.18	0.18	0.17	0.17	0.17	0.17	0.17	0.17
0.00	0.20	0.18	0.17	0.16	0.15	0.14	0.15	0.15	0.14	0.14	0.15
0.20	0.18	0.16	0.15	0.14	0.13	0.12	0.13	0.12	0.12	0.12	0.12
0.40	0.16	0.14	0.12	0.12	0.11	0.11	0.10	0.10	0.10	0.10	0.10
0.60	0.14	0.12	0.11	0.10	0.09	0.09	0.09	0.09	0.08	0.09	0.08
0.80	0.13	0.11	0.10	0.08	0.08	0.07	0.07	0.06	0.06	0.06	0.06
1.00	0.12	0.10	0.08	0.07	0.06	0.06	0.05	0.05	0.05	0.05	0.05
1.20	0.11	0.09	0.08	0.07	0.06	0.05	0.05	0.04	0.04	0.04	0.04
1.40	0.11	0.09	0.07	0.06	0.05	0.04	0.04	0.03	0.03	0.03	0.03
1.60	0.10	0.08	0.07	0.05	0.04	0.03	0.04	0.03	0.03	0.02	0.02
1.80	0.10	0.08	0.07	0.05	0.04	0.04	0.03	0.03	0.02	0.02	0.02
2.00	0.10	0.08	0.06	0.05	0.04	0.04	0.03	0.03	0.02	0.02	0.02

cases the mass of the distant body is ≥ 10 times larger than the mass of the inner binary.

Table 3 shows a good agreement (within 5%) between the numerical and analytical criteria for $\beta \leq -1$. Usually, the analytical ratios X_0^{cr} are significantly overestimated with respect to X_0^{min}. The maximum of difference $X_0^{cr} - X_0^{min}$ is achieved at $\alpha = 0.8$ and $\beta = 0.2$ and is $\simeq 1.7$. On the other hand, there are two regions of underestimation when $X_0^{cr} < X_0^{min}$. These regions are shown by boldface in Table 3.

We have plotted the trajectories for both the regions. In both cases, the escape of the light distant body takes place without exchanges and close triple approaches. Therefore, the c^2H-criterion with respect to exchange does work in this situation; however it says nothing about escape. Some additional condition for dynamical stability is necessary in this case.

Now let us consider the zones of 'disruptive resonances' found in KEA inside the area of stability. These resonances may be associated (but not very clearly) with 4:1 and 3:1 commensurability in $\overline{X}$. We can distinguish two kinds of resonances. In the first case ($\alpha \leq 0.2$), the resonance values $X_0^{res} > X_0^{cr}$ (see Table 1 and Table 3). Therefore, such triple systems must be stable in the Hill sense. However, the escape takes place without an exchange or a close triple approach in these cases. The ejection of the distant body with an increasing amplitude is observed. In the second case ($\alpha \geq 0.4$), the inequality $X_0^{res} < X_{cr}$ is satisfied. Therefore, the Hill-type stability criterion is violated and escape is also possible. The exchange of the components is observed after a few revolutions of the outer binary. Then one body has a series of ejections with return. These ejections are sometimes followed by capture to bound orbits with either of the other bodies. The evolution may end by escape of one body.

In our N-body simulations the escape of the distant component without exchange for the 'upper' boundary of stability is observed in all cases of resonance with $\alpha \leq 0.8$. For the 'lower' boundary (below the resonance) we also have a number of escapes without exchange.

Between the resonance zone and the area of instability there is a region of stability. Similar zones are also seen in Table 1 in Ge's paper (1991) for the initially elliptic three-body problem. Inside the region of stability $X_0 < X_{cr}$, therefore the c^2H-criterion is not satisfied and exchange is possible in principle. However, exchange is not observed during the time of calculations (upto 10000 periods of the initial outer binary in some cases). There are two possibilities for this behaviour:

1. these regions are actually stable and the regions of orbit are closed, in spite of intersecting the regions of possible motions;

2. the systems are unstable but have a very long life-time, the channel between two lobes is very narrow and the probability of exchange is very small.

The clarification of this situation is a subject of further investigations.

Considering a significant number of typical trajectories in different areas on the (α, β) plane, we suppose that the type of instability in triple systems near the limit of stability depends on the mass ratio. Near the boundary of stability the escapes of the third body without exchange and without close triple approaches takes place only for $\beta \geq 0.4$ and some sets of α. For other pairs (α, β) some exchanges of the bodies are observed. Sometimes we can see only the short ejections and interplay without escape. The last states are typical for $\alpha \geq 1.0$ and $\beta \leq 0.0$. In other cases, the exchanges are followed by long-term distant ejections and finally escapes of the bodies. The series of exchanges and ejections, as a rule, takes place at $\alpha \leq 1.0$ and $\beta \leq 0$ and at $\alpha \geq 1$ and $\beta \geq 1$. For $0.2 < \beta < 1.4$, there is a transient region where we have two zones of stability, separated by disruptive resonances. The resonances may result in both escape without exchange ($\alpha \leq 1.0$) and a series of exchanges ($\alpha > 1$). The lower boundary of the second zone of stability corresponds to escapes or exchanges as well. Indeed, the picture described has a rather preliminary character, and the actual zones of instability may be even more complicated. Their study is the subject of further work.

CONCLUSIONS AND PROSPECTS

Our study of dynamical stability in the triple star systems with initially circular orbits and mass ratios between 1:1 and 100:1 for the inner and outer pairs shows the following:

1. The analytical c^2H-criterion of the Hill-type stability is in good agreement with numerical results of KEA for the mass ratios $m_3/(m_1 + m_2) \geq 10$ at any ratio m_1/m_2.

2. For all other cases under study (except two zones), the c^2H-criterion overestimates significantly the limiting period ratio X for dynamical stability.

3. In the plane (α, β) there are two zones where the c^2H-criterion underestimates the limiting ratio X.

4. Both above zones correspond to escape of the distant body without exchange and close triple approaches.

Thus, in most cases under study the analytical c^2H criterion of Hill-type stability gives an upper estimation of the critical period ratio for the dynamical stability. In further studies we are going to investigate the dynamics of the triple systems inside the zone of 'disruptive resonances' in order to understand the reasons for dynamical instability in such cases. We also intend to make a more accurate numerical determination of the dynamical instability in such cases. We also intend to make a more accurate numerical determination of the dynamical stability, especially in the case when the distant body has a small mass ($\beta \geq 0.4$).

REFERENCES

1. Aarseth, S.J. and Zare, K. 1974, *Celest. Mech.,* **10**, 185
2. Bozis, G. 1976, *Astrophys. Space. Sci.*, **43**, 355
3. Ge, Y-C., 1991, in *Predictability, Stability and Chaos in N-Body Dynamical Systems* (ed. A.E. Roy), Plenum Press, New York
4. Golubev, V.G. 1967, *Doklady Acad. Nauk USSR*, **12**, 529
5. Golubev, V.G. 1968, *Doklady Acad, Nauk USSR*, **13**, 373
6. Golubev, V.G. and Grebenikov, E.A., 1985, *The Three-body Problem in Celestial Mechanics*, Moscow State Univ.
7. Kiseleva, L.G., Eggleton, P.P. and Anosova, J.P., 1993, *MNRAS*, in press (KEA)
8. Marchal, C. 1990, *The Three-body Problem*, Elsevier Publishers, Amsterdam
9. Marchal, C., Bozis, G. 1982, *Celest. Mec.*, **26**, 311
10. Szebehely, V. and Zare, K., 1977, *AA*, **58**,145

ON THE ROLE OF THE MOMENT OF INERTIA IN THREE-BODY SCATTERING

Roger A. Broucke
Department of Aerospace Engineering
University of Texas
Austin, TX 78712

ABSTRACT

We made some numerical simulations of the planar general three-body problem, in order to understand scattering and disruption of triple stellar systems. We introduce a new classification principle of the solutions with negative energy. Most of these solutions are Hyperbolic-Elliptic, according to the terminology of J. Chazy. We classify them according to the number n of variations (maxima or minima) of their moment of Inertia $I(t)$. These classes with a given n have clearly defined boundaries: Chazy's so-called Parabolic - Elliptic solutions. These families of Parabolic - Elliptic solutions usually terminate at isolated triple collision solutions. We show a few examples. This classification also gives us a quantitative measure of the complexity of the solutions, (the amount of interplay).

In order to illustrate the new classification scheme, we integrated a two-parameter collection of about 1 million solutions with zero initial velocities. The numerical integrations are all performed with a Steffensen Recurrent Power Series Method, with an automatic step selection, with an order of 14 to 16 and with a 2-dimensional Levi-Civita Regularization.

INTRODUCTION

In the last 30 years a very large number of numerical simulations of the general three-body problem have been made, in order to understand scattering and disruption of triple stellar systems for which the usual outcome is the escape of a single body in one direction and a binary escaping in the other direction. These are Chazy's so-called Hyperbolic-Elliptic solutions.

Among the early important large-scale numerical simulations, we must mention the works of Valtonen(1975), Monaghan (1976) as well as Valtonen and Heggie (1979).

Some of the most important of these studies are by Graziani (1981) and Black (1982), who developed a numerical criterion for the separation in regions of bounded and separating solutions.

From Newton to Chaos, Edited by A.E. Roy
and B.A. Steves, Plenum Press, New York, 1995

Anosova(1986) and Anosova and Orlov (1992) try to estimate the lifetime of the system before an escape occurs, due to a triple close approach. Kiseleva and Orlov (1993) as well as Kiseleva, Eggleton and Anosova (1994) correlate their numerical experiments with the known observational data on galaxies.

Piet Hut and Bahcall (1983) and Piet Hut (1983) also made extensive numerical simulations to determine regions of escaping versus temporary bounded systems. They especially use the concept of scattering cross section that is often used in particle physics. The same ideas are widely used in the papers by Benz and Hills (1992) and many other papers by Hills (1991). While we are using a purely point-mass approach, his group frequently used a three-dimensional Hydrodynamics Code(Davies, Benz and Hills, 1992).

Some attempts have also been made to classify the solutions but no strong general classification principles seem to have been found, since the work of Chazy. Some classifications were made however, by J. Anosova (1991) and by Piet Hut (1983) who describes the complexity of the solutions as streaming rivers and watersheds (his page 321).

The work of Greenberg, Carusi and Valsecchi (1988)(1990) is rather different in the sense that Geometrical-Analytical reasoning is used instead of pure numerical simulations. They expand on some basic ideas from Opik's "Interplanetary Encounters" (1976).

All this previous work could be basically summarized in two ideas: first of all the attempts to separate the solutions in two categories, escaping versus bounded. Secondly the major role of the triple close approaches is emphasized, (Szebehely,1974).

Our present work introduces a new classification principle for solutions with negative energy. Most of these solutions are Hyperbolic- Elliptic. We classify them according to the number n of variations (maxima) of their moment of inertia $I(t)$. As for the Theory of the moment of inertia in the 3-body problem, we refer to the last chapter of the book by Birkhoff, (1927) or Marchal (1990), as well as Zare (1981). These classes with a given n also have clearly defined boundaries: Chazy's so-called Parabolic - Elliptic solutions. In Figure 1 we show an example of a solution with $n = 1$: $I(t)$ has only one minimum. It is a rather simple solution: the smallest mass m_3 passes almost straight in between the two other masses before escaping. Figure 2 shows a solution that is quite a bit more complicated:$I(t)$ has 3 minima before one of the bodies escapes, as can be seen in the graph of $I(t)$ shown in figure 3.

The permanently bounded solutions which are also one of Chazy's classes (in the case of negative energy) appear in our classification as a subclass of solutions with an infinite number of maxima of the moment of inertia $I(t)$.

Our classification principle applies to the solutions with all angular momentum values although the illustrations in the present paper are with zero angular momentum. We show that, in this last case, the triple collision solutions play an extremely important role in the classification. The existence of a large number of triple collision solutions greatly increases the complexity of the solutions manifold, and its subdivisions in classes with a fixed number of maxima of $I(t)$.

In the past, several authors have suggested that the strong three-particle interaction is chaotic or quasi-ergodic. In particular, the complexity of triple close approaches is being thought of as a form of chaos. However, with our new classification principle, based on the variations of the moment of inertia $I(t)$, the impression of chaos weakens and some order begins to appear in this complex phenomenon of three-body scattering.

We do not give the equations of motion here because they are too well known. We

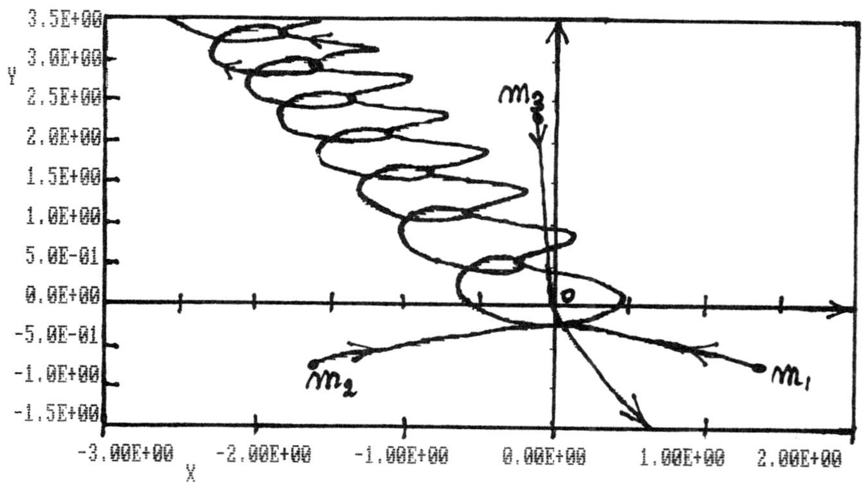

Figure 1. An example of one of the most simple Hyperbolic-Elliptic solutions, of Complexity $C_3^{(1)}$. The heliocentric initial conditions are (-1.5, 3.0) for m_3. with the standard values (-3.0, 0.0) for m_2, the largest mass m_1 being at the origin. The initial velocities are all zero.

refer to the book by C. Marchal (1990). In the absence of close approaches we integrate in heliocentric coordinates, with a system of ODE's of order 8, with the largest of the 3 masses at the origin. However, if one side of the triangle is less than 15 percent of the perimeter, we perform a standard Levi-Civita regularization with the algorithm that we developed about 20 years ago, (Davoust and Broucke, 1982).

The numerical integrations are all performed with a Steffensen Recurrent Power Series Method, with an automatic step selection, and with a maximum order of 30 . Most of the Integrations were done with power series of order of 14 to 16. This method is appreciably more accurate than the Runge Kutta Fehlberg method (of order 8) that we also used frequently in the past.

THE BASIC CHARACTERISTICS OF OUR SOLUTIONS

We will investigate a class of planar solutions of the general three-body problem with the following characteristics, (which are not necessarily independent). They all have zero initial velocities, zero angular momentum and negative energy. The three masses will be fixed ($m_1 = 5; m_2 = 4; m_3 = 3$) and the constant of gravity is 1.

The solutions are symmetric in time with respect to the origin t=0. For instance the moment of inertia $I(t)$ also has the same symmetry: $I(t) = I(-t)$. Therefore, we only study the solutions for positive times. The moment of inertia $I(t)$ has a local maximum $I(0)$ at $t = 0$.

Our problem depends on only two essential parameters. We will take them to be the initial positions $[x_3(0), y_3(0)]$ of m_3 , (section 3 below). In our numerical integrations we will put the large mass $m_1(= 5)$ at the origin and we put the second mass m_2 at $(-3, 0)$, so that the base of the triangle formed by the three masses is 3. We will only study the solutions with $y_3(0) > 0$.

In the initial numerical explorations, $x_3(0)$ was varied by steps of 0.01 from -7.5 to $+7.5$ and $y_3(0)$ from 0.0 to $+7.5$ with the same step, essentially a million orbits. This took 3 months of IBM-PC CPU-time, on an Intel - 80386 Processor, roughly at the rate

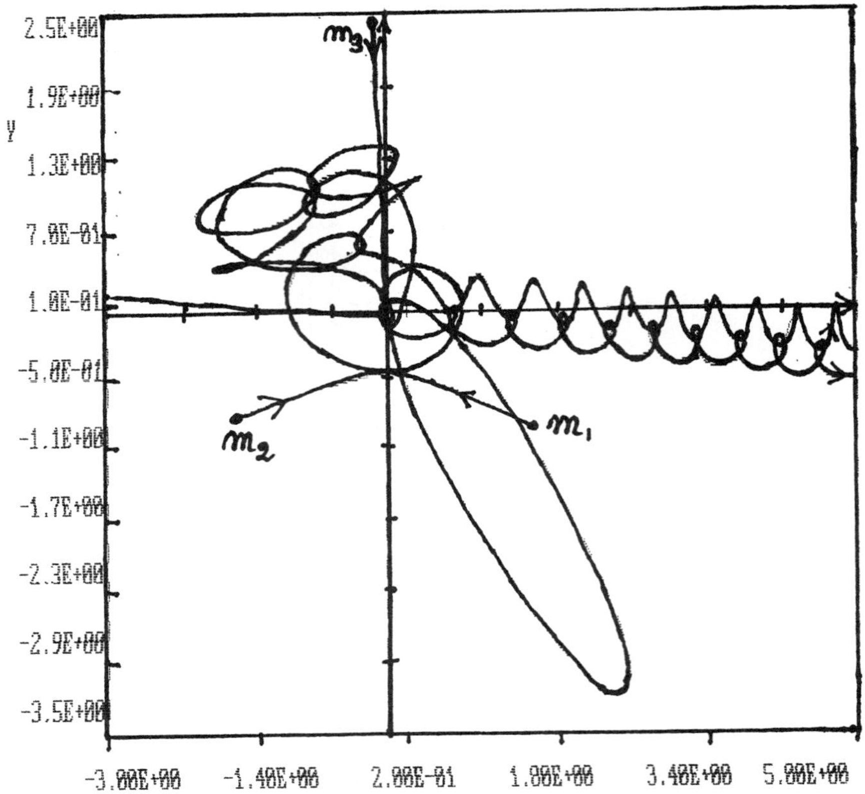

Figure 2. An example of a more complicated Hyperbolic-Elliptic solution, of Complexity $C_k^{(3)}$. The heliocentric initial conditions are (-1.5, 3.5), so that this point falls in the white area of the histogram in figure 6.

of 5 seconds per orbit. We generally used the Standish (1971) escape condition to stop an integration. We are aware of a large number of escape tests, by Birkhoff, Tevzadze, Griffith and North, Yoshida, Marchal, Saari. For a review, we refer to Marchal's book (1990) or the article by Anosova (1986).

All our numerical integrations are done in an eight-order heliocentric formulation when there are no binary close approaches and with an eight-order Jacobian formulation, regularized with a Levi-Civita transformation when there is a binary close approach. We define a binary close approach as a situation where one side of the triangle of the three masses is less than 15 percent of the perimeter. The constant 15 is actually arbitrary but we have made experiments with numbers ranging from 5 to 30 and 15 has been found to be a reasonable compromise between speed and precision.

We are aware of the sixth-order formulation of the general three-body problem with zero angular momentum: the so-called Wintner-Van Kampen symmetric formulation which uses the sides of the triangle as the three principal variables. We did extensive work with these variables, in particular a Lemaitre regularization (Lemaitre, 19) and a MacGehee Transformation, (MacGehee, 19..), but we will describe the results in a later publication.

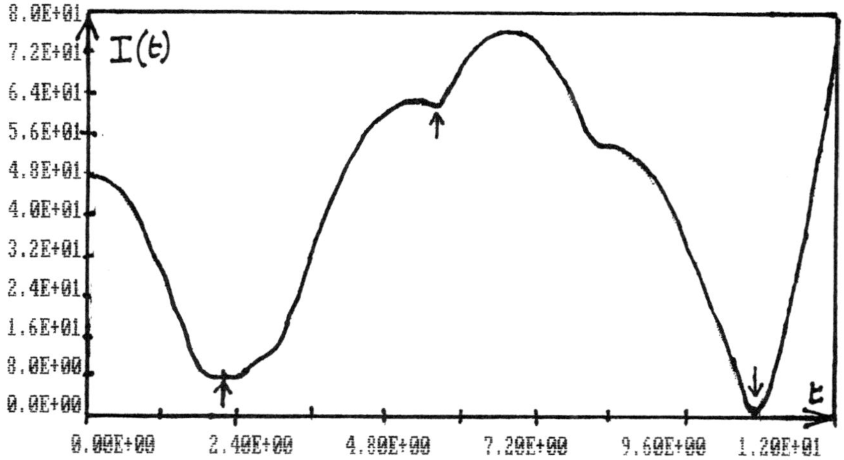

Figure 3. The moment of inertia $I(t)$ of the solution of figure 2. We see that the escape occurs at about $t = 12.0$, following a very close triple approach. There are 3 maxima and 3 minima.

THE BASIC INVARIANCE OF THE PROBLEM

We know that the formulation of the general three-body problem is invariant under several elementary transformations. We will take advantage of these invariances in order to reduce the number of initial values that need to be varied without any loss of generality on the totality of solutions that we want to study.

The basic rule is that the solutions depend only on the initial shape of the triangle but not on the position or orientation or size of the triangle in the xy-plane. These invariances are essentially related to the existence of the various first integrals, as we know from the theory of the Noether theorem. The translational position of the triangle is irrelevant because of the integrals of the center of mass. We avoid the corresponding redundancy by selecting a given point as a reference. In the integrations we use m_1 as the origin $(0, 0)$ of the coordinate system, while in the different plots of orbits in the present article, we use the center of mass as origin.

As for the rotational invariance, which corresponds to the integral of angular momentum, we introduce the convention that the side $m_2 - m_1$ of the triangle shall be parallel to the x-axis. As a matter of fact in the heliocentric frame, this side $m_2 - m_1$ will be assumed to have fixed length. This could be done without any loss of generality to the solutions of the problem because of the homogeneity of the potential function. We also note that this invariance can easily be taken care of by fixing the energy constant.

Finally, there is a last invariance that should be mentioned: to each solution starting on a direct triangle $m_1 m_2 m_3$ corresponds a mirror image solution with an initial similar but retrograde triangle. The two solutions correspond to changing the sign of $y_3(0)$. We avoid this ambiguity by restricting our work to positive values of $y_3(0)$ only.

In summary we are left with only two variable parameters $x_3(0), y_3(0)$, that are required to describe the solutions of the general three-body problem, which have all zero initial velocities. Among those two parameters, the second one is non-negative. It will thus be very convenient to discuss the solutions and their properties in the $[x_3(0), y_3(0)]$ - half-plane.

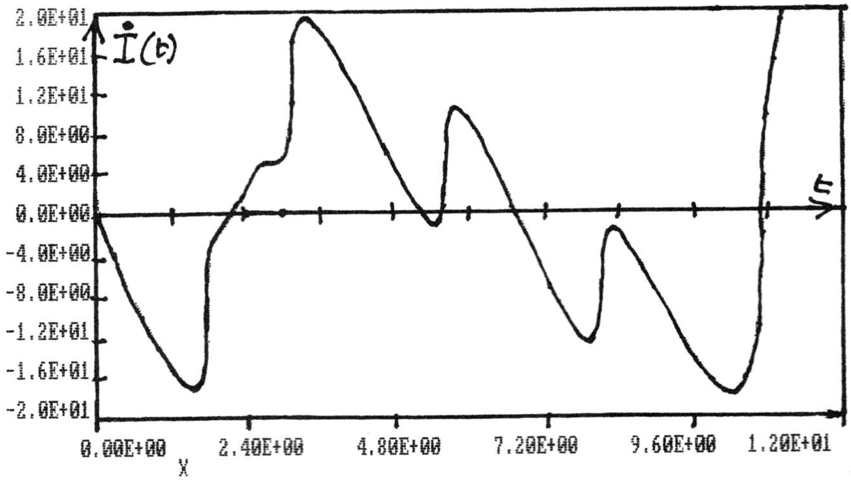

Figure 4. The first derivative of the moment of inertia $I(t)$ of the solution of figure 2. We clearly see the six roots, corresponding to the 6 extrema of $I(t)$.

In relation to the scaling of the solutions, let us mention that we could have scaled the solutions with the energy. For instance, the energy could have been taken equal to $-1/2$, which is the standard value corresponding to a semi-major axis $a = 1$ and a unit total mass, (Marchal, 1990). In fact, instead of using the masses $(5, 4, 3)$, we could have scaled them in such a way that $m_1 + m_2 + m_3 = 1$, which is the standard total mass used in the two-body problem and in the restricted three-body problem. Scaling the masses changes only the time-scale but not the form of the solutions.

THE CHAZY CLASSIFICATION OF THE FINAL TERMINATIONS

We recall here the basic classification of the possible final terminations in the 3-body problem, according to Chazy (1922), (1929). The detailed theory can also be found in Marchal (1990). In the present research we only need the case with negative Energy, but for completeness we quickly summarize all three cases.

1. The Positive Energy Orbits. There are basically three types of final evolutions; (the first 2 types form sets of measure 0). There are no bounded motions.

a) The Tri-Hyperbolic Orbits: All three particles escape on separate hyperbolic orbits and the three distances from the center of mass finally increase proportional to t.

b) The Hyperbolic-Parabolic Orbits: Two of the particles escape on parabolic orbits: their distance increases proportional to $t^{2/3}$ only. The third particle escapes on a hyperbolic orbit and its distance increases as t.

c) The Hyperbolic-Elliptic Orbits: A small binary system forms and the third particle escapes from it on a hyperbolic orbit: two sides of the triangle increase as t and the third one remains bounded. In a statistical sense, this is the generic case.

2. The Zero-Energy Orbits. Again there are no permanently bounded motions in this case: all bodies always go to infinity. There are two types of final motion.

a) The Hyperbolic Elliptic Orbits form the generic case: one binary system and a third mass escaping alone on a hyperbolic orbit.

b) The Tri-Parabolic Orbits: All three particles are on parabolic orbits. The sides of the triangle increase asymptotically as $t^{2/3}$ These solutions always tend to one of the

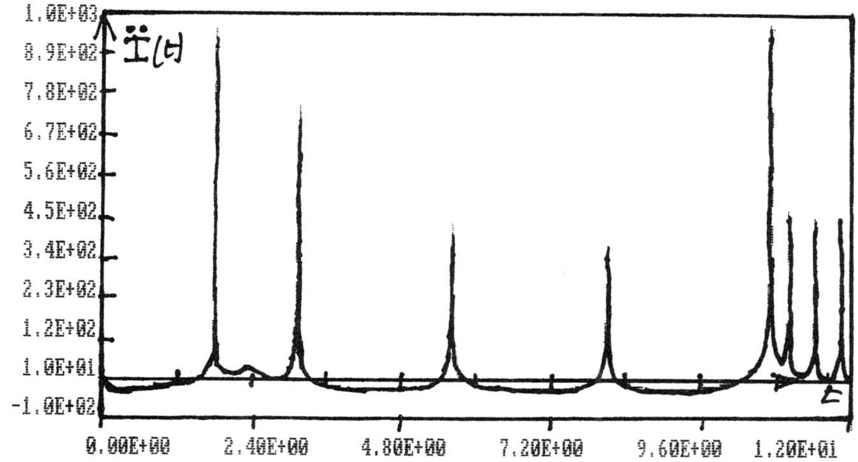

Figure 5. The second derivative of the moment of inertia $I(t)$ of the solution of figure 2. This is also the graph of the potential function $U(t)$. The vertical spikes correspond to all the close approaches, (binary as well as triple).

two central configurations: the Equilateral Triangle (Lagrangian case) or the collinear configuration (Eulerian case) where the distances satisfy the standard quintic equation.

3. The Negative Energy Orbits. We have essentially four types of solutions:

a) Bounded Motions are possible in the case of negative energy. This includes in particular the periodic solutions and the relative equilibria of the Lagrangian type (Equilateral Triangle) or the Eulerian type (collinear solutions satisfying the quintic equation).

b) The Hyperbolic-Elliptic Orbits form again the generic case. Our initial numerical explorations clearly confirm that, at least for our set of masses, the Hyperbolic-Elliptic orbits are by far the most abundant.

c) The Parabolic-Elliptic Solutions are solutions where two masses form a binary pair while the third one escapes from the pair along a parabolic orbit, such that the distance increases like $t^{2/3}$.

d) The hypothetical oscillatory solutions will not be considered here.

The last of Chazy's articles (1929) on the classification of the final evolutions relates to the solutions with negative total energy. In the wording of his article there exist 4 important classes; he uses the word "continuum". The inner continuum contains the bounded solutions.

This "continuum" is surrounded by a surface of P.E. Solutions which separates it from the three important continua of H.E. solutions with the same escaping body. In the same article, he says that the continua have no holes in them, (page 361).

Chazy does not mention the variations of the moment of inertia. He does not consider the case of different classes of H.E.-Orbits, of Complexity $M_k^{(n)}$ and $M_{k'}^{(n+1)}$, separated by a boundary of P.E.-Orbits of complexity $M_k^{(m)}$. Note that k and k' may be different integers, (1, 2 or 3). In other words, the escaping body is not necessarily the same.

The H.E.-solutions can be characterized by the energy of the escaping body, which is positive. However, towards a boundary of a class, this energy decreases and goes through zero at the exact boundary, where we have the Parabolic-Elliptic Orbits.

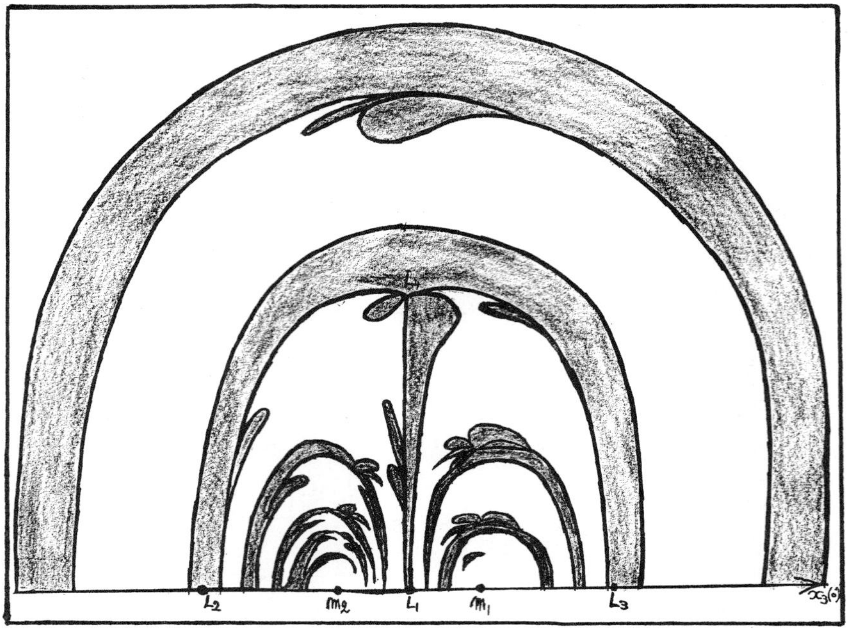

Figure 6. The principal histogram summarising a classification of a million solutions of the general 3-body problem. The grey areas correspond to the most simple Hyperbolic-Elliptic solutions, with only one minimum of the moment of inertia $I(t)$, before the escape of one of the 3 particles. The white areas correspond to more complex solutions with more minima. The boundaries, (black lines), correspond to Parabolic-Elliptic solutions. The horizontal axis corresponds to $x_3(0)$. The exploration was done from -7.0 to $+7.0$, with a step of 0.01. The vertical axis is $y_3(0)$, from 0 to $+7.0$, with the same step.

SUMMARY OF THE NEW CLASSIFICATION PRINCIPLES

The principal goal of our study is to classify the solutions, principally the Hyperbolic-Elliptic solutions. We discovered that indeed, they can be classified in well-defined connected groups, (with no holes in them, as Chazy writes), in the $[x_3(0), y_3(0)]$-plane, according to their complexity.

We quantify the complexity, or the amount of interplay, by the number of minima of the moment of inertia of the system, before an escape occurs. We say that a solution is of complexity $C_k^{(n)}$, if the moment of inertia has n minima before the escape of a body occurs. There are of course three such classes (or groups) according to which of the three bodies is the isolated escaper, as indicated by the subscript $k = 1$, 2 or 3.

Generally, the boundaries of these classes of Hyperbolic - Elliptic solutions of complexity $C_k^{(n)}$ are lines with Parabolic-Elliptic Orbits, with the same complexity. The important fact is that in the initial condition plane, the Parabolic - Elliptic solutions form lines, i.e one-parameter sets.

The next important question is the termination of the lines with Parabolic-Elliptic solutions. It was found that there are generally two possible answers: in a few cases, these lines turn around and form closed curves. The second possibility is in the Triple Collision Solutions. There exists in the initial condition plane many isolated points which result in triple collision solutions.

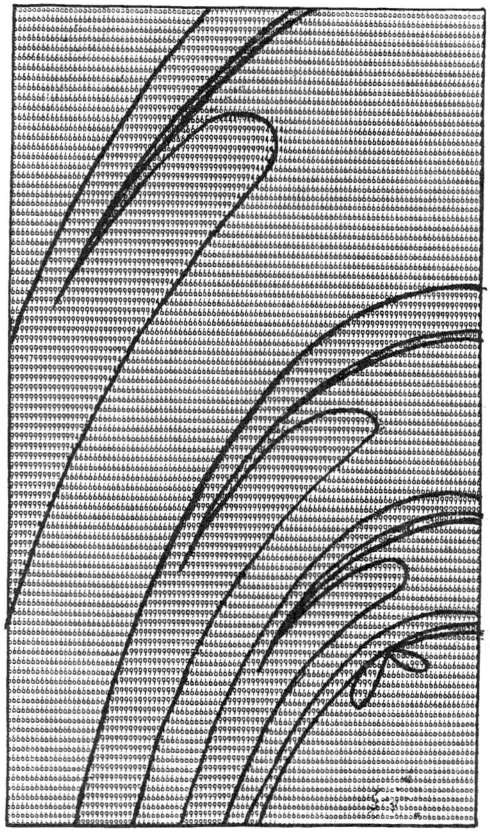

Figure 7. A more detailed histogram summarising a classification of some of our solutions of the general 3-body problem. The most simple solutions correspond to the codes 7, 8 or 9. The more complicated solutions have code 6. The boundaries, (black lines), correspond again to the Parabolic-Elliptic solutions. The convergence points of these lines correspond to triple collision solutions. The boundaries of this figure are $x_3(0) = -4.0$ to -3.0, (horizontal), while the vertical boundaries are $y_3(0) = +0.37$ to $+1.23$.

In a nutshell, we can thus state that the $[x_3(0), y_3(0)]$-plane is covered with many points leading to triple collision solutions. These points are connected with lines of Parabolic-Elliptic solutions which in turn are boundaries between classes of similar Hyperbolic-Elliptic solutions of the same complexity .

As for the practical implementation of this classification, we first integrate all the orbits on the computer with the maximum number of minima of the moment of inertia $I(t)$ limited to 2. In other words, the integration of each orbit is stopped when the second minimum is reached. This approach immediately gives us four classes of solutions: the three classes where one of the three masses escapes after the first minimum is reached. The fourth class contains the solutions of higher complexity, which truly reach the second minimum.

Next we continue the integration of all the solutions of this fourth class, up to the third minimum of the moment of inertia. This gives us a new subdivision of this class in four new classes: three classes with an escaping body and a fourth class with solutions

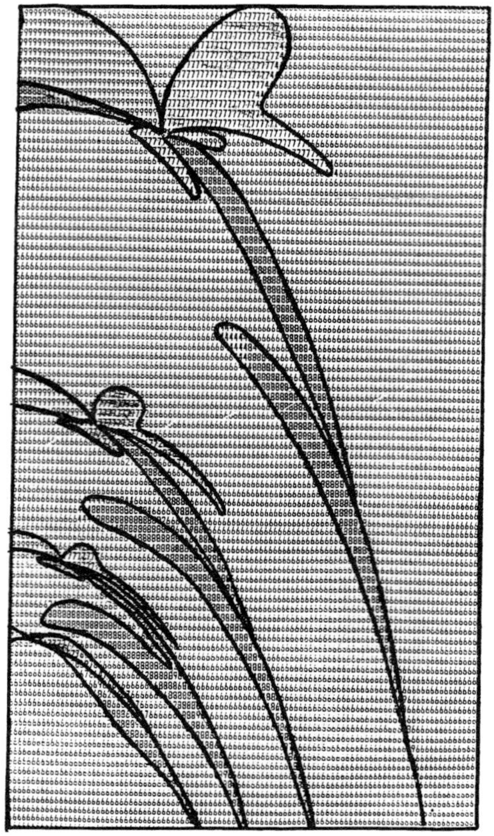

Figure 8. Another detailed histogram summarising a classification of some of our solutions of the general 3-body problem. The horizontal boundaries of this graph are here $x_3(0) = -3.0$ to -2.0, while the vertical boundaries are $y_3(0) = +0.357$ to $+1.388$.

of higher complexity. Then we continue this process further, giving us finer and finer subdivisions in smaller classes. In theory, we should finally reach classes of bounded solutions which contain stable periodic and quasi-periodic solutions.

In the present study, with masses $(5, 4, 3)$, we did not find any classes of bounded orbits. However, in the isosceles problem, we did find a large group of bounded stable quasi-periodic solutions with zero-angular momentum, (with two equal masses $m_1 = m_2$). The real existence of these stable bounded solutions is thus not in question.

THE NUMERICAL IMPLEMENTATION OF THE CLASSIFICATION

As we have explained above, the initial condition plane $[x_3(0), y_3(0)]$ can be subdivided in regions with various complexities $C_k^{(n)}$, with $n = 1, 2, \ldots$. In order to detect the location of some of these regions, we computed a large number of solutions as an initial exploration of the problem. After some experimenting we settled on the limits $-7.0 < x_3(0) < +7.0$ and $0.0 < y_3(0) < 7.0$ with an interval of 0.01, which corresponds thus to $1400 \times 700 = 980000$ solutions.

Our computer program was set up to compute $\dot{I}(t)$ at every integration step and

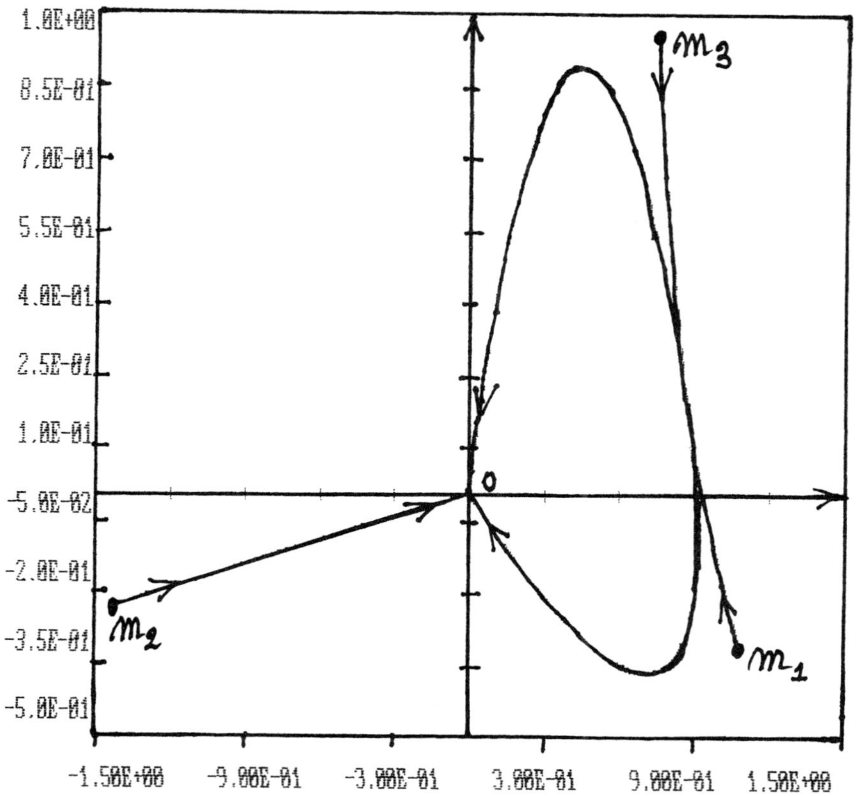

Figure 9. An example of a near triple-collision solution. the 2 masses m_1 and m_3 have one near binary collision. The heliocentric initial position of m_3 is $(-0.334263708, +1.26552838)$.

to monitor and count its changes of sign and roots of both kinds. Figure 4 shows the graph of $\dot{I}(t)$, for the example that was given in figure 2. We see the 6 roots, (including the initial point $t = 0$). Only 3 of these roots correspond to the minima of $I(t)$, visible on figure 3. Figure 5 shows the graph of $\ddot{I}(t)$, for the same solution. The vertical spikes correspond to the binary as well as the triple close approaches. This essentially shows the variations of the potential function, (because of the Lagrange-Jacobi Identity), but we did not use $\ddot{I}(t)$ in the present classification principle.

The termination conditions of each solutions were saved, especially the degree n of the complexity $C_k^{(n)}$. In the initial explorations we set the maximum n equal to 2. We designed our computer program to terminate the integration when certain termination conditions are reached. For the present discussion, the following five termination codes are the most relevant.

$N = 2$: Maximum Number of Integration Steps reached; Integration aborted. These solutions will be studied in more detail later.

$N = 6$: the allowed maximum number of positive going roots of $\dot{I}(t)$ has been reached. In our initial exploration, this value was 2. These are thus orbits with a higher degree of interplay and complexity. Some of them could be permanently bounded orbits.

$N = 7$: the Escape condition is satisfied and m_1 escapes; we have a termination in a Hyperbolic-Elliptic orbit.

$N = 8$: same as $N = 7$, but the escaping body is m_2.

$N = 9$: same as $N = 7$, but the escaping body is m_3.

It was observed that the plane of the two initial parameters $[x_3(0), y_3(0)]$ is then subdivided in large distinct connected regions defined by their termination code $N = 6, 7, 8$ *or* 9. The largest region corresponds to $N = 6$:these are the orbits with the most interplay; they reach a point where $I(t)$ is minimum for the second time. They are thus of class $C_k^{(2)}$ or higher, (the white area in figure 6).

The other three regions have the termination codes $N = 7, 8$ *or* 9. They are all of complexity $C_k^{(1)}$, one of the standard escape conditions being satisfied. They are thus the most simple Hyperbolic-Elliptic solutions in this problem: they begin with a maximum of $I(t)$ at $t = 0$ (which is the only maximum) and they have exactly one and only one minimum of $I(t)$ (for $t > 0$) before the dissociation of the system in a binary and single hyperbolically escaping particle. These solutions of complexity $C_k^{(1)}$ form the grey area in figure 6. The white area corresponds to the more complicated orbits with $N = 6$. The grey area corresponds to the simple Hyperbolic-Elliptic Orbits of class $C_k^{(1)}$, of the three types, $k = 1, 2$ *or* 3, (the escaping particle). The typical solution shown in figure 2 corresponds to a point in the white area.

Myles Standish (1976) has already published four orbits of this type. This curved region of $C_k^{(1)}$-orbits, (the Standish column!), can be recognized as a small subset in our figure 6.

In order to further classify the orbits with termination code $N = 6$, we would have to continue the numerical integration with the maximum number of minima of $I(t)$ set equal to 3. We would then obtain a new division in 4 classes: the classes with $N = 7, 8$ *or* 9 which are Hyperbolic-Elliptic of Complexity $C_k^{(2)}$, as well as a new class $N = 6$, which are all orbits with more interplay, of complexity $C_k^{(3)}$ (or higher). These orbits could in turn be subdivided in classes by making longer numerical integrations. These classes of similar solutions form of course two-parameter sets in the $[x_3(0), y_3(0)]$-plane.

DETAILED DESCRIPTION OF THE HISTOGRAMS

Figures 7 and 8 give finer details of the global histogram that was given in figure 6. Each digit (2,6,7,8 or 9) corresponds to a solution with a given $x_3(0)$ and $y_3(0)$. The horizontal axis corresponds to the value of $x_3(0)$ and the vertical axis to $y_3(0)$. The meaning of the letter codes $2, 6, 7, 8, 9$ was described before.

The 4 classes (with code $6, 7, 8, 9$) form clearly separated distinct regions. The lines separating the regions correspond to Parabolic - Elliptic Solutions. We also see very clearly that many of these lines converge to common points; these points correspond to triple collision solutions. Figures 9 and 10 show two of the many approximate triple collisions that were discovered. For both of these solutions, the perimeter of the triangle is about 0.00005 at the final point, (which is the closest approach point, when $I(t)$ reaches its first minimum).

As we have said before, the solutions with code 6 are those with the most interplay. They have two (or more) minima of the moment of inertia. We could now integrate all these code-6 solutions again in order to find the three subclasses for which m_1 or m_2 or m_3 escapes after the second minimum of $I(t)$. As was explained above, a finer and

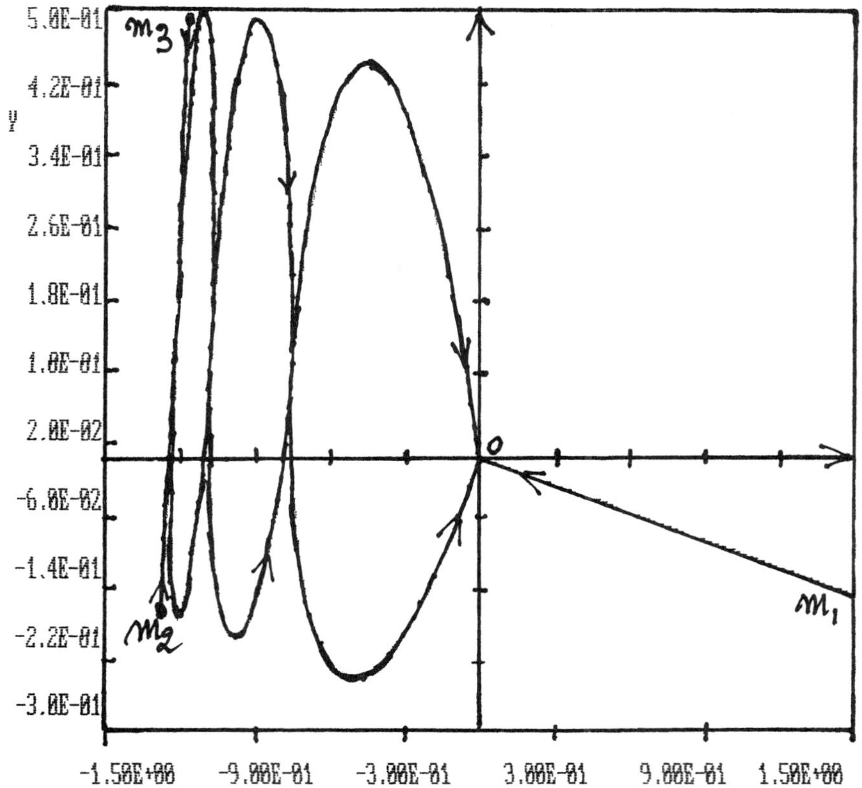

Figure 10. An example of a near triple-collision solution. The 2 masses m_2 and m_3 have three near binary collisions before the final triple collapse. The heliocentric initial position of m_3 is $(-2.89740644, +0.6911152888)$.

finer classification can be made in this way, by making longer and longer integrations of the code-6 solutions. At this time, we have this process completely automated.

We also want to mention that a finer classification of the solutions could be made, even if we restrict ourselves to a small number of minima of $I(t)$, just by using a finer grid in the $x_3(0)$ and $y_3(0)$ values, especially near the triple collision points where the complexity of the forms of solutions is extreme.

On each of the figures 7 and 8 the number of $x_3(0)$-values (on the horizontal axis) is 100 and the number of $y_3(0)$-values (on the vertical axis) is 75, corresponding thus to a total of 7500 solutions represented on each figure. On figure 7, the horizontal boundaries are $x_3(0) = -4.0$ to -3.0, while the vertical boundaries are $y_3(0) = 0.37$ to 1.23 . On figure 8, the horizontal boundaries are $x_3(0) = -3.0$ to -2.0, while the vertical boundaries are $y_3(0) = 0.357$ to 1.388.

REFERENCES

1. Anosova, J.P. Orlov, V.V., 1992, "The Types of Motion in Hierarchical and Non-Hierarchical Triple Systems Numerical Experiments", Astronomy and Astrophysics, Vol. 260, p. 473-484.

2. Anosova, J., 1986, "Dynamical Evolution of Triple Systems", Astrophysics and Space Science. Vol. 124, p. 217-241.

3. Benz, W., Hills, J.G., 1992, "Three-Dimensional Hydrodynamical Simulations of Colliding Stars"., The Astrophysical Journal, Vol. 389, p. 546-557.

4. Birkhoff, G. D. 1927, "Dynamical Systems", Am. Math. Soc. Publ., Providence, R. I.

5. Black, D. C., 1982, "A Simple Criterion for Determining the Dynamical Stability of Three-Body Systems", The Astronomical Journal, Vol. 87, Nr. 9, p. 1333-1337.

6. Carusi, A., Valsechi, G.B., 1990, "Planetary Close Encounters: Geometry of Approach and Post-Encounter Orbital Parameters", Celestial Mechanics and Dynamical Astronomy, Vol. 49, p. 111-131.

7. Chazy, J., 1922, Annales de l'ecole Normale, 3eme serie,Vol 39, 29-130.

8. Chazy, J., 1929, Journal de mathematiques pures et appliquees, Vol. 8, p.353-380.

9. Davies, M.B., Benz, W., Hills, J.G., 1992 "Stellar Encounters Involving Neutron Stars in Globular Cluster Cores", The Astrophysical Journal, Vol. 401, Nr.1, Part 1, p. 246-259.

10. Davoust, E. and Broucke, R., 1982, "Families of periodic orbits in the general three-body problem", Astronomy and Astrophysics, Vol. 112, p.305-320.

11. Graziani, F., Black, D.C., 1981, "Orbital Stability Constraints on the Nature of Planetary Systems", The Astrophysical Journal, Vol. 251, p. 337-341.

12. Greenberg, R., Carusi, A., Valsecchi, G.B., 1988, "Outcomes of Planetary Close Encounters: A Systematic Comparison of Methodologies". Icarus, Vol. 75, p. 1-29.

13. Hills, J.G., 1991, "Computer Simulations of Encounters Between Massive Black Holes and Binaries"., The Astronomical Journal, Vol.102, Nr.2 p. 704-715.

14. Hut, P., Bahcall, J., 1983, "Binary Single Star Scattering. I. Numerical Experiments for Equal Masses", The Astrophysical Journal, Vol. 268, p.319-341.

15. Hut, P., 1983, "The Topology of Three-Body Scattering", The Astronomical Journal, Vol. 88, Nr. 10, p. 1549-1559.

16. Kiseleva, L.D., Orlov, V.V., 1993, "Dynamics of Galaxy Groups: Computer Simulations Versus Observations", Vistas in Astronomy, Vol.36, p.1-30.

17. Kiseleva, L.G., Eggleton, P.P., Anosova, J.P.,1994, "A Note on the Stability of Hierarchical Triple Stars with Initially Circular Orbits", R. Astron. Soc. Vol. 267, p. 161-166.

18. Marchal, C., 1990, "The Three Body Problem", Elsevier Science Publishers, Amsterdam, New York.

19. Monaghan, J.J., 1976, "A Statistical Theory of the Disruption of Three-Body Systems-1", Mon. Not. R. Astr. Soc., Vol. 176, p. 63-72.

20. Opik, E.J., 1976, "Interplanetary Encounters", Elsevier, Amsterdam.

21. Standish, E. M., 1971, "Sufficient Conditions for Escape in the Three-Body Problem", Celestial Mechanics, Vol. 4, p. 44-48.

22. Standish, E. M., 1976, "On the Stability of the Solutions of the General Three-Body Problem", Celestial Mechanics, Vol. 14, p. 493-498.

23. Szebehely, V., 1974, "Analysis of a one-parameter family of triple close approaches occuring in stellar systems", The Astronomical Journal, Vol. 79, No 9, p. 981-983.

24. Valtonen, M., 1975, "A Gravitational Three-Body Scattering Experiment", Mem. R. Astr. Soc. Vol. 80, p. 61-75.

25. Valtonen, M., Heggie, D.C., 1979, "Three-Body Gravitational Scattering: Comparison Between Theory and Experiment", Celestial Mechanics, Vol. 19, p. 53-58.

26. Zare, K., 1981, "Properties of the Moment of Inertia in the Problem of Three Bodies", Celestial Mechanics, Vol.24, p. 345-354.

THE BICIRCULAR MODEL NEAR THE TRIANGULAR LIBRATION POINTS OF THE RTBP

Carles Simó,[1] Gerard Gómez,[1] Àngel Jorba, [2] and Josep Masdemont [2]

[1]Departament de Matemàtica Aplicada i Anàlisi
Universitat de Barcelona
Gran Via 585, 08007 Barcelona (Spain)
[2]Departament de Matemàtica Aplicada I
Universitat Politècnica de Catalunya
Diagonal 647, 08028 Barcelona (Spain)

Abstract

We present a study of a simplified model of the Restricted Four Body Problem consisting of Earth, Moon, Sun and a massless particle, as a model of the dynamics of a spacecraft. The region where we look for the motion is a vicinity of the triangular libration points of the Restricted Three Body Problem. The model we discuss here is the so called Bicircular Problem. The main question is the existence of zones where the motion has good stability properties. The answer is positive, but the stable motions can not be confined to a small distance of the ecliptic plane. Both numerical simulations and analytical results are presented. Some tentative explanations offer a possible way to study many other kinds of problems. Some applications to space missions are mentioned.

INTRODUCTION

The objective of the work is to understand the kinds of motion which appear on a vicinity (may be a large vicinity) of the geometrically defined equilateral points of the Earth-Moon system when the bicircular model is considered. In this Introduction we summarize the present knowledge, some of the achievements of this work and we discuss the main difficulties of the problem.

The simplest model for the motion near the triangular libration points is, of course, the spatial Restricted Three Body Problem (RTBP). With respect to the RTBP the main perturbations are due to the presence of the Sun as it is shown in Gómez et al.[5]. This is the reason to devote our attention to the bicircular problem.

From Newton to Chaos, Edited by A.E. Roy
and B.A. Steves, Plenum Press, New York, 1995

Definition of the model and equations of motion

The bicircular problem is a simplified version of a restricted four body problem. The objective is to describe the motion of a massless particle under the gravitational atraction of Earth, Moon and Sun. In this model we suppose that the Earth and the Moon are revolving in circular orbits around their centre of mass, B, and the Earth-Moon barycentre is moving in a circular orbit around the centre of masses of the Sun-Earth-Moon system. We remark that, with these assumptions, the motion of these three bodies is not coherent. That is, the assumed motions do not satisfy Newton's equations. However the model is extremely useful for some kind of orbits.

Let μ be the mass of the Moon, $1-\mu$ the mass of the Earth and m_S the mass of the Sun. Let the distance from the Earth to the Moon be taken as unity. Then the distance from B to the Sun is a_S. We use synodic coordinates with respect to the Earth-Moon system, so that the positions of Earth and Moon are fixed at $(\mu, 0, 0)$ and $(\mu-1, 0, 0)$, respectively. The mean angular velocity of the Sun in these synodic coordinates is denoted by ω_S.

To keep a Hamiltonian form we use synodic coordinates x, y, z, for the position of the massless body, but instead of the velocities we use momenta p_x, p_y, p_z, defined by $p_x = \dot{x} - y$, $p_y = \dot{y} + x$, $p_z = \dot{z}$.

In this way the equations of the bicircular problem are:

$$
\begin{aligned}
\dot{x} &= p_x + y, \\
\dot{y} &= p_y - x, \\
\dot{z} &= p_z, \\
\dot{p}_x &= p_y - \frac{1-\mu}{r_{PE}^3}(x-\mu) - \frac{\mu}{r_{PM}^3}(x-\mu+1) - \frac{m_S}{r_{PS}^3}(x-x_S) - \varepsilon_S \cos\theta, \\
\dot{p}_y &= -p_x - \left(\frac{1-\mu}{r_{PE}^3} + \frac{\mu}{r_{PM}^3}\right) y - \frac{m_S}{r_{PS}^3}(y-y_S) + \varepsilon_S \cos\theta, \\
\dot{p}_z &= -\left(\frac{1-\mu}{r_{PE}^3} + \frac{\mu}{r_{PM}^3} + \frac{m_S}{r_{PS}^3}\right) z,
\end{aligned}
$$

where $r_{PE}^2 = (x-\mu)^2 + y^2 + z^2$, $r_{PM}^2 = (x-\mu+1)^2 + y^2 + z^2$ and $r_{PS}^2 = (x-x_S)^2 + (y-y_S)^2 + z^2$, with $x_S = a_S \cos\theta$, $y_S = -a_S \sin\theta$, $\theta = \omega_S t + \theta_0$, and θ_0 is some initial phase of the Sun. For shortness we have used the perturbation parameter of the Sun $\varepsilon_S = \frac{m_S}{a_S^2}$.

For further use we give here the related Hamiltonian

$$
H = \frac{1}{2}(p_x^2 + p_y^2 + p_z^2) + y p_x - x p_y - \frac{1-\mu}{r_{PE}} - \frac{\mu}{r_{PM}} - \frac{m_S}{r_{PS}} - \frac{m_S}{a_S^2}(y \sin\theta - x \cos\theta).
$$

The values of the parameters that we have used for the symbolic and numerical computations are as follows:

Mass parameter for the Earth-Moon system:

$$
\mu = \frac{1}{82.300587} \approx 0.012150582.
$$

Sun mass (1 unit = Earth+Moon mass):

$$
m_S = \frac{0.29591220828559 \times 10^{-3}}{0.89970116585573 \times 10^{-9}}.
$$

Mean angular velocity of the Sun in synodic coordinates:

$$\omega_S = 1 - \frac{129602770.31}{1732564371.15}.$$

Semimajor axis of the Sun (1 unit = Earth-Moon distance):

$$a_S = \left(\frac{1 + m_S}{(1 - \omega_S)^2}\right)^{1/3}.$$

Known results near $L_{4,5}$

The bicircular problem can be obtained from the circular RTBP by continuation with respect to the mass of the Sun or by means of some intermediate Hill's problem (see Gómez et al.[4] and references therein).

There are three simple periodic orbits (two of them linearly stable and another one small and slightly unstable) with period equal to the synodic period of the Sun in the Earth-Moon system: $T_S = 6.79117$. The two stable orbits lie relatively far away from the triangular libration points of the RTBP and, when the full set of perturbations is included, they become slightly unstable (see Gómez et al.[4]). One can use a continuation method to pass from the RTBP to the bicircular model by introducing a parameter ε instead of ε_S in the equations, and then by increasing it from zero to ε_S. At the zero value we can take the L_5 point (or, equivalently, the L_4 one) as initial condition. Of course, it can be considered as a T_S periodic orbit. For small values of ε a periodic orbit is obtained, close to a small loop travelled twice around L_5. By increasing ε till ε_S, the orbit we reach is one of the big periodic orbits mentioned above. The other big orbit is essentially identical but with a change in the phase. Starting at this orbit and decreasing now ε, a turning point is found. When crossing the turning point, the stability of the periodic orbit along the family changes. From that point on ε increases again, and when it reaches the value ε_S the small unstable periodic orbit is obtained. The characteristic curve can be seen as obtained from a pitchfork by symmetry breaking. From now on we shall consider these small periodic orbits as the substitute of the triangular libration points in the case of the bicircular model, despite they can not be obtained by direct continuation (see Figure 1).

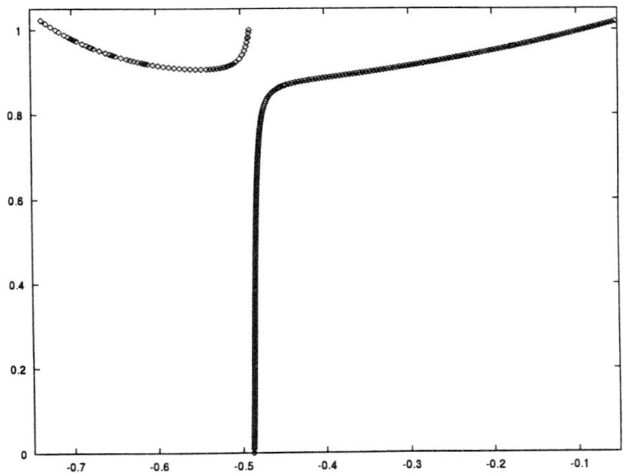

Figure 1

We remark that for other natural 4-body problems in the Solar System, where the bicircular model can be used as an approximation near some triangular points of a RTBP, the behaviour can be different. For instance, consider as RTBP the system Sun-Jupiter-asteroid and take Saturn as fourth body. In a bicircular model one finds that, under the influence of Saturn, the L_4 and L_5 points are replaced by small stable periodic orbits. The same is true if we consider the L_4, L_5 points of the Sun-Saturn system when perturbed by Jupiter.

No fixed points, nor autonomous first integral exist for the bicircular problem. The existence of a "stable domain" in some extended vicinity of the small periodic orbit is, up to here, an open question. We will see later on that, according to our numerical simulations, these stable domains exist. In fact similar domains, with good stability properties, exist also for the full problem (i.e., when all the solar system is taken into account) and are close to the ones found for the bicircular model (see Gómez et al.[5]).

Main difficulties of the problem

We can summarize the difficulties to be found in our problem after the previous results are known:

1. An external frequency appears together with the (amplitude varying) inner frequencies. These 4 frequencies lead to many resonances.

2. The effect of the Sun is uniformly large in the vicinity of $L_{4,5}$.

3. At the geometrical $L_{4,5}$ the frequencies of: the proper short period, the vertical oscillations and the synodic frequency of the Sun are rather close.

In the circular RTBP we are faced to a three degrees of freedom autonomous Hamiltonian. In the bicircular case the system is periodic instead of autonomous. Even in the 3-D autonomous case, to describe to some extent the dynamical behaviour in big regions of the phase space can be a formidable task.

Starting at some integrable problem and adding a perturbation with increasing parameter we find, successively:

a) For the integrable problem the phase space is completely foliated by invariant manifolds, mainly 3-D tori.

b) For small perturbations (or, equivalently, when we look very close to a totally elliptic equilibrium point) most of the 3-D tori subsist, as assured by the celebrated KAM theorem (see Arnol'd and Avez[1]), and very small zones of chaotic motion appear. They are hardly seen in practice if the perturbation is small because of the exponentially small character of the splitting of separatrices (see Lazutkin et al.[7], Fontich and Simó[2], Simó[9]. As the 3-D tori do not separate the levels of energy (which are 5-dimensional), Arnol'd diffusion can appear as a wandering motion between the tori. This prevents, in general, the existence of true barriers for the motion of the momenta. However, this motion is, at most, extremely slow, as assured by the results of Nekhorosev[8].

c) When the perturbation is increased (or, equivalently, when we look at a relatively large distance from a totally elliptic equilibrium point), the 3-D tori are destroyed by some not yet fully understood mechanism. However, cantorian families of normally hyperbolic tori still subsist. They constitute a kind of skeleton of the motion. In some sense, one can consider the motion as a sequence of passages near lower dimensional tori, where they stay for some time interval before reaching the vicinity of next tori, following closely a heteroclinic orbit. The observed behaviour in the Solar System simulations for very long time intervals seems to be of this type (see Laskar[6]).

In the present case we have, as an additional difficulty, the fact that the system is not autonomous but periodic. In principle, we can assume that the system is reduced to autonomous form by means of time-dependent canonical changes. But this is purely formal and, furthermore, we can be faced to resonances in this process. If no resonances appear and we are satisfied with a study for moderate time intervals, we can describe the motion as the one of the autonomous system obtained after the canonical changes, which is then shaken by the periodic change of variables.

Contents of the work

In the second Section we describe numerical simulations carried out for the bicircular model. We present the results concerning sets of points which subsist, in an extended neighbourhood of the former triangular equilibrium points, for a long time span. Furthermore frequency analysis of several orbits have been carried out. The Section is ended with a discussion of the results. These ones suggest some reasonable explanations. Just looking at the frequency analysis we can detect the key role played by the inner frequencies on the system (that is, what can be seen as a generalization of the vertical, short period and long period modes of the circular RTBP), together with the solar synodic frequency. A sketch of the geometrical and analytical explanations of the zone of stable motion is presented.

The third Section describes the reduction of the bicircular model to some Normal Form. The essential part is to convert the periodically dependent Hamiltonian in an autonomous one. This is done around the small unstable periodic orbit described above. Then one obtains a clear explanation for the instability found in the numerical simulations of that problem for small values of the z amplitude. The source of the instability is the 1-1 resonance between the short period frequency and the solar one. An analytical description and some illustrations are given for the 2-D unstable tori which emerge from the small unstable periodic orbit when the vertical mode is added.

The work is ended with some discussion and conclusions concerning the applicability of the results to space missions.

NUMERICAL SIMULATIONS

Description of the massive simulations

A typical computation, for the determination of the stability zones, starts as follows. Give values ρ, α, z. Then select a point x, y, z by $x = (1+\rho)\cos(2\pi\alpha) + \mu$, $y = (1+\rho)\sin(2\pi\alpha)$. We shall use α between zero and 1/2. So we are doing the computations near the position of L_5 in the RTBP, the case of L_4 being symmetric. Take zero initial velocity in synodic coordinates ($\dot{x} = \dot{y} = \dot{z} = 0$). Perform an integration for a time span up to some prescribed t_f. Stop the computation if $y < 0$ at some value of t, $t \leq t_f$. If we reach t_f the position (ρ, α, z) is stored as belonging to the "stable domain" in some weak sense.

In the bicircular problem the triangular libration points do not longer exist. In its place there are unstable periodic orbits (see Introduction, Gómez et al.[4,5]). So, locally near that orbit, the points escape. As the system is not autonomous, we have to take into account the initial phase of the Sun, θ_0. This initial phase plays an important role in the stable domain. The effect is almost the same when the initial phase in increased by π. To see the effect for different initial phases of the Sun we refer to Gómez et al.[5]. Here, for shortness we consider $\theta_0 = 0$ in all the simulations.

At this moment some considerations must be done. They apply to the bicircular problem, but are also valid for the RTBP, both in the circular and the elliptic cases. The first one concerns the set of initial data. Taking zero initial (synodic) velocity seems to be a strong restriction. Consider, however, the case of a totally elliptic fixed point of a Hamiltonian system, and assume that the frequencies are rationally independent. In a small vicinity of that point the motion takes place mainly on tori. Let q_j and p_j denote some local coordinates, reducing the quadratic part of the Hamiltonian to diagonal form. The motion is close to $q_j = A_j \cos(\omega_j t + \phi_j)$, $p_j = -A_j\omega_j \sin(\omega_j t + \phi_j)$, for some frequencies, ω_j, and phases ϕ_j. Assume p_j are the velocities (otherwise, it is similar). By the density of the orbits in the tori, there exists some time such that all the velocities are so close to zero as desired. It is true that, if at some distance of the fixed point there is a linearly stable periodic orbit, surrounded by some tori, the motion on them is stable and the velocity on these tori is never zero. Another consideration is that, if most of the motion takes place on tori (or close to them), taking an initial generic 3-D set (as it seems to be the set with zero synodic velocity) the flow will span a 6-D domain.

The second consideration concerns the stability criterion. To choose $y < 0$ as an indicator of instability seems reasonable. It has been seen that, as soon as the orbit enters in that region, a wild behaviour shows up. Either passages near collision with the Earth or the Moon occur, or the motion is confined for some time near the region of the other triangular point, with an eventual escape of the vicinity of the Earth-Moon system. The criterion can be too strict for orbits reaching big values of z (say, larger than 0.8). May be orbits showing good stability properties (they seem to be in tori passing close to the L_5 point, in the position coordinates) and oscillate up and down, enter the $y < 0$ region near the extreme values of z. In any case this seems to affect only a small fraction of the initial conditions. The next question deals with the value of t_f. Usually points which escape, according to our criterion, go away in a relatively short time (say, 100 lunar periods). On the other hand, among the points subsisting for 10000 periods the escape takes place at a small rate. See the next subsections for more qualitative and quantitative information, and for the behaviour of the escaping rate.

The integrations have been carried out using an RKF78 method with automatic stepsize control and maximum local error equal to 10^{-13}.

Results of the simulations

For values of z ranging from 0 to 1 with step 0.05, we have taken values of ρ, α in the ranges [-0.250,0.025], [0.100,0.450], respectively, with stepsize 0.001 in both directions. The final time has been set to $t_f = 10000 \times 2\pi$, and all the points subsisting up to a time at least equal to $100 \times 2\pi$ have been recorded. The Table 1 presents a summary of results.

The Figure 2 displays the results for values of z equal to 0.25, 0.35, ..., 0.85 from top to bottom. The stable region after 10000 lunar synodic revolutions is shown. The

Table 1. Points which subsist after 10000 revolutions of the Moon for the bicircular problem. In the first row the value of z is given. From the second to the last row we display, respectively: Points subsisting for 100, 1000 and 10000 lunar revolutions, minima and maxima of $1000 \times \alpha$ and $1000 \times \rho$ for the points which subsist for 10000 revolutions. In all cases we have taken $\theta_0 = 0$.

.00	.05	.10	.15	.20	.25	.30	.35	.40
74	69	75	58	80	252	538	1189	2214
15	13	11	8	12	153	397	820	1537
12	10	9	5	2	139	381	720	1225
333	333	334	334	336	323	312	302	290
339	338	339	340	337	350	358	365	377
-3	-3	-5	-7	-12	-20	-28	-38	-50
-1	-2	-4	-6	-12	-14	-18	-22	-27
.45	.50	.55	.60	.65	.70	.75	.80	.85
2371	3214	3158	2630	2258	2099	1548	743	787
1484	1588	2224	1391	824	690	627	163	205
1279	1062	1748	847	366	228	397	69	115
295	283	290	289	313	323	310	331	328
376	371	378	372	366	373	360	346	348
-63	-74	-87	-98	-108	-122	-138	-162	-175
-37	-45	-50	-65	-82	-99	-123	-151	-166

horizontal variable is α and the vertical one is ρ. For α we have taken a range going from $\alpha = 0.38$ (to the left) to $\alpha = 0.285$ (to the right). The windows for ρ depend on z but all the ranges in ρ have amplitude equal to 0.04. The lower values of ρ in the windows are -0.035, -0.050, -0.070, -0.088, -0.115, -0.150 and -0.190, respectively. The divisions inside each plot correspond to multiples of 0.025, both in ρ and α.

The behaviour is quite different from the one found for the circular and elliptic RTBP (see Gómez et al.[5]). For z less than 0.25, almost all points disappear after 10000 revolutions. But from $z = 0.25$ till $z = 0.55$ the size of the stable region increases, to decrease again when higher values of z are taken. For some values of θ_0 some points subsist even for $z = 0.95$. Averaging over θ_0, the number of subsisting points for $z = 0.40, \ldots, 0.60$ is comparable to the RTBP case (compare Gómez et al.[5]).

Let us present an explanation for the stable points found till now. At L_5 point the RTBP, considering the Hamiltonian up to second degree, behaves like three harmonic oscillators. One of them, with frequency equal to 1, corresponds to vertical oscillations. When the full system is considered, these vertical oscillations give rise to a family of "vertical" periodic orbits. These orbits, in fact, project on the (x, y) plane close to an ellipse travelled twice and, on the (x, z) and (y, z) planes, as figure eight curves. The initial conditions can be chosen with $z = 0$. In the bicircular model, L_5 has been replaced by the small periodic orbit. In a similar way, the RTBP "vertical" periodic orbits are replaced by 2-D tori, unstable for small vertical amplitudes (see next Section). It seems that for larger z these 2-D tori become linearly stable. One can also consider them as a periodic perturbation of "big" orbits of the vertical family of the RTBP. Hence, motion near these 2-D stable tori must be, for a set of points of (big) positive measure on 4-D tori. Some of those points can sit on the 4-D manifold of zero synodical velocity (now we have 7 dimensions: 6 spatial and the time).

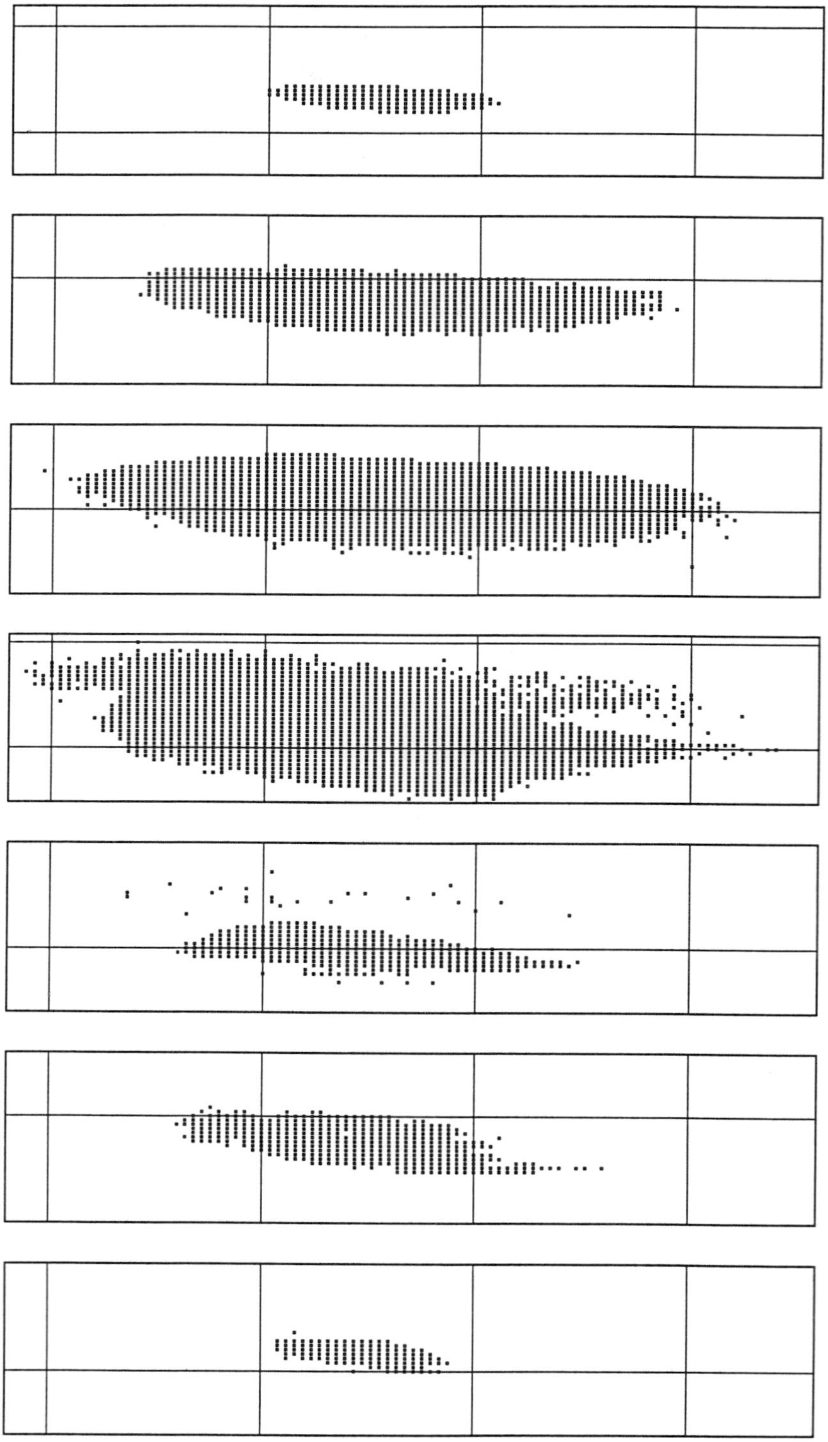

Figure 2. Representation of the stable region for several values of z. See the text for the concrete values and windows.

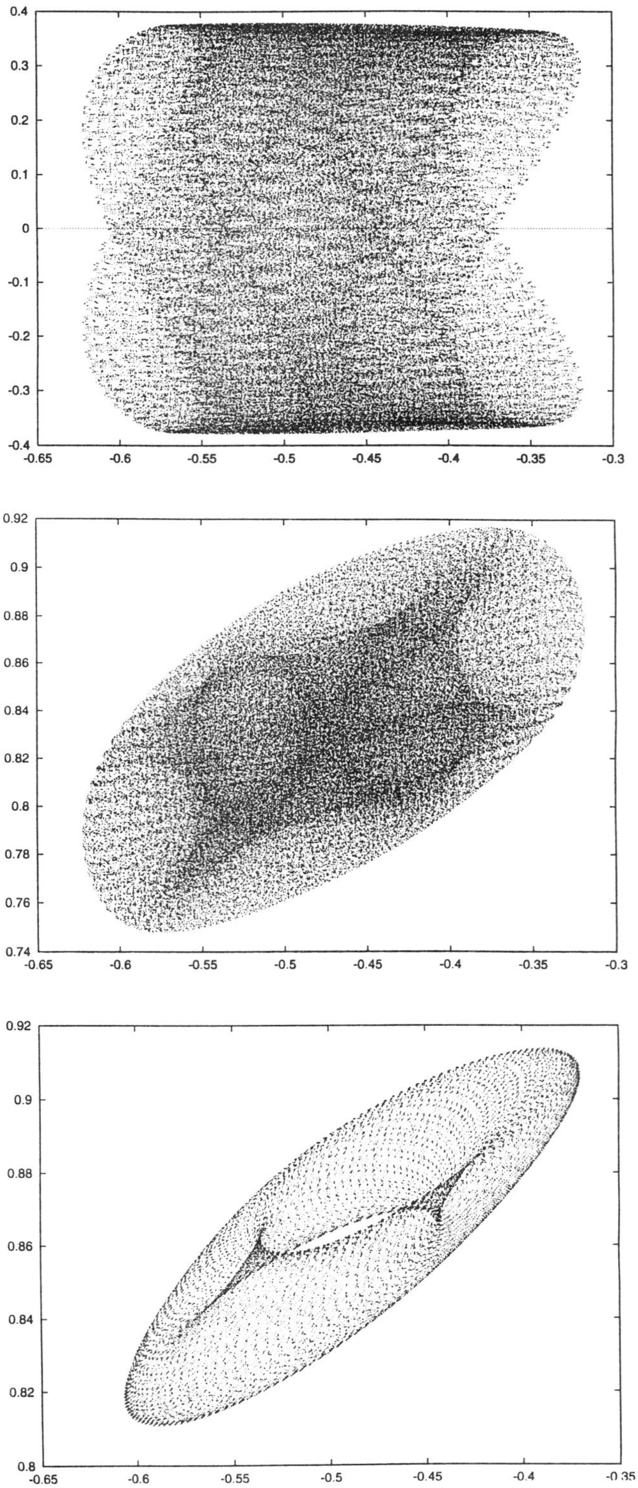

Figure 3. Projections of 3-D and 2-D tori for $\dot{z}_0 = 0.36$. See the text for additional explanation.

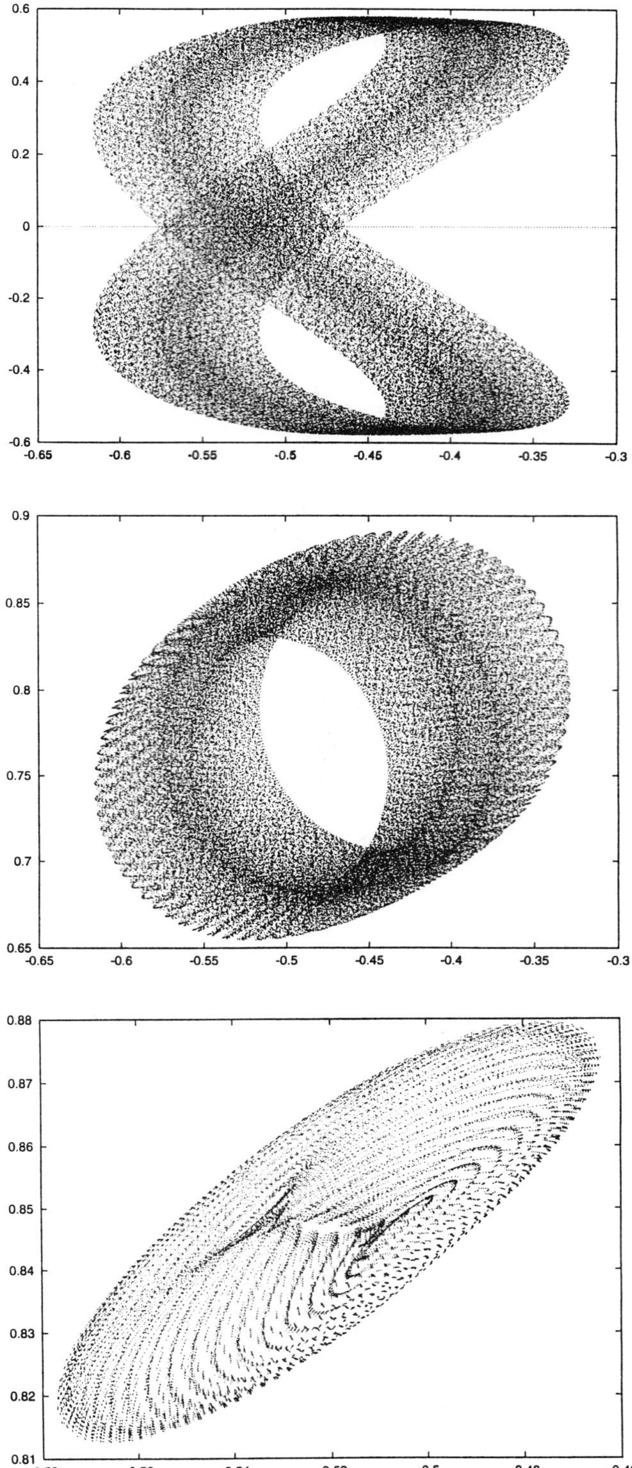

Figure 4. Projections of 3-D and 2-D tori for $\dot{z}_0 = 0.56$. See the text for additional explanation.

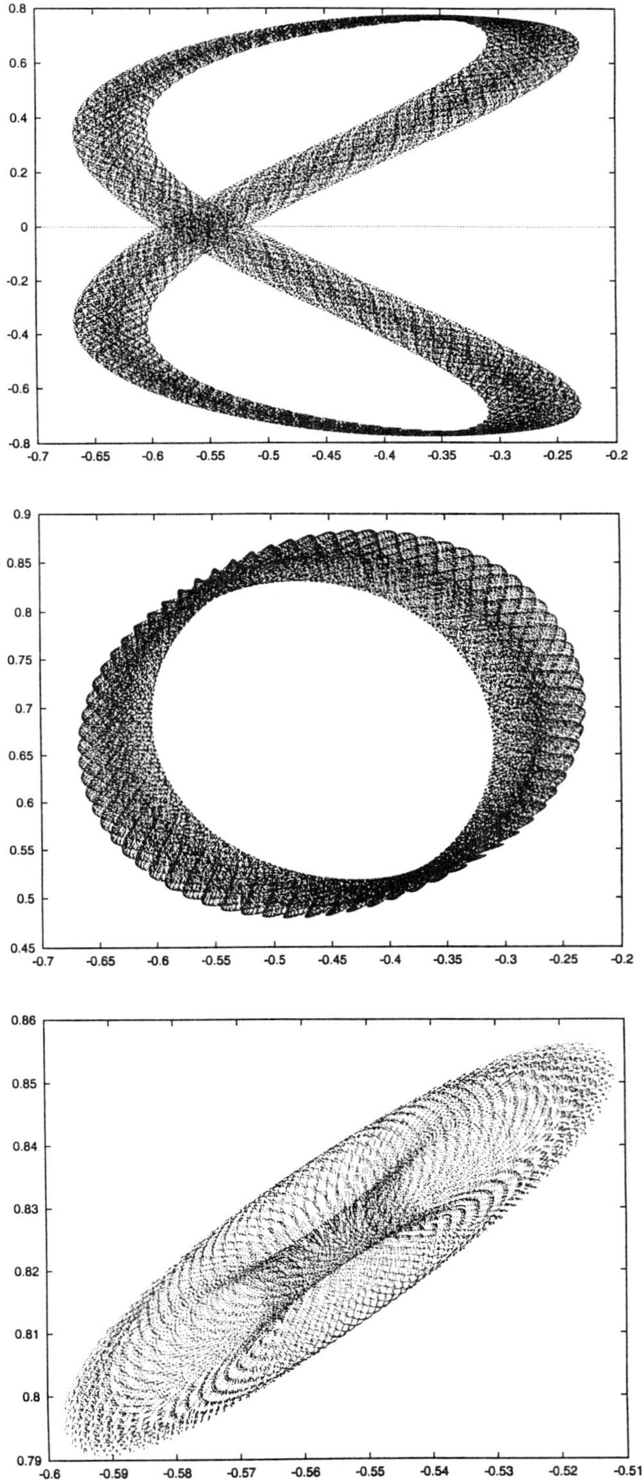

Figure 5. Projections of 3-D and 2-D tori for $\dot{z}_0 = 0.76$. See the text for additional explanation.

Some simulations with non zero initial synodic velocity

We can try another kind of simulations. Assume that we take as initial conditions the ones of the vertical family of periodic orbits of the RTBP, with $z = 0$ at $t = 0$ and also with zero initial phase of the Sun. We shall use as label for the orbit the initial value, $\dot{z}_0$, of $\dot{z}$ (this value is close to the maximum z-amplitude of the periodic orbit). For small z-amplitudes they must behave in an unstable way, but for larger amplitudes they can be stable, and even in some 4-D torus, typically, because of the preceding reasoning. We have tested this approach. To simplify the visualization we have used the "time the synodic period of the Sun"-map, to be denoted by T_{SS}. This is a 6-D canonical map and the invariant tori of maximal dimension, if any, must be 3-D. Even the 2-D projections of these 3-D tori can be difficult to recognize. To make things simpler one can take "slices" of the 3-D tori. The slices are defined by $|z| < 0.01 \times \dot{z}_0$. They should be close to 2-D tori, and their 2-D projections can be better identified as corresponding to 2-D tori.

The Figures 3,4 and 5 present the (x, z) and (x, y) projections of the 3-D tori and the (x, y) projection of the slice, from top to bottom. The values of $\dot{z}_0$ are taken equal to 0.36, 0.56 and 0.76, respectively. They have been chosen to have a representative sample (and, may be, by subjective aesthetic reasons). In all cases 4×10^6 iterates of T_{SS} have been done. The projections of the 3-D tori show only the 10^5 first iterates. For the slices we have considered all the iterates satisfying the slice condition. The respective number of points in the slices displayed are 24979, 25298 and 25792. To get a better feeling about the behaviour of the dynamics on the 3-D tori, one can look at how they appear on a computer screen when displayed at several rates of points/second. This is difficult to transmit in written form, so that we encourage the reader to try by himself. One can see that the points on the (x, z) projection travel along the "figure eight" shape of the set (this is very clear for $\dot{z}_0 = 0.76$) and the amplitudes behave in a pulsating way. These facts give a good evidence that the points sit in 3-D tori.

Frequency analysis

One can perform a frequency analysis of the x, y and z coordinates, as functions of time, for the orbits with initial conditions the ones of the vertical periodic orbits of the RTBP and zero initial phase of the Sun, as before. A refined Fourier analysis allows for a precise determination of the basic frequencies (see Gómez et al.[5]). The method is similar, but not equal, to the one used by Laskar[6]. The selected orbits have $\dot{z}_0$ between 0.24 and 0.92. Other values outside this interval lead to escape. The time span has been 1024 revolutions of the Moon. The Figure 6 shows the evolution of the four basic frequencies as a function of $\dot{z}_0$. In the figure the number of the orbit, k, is associated to $\dot{z}_0$ by $k = 50 \times \dot{z}_0$. The basic frequencies 1 to 4 correspond to the analogous of the long period frequency of the RTBP, the synodic Sun frequency, the short period and the vertical period, respectively. The solar frequency is constant, of course. It stops their appearition at orbit number 23 because the maximum amplitude (in any of the x, y or z coordinates) is less than 10^{-3}. Some remarks are relevant.

The long period frequency decreases (in absolute value) with the vertical amplitude. This is in good agreement with the RTBP behaviour, where a Normal Form computation around L_5 (see Gómez et al.[5]) gives, for the Earth-Moon problem, the expansion

$$\begin{aligned}\omega_l &= -0.298208 + 0.225319 I_v - 0.178982 I_v^2 + 0.071211 I_v^3 + 0.051720 I_v^4 + \\ &\quad +0.000197 I_v^5 - 0.106373 I_v^6 + 0.045702 I_v^7 + \ldots,\end{aligned}$$

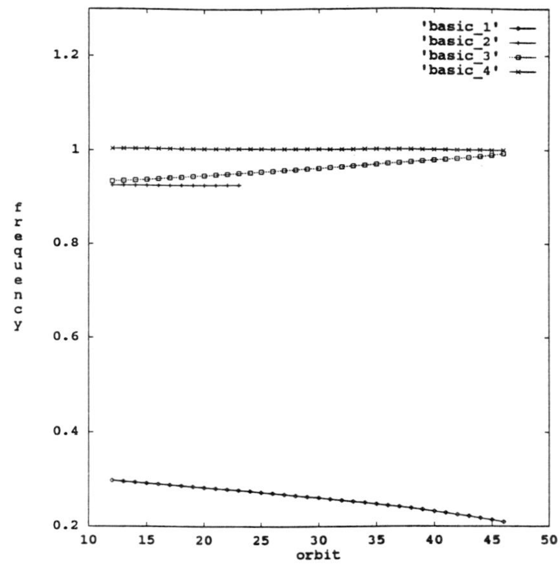

Figure 6. Evolution of the four basic frequencies along the family of orbits explored.

where the Hamiltonian is expanded in power series of three actions, I_s, I_l, I_v, associated to the short period, long period and vertical modes, respectively. The expansion above is just $\frac{\partial H}{\partial I_l}(0,0,I_v)$. For I_v one has $I_v = \frac{1}{2}(q_v^2 + p_v^2)$ (and similar for I_s, I_l) and q_v, p_v are rather close to z, $\dot{z}$. For completeness we give also ω_s and ω_v from the Normal Form of the RTBP as a function of the "vertical" action I_v (and $I_s = I_l = 0$):

$$\begin{aligned}
\omega_s &= 0.964501 + 0.089131 I_v - 0.055905 I_v^2 + 0.035475 I_v^3 - 0.004256 I_v^4 + \\
&\quad +0.042148 I_v^5 - 0.010521 I_v^6 + 0.048638 I_v^7 + \ldots, \\
\omega_v &= 1 - 0.004471 I_v - 0.000154 I_v^2 + 0.002088 I_v^3 + 0.000241 I_v^4 - 0.001846 I_v^5 \\
&\quad -0.000379 I_v^6 + 0.002006 I_v^7 + \ldots \ .
\end{aligned}$$

It follows that, in the RTBP and along the "vertical" family, the short period frequency increases and the vertical one decreases very slowly. This also happens in the bicircular model along the family used in the simulations, as can be checked in Figure 6. An important effect of the Sun is that it increases the value of the vertical frequency and decreases the short period frequency, with respect to what gives the RTBP. When the z-amplitude is of medium size (say, around $\dot{z}_0 = 0.22$, see next Section) there is a 1 to 1 resonance between the short period frequency and the solar one, leading to stability. On the other hand, when the z-amplitude increases (near $\dot{z}_0 = 0.93$) the vertical frequency and the short period one come into 1 to 1 resonance, leading again to instability. In this upper part the effect of the Sun (due to the variation of the frequencies with respect to the ones of the RTBP mentioned above) produces a delay in the instabilization. Indeed, in the RTBP the vertical family becomes unstable near $\dot{z}_0 = 0.89$.

As one can check with the expression of ω_v above, the dependence of ω_v with respect to I_v is quite mild. This also happens in the bicircular model. Vertically the motion is rather close to an harmonic oscillator. This can lead to a strong magnification of the effect of perturbations with frequencies close to resonance with the vertical one. The "twist condition" is very weak, say. The lack of robustness of the RTBP near L_4

and L_5 in front of the perturbations, for the Earth-Moon mass ratio, can also be seen from the torsion matrix (that is, the matrix of second derivatives of the Hamiltonian with respect to the actions). Using the Normal Form around L_5 one has that the torsion matrix is non definite and the torsion is small. The eigenvalues are 2.19621, -1.28718 and -0.02578, approximately.

Discussion and tentative explanations

We summarize first the results obtained so far. As presented in Gómez et al.[5], in the circular RTBP there is a big set of points which display stable motion around L_5. If, as we did, we take zero initial synodic velocity, this set of points is a kind of thick shell (which is close, but definitely departs, from part of a spherical surface). The form of this shell can be useful to try to derive "adapted" variables to carry out analytical studies. For the elliptic RTBP the results are not too far from the previous ones.

However the bicircular problem shows a quite different behaviour. Starting with zero initial velocity almost no point subsist until $z = 0.25$. But for z near 0.50 (i.e. orbits which start with zero velocity in synodical coordinates at a distance from the Earth-Moon plane which is of the order of magnitude of one half of the Earth-Moon distance) a big amount of points subsists. Near the periodic solution which replaces L_5 the behaviour is unstable. The fact that for higher values of z the motion seems to be on a 3-D torus which in turn is moving periodically with the period of the Sun, becomes more evident after a frequency analysis.

A part of the points which subsist in the bicircular problem do also subsist in the real one. A few points have been also seen to subsist starting on $z = 0$ in adimensional coordinates (see Gómez et al.[5]).

Now we present tentative explanations concerning the escape mechanism, the nature of the "boundary" of the set of remaining points and the rate of escape. These explanations are of different kind: geometrical and analytical.

Geometrical considerations. First we present these considerations for the RTBP. It seems that the local boundary of the stable set is related to resonances. May be these resonances are of high order (order= $\sum_{j=1}^{3} |k_j|$). Orders as 10, 12 or 17 are common in the Earth-Moon case. A complete description of the involved resonances is a hard task and it seems hopeless to try just analytical methods. A strong numerical effort is required to have a detailed description of those resonances.

Furthermore we have detected zones of stable motion which do not look as quasi-periodic, but chaotic close to resonance. This means that local resonance alone is not enough to produce escape. Let us try to explain this. The local resonance is due, in the simplest case to the appearition of some 2-D tori possessing stable and unstable manifolds. But to go out from the stable zone it is necessary that the stable manifold be connected to regions far away from L_5. These connections seem to be produced by a heteroclinic intersection with the stable manifold of some invariant object. Let us describe this object. We recall that the point L_3 is of saddle×centre×centre type. In particular it has a 4-D centre manifold. This manifold contains a Lyapunov orbit, halo orbits, etc. and also a Cantor set (with big fractal dimension) of 2-D tori in each level of energy. All these objects (periodic orbits and tori) are normally hyperbolic in their level of energy. It seems that the global stable manifold of this centre manifold ($W^s(W^c_{L_3})$), which is a manifold of codimension 1, acts as the effective boundary of the stable set.

Some numerical computations of intersections of this 5-D manifold with the set of the zero initial velocity, have been obtained. Generically this is a surface in the (x, y, z) variables. That surface can be plotted together with the set of stable points from direct numerical simulations (for instance using sections through z =ctant) to see if the suspected relation holds (see Gómez et al.[5]). For small values of μ the coincidence, between the 5-dimensional manifold and the boundary of the stable set, becomes more apparent.

Now we return to the bicircular model. We have already commented on the fact that the orbit which replaces the libration point has a hyperbolic behaviour in some directions. A transformation which (at least formally) skips the time dependence, if this is possible (see the third Section), produces an autonomous Hamiltonian with a fixed point of saddle × centre × centre type instead of the points L_4 or L_5. Hence, these fixed points also have codimension 1 invariant manifolds. We suspect that the "stability" domains are "confined" by the codimension 1 manifolds (of centre-stable and centre-unstable type) associated to the periodic (in the bicircular model) or quasiperiodic (in better models) orbits which take the place of L_3, L_4 and L_5.

We want to remark that these "rough boundaries" look very close to a "true" boundary when the mass parameter decreases (e.g., in the RTBP Sun-Jupiter case). Similar objects appear in many other problems when one looks at the stability zones around a general totally elliptic point (for instance, in symplectic maps in dimension four). In our case they come from the centre-unstable and centre-stable manifolds associated to the libration point L_3. In a similar way, the stability zones around the Moon, for instance, can be seen to have a rough boundary made of the centre-unstable and centre-stable manifolds associated to the libration points L_1 and L_2 of the Earth-Moon system. Some interesting transfers to halo orbits around the Moon or to highly elliptic lunar orbits, can use the fact that the manifolds for the L_1 Sun-Earth problem and the corresponding ones for the L_1 and L_2 Earth-Moon problems, do intersect for suitable epochs. But this topic is outside the scope of the present work.

Analytical considerations on the rate of escape. We shall consider the rate of escape of the points, according to the numerical simulations described before. It is readily seen that, for the bicircular model and the set of initial data we have taken, most of the points do not satisfy the stability criterion even after just 1 period of the Moon. We want to look at the ultimate rate of escape, after a transient (probably as long as a few hundreds of lunar periods, in our case).

First we look at a model problem. Consider the vicinity of a totally elliptic point of a Hamiltonian system and assume that the frequencies at that point satisfy a non-resonant diophantine condition (for some "admissible" resonances we refer to Giorgilli et al.[3]). Then the rate of diffusion of the points at a distance R of the fixed point is of the form $\exp(-\frac{c}{R^\alpha})$ (see Giorgilli et al.[3]) for some positive values of c and α, as it should be, in good agreement with the Nekhorosev[8] estimates. This is a bound on the time derivative of the radius, where the radius, R, is the square root of the sum of the actions (or of some positive definite linear combination of them). On the other hand we know that some power of the radius can be considered as a measure of the perturbation with respect to an integrable model, for instance, a Normal Form up to some degree. As it is well know, the set of points which are not in invariant tori is, in turn, a power of the perturbation parameter. Furthermore, we shall make the simplificative assumption that all the points which can go away from the fixed point, do really go away and with the maximum speed given by the previous bound on the speed of diffusion.

Summarizing, we consider, that for some radius, R_1, there are $R_1^s dR_1$ particles travelling away from 0 with a velocity equal to $\exp(-\frac{c}{R^\alpha})$, where s is some positive exponent. We consider that the particles escape from the neighbourhood of the elliptic point when they reach a value of the radius equal to R_0. Then it is possible to compute two values, $R_{1,n-1}$ and $R_{1,n}$ such that for R_1 between them, the escaping time is between $n-1$ and n. We define the set of points escaping in this time interval as e_n, and the set of points which remain at time n as r_n. Obviously $r_n = r_m - \sum_{j=m+1}^{n} e_j$ for $n > m$. After carrying out all the computations we obtain

$$r_n = L + \frac{A}{(\log n)^\beta},$$

where L (eventually zero) are the points which subsist for all time and A, β are positive quantities. Of course, β is related to the previous exponents α and s. We must remark that the formula above is a first approximation, more accurate formulas requiring corrections of the form $\log(\log n)$ to be added to the denominator. For moderate values of n this correction can be important. Furthermore, the "experimental" values of e_n, i.e., obtained from some numerical simulation, seem to behave in an irregular way. Hence, one has to consider successive values of r_{n_k} for $k = 1, 2, 3, \ldots$ not too close. To have values of n_k such that the fractions

$$\frac{r_{n_k} - r_{n_{k+1}}}{r_{n_{k-1}} - r_{n_k}}$$

have a limit different from 1, and has to take $n_{k+1} = n_k^\gamma$, for some γ greater than 1.

We have tested the approach above for the results of the numerical simulations, concerning the points which remain after n lunar revolutions in the bicircular model. To this end we have considered all the points which remain for more than 100 lunar revolutions (and up to 10000 revolutions due to the severe CPU time requirements). We have taken 64 time intervals, i.e., $n_0 = 100$ and $n_{64} = 10000$. Using the previous relation between successive values of n_k one has to take $\gamma = 1.010889$. Then we have obtained the best fit (by minimum squares, after taking logarithms in the formula above) for L, A and β by using the values of n with indices ranging from 32 to 64. The values obtained for the parameters are

$$L = 4636.3, A = 521306, \beta = 2.19437.$$

With those data the relative differences between the values given by the formula above and the observed data are less than 0.0022 for almost all the points (and certainly for the second half of the set of data), reaching 0.0211 for $n = 100$. We remark that, close to $n = 100$ the selected values of n are too close and the irregularity in the observed behaviour seems to give a good evidence that the initial transient should be considered till $n = 400$ or $n = 500$. The Figure 7 displays the data r_{n_k} for $k = 0, 1, \ldots, 64$ versus $\log(\log n)$ and the fit with the coefficients given above.

A natural question shows up. We have derived the formula for r_n, under some simplifying assumptions, for an elementary model. Why it seems to work in the present, harder case, of the bicircular model? An heuristic explanation can be as follows. The role played by the elliptic fixed point in the elementary model, is played by different invariant objects, mainly invariant tori. Instead of considering points in spherical shells which travel with some velocity, one has to consider subsets of toroidal annuli. The estimates for the speed of diffusion must also be of the Nekhorosev type, where now

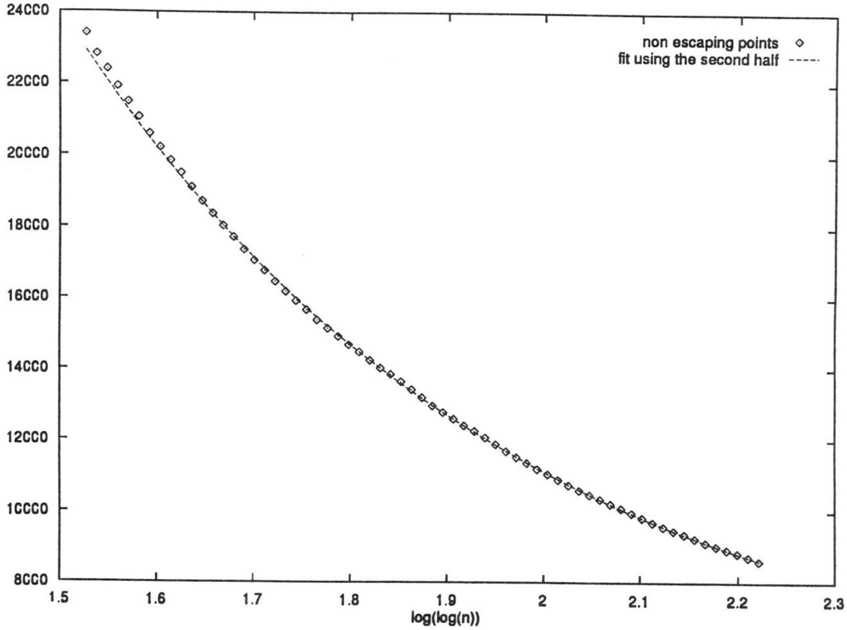

Figure 7

the meaning of R is the distance to some invariant object. Then one has to average to take into account different contributions, but if most of them are of the same kind, it is not a surprise that the general formula be of the same type as before.

SOME ANALYTIC COMPUTATIONS AND THE BOUNDARY OF INSTABILITY NEAR THE SMALL PERIODIC ORBIT

In this Section we shall consider different analytic tools to obtain an approximation of the boundary where the transition from instability to stability occurs in the bicircular model. We start at $z = 0$ near the small unstable periodic solution which replaces L_5, and then the vertical component is increased.

Normal Form

To obtain a Normal Form, suitable for the analysis in a vicinity of the small periodic orbit, first we introduce a new action variable, p_θ, canonically conjugate to the Sun's angle θ. Then the Hamiltonian becomes

$$H = \omega_S p_\theta + \frac{1}{2}(p_x^2 + p_y^2 + p_z^2) + yp_x - xp_y - \frac{1-\mu}{r_{PE}} - \frac{\mu}{r_{PM}} - \frac{m_S}{r_{PS}} - \frac{m_S}{a_S^2}(y \sin\theta - x\cos\theta),$$

without explicit appearition of the time. For shortness we just describe the steps to be carried out for the computation of the desired Normal Form, refering to Gómez et al.[5] for the details. We shall keep the names θ and p_θ for the Sun's angle and its related momentum through all the transformations. In fact θ remains unchanged, while p_θ changes because the initial Hamiltonian is not autonomous.

1. The first step is to shift the origin of coordinates to the periodic solution. One needs the Fourier expansion of the periodic orbit. This can be obtained by some symbolic computation or, in a more expeditive way, by computing both the periodic solution and the corresponding Fourier analysis numerically.

2. At this moment the Hamiltonian has no terms of first degree in the coordinates. All the other terms are monomials in the space coordinates and momenta, with coefficients depending periodically on θ. First we want to make the terms of degree 2 of the Hamiltonian, H_2, independent of θ. This is possible by Floquet's Theorem. Furthermore, it is easy to show that the transformation can be chosen to be canonical. One needs a representation of the (periodic) matrix of change of coordinates as a trigonometric series. Again one can use either a symbolic computation of the coefficients or a numerical Fourier analysis. In this way the actual terms of degree 2, H_2, (we keep H to denote the Hamiltonian in all the sequence of transformations) are independent of θ.

3. The next step is to put the terms of degree 2 in some suitable "diagonal" form:

$$H_2 = \lambda_1 Q_1 P_1 + \frac{1}{2}\omega_2(Q_2^2 + P_2^2) + \frac{1}{2}\omega_3(Q_3^2 + P_3^2),$$

   where ω_2 and ω_3 are close to the frequencies of the RTBP ω_l and ω_v. The frequency ω_s is the one which has dissappeared. In its place the behaviour is hyperbolic with eigenvalue λ_1. We denote by Q_j and P_j for $j = 1, 2, 3$, the current variables.

4. For further use it is convenient to introduce complex coordinates by means of

$$X_j = \frac{Q_j + \sqrt{-1}P_j}{\sqrt{2}}, \qquad Y_j = \frac{\sqrt{-1}Q_j + P_j}{\sqrt{2}}, \qquad j = 2, 3,$$

   and we put $X_1 = Q_1, Y_1 = P_1$. Then we have

$$H_2(X, Y) = \lambda_1 X_1 Y_1 + \sqrt{-1}\omega_2 X_2 Y_2 + \sqrt{-1}\omega_3 X_3 Y_3.$$

5. At this point we have to take the full Hamiltonian and perform all the changes described so far. To do this is quite convenient to put all the changes together. One obtains an affine transformation depending periodically on the angle θ. Then, as the inverses of the distances of the particle to Earth, Moon and Sun, r_{PE}, r_{PM} and r_{PS}, respectively, can be expanded by using the recurrence of the Legendre polynomials, it is easy to obtain the expansion of the Hamiltonian in the X_j and Y_j, $j = 1, 2, 3$ variables. The H_2 is already in Normal Form.

6. Now we have to show how to reduce the higher order terms of the Hamiltonian. The actual Hamiltonian is of the form

$$H(X, Y, \theta, p_\theta) = \omega_S p_\theta + H_2(X, Y) + \sum_{r>2} H_r(X, Y, \theta),$$

   with H_2 as before and

$$H_r(X, Y, \theta) = \sum_{|k|=r} h_r^k(\theta) X^{k^1} Y^{k^2},$$

   where $k = (k_1, \ldots, k_6)$, $k^1 = (k_1, k_2, k_3)$ and $k^2 = (k_4, k_5, k_6)$.

The purpose is to obtain a (periodic) change of variables transforming this Hamiltonian to a simpler one. That is, we will try to eliminate the dependence with respect to θ and, at the same time, to cancel the maximum possible number of monomials. Let us describe the method to be used. The main point is contained in the following proposition, which is presented in a general setting.

Proposition 1 *Let us consider the Hamiltonian*

$$H = \omega_S p_\theta + H_2(X,Y) + \sum_{i=3}^{r-1} H_i(X,Y) + H_r(X,Y,\theta) + H_{r+1}(X,Y,\theta) + \cdots,$$

where $r > 2$, $H_2(x,y) = \sum \omega_i X_i Y_i$ $(\omega_i \in \mathbb{C})$, $H_i(X,Y) = \sum_{|k|=i} h_i^k X^{k^1} Y^{k^2}$,

$H_r(X,Y,\theta) = \sum_{|k|=r} h_r^k(\theta) X^{k^1} Y^{k^2}$ *and* $h_r^k(\theta) = \sum_j h_{r,j}^k \exp(j\theta\sqrt{-1})$.

Let $G_r(X,Y,\theta) = \sum_{|k|=r} g_r^k(\theta) X^{k^1} Y^{k^2}$ *be a generating function defined as follows:*

(a) If $k^1 \neq k^2$ *then:*

$$g_r^k(\theta) = \frac{c_k - h_{r,0}^k}{< \omega, k^2 - k^1 >} + \sum_{j\neq 0} \frac{h_{r,j}^k}{j\omega_S\sqrt{-1} - < \omega, k^2 - k^1 >} \exp(j\theta\sqrt{-1}).$$

(b) If $k^1 = k^2$ *then:*

$$g_r^k(\theta) = \sum_{j\neq 0} \frac{h_{r,j}^k}{j\omega_S\sqrt{-1}} \exp(j\theta\sqrt{-1}).$$

Then, the new Hamiltonian, H', obtained from H by means of the change of variables given by the generating function G_r,

$$H' = H + \{H, G_r\} + \frac{1}{2!}\{\{H, G_r\}, G_r\} + \cdots,$$

satisfies that

$$H' = \omega_S p_\theta + H_2(x,y) + \sum_{i=3}^{r-1} H_i(x,y) + H_r'(x,y) + H_{r+1}'(x,y,\theta) + \cdots,$$

where $H_r'(x,y) = \sum (h')_r^k x^{k^1} y^{k^2}$,

$$(h')_r^k = \begin{cases} c_k & \text{if } k^1 \neq k^2 \\ h_{r,0}^k & \text{if } k^1 = k^2 \end{cases}$$

and x_j, y_j, $j = 1,2,3$ *denote the new variables after the change.*

We refer to Gómez et al.[5] for the proof.

Remark: The value c_k that appears in the function G_r is arbitrary. Usually it will be selected equal to zero, unless the divisors $< \omega, k^2 - k^1 >$ were small, in which case it would be chosen equal to $h_{r,0}^k$. To have the final Hamiltonian as simple as possible, it would be desirable that all the c_k were chosen equal to 0 (except, of course, the ones corresponding to the resonant terms $k^1 = k^2$). But doing that, the expressions of the generating functions contain the divisors $< \omega, k^2 - k^1 >$, that can be very small and produce convergence problems. For this reason, in practice, one can use a threshold ε and select $c_k = 0$ only when the corresponding divisor is greater than ε. Otherwise it is chosen to be equal to $h_{r,0}^k$. With this, one can expect to obtain a final Hamiltonian with only a few monomials, being a useful approximation over a relevant region of the phase space.

The results of the computation of the Normal Form

The steps described above have been implemented and applied to the case of the bicircular model. The computations have been done, in fact, for the periodic orbit replacing L_4. They are similar for L_5. Using a threshold $\varepsilon = 10^{-2}$ for the small divisors, we have computed the Normal Form up to degree 12 (in the cartesian coordinates, and, hence, up to degree 6 in the actions). This expansion contains only the exactly resonant terms (this is because there is not any small divisor less than 10^{-2} up to order 12), and this implies that the Normal Form is integrable.

For the analysis it is convenient to translate the final Hamiltonian to real coordinates (we are interested on the behaviour of the system for real values of the coordinates). We use the inverse of the change of variables used when complexifying:

$$u_i = \frac{x_i - \sqrt{-1}y_i}{\sqrt{2}}, \quad v_i = \frac{-\sqrt{-1}x_i + y_i}{\sqrt{2}}, \; i = 2, 3,$$

and (as before) $u_1 = x_1$ and $v_1 = y_1$.

We can write this Hamiltonian in an simpler form by defining the actions

$$I_1 = u_1 v_1, \quad I_2 = \frac{1}{2}(u_2^2 + v_2^2), \quad I_3 = \frac{1}{2}(u_3^2 + v_3^2).$$

We note that the variables u_3 and v_3 are rather close to z and $\dot{z}$.

We give that Hamiltonian in Table 2, where the three first columns contain the exponents of I_1, I_2 and I_3, and the fourth one contains the corresponding coefficient.

The Normal Form obtained so far shows that, around the equilibrium point which takes the place of the periodic orbit, the Hamiltonian behaves like a saddle $\times$ centre $\times$ centre. The action variable I_1 is related to the saddle. The local formal centre manifold is defined by $I_1 = 0$. For any couple of (small enough) positive values of I_2 and I_3, we obtain an approximation of a 2-D torus, the frequencies being similar to the long period one and the vertical one of the RTBP. Going back through the changes of variables, we obtain 3-D tori, the new frequency being the solar synodic one. Along the full centre manifold we can compute the normal hyperbolicity. This is given by the function $\lambda = \frac{\partial H}{\partial I_1}(0, I_2, I_3)$. Of course, at the origin λ is the coefficient of I_1, that is, λ_1.

We can consider the part of the centre manifold restricted to $I_2 = 0$ having I_3 as unique parameter. This describes a family of unstable periodic orbits of the Normal Form, corresponding to 2-D tori of the bicircular problem. We want to know the value of I_3 for which the stability changes (and so does the dimension of the centre manifold). From Table 2 we obtain the following expression for λ, now being simply $\lambda(I_3)$:

$$\begin{aligned}\lambda(I_3) &= -0.013852 + 0.257784 I_3 + 2.607311 I_3^2 + 50.49658 I_3^3 + \\ &\quad +1299.415 I_3^4 + 38773.19 I_3^5 + \dots \; .\end{aligned}$$

Of course, the Normal Form of the kind considered here can not be convergent when the desired bifurcation occurs (letting aside the small divisors problems that, in fact, prevent from convergence in any open set). We can set $\lambda(I_3) = 0$ with the expression given here to obtain a critical value of I_3. We find $I_3 = 0.030481$. This gives a value for, say v_3 if $u_3 = 0$, of 0.246905. The observation concerning u_3 and v_3 tells us that this gives an idea of the suitable value of $\dot{z}$ if $z = 0$. One can try to improve a little bit

Table 2. Hamiltonian in Normal Form around the small unstable periodic orbit of the bicircular problem.

1	0	0	-.1385220057080626D-01	2	1	2	-.1864225143531067D+07
0	1	0	-.3005039252506557D+00	2	0	3	.5304991639454239D+05
0	0	1	.1004006523604956D+01	1	4	0	-.4052415718452325D+08
2	0	0	.8515077021376744D+00	1	3	1	-.1360849493836872D+08
1	1	0	-.3324999040181953D+01	1	2	2	.5520658628803922D+05
1	0	1	.2577837871812799D+00	1	1	3	-.5699313229911387D+05
0	2	0	.3270045540584760D+00	1	0	4	.1299414585487088D+04
0	1	1	.2307565086566667D+00	0	5	0	-.2448924023843400D+07
0	0	2	-.4032569929542033D-02	0	4	1	.1501613346456005D+07
3	0	0	.1037922842258645D+03	0	3	2	.1711773835404635D+06
2	1	0	-.4565506222871429D+03	0	2	3	.1632111440703841D+04
2	0	1	.3321899325302790D+02	0	1	4	-.9347880155609670D+02
1	2	0	-.2146275456385827D+04	0	0	5	.1551444240916563D+01
1	1	1	-.4784863921619558D+02	6	0	0	.2987999377729337D+10
1	0	2	.2607311336225566D+01	5	1	0	-.2411838508681229D+11
0	3	0	.3096780901433248D+02	5	0	1	.1782373214959049D+10
0	2	1	.2494045964589006D+01	4	2	0	-.8672410442035261D+12
0	1	2	-.2270782098281886D+00	4	1	1	-.1052914435151395D+11
0	0	3	.9052424722400662D-03	4	0	2	.4078275712719006D+09
4	0	0	.2499474130723898D+05	3	3	0	-.1116963084432885D+12
3	1	0	-.1379292451759401D+06	3	2	1	-.2505479615606837D+11
3	0	1	.1018100448948390D+05	3	1	2	-.1622067585088890D+10
2	2	0	-.1748183316981774D+07	3	0	3	.4405125443309778D+08
2	1	1	-.3120287556473054D+05	2	4	0	.1858745324134717D+13
2	0	2	.1318002728888036D+04	2	3	1	.5250659284283364D+11
1	3	0	.1288209631845397D+07	2	2	2	.4716462061283490D+09
1	2	1	.4931293498255790D+05	2	1	3	-.1020454848295031D+09
1	1	2	-.1616391528463035D+04	2	0	4	.2189796196165091D+07
1	0	3	.5049658419448990D+02	1	5	0	-.4715529712174978D+12
0	4	0	-.2709111320415782D+05	1	4	1	-.2821114699272034D+11
0	3	1	-.3353580692587518D+04	1	3	2	.9001096107686855D+09
0	2	2	.6040532947455749D+01	1	2	3	.8144952535614336D+07
0	1	3	-.1944360487197618D+01	1	1	4	-.2160288537221536D+07
0	0	4	.4157193549100161D-01	1	0	5	.3877318738835637D+05
5	0	0	.7974901100840953D+07	0	6	0	.8912688288853617D+10
4	1	0	-.5407213341839783D+08	0	5	1	.5578588877780112D+09
4	0	1	.3996634109602074D+07	0	4	2	-.1075748940803764D+09
3	2	0	-.1253144279500538D+10	0	3	3	-.3296022936390645D+07
3	1	1	-.1807321686550926D+08	0	2	4	.8616212602968118D+05
3	0	2	.7164000260814325D+06	0	1	5	-.4191316945624839D+04
2	3	0	.8247843021482749D+09	0	0	6	.5876577678574162D+02
2	2	1	.1380185818563243D+08				

the estimate by proceding in an heuristic way. In the formula for $\lambda(I_3)$ one can assume that the coefficients of higher order increase at a given rate, similar or slightly higher than the quotient of the coefficients of degree 5 and 4. Assuming an increase by a factor 30 one gets a critical value for $\dot{z}$ of 0.233358, and if the factor is 40, one obtains 0.217431. However, one can not rely very much on this result, the main problem being the low order at which we have computed the Normal Form. To increase the order in a substantial way seems to be a formidable task. We better proceed to produce estimates using a different analytic approach, which can lead to a higher order expansion with a similar computational effort.

The equations and the algorithm for the computation of unstable 2-D tori

Vertical periodic orbits of the circular 3-D RTBP give rise, in the case of the bicircular problem, to 2-D tori. One of the frequencies is the vertical one, the other being the solar synodic frequency. The Normal Form procedure gives good results only for small z amplitude. We proceed to the direct computation of these tori.

Let ξ, η, ζ be the coordinates with respect to the L_5 point in the synodic system (for the L_4 case is similar). Then the equations of motion are ($' = d/dt$)

$$\begin{aligned}
\xi'' &= 2\eta' + \frac{3}{4}\xi - a\eta + \frac{\partial}{\partial \xi}V, \\
\eta'' &= -2\xi' - a\xi + \frac{9}{4}\eta + \frac{\partial}{\partial \eta}V, \\
\zeta'' &= -\zeta + \frac{\partial}{\partial \zeta}V,
\end{aligned}$$

where $a = (\sqrt{27}/4)(1-2\mu)$ and V is the potential accounting for the higher order terms of the RTBP plus the solar effect:

$$\begin{aligned}
V &= (1-\mu)\sum_{n\geq 3} P_n\left(\frac{(1/2)\xi - (\sqrt{3}/2)\eta}{\rho}\right)\rho^n + \mu\sum_{n\geq 3} P_n\left(\frac{-(1/2)\xi - (\sqrt{3}/2)\eta}{\rho}\right)\rho^n \\
&\quad + \frac{m_S}{a_S}\sum_{n\geq 2} P_n\left(\frac{(\xi+\mu-1/2)\cos(\omega_S t) - (\eta+\sqrt{3}/2)\sin(\omega_S t)}{R}\right)\frac{R^n}{a_S^n}.
\end{aligned}$$

In the last expression one has $\rho^2 = \xi^2+\eta^2+\zeta^2$, $R^2 = (\xi+\mu-1/2)^2+(\eta+\sqrt{3}/2)^2+\zeta^2$, and a_S, m_S, ω_S denote the radius of the orbit, the mass and frequency of the Sun.

We write first $\xi = \xi_0 + \overline{\xi}$, $\eta = \eta_0 + \overline{\eta}$, $\zeta = \zeta_0 + \overline{\zeta}$, where ξ_0, η_0 ($\zeta_0 = 0$) are the coordinates of the periodic solution, with frequency ω_S, which replaces L_5 in the bicircular problem. Let $\psi = \exp(i\omega_S t)$. Then $\xi_0 = \sum_k \xi_{00k}\psi^k$, $\eta_0 = \sum_k \eta_{00k}\psi^k$. The coefficients are known from the numerical solution and numerical Fourier analysis. We look for a solution of the form (with complex coefficients)

$$\xi = \sum_{l,j,k} \xi_{ljk}\alpha^l\phi^j\psi^k, \eta = \sum_{l,j,k} \eta_{ljk}\alpha^l\phi^j\psi^k, \zeta = \sum_{l,j,k} \zeta_{ljk}\alpha^l\phi^j\psi^k,$$

with $l \geq 0$, $j, k \in Z$, where $\phi = \exp(i\omega t)$, α is some (vertical) amplitude, which is being considered as a parameter, and ω is an amplitude depending (vertical) frequency, of the form $\omega = \omega_0 + \omega_2\alpha^2 + \omega_4\alpha^4 + \ldots$. As normalizing condition on α we choose $\zeta_{1,1,0} = 1/2$, $\zeta_{l,1,0} = 0$, $\forall l > 1$. As the solutions are real we have $\xi_{l,j,k} = \overline{\xi}_{l,-j,-k}$ and similar relations hold for the η and ζ coefficients. Furthermore, the symmetries of the problem imply that in ξ, η, the indices l and j are even, in the ζ they are odd, and in all the cases $|j| \leq l$.

Let us denote by ξ_l the sum $\sum_{j,k}\xi_{ljk}\phi^j\psi^k$. So $\overline{\xi} = \alpha\xi_1 + \alpha^2\xi_2 + \ldots$. Similar representations are used for $\overline{\eta}$, $\overline{\zeta}$. First we discuss how we have computed the terms in ζ_1 (ξ_1 and η_1 being zero) and then how to proceed, in a different way, to compute ξ_n, η_n, ζ_n, $n \geq 2$.

The standing equation is $\zeta'' = -\zeta + (\partial/\partial\zeta)V$, and this, to order 1 in α, reads $\zeta_1'' = -\zeta_1 + \frac{\partial^2}{\partial\zeta^2}V(\xi_0, \eta_0, 0)\zeta_1$. Furthermore we keep $\zeta_{110} = 1/2$, and we have to determine ω_0, where $(\zeta_{1jk}\phi^j\psi^k)'' = \zeta_{1jk}\phi^j\psi^k[-(j\omega_0 + k\omega_S)^2]$. By equating terms in $\phi^j\psi^k$ in the previous equation for ζ_1 we get a linear system (but the problem is nonlinear because of ω_0) in the ζ_{1jk} coefficients $|j| \le j_{max}$, $0 \le k \le k_{max}$. As the system is homogeneous one has to take the right value of ω_0 to make the solution possible and some normalization, as it has been said. There are several possible ways to solve the problem. We can get ω_0 from the numerical integration of the variational equations along the periodic solution $(\xi_0, \eta_0, 0)$. Another possibility is to use an iterative process.

We write $(\zeta_1'' + \zeta_1)^{(k+1)} = \frac{\partial^2}{\partial\zeta^2}V(\xi_0, \eta_0, 0)(\zeta_1)^{(k)}$ and we start with the initial guess $\zeta_1^0 = (1/2)\phi$ (notice that we had to add to the representation used the complex conjugate and that if $k = 0$ then the range of j in the representation is $0 \le j \le j_{max}$). Let us denote by $c_1^{(k)}(\phi, \psi)$ the product $\frac{\partial^2}{\partial\zeta^2}V(\xi_0, \eta_0, 0)(\zeta_1)^{(k)}$. As we want to keep $\zeta_{1,1,0} = 1/2$ this gives immediately the value of $\omega_0^{(k+1)}$ by equating the terms in $\phi^1\psi^0$: $[1 - (\omega_0^{(k+1)})^2](1/2) = c_{1,1,0}^{(k)}$. Then the terms $\zeta_{1,j,k}^{(k+1)}$, $(j, k) \ne (1, 0)$ are given by $\zeta_{1,j,k}^{(k+1)} = c_{1,j,k}^{(k)}/(1 - (j\omega_0^{(k+1)} + k\omega_S)^2)$. When $\zeta^{(k+1)}$ is available we substitute to get the new $c_1^{(k+1)}$ and the process is repeated.

Unfortunately, that process, as it is described, fails to converge. However, it has been seen that it has just one unstable direction. This suggests to use an extrapolation method (Aitken) that then leads to convergence. The value obtained for ω_0 is 1.0040065236..., in perfect agreement with the monodromy matrix.

When ζ_1 is available we could continue the iterative- extrapolation process. However the number of unstable directions increases when we look for (ξ_n, η_n, ζ_n) (remark that if n is even we had to solve only for (ξ_n, η_n) and if n is odd, only for ζ_n). Hence we need a different method. We describe first the procedure for (ξ_n, η_n).

We write the equations as

$$\xi_n'' - 2\eta_n' - \left(\frac{3}{4}\xi_n - a\eta_n + \frac{\partial^2 V}{\partial\xi^2}(\xi_0, \eta_0, 0)\xi_n + \frac{\partial^2 V}{\partial\xi\partial\eta}(\xi_0, \eta_0, 0)\eta_n\right) =$$

$$= \left[\frac{\partial V}{\partial\xi}(\overline{\xi}_{n-1}, \overline{\eta}_{n-1}, \overline{\zeta}_{n-1}) - \overline{\xi}''_{n-1} + 2\overline{\eta}'_{n-1}\right]\Big|_n ,$$

$$\eta_n'' + 2\xi_n' - \left(a\xi_n + \frac{9}{4}\eta_n + \frac{\partial^2 V}{\partial\xi\partial\eta}(\xi_0, \eta_0, 0)\xi_n + \frac{\partial^2 V}{\partial\xi^2}(\xi_0, \eta_0, 0)\eta_n\right) =$$

$$= \left[\frac{\partial V}{\partial\eta}(\overline{\xi}_{n-1}, \overline{\eta}_{n-1}, \overline{\zeta}_{n-1}) - \overline{\eta}''_{n-1} - 2\overline{\xi}'_{n-1}\right]\Big|_n ,$$

where the notation $\overline{\xi}_{n-1}$ means $\xi_0 + \alpha\xi_1 + \ldots \alpha^{n-1}\xi_{n-1}$ (and similar for η, ζ) and $[\,]|_n$ denotes the terms in α^n. Notice that in the $[\,]|_n$ terms, there is a contribution in α^n when ξ, η, ζ are known to order $n-1$ in α because of the nonlinearities in the derivatives of V and because of the powers of α appearing in ω in the time derivatives.

For ζ_n (n odd) the procedure is similar, but keeping $\xi_{n,1,0} = 0$ allows the determination of ω_{n-1} when equating the terms in $\alpha^n\phi^1\psi^0$.

We remark that the systems of linear equations which appear to get (ξ_n, η_n) for $n = 2, 4, \ldots$ have the same matrix. The same thing happens for ζ_n, $n = 3, 5, \ldots$. Furthermore the systems for different values of j are uncoupled. Some caution has to be taken with the poorly conditioned character of the matrices, specially in the ζ_n case, $j = 1$. It has been found very useful, to decrease the numerical errors, to work in quadruple precision complex variables (or simulate it, depending on the software). The computation of the derivatives of V is shortened by the use of recurrences derived from the ones of Legendre polynomials.

Truncated power series results

The previous algorithm has been implemented with l_{max} (and hence, j_{max}) $= 41$ and $k_{max} = 18$. From the results of the second Section, we know that the previous 2-D unstable tori should be convergent up to moderate values of α (and, definitely, less than 0.25). In fact, the coefficients of ω seem to give a radius of convergence ($\lim_{n\to\infty, n\ even} \omega_n^{-1/n}$) close to 0.219. If, instead, we consider the norm of ξ_n as $|\xi_n| = \sum_{j,k} |\xi_{n,j,k}|$, we can also get an idea of the radius of convergence of the expansions with respect to α. The same is true for the $|\eta_n|$, $|\zeta_n|$. All of them give essentially the same radius of convergence as ω. Notice that for n large the numerical errors propagated by the successive solutions at the previous orders, start to show up, giving a slightly irregular behavior of the components (see Table 3).

A test has been passed as follows. Consider a value of α and look at the approximate quasiperiodic solution provided by the formulas above. For a given value of t one can compute $\xi, \eta, \zeta, \xi', \eta', \zeta', \xi'', \eta'', \zeta''$. Then, from the positions and velocities, one obtains the acceleration from the direct equations of motion for the bicircular problem. The difference with respect to the values computed from the series is the residual acceleration. Table 4 shows some results for a time span of 100 adimensional units with time step 0.1. There is a remarkable agreement till $\alpha = 0.16$ and still acceptable till $\alpha = 0.2$. Figures 8 and 9 show the (x, z) and (x, y) projections of the analytic orbits computed for $\alpha = 0.10$ and $\alpha = 0.18$. In both Figures we plot just one dot for each value of t, these values being of the form $k \times 0.1$ for $k = 0, 1, \ldots, 100000$, without joining successive points in the orbit. One can suspect that in the case of Figure 9 we are close to a resonance. The frequencies turn out to be $\omega_S = 0.9251959855$ and $\omega_v = 1.0038778841$, not too close to produce a low order resonance. The behaviour seen in the Figure 9 is due to the time interval. Indeed, $18 \times \omega_S \times 0.1 + 46 \times \omega_v \times 0.1 - 2\pi$ is less than 3×10^{-6}. Taking 1.1×10^6 points (with the same time step) shows a very good equidistribution of points.

Discussion and Conclusions

Invariant 2-D unstable tori have been found by a symbolic manipulation implementation of a version of the Lindstedt-Poincaré method. By a suitable modification and an additional computational effort, one can compute also 3-D unstable tori. The frequency to be added is equivalent to the long period frequency of the RTBP. We also recall that the action of the Sun produces an increase of the vertical frequency (whose limit is 1 in the RTBP when the vertical amplitudes goes to zero). However, the behaviour of the "vertical" frequency when the amplitude increases is preserved. An increase of the amplitude produces a reduction of the frequency. The boundary of convergence is associated to a bifurcation (as was already outlined in the previous Section): The tori lose their hyperbolic character, becoming stable (in some weak sense). They can be considered as "big" KAM tori of the bicircular problem. As sketched in the second Section, the lose of stability, for z amplitudes around 0.22, is due to a resonance between the "short period frequency" and the solar one.

The existence of stable zones for the bicircular model (and, in fact for the real model, this one being a slight perturbation of the previous one, in the region of the study) opens the way to a new series of space missions. Indeed, the stability of the orbits implies that no station keeping is needed for the orbit dynamics. Furthermore is has been seen that it is relatively cheap to reach those zones from a GTO or a parking

Table 3. For the variables ω, ξ, η and ζ the non zero norms of the coefficients for orders of α from 16 up to 41 are given, and also the estimated radii of convergence.

ω			ξ		
16	.125580D+06	.214260D+00	16	.650326D+08	.226198D+00
18	.270934D+07	.215292D+00	18	.127693D+10	.225673D+00
20	.580404D+08	.216056D+00	20	.251957D+11	.225123D+00
22	.123691D+10	.216618D+00	22	.499663D+12	.224556D+00
24	.262616D+11	.217024D+00	24	.996014D+13	.223978D+00
26	.556124D+12	.217307D+00	26	.199577D+15	.223396D+00
28	.117566D+14	.217492D+00	28	.401983D+16	.222818D+00
30	.248292D+15	.217600D+00	30	.813808D+17	.222250D+00
32	.524165D+16	.217644D+00	32	.165577D+19	.221697D+00
34	.110661D+18	.217639D+00	34	.338509D+20	.221164D+00
36	.233726D+19	.217592D+00	36	.695257D+21	.220654D+00
38	.494006D+20	.217514D+00	38	.143426D+23	.220170D+00
40	.104511D+22	.217412D+00	40	.297108D+24	.219713D+00
η			ζ		
16	.471914D+08	.226247D+00	17	.547271D+08	.226223D+00
18	.926194D+09	.225725D+00	19	.107433D+10	.225699D+00
20	.182662D+11	.225178D+00	21	.211934D+11	.225148D+00
22	.362062D+12	.224612D+00	23	.420179D+12	.224586D+00
24	.721361D+13	.224034D+00	25	.837397D+13	.224001D+00
26	.144470D+15	.223453D+00	27	.167751D+15	.223425D+00
28	.290843D+16	.222874D+00	29	.337852D+16	.222828D+00
30	.588521D+17	.222304D+00	31	.683700D+17	.222295D+00
32	.119684D+19	.221749D+00	33	.139109D+19	.221694D+00
34	.244575D+20	.221213D+00	35	.284283D+20	.221208D+00
36	.502114D+21	.220701D+00	37	.583760D+21	.220677D+00
38	.103540D+23	.220214D+00	39	.120527D+23	.220077D+00
40	.214405D+24	.219754D+00	41	.249632D+24	.219731D+00

Table 4. Test of errors for the unstable 2-tori obtained by the Lindstedt-Poincaré method. The first column gives the amplitude. 1he other three columns give the residual accelerations in x, y and z, respectively.

.11	.118048D−13	.124831D−13	.181799D−14
.12	.152708D−13	.175892D−13	.262290D−14
.13	.148928D−12	.204631D−12	.291295D−13
.14	.337156D−11	.490744D−11	.710931D−12
.15	.669266D−10	.957271D−10	.148328D−10
.16	.112226D−08	.156734D−08	.259666D−09
.17	.162949D−07	.221698D−07	.391360D−08
.18	.210096D−06	.277842D−06	.532121D−07
.19	.246850D−05	.319889D−05	.660951D−06
.20	.272526D−04	.352209D−04	.769544D−05
.21	.292881D−03	.380884D−03	.873664D−04

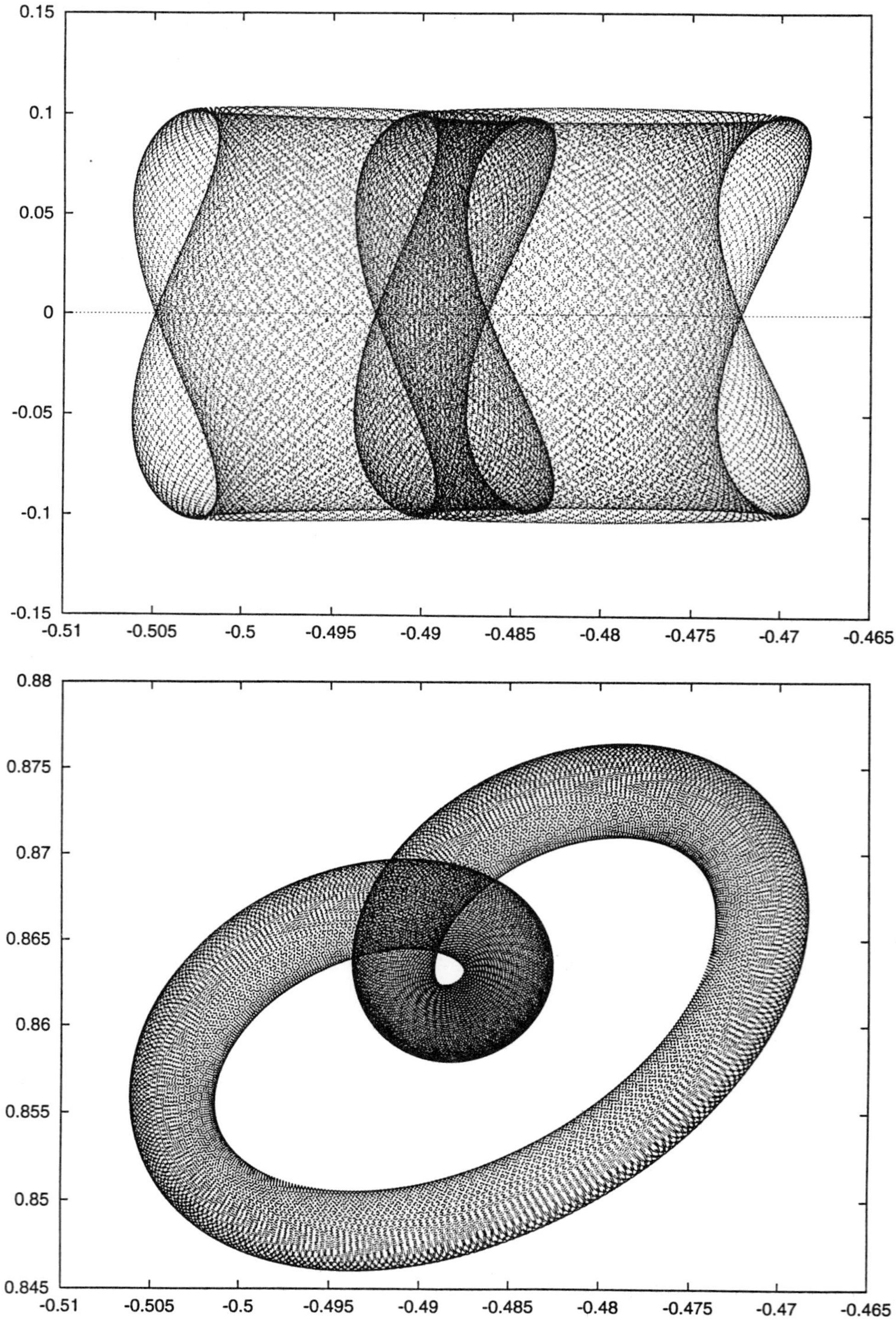

Figure 8. Projections (x, z) (top) and (x, y) (bottom) of the 2-torus of z amplitude equal to 0.10. The final time is 10000 units (recall 2π is equivalent to the lunar period). The time step is 0.1 time units. The distance unit is the Earth-Moon distance.

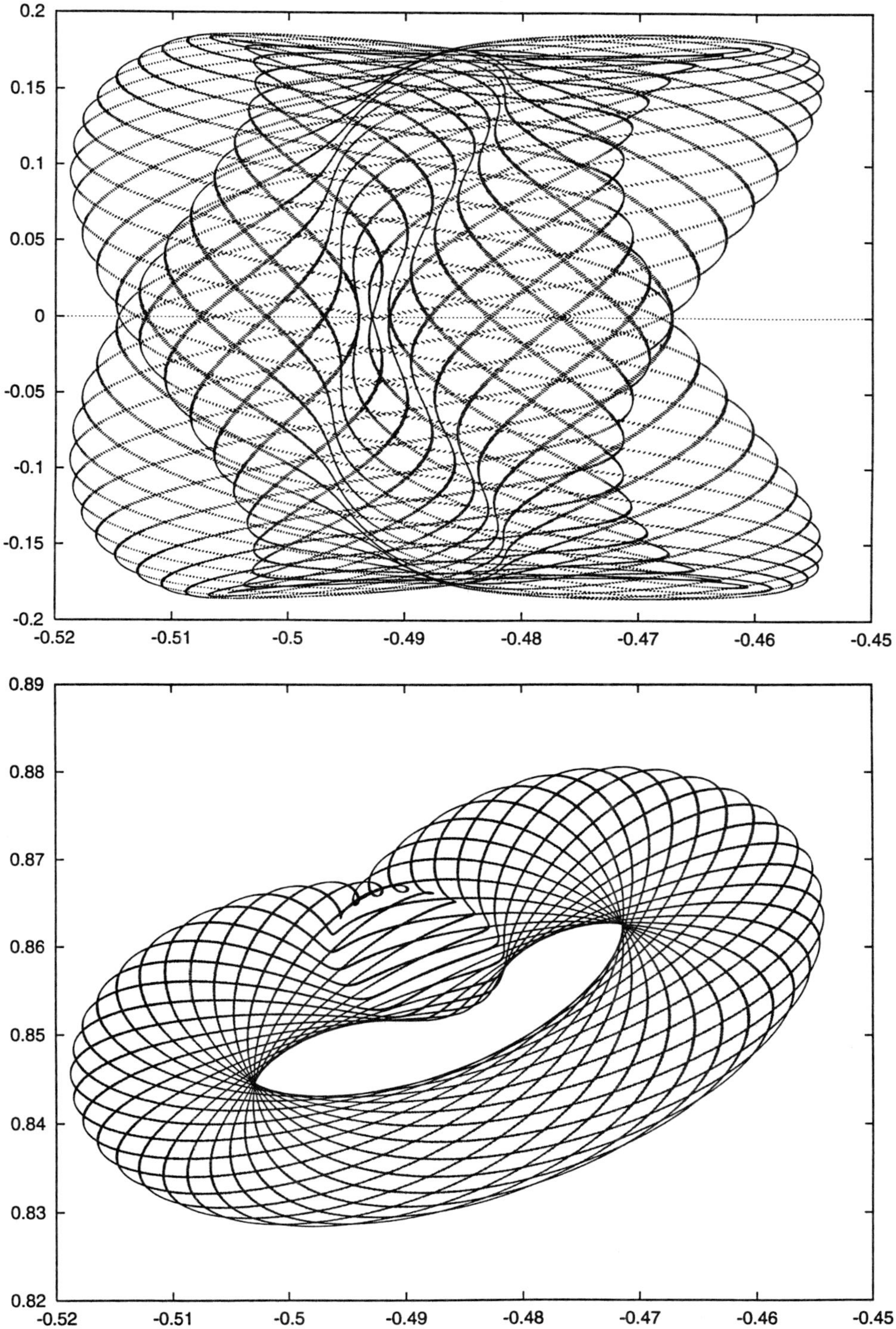

Figure 9. Projections (x, z) (top) and (x, y) (bottom) of the 2-torus of z amplitude equal to 0.18. The final time is 10000 units. All the units as in the previous Figure.

orbit around the Earth (see Gómez et al.[5]). Possible space missions using this kind of orbits are: Parking orbits for micro spacecrafts, ready to travel to asteroids approaching the Earth-Moon system (possibly with a lunar gravity assisted manoeuvre); Deep space astronomy, close enough to the Earth to realize easely the observational commands, and far enough to have low noise level.

Acknowledgements

The authors have been supported by a CICYT Grant ESP91-0403 from the Spanish Ministery of Education. They are indebted to the staff of ESOC, specially to Dr. J. Rodríguez Canabal, for his encouragement during the realization of the ESOC Contract 9711/91/D/IM(SC), where applications of the present work can be found in detail.

REFERENCES

1. V.I. Arnol'd and A. Avez. "Problèmes Ergodiques de la Mécanique Classique," Gauthier-Villars, Paris (1967).

2. E. Fontich and C. Simó,The splitting of separatrices for analytic diffeomorphisms, *Ergodic Theory and Dynamical Systems* **10**, 295-318 (1990).

3. A. Giorgilli, A. Delshams, E. Fontich, L. Galgani and C. Simó, Effective stability for a Hamiltonian system near an elliptic equilibrium point, with an application to the restricted three body problem, *J. Differential Equations* **77**, 167-198 (1989).

4. G. Gómez, J. Llibre, R. Martínez and C. Simó. "Study on Orbits near the Triangular Libration Points in the Perturbed Restricted Three Body Problem," ESOC Contract 6139/84/D/JS(SC), Final Report (1987).

5. G. Gómez, A. Jorba, J. Masdemont and C. Simó. "Study of Poincaré Maps for Orbits near Lagrangian Points," ESOC Contract 9711/91/D/IM(SC), Final Report (1993).

6. J. Laskar, A numerical experiment on the chaotic behaviour of the solar system, *Nature* **338**, 237-238 (1989).

7. V.F. Lazutkin, I.G. Schachmanski and M.B. Tabanov, Splitting of separatrices for standard and semistandard mappings, *Physica* **D40**, 235-248 (1989).

8. N.N. Nekhoroshev, An exponential estimate of the time of stability of nearly-integrable Hamiltonian systems, *Russ. Math. Surveys* **32**, No 6, 1-65 (1977).

9. C. Simó, Averaging under fast quasiperiodic forcing, University of Barcelona Preprint (1993).

NON-LINEAR STABILITY ZONES AROUND THE TRIANGULAR LAGRANGIAN POINTS

V. V. Markellos, K. E. Papadakis and E. A. Perdios

Department of Engineering Science
Division of Applied Mathematics and Mechanics
University of Patras, Greece

ABSTRACT

Non-linear stability zones of the triangular Lagrangian points are determined numerically and the effect of the parameters of mass distribution and eccentricity of primaries are considered within the framework of the elliptic restricted three-body problem. It is found that both parameters have a strong effect reducing the stability zones to negligible size for some parameter values within the linear stability regions. The effect is identified to be due to the non-linearly unstable resonant cases and the associated curves in the μ-e parameter space. It is thus concluded that the classical μ-e linear stability diagram is only partially valid in terms of the more practical concept of non-linear stability.

NON-LINEAR STABILITY ZONES IN THE CONFIGURATION SPACE AND EXTENT-OF-STABILITY REGIONS IN THE PARAMETER SPACE

The stability of the triangular equilibrium points in the elliptic restricted three-body problems has been a popular subject in the literature due to the practical importance of the corresponding configurations (e.g. Trojan asteroids), the classical contribution being Danby (1964) (also Bennett, 1965), culminating in the μ-e linear stability diagram reproduced here in Figure 1.

More recently McKenzie and Szebehely (1981) considered the stability of the triangular equilibrium points (TEP) in more practical terms, namely in terms of the extent of the zone around the TEP for which stability of non-equilibrium orbital motion is valid in the sense of persisting librations, or non-escaping motions, over long times. The time intervals to be adopted in such computations were discussed by these authors and can be chosen on the basis of certain criteria as transpiring e.g. from the work of Erdi (1979) regarding the time of variation of long librations around the TEP.

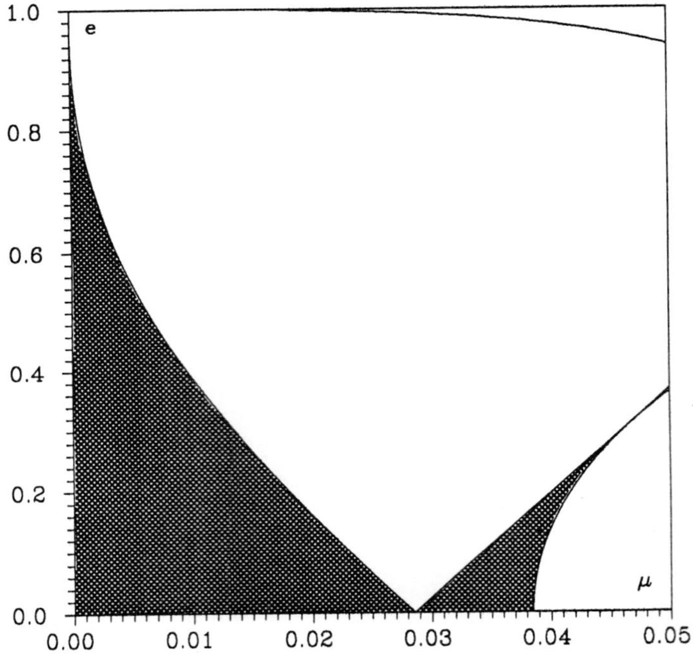

Fig. 1. The region of linear stability in the parameter space μ-e (Danby 1964, Bennett 1965).

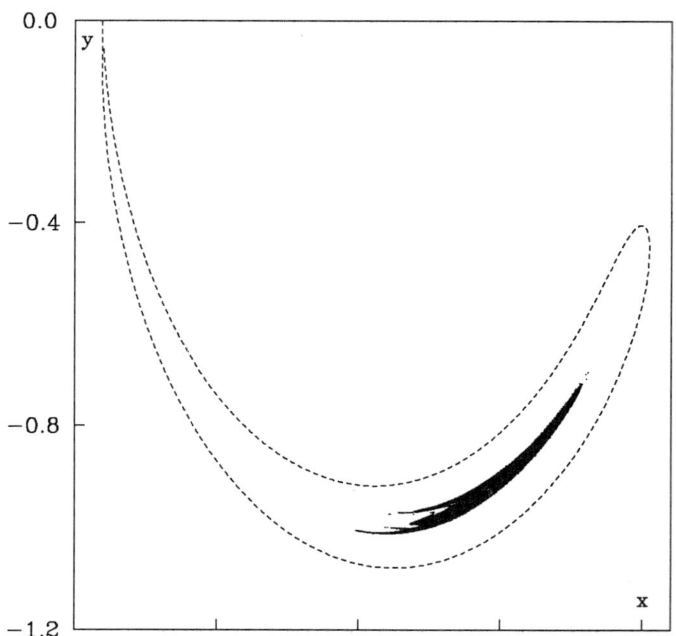

Fig. 2. The non-linear stability zone in the configuration space for the Earth-Moon case and zero eccentricity (McKenzie and Szebehely, 1981) and its maximum boundary (Whipple 1983).

In our calculations below we used mostly the final integration time,

$$t = 20\ T_{long},$$

where T_{long} is the period of the small long-periodic orbits around the equilibrium. However, the question of the length of time or the number of revolutions to be used in such studies (in order to decide after which point in time a change of orbital behaviour is not expected) is not crucial. A change of orbital behaviour, from librational to non-librational orbit, will usually occur early or not at all. The criterion adopted here for librational motion was that after t has elapsed the particle has not crossed the axis of the primaries (as in Dvorak and Lohinger, 1991).

McKenzie and Szebehely (1981) considered a particular value for the mass parameter (mass distribution among the primaries), namely the Earth-Moon case $\mu \cong 0.01213$, and a particular value of the eccentricity, namely the circular case $e = 0$, and found the stability zone in the configuration space around the TEP given in Figure 4 of their paper.

Whipple (1983) discussed the zero-velocity curve through the outer collinear Lagrangian point L_3 as a maximum boundary of the above stability zone in the configuration space. The above results of McKenzie and Szebehely (1981) and Whipple (1983) are summarised here in the diagram of Figure 2.

Györgyey (1985), extended the results of McKenzie and Szebehely (1981) by considering the effect of eccentricity on the size of the stability zone in the Earth-Moon case and found that the stability zone steadily decreases for increasing eccentricity of primaries up to the value $e = 0.15$.

Other values of the mass parameter or the eccentricity were not considered by Györgyey. However, Dvorak and Lohinger (1991) considered the effect of the mass parameter μ on the size of the stability zone in the configuration space and found a drastic reduction of the zone at the critical values of the μ corresponding to non-linear instability of the TEP in the circular case as established by Deprit and Deprit Bartholomé (1967): $\mu = 0.013516\ldots$, $\mu = 0.024293\ldots$.

In the present work we firstly considered the Earth-Moon case $\mu \cong 0.01213$ and extended Györgyey's results to include a fuller scan of eccentricity values extending up to the limiting value at which terminates the linear stability in the μ-e diagram of Figure 1 ($e \cong 0.33$). We found that the stability zone initially decreases as found by Györgyey (1985), practically disappearing at about $e = 0.2$. Surprisingly, however, the zone increases again in size for values of the eccentricity higher than $e = 0.2$ and up to about $e = 0.3$. Finally it reverts to decreasing again, for higher values of the eccentricity, until the critical value $e \cong 0.33$ has been reached, at the boundary of the instability region in the μ-e parameter space. These results are illustrated here in Figure 3 exhibiting clearly the variation of the size of the stability zone with increasing e, and also in Figure 4, frames a to h, showing the stability zones themselves as they evolve with varying e.

In order to resolve the problem more globally we considered here the effects of both the mass parameter μ and the eccentricity on the size of the non-linear stability zone around the TEP, simultaneously. This we did by performing extended calcula-

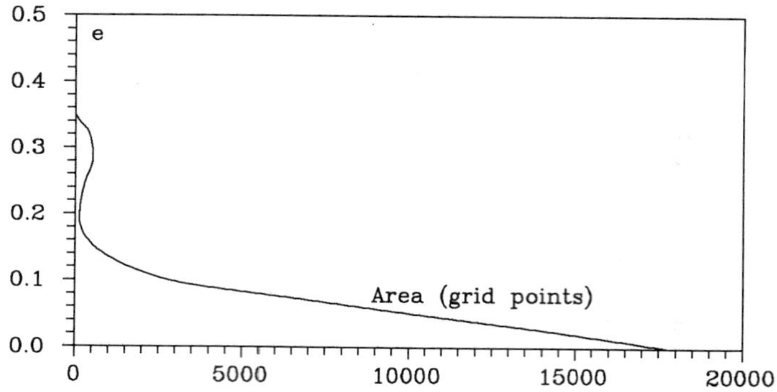

Fig. 3. The size (area measure) of the zone of non-linear stability in the configuration space as a function of the eccentricity, for the Earth-Moon mass distribution.

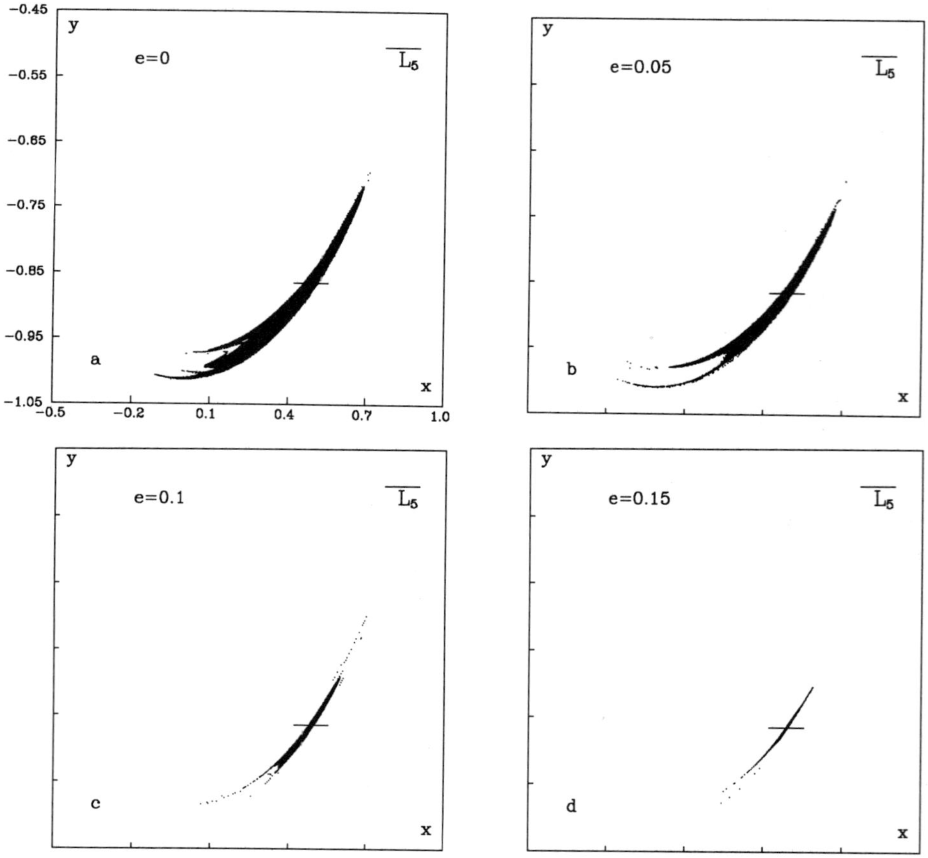

Fig. 4. Frames a to d. The zone of non-linear stability in the configuration space as a function of the eccentricity, for the Earth-Moon mass distribution.

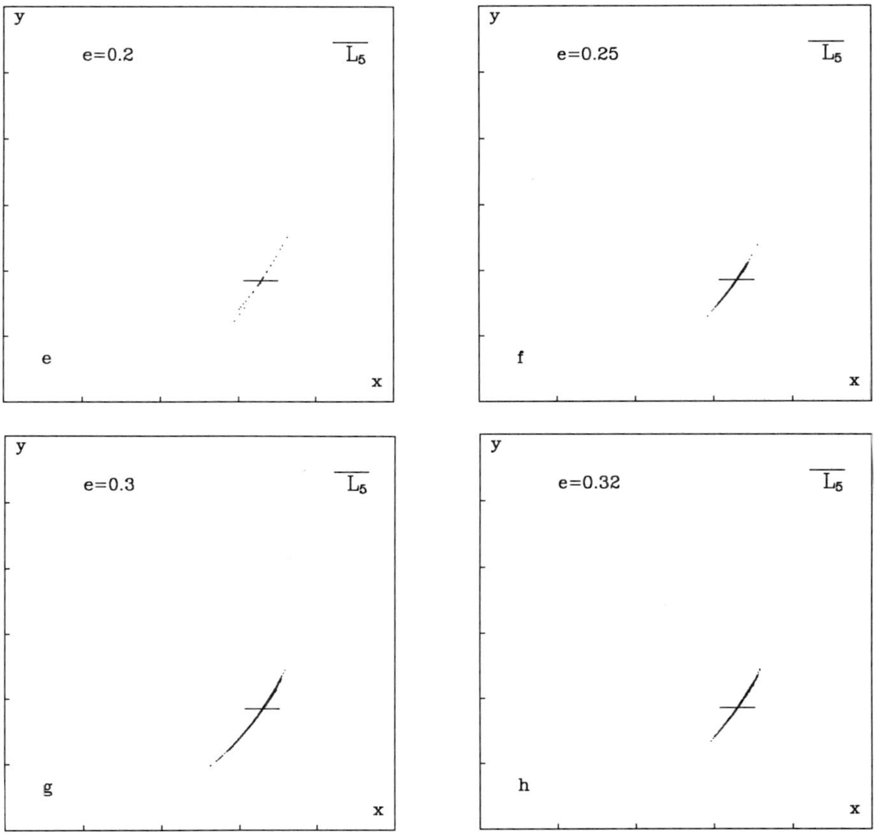

Fig. 4. Frames e to h. The zone of non-linear stability in the configuration space as a function of the eccentricity, for the Earth-Moon mass distribution.

tions of the size and shape of the stability regions for many hundreds of combinations of μ and e values, covering densely the linear stability region of the μ-e parameter diagram of Figure 1. We introduced a measure of the extent of non-linear stability in terms of the size or area of the stability zone in the configuration space and represented this measure in the parameter space μ-e by points of varying darkness, the darker points representing larger stability zones for the corresponding pair of μ-e values (in a more or less arbitrary scale).

The result is shown in Figure 5-a where the effect of both parameters μ and e of the problem, in reducing the area of the non-linear stability zone around the equilibrium points, is clear. Further, the effect is identified to be due to the non-linearly unstable resonant cases and the associated curves in the μ-e parameter space which we calculated numerically and present here in Figure 5-b superimposed on the results of Figure 5-a. These curves represent an extension into the elliptic case of the Deprit and Deprit-Bartholomé (1967) results and have also been presented in a different context in Figure 7 of Marchal (1991).

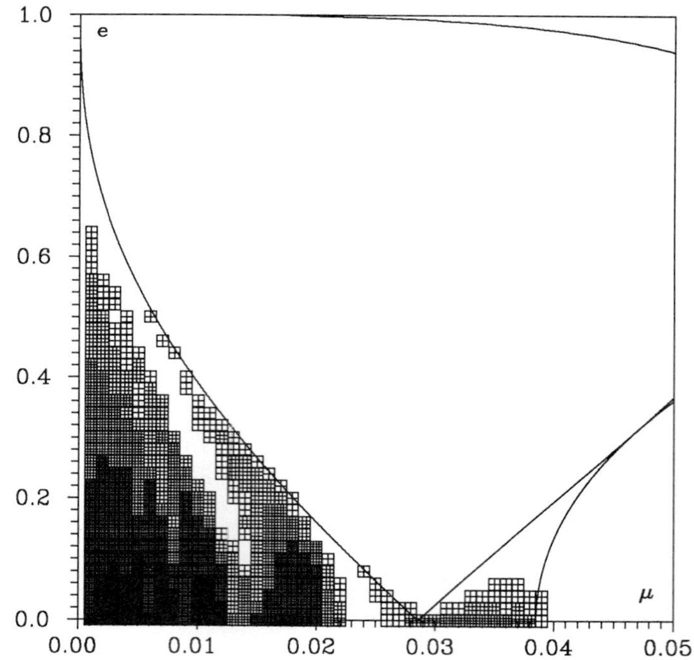

Fig. 5-a. The extent of non-linear stability zones as represented in the parameter space μ-e.

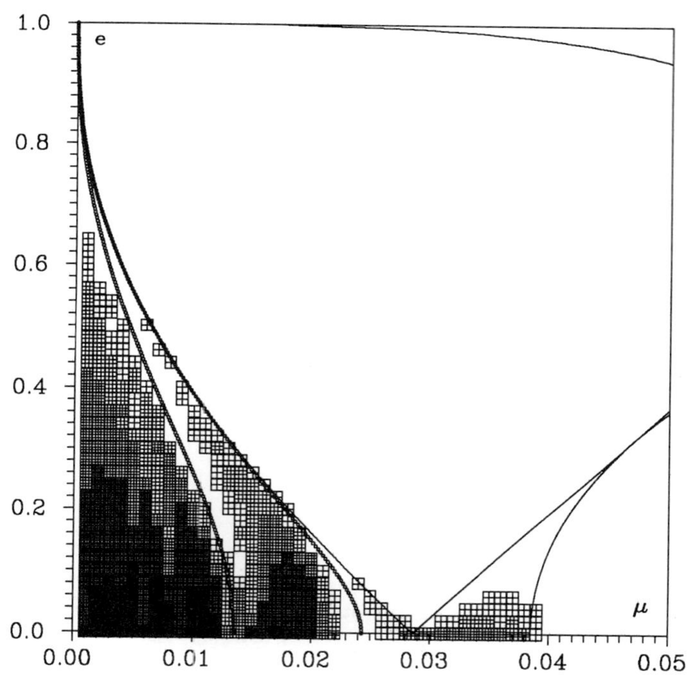

Fig. 5-b. The results of Figure 5 with the associated curves of non-linearly unstable resonant cases superimposed.

As a conclusion, it follows from the present results that the classical μ-e linear stability diagram (Figure 1) is to be considered with caution for practical applications and that with respect to the more practical concept of non-linear stability and considering the extent of the non-linear stability zones the present diagram of Figure 5-a provides a more realistic representation of the stability properties of the triangular equilibrium points and the associated librations around them.

REFERENCES

Bennett, A.: 1965, *Icarus* **4**, 177.

Danby, J.M.A.: 1964, *Astron. J.* **69**, 165.

Deprit A. and Deprit-Bartholomé A. (1967), *Astron. J.* **72**, 173.

Dvorak, R. and Lohinger, E.: 1991, in A. E. Roy (ed.), *Predictability, Stability and Chaos in N-Body Dynamical Systems*, Plenum Press, New York, 439.

Erdi, B.: 1979, *Celes. Mech.* **20**, 59.

Györgyey, J.: 1985, *Celes. Mech.* **36**, 281.

Marchal C.: 1991, in Roy A.E. (ed.), *Predictability, Stability and Chaos in N-Body Dynamical Systems*, Plenum Press, New York, 73.

McKenzie, R. and Szebehely, V.: 1981, *Celes. Mech.* **23**, 223.

Whipple, A.: 1983, *Celes. Mech.* **30**, 385.

AN INVESTIGATION OF ORBITS AROUND THE TRIANGULAR LAGRANGIAN POINTS OF SATURN

Colette M. de la Barre and William Kaula

University of California
Department of Earth and Space Sciences
Los Angeles, CA 90024

ABSTRACT

We numerically integrated orbits around the L4 and L5 Lagrangian points of Jupiter (Trojans), and Saturn ('Bruins'): 40 Trojans and 350 Bruins, all of inclination less than 12°. Four Bruins remained stable until the integration was stopped at 412 Myrs. Properties of these stable orbits were: (1) proper eccentricities less than 0.028; (2) librations in true longitude about L4 and L5 of more than 80°; and (3) librations of the longitude of perihelion (ϖ) relative to Saturn's perihelion, such that they were never close when the forced eccentricity was near 0.08. Orbits with librations in true longitude less than 80° were unstable, the time to instability being correlated with libration angle. No unstable region was found around Jupiter's Lagrangian points, and stable Trojans may have longitudes of perihelion that either circulate or librate with respect to jupiter's.

INTRODUCTION

We numerically integrated the orbits of more than 350 low inclination Bruins, and more than 40 low inclination Trojans on a SPARC-2 workstation searching for stable orbits. We define a stable orbit as one that remains a tadpole or horseshoe orbit in a coordinate frame rotating with the primary planet for the duration of the numerical integration. We used a Cowell-Stormer, 12th-order, numerical integrator, and a model where the terrestrial planets are thrown into the Sun, and the four major planets are numerically integrated. Orbital parameters were chosen at random within ranges based on the observed Trojan population.

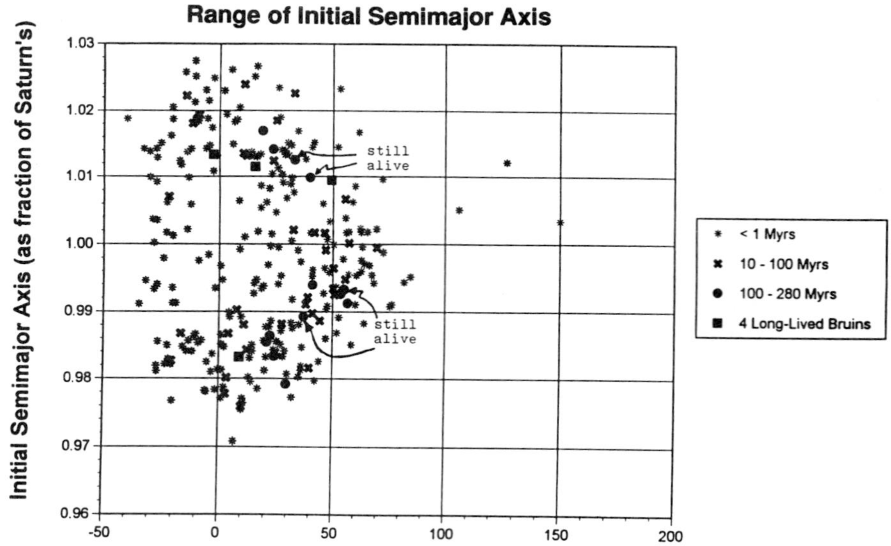

Figure 1a. Bruin longevity

BRUINS

We discovered four Bruins that are stable for more than 412 Myrs, and a region about Saturn's Lagrangian points where no stable orbits exist; some orbits having longevity less than one Myrs. A plot of initial semimajor axis versus initial angle from the Lagrangian point (Fig. 1a) shows that if a Bruin orbit is to be stable, it must not reside too close to Saturn's Lagrangian points. M. Holman and J. Wisdom found similar results for numerical integrations lasting 20 Myrs [1]. Figure 1a indicates that although it is necessary for an orbit to reside outside the unstable region for stability, it is not a sufficient condition. A similar plot for Trojans (Fig. 1b) does not show this unstable region.

A plot of maximum libration angle versus time for the four long-lived Bruins shows they generally librate $80° - 90°$, although one librated as much as $111°$, about their Lagrangian points (libration angle is measured from head to tail of the tadpole orbit) keeping them well away from the unstable region (Fig. 2).

Figure 3 shows eccentricity versus time for two Bruins with the same initial orbital elements, except their true longitude. The Bruin initially placed at L4 goes unstable within 650 centuries, whereas the Bruin initially 23° from L4 is stable.

The proper eccentricity and proper longitude of perihelion were calculated for the stable Bruin orbits by Fourier analysing their discrete time series in $h = e \times sin(\varpi)$ and $k = e \times cos(\varpi)$ over time spans of one million years each.The contributions to the time series from the perihelia of Jupiter (g5), Saturn (g6), and Uranus (g7) were removed, along with the largest nonlinear forcing term (affected by the Great Inequality) between Jupiter and Saturn ($2 \times g6 - g5$.) The residuals were Fourier analysed to determine the Bruin's proper longitude of perihelion rate (B1), and the amplitude or proper eccentricity. The contributions to the time series from other nonlinear forcing terms did not significantly alter these proper elements.

Figure 4 is a spectral analysis for long-lived Bruin #14 compared to that of Saturn,

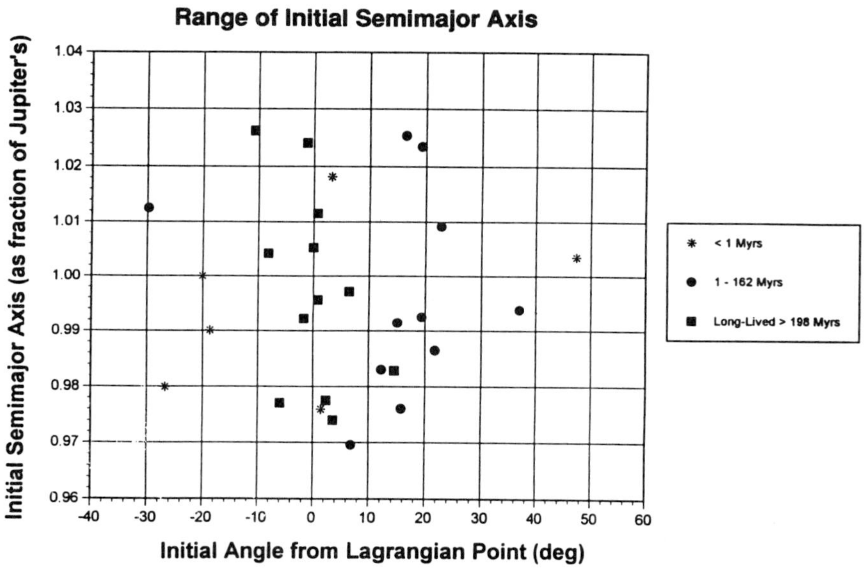

Figure 1b. Trojan longevity

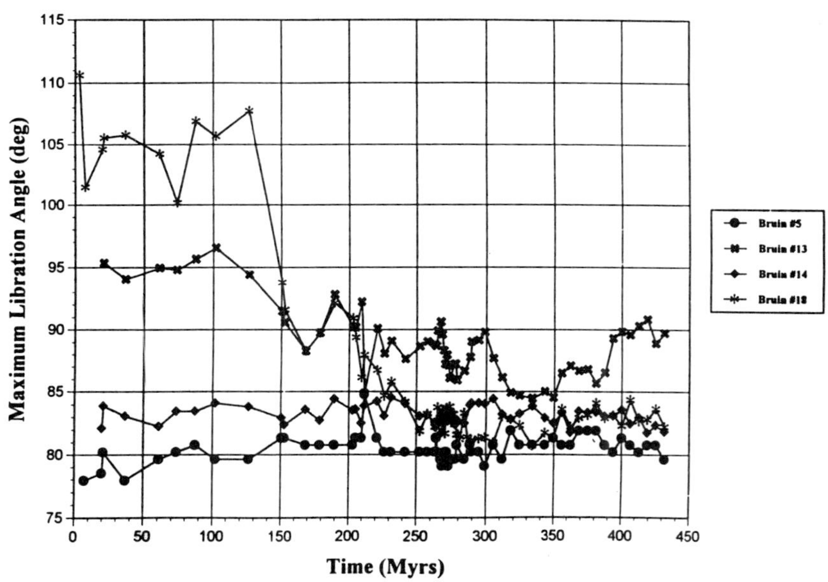

Figure 2. Maximum libration angle for four long-lived Bruins.

and shows clearly the Bruin's free or proper motion. It also shows that the Bruin's perihelion precesses at a rate of g6, with oscillations about this rate due to B1. Figures 5 and 6 show little variation in the period of proper longitude of perihelion and proper eccentricity for the four long-lived Bruins for the duration of the numerical integration.The proper eccentricities of the Bruins (less than 0.028) are an order of magnitude less than the largest proper eccentricity of the observed Trojans determined by A. Milani [2], which keeps the Bruins from coming too close to Jupiter or Saturn.

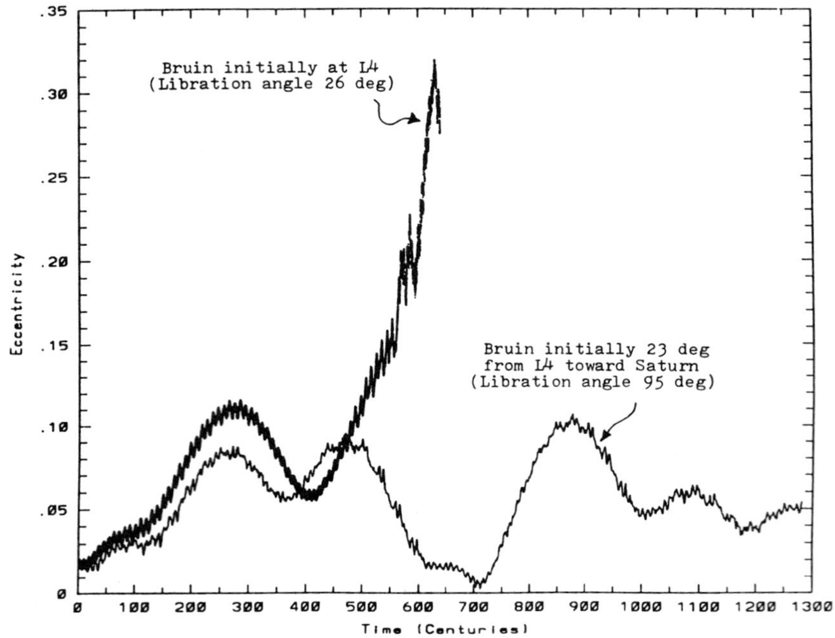

Figure 3. Stable and unstable Bruins.

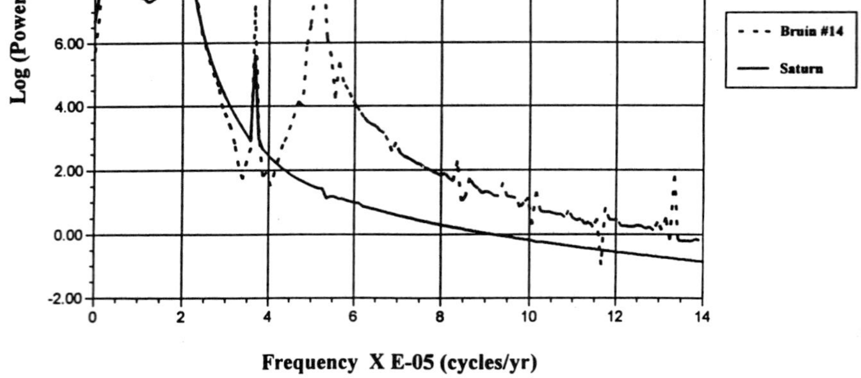

Figure 4. Spectrum of variations in h and k for Bruin #14 and Saturn

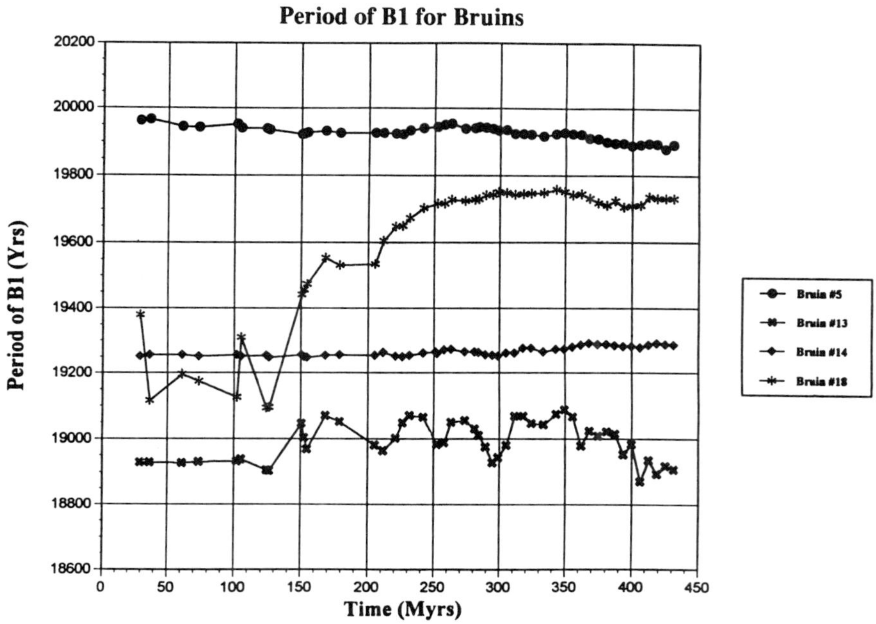

Figure 5. Period of proper longitude of perihelion for four long-lived Bruins.

The four long-lived Bruins have longitudes of perihelion that librate with respect to Saturn's longitude of perihelion such that the perihelia are never close when the Bruin's eccentricity is at its maximum, and Saturn's maximum eccentricity ($eS = 0.085$) doesn't occur with the Bruin's maximum eccentricity ($eB = 0.1$.) Figure 7 shows the longitude of perihelion of Bruin #14 (L4 Bruin) librating about a centre 45 degrees ahead of Saturn's longitude of perihelion. These constraints on the Bruin's perihelion and eccentricity reduce the perturbations by Saturn, which helps to keep the Bruin's eccentricity low.

TROJANS

We found 12 Trojan orbits that were stable for more than 198 Myrs, and no unstable region near Jupiter's Lagrangian points. A spectral analysis was made of the discrete time series in parameters h and k for Jupiter and two long-lived Trojans. Figure 8 shows Trojan #11 whose perihelion precesses at a rate of g5, with smaller contributions due to its proper motion (T1.) Therefore, the longitude of perihelion of Trojan #11 (L5 Trojan) librates relative to Jupiter's longitude of perihelion, with a libration centre about 45 degrees behind Jupiter's perihelion (Fig.9.) This is similar to what we observed for the four long-lived Bruins, however, it differs from that for long-lived Trojan #10. Figure 10 shows the perihelion of Trojan #10 precessing at its proper rate, with smaller contributions by g5. Since this Trojan's longitude of perihelion circulates relative to Jupiter's longitude of perihelion, its maximum eccentricity ($eT = 0.16$) is much larger than that of Trojan #11 ($eT = 0.075$), its maximum eccentricity can occur when its perihelion and Jupiter's are close, or during Jupiter's maximum eccentricity ($eJ = 0.06$). Since Saturn's perturbation of a Trojan is not as large as Jupiter's perturbation of a Bruin, even a Trojan with maximum eccentricity as large as 0.16 may be stable. Bruins with circulating longitudes of perihelion relative to Saturn's, resulting in large eccentricity are short-lived. B. Erdi (among others) found asteroids in the ob-

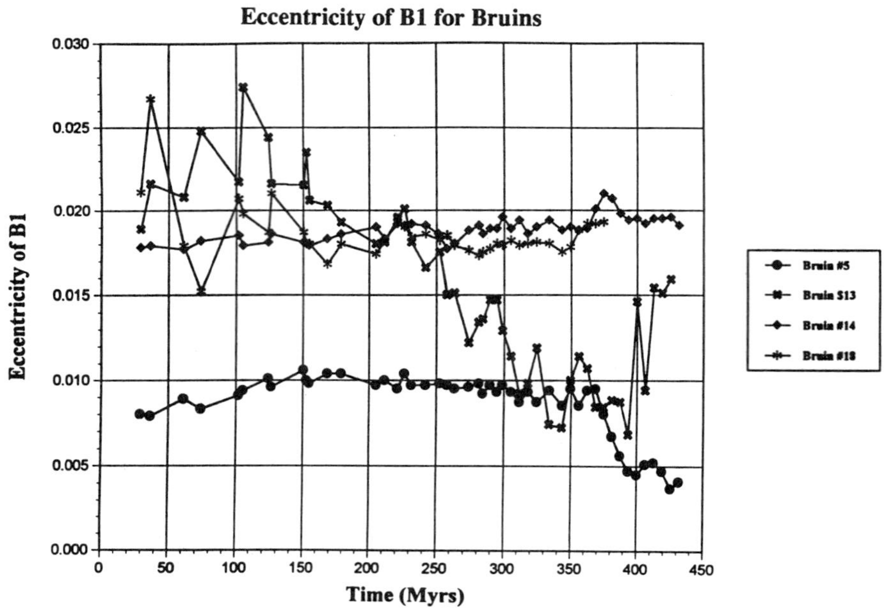

Figure 6. Proper eccentricity for four long-lived Bruins

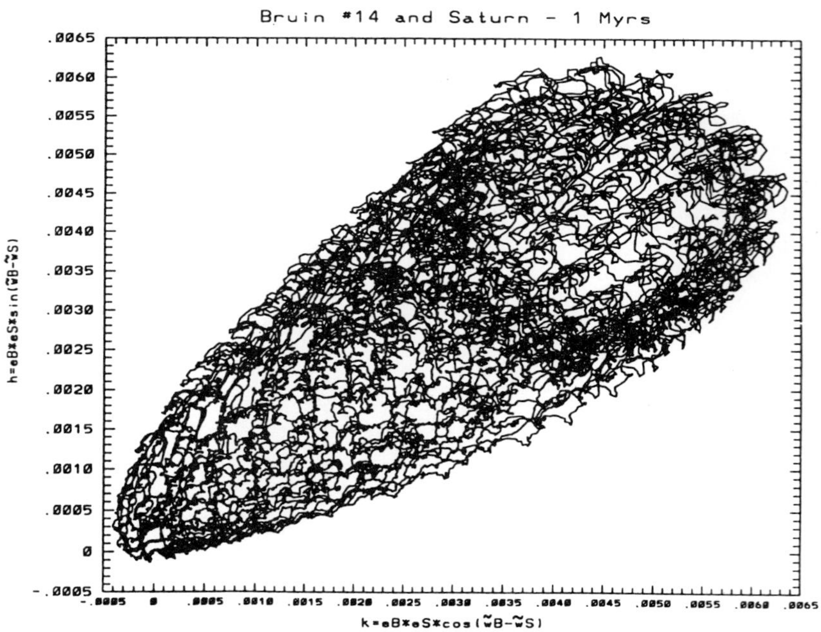

Figure 7. Expanded h and k for Bruin #14

served Trojan population with circulating longitudes of perihelion relative to Jupiter's longitude of perihelion, as well as librating longitudes of perihelion [3].

CONCLUSIONS

Bruins with small proper eccentricity and librating longitude of perihelion relative to Saturn's, such that the perihelia are never close when the Bruin's eccentricity is large

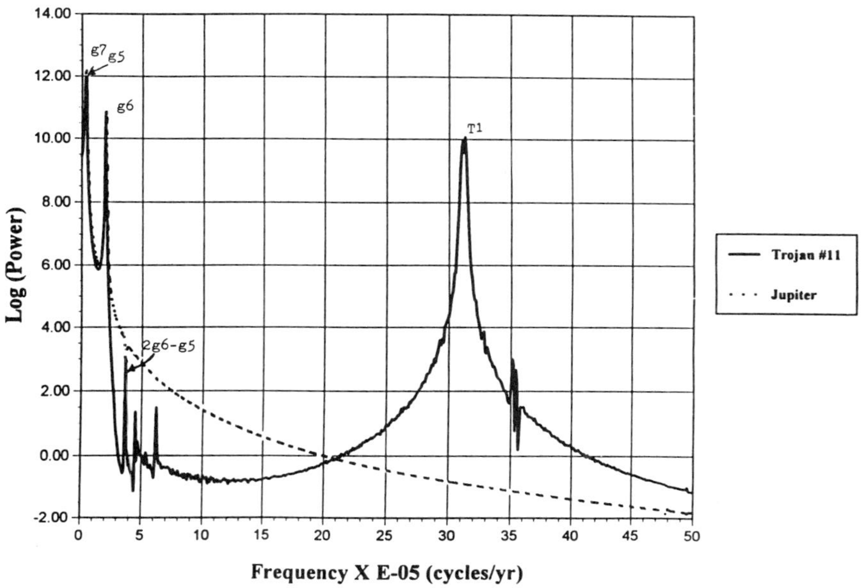

Figure 8. Spectrum of variations in h and k for Trojan #11 and Jupiter.

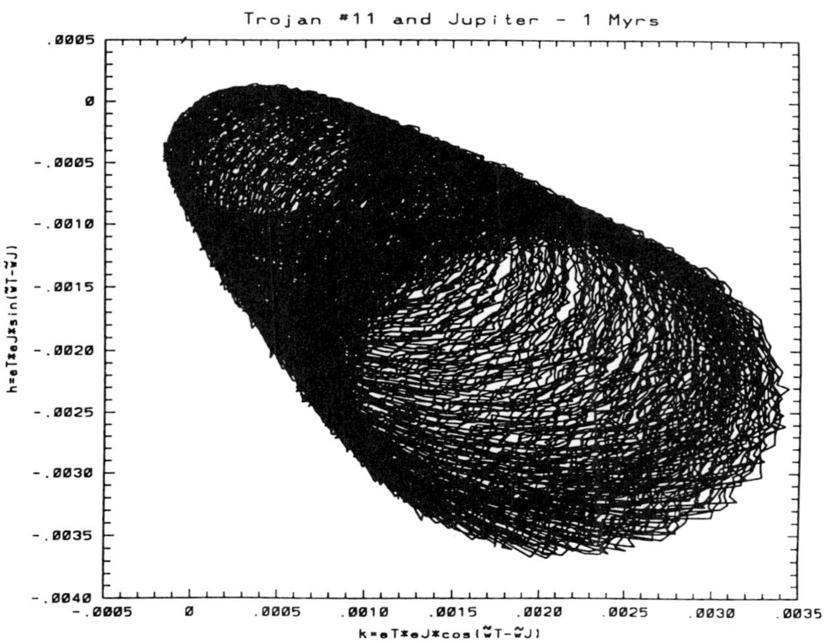

Figure 9. Expanded h and k for Trojan #11.

are candidates for long-term stability. These constraints need not be imposed on Trojan asteroids, since Saturn's threat to Trojan stability is not as great as Jupiter's threat to Bruin stability.

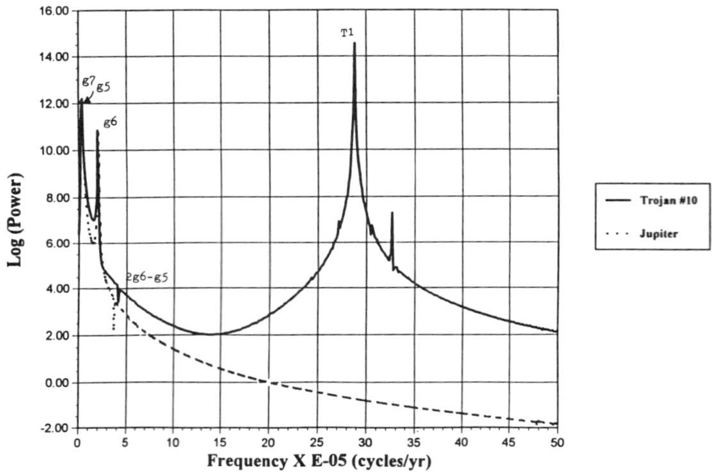

Figure 10. Spectrum of variations in h and k for Trojan #10 and Jupiter.

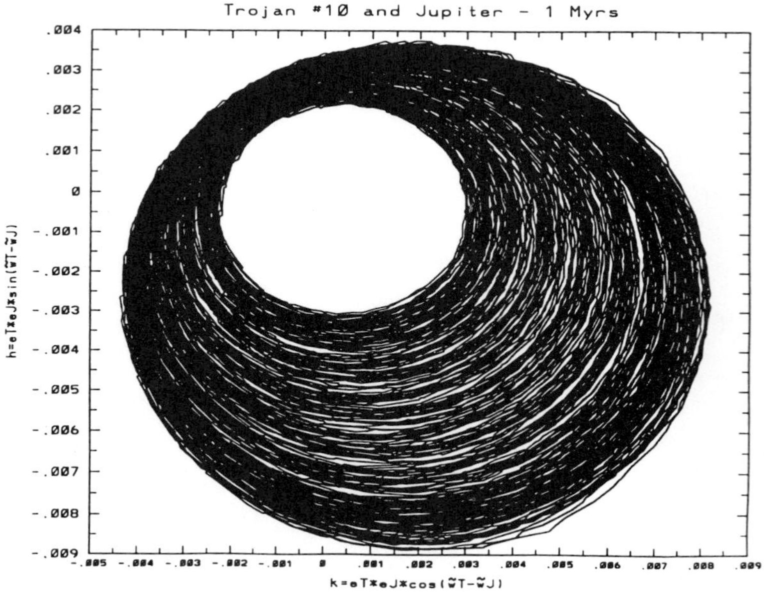

Figure 11. Expanded h and k for Trojan #10.

REFERENCES

1. M. Holman and J. Wisdom, "Dynamical Stability in the Outer Solar System and the Delivery of Short Period Comets", The Astronomical Journal, Vol. 105, No. 5, May 1993, pp. 1987-1999.

2. A. Milani. "The Trojan Asteroid Belt: Proper Elements, Stability, Chaos and Families", Celestial Mechanics, 20 June 1992, in press.

3. B. Erdi. "Orbital Evolution of Trojan Asteroids", Dynamical Trapping and Evolution in the Solar System. V.V. Markellos and Y. Kozai (eds.), D. Reidel Publishing Company, 1983, pp. 165-176.

COLLISION ORBITS IN THE ISOSCELES RECTILINEAR RESTRICTED PROBLEM

J. Martínez Alfaro[1] and Rosa B. Orellana[2]

[1]Departament de Matemàtica Aplicada i Astronomia.
Facultat de Matemàtiques. Universitat de València. Spain
[2]CONICET.Observatorio Astronómico de La Plata. Argentina

INTRODUCTION

In the study of the Collinear Three-Body Problem, McGehee (1974) introduced a new set of coordinates which had the effect of blowing up the triple collision singularity. Subsequently, his method has been used to analyze some other collision or singularities. Recently, Wang (1986) introduced another transformation which differs from the McGehee´s coordinates in the fact that the blowing-up factor is now the potential function, U, instead of the moment of inertia, I. Meyer and Wang (1993) have applied this method to the Restricted Isosceles Three-body Problem with positive energy and Cors and Llibre (1994) to the hyperbolic restricted three-body problem. In this paper we study the singularities of this problem (IRRP), due to collisions, in the case of negative energy by means of a modified version of McGehee's coordinates. Moreover, we remove the singularities due to a velocity not bounded.

The IRRP can be considered as the Sitnikov's Problem with eccentricity one, and has received different denominations according to the authors, (Broucke, 1971, 1979; Meyer and Wang, 1993; Puel, 1979; Waldvogel, 1973). It can be also, seen as the limit of the Isosceles Three-Body Problem when the unequal mass tends to zero. Then, the IRRP can be considered as a bridge through the Sitnikov's Problem and the Isosceles Three-Body Problem.

In this restricted problem, the primaries are two equal masses (M) with symmetric initial conditions and with negative energy; they move in an oscillatory fashion on a straight line under the action of their Newtonian attraction and have thus, consecutive collisions. Let r be the distance of each primary to the center of mass; we normalize the units so that the maximum value or r will be 1, and the period 2π. We want to describe the motion of a third massless particle equidistant of both primaries and in the same plane (Fig. 1).

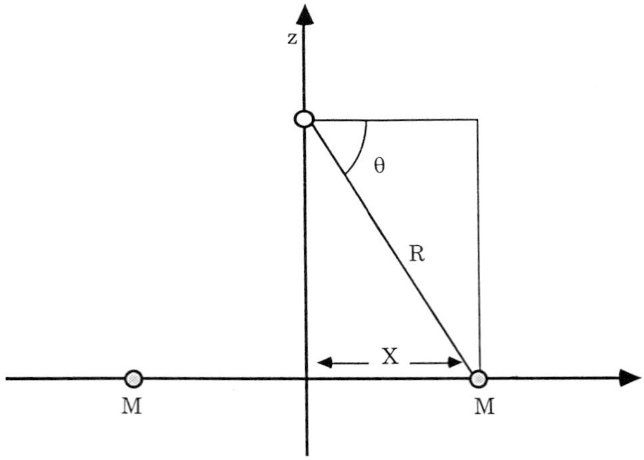

Figure 1. The IRRP

Let z and v denote its coordinate and velocity. Then, the differential equation is (Moser, 1973):

$$\frac{dz}{dt} = v \quad ,$$

$$\frac{dv}{dt} = - \frac{z}{(z^2+r(t)^2)^{3/2}} \quad . \tag{1}$$

In order to have r in a closed form we introduce the eccentric anomaly E as a new temporal variable as in Waldvogel (1973). According to the Kepler's equation:

$$t = E - \sin E \quad ,$$

$$\frac{dt}{dE} = 1 - \cos E = 2r \quad . \tag{2}$$

The system becomes:

$$\frac{dz}{dE} = 2\, r(E)\, v \ ,$$

$$\frac{dv}{dE} = - \frac{2\, r(E)\, z}{(z^2+r(E)^2)^{\frac{3}{2}}} \quad . \tag{3}$$

We consider E modulus 2π, as an angular variable, then the extended phase space is diffeomorphic to a solid torus with E as the longitudinal coordinate, and in each section E = constant, z and v define a plane or equivalently a two-dimensional disc. The system (3) is singular since z = 0, E = 0 and corresponds to a collision of the three masses.

THE TRIPLE COLLISION MANIFOLD

To blow-up the triple collision, we consider the polar coordinates, R, θ:

$$z = R\sin\theta,$$

$$X = R\cos\theta, \qquad (4)$$

where X is the abscissa of the right primary (Fig. 1).

The equations, with R, θ and v are:

$$\frac{dR}{dE} = 2Rvs|c| - c\sqrt{R|c|(1-R|c|)},$$

$$\frac{d\theta}{dE} = 2vc|c| + s\sqrt{\frac{|c|(1-R|c|)}{R}}, \qquad (5)$$

$$\frac{dv}{dE} = -\frac{2s|c|}{R},$$

where $s = \sin\theta$, $c = \cos\theta$.

Then, we introduce a new scaled velocity and time:

$$\omega = v\sqrt{R}, \qquad \frac{dE}{dL} = \sqrt{R/|c|}. \qquad (6)$$

The system goes over to:

$$\frac{dR}{dL} = R\left[2\omega s|c|^{1/2} - c\sqrt{1-R|c|}\right],$$

$$\frac{d\theta}{dL} = 2\omega c|c|^{1/2} + s\sqrt{1-R|c|}, \qquad (7)$$

$$\frac{d\omega}{dL} = s|c|^{1/2}(\omega^2-2) - \frac{\omega c}{2}\sqrt{1-R|c|}.$$

Then, R = 0 becomes an invariant manifold, a cylinder. The triple collision manifold and the system are now:

$$\frac{d\theta}{dL} = 2\omega c|c|^{1/2} + s,$$

$$\frac{d\omega}{dL} = s|c|^{1/2}(\omega^2-2) - \frac{\omega c}{2}. \qquad (8)$$

The equilibrium points and their associated eigenvalues are :

Notation	Coordinates	Eigenvalues
E^+	$\theta = \pi \quad , \omega = 0$	$\frac{1}{4}(-1 \pm i\sqrt{55})$
E^-	$\theta = 0 \quad , \omega = 0$	$\frac{1}{4}(1 \pm i\sqrt{55})$
L_4^+	$\theta = \frac{2\pi}{3}, \omega = \sqrt{\frac{3}{2}}$	$\frac{1}{2}(-1 \pm \sqrt{19})$
L_4^-	$\theta = \frac{\pi}{3}, \omega = -\sqrt{\frac{3}{2}}$	$\frac{1}{2}(1 \pm \sqrt{19})$
L_5^+	$\theta = \frac{4\pi}{3}, \omega = -\sqrt{\frac{3}{2}}$	$\frac{1}{2}(-1 \pm \sqrt{19})$
L_5^-	$\theta = -\frac{\pi}{3}, \omega = \sqrt{\frac{3}{2}}$	$\frac{1}{2}(1 \pm \sqrt{19})$.

To understand the flow on the collision manifold it is necessary to know the connection between the invariant manifolds of the equilibrium points. The IRRP is the limit of the isosceles problem when the unequal mass, m, tends to zero. Therefore, these connections can be obtained directly from Simó (1981). In particular, by the symmetry of the problem, if there exists a connection between the Lagrangian points, it must pass through the point $\theta = \pi/2$, $\omega = 0$. The minimum of the distance between this invariant connection and the point is larger than a positive constant as m tends to 0. We sketch them in Figure 2.

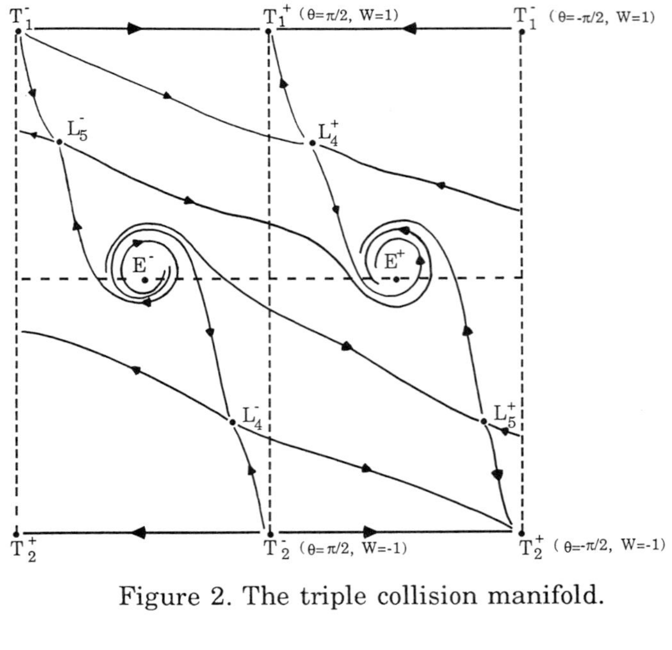

Figure 2. The triple collision manifold.

To study the flow as $|\omega|$ tends to ∞, we introduce a new scaled velocity and time:

$$W = \frac{\omega}{1 + |\omega|} \quad , \frac{dL}{d\tau} = 1 - |W| . \tag{9}$$

We get the system:

$$\frac{d\theta}{d\tau} = 2Wc|c|^{1/2} + s(1 - |W|) ,$$

$$\frac{dW}{d\tau} = (|W| - 1)[s|c|^{1/2}(W^2 - 4|W| + 2) + \frac{W(1 - |W|)c}{2}] . \tag{10}$$

Now, we have four new equilibrium points (Fig. 2):

Notation	Coordinates	Eigenvalues
T_1^+	$\theta = \frac{\pi}{2}, W = 1$	0
T_1^-	$\theta = -\frac{\pi}{2}, W = 1$	0
T_2^+	$\theta = -\frac{\pi}{2}, W = -1$	0
T_2^-	$\theta = \frac{\pi}{2}, W = -1$	0

Property 1.- *T_i^+ are attractors and T_i^- repellors.*
Proof: Consider the point T_1^+ (Fig. 3).
W = 1 is an invariant manifold, the differential equation on it is:

$$\frac{d\theta}{d\tau} = 2c|c|^{1/2} . \tag{11}$$

The flow consists of the two equilibrium points T_1^+ and T_1^- and two trajectories γ_1 and γ_2 attracted by T_1^+.
Now, we consider the variation of W near T_1^+.
If $c \neq 0$, the second equation of (10) can be written:

$$\frac{dW}{d\tau} = (|W| - 1)|c|^{1/2} [s(W^2 - 4|W| + 2) + \frac{W(1 - |W|)}{2} \frac{c}{|c|} |c|^{1/2}].$$

The first factor of the derivative is negative. The second one is also negative because the first term is $(-1 + \varepsilon)$, and the second term is as small as we want. Consequently, the derivative of W is positive. W it is an increasing function.

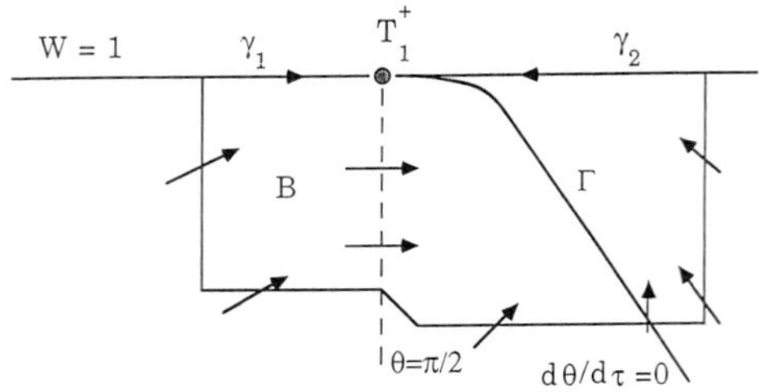

Figure 3. The flow near T_1^+.

Now, let us consider a box, B, near the point T_1^+, like the box in Figure 3.

The curve Γ is defined by $d\theta / d\tau = 0$. Therefore θ increases at the left of Γ and decreases at the right.

A trajectory that arrives to the border B comes into B. Moreover, orbits on the left side of the line $\theta = \pi/2$, crosses transversally this line, since:

$$\frac{d\theta}{d\tau}(W, \pi/2) = 1 - |W| > 0,$$

$$\frac{dW}{d\tau}(W, \pi/2) = 0. \tag{12}$$

Once on the right side of $\theta = \pi/2$, W is an increasing function, and therefore all orbit goes to the point T_1^+.

The prove for the other equilibrium points it is similar.

Finally, next properties explain the behavior of the orbits near triple collision.

Property 2.- *The collision orbits of the IRRP consist of two 2-dimensional manifolds associated to each of the points,* L_4^-, L_5^-, *and the orbit defined by* $\theta = 0$, $W = 0$. *The ejection orbits of the IRRP consist of two 2-dimensional manifolds associated to each of the points,* L_4^+, L_5^+, *and the orbit defined by* $\theta = \pi$, $W = 0$.

Proof: It is immediate having into account the eigenvalues of the equilibrium points outside the t.c.m. From system (7) we have:

Point	Eigenvalue	Point	Eigenvalue
E^+	1	L_4^-	-2
E^-	-1	L_5^+	2
L_4^+	2	L_5^-	-2.

E^+ is connected with E^- by means of their common invariant manifold, $W^u(E^+) = W^s(E^-)$. In particular, if $\theta = 0$, $W = 0$, the system (7) goes over to:

$$\frac{dR}{dL} = - R\sqrt{1 - R}. \qquad (13)$$

Therefore, R decreases and we have a collision orbit. Similarly, $\theta = \pi$, $W = 0$, is an ejection orbit.

Property 3.- *All orbit that arrives close enough to the triple collision manifold, it ends in a collision or it escapes near the equilibrium points T_1^+ and T_2^+.*

Proof: The only critical points of the system are the equilibrium points. We introduce a partial order into them saying that $p_1 < p_2$ if the unstable manifold of p_1 on the t.c.m. intersect the stable manifold of p_2 .Then, the extremum points in this order are.

initials: T_1^-, E^-, T_2^-. finals: T_1^+, E^+, T_2^+.

All orbit close enough to the t.c.m. falls on the stable manifolds of the equilibrium points and ends in a collision or follows closely the trajectories on the t.c.m. The order of critical points implies that it must leave the t.c.m. trough T_i^+ or E^+. Moreover, E^+ is connected with E^- by means of their common invariant manifold, $W^u(E^+) = W^s(E^-)$; then the point comes back to t.c.m.. The t.c.m. and $W^u(E^+)$ form a new invariant manifold.

The points of escape from this manifold are T_i^+ $i = 1, 2$; the property is immediate from this fact.

REFERENCES

1. R. Broucke, Periodic orbits in the rectilinear restricted three-body problem. *Journal de Mécanique*, 10, nº 3, 449:465 (1971).
2. R. Broucke, On the isosceles triangle configuration in the planar general three body problem. *Astronomy and Astrophysics*, 73, 303:313 (1979).
3. J. Cors, J. Llibre, The global flow of the hyperbolic restricted three-body problem . See this volume (1994).
4. R. McGehee, Triple collision in the collinear three-body problem. *Inventions Mathematiques*, 27, 191:227 (1974).
5. K. Meyer, Q. Wang, Global phase structure of the restricted isosceles three-body problem with positive energy. *Transaction of the A.M.S.* 339, 311:336 (1993).
6. J. K. Moser. "Stable and Random Motions in Dynamical Systems". Princeton Univ. Press. (1973).
7. F. Puel, Numerical study of the rectilinear isosceles restricted problem. *Celestial Mech.* 20, 105:123 (1979).
8. C. Simó, Analysis of triple collision in the isosceles problem, in "Classical Mechanics and Dynamical Systems", Marcel Dekker, 203:224 (1981).
9. J. Waldvogel,The rectilinear restricted problem of three bodies. *Celestial Mechanics*, 8, 189:198 (1973).
10. Q. D. Wang, Qualitative study of n-body problem: Unitized momentum transformation and its application, "Space Dynamics and Celestial Mechanics". Reidel. 61:69 (1986).

ASYMPTOTIC ORBITS AS TERMINATIONS OF FAMILIES OF PERIODIC ORBITS IN THE COPENHAGEN PROBLEM WITH AND WITHOUT RADIATION PRESSURE

V. V. Markellos, E. A. Perdios and K. E. Papadakis

Department of Engineering Science
Division of Applied Mathematics and Mechanics
University of Patras, Greece

ABSTRACT

Heteroclinic asymptotic orbits of the Copenhagen problem are important as they can be combined in pairs to form infinite period terminations of families of periodic orbits. These families and their terminations are considered here as limiting cases of the photogravitational restricted problem with equal masses and radiation factors of primaries. The terminating asymptotic orbits, and the corresponding families, are first computed as homoclinic spirals at the inner collinear equilibrium point in the case of strong radiation and then continued numerically to reach the gravitational Copenhagen problem in which these terminating orbits have evolved to heteroclinic orbits at the triangular points.

INTRODUCTION

In the photogravitational circular restricted three-body problem (for background information see Simmons et al. 1985), and for certain values of the radiation factors, the triangular equilibrium points approach the inner collinear point L_1 and finally disappear by coalescing at L_1. In order to study the heteroclinic asymptotic orbits of the gravitational Copenhagen problem as elements of terminations of families of periodic orbits, according to the results of Strömgren (see Szebehely 1967, and for more recent work: Gomez et al. 1988 and Markellos 1991), we consider here the photogravitational restricted problem with equal masses ($\mu = 1/2$) and equal radiation factors ($q_1 = q_2 = q$) of primaries. In this, generalised, Copenhagen problem the triangular equilibrium points are at:

$$x_{L_{4,5}} = \frac{1}{2}, \qquad y_{L_{4,5}} = \pm\sqrt{q^{2/3} - \frac{1}{4}}, \tag{1}$$

and it is seen that in this case the triangular points disappear for $q = 1/8$. The stability properties of the triangular points $L_{4,5}$ are ultimately transferred to the

From Newton to Chaos, Edited by A.E. Roy
and B.A. Steves, Plenum Press, New York, 1995

inner collinear equilibrium point L_1 which is linearly stable for $1/9<q<1/8$ and unstable for $q<1/9$. In fact the points $L_{4,5}$ for $\mu=1/2$, become linearly stable for q less than the critical value:

$$q_{crit.}=\left[3(3-2\sqrt{2})/2\right]^{3/2}\cong 0.130560, \tag{2}$$

following from Equation (10) of Markellos et al. (1993a). For more information on the equilibrium points of the photogravitational restricted problem and their stability properties see Markellos et al. 1993a, 1993b. Note, in particular, that the instability of the collinear point L_1 for $q<1/9$ is of the same type as the instability of the triangular point L_4 in the classical Copenhagen case (q=1), i.e. the type allowing asymptotic spiralling solutions of the linearised equations of motion around the equilibrium.

The important difference, however, is that in the case of the point L_1 due to the symmetry properties of the problem with respect to the axis of the primaries, and the fact that L_1 is collinear, the outgoing spirals are mirror images of the incoming spirals at the same point L_1 (homoclinic orbits). This is in contrast to the Strömgren outgoing spirals at L_4 which are symmetric to the incoming spirals not at L_4 itself but at L_5 (heteroclinic orbits). The difference is important in that having found the outgoing spirals at L_1, and by symmetry the incoming spirals as well, we immediately obtain the full infinite period orbit closing on itself and not just an open segment of it as in the case of the Strömgren spirals at L_4. This means that it is feasible to identify the asymptotic orbit as the termination of a family of periodic orbits and proceed, if with difficulty, to extract from the spiralling doubly-asymptotic orbit at L_1 the entire family to which it is the terminating orbit.

In the present work we compute spiralling homoclinic asymptotic orbits at L_1, and corresponding families of periodic orbits, for $q=1/10<1/9$. We then recompute (numerically continue) the families for $q=0.12$, for which L_1 is stable and the families cannot terminate with spiralling asymptotic orbits at it. Finally we numerically continue the families for $q=1$, i.e. for the classical Copenhagen problem, and thus determine families of periodic orbits in the classical case and their terminations as spiralling asymptotic orbits at the triangular points.

This deviation into the photogravitational case serves to provide the spiralling asymptotic periodic orbits at the triangular points, i.e. the heteroclinic termination segments of the Copenhagen problem in the correct pairing, as final evolution stages of the spiralling asymptotic orbits at the collinear point L_1, i.e. the homoclinic orbits of the photogravitational case.

SPIRALLING HOMOCLINIC ORBITS AT L_1

A spiralling homoclinic orbit at the inner collinear equilibrium point L_1 is determined numerically by following the outgoing spiral from L_1 to a distant perpendicular crossing of the x-axis. As the orbit crosses the axis infinitely many times near L_1, we start the numerical integration at a distance r away from L_1, with r sufficiently large to avoid close-by crossings.

In the immediate neighborhood of L_1, i.e inside the circle centered at (0,0) with radius r, the orbit is described by analytical approximations and r must be sufficiently small to ensure validity of these approximations. We have used the symbolic computation package "*Mathematica*" to determine expansions of the outgoing spirals, in the case $\mu = 1/2$ and $q = 1/10$, accurate to the 6-th order in the small orbital parameter r and we give below the initial conditions for the outgoing orbit to 4-th order terms in r (only odd-order terms are present), which can be adopted as sufficiently accurate for the value of the radius $r = 0.05$ used in our calculations:

$$x_o = C\,r + \left(17.741\,826\,C^3 + 7.787\,601\,C^2 S + 7.605\,269\,C\,S^2 + 1.460\,736\,S^3\right) r^3 + \cdots,$$

$$y_o = S\,r + \left(-16.670\,463\,C^3 + 6.507\,999\,C^2 S - 2.238\,289\,C\,S^2 + 2.013\,350\,S^3\right) r^3 + \cdots,$$

$$\dot{x}_o = (0.385\,297\,C + 0.434\,259\,S)\,r + \left(8.314\,241\,C^3 + 6.131\,126\,C^2 S + 4.457\,975\,C\,S^2 + 3.770\,945\,S^3\right) r^3 + \cdots, \tag{3}$$

$$\dot{y}_o = (-1.565\,741\,C + 0.106\,862\,S)\,r + \left(-29.459\,076\,C^3 - 8.998\,227\,C^2 S - 5.145\,633\,C\,S^2 - 0.326\,544\,S^3\right) r^3 + \cdots,$$

where we have abbreviated: $C = \cos\theta$, $S = \sin\theta$, and θ is the angle giving the point of exit on the circle of radius r.

RESULTS

We give here, in Figure 2, six typical asymptotic homoclinic orbis at L_1. Also in Figures 3, 4 and 5, we give the family associated with orbit f_2 for three different values of the radiation factor: $q = 0.1$, 0.12 and 1, and similarly in Figures 6, 7 and 8, the family associated with orbit f_3 for the same three values of q.

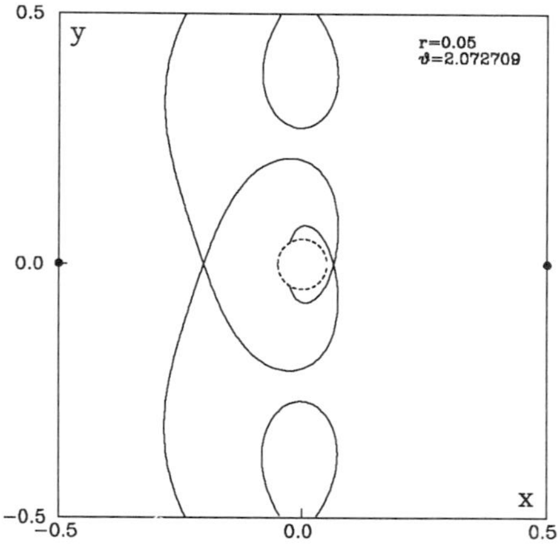

Fig. 1. The analytical solution giving the initial conditions (3) is valid inside the small circle of radius r = 0.05. Both the outgoing spiralling orbit and its mirror image (the incoming orbit) are shown. The collinear equilibrium point L_1 is located at (0,0).

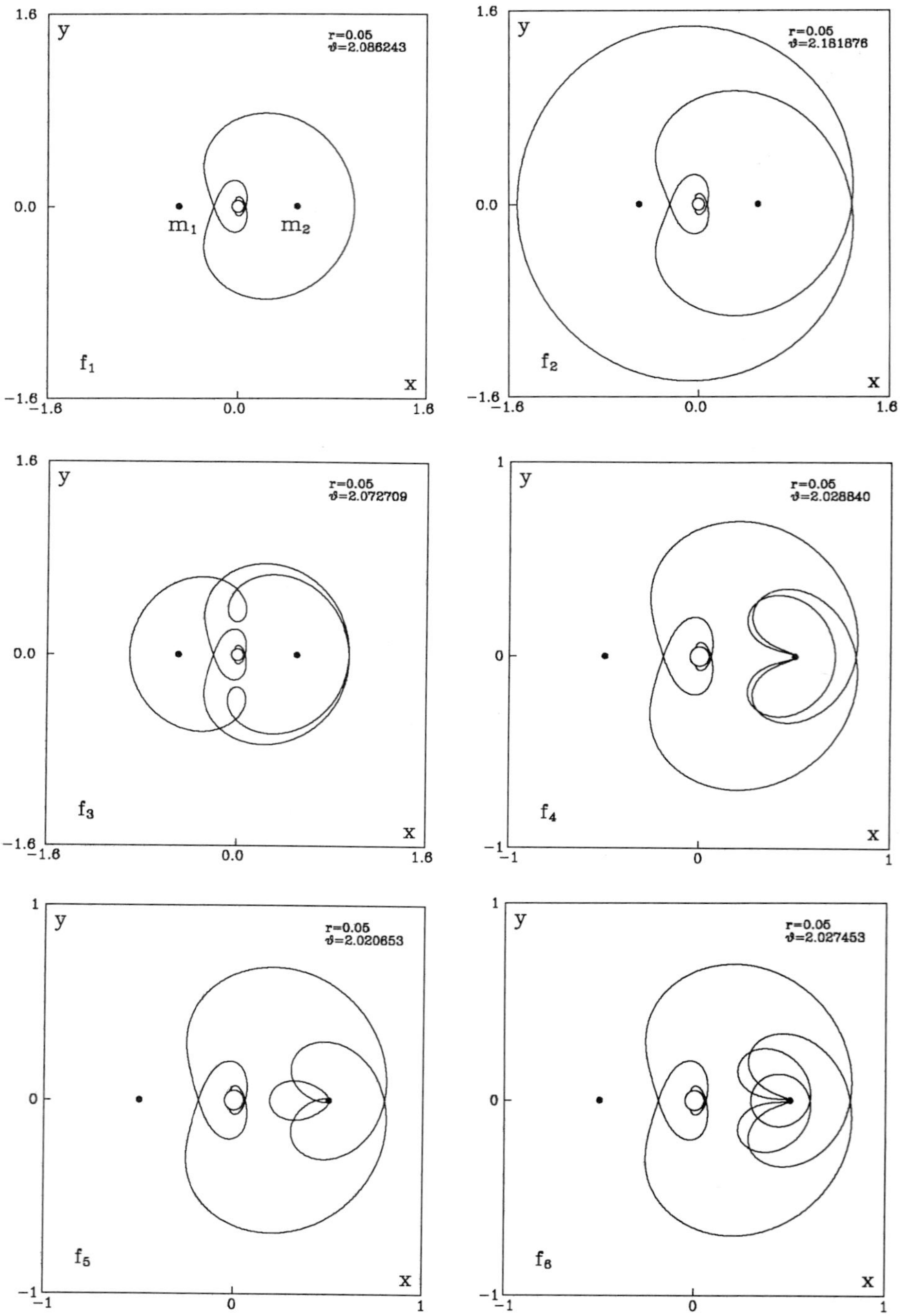

Fig. 2. Six homoclinic spiralling asymptotic orbits at L_1 (f_1 to f_6) for $\mu = 1/2$ and $q_1 = q_2 = q = 1/10$.

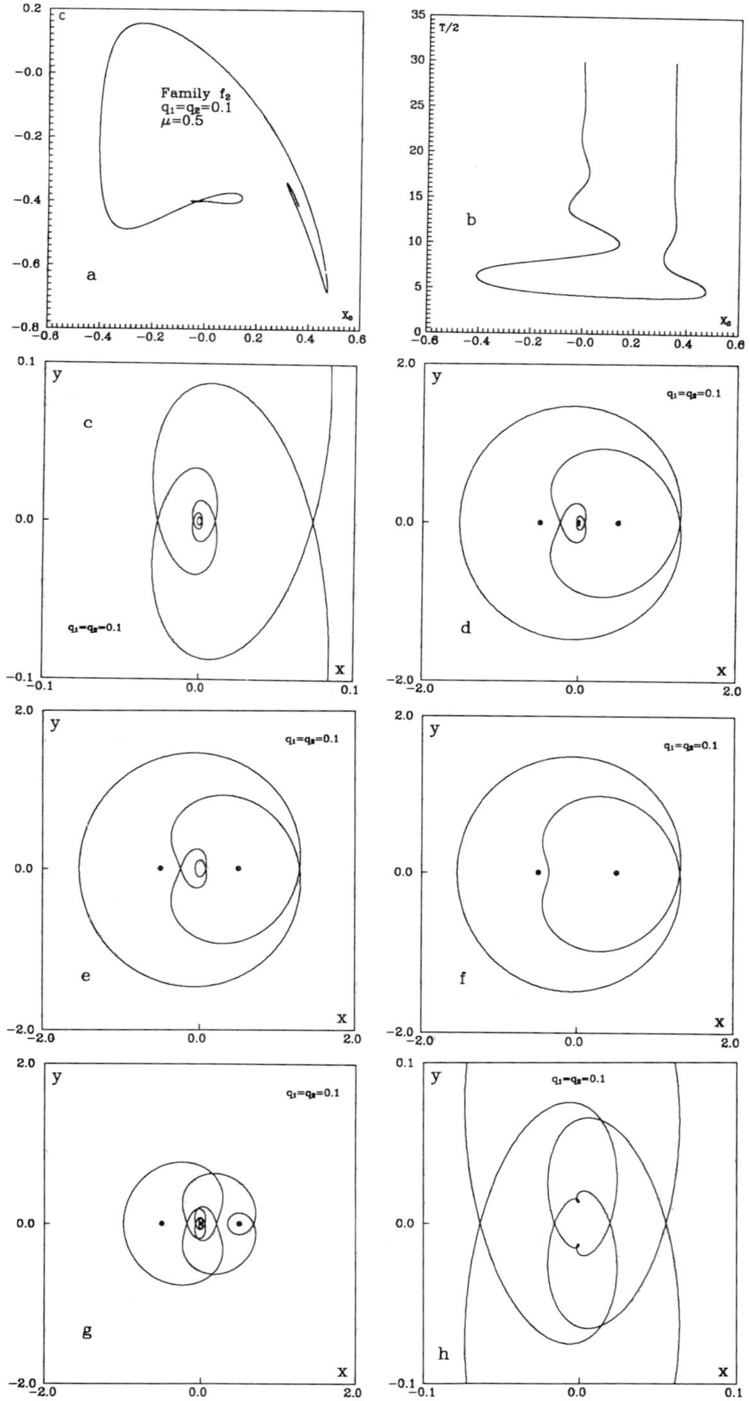

Fig. 3. Frames a to h. The family of periodic orbits f_2 extracted from the asymptotic orbit f_2 for q = 1/10. The family terminates at both ends with homoclinic asymptotic orbits at L_1. a: The family characteristic in terms of the Jacobi constant C as a function of x_0. b: The half period T/2 of the family members as a function of x_0. c to h: Typical orbits of the family. c is a magnification of the termination orbit d (f_2) and h is a magnification of the termination orbit g.

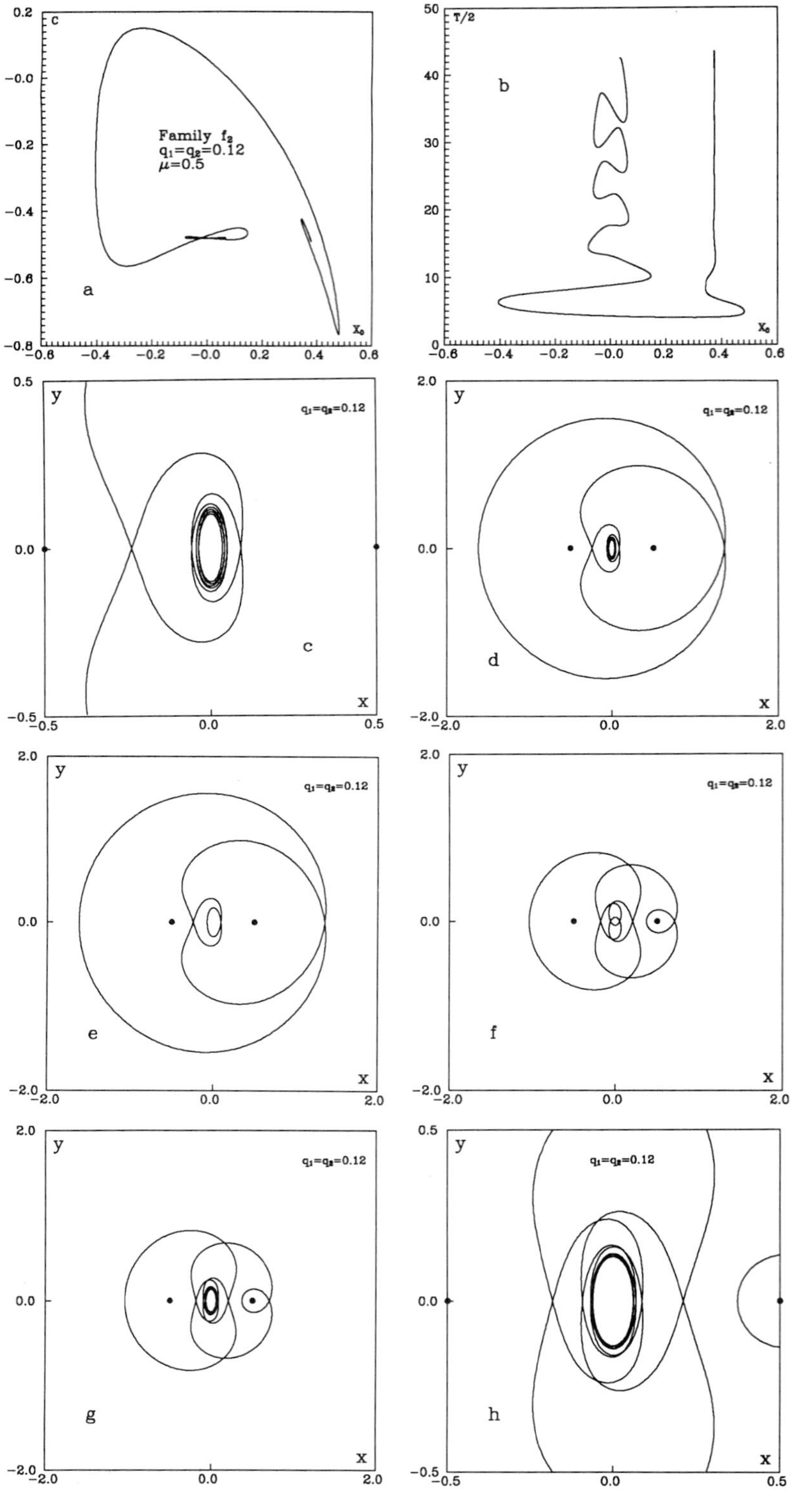

Fig.4. Frames a to h. The family of periodic orbits f_2 for q = 0.12. The family now terminates at both ends with asymptotic orbits at small periodic orbits around L_1. This is shown clearly in the magnifications c and h.

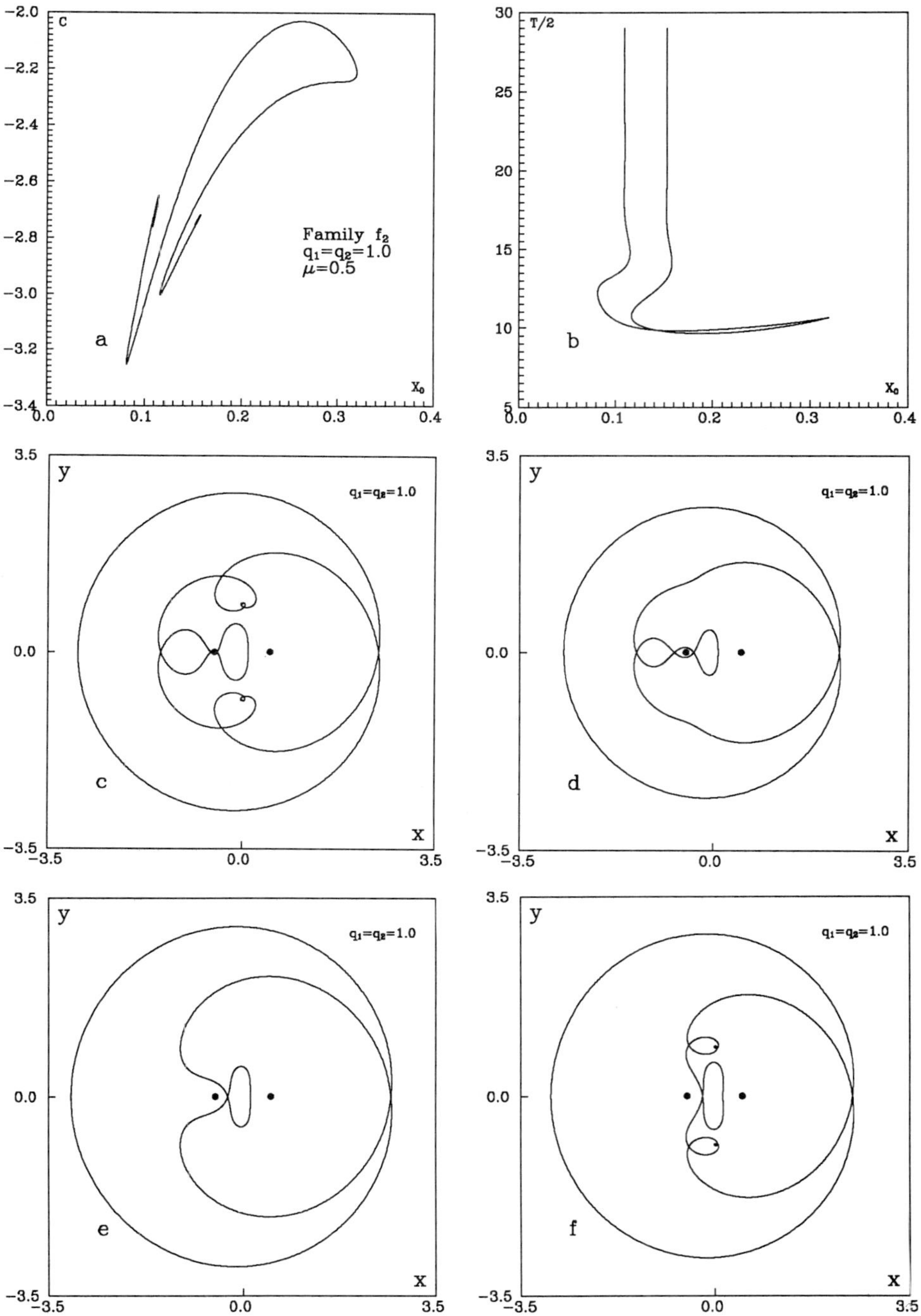

Fig. 5. Frames a to f. The family of periodic orbits f_2 for $q = 1$ (the gravitational Copenhagen problem). The family now terminates at both ends (frames c and f) with combinations of heteroclinic asymptotic orbits at L_4 and L_5.

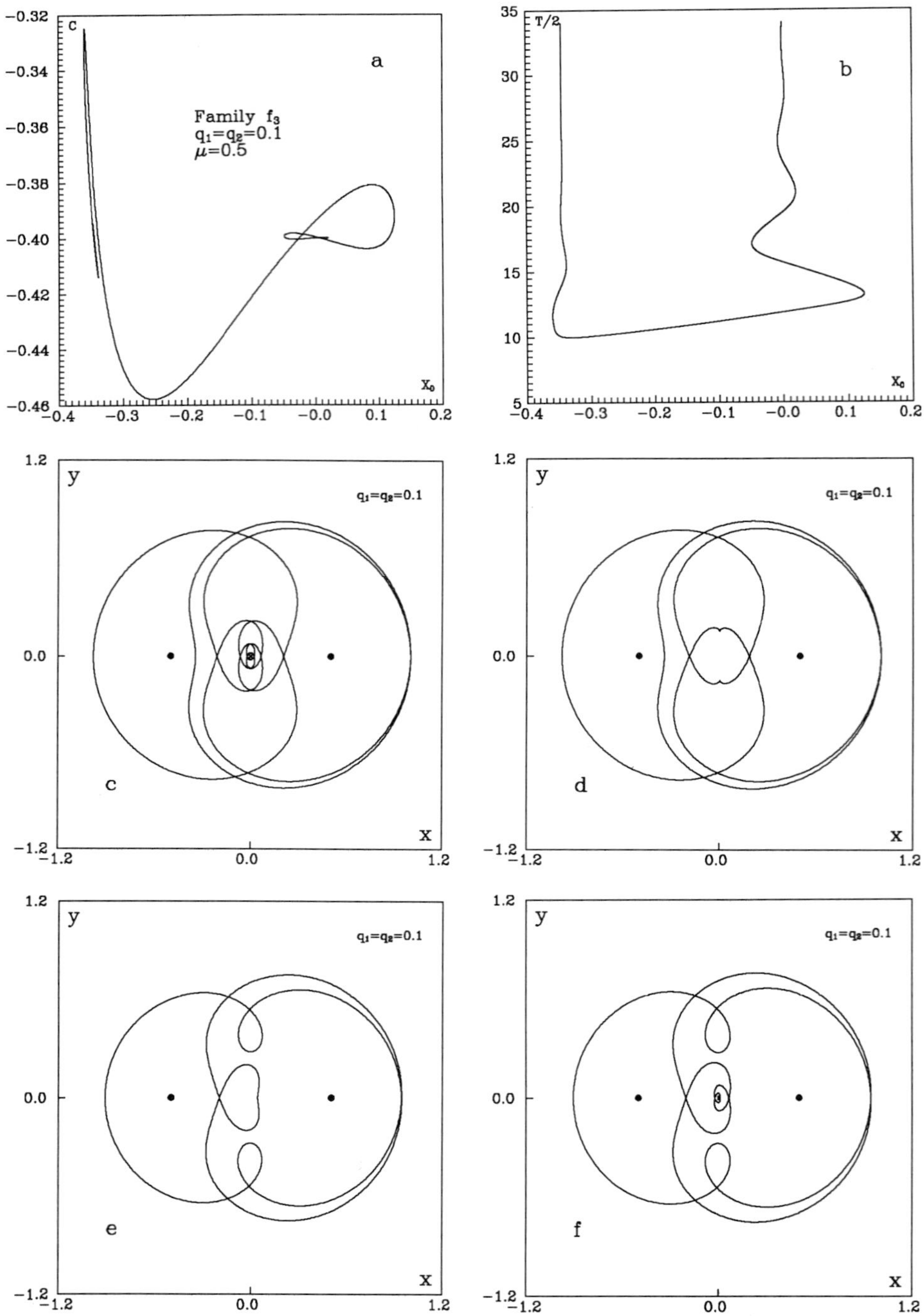

Fig. 6. Frames a to f. The family of periodic orbits f_3 extracted from the asymptotic orbit f_3 for q = 1/10. The family terminates at both ends with homoclinic asymptotic orbits at L_1. a: The family characteristic in terms of the Jacobi constant C as a function of x_0. b: The half period T/2 of the family members as a function of x_0. c to f: Typical orbits of the family.

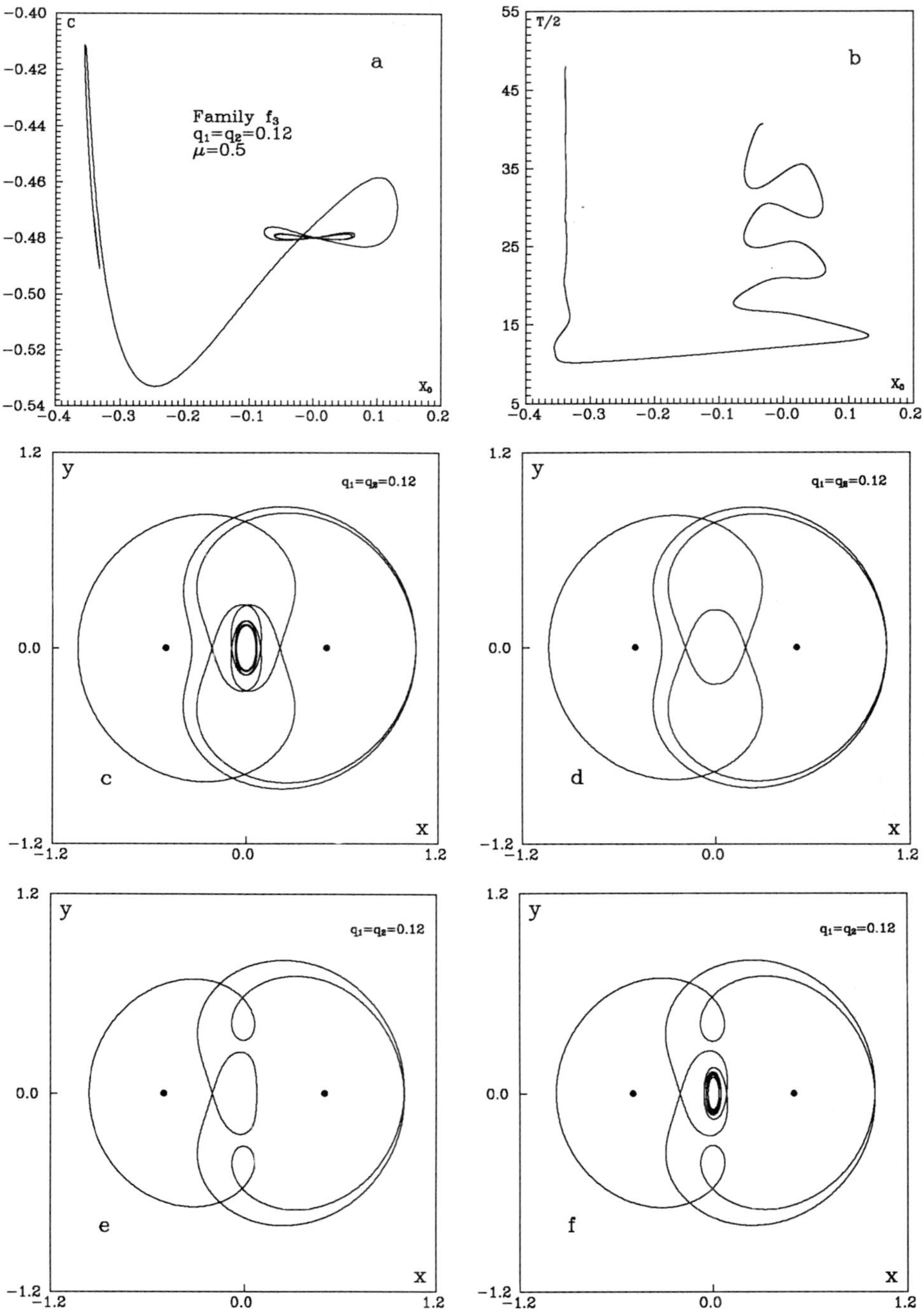

Fig.7. Frames a to f. The family of periodic orbits f_3 for q = 0.12. The family now terminates at both ends with asymptotic orbits at small periodic orbits around L_1.

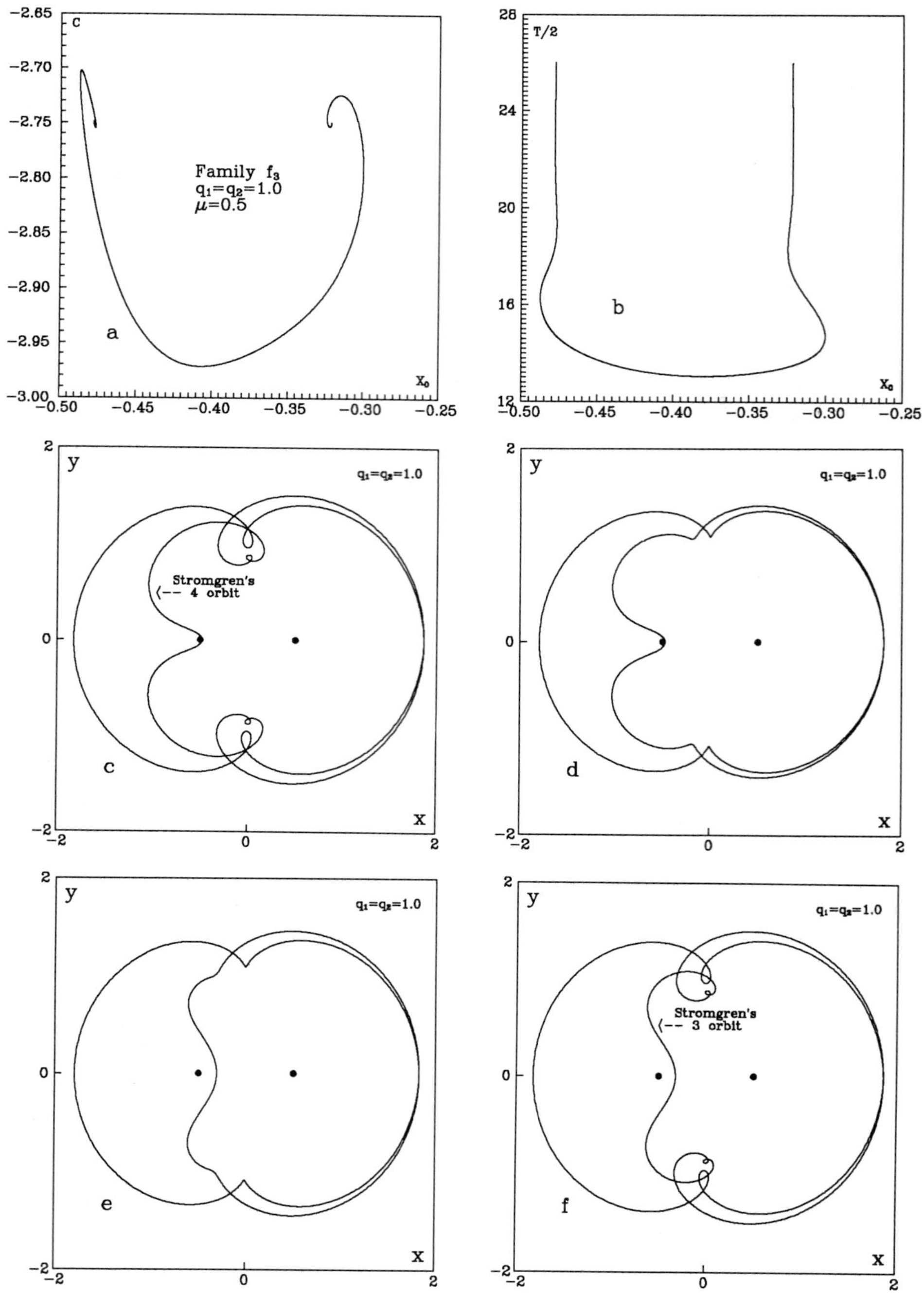

Fig. 8. Frames a to f. The family of periodic orbits f_3 for $q = 1$ (the gravitational Copenhagen problem). The family now terminates at both ends (frames c and f) with combinations of heteroclinic asymptotic orbits at L_4 and L_5.
In this case the classical orbits 3 and 4 of Strömgren (see Szebehely 1967) are seen to combine with other heteroclinic orbits to form the terminations.

It is interesting to note that in the case q=0.12 the point L_1 is stable and the families cannot terminate with spiralling orbits asymptotic at it. However, the families behave similarly except that they terminate at both ends with orbits asymptotic at small periodic orbits around the equilibrium instead of at the equilibrium itself (see frames d and g of Figure 4 and frames c and f Figure 7).

REFERENCES

Gomez, G., Llibre, J. and Masdemont, J.: 1988, *Celes.Mech.* **44**, 239.

Markellos, V.V., Perdios, E. and Georghiou, K.: 1993a, *Astrophys. Space Sci.* **199**, 23.

Markellos, V.V., Perdios, E. and Papadakis, K.: 1993b, *Astrophys. Space Sci.* **199**, 139.

Markellos, V.V.: 1991, in Roy, A.E. (ed.), *Predictability, Stability and Chaos in N-Body Dynamical Systems*, Plenum Press, New York, 413.

Szebehely, V.: 1967, *Theory of Orbits*, Academic Press, New York.

Simmons, J.F.L., McDonald, A.J.C., Brown, J.C.: 1985, *Celes.Mech.* **35**, 145.

A TYPICAL FAMILY FII OF SYMMETRIC AND PERIODIC ORBITS OF CHARGED PARTICLES MOVING IN THE PLANE OF MOTION OF TWO PARALLEL ROTATING MAGNETIC DIPOLES

Maria Leftaki

Department of Mathematics
University of Patras
26110 Patras, Greece

INTRODUCTION

This work is a contribution to the study of symmetric and periodic orbits of charged particles moving under the action of Lorentzian forces in the plane of motion of two parallel magnetic dipoles rotating about the barycenter of their carrier stars.

The Newtonian forces are neglected on account of their small measure in comparison with the Lorentzian forces.

THE EQUATIONS OF MOTION AND VARIATIONS

The frame of reference adopted is obviously rotating synodical barycentric (see Figure 1) and the system of units is explained in detail by Goudas et al (1985) or by Mavraganis (1978).

The method used is numerical and followed the steps given below.

Assuming $x_1 = x, x_2 = y, x_3 = \dot{x}$ and $x_4 = \dot{y}$ the system of the differential equations of motion becomes

$$\begin{aligned} \dot{x}_1 &= x_3 = f_1 \\ \dot{x}_2 &= x_4 = f_2 \\ \dot{x}_3 &= x_4 r_3 - \frac{\partial V}{\partial x_1} = f_3 \\ \dot{x}_4 &= -x_3 r_3 - \frac{\partial V}{\partial x_2} = f_4 \end{aligned} \tag{1}$$

From Newton to Chaos, Edited by A.E. Roy
and B.A. Steves, Plenum Press, New York, 1995

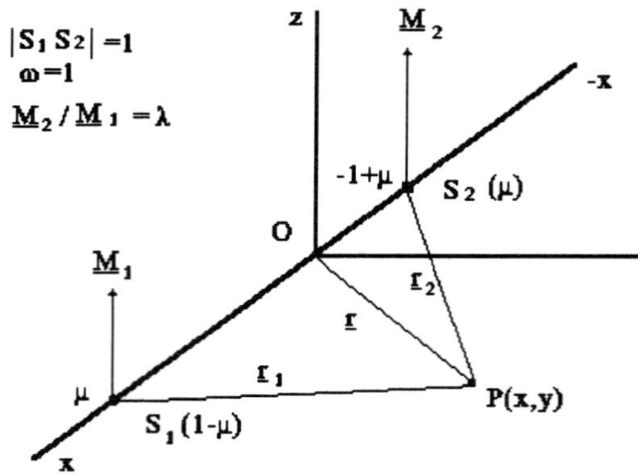

Figure 1. Geometrical configuration of the two dipole problem

with

$$r_1 = ((x_1 - \mu)^2 + x_2^2)^{1/2}$$
$$r_2 = ((x_1 + 1 - \mu)^2 + x_2^2)^{1/2}$$
$$r_3 = 2 - \frac{1}{r_1^3} - \frac{\lambda}{r_2^3}$$
$$V = -(x_1^2 + x_2^2)(\frac{1}{2} + \frac{1}{r_1^3} + \frac{\lambda}{r_2^3}) + x_1(\frac{\mu}{r_1^3} - \frac{\lambda(1-\mu)}{r_2^3}) \tag{2}$$

where $\lambda = \frac{M2}{M1}$ the ratio of magnetic moments, $\mu = m_{S_2}$, $1\text{-}\mu = m_{S_1}$.

Moreover, the Jacobi integral has the form

$$\frac{1}{2}(x_3^2 + x_4^2) + V = c \tag{3}$$

The system of differential equations (1), which is autonomous ($\frac{\partial f_i}{\partial t} = 0, i = 1,2,3,4$) in vector form becomes

$$\frac{d\underline{x}}{dt} = \underline{f}(\underline{x}) \tag{4}$$

with

$$\underline{x} = (x_1, x_2, x_3, x_4) \text{ and } \underline{f}(\underline{x}) = (f_1(\underline{x}), f_2(\underline{x}), f_3(\underline{x}), f_4(\underline{x})) \tag{5}$$

Its solutions

$$\underline{x} = \underline{x}(x_0; t) \tag{6}$$

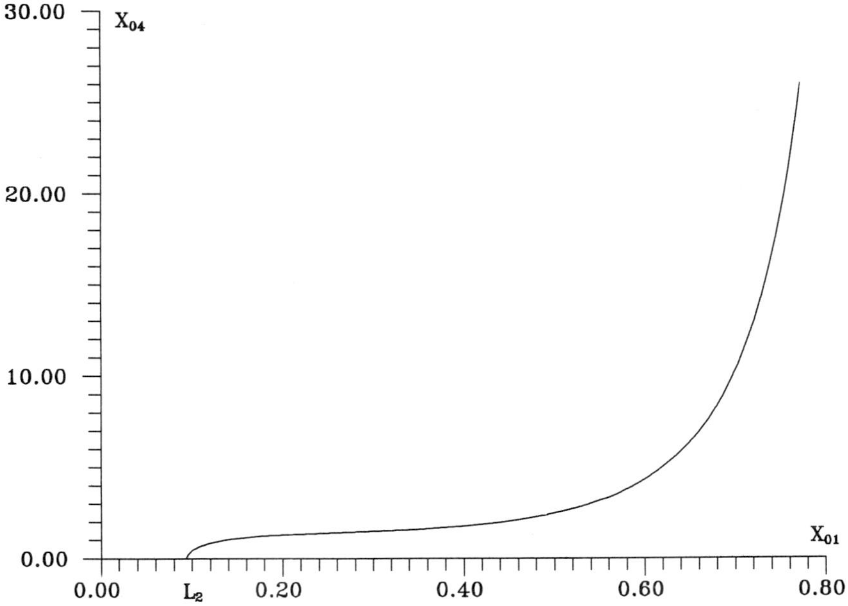

Figure 2. Graphical representation of the family FII of periodic and symmetric orbits. The values (x_{01}, x_{04}) correspond to the initial conditions listed in Table 1.

are functions of the initial conditions

$$\underline{x}_0 = (x_{01}, x_{02}, x_{03}, x_{04}) \tag{7}$$

and the time t, not depending on the initial time which is selected to be zero.

The system of the differential equations of the first order variations is then:

$$\begin{aligned}
\frac{d}{dt}\left(\frac{\partial x_1}{\partial x_{0j}}\right) &= \frac{\partial f_1}{\partial x_{0j}} = \sum_{k=1}^{4} \frac{\partial f_1}{\partial x_k}\frac{\partial x_k}{\partial x_{0j}} \\
\frac{d}{dt}\left(\frac{\partial x_2}{\partial x_{0j}}\right) &= \frac{\partial f_2}{\partial x_{0j}} = \sum_{k=1}^{4} \frac{\partial f_2}{\partial x_k}\frac{\partial x_k}{\partial x_{0j}} \\
\frac{d}{dt}\left(\frac{\partial x_3}{\partial x_{0j}}\right) &= \frac{\partial f_3}{\partial x_{0j}} = \sum_{k=1}^{4} \frac{\partial f_3}{\partial x_k}\frac{\partial x_k}{\partial x_{0j}} \\
\frac{d}{dt}\left(\frac{\partial x_4}{\partial x_{0j}}\right) &= \frac{\partial f_4}{\partial x_{0j}} = \sum_{k=1}^{4} \frac{\partial f_4}{\partial x_k}\frac{\partial x_k}{\partial x_{0j}}
\end{aligned} \tag{8}$$

for $j = 1, 2, 3, 4$
and in matrix form

$$\frac{d\Delta}{dt} = P\Delta \tag{9}$$

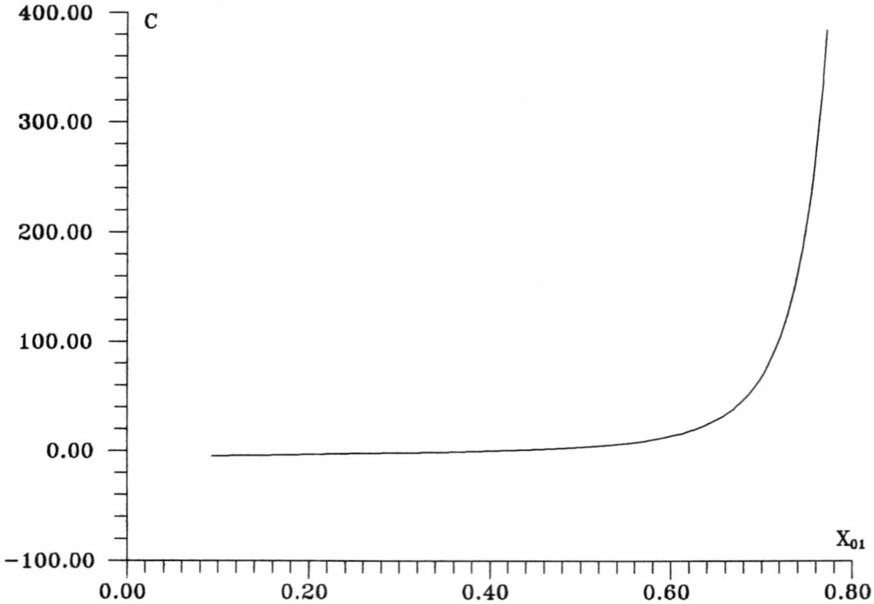

Figure 3. Graphical representation of points (x_{01}, c) for the family FII of symmetric and periodic orbits.

where

$$\Delta = \begin{bmatrix} \frac{\partial x_1}{\partial x_{01}} & \frac{\partial x_1}{\partial x_{02}} & \frac{\partial x_1}{\partial x_{03}} & \frac{\partial x_1}{\partial x_{04}} \\ \frac{\partial x_2}{\partial x_{01}} & \frac{\partial x_2}{\partial x_{02}} & \frac{\partial x_2}{\partial x_{03}} & \frac{\partial x_2}{\partial x_{04}} \\ \frac{\partial x_3}{\partial x_{01}} & \frac{\partial x_3}{\partial x_{02}} & \frac{\partial x_3}{\partial x_{03}} & \frac{\partial x_3}{\partial x_{04}} \\ \frac{\partial x_4}{\partial x_{01}} & \frac{\partial x_4}{\partial x_{02}} & \frac{\partial x_4}{\partial x_{03}} & \frac{\partial x_4}{\partial x_{04}} \end{bmatrix} \quad \text{and} \quad P = \begin{bmatrix} 0 & 0 & 1 & 0 \\ 0 & 0 & 0 & 1 \\ \frac{\partial f_3}{\partial x_1} & \frac{\partial f_3}{\partial x_2} & 0 & r_3 \\ \frac{\partial f_4}{\partial x_1} & \frac{\partial f_4}{\partial x_2} & -r_3 & 0 \end{bmatrix} \tag{10}$$

$\mid \Delta \mid$ is the Jacobian of the solution with respect to the initial conditions $\underline{x}_0$.

$$\mid \Delta \mid = \mid \frac{\partial(x_1, x_2, x_3, x_4)}{\partial(x_{01}, x_{02}, x_{03}, x_{04})} \mid = 1 \tag{11}$$

Proof:

$$\frac{d \mid \Delta \mid}{dt} = \mid \Delta \mid tr(P) = \mid \Delta \mid \sum_{k=1}^{4} \frac{\partial f_k}{\partial x_k} = \mid \Delta \mid 0 = 0$$

Consequently $\mid \Delta \mid = \text{constant} = \mid \Delta(t=0) \mid = \mid I \mid = 1$
Thus, the condition (11) insures the validity of the principle of conservation of volumes in the phase space.

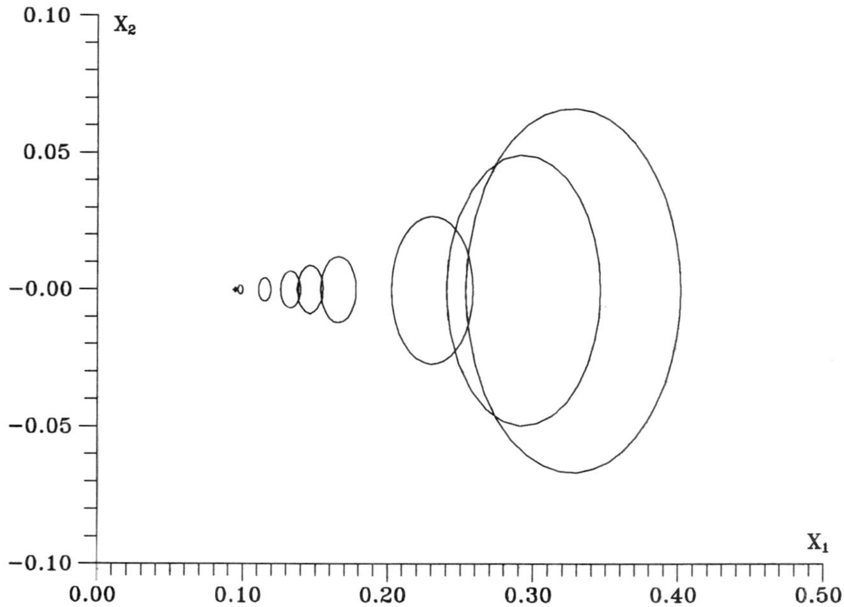

Figure 4. Graphs of some symmetric and periodic orbits obtained from Table 1 with c ranging from -4.8546 to -0.0832

In order to start with the simplest orbit that may exist, namely symmetric periodic orbit of simple symmetry, intersecting the $0x_1$ axis only twice in every period, we select arbitrary initial conditions of the form

$$(x_{01}, 0, 0, x_{04}), \quad t = 0 \tag{12}$$

and integrate the equations of motion and of variations (4) and (9) until the condition of symmetry

$$x_2(x_{01}, x_{04}; T) = 0, \tag{13}$$

at half period T, is satisfied.

In practice an orbit is considered symmetric if

$$\mid x_2(x_{01}, x_{04}; T) \mid < \epsilon = 10^{-6} \tag{13a}$$

We stop the integration to the first point where $x_2(x_{01}, x_{04}; t)$ changes sign. We then continue the integration with a step in time

$$h = \frac{-x_2(x_{01}, x_{04}; t)}{x_4(x_{01}, x_{04}; t)}$$

and after the integration we check the absolute value of x_2. This is repeated until the condition $(13a)$ is satisfied.

$$\text{If} \qquad x_3(x_{01}, x_{04}; T) = 0 \tag{14}$$

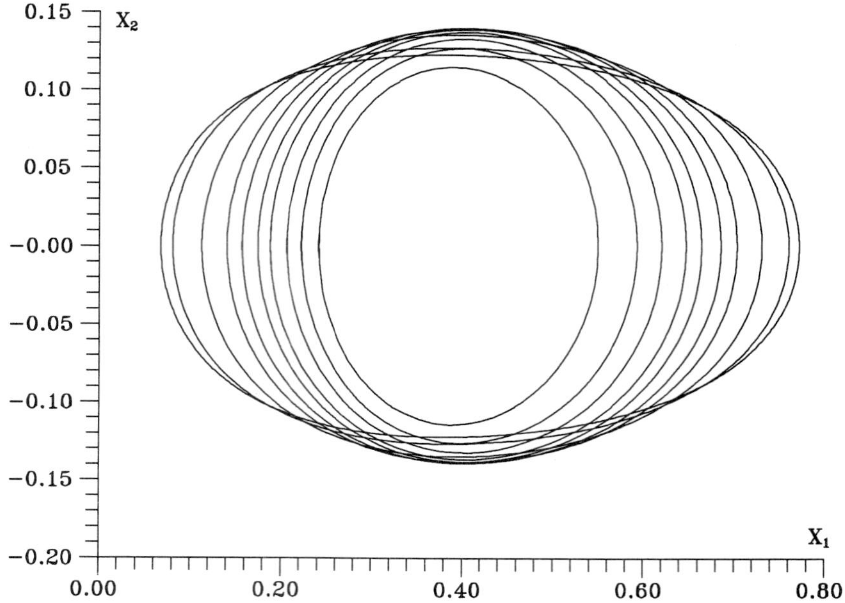

Figure 5. Graphs of some symmetric and periodic orbits listed in Table 1 with c ranging from 0.8929 to 383.8.

at half period, the orbit is said to be periodic of period 2T. In practice it suffices

$$| x_3(x_{01}, x_{04}; T) | < \epsilon = 10^{-6} \tag{14a}$$

In general, however, for initial conditions of the type (12), selected at random the condition (14) is not satisfied and a correction $(\delta x_{01}, \delta x_{04}; \delta T)$ is needed in order that at $t = T + \delta T$ the conditions of symmetry and periodicity

$$\begin{aligned} x_2(x_{01} + \delta x_{01}, x_{04} + \delta x_{04}; T + \delta T) = 0 \\ x_3(x_{01} + \delta x_{01}, x_{04} + \delta x_{04}; T + \delta T) = 0 \end{aligned} \tag{15}$$

are satisfied simultaneously.

On expanding the conditions (15) in Taylor series of the corrections and on retaining only the first-order terms, we come to the linear system

$$\frac{\partial x_2}{\partial x_{01}} \delta x_{01} + \frac{\partial x_2}{\partial x_{04}} \delta x_{04} + \frac{\partial x_2}{\partial T} \delta T = 0$$

$$\frac{\partial x_3}{\partial x_{01}} \delta x_{01} + \frac{\partial x_3}{\partial x_{04}} \delta x_{04} + \frac{\partial x_3}{\partial T} \delta T = -x_3(x_{01}, x_{04}; T) \neq 0$$

to which we add the equation

$$\begin{vmatrix} \delta x_{01} & \delta x_{04} & \delta T \\ \frac{\partial x_2}{\partial x_{01}} & \frac{\partial x_2}{\partial x_{04}} & \frac{\partial x_2}{\partial T} \\ \frac{\partial x_3}{\partial x_{01}} & \frac{\partial x_3}{\partial x_{04}} & \frac{\partial x_3}{\partial T} \end{vmatrix} = 0 \tag{16}$$

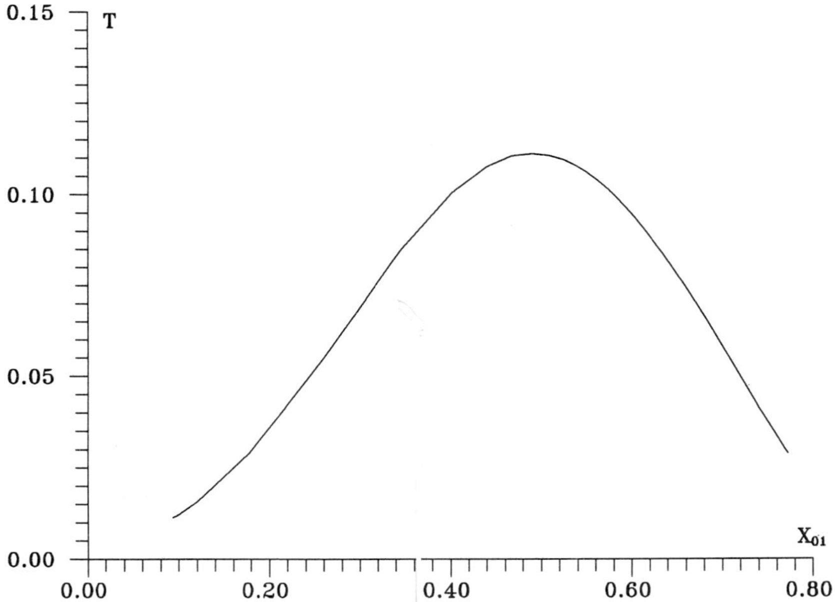

Figure 6. Graphical representation of the variation of the half period T as a function of x_{01}. An interesting finding: There exist symmetric and periodic orbits of equal period within the same family FII.

The third equation of the system (16) secures the steepest descent to the nearest initial conditions of a periodic orbit.

After computing the predictor corrections δx_{01},δx_{04},δT we apply the same procedure starting with the new initial conditions

$$x_{01}^* = x_{01} + \delta x_{01}, \qquad x_{04}^* = x_{04} + \delta x_{04}$$

until the conditions (13)-(14) are satisfied with the desired accuracy.

All points (x_{01}, x_{04}) satisfying the conditions (13)-(14) rest upon a curve that constitutes the family of periodic orbits. Conditions (13)-(14) can be written equivalently in the form

$$\left\{ \begin{array}{l} x_{01} = \phi_1(T) \\ x_{04} = \phi_2(T) \end{array} \right\} \quad \text{if} \quad \frac{\partial(x_2, x_3)}{\partial(x_{01}, x_{04})} \neq 0$$

Once one member of a family is found the next member of the same family is provided by the homogeneous simultaneous equations

$$\begin{aligned} x_2(x_{01} + \delta x_{01}, x_{04} + \delta x_{04}; T + \delta T) = 0 \\ x_3(x_{01} + \delta x_{01}, x_{04} + \delta x_{04}; T + \delta T) = 0 \end{aligned} \tag{17}$$

The system (17) linearized with respect to δx_{01}, δx_{04}, δT takes the form

$$\frac{\partial x_2}{\partial x_{01}} \delta x_{01} + \frac{\partial x_2}{\partial x_{04}} \delta x_{04} + \frac{\partial x_2}{\partial T} \delta T = 0$$

$$\frac{\partial x_3}{\partial x_{01}}\delta x_{01} + \frac{\partial x_3}{\partial x_{04}}\delta x_{04} + \frac{\partial x_3}{\partial T}\delta T = 0 \tag{18}$$

in association with the tangential predictor

$$\delta x_{01}^2 + \delta x_{04}^2 = l^2$$

where l is an arbitrary tangential segment to the curve of the family; consequently δT is taken arbitrarily.

The choice of the value of l depends on the density of members of the family and on the curvature of the family at each point. We proceed with small l if the curvature is large and with larger l if the curvature is small.

We integrate then the system of motion and variations (4) and (9) with initial conditions

$$x_{01}^* = x_{01} + \delta x_{01}, \qquad x_{04}^* = x_{04} + \delta x_{04}$$

located on the tangent of the family and we repeat the steps (13)-(16) of the numerical technique explained above (steepest descent etc.).

All the orbits we have studied are unstable. They do not satisfy the condition of stability for periodic orbits which is

$$0 \leq tr\Delta(2T) \leq 4 \tag{19}$$

The matrix $\Delta(2T)$ is computed by means of the relation

$$\Delta(2T) = L\Delta^{-1}(T)L\Delta(T) \tag{20}$$

for

$$L = \begin{bmatrix} 1 & 0 & 0 & 0 \\ 0 & -1 & 0 & 0 \\ 0 & 0 & -1 & 0 \\ 0 & 0 & 0 & 1 \end{bmatrix}$$

which is the matrix appearing in the transformation $\underline{Y} = L\underline{X}$ where t is substituted by -t.

We know that

$$tr\Delta(\underline{x}_0; 2T) = \sum_{i=1}^{4} \lambda_i$$

Also, that two eigenvalues of the matrix $\Delta(2T)$, say λ_1 and λ_2 are equal to unity ($\lambda_1 = \lambda_2 = 1$). Consequently the λ_3,λ_4 eigenvalues are computed by the algebraic equation of second degree

$$\lambda^2 - (tr\Delta(\underline{x}_0; 2T) - 2)\lambda + 1 = 0$$

because $\lambda_3 + \lambda_4 = tr\Delta(\underline{x}_0\lambda; 2T) - 2$ and $\lambda_3\lambda_4 = 1$.

SOME NUMERICAL RESULTS

Using the above algorithm we treated the case μ =0.9 and λ =2 and found a family of planar symmetric and periodic orbits that is associated with the equilibrium point L_2 between the two dipoles. We note that $x_{S_1} = 0.9, x_{S_2} = -0.1$.

Some numerical values of the initial and final conditions, the constant of energy and the half of the period of a representative group of members we list in the Table 1

Table 1. Initial and final conditions, constant of energy and half of the period of symmetric and periodic orbits of a typical family FII computed for $\mu = .9$ and λ =2.

x_{01}	x_{04}	x_1(T)	x_4(T)	c	T
+0.0936	+0.0259	+0.0934	-0.0259	-4.8546	+0.0114
+0.1000	+0.4258	+0.0966	-0.4367	-4.7581	+0.0122
+0.1194	+0.8253	+0.1109	-0.8752	-4.4318	+0.0156
+0.1779	+1.2198	+0.1535	-1.3971	-3.5380	+0.0291
+0.3463	+1.5792	+0.2403	-2.2212	-1.1602	+0.0851
+0.4396	+1.9672	+0.2574	-2.9439	+0.8929	+0.1078
+0.4900	+2.3639	+0.2552	-3.5323	+2.7743	+0.1114
+0.5241	+2.7624	+0.2487	-4.0574	+4.6959	+0.1097
+0.5496	+3.1616	+0.2410	-4.5470	+6.7185	+0.1064
+0.5699	+3.5611	+0.2331	-5.0144	+8.8673	+0.1025
+0.5820	+3.8433	+0.2276	-5.3350	+10.4677	+0.0997
+0.6034	+4.4429	+0.2165	-5.9986	+14.1088	+0.0940
+0.6207	+5.0426	+0.2061	-6.6456	+18.0897	+0.0886
+0.6351	+5.6425	+0.1966	-7.2816	+22.4188	+0.0838
+0.6474	+6.2424	+0.1878	-7.9098	+27.1008	+0.0793
+0.6580	+6.8422	+0.1797	-8.5325	+32.1385	+0.0754
+0.6702	+7.6422	+0.1699	-9.3566	+39.4123	+0.0706
+0.6854	+8.8421	+0.1569	-10.5834	+51.5209	+0.0645
+0.6979	+10.0420	+0.1455	-11.8029	+65.0710	+0.0593
+0.7130	+11.8420	+0.1310	-13.6236	+88.1045	+0.0530
+0.7287	+14.2419	+0.1151	-16.0418	+123.8743	+0.0464
+0.7454	+17.6419	+0.0974	-19.4576	+184.4470	+0.0395
+0.7586	+21.2418	+0.0827	-23.0679	+261.2285	+0.0342
+0.7679	+24.4418	+0.0721	-26.2739	+340.3925	+0.0305
+0.7719	+26.0418	+0.0674	-27.8762	+383.8261	+0.0290

There are some symmetric periodic orbits of very short periods and very large velocities corresponding to large values of c. An interesting finding is the existence of orbits of equal period within the same family, namely there exist orbits of very small size and orbits of relatively large size, both having the same period.

In Figure 2 we present the members (x_{01}, x_{04}) of this *family*. We observe that to the left the family terminates at L_2 with position $x_{L_2} = 0.0934805$ and energy $c_{L_2} = -4.854981$ while to the right $c \to +\infty$.

In Figure 3 we give the variation of c as a function of x_{01}. For values of x_{01} approaching 0.9 we see that $c \to +\infty$.

Finally in Figures 4 and 5 we give some orbits appearing in Table 1 and show their paths as well as the overall appearance of this family.

As the energy increases we observe that the second point of intersection of each orbit moves away from L_2. From certain value of c this point approaches L_2 and moves away again towards the opposite side of Ox_1 axis.

Moreover the shape of the orbit is compressed along Ox_2 axis as the constant of c exceeds a certain value (72.38).

As a conclusion we can say that the family FII is well behaved. As the periodic orbits are unstable we don't have trapping areas in the region between the two dipoles.

An application of this problem refers to the motions of charged particles in the field of double magnetic stars. We have found the periodic orbits and we plan in future to study the distribution of non periodic orbits, and as this problem is not integrable we can find chaos.

ACKNOWLEDGEMENTS

I wish to express my appreciation to Prof. Goudas who suggested this piece of research and to Mr. E. Valaris who helped me a lot to realize it.

REFERENCES

Goudas C.L., Leftaki M., and Petsagourakis E., 1985, Motions in the field of two rotating magnetic dipoles. I: Equilibrium points, Celestial Mechanics 37: 127.

Mavraganis, A., 1978, The magnetic-binaries problem: Motion of a charged particle in the region of a magnetic-binary system, Astrophys. Space Sci. 54: 305.

INTERLUDE

FRACTALS, CHAOS AND QUATERNIONS

John F. Doyle, Bonnie Steves, and Jagannathan Gomatam

Department of Mathematics
Glasgow Caledonian University
Glasgow, UK. G4 0BA

ABSTRACT

Fractals characterise the geometry of Chaos. To understand this geometry it is necessary to formulate the concept of non integral dimensions, a continuum of dimension from zero upwards. One of the signatures of Chaos theory is the Mandelbrot Set which is embedded in a two dimensional plane. This expository talk is aimed at illustrating the notions of fractals, chaos and associated generalisations. Examples will be developed which show the need for seeking extension of the Mandelbrot set from the complex plane to division rings of quaternions. Since this conference is a celebration of Poincaré, we have chosen to develop our theme with a hint of history and a glint of shifting paradigm.

1. INTRODUCTION

> "I know that most men, including those at ease with problems of the greatest complexity, can seldom accept even the simplest and most obvious truth if it be such as would oblige them to admit the falsity of conclusions they reached, perhaps with great difficulty, conclusions which they have delighted in explaining to colleagues, which they have proudly taught to others, and which they have woven, thread by thread, into the fabric of their lives."
>
> Leo Tolstoy (Lukacs, 1972)

The theory of Mathematical Chaos, embedded as it is in non-linear systems analysis, threatens the whole fabric of classical determinism which has traditionally treated non-linearities as mere perturbations. If more problems have been solved using linear theories, it is not because such problems are more numerous or because non–linear methods are appropriate for 'unusual' problems, but rather because our understanding of non–linear methods has lagged behind that of linear methods. It is no accident that the recent surge of interest in the dynamics of chaos has been matched by a widening field of applications.

One of the paradoxes of chaos is that it arises in problems that are modelled using traditional differential equations. But calculus is based on the concepts of smoothness and continuity, while chaos theory is based on the inherent lack of smoothness in most problems. Perrin, in 1906, said "Curves that have no tangent are the rule and regular curves such as the circle, are interesting but quite special" (Schroeder, 1991). The word curve conveys the impression of smoothness and continuity. Perhaps a wider definition should be: a curve is a collection of points obeying some rule of association.

In his Nobel speech, Albert Einstein, hinted at the problems to come with chaos theory when he said, "As far as the laws of mathematics refer to reality, they are not certain; and as far as they are certain, they do not refer to reality" (Schroeder, 1991).

2. EXAMPLES OF CHAOTIC BEHAVIOR AS A RESULT OF ITERATION

One of the methods often used for solving problems is iteration. In order to solve the equation $F(z) = 0$, it can be re-written in the form:

$$z_{n+1} = G(z_n)$$

For example:

$$y_{n+1} = \frac{y_n}{2} - \frac{1}{2y_n}$$

Starting the iteration with two very similar initial conditions $y_0 = 0.5$ and $y_0 = 0.500001$ produces the same values for the first twelve iterations. The divergence of the two evolutions from each other thereafter, the one giving 0.64 and the other 0.15 at the 20th iteration , illustrates one of the characteristics of Chaos: sensitivity to initial conditions.

Newton's Method, a very common method for solving equations of the form $F(z) = 0$, also involves the use of iteration:

$$z_{n+1} = z_n - \frac{F(z_n)}{F'(z_n)}$$

In applying Newton's method to one of the simplest non-linear problems:

$$z^2 - 1 = 0 \,, \qquad z = x + iy$$

it can be seen that two very similar initial conditions can develop chaos within 20 iterations. This can be visualised geometrically by noting that if the iteration begins with a number on the y-axis exactly equi-distant from the two roots, successive iterations jump chaotically up and down the y-axis, but cannot escape from it. Iterations beginning with any point to the right of the y-axis give convergence to the root $z = 1$, while iterations beginning with any point to the left give convergence to the root $z = -1$. Therefore two points which produce a different behavior of the system can be arbitrarily close together – a characteristic of chaotic systems. The z-plane can be thought of as consisting of two regions called Basins of Attraction separated by the y-axis. The boundary, ie the y-axis, is where chaos reigns.

Zare and Szebehely (1994) give illustrations of the regions of escape for the three body problem which are similar to the patterns of chaos resulting from Newton's method applied to the equation $z^3 - 1 = 0$. Here there are three roots and, in the neighbourhood of each root, there is convergence to that root. Each root has its own basin of attraction and one might expect that the plane is divided neatly into three basins of attraction by the lines of points equidistant from any two roots. However, there is chaos, not only along these lines but within the regions to either side of them. In these chaotic regions, the iteration may lead to any one of the three roots.

A similar behaviour is illustrated more dramatically by a magnetic pendulum with three magnets. It is not possible to predict the individual magnet to which the magnetic pendulum will eventually be drawn permanently. Those points from which the magnetic pendulum begins its motion are located in the basin of attraction of the individual magnet to which it is ultimately drawn. The three basins are found to be inextricably interwoven.

In all these examples of chaotic behavior, the systems can be precisely modelled mathematically, yet the mathematics cannot predict the behaviour precisely because of sensitivity to initial conditions. A more detailed discussion of examples of chaotic behavior can be found in Doyle (1992).

3. FRACTALS

In Mathematics the word fractal has the precise definition: the boundary of a basin of attraction. Such boundaries are generally difficult to measure. They are like a geographical coastline in that, the length of the coastline increases as the accuracy of the measuring stick increases. On a small scale map a coastline loses much of its detail. As the scale increases, the more refined detail leads to ever greater lengths. It has been found that by magnifying the scale, the length of a coastline can be increased indefinitely.

Length can be expressed as a number N of units of a fundamental length r. For example a rope which is 2 metres (ie N=2, r=1m) is also 200 cm (ie N=200, r=1cm). Thus the length L can be expressed by the formula: $L = Nr$. This can be ammended to give: $L = Nr^D$ where D is the Hausdorff dimension. The well known Koch curve has dimension equal to 1.2618.... Hilbert constructed a curve which fills a two-dimensional space, which not surprisingly has a Hausdorff dimension of 2. By using variations of the Cantor set one can construct curves of any dimension between 0 and 1, while variations of the Koch curve give dimensions between 1 and 2. The coastline of Britain, for example, has a Hausdorff dimension of 1.26.

In principle the notion of fractional dimensions could be extended to fractional surfaces 'embedded' in higher dimensional spaces. Therefore it is natural to inquire, just as a Mandelbrot set is defined on a two dimensional plane, whether there is a generalised Mandelbrot set for a suitably generalised complex plane? The answer is yes, and this brings with it the concept of quaternions invented by W. R. Hamilton (1843).

4. THE 4-D MANDELBROT SET

The construction of the famous Mandelbrot Set comes from identifying those points C in the complex plane which lead to convergence in the iterative formula:

$$z_{n+1} = z_n^2 + C$$

where z_n and C are complex numbers. The logical extension of the above iteration is not to three dimensions as one might expect, but rather to four. The 4-D Mandelbrot Set and the stability of its multiple cycles is discussed in Gomatam et al (1994).

It is interesting to retrace the historical steps imprinted in the development of quaternions. The four dimensional space was first proposed by Hamilton in 1843 (Ashurst, 1982) using quaternions, an ordered quadruple of numbers expressible as $Q = a + bi + cj + dk$. If the complex plane is produced by the real line and another line orthogonal to it, generated by multiplication by the square root of -1 (i), so a four dimensional space can be represented by the complex plane and another complex plane orthogonal to it, generated by multiplication by the square root of -1 (j). Thus a quaternion is:

$$x + yi + (u + vi)j = x + yi + uj + vk$$

where k replaces ij and $i^2 = j^2 = k^2 = ijk = -1$.

Hamilton had difficulty accepting this as it gives rise to a non-commutative algebra. He began with the idea that all real numbers can be represented geometrically by a line. Negative numbers could be included by adding the concept of direction and magnitude. Hence a vector $\underline{OP}$ could become an anti–parallel vector through multiplication by -1. A second multiplication by -1 would rotate the vector back to its original position. Hence $--\underline{x} = \underline{x}$. Likewise, he saw that complex numbers could be represented geometrically by a plane where a second line containing the imaginary numbers $i = \sqrt{-1}$ was placed orthogonal to the real number line and multiplication by i could be thought of as a rotation of the vector $\underline{OP}$ through 90 degrees. Hence $i^4\ \underline{x} = \underline{x}$.

The question he then posed was: What kind of 'number' represented a three-dimensional space? For ten years he struggled with the possibility of an ordered triplet, (a,b,c) = a+bi+cj where j represented a third orthogonal line forming a 3 dimensional space and $i^2 = j^2 = -1$.

What he didn't realise was that rotation and contraction of a 3 dimensional vector requires an operator that has four parameters, not three. Two parameters are needed to specify the orientation of the plane of rotation (ie the angular distance Ω to the line of nodes $\underline{NN'}$ from a reference direction and the inclination i of the plane). Within the plane, a third parameter is needed to specify the rotation θ of the vector from the line of nodes and a fourth to specify the amount of magnification or contraction r of the vector (See Figure 1).

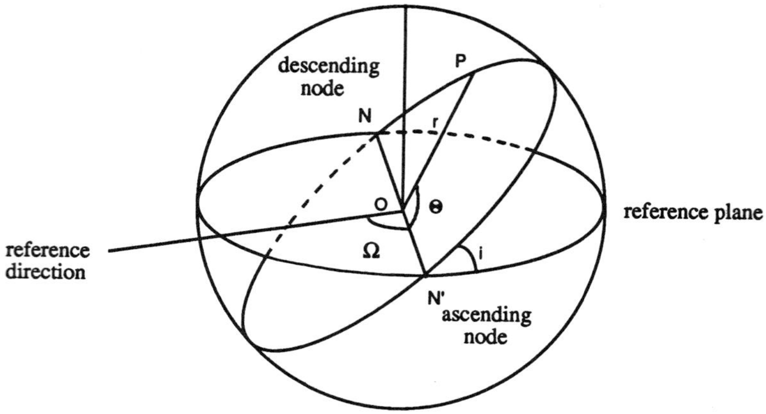

Figure 1. Four parameters are needed to describe the rotation and contraction of a 3-dimensional vector $\underline{OP}$: Ω, i, θ and r.

His difficulty with ordered triplets arose when applying the fundamental laws of algebra. Multiplication of a 3 dimensional vector by another 3 dimensional vector leads to an unknown quantity involving ij and ji.

$$(a+bi+cj)(a'+b'i+c'j) = aa' - bb' - cc' + (ba'+ab')i + (ca'+ac')j + bc'ij + cb'ji$$

Hence multiplication by an ordered triplet does not transform one 3 dimensional vector into another. In a flash of inspiration, he saw that a quadruplet of numbers would be sufficient, as long as he could abandon the commutative law of multiplication and hence stipulate that $ij = -ji = k$, where $i^2 = j^2 = k^2 = ijk = -1$. The quadruplet or quaternion could then be written as $Q = a + bi + cj + dk$ and a vector $\underline{x} = xi + yj + zk$ multiplied by it would be transformed into another vector $\underline{u} = ui + wj + zk$.

Today Hamilton's genius is not that he discovered quaternions but that he was the first person to realise that one could describe a logical, useful algebra that did not require all the fundamental laws of elementary algebra to hold. Hamilton spent the last 20 years of his life studying quaternions. For a time they were fashionable and Hamilton was feted as one of the great scientists of his time. But even during Hamilton's lifetime a fierce debate was waged between the proponents of quaternions and the proponents of James Clerk Maxwell's new algebra of vectors. Eventually the algebra of vectors began to dominate mathematical analysis and quaternions became a mathematical dinosaur.

Interest was renewed in quaternions with the study of quantum mechanics. In particular, the rotation of a vector (Goldstein, 1980) was expressed by Pauli (1927) in his quantum theory , using Pauli Spin matrices:

$$\sigma_1 = \begin{bmatrix} 0 & 1 \\ 1 & 0 \end{bmatrix} \qquad \sigma_2 = \begin{bmatrix} 0 & -i \\ i & 0 \end{bmatrix} \qquad \sigma_3 = \begin{bmatrix} 1 & 0 \\ 0 & -1 \end{bmatrix}$$

Each spin matrix is associated with rotation about one particular axis x, y or z through a particular Euler angle. If the spin matrices are complemented by the identity matrix I, one obtains a matrix quaternion which gives the transformation matrix for rotation through the three Euler angles.

$$Q = Ix + yi\sigma_1 + zi\sigma_2 + wi\sigma_3$$

The matrix quaternion satisfies all the properties of the original quaternion first described by Hamilton, and is the representation used to extend the Mandelbrot set to four dimensions in Gomatam et al (1994) starting with the map $Q \rightarrow Q^2 + C$.

Hamilton spent the rest of his life studying quaternions, convinced that he had discovered something imbued with cosmic significance. He became more and more of a recluse, concentrating on his work. In 1865 at the age of 60, he died from a severe attack of gout brought on by his longstanding overindulgence in alcohol. But he left us with a doorway into a whole realm of algebras which progressively relinquish standard laws only to gain greater generality and structure.

ACKNOWLEDGEMENTS

We (J. Gomatam and J. F. Doyle) would like to thank the Department of Mathematics, Glasgow Caledonian University and the Carnegie Foundation, respectively, for travel grants.

REFERENCES

Ashurst, G., 1982, "Founders of Modern Mathematics", Frederick Muller Ltd, London, 15.

Doyle, J.F., 1992, An introduction to fractals and chaos, *Teaching Mathematics and its Applications*, Vol.11, No.4, 166.

Goldstein, H., 1980, "Classical Mechanics", II Edition, Addison-Wesley, Reading, Massachusetts, 155.

Gomatam, J., Doyle, J.F., and Steves, B., 1994, Quaternionic generalisation of the Mandelbrot set, *in*: "From Newton to Chaos: Modern Techniques for Understanding and Coping with Chaos in N-Body Dynamical Systems", A.E. Roy and B. Steves, eds., Plenum, London.

Lukacs, Gyorgy, 1972, "Studies in European Realism", Merlin Press, London.

Schroeder, M., 1991, "Fractals, Chaos, Power Laws: Minutes from an Infinite Paradise", W.H. Freeman, New York, 8 and 139.

Zare, K. and Szebehely, V., 1994, Order out of chaos in the three body problem: regions of escape, *in*: "From Newton to Chaos: Modern Techniques for Understanding and Coping with Chaos in N-Body Dynamical Systems," A.E. Roy and B. Steves, eds., Plenum, London.

PART FIVE

SELECTED TOPICS IN DYNAMICS

ORDER IN CHAOS

G. Contopoulos

Department of Astronomy
University of Athens
Panepistimiopolis
GR 157 83-Athens
Greece

and

Department of Astronomy
University of Florida
Gainesville, Fl. 32611
U.S.A.

I. INTRODUCTION

Deterministic chaos is different from randomness. In the case of chaos we can make <u>some</u> predictions about the behaviour of individual orbits, while in the case of randomness only an average behaviour of the system can be estimated.

We consider here some examples of order in chaotic regions in various systems. In sections II and III we deal with escapes, and their statistics. In section IV we describe the structure of the homoclinic tangle. In section V we describe systems without escapes.

Finally in section IV we deal with a genuinely 3-D problem, the Mixmaster model of the Universe. This model was assumed to be completely chaotic. However we will provide evidence that it is an integrable system, therefore it does not have any chaos at all.

II. ESCAPES

A dynamical system in which escapes play a most important role is the problem of two fixed black holes in General Relativity (Contopoulos 1990a, 1991). The orbits of photons (particles of zero rest mass) around two black holes are almost all escaping, either to one of the black holes, or to infinity. The non escaping orbits form a set of measure zero.

The line element in this system is the so-called Majumdar-Papapetrou solution of the Einstein field equations (Chandrasekhar 1989) namely

$$ds^2 = \frac{dt^2}{U^2} - U^2(dx^2+dy^2+dz^2), \tag{1}$$

where

$$U = 1 + \frac{M_1}{r_1} + \frac{M_2}{r_2}\ , \tag{2}$$

is the classical potential with r_1, r_2 the distances of a particle from the black holes M_1 and M_2, which are located at the points $z=\pm1$. The only non-escaping orbits are the periodic orbits, which are all unstable. All other orbits escape either to the black holes M_1 and M_2 (types of orbits I and II), or to infinity (type of orbits III).

The periodic orbits separate the nonperiodic orbits of types I, II, and III. Orbits of all three types are shown in Fig. 1, where a beam of photons is coming from infinity and is scattered by the black holes M_1 and M_2.

Simple topological considerations show that between an orbit of type I and an orbit of type II there are orbits of type III. Similarly between an orbit of type I and an orbit of type III there are orbits of type II, and so on.

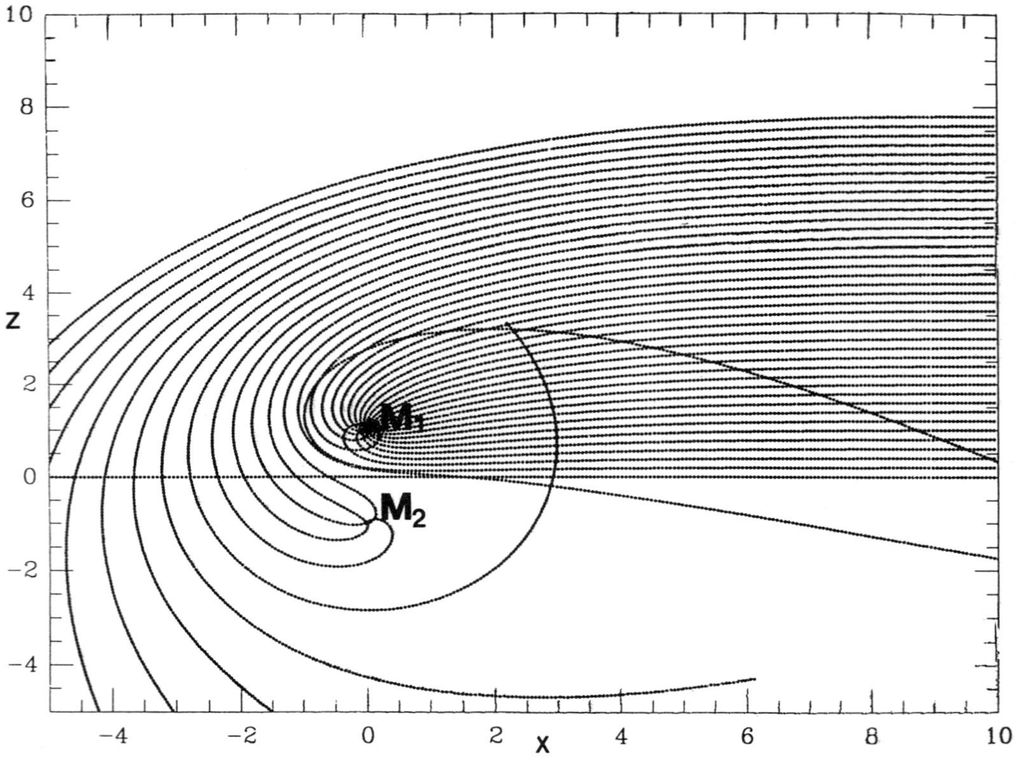

Figure 1. A set of photons coming from infinity along directions parallel to the axis-x are scattered by the black holes M_1 and M_2 located at the points (x=0, z=1) and (x=0, z=−1). Only the orbits with initial z>0 are shown.

This property ensures that the sets of orbits of types I, II, and III are Cantor sets of non zero measure (their limits are Cantor sets of zero measure).

This case is a typical example of chaotic scattering. Similar problems appear in dynamical systems with more than one escape windows, e.g. the dissociation of complex molecules (Noid et al. 1986), or the interaction of asteroids (Petit and Henon 1986).

The problem of photons in the case of two fixed black holes is relatively easy, because the set of non-escaping orbits has measure zero. In more general cases there are also orbits that do not escape. Such is the case of test particles with non-zero rest mass moving around the two fixed black holes. In such a case there are stable periodic orbits around one or both black holes. Orbits close to these periodic orbits remain trapped for ever and do not escape (Contopoulos 1991).

We will consider now the cases of some simple Hamiltonians that contain both escaping and nonescaping orbits

$$H \equiv \frac{1}{2}(\dot{x}^2+\dot{y}^2+Ax^2+By^2) - \varepsilon x^2 y = h \quad , \tag{3}$$

$$H \equiv \frac{1}{2}(\dot{x}^2+\dot{y}^2+x^2+y^2) - \varepsilon x^2 y^2 = h \quad , \tag{4}$$

(Contopoulos 1960b, Contopoulos and Kaufmann 1992, Contopoulos and Polymilis 1993) with $\varepsilon>0$.

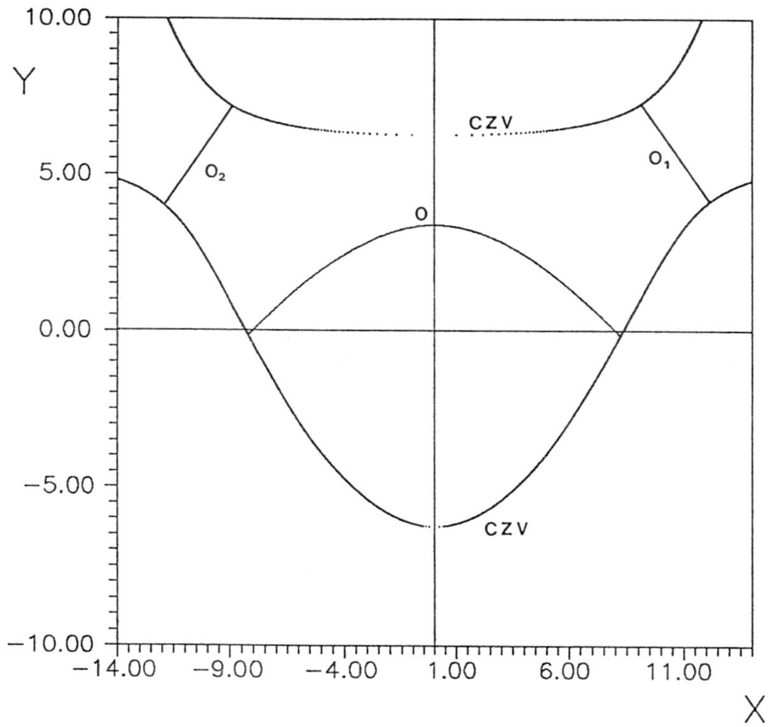

Figure 2. The curves of zero velocity (CZV) in the Hamiltonian (3) for A=0.9, B=1.6, ε=0.08 and h=31.3 (above the escape energy h_{esc}=25.31). The Lyapunov orbits O_1 and O_2 cross the openings of the CZV. The orbit O is a simple unstable periodic orbit crossing the x-axis perpendicularly (y=3.3698, $\dot{y}$=0).

If the energy h exceeds the escape energy h_{esc} the curves of zero velocity (CZV) are open (Figs 2 and 3) and there are orbits that escape to infinity. At each escape channel of the CZV there is an unstable periodic orbit across the opening of the channel, called a Lyapunov orbit. It is known (Rod 1973, Churchill et al. 1975) that orbits crossing a Lyapunov orbit outwards escape from the system and rever return close to the origin. Therefore the limits of the sets of escaping orbits are asymptotic orbits to the Lyapunov orbits. In fact, small deviations from these asymptotic orbits outwards lead to escapes.

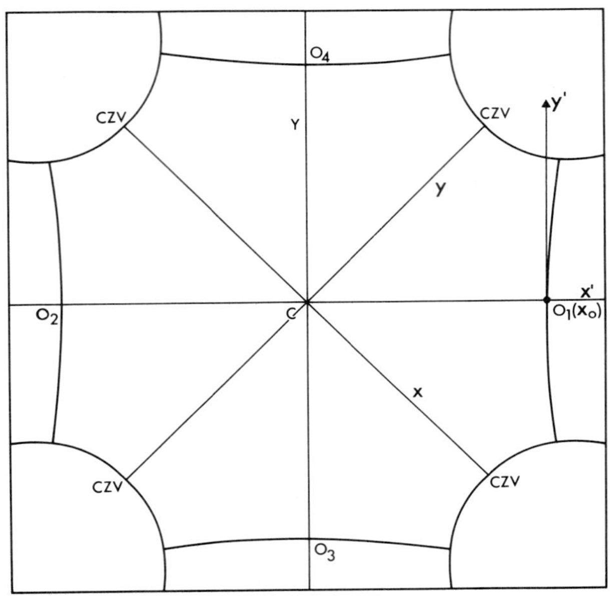

Figure 3. The same as in Fig. 2 for the Hamiltonian (4) (ε=5.26, h=0.12). In this case there are 4 Lyapunov orbits O_1, O_2, O_3 and O_4. The axes x',y' start at the Lyapunov orbit O_1 and form an angle 45° with the original axes (x,y). The Lyapunov orbit O_1 starts at a distance x_o=0.467760... from the center C.

In order to define the escaping sets we must find the asymptotic manifolds of the Lyapunov orbits in phase space. Such a manifold consists of orbits which are asymptotic to a particular Lyapunov orbit. In a problem of two degrees of freedom and a fixed value of the energy the phase space is 3-dimensional and the asymptotic manifold of each Lyapunov orbit is a 2-dimensional surface. This surface close to the Lyapunov orbit is rather simple, like a cylinder (Fig. 4). The intersection of this surface by a transverse Poincare surface of section is a closed curve called a "limiting asymptotic curve".

Any orbit starting inside this limiting asymptotic curve escapes to infinity without intersecting again the Poincare surface of section.

If we extend further the asymptotic manifold we see that it has an extremely complicated structure (Contopoulos and Kaufmann 1992).

In the case of the Hamiltonian (3) we have studied the escapes for A=0.9, B=1.6, ε=0.08 and various values of h. We have considered, in particular, orbits starting asymptotically from the neighbourhood of an orbit O perpendicular to the y-axis, as the one shown in Fig. 2.

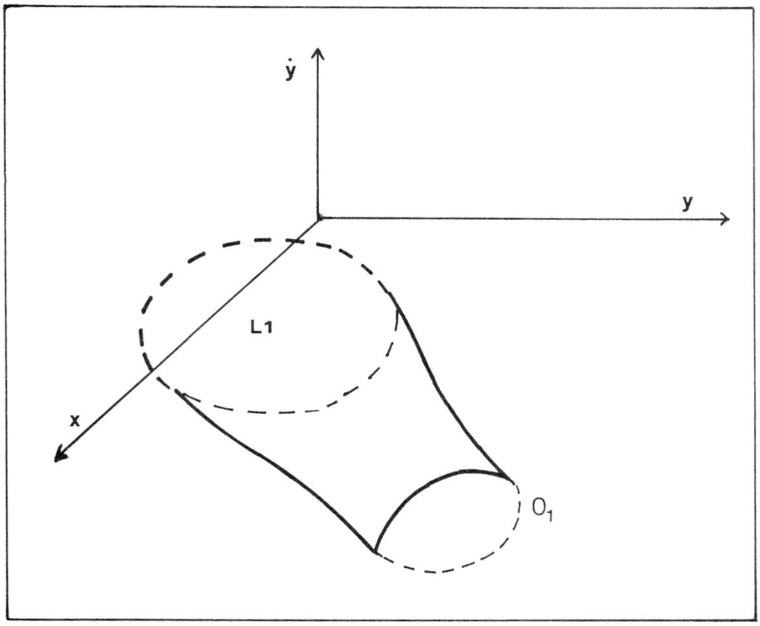

Figure 4. The stable asymptotic manifold of the Lyapunov orbit O_1 of Fig. 2 in the phase space $(x,y,\dot{y})$ (schematically). This roughly cylindrical surface intersects the Poincare surface of section $(y,\dot{y})$ along a "limiting asymptotic curve".

In Fig. 5 we have marked on a Poincare surface of section two asymptotic curves from the unstable periodic orbit O for a value of the energy h=31.30, above the escape energy, which is equal to h_{esc}=25.31. One of these asymptotic curves is unstable (U) and the Poincare consequents of its points recede away from O. The other asymptotic curve (S) is stable and the Poincare consequents of its points approach O. The asymptotic curve U is generated numerically by finding the successive Poincare consequents of a large number of points on a small arc very close to the point O, along the unstable eigenvector of the periodic orbit O.

There are two more asymptotic curves UU (unstable) and SS (stable), symmetric to S and U, respectively, whose beginnings are marked in Fig. 5.

The asymptotic curve U makes infinite spiral rotations around the "limiting asymptotic curve" L1 and tends to L1 asymptotically. Any orbit with initial conditions $x=0$ and $(y,\dot{y})$ inside this limiting asymptotic curves escapes to infinity as $t \longrightarrow +\infty$ from the opening at O_1 $(x>0,\ y>0)$, of Fig. 2 without intersecting again the Poincare surface of section.

On the Poincare surface of section the arc e_0e_1 (Fig. 6) along the unstable asymptotic curve U is mapped on the whole curve that starts at e_1 and makes infinite rotations around L1. The length of this curve is infinite. A point a little beyond e_1 has no further Poincare consequents, because the orbit starting at this point escapes to infinity through the Lyapunov orbit O_1.

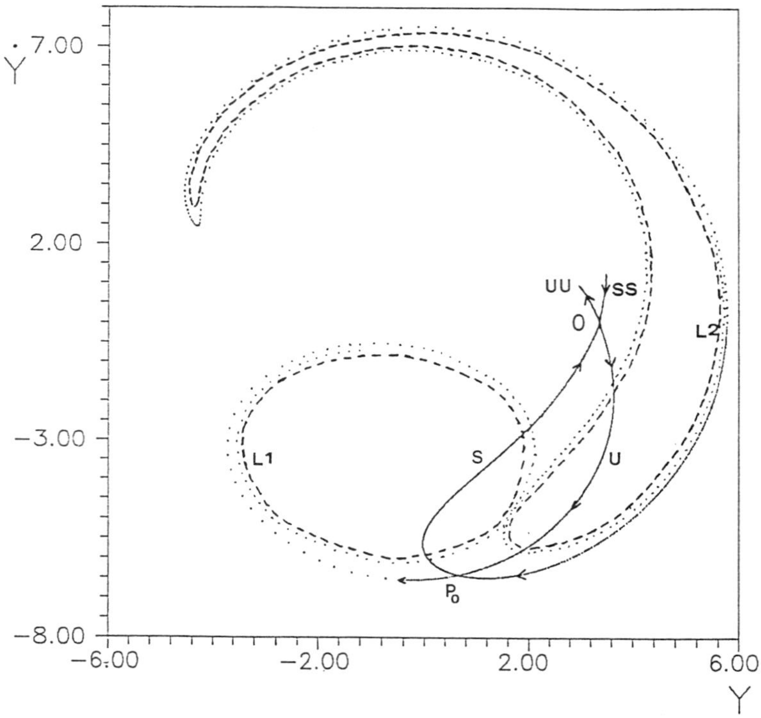

Figure 5. The asymptotic curves U and S in the case $h=31.3$ make infinite rotations, approaching asymptotically the limiting asymptotic curves L1 and L2. P_0 is a homoclinic point.

Similarly the asymptotic curve S tends to another "limiting asymptotic curve" L2 after infinite rotations (Fig. 5). The curve S is found by calculating orbits for $t<0$. Orbits inside L2 escape to infinity as $t \longrightarrow -\infty$ from the opening O_2 at $(x<0,\ y>0)$, also without intersecting the Poincare surface of section.

There are two more limiting asymptotic curves L3 and L4, symmetric to L1 and L2 with respect to the axis $\dot{y}=0$. These curves are the limits of the curves SS and UU and lead to escapes for $t \longrightarrow -\infty$ through O_1 $(x>0,\ y>0)$ and $t \longrightarrow +\infty$ through O_2 $(x<0,\ y>0)$ respectively.

There are more orbits escaping to infinity. In fact the points on the asymptotic curve a little beyond e_2 (Fig. 6) have Poincare consequents that form an arc starting after infinite rotations around the limiting asymptotic curve L1 outwards in a countercclockwise direction. This arc has further intersections (homoclinic points) with the curve S. After the intersections P_1',P_1, P_2',P_2 it forms the arc U_3', which makes infinite spiral rotations around the curve L4, which is symmetric to the curve L2 with respect to the $\dot{y}=0$ axis.

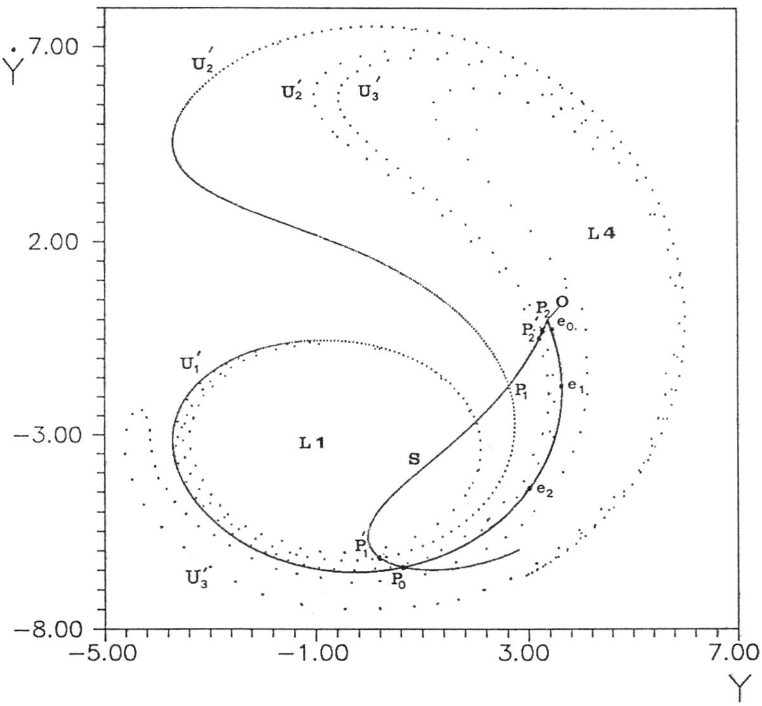

Figure 6. The asymptotic curve U has further arcs beyond the one reaching asymptotically the curve L1 of Fig. 5 (for h=31.3). A second arc starts asymptotically from L1 and, after some oscillations, reaches asymptotically the curve L4.

In the same way there are infinite arcs of the curve U that lead to escapes after infinite spirals around L1 or L4. This phenomenon was found already in the Hamiltonian (4) (Contopoulos 1990b).

Also the asymptotic curves of other unstable periodic orbits that lead to escapes spiral asymptotically to the same limiting asymptotic curves (Contopoulos 1990b). In fact we may state the general conclusion that the asymptotic curves from any unstable periodic orbit that lead to escapes must approach in a spiral way, after infinite rotations, the limiting asymptotic curves L1,L4 for orbits escaping for $t \longrightarrow \infty$, or the curves L2, L3 if they escape for $t \longrightarrow -\infty$. The reason is that the curves L1,L2,L3, L4 are the only sets of initial conditions that lead to escapes through one or the other Lyapunov orbit, as $t \longrightarrow \infty$, or $t \longrightarrow -\infty$, without any further intersections with the Poincare surface of section.

If the energy h is smaller than h=31.30, but above the escape energy h_{esc}=25.31, the asymptotic curves U and S may not make infinite spiral rotations around L1, or L2. This is the case for h=27.80 (Fig. 7).

In this case the asymptotic curve U does not reach L1 asymptotically, but makes only one rotation clockwise alound the limiting asymptotic curve L1 and then returns in a counterclockwise direction. The curve U intersects again the stable asymptotic curve S. The arc between the intersections P_0 and P_1' is called U_1'. This arc, together with the arc P_0P_1' along the curve S, form the lobe U_1'. As the curve U continues, it forms the inner arc U_1 (inside the resonance region defined by the arcs OP_0 along U and P_0O along S), the outer arc U_2', the inner arc U_2, and then continues along the arc U_3'. This arc terminates after infinite rotations around the limiting asymptotic curve L4.

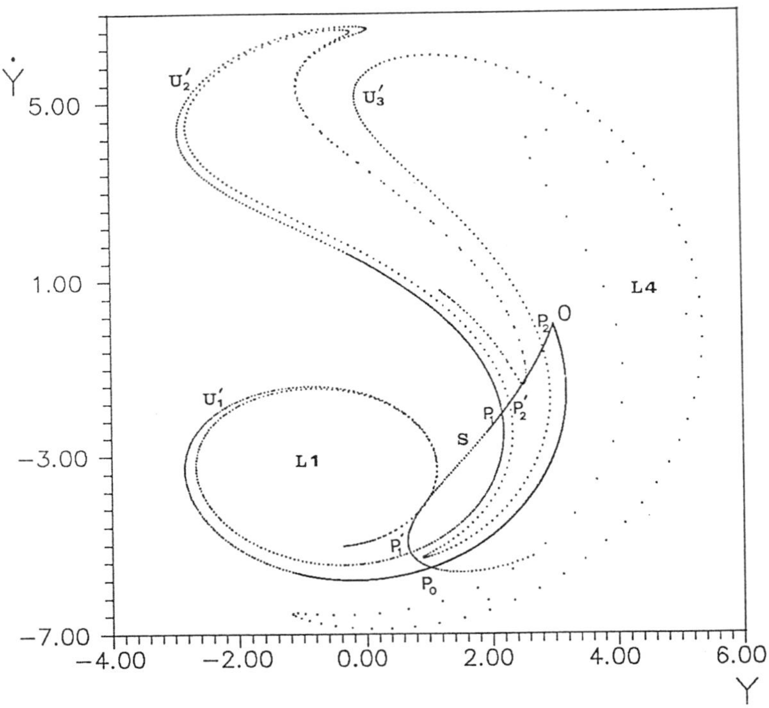

Figure 7. The asymptotic curve U for the Hamiltonian (3) with h=27.80. This curve forms an outer lobe U_1' that makes one rotation around L1 but does not tend asymptotically to L1 as in Fig. 5, then an inner lobe U_1, an outer lobe U_2', an inner lobe U_2, and then it spirals around the limiting asymptotic curve L4, tending to it after infinite rotations.

More generally if some orbits, starting asymptotically from the unstable periodic orbit O, escape to infinity, they do so while the asymptotic curve U from O reaches one of the limiting asymptotic curves, L1 or L4, after infinite rotations around it. Before finishing in this way the asymptotic curve U makes an arbitrary number of lobes, approaching the curves L1 and L4 but not very close.

In the same way we may have escapes from the curve UU as $t \longrightarrow \infty$, and from the curves S or SS as $t \longrightarrow -\infty$.

We stress here the word <u>may</u>. Because it is possible that for energies slightly above the escape energy h=25.31 the asymptotic curves of the orbit O do not lead to escapes, but remain bounded for all times. What we discussed above is the mechanism of escape, <u>if</u> we have orbits asymptotic to O escaping to infinity.

III. STATISTICS OF ESCAPES

Another way of approaching the problem of escapes is by finding the proportion of particles that escape within a certain time interval. We studied this problem in the case of the Hamiltonian (4) with h=1 and ε as a parameter (Contopoulos et al 1993).

In this case the escape perturbation is $\varepsilon_1=2.08$. For $\varepsilon>\varepsilon_1$ the curves of zero velocity are open and some orbits escape to infinity. But there may be orbits that never escape, especially if ε is not much larger than ε_1.

In the present study we consider all initial conditions, and not only those along the asymptotic curves. In total we calculated several hundreds of thousands of orbits, in order to find the statistics of the escapes.

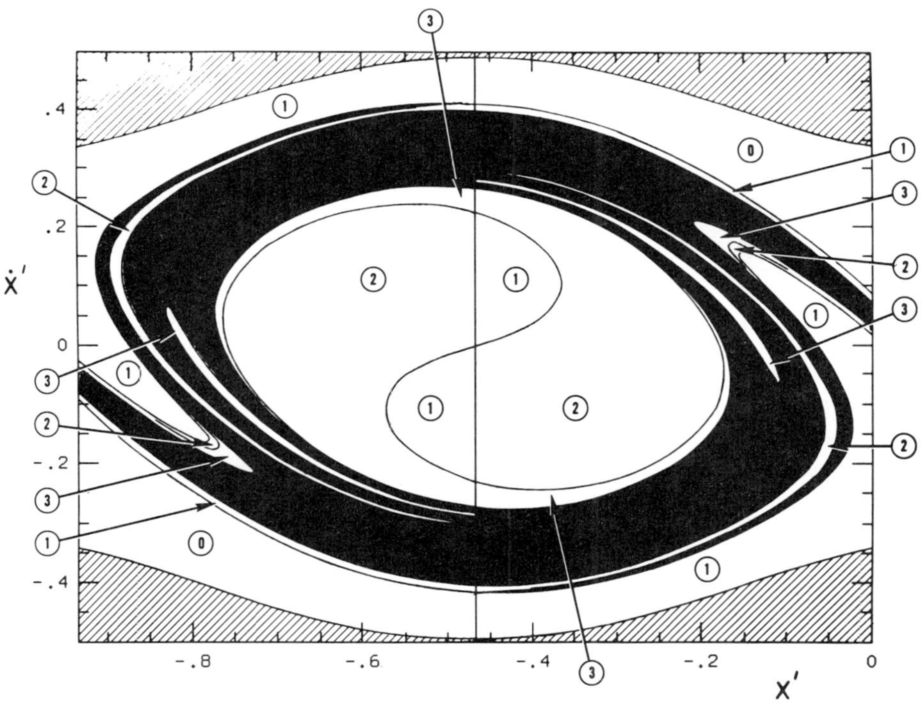

Figure 8. Regions of escape on a Poincare surface of sec tion $(x',\dot{x}')$ for $\varepsilon=3.9$. The number in circles represent the intersections of the central axis $(x'=-x_o)$, drawn as a vertical line in the figure, before escape. The black region represents nonescaping orbits. The shaded regions close to the top and the bottom are forbidden (no orbits are possible there).

In Fig. 8 we mark the orbits that escape after 0,1,2 or 3 intersections with the central axis (the origin of the system (x',y') is moved by x_o from the center, where x_o is the position of the Lyapunov orbit; in the system $(x',\dot{x}')$ the axis $x'=-x_o$, passes through the center) for $\varepsilon=3.9$. The escape regions of higher order are extremely small. The dark regions represent nonescaping orbits. In these regions there are stable periodic orbits that trap orbits around them for all times.

The escape regions for $\varepsilon=5.26$ are shown in Fig. 9. In this case we did not find any stable periodic orbits of low multiplicity, but we expect that some small regions of nonescaping orbits do exist.

We notice that the orbits either escape fast, as a bulk (large regions of escape after 0,1,2 iterations in Figs. 8 and 9), or the escape regions are very elongated spiral threads, which are thinner as the order of escape increases.

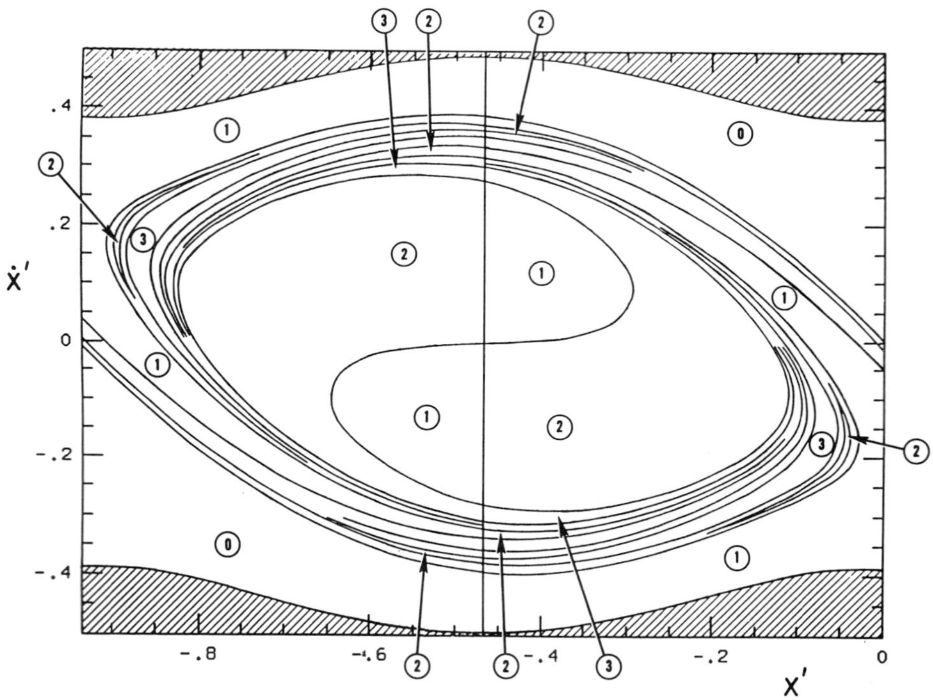

Figure 9. The same as in Fig. 7 for $\varepsilon=5.26$.

We found the escape probability in the regions of complicated structure by calculating large numbers of trajectories starting in various cells inside these regions. The escape probability is given by the formula

$$p(t) = \frac{N_{esc}}{N_{tot}}, \tag{5}$$

where N_{tot} is the number of trajectories from one cell that have $(t-1)$ intersections with the central line, and N_{esc} is the number of escapers through some Lyapunov orbit between times (t-1) and t where t is an integer "time". Thus the N_{esc} particles do not have a tth intersection with the Poincare surface of section.

In order to have good statistics we calculated some tens of thousands of orbits in each cell of size 0.05x0.05 in the units of Fig. 8. In Fig. 10 we see an example of the variation of the escape probability p(t) as a function of t for 22 different cells. We notice that the numbers p(t) for t≥5 tend to the same limiting probability $p_\infty \approx 0.51 \pm 0.02$.

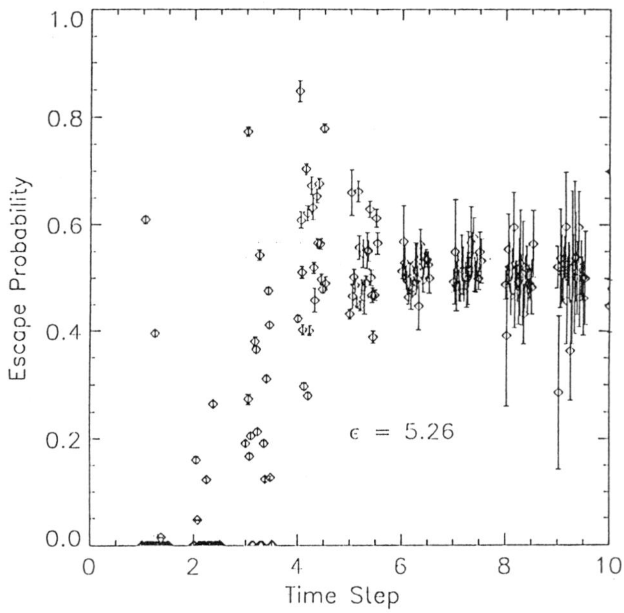

Figure 10. The escape probability as a function of the time t for 22 cells, each containing some tens of thousants of trajectories. The data for different cells are slightly displaced in t. The uncertainties are proportional to $N_{esc}^{1/2}$. The values of p(t) for large t converge to a value p_∞, independent of t.

If we consider that p_∞ is independent of t then the number of trajectories remaining after a long time t is given by an exponential formula

$$N=N_o e^{-\lambda t} \tag{6}$$

where

$$\lambda= -\ln(1-p_\infty) . \tag{7}$$

We have repeated the same experiments for several values of ε and found the following rather astonishing result. The limiting values of the escape probability, considered as a function of the perturbation ε, follow a power law

$$p_\infty \propto (\varepsilon-\varepsilon_2)^\alpha \ , \tag{8}$$

where $\varepsilon_2 \approx 4.90 \pm 0.01$ and $\alpha=0.49\pm0.05$. The critical perturbation, ε_2, is almost 2.5 times larger than the escape perturbation $\varepsilon_1=2.08$.

For ε between ε_1 and ε_2 the values of p(t) seem to tend to zero, and no power law like (8) seems to exist.

The value of p_∞ is independent of the size of the cells. We have checked that by reducing the cell size from 0.05x0.05 to 0.005x0.005 and to 0.0005x0.0005. Furthermore the structure of the regions with different escape times in cells of smaller and smaller size, are qualitatively the same, indicating a fractal structure (Contopoulos et al 1993).

The time T needed to reach approximately the asymptotic value p_∞ is shorter as ε increases. Using two different definitions for this time T we found essentially the same numerical results. The value of T as a function of ε is given by a scaling law

$$T(\varepsilon) \propto (\varepsilon-\varepsilon_2)^{-\beta} \ , \tag{9}$$

where ε_2 is the same as in Eq. (8), and $\beta=0.39^{+0.14}_{-0.06}$.

Finally we found that the time T for a fixed value of ε depends on the size of the cell r.

If r is small the time needed for p(t) to reach the limiting value p_∞ is longer. The scaling law is

$$T \propto r^{-\delta} \ , \tag{10}$$

where $\delta=0.08\pm0.02$.

These scaling laws that occur for ε above the critical value ε_2 indicate that at $\varepsilon=\varepsilon_2$ a phase transition occurs in our system.

A phase transition in a dynamical system that represents the central parts of a deformed galaxy is quite remarkable. Probably this is the first case where such a phenomenon appears in a gravitational system of astronomical importance.

This result brings about the question what is the physical mechanism that producces this phase transition. The most obvious qualitative change that occurs near $\varepsilon=\varepsilon_2$ is the (approximate) disappearance of KAM islands for ε larger than this value.

If for $\varepsilon<\varepsilon_2$ we have islands of stability in every cell examined, then the number of remaining trajectories at time t, N_{tot}, divided by the original number of trajectories, N_o, tends to the ratio of the size of the islands to the total size of the cell. Thus N_{tot}/N_o is finite, while $N_{esc} \longrightarrow 0$, therefore the probability p(t) tends to zero, as $t \longrightarrow \infty$. This tendency was observed in our numerical calculations.

However this is not the whole story. In fact we found some stable periodic orbits for $\varepsilon>\varepsilon_2$, which are surrounded by small islands of stability. In such a case for any cell containing an island of stability the value of p(t) should tend to zero after a very long time t. We believe that for $\varepsilon>\varepsilon_2$ the size of the islands is extremely small and the value of p(t) should be quasi-stationary for a long time, before going finally to zero.

It is obvious that this subject needs a more detailed quantitative study. However it is remarkable that phenomena like phase transitions can occur in very simple dynamical systems, like (3) and (4), that are of interest in an astronomical context.

IV. THE STRUCTURE OF THE HOMOCLINIC TANGLE

The orbit O (Fig. 2) is stable if $h<h_c=22.17$, and unstable for $h>h_c$. In the latter case, the asymptotic curves on a Poincare surface of section form elongated lobes that intersect in an intricate way, generating what is called the homoclinic tangle (Guckenheimer and Holmes 1983; Wiggins 1990).

If the energy h is smaller than the escape energy h_{esc} there are no "limiting asymptotic curves", therefore the arcs of the asymptotic curves between successive homoclinic points are finite. But the lengths of successive lobes increase by a factor of order λ, where λ is the larger eigenvalue of the unstable orbit O (Contopoulos and Polymilis 1993).

The structure of the lobes and their intersections follow certain rules that put an order in the seemingly random structure of the homoclinic tangle.

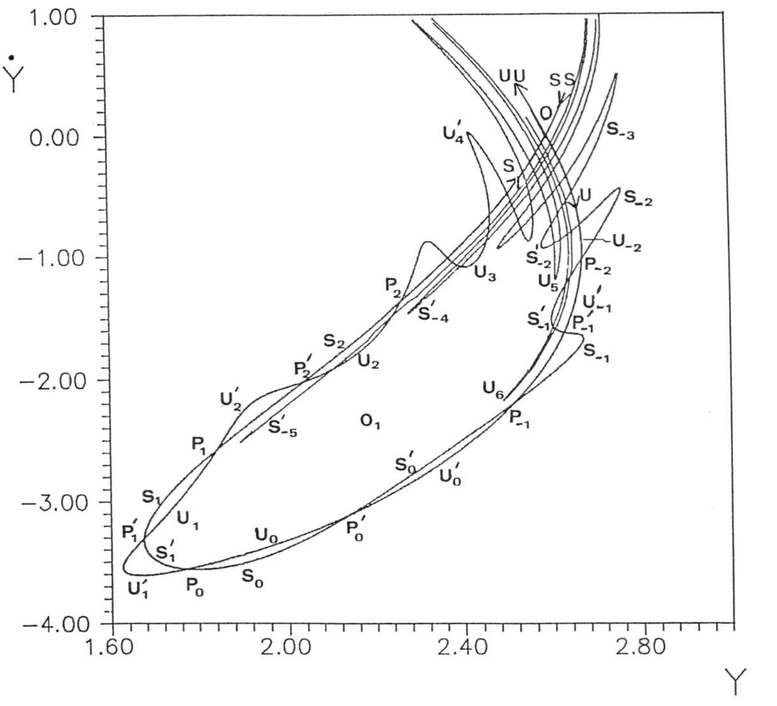

Figure 11. The asymptotic curves U (unstable) and S (stable) from the unstable invariant point (periodic orbit) O. These curves intersect at infinite homoclinic points, forming infinite lobes of equal area. The arcs OP_o along U and P_oO along S define the resonant region O_1. Inside this region there is a stable periodic orbit O_1. There are two more asymptotic curves, UU (unstable), and SS(stable), symmetric to S and U, with respect to the axis $\dot{y}=0$.

The forms of the low order lobes can be seen in Fig. 11 for energy h=24. The unstable and stable asymptotic curves U and S, intersect along infinite homoclinic points. The point P_o is roughly at equal distances from O along U and S, and the arcs OP_o along U and P_oO along S define the resonant region O_1, which contains a stable periodic orbit O_1, surrounded by closed invariant curves. The orbit O_1, and the symmetric orbit O_2, above O, were produced by bifurcation from the orbit O, when this became unstable as h increased above h_c. At the point P_o, the curve U intersects S outwards from the resonance region O_1. The curve U intersects S outwards also at the points $....P_{-2},P_{-1},P_1,P_2...$ and inwards at the points $...P_{-1}',P_o',P_1',P_2'...$.

The lobes are defined as follows. A lobe is formed by the arcs U_n' and S_n' joining the homoclinic points P_{n-1} and P_n' and is called U_n' or S_n'. Similarly a lobe U_n is formed by the arcs U_n and S_n joining the homoclinic points P_n' and P_n. The lobes U_n' and U_n with n>0 are called outer and inner lobes respectively. Also the lobes S_n and S_n' with $n\leq 0$ are outer and inner lobes respectively (Fig. 11).

The areas of the various lobes are equal because of the conservation of areas in the Poincare map, and the symmetry of the problem. The curve U does not intersect itself, or UU. Similarly the curve S does not intersect S or SS. But the various lobes intersect each other and the areas of intersection are conserved. E.g. as the lobes U_5 and S_{-2}' intersect (Fig. 11) the lobes U_6 and S_{-1}' also intersect and in general the lobes U_n and $S_{n-7}'(=U_{n-7}')$ and the areas of intersection are all equal. We have a similar behaviour as regards the intersections of the lobes U with SS, or the intersections of UU with S, or SS.

We can then write the following rules:

Rule I. If two lobes U_m and $S_n'(=U_n')$, or $S_n(=U_n)$, intersect, then the lobes U_{m+q} and $S_{n+q}'(=U_{n+q}')$, or S_{n+q} $(=U_{n+q})$ respectively, also intersect, and the areas of intersection are equal.

Rule II. If two lobes U_m' and $SS_n(=UU_n)$, or $SS_n'(=UU_n')$, intersect, then the lobes U_{m+q}' and $SS_{n-q}(=UU_{n-q})$, or SS_{n-q}' $(=UU_{n-q}')$ respectively, also intersect, and the areas of intersection are equal.

These rules allow the prediction of the properties of the higher order lobes from the properties of the lower order lobes.

As an example we give in Figs. 12 and 13 the outer lobes U_{10}' and U_{11}', together with a background of the curves U,S, UU and SS.

In the same way we can predict the properties of the higher order lobes.

If we superimpose the lobes U', up to U_{26}' we have the form of Fig. 14. In this figure we notice that the higher order lobes extend to large distances from the unstable point O (notice the difference in scale of this figure and Figs 12 and 13).

However the various lobes make many clockwise and counterclockwise rotations around the resonant regions O_1 and O_2 and they produce a large concentration of points around these regions (black areas in Fig. 14). Thus if we have many points in an original lobe, say U_1', its images (**a**) have great difficulty to leave the close neighbourhood of the resonant regions O_1 and O_2, and (**b**) even the higher order lobes, like U_{25}' and U_{26}' that go far from O_1 and O_2 have their greatest part (a large proportion of their points) close to O_1 and O_2. This phenomenon shows the "stickiness" of the resonant regions O_1 and O_2. The regions O_1 and O_2 are filled with closed invariant curves (KAM curves). If we try to find the outermost KAM curve around O_1 and O_2, by starting orbits a little outside, we find that the Poincare consequents stay for a long time close to O_1 and O_2 before going further away. This phenomenon was observed numerically by Contopoulos (1971) and was emphasized by Shirts and Reinhardt (1982), Menyuk (1985), and others.

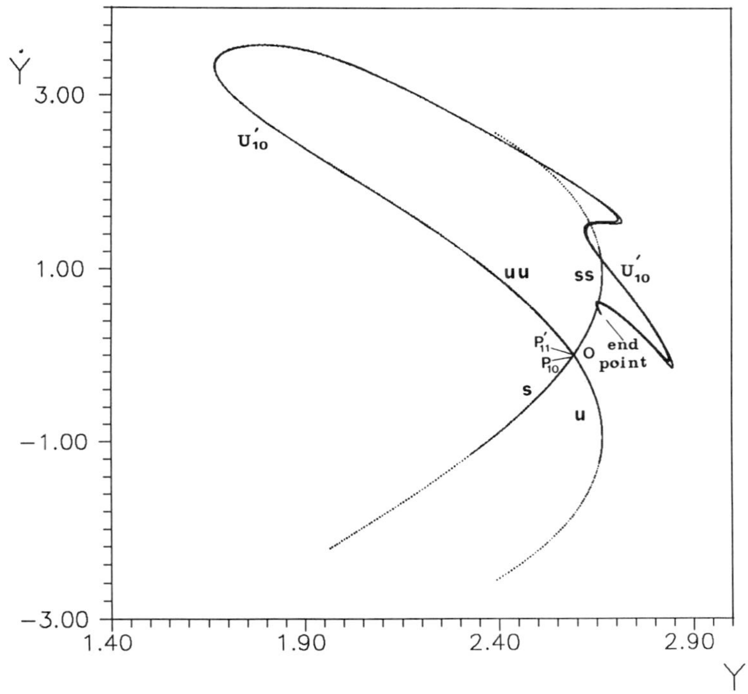

Figure 12. The arc U'_{10} of the asymptotic curve U starts at the point P_{10}, reaches an "end point" and then returns very close to the original arc, terminating at the point P_{11}', very close to O. This elongated arc, together with the segment $P_{10}P_{11}'$ on the curve S define the lobe U_{10}'. The beginnings of the arcs U,S,UU,SS are given here as a background.

As a consequence, although the chaotic region around O fills almost the whole frame of Fig. 14, the density of the points is much larger close to O_1 and O_2 than further away. It should take an exceedingly long time to derive a more or less smooth distribution of points in the whole chaotic region. This phenomenon is related to the diffusion of the points in the chaotic regions. Any diffusion mechanism takes an extremely long time to become operative.

If we increase the energy, the stickiness phenomenon becomes less pronounced, and the diffusion takes place faster. E.g. if h=25 we have Fig. 15. In this case thel outer lobes go away from the resonance regions O_1 and O_2 much faster than when h=24. We see that the lobes U_5',U_6' in Fig. 15 deviate from O as much as U_{25}',U_{26}', in Fig. 14.

In this particular problem we cannot explore much larger values of h without encountering escapes. In fact the escape energy is h_{esc}=25.31, and if we go well above h_{esc} most of the orbits escape to infinity.

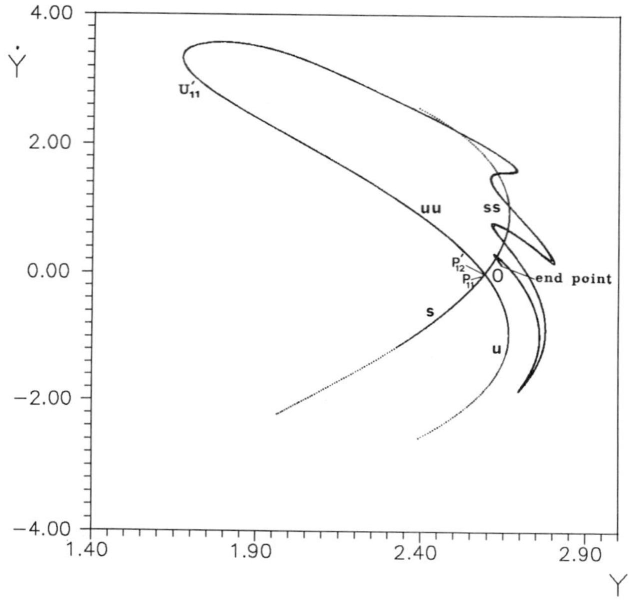

Figure 13. The same as in Fig. 12 for the lobe U_{11}'.

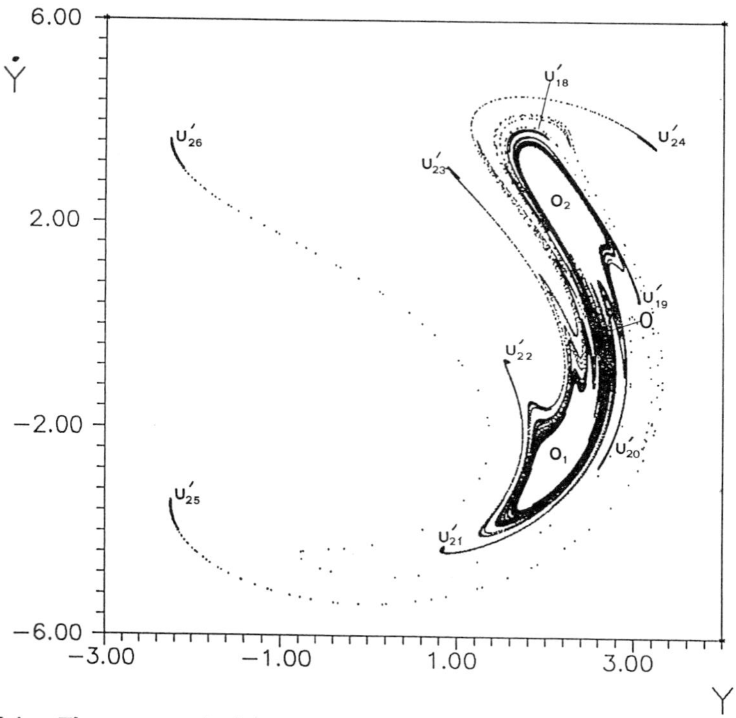

Figure 14. The asymptotic curve U all the way to the arc U_{26}' for h=24. The low order lobes remain close to the resonant regions O_1 and O_2 (stickiness phenomenon), but the high order lobes extend to large distances from these regions.

V. SYSTEMS WITHOUT ESCAPES

In order to study the behaviour of the homoclinic tangle for very large energies we studied the Hamiltonian

$$H \equiv \frac{1}{2}(\dot{x}^2+\dot{y}^2+x^2+y^2)+\varepsilon x^2 y^2=1 \tag{11}$$

with $\varepsilon>0$. This is similar to the Hamiltonian (4), but the sign of ε is different, and it has no escapes, however large ε is. In order to compare the various cases we keep $h=1$=constant and vary ε (a change of scale can make ε=const and h varying).

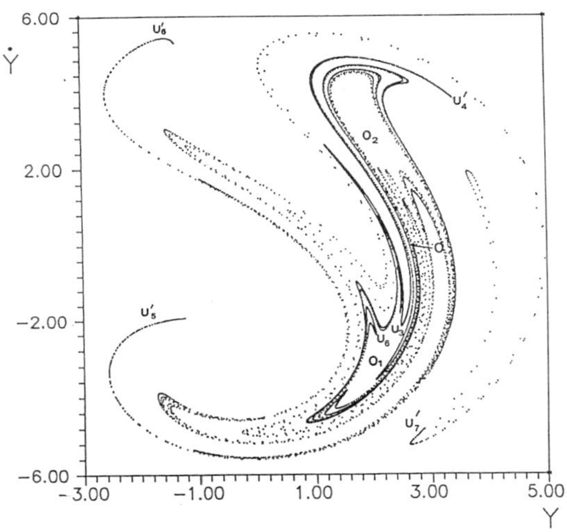

Figure 15. The asymptotic curve U up to the arc U_6' for h=25. The stickiness in this case is less than in Fig. 14.

If ε is small the Poincare surface of section $(y,\dot{y})$ is mostly filled with invariant (KAM) curves ($\varepsilon=0.12$; Fig. 16). There are 4 stable periodic orbits: (1) the origin O, corresponding to the orbit y=0 in the configuration space, (2) the boundary $y^2+\dot{y}^2=2$, corresponding to the orbit x=0, (3) the orbit B_1 ($y=0$, $\dot{y}=1$), corresponding to the orbit x=y, and (4) the orbit B_2($y=0$, $\dot{y}=-1$), corresponding to the orbit x=-y. There are also two unstable orbits, C1 and C2, corresponding to an almost rectangular orbit in the configuration space (Contopoulos 1993), described in two opposite directions.

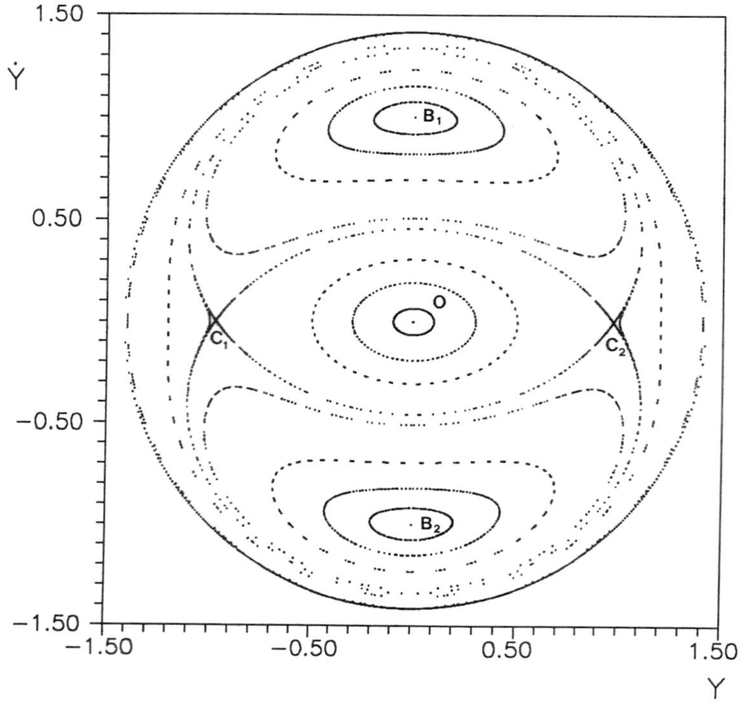

Figure 16. The Poincare surface of section $(y,\dot{y})$ in the case of the Hamiltonian (11) for $\varepsilon=0.12$.

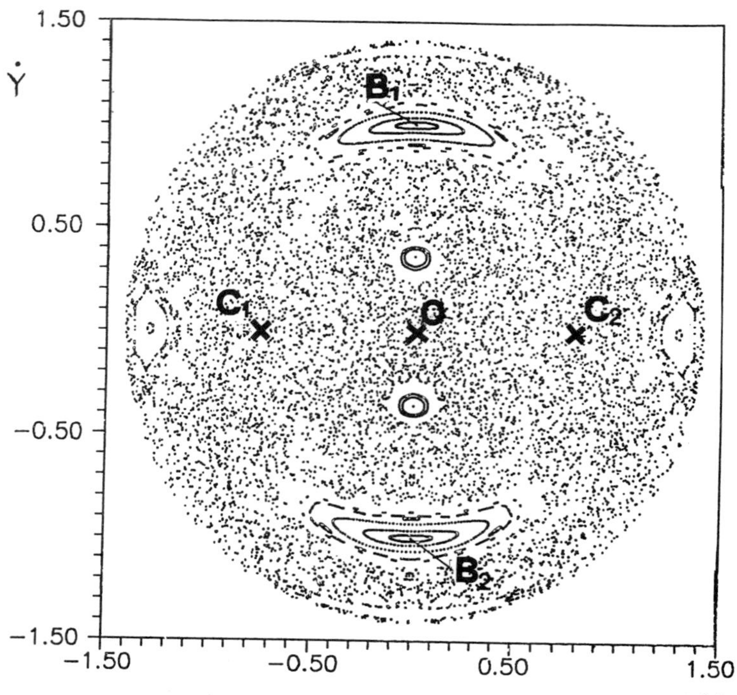

Figure 17. As in Fig. 16 for $\varepsilon=1.92$.

In this case there is only a little chaos around the unstable invariant points C1 and C2. However, as ε increases the chaotic regions around C1 and C2 increase and for ε=1.92 they cover most of the phase space (Fig. 17). In this case the origin O is unstable, and the bounary $y^2+\dot{y}^2=2$ is also unstable. At the transition to instability two stable orbits bifurcate from O represented by two points, above and below O, surrounded by closed invariant curves (islands). At the same transition value of ε bifurcate two stable orbits from the orbit x=0 (represented by $y^2+\dot{y}^2=2$ in Fig. 17), which are surrounded by islands close to the left and the right boundary on the $\dot{y}=0$ axis. For this value of ε the orbits B1 and B2 are stable.

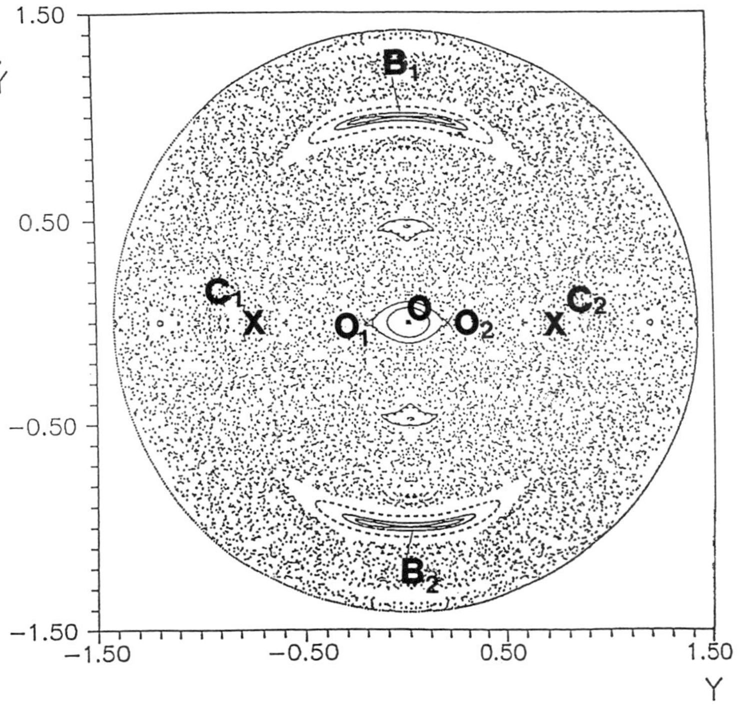

Figure 18. As in Fig. 16 for ε=2.5.

As ε increases further the orbit O becomes again stable (Fig. 18, for ε=2.5). At the transition to stability the family O generates, by bifurcation, two unstable orbits O_1 and O_2.

At the same time the asymptotic curves of O are duplicated (Fig. 19) and 4 of them (U^1,S^1,UU^1,SS^1) emanate from O_1, while another 4 (U^2,S^2,UU^2,SS^2) emanate from O_2. The outer curves $(S^1,UU^1,U^2$ and $SS^2)$ far from O are very close to the asymptotic curves from O (S,UU,U and SS), before O became stable. The only difference is in the nonenclature, because the intersections of (S^1,U^2) and of (UU^1,SS^2) are now called heteroclinic, while they were called homoclinic when the curves S,UU,U and SS emanated from O.

On the other hand the inner curves (U^1,S^2) and (SS^1,UU^2) are almost like separatrices, i.e. they form very small oscillations when they intersect and the chaotic regions close to them are very small.

As ε increases the central orbit O is alternatively stable and unstable (Fig. 20). It has been proven (Sohos et al. 1989) that the variational equation for the orbit O is a Mathieu equation that has infinite intervals of stability and instability. Thus there is no maximum value of ε beyond which O is unstable. Beyond any given ε there are infinite intervals of stability of the orbit O. Thus chaos cannot be complete near the center O for arbitrarily large ε.

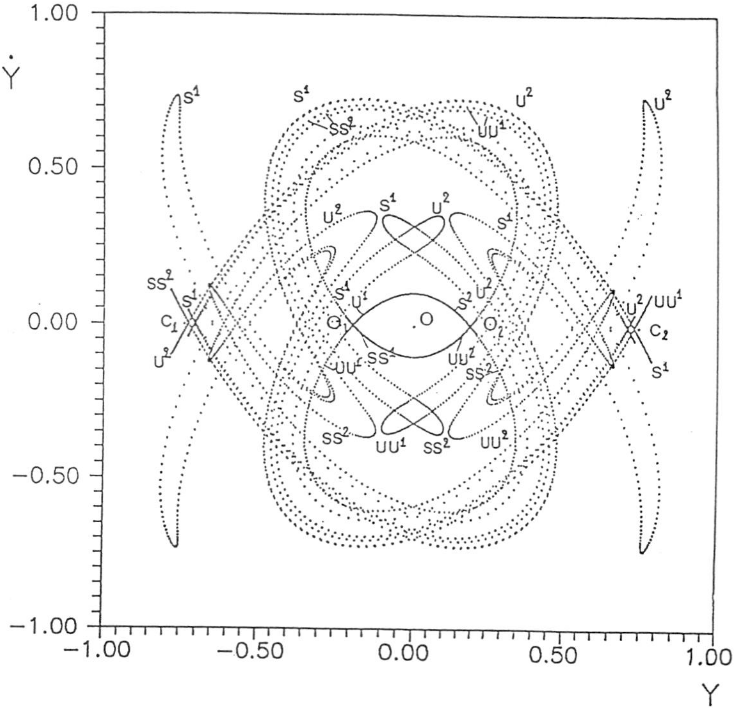

Figure 19. The asymptotic curves from O_1 and O_2 for $\varepsilon=2.5$, when O has become again stable.

When $\varepsilon>2.367$ the diagonal orbits $x=\pm y$, which are represented by the points B1,B2 ($y=0$, $\dot{y}=\pm 1$) in Figs 16-18, are unstable. For even larger ε the asymptotic curves of B1 and B2 cover most of the phase space, intersecting along homoclinic and heteroclinic points, and also along heteroclinic points with the asymptotic curves from C1,C2, and possibly also O (when this is unstable). As ε increases the lobes of the orbits B1,B2 become longer. However, as there are no escapes to infinity, there are no "limiting asymptotic curves", which the lobes of B1,B2 could approach. Thus the lobes become longer and longer, spiralling clockwise and counterclockwise. An example of such spiral lobes is shown in Fig. 21 for $\varepsilon=40$.

As ε increases, and the spirals become longer, there are more and more homoclinic and heteroclinic intersections between these lobes. Before new intersections appear the lobes become tangent to each other. Then, according to the Newhouse (1977, 1983) theorem, there are nearby stable periodic orbits, surrounded by closed invariant curves (islands). As ε increases all these stable orbits become unstable after infinite period doubling bifurcations. But for larger ε new stable orbits appear, close to new tangencies of the lobes of the asymptotic curves of the various unstable periodic orbits.

Thus we conclude that chaos is not complate for large ε, not only close to the center O, but also near all the tangencies of the lobes of the various unstable periodic orbits, which may occur anywhere on the Poincare surface of section.

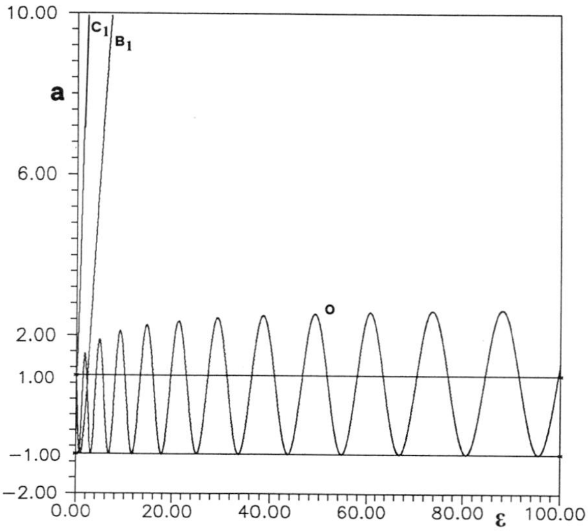

Figure 20. The stability parameter a of the periodic orbits O, B_1 and C_1 as a function of ε. When $|a|<1$ the orbit is stable and when $|a|>1$ it is unstable. This orbit has infinite transitions from stability to instability and vice-versa.

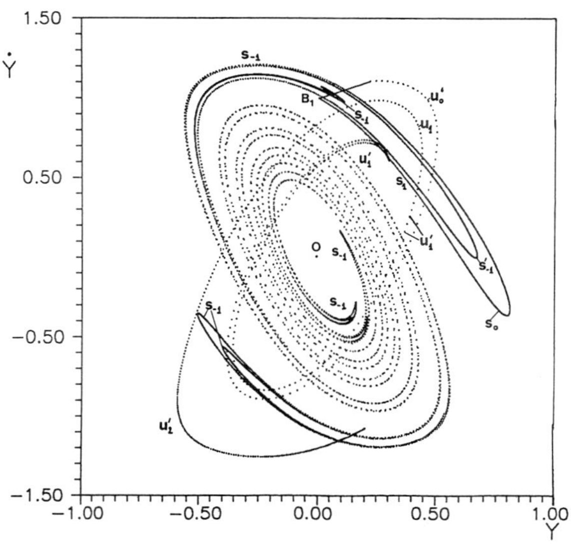

Figure 21. The asymptotic curves U and S from the unstable orbit B_1 for ε=40. These curves make many spiral rotations and intersect at a large number of homoclinic points.

VI. THE MIXMASTER UNIVERSE

The Mixmaster Model of the Universe was introduced by Belinsky, Khalatnikov and Lifshitz (1969, 1970) and independently by Misner (1969). It is a particular solution of Einstein field equations, known as Bianchi Type IX model. It has been called "Mixmaster" by Misner, because of its assumed mixing properties (Misner et al. 1977). The line element of this model is

$$ds^2 = dt^2 - d\sigma^2 , \tag{12}$$

where

$$d\sigma^2 = a^2 dx_1^2 + b^2 dx_2^2 + c^2 dx_3^2 . \tag{13}$$

The scale factors a,b,c satisfy the equation

$$2abc \frac{d}{dt} (\frac{da}{dt} bc) = (b^2 - c^2)^2 - a^4 \tag{14}$$

and its cyclic permutations. At any time t two of the scale factors represent expansion, while the third represents contraction of the Universe. However the directions of expresion and contraction change in a presumably chaotic way as t goes to zero (big bang). This is why this model was called also a Chaotic Universe. In recent years this model has drawn much attention because it was supposed to produce a chaotic mixing of the various parts of the early universe, thus avoiding the cosmological "causality problem", without introducing inflation.

If we introduce a new time (Belinsky et al. 1970)

$$\tau = -\ln t \tag{15}$$

and write

$$a=e^{\alpha}, \ b=e^{\beta}, \ c=e^{\gamma} \tag{16}$$

the big bang (t=0) coresponds to $\tau=+H$ and the equations of motion take the form

$$2\ddot{\alpha} = (e^{2\beta} - e^{2\gamma})^2 - e^{-4\alpha} \tag{17}$$

and its cyclic permutations, where $.=d/d\tau$.

These equations can be derived from the Hamiltonian

$$H = \frac{1}{16}(p_\alpha^2 + p_\beta^2 + p_\gamma^2 - 2p_\alpha p_\beta - 2p_\beta p_\gamma - 2p_\gamma p_\alpha) + e^{4\alpha} + 4^{4\beta} + e^{4\gamma}$$
$$-2e^{2(\alpha+\beta)} - 2e^{2(\beta+\gamma)} - 2e^{(2\gamma+\alpha)} . \tag{18}$$

The numerical value of H is known to be zero. This is a genuine system of 3 degrees of freedom.

In order to check the chaotic character of the Mixamster Universe several people calculated the maximal Lyapunov characteristic number (LCN) along several orbits. The first rough calculations indicated positive LCN's; however more accurate calculations (Hobill et al 1989, Burd et al. 1990,

Hobill 1991, Berger 1991) have shown that the LCN is probably zero. Therefore the Mixmaster Universe does not seem to be chaotic.

In order to avoid this conclusion some people have redefined the time in order to derive a positive LCN. However such a redefinition of time may make even an integrable system look chaotic. A trivial example is the following. Consider a linear deviation ξ from a given orbit, of the form

$$\xi = \xi_o t_1 \ . \tag{19}$$

In time t_1 the Lyapunov characteristic number

$$LCN_1 = \lim_{t_1 \to \infty} \frac{\ln(\xi/\xi_o)}{t_1} = \lim_{t_1 \to \infty} \frac{\ln t_1}{t_1} = 0 \ . \tag{20}$$

However, if we introduce a new time t_2 by the relation

$$t_1 = e^{qt_2} \tag{21}$$

with $q>0$, we find

$$\xi = \xi_o e^{qt_2} \tag{22}$$

and the Lyapunov characteristic number in time t_2 is

$$LCN_2 = q>0 \ . \tag{23}$$

The question is now what time is physically acceptable. A physically appropriate time is the length of an orbit, or a similar quantity, because then the deviations from an orbit are measured with respect to the length of the orbit itself. E.g. such is the Ω-time introduced by Misner et al (1977) that is equal to

$$\Omega = \frac{1}{6} (x+y+z) \ . \tag{24}$$

In such a time the Lyapunov characteristic number is zero.

A quite different check of the chaotic character of the Mixmaster Universe is by finding whether this model is integrable. If this is so, then the Mixmaster Universe is definitely non chaotic. We have shown recently (Contopoulos, Grammaticos and Ramani 1993) that the Mixmaster Universe has the Painleve property, which is a strong indication of integrability (Bountis et al. 1982, Ramani et al. 1989). Namely we have found that a generic solution of the equations of motion, depending on 6 arbitrary constants, has only poles as (movable) singularities in complex τ-time. Using the ARS method (Ablowitz, Ramani and Segur 1980) we derived the general solution in the form:

$$e^{2\alpha} = \pm\frac{i}{s} + \gamma_1 s + \ldots$$

$$e^{2\beta} = x_2 s + \ldots$$

$$e^{2\gamma} = x_3 s + \ldots$$

$$p_\alpha = -\frac{4}{s} + 2(p_2+p_3)\pm 4i(x_2+x_3-2\gamma_1)s+\ldots \qquad (25)$$

$$p_\beta = 2p_2 \pm 4ix_2 s + \ldots$$

$$p_\gamma = 2p_3 \pm 4ix_3 s + \ldots$$

where $s=\tau-\tau_o$, and the 6 arbitrary constants are $\tau_o,\gamma_1,x_2,x_3, p_2,p_3$. The dots represent higher order terms in s, which are found, step by step, by using the ARS method. There are also special solutions, depending on fewer than 6 arbitrary coefficients, which are Laurent series in the "time" $s=\tau-\tau_o$.

Up to now many systems that have the Painleve property were proved to be integrable by explicitely finding the integrals of motion. There is no counterexample of a system that has the Painleve property and is not integrable.

In the present case we have not yet found the integrals of motion appropriate for the Mixmaster Universe, despite some efforts using computer algebra. However the fact that this system has the Painleve property is a strong indication that it is integrable, therefore it is not chaotic.

On the other hand one may introduce perturbations in the Mixmaster Universe that may make it chaotic. Therefore the possibility of the existence of chaotic models of the Universe not only persists but it is quite probable.

ACKNOWLEDGEMENTS

This research was supported in part by the Greek Ministry of Industry Energy and Technology under contract No 9013502. I thank Dr. C. Polymilis and Miss H. Papadaki for many calculations in preparing some figures.

REFERENCES

Ablowitz, M.J., Ramani, A. and Segur, H.,1980, J. Math. Phys. 21,715.

Berger, B.K.,1991, Class.Quant.Grav. 7,203.

Belinski, V.A. and Khalatnikov, I.M.,1969, Sov.Phys.JETP 29, 911; 30,1174.

Belinski, V.A., Khlatnikov, I.M. and Lifshitz, E.M.,1970, Adv. in Phys. 19, 525.

Bountis, A., Segur, H. and Vivaldi, F.,1982, Phys.Rev. A25, 1257.

Burd, A.B., Buric, N. and Ellis, G.F.R.,1990, Gen.Rel.Grav. 22,349.

Chandrasekhar, S.,1989, Proc.R.Soc.Lond A421, 227.

Churchill, R.C., Pecelli, G. and Rod, D.L.,1975, J.Diff.Equ. 17,329.

Contopoulos, G.,1971, Astron.J. 76, 147.
Contopoulos, G.,1990a, Proc.R.Soc.Lond. A431,183.
Contopoulos, G.,1990b, Astron.Astrophys. 231,41.
Contopoulos, G.,1991, Proc.R.Soc.Lond. A435,551.
Contopoulos, G. and Kaufmann, D.,1992, Astron.Astrophys. 253,379.
Contopoulos, G.,1993, in Seimenis, J., (ed) "Hamiltonian Mechanics. Integrability and Chaotic Behaviour", Plenum Press (in press).
Contopoulos, G. and Polymilis, C.,1993, Phys.Rev.E 47,1546.
Contopoulos, G., Kandrup, H.E. and Kaufmann, D.,1993, Physica D64,310.
Contopoulos, G., Grammaticos, B. and Ramani, A., 1993, J. Phys. A. (in press).
Guckenheimer, J. and Holmes, P.,1993, Nonlinear Oscillations Dynamical Systems, and Bifurcations of Vector Fields, Springer Verlag, N. York.
Hobill, D.,1991,in Detweiler,S.L. and Ipser,J.R.,(eds.) "Nonlinear problems in Relativity and Cosmology", N. York Acad.Sci. 631,15.
Hobill, D., Bernstein, D., Simpkins, D. and Welge, M.,1989, in Proc. 12th Int.Conf.Gen.Rel.Grav., Boulder, Univ. of Colorado, Abstracts, p. 337.
Menyuk, C.R.,1985, Phys.Rev.A. 31, 3282.
Misner, C.M.,1969, Phys.Rev.Lett. 22, 1071.
Misner, C.M., Thorne, K. and Wheeler, J.A.,1977, Gravitation, Freeman, San Francisco.
Newhouse, S.E.,1977, Am.J.Math. 99, 1061.
Newhouse, S.E.,1983 in Iooss, G., Helleman, R.H.G. and Stora, R. (eds) Chaotic Behaviour in Deterministic Systems, North Holland, Amsterdam, p. 443.
Noid, D.W., Gray, S.K. and Rice, S.A.,1986, J.Chem.Phys. 84, 2649.
Petit, J.M. and Henon, M.,1986, Icarus 66, 536.
Ramani, A., Grammaticos, B. and Bountis, A.,1989, Phys.Rep. 180,159.
Rod, D.L.,1973, J.Diff.Equ. 14,129.
Shirts, R.B. and Reinhardt, W.P.,1982, J.Chem.Phys. 77, 5204.
Sohos, G., Bountis, T. and Polymilis, H., 1989, Nuovo Cim. 104B, 339.
Wiggins, S.,1990, Introduction to Applied Nonlinear Dynamical Systems and Chaos, Springer Verlag, N. York.

CHAOS AS THE TRUE SOURCE OF THE IRREVERSIBILITY OF TIME

C. Marchal

Des-Onera
92320 Châtillon
France

ABSTRACT

After a first part about the various definitions of time, the main examples of irreversibilities are presented.

The classical irreversibilities have some surprising connections.

The paradox of reversible physical laws associated with irreversible phenomena is related to the existence of systems of chaotic character with a very large number of parameters.

A simplified model helps to understand and resolve the paradox.

INTRODUCTION

Many phenomena of day-to-day life are, or at least seem, irreversible: we remember the past and ignore the future , we are growing old, the heat always goes from hot bodies to cold ones and the sugar disappears into the coffee... However, the known physical laws are reversible.

This paradox has led to many researches and many controversies at the limit of philosophy, the classical answers are incompatible with each other and neglect or underestimate the importance of chaos that is certainly the essential reason of existence of physical irreversibility.

LIST OF SYMBOLS

In this paper with many very large and very small numbers, we will use the notations "by figures and sizes" with the letter p for "positive power of ten" and the letter n for "negative power of ten".

Hence, for instance:

$$6.02p23 = \text{ Avogadro Number } = 6.02 \times 10^{23}$$

From Newton to Chaos, Edited by A.E. Roy
and B.A. Steves, Plenum Press, New York, 1995

$$1.66n24 = \text{ reciprocal of Avogadro Number } = 1.66 \times 10^{-24}$$

Notice that the "figures", i.e. here 6.02 and 1.66 are always between 1 and 10. This gives an unambiguous definition to the "sizes" (here p23 and n24). The size is the main element of very large and very small quantities, it is even very often their only known element.

FIRST PART : THE TIME

Let us consider for a moment an old Babylonian tablet of the upper period. It shows for instance a mathematical demonstration or the computation of the square root of 2 with five or six digits...

What moves us? These scientific results so small at our scale? No, it is the time that moves us... the fact that these stammerings of human intelligence occured 4000 years ago.

Time is something very mysterious, it is extremely subjective with one meaning per profession and almost one meaning per people.

We find for instance (from Tables 1, 2 and 3):

Table 1. The time of cosmologists (in the standard model).

Time	Events
0	Big bang
n43 second	Planck's wall
n35 to n32 s	Inflation
n6 to n4 s	Hadronic era (protons, neutrons)
n4 second	Leptonic era (electrons, photons, neutrinos)
0.5 second	→ Escape of neutrinos
1 second	
100 s	First atomic nucleus, primordial nucleo-synthesis
300 000 years	Escape of photons
	Formation of atoms (H, He, ...)
	Quasars, formation of galaxies
4 p9 years	Successive generations of stars
p10 years	Birth of the Sun and the Solar System
1.5 p10 years	To-day

It is well-known that the time of cosmologists is a mixture of space and time, but this is also true (in another way) for the time of geologists. For instance the "cretaceous-tertiary transition" is given by the disappearance of dinosaurs that has perhaps required tens of thousands of years to go from India to Mexico.

It must be noticed that geologists recognize the periods of sedimentary strata by the presence of various types of fossils. Very few fossils exist in pre-cambrian strata and at the beginning of the Cambrian period occured literally an "explosion of life".

Since then, the differences are essentially given by the mass extinctions (Fig. 1). If sometimes species seem to appear here or there, as mammals at the beginning of tertiary, this is mainly because before these periods these species were confined by their enemies to few varieties and remote areas.

The true motor of evolution is not yet known and even if Darwinism is true it is insufficient. It explains very well a small evolution (change of scale, apparition of

Table 2. The time of geologists

Approximated dates (millions of years)	Events, era and periods		
4600	Formation of earth		
	Pre-cambrian time		
600			
	Cambrian		Paleozoic era or primary
500			
	Ordovician		
440			
	Silurian		
400			
	Devonian		
350			
	Carboniferous		
270			
	Permian		
225			
	Triassic		Mesozoic era or secondary
180			
	Jurassic		
135			
	Cretaceous		
65			
	Paleocene	Tertiary	Cenozoic era
55			
	Eocene		
38			
	Oligocene		
25			
	Miocene		
5.1			
	Pliocene		
2.0			
	Pleistocene	Quaternary	
0.01			
	Holocene		
0			

a color) or a regressive evolution (loss of an organ), but the billions and billions of mutations of geologic times are, by far, not numerous enough to explain a progressive evolution such as the gain of a new organ.

Notice two opposite effects: the time of cosmologists seems to run very fast at the beginning with many events occuring in a very small period, and it progressively slows down. This is contrary to geologic times with seven-eighths occuring in pre-cambrian times and with always smaller and smaller eras.

This acceleration is partly an effect of perspective: recent events and recent fossils have a larger chance to have remains in our time, but it is also partly a real effect that is even more visible in the time of prehistorians and archeologists.

The paleolithic scale depicted on Table 3, in which glaciations have German names and periods have mostly French names, has been defined for Europe. The different paleolithic scales (Europe, Africa, Middle-East, Far-East, America) have still uncertain and approximate correspondences.

Notice Adam and Eve at the beginning of middle paleolithic: these names have been given by the prehistorians to the first men that buried their deads... the awakening of human consciousness.

The acceleration encountered in geological and archeological times is even stronger in historical times (Antiquity, Middle Age, Modern Times) and events seem to appear at an always faster and faster rate.

Table 3: The time of prehistorians and archeologists

3.1. Paleolithic age (hunters, cavern artists).

Dates (years)	Men	GLACIATIONS	Periods	
- 3 p6*	Australopithecus (Homo habilis)			Lower paleolithic
- 1 p6	Pithecanthropus			
- 500 000	Sinanthropus (Homo erectus)	GÜNZ		
		MINDEL	Abbevillian Clactonian	
		RISS	Chellean Premousterian	
- 100 000	(Adam and Eve)	WÜRM (1)	Mousterian (Pontinian Altmühlian	Middle paleolithic
	Neanderthal		Jabroudian)	
- 35 000		WÜRM (2)	Perigourdian	Upper paleolithic
	First Amerindians Cro-Magnon		Aurignacian Solutrian	
	(Homo sapiens)	WÜRM (3)	Magdalenian Azilian	
- 9000				

* - 3 p6 years = 3 millions of years B.C., see the list of symbols.

3.2. Neolithic age (shepherds, farmers, builders).

Dates (B.C.)	Domestic animals	and cereals	Events
9000	Dogs		
	Bulls and cows		Villages with round houses
8000	Horses		
	Goats,		Villages with square houses
7000	Sheeps	Corn	Jericho
	Pigs		Baked clay
6000		Maize (U.S. corn)	Catal Hoyük (Turkey)
5000		Rice	Megaliths

3300 --

3.3. Bronze age — PREHISTORY

1200 --

3.4. Iron age — HISTORY

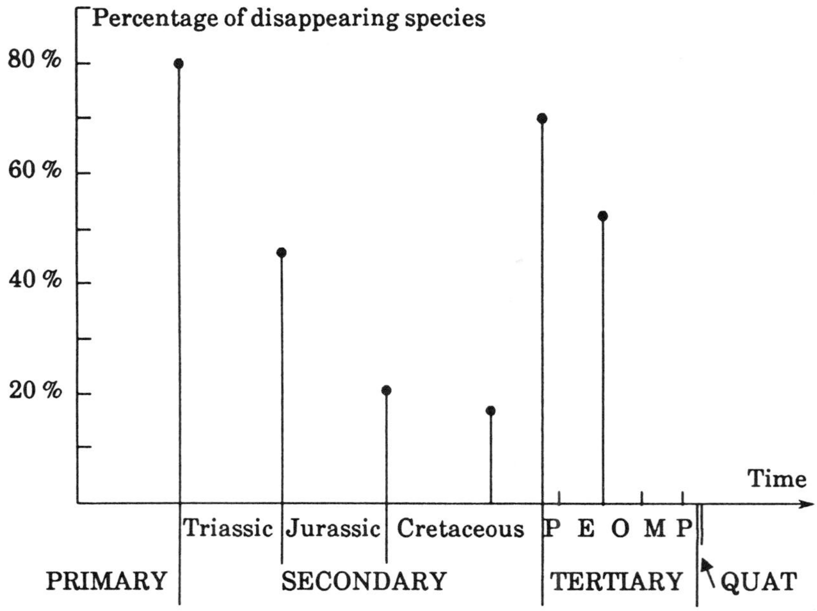

Fig. 1 – The mass extinction record

This obvious property of lived time is in strong contrast with the regular times defined by scientists:

The time of mechanicians: $\vec{F} = m\frac{d^2\vec{r}}{dt^2}$
The time of quantum physicists: $\Delta E.\Delta t \geq h$
The time of biologists: cycles, rhythms, yearly hibernation, prolongated hibernation.

Similarly, we find the time of astronomers, seamen, aviators, watchmakers, musicians, etc...

Every profession has its own time with its own definition and its own accuracy. For instance, seamen use essentially the U.T.C. (Universal Time Coordinated) that is defined by the slightly irregular rotation of Earth and that gives them an easy way for finding longitudes.

The different times used by astronomers are called: true local solar time, mean local solar time, sidereal local time, universal time, international atomic time, ephemeris time, terrestrial dynamic time, barycentric time, etc...

Fortunately, there is at least a time that is common to all mankind in spite of its variability, it is the psychological time:

"Waiting half an hour for Dad when he is going to arrive at the station... It's very long!

Playing half an hour before sleeping... It's very short! I have to go to bed already! It's unfair!"

My grand-daughter Eleonor was only five years old when she said these words (in French of course) and thus a five year old child already knows that the psychological time is not uniform. Furthermore, it has the unpleasant but universal property of being slow during unpleasant periods and fast during pleasant ones.

Unfortunately my grand-daughter died suddenly a few weeks later and this is the strongest and most painful evidence of the irreversibility of time.

SECOND PART : THE PARADOX OF TIME

The second principle of thermodynamics is the basis of physical irreversibility, but it is generally considered as a "principle" and not as a "law". This linguistic subtlety allows us to write that all known scientific laws can always be expressed in a reversible form, a form in which past and future are symmetrical.

This leads to well-known reversible properties. For instance if, in the Solar System, we reverse the velocities of all planets and satellites, the orbits will remain the same and they will be described in the other direction.

However, this beautiful symmetry contradicts many phenomena of ordinary life.

The most known irreversible phenomena are the following:

1) The memory: we remember the past and ignore the future.

2) The second principle of thermodynamics. The entropy of an isolated system, the measure of its disorder, is always increasing.

3) The expansion of the Universe.

4) The black holes: light and matter fall into black holes, they never escape from them.

5) The propagation of light: by diverging waves and not by converging ones.

There are some other phenomena considered as irreversible in quantum mechanics and in radio-active disintegration.

Remarkable connections between these irreversibilities exist.

Consider for instance a lake in a cold country. It freezes each autumn and thaws each spring.

Ice is much more organized than water, it is then during freezing that the entropy of the lake decreases.

What happens then? The lake cannot be considered as an isolated system and, during the long and starry nights of November and December, it sends towards space a huge quantity of infra-red photons that carry away its heat and its entropy.

What happens to these photons? Because of the expansion of the Universe, most of them will never arrive anywhere, they will wander forever in an always emptier space.

If the Universe was static, the number of arriving photons would equal the number of departing photons and the lake could not freeze.

There is however a question: if the Universe is dense enough, it will not expand forever and in some tens of billions of years it will begin to contract. What will then happen? Some theoreticians think that then the time will reverse, we will escape from the grave and return to childhood! This is really very surprising and it is more probable that the black holes, so rare today, will then be very numerous. They will take the place of expansion and most wandering photons will fall into a black hole and disappear forever. The lakes will continue to freeze each autumn...

A CONCRETE EXAMPLE

Let us open the communication between two neighbouring closed vessels full of gas. The brownian motion will equalize the temperatures, the pressures and the compositions while the opposite evolution never appears.

However:

A) The brownian motion and the kinetic theory of gas are conservative and reversible, as conservative and reversible as the Celestial Mechanics itself.

B) Henri Poincaré[1] has demonstrated that for bounded and conservative systems, almost all initial conditions lead to an infinite number of returns in the vicinity of these initial conditions (the mathematicians specify: "in any vicinity of the initial conditions").

These returns to the vicinity of initial conditions are of course contradictory with the equalization of temperatures, pressures and compositions.

1. CLASSICAL AND UNSATISFACTORY ANSWERS

A) "There exist perhaps some very small, irreversible and dissipative hidden phenomena that forbid the application of Poincaré results..."

This rejection of a major symmetry of nature is not justified and our present knowledge is sufficient for the resolution of the observed contradiction.

B) "The notion of trajectory remains accurate for only some "Liapounov times" i.e. much less than the return time of Poincaré that has never been observed in this type of experiment..."

This answer is true but insufficient. The impossibility of accurate long-term computation of the future evolution doesn't resolve the contradiction.

C) "In principle Poincaré is right and for strictly isolated systems there is indeed this mysterious correlation between initial and final conditions (after the return time of Poincaré). But our systems are not isolated and even very small perturbations, such as the attraction of planets, destroy this correlation..."

These "mysterious correlations" are imaginary and it is in a natural fashion that the system returns toward all states attainable from the given initial conditions. The "very small pertubations" will not modify the order of magnitude of the return time of Poincaré, even if it is true that they can modify very much the evolution in a relatively short interval of time (a few "Liapounov times") and thus contribute to the disparition of correlations.

2. THE TRUE ANSWER

It is because a system is "sensible to initial conditions" and because it depends on billions of parameters, while we measure only a few of them, that we ascertain almost no correlation between successive states at large intervals and that the Poincaré return time is very large, much larger than the age of the Universe.

We thus reach the physical irreversibilty of our experiments in spite of reversible and conservative laws.

Notice that for non-chaotic evolutions, for instance for periodic or quasi- periodic evolutions, the deterministic predictions can be excellent even if the knowledge of initial conditions is weak. A solution of these types has a natural reversibility and remains in a very small part of phase space, a part much smaller than that corresponding to a chaotic motion.

The chaotic evolutions compensate their impossibility of long-term deterministic predictions by excellent long-term statistic predictions (notice the similarity with quantum mechanics). This excellency is related to the chaos itself that introduces randomness permanently and, even if it is impossible to predict the future motion of a molecule in the brownian motion, we can model very well the statistic elements as the temperature and the pressure.

The following extremely simplified model, with only didactic purposes, will help to understand these questions.

Notice that this model satisfies practically the Boltzmann hypothesis of "molecular chaos" (no correlation between successive variations) but reaches opposite conclusions: the molecular chaos doesn't forbid the Poincaré return.

3. A SIMPLIFIED MODEL

Let us consider one billion billion molecules (that is p18 molecules with the notations of the above list of symbols). This number is the number of molecules in 37mm^3 of air in "normal conditions", which is a very small volume. In most experiments, the effects will be even larger.

These p18 molecules will be put in the two identical vessels A and B and will be numbered from 1 to p18.

The evolution will be: at each step an integer number between 1 and p18 will be chosen and the corresponding molecule will be transfered from its present vessel to the other one.

The rate of these exchanges can be, for instance, one million billion (that is p15) per second.

We will consider the temperature as constant and we will measure only the number of molecules in the two vessels A and B, i.e. the local pressure. We can for instance start with the following initial pressures:

$$P_A(0) = 1.4 \text{ bar};\ P_B(0) = 0.6 \text{ bar} \tag{1}$$

what will be the evolution?

4. EVOLUTION OF THE PRESSURES P_A AND P_B

The total number of molecules is constant and thus at any time:

$$P_A(t) + P_B(t) = P_A(0) + P_B(0) = 2 \text{ bars} \tag{2}$$

It is then sufficient to consider $P_A(t)$.

An essential question is the mode of choice of the successive 18 digit integer numbers.

We can consider a purely random choice (it is the "molecular chaos" of Boltzmann), but, because of the usual philosophical objections, we will also consider the following deterministic choice: the kth choice will be given by the decimals of rank (18k - 17) to 18k of a given real number x.

For instance, with x = π, that is:

$$x = 3.141\ 592\ 653\ 589\ 793\ 238\ 462\ 643\ \ldots \tag{3}$$

the first choice will be: 141 592 653 589 793 238.

The purely random choice leads to a simple analysis. With P_A and its variations δP_A expressed in bars we obtain: at each step we have $\delta P_A = \pm 2n18$ (that is 2×10^{-18}) with the probability $P_A/2$ for $\delta P_A = -2n18$ and the probability $1-(P_A/2)$ for $\delta P_A = +2n18$.

Hence the average evolution $P_{AM}(t)$ is given at each step by:

$$\delta P_{AM} = (1 - P_{AM}) \times 2n18 \tag{4}$$

that is, after k steps:

$$P_{AM.k} = 1 + [P_A(0) - 1][1 - 2n18]^k \tag{5}$$

and, with $P_A(0) = 1.4bar$ and p15 steps per second:

$$\left\{ \begin{array}{l} P_{AM}(t) = [1 + exp\{-0.002t\} \times 0.4] \text{ bar} \\ \text{with t expressed in seconds} \end{array} \right\} \tag{6}$$

$$\left. \begin{array}{l} P_{AM}(500s) = 1.1471 \text{ bar}; P_{AM}(1000s) = 1.0541 \text{ bar} \\ P_{AM}(1 \text{ day}) = (1 + 4n76) \text{ bar; practically 1 bar} \end{array} \right\} \tag{7}$$

This converging exponential evolution of the average value of the pressure $P_A(t)$ must be accompanied by that of its variance V(t).

The $(k + 1)^{th}$ step gives:

$$V_{k+1} = (1 - 4n18)V_k + 4n36[1 - (P_{AM.k} - 1)^2] \tag{8}$$

Hence, with (5), $P_A(0) = 1.4$ bar and $V_0 = 0$:

$$V_k = [n18(1 - (1 - 4n18)^k) - 0.16\{(1 - 2n18)^{2k} - (1 - 4n18)^k\}] \text{ bar}^2 \tag{9}$$

In terms of the time t (expressed in seconds) the exact expression (9) gives almost:

$$V(t) = [1 - (1 + 0.00064t)exp\{-0.004t\}]n18 \text{ bar}^2 \tag{10}$$

The variance V(t) has a monotonic evolution, it increases from zero to n18 bar^2 and at t = 1000 s, it is already at 97% of its final value.

The main result is that the variance V(t) remains forever very small.

The standard deviation $\sigma(t)$, the square root of the variance, will also remain forever very small, its maximum is n9 bar, that is one billionth of a bar or one deci-milli-pascal (one pascal = 1 Pa = 1 N/m^2 = n5 bar).

If we measure the pressure with the excellent accuracy of one millipascal, i.e. ten standard deviations, we will notice from time to time a fluctuation with respect to the average evolution. For instance flucuations larger than 7.94 standard deviations have an average frequency of one per two years and an average duration of only 0.127 microsecond.

If we measure the pressure with the accuracy of 5 millipascals, i.e. fifty standard deviations, we will never notice a fluctuation and the evolution will appear as irreversible. Indeed, there is only the probability n200 (that is 10^{-200}) that the first fluctuation of 5 millipascals appears before the time t = 4.625 p329 seconds, that is t = 1.465 p322 years.

In these conditions, the return time of Poincaré is extremely large and purely theoretical but it can be computed: if we neglect a probability of n200 (which corresponds to the "threshold of certainty of observable Universe"); the first Poincaré return to the pressure 1.4 bar will occur after 10^Q seconds with:

$$3.57349p16 < \text{ Q } < 3.57351p16 \tag{11}$$

These results correspond to the random choice of successive exchanges and we have also to consider the case of deterministic choices as explicited with equation (3). The computations (4) - (11) give then that the values of x that satisfy $0 \leq x \leq 1$ and that don't satisfy (11) have a total measure smaller than n200. This set of values is then completely negligible even if many remarkable values of x, such as x = 0.5, belong to that set.

Remark

The equations (6) and (7) of evolution of the average pressure $P_{AM}(t)$ show clearly an irreversible phenomena but can also give the false impression of an essential dissymmetry between the past (far from equilibrium) and the future (near to equilibrium).

On the contrary there is a strict past-future symmetry and if we only know that at t_1 and t_2 the pressure was P_{A1} and P_{A2} the evolution of the average pressure $P_{AM}(t)$ for $t_1 < t < t_2$ is given by:

$$\left.\begin{array}{l} P_{AM}(t) = [1 + P_1' e^{0.002(t_1 - t)} + P_2' e^{0.002(t - t_2)}] bar \\ t \text{ given in seconds; } t_1 < t < t_2 \\ P_1' \text{ and } P_2' \text{ given by: } P_{AM}(t_1) = P_{A1}; P_{AM}(t_2) = P_{A2} \end{array}\right\} \quad (12)$$

Thus even if the evolution (6), (7) is irreversible the past-future symmetry is conserved. Look for instance to the arrival on a Poincaré return with t_2 arbitrary and t_1 negative and extremely far.

CONCLUSION

The irreversibility of the second principle of thermodynamics agree fully with our experiments and our measures that are by far neither long nor numerous enough to lead to a contradiction. However, some very small temporary fluctuations appear from time to time in very accurate experiments.

Thus, the paradox of reversible physical laws associated with irreversible phenomena can be explained without "perfect isolation", "hidden correlations" and/or "small hidden irreversibilities". The main reasons of irreversibility are the chaotic character and the very large number of parameters of irreversible systems.

The Boltzmann's hypothesis of "molecular chaos" is excellent and allows very accurate computations. The correlations will not increase slowly and insidiously after a very long time and we can almost write that the return of Poincaré occurs by chance which usually requires such a large delay, much larger than the age of the Universe, that the corresponding decrease of entropy never appears in our experiments.

The return time of Poincaré is exponentially related to the number of independent parameters of the system of interest and we can thus write:

"If after the usual mathematical simplification (integral of motions, decomposability, etc.) a system:

A) remains with N independent parameters where N > 10 000.

B) is sensible to the initial conditions (chaotic system).

C) is analysed through statistical parameters as temperature and pressure.

Then its evolution will physically appear as irreversible for measures of relative accuracy worse than $(50/\sqrt{N})$ even if its laws are mathematically reversible and conservative."

REFERENCE

1. H. Poincaré. "Les méthodes nouvelles de la Mécanique Céleste," Dover Publication, New York, Vol. 3, P. 140 - 174 (1957).

REPRESENTATION AND CLASSIFICATION OF DYNAMICAL SYSTEMS

K. Zare

The Department of Aerospace Engineering
and Engineering Mechanics
The University of Texas at Austin
Austin, TX 78712

Abstract

It is shown that any dynamical system may be represented by a system of first order differential equations of the form $\dot{x}_i = a_{ij}\partial Z/\partial x_j$, where on the right side the $(n \times n)$ matrix-elements, a_{ij} and the scalar potential Z depend on the coordinates $x_1, x_2 \dots x_n$ and the summation convention is applied with respect to the repeated subscript j. Using direct notation the equation is written as $\dot{\boldsymbol{x}} = A(\boldsymbol{x})\partial Z/\partial \boldsymbol{x}$. Invariant properties of the system under transformations are studied and the classifications of the systems are established depending on the properties of the matrix $A(\boldsymbol{x})$. With this representation, many of the classical as well as the modern results may be shown using only linear algebra. In particular for a constant skew-symmetric matrix A, it is shown that many of the known properties of the motion for Hamiltonian systems such as form preservation, volume preservation, and variational principle remain valid.

1. INTRODUCTION

The equations of motion of dynamical systems can be represented in many different forms. The primary requirements of analytical representations are their general applicability, simplicity, relation to transformations, etc. One of the purposes of the present paper is to establish a representation which allows an immediate and simple classification of dynamical systems.

As it will be shown, the general form used in this paper allows straight forward easy classification and easy demonstration of many of the classical (e.g. Poincaré (1892), Birkholf (1927), Wittaker (1959), Pars (1965), etc.) as well as the modern results (e.g. Abraham and Marsden (1978), Smale (1980), Arnold (1988), etc.), using only linear algebra.

The use of direct matrix notation, instead of subscripts was selected to simplify and shorten the analytical manipulations.

From Newton to Chaos, Edited by A.E. Roy
and B.A. Steves, Plenum Press, New York, 1995

2. FORMULATION

The general representation of a dynamical system may be described by the set of differential equations,

$$\dot{\boldsymbol{x}} = A(\boldsymbol{x})\frac{\partial Z(\boldsymbol{x})}{\partial \boldsymbol{x}} , \tag{1}$$

where $\boldsymbol{x}$ is an n-dimensional state vector, $A(\boldsymbol{x})$ is a matrix whose elements are continuously differentiable functions of $\boldsymbol{x}$, and $Z(\boldsymbol{x})$ is a twice continuously differentiable function. The differentiability conditions are imposed to insure the existence and uniqueness of the solutions. Geometrically, we may think of $Z(\boldsymbol{x})$ as an n-dimensional manifold embedded in the $(n+1)$-dimensional Euclidean space, and Eq. (1) as a vector field on this manifold. This vector field is not restricted to the gradient direction, but by the action of $A(\boldsymbol{x})$, it may have any direction in the tangent space.

Remarks:

1. A vector field may be identified by its matrix $A(\boldsymbol{x})$. For instance, if $A(\boldsymbol{x})$ is the identity matrix, (i.e., $A(\boldsymbol{x}) = I$), Eq. (1) reduces to the potential flow with $Z(\boldsymbol{x})$ as a potential function, and if $A(\boldsymbol{x})$ is symplectic, $A(\boldsymbol{x}) = \begin{pmatrix} 0 & I \\ -I & 0 \end{pmatrix}$, Eq. (1) reduces to the Hamiltonian system with $Z(\boldsymbol{x})$ as a Hamiltonian function. Another example in celestial mechanics is the Lagrange planetary equations where $\boldsymbol{x}$ represents a set of orbital elements, $A(\boldsymbol{x})$ is a skew-symmetric matrix, and $Z(\boldsymbol{x})$ is the disturbing function.

2. The general differential equations $\dot{\boldsymbol{x}} = \boldsymbol{F}(\boldsymbol{x})$ may be reduced to the form given by Eq. (1). This reduction is not unique and it depends on the selection of the function Z. As an example, let $Z(\boldsymbol{x}) = \boldsymbol{F}^T(\boldsymbol{x})\boldsymbol{F}(\boldsymbol{x})/2$ and assume $\det(\partial \boldsymbol{F}(\boldsymbol{x})/\partial \boldsymbol{x}) \neq 0$, then $A(\boldsymbol{x}) = (\partial \boldsymbol{F}(\boldsymbol{x})/\partial \boldsymbol{x})^{-T}$. Once again the generality of Eq. (1) should be emphasized by this example.

3. The variation of $Z(\boldsymbol{x})$ along the trajectories of Eq. (1) is given by

$$\dot{Z}(\boldsymbol{x}) = \left(\frac{\partial Z(\boldsymbol{x})}{\partial \boldsymbol{x}}\right)^T \dot{\boldsymbol{x}} = \left(\frac{\partial Z(\boldsymbol{x})}{\partial \boldsymbol{x}}\right)^T A(\boldsymbol{x}) \left(\frac{\partial Z(\boldsymbol{x})}{\partial \boldsymbol{x}}\right) = g(\boldsymbol{x}) . \tag{2}$$

   Therefore $Z(\boldsymbol{x})$ is increasing or decreasing according to the sign of $g(\boldsymbol{x})$.

3. INVARIANT PROPERTIES

To study the properties of Eq. (1) under coordinate transformations, let

$$\boldsymbol{x} = \boldsymbol{f}(\boldsymbol{y}) , \qquad J(\boldsymbol{y}) = \left(\frac{\partial \boldsymbol{f}(\boldsymbol{y})}{\partial \boldsymbol{y}}\right) , \tag{3}$$

be a coordinate transformation and its corresponding Jacobian matrix with a non-vanishing determinant (i.e., $\det J(\boldsymbol{y}) \neq 0$). The differentiation of Eq. (3) yields $\dot{\boldsymbol{x}} = J(\boldsymbol{y})\dot{\boldsymbol{y}}$. Furthermore, by application of the chain rule and the rule for differentiation of composite functions, the following identity may be obtained

$$\left[\frac{\partial Z(\boldsymbol{x})}{\partial \boldsymbol{x}}\right]_{\boldsymbol{x}=\boldsymbol{f}(\boldsymbol{y})} = J^{-T}(\boldsymbol{y}) \left[\frac{\partial Z(\boldsymbol{f}(\boldsymbol{y}))}{\partial \boldsymbol{y}}\right] .$$

Substitution in Eq. (1) leads to

$$\dot{\boldsymbol{y}} = B(\boldsymbol{y})\frac{\partial Z(\boldsymbol{f}(\boldsymbol{y}))}{\partial \boldsymbol{y}}, \qquad B(\boldsymbol{y}) = J^{-1}(\boldsymbol{y})A(\boldsymbol{f}(\boldsymbol{y}))J^{-T}(\boldsymbol{y}). \tag{4}$$

Notice that the above matrix is transformed according to a congruence transformation. Therefore, the invariant (i.e., coordinate free) properties of a vector field are those properties of $A(\boldsymbol{x})$ which are invariant under congruence transformations. For example, we may show that definiteness, symmetry, and skew-symmetry are invariant properties.

1. If A is definite (positive or negative), then $\boldsymbol{u}^T A\boldsymbol{u} \geq 0$ for all $\boldsymbol{u}$. Let $\boldsymbol{u} = J^{-T}\boldsymbol{v}$, then $\boldsymbol{v}^T J^{-1}AJ^{-T}\boldsymbol{v} = \boldsymbol{v}^T B\boldsymbol{v} \geq 0$ for all $\boldsymbol{v}$.

2. If A is symmetric, then $B = J^{-1}AJ^{-T} = J^{-1}A^T J^{-T} = B^T$.

3. If A is skew-symmetric, then $B = J^{-1}AJ^{-T} = -J^{-1}A^T J^{-T} = -B^T$. Since $2\boldsymbol{u}^T A\boldsymbol{u} = \boldsymbol{u}^T A\boldsymbol{u} + \boldsymbol{u}^T A^T\boldsymbol{u} = \boldsymbol{u}^T(A + A^T)\boldsymbol{u} = 0$, A being skew-symmetric is equivalent to $\boldsymbol{u}^T A\boldsymbol{u} = 0$ for all $\boldsymbol{u}$.

4. CLASSIFICATION BASED ON THE PROPERTIES OF $A(\boldsymbol{x})$

Class I. A is definite, $\boldsymbol{u}^T A\boldsymbol{u} \geq 0$ for all $\boldsymbol{u}$. For this class, $g(\boldsymbol{x})$ is a positive or negative function (independent of Z), and $Z(\boldsymbol{x})$ is monotonic along the trajectories of Eq. (1) as it is shown by Eq. (2). The equilibrium points for these dynamical systems are the critical points of $Z(\boldsymbol{x})$, i.e., $(\partial Z(\boldsymbol{x})/\partial \boldsymbol{x}) = 0$, and the other trajectories of Eq. (1) approach these equilibrium points if $Z(\boldsymbol{x})$ remains bounded. An invariant subclass is defined by symmetric definite matrices. This subclass includes the potential flow ($A = I$) as a subset. The subset defining the potential flow is not invariant under Eq. (4). Since in the set of all $n \times n$ matrices, the set of positive (negative) definite matrices is an open set, Class I defines an open set in the space of all vector fields given by Eq. (1). As a consequence if a vector-field is in Class I, so is any nearby vector-field. The subclass defined by symemtric definite matrices is not an open set.

Class II. A is skew-symmetric, $\boldsymbol{u}^T A\boldsymbol{u} = 0$ for all $\boldsymbol{u}$. For this class, $g(\boldsymbol{x}) = 0$ and $Z(\boldsymbol{x})$ is constant along the trajectories of Eq. (1) as indicated by Eq. (2). The trajectories for these dynamical systems are restricted to the invariant submanifolds defined by $Z(\boldsymbol{x}) =$ constant. This class includes the Hamiltonian system as a subset. The subset defining the Hamiltonian systems is not invariant under Eq. (4). Since the set defined by skew-symmetric matrices is not an open set, Class II does not define an open set in the space of all vector fields given by Eq. (1). This implies that in any neighborhood (no matter how small) of a vector field in Class II, there are vector fields which do not belong to Class II. Therefore Class II does not share the stability characteristic of Class I.

Class III. A is non-definite, $\boldsymbol{u}^T A\boldsymbol{u}$ does not have the same sign for all $\boldsymbol{u}$. For this class, the $(n-1)$-dimensional submanifold defined by $g(\boldsymbol{x}) = 0$ divides the domain of $Z(\boldsymbol{x})$ into two regions: $P = [\boldsymbol{x}, g(\boldsymbol{x}) \geq 0]$, and $N = [\boldsymbol{x}, g(\boldsymbol{x}) < 0]$. If for all $t > T$ a trajectory does not enter P or N and $Z(\boldsymbol{x})$ remains bounded, then this trajectory will approach one of the equilibrium points. These solutions are similar to the solutions in Class I. However, in addition to these solutions, Class III may contain trajectories which cross the submanifold defined by $g(\boldsymbol{x}) = 0$ infinitely often and enter both P and N. Examples of such trajectories are periodic and quasi-periodic solutions. Therefore for Class III, $g(\boldsymbol{x}) = 0$ defines a Poincaré section. Since the set of non-definite matrices is an open set, Class III defines an open set in the space of all vector fields given by Eq. (1). Therefore, the stability characteristic of Class I is also true for Class III. However the phase portrait (i.e., the totality of solutions) for a member of Class III is more complex than the phase portrait for a member of Class I. An invariant subclass is defined by symmetric non-definite matrices.

Class IV. A is not skew-symmetric, $\boldsymbol{u}^T A \boldsymbol{u}$ is not zero for all $\boldsymbol{u}$. This is the complement set of Class II or equivalently the union of the open sets defined by Class I and Class III. The matrix $A(\boldsymbol{x})$ for a typical member of this class is definite in some regions and non-definite in the other regions. Since the set of non-skew-symmetric matrices is both open and dense, Class IV is defined by a generic property. This implies that the vector fields in Class IV are typical and the vector fields in Class II (i.e., its complement) are exceptional.

In summary we may present the above classification in the following hierarchical diagram:

Set of All Vector Fields $A(\boldsymbol{x})\dfrac{\partial Z(\boldsymbol{x})}{\partial \boldsymbol{x}}$

↓ Class IV (generic)
$A(\boldsymbol{x})$ is not skew-symmetric;
invariant set (open and dense)

↘ Class II
$A(\boldsymbol{x})$ is skew-symmetric;
invariant set

Class IV ↓ Class I
$A(\boldsymbol{x})$ is definite;
invariant set (open)

↓ Invariant Sub-Class
$A(\boldsymbol{x})$ is symmetric
definite

↓ Potential flow
$A(\boldsymbol{x})$ is identity

Class IV ↓ Class III
$A(\boldsymbol{x})$ is non-definite;
invariant set (open)

↓ Invariant Sub-Class
$A(\boldsymbol{x})$ is symmetric
non-definite

Class II ↓ Hamiltonian Systems
$A(\boldsymbol{x})$ is symplectic

5. FORM PRESERVING TRANSFORMATIONS

The invariant properties discussed in the above sections are true for any coordinate transformation. However, the form of Eq. (1) is invariant if the transformations are restricted to the group which preserves $A(\boldsymbol{x})$ under congruence transformation. This group may be identified by

$$J(\boldsymbol{y}) A J^T(\boldsymbol{y}) = A \, . \tag{5}$$

Examples:

1. For potential flow A is identity and the form is preserved by the group of orthogonal transformations, $J(\boldsymbol{y}) J^T(\boldsymbol{y}) = I$.

2. For the Hamiltonian systems A is simplectic and the form is preserved by the group of canonical transformations, $J(\boldsymbol{y})\begin{bmatrix} 0 & I \\ -I & 0 \end{bmatrix} J^T(\boldsymbol{y}) = \begin{bmatrix} 0 & I \\ -I & 0 \end{bmatrix}$.

By taking the determinant of Eq. (5), it may be shown that $\det J(\boldsymbol{y}) = 1$ is a property of the form preserving transformations. If $\det J(\boldsymbol{y}) = 1$, then the volume given by $\int dx_1 dx_2 \ldots dx_n = \int \det J(\boldsymbol{y}) dy_1 dy_2 \ldots dy_n$ is invariant and the transformation preserves the volume. This implies that the group of form preserving transformations is a sub-group of the volume preserving transformations.

We shall now consider the infinitesimal (near identity) transformation

$$\boldsymbol{x} = \boldsymbol{y} + \varepsilon A \frac{\partial Z(\boldsymbol{y})}{\partial \boldsymbol{y}} , \tag{6}$$

which is a transformation from the value of the state vector $(\boldsymbol{y})$ at time t to the value of the state vector $(\boldsymbol{x})$ at time $t+\varepsilon$. Here ε represents an arbitrary infinitesimal constant. Notice that Eq. (6) expresses a one parameter group of infinitesimal transformations representing the solution of Eq. (1) with time regarded merely as a parameter of the transformation. Assuming A is constant, the Jacobian matrix for this transformation is given by $J(\boldsymbol{y}) = I + \varepsilon A H(\boldsymbol{y})$, where $H(\boldsymbol{y}) = \left(\frac{\partial^2 Z(\boldsymbol{y})}{\partial^2 \boldsymbol{y}}\right)$ is the Hessian matrix of Z. For this transformation we have

$$J(\boldsymbol{y}) A J^T(\boldsymbol{y}) = (I + \varepsilon A H(\boldsymbol{y})) A (I + \varepsilon H(\boldsymbol{y}) A^T) = A + \varepsilon A H(\boldsymbol{y})(A + A^T) + O(\varepsilon^2) .$$

Therefore for a vector field defined by a constant skew-symmetric matrix A, the infinitesimal transformation (6) is a form preserving transformation. Since form preservation implies volume preservation, the infinitesimal transformation (6) is also volume preserving if A is a constant skew-symmetric matrix. For these vector fields, the motion itself can be regarded as a sequence of the form preserving (as well as the volume preserving) infinitesimal transformations. The vector fields defined by the constant skew-symmetric matrices are members of Class II and they include the Hamiltonian systems.

Remark: Using the form preserving property of the motion, we may generate the near identity form preserving transformations for any constant skew-symmetric matrix A. This leads to

$$\boldsymbol{x} = \boldsymbol{y} + \varepsilon A \frac{\partial S_0}{\partial \boldsymbol{y}} + \varepsilon^2 \left(A \frac{\partial S_1}{\partial \boldsymbol{y}} + \tfrac{1}{2}\, A \frac{\partial^2 S_0}{\partial \boldsymbol{y}^2} A \frac{\partial S_0}{\partial \boldsymbol{y}} \right) + O(\varepsilon^3) , \tag{7}$$

where $S(\boldsymbol{y}) = S_0(\boldsymbol{y}) + \varepsilon S_1(\boldsymbol{y}) + O(\varepsilon^2)$ is an arbitrary generating function. This is the generalization of the canonical transformation obtained by the method of Lie-series, for example, see Hori (1966).

6. VARIATIONAL PRINCIPLE

Assuming a constant skew-symmetric matrix A, let us evaluate the following integral

$$I = \int_{t_1}^{t_2} \left[\tfrac{1}{2}\, \boldsymbol{x}^T A^{-1} \dot{\boldsymbol{x}} - Z(\boldsymbol{x}) \right] dt , \tag{8}$$

along a one parameter family of admissible curves connecting two fixed points in the phase space, namely $\boldsymbol{x}_1 = \boldsymbol{x}(t_1)$ and $\boldsymbol{x}_2 = \boldsymbol{x}(t_2)$. Note that each curve represents a solution corresponding to a different vector field. Furthermore, we assume the vector field $\dot{\boldsymbol{x}}(t,\alpha)$, the state vector $\boldsymbol{x}(t,\alpha)$, and therefore the above integral $I(\alpha)$ are continuously differentiable functions of α, where α represents the parameter of the family of admissible curves. The boundary conditions may be stated as $\frac{\partial}{\partial\alpha}\boldsymbol{x}(t_1,\alpha) = \frac{\partial}{\partial\alpha}\boldsymbol{x}(t_2,\alpha) = 0$.

The critical values of the function $I(\alpha)$ may be obtained from $\frac{dI}{d\alpha} = 0$. Using lemmas from calculus, this may be written as

$$\int_{t_1}^{t_2} \left[\tfrac{1}{2} \left(\frac{\partial \boldsymbol{x}}{\partial \alpha}\right)^T A^{-1}\dot{\boldsymbol{x}} + \tfrac{1}{2}\, \boldsymbol{x}^T A^{-1} \frac{d}{dt}\left(\frac{\partial \boldsymbol{x}}{\partial \alpha}\right) - \left(\frac{\partial \boldsymbol{x}}{\partial \alpha}\right)^T \frac{\partial Z}{\partial \boldsymbol{x}} \right] dt = 0 \; .$$

Integration by part, using the boundary conditions, and noting that A is skew-symmetric leads to

$$\int_{t_1}^{t_2} \left(\frac{\partial \boldsymbol{x}}{\partial \alpha}\right)^T \left[A^{-1}\dot{\boldsymbol{x}} - \frac{\partial Z}{\partial \boldsymbol{x}} \right] dt = 0 \; .$$

This requires that the expression in the bracket to vanish identically (i.e. $\dot{\boldsymbol{x}} = A\frac{\partial Z}{\partial \boldsymbol{x}}$). Therefore for a constant skew-symmetric matrix A, Eq. (1) represents the vector field which results in a critical value of the integral (8), namely

$$\int_{t_1}^{t_2} \left[\tfrac{1}{2}\, \boldsymbol{x}^T \frac{\partial Z}{\partial \boldsymbol{x}} - Z(\boldsymbol{x}) \right] dt \; .$$

Remark: If A is symplectic, Eq. (1) is the Hamiltonian equations and the integral (8) reduces to Hamilton's integral. However it is important to point out the differences between the above formulation and the classical Hamilton's principle. In the classical version, the admissible curves are in the configuration space. The resulting second order differential equations are the Euler-Lagrange equations, and the Hamiltonian equations are obtained by applying the Legendre transformation. In the above formulation, the admissible curves are in the phase space, and the Hamiltonian equations are obtained directly.

7. VARIATIONAL EQUATION

Let us assume that a particular solution of Eq. (1) is given by $\boldsymbol{x} = \varphi(t)$, and we are interested in the neighboring trajectories of this reference solution. If $\boldsymbol{u}(t)$ represents a small variation from this reference trajectory, we may define $\boldsymbol{x} = \varphi(t) + \boldsymbol{u}(t)$. Substitution into Eq. (1), defining $\boldsymbol{F}(\boldsymbol{x}) = A(\boldsymbol{x})\frac{\partial Z(\boldsymbol{x})}{\partial \boldsymbol{x}}$, and expanding in Taylor's formula about the reference solution leads to

$$\dot{\boldsymbol{u}}(t) = C(t)\boldsymbol{u}(t) \; , \quad C(t) = \left(\frac{\partial \boldsymbol{F}(\boldsymbol{x})}{\partial \boldsymbol{x}}\right)_{\boldsymbol{x}=\varphi(t)} \; . \tag{9}$$

This is the variational equation belonging to Eq. (1) along the reference solution $\varphi(t)$. The general solution of Eq. (9) may be written as $\boldsymbol{u}(t) = \Phi(t,0)\boldsymbol{u}(0)$, where $\Phi(t,0)$ is called the state transition matrix. The differential equation for the state transition matrix may be written as $\dot{\Phi}(t,0) = C(t)\Phi(t,0)$, $\Phi(0,0) = I$.

Remarks

1. A particular solution of the variational equation (9) is given along the velocity vector of the reference solution $\boldsymbol{u}(t) = \varepsilon\dot{\varphi}(t) = \varepsilon \boldsymbol{F}(\varphi(t))$. This may be shown by differentiation $\dot{\boldsymbol{u}}(t) = \varepsilon \left(\frac{\partial \boldsymbol{F}(\boldsymbol{x})}{\partial \boldsymbol{x}}\right)_{\boldsymbol{x}=\varphi(t)} \dot{\varphi}(t) = C(t)\boldsymbol{u}(t)$. The existence of this particular solution is a consequence of the fact that the vector field represented by Eq. (1) is autonomous (i.e., $\boldsymbol{F}(\boldsymbol{x})$ is not an explicit function of t). Since $\boldsymbol{x} = \varphi(t) + \boldsymbol{u}(t) = \varphi(t) + \varepsilon\dot{\varphi}(t) + \ldots = \varphi(t+\varepsilon)$, this solution corresponds to a constant shift in time along the reference solution.

2. The gradient of a first integral is a particular solution of the differential equation dual to the variational equation. To show this, let us assume that a first integral of Eq. (1) exists, i.e., $G(\varphi(t) + \boldsymbol{u}(t)) = G(\varphi(0) + \boldsymbol{u}(0))$ for all t. Expanding this equation in Taylor's formula about the reference solution $\varphi(t)$ leads to

$$\Phi^T(t,0)\left(\frac{\partial G(\boldsymbol{x})}{\partial \boldsymbol{x}}\right)_{\boldsymbol{x}=\varphi(t)} = \left(\frac{\partial G(\boldsymbol{x})}{\partial \boldsymbol{x}}\right)_{\boldsymbol{x}=\varphi(0)} .$$

This is a necessary (but not sufficient) condition for the existence of a first integral. The differential equation dual to variational equation is defined by $\dot{\boldsymbol{v}}(t) = -C^T(t)\boldsymbol{v}(t)$. The general solution of this equation is given by $\boldsymbol{v}(t) = \Psi(t,0)\boldsymbol{v}(0)$, where $\Psi(t,0)$ satisfies the following differential equation $\dot{\Psi}(t,0) = -C^T(t)\Psi(t,0), \Psi(0,0) = I$. Since $\frac{d}{dt}[\Phi^T(t,0)\Psi(t,0)] = 0$ the following relation holds $\Phi^T(t,0)\Psi(t,0) = I$. Notice that the basis defined by the columns of $\Psi(t,0)$ is dual to the basis defined by the columns of $\Phi(t,0)$. Therefore, the vector spaces defined by these matrices are dual vector spaces. This leads to

$$\left(\frac{\partial G(\boldsymbol{x})}{\partial \boldsymbol{x}}\right)_{\boldsymbol{x}=\varphi(t)} = \Psi(t,0)\left(\frac{\partial G(\boldsymbol{x})}{\partial \boldsymbol{x}}\right)_{\boldsymbol{x}=\varphi(0)} . \tag{10}$$

While any solution of the variational equation belongs to the vector space defined by $\Phi(t,0)$, the gradient of any first integral of the system belongs to the dual vector space defined by $\Psi(t,0)$.

7.1 Liouville's Theorem

The determinant of the state transition matrix is given by

$$\det \Phi(t,0) = \exp \int_0^t trC(S)ds . \tag{11}$$

This follows from Liouville's Theorem. To show this, let us consider the infinitesimal transformation $\Phi(t+\varepsilon,0) = \Phi(t,0) + \varepsilon\dot{\Phi}(t,0) = [I + \varepsilon C(t)]\Phi(t,0)$, which is a transformation of the state transition matrix at time t to the state transition matrix at time $t+\varepsilon$. Taking the determinant leads to $\det\Phi(t+\varepsilon,0) = \det[I+\varepsilon C(t)]\det\Phi(t,0)$. The determinant of the matrix $[I+\varepsilon C(t)]$ is equal to the product of its eigenvalues. Since the eigenvalues of $[I+\varepsilon C(t)]$ are equal to $1+\varepsilon\lambda_i$, where λ_i are the eigenvalues of $C(t)$, we may write $\det[I+\varepsilon C(t)] = \Pi(1+\varepsilon\lambda_i) = 1+\varepsilon tr\, C(t) + O(\varepsilon^2)$. Using this identity and taking the limit as $\varepsilon \to 0$ yields $\frac{d}{dt}[\det\Phi(t,0)] = tr\; C(t)\det\Phi(t,0)$. This is Liouville's theorem, from which Eq. (11) follows by integration.

Remarks

1. If $tr\ C(t) = 0$, then $\det \Phi(t,0) = \exp \int_0^t tr\ C(s)ds = e^0 = 1$. This implies that the phase flow of the variational equation is volume preserving. Now, assuming A is a constant skew-symmetric matrix, we have $C(t) = AH(\varphi(t))$, where $H(\boldsymbol{x}) = \left(\dfrac{\partial^2 Z(\boldsymbol{x})}{\partial \boldsymbol{x}^2}\right)$ is the Hessian matrix of Z. Using symmetry of H and skew-symmetry of A, it follows that $trC(t) = tr(AH) = tr(AH)^T = tr(HA^T) = -tr(HA) = -tr(AH) = -trC(t)$. Therefore, for a vector field defined by a constant skew-symmetric matrix A, $tr\ C(t) = 0$ and the corresponding motion is volume preserving. Notice that the volume preserving property of these vector fields is already shown by a different method in Section 5.

2. The divergence of a vector field is defined by $\text{div}\ (\boldsymbol{F}(\boldsymbol{x})) = \sum \dfrac{\partial F_i(\boldsymbol{x})}{\partial x_i}$. From the definition of $C(t)$, it follows that $\text{div}\ (\boldsymbol{F}(\boldsymbol{x})) = trC(t)$. Therefore, $\text{div}\ (\boldsymbol{F}(\boldsymbol{x})) = 0$ implies the volume preserving property. To study the properties of the divergence of a vector field under coordinate transformations, assume that the vector field $\dot{\boldsymbol{x}} = \boldsymbol{F}(\boldsymbol{x})$ is transformed into $\dot{\boldsymbol{y}} = \boldsymbol{G}(\boldsymbol{y})$ using the coordinate transformation (3). Then it follows that

$$\left(\frac{\partial \boldsymbol{F}(\boldsymbol{x})}{\partial \boldsymbol{x}}\right)_{\boldsymbol{x}=\boldsymbol{f}(\boldsymbol{y})} = J(\boldsymbol{y}) \left(\frac{\partial \boldsymbol{G}(\boldsymbol{y})}{\partial \boldsymbol{y}}\right) J^{-1}(\boldsymbol{y}) + S(\boldsymbol{y})J^{-1}(\boldsymbol{y}) ,$$

where the elements of matrix $S(\boldsymbol{y})$ are given by $S_{ij} = \sum_{k=1}^{n} \left(\dfrac{\partial J_{ik}}{\partial y_j}\right) G_k$. Applying the trace operation to both sides of this equation leads to

$$tr \left(\frac{\partial \boldsymbol{F}(\boldsymbol{x})}{\partial \boldsymbol{x}}\right)_{\boldsymbol{x}=\boldsymbol{f}(\boldsymbol{y})} = tr \left(\frac{\partial \boldsymbol{G}(\boldsymbol{y})}{\partial \boldsymbol{y}}\right) + tr(S(\boldsymbol{y})J^{-1}(\boldsymbol{y})) .$$

Using the properties of the trace and the determinant, a computation shows that

$$tr(S(\boldsymbol{y})J^{-1}(\boldsymbol{y})) = (\det J(\boldsymbol{y}))^{-1} \sum_{k=1}^{n} \left(\frac{\partial \det J(\boldsymbol{y})}{\partial y_k}\right) G_k .$$

Use of this identity and the equivalence of the divergence and the trace yields

$$\text{div}\ (\boldsymbol{F}(\boldsymbol{x}))_{\boldsymbol{x}=\boldsymbol{f}(\boldsymbol{y})} = (\det J(\boldsymbol{y}))^{-1} \text{div}\ (\det J(\boldsymbol{y}) \boldsymbol{G}(\boldsymbol{y})) . \tag{12}$$

If $\det J = 1$, then $\text{div}\ (\boldsymbol{F}(\boldsymbol{x}))_{\boldsymbol{x}=\boldsymbol{f}(\boldsymbol{y})} = \text{div}\ (\boldsymbol{G}(\boldsymbol{y}))$ which implies that under the group of volume preserving transformations, the divergence of a vector field is invariant. In particular when $\text{div}\ (\boldsymbol{F}(\boldsymbol{x})) = 0$, the motion itself is a sequence of volume preserving transformations and, therefore, the general properties of the divergence free vector fields are equivalent to the group properties of the volume preserving transformations. However, it is important to note that the divergence of a vector field is not invariant under a general transformation as shown by Eq. (12), and, therefore, it is not an invariant property.

7.2 Eigenanalysis

We shall now proceed to the eigenanalysis of the vector space defined by $\Phi(t,0)$. Here we show under what conditions the eigenvalues are independent of the coordinates.

Only in these cases will the eigenvalues reveal the intrinsic properties of a reference orbit.

Let us consider the vector field $\dot{\boldsymbol{x}} = \boldsymbol{F}(\boldsymbol{x})$, and its corresponding variation $\boldsymbol{u}(t) = \Phi_x(t,0)\boldsymbol{u}(0)$, $\boldsymbol{x}(t) = \varphi(t)+\boldsymbol{u}(t)$, along the reference solution $\varphi(t)$. Using the coordinate transformation (3), the vector field is transformed into $\dot{\boldsymbol{y}} = \boldsymbol{G}(\boldsymbol{y})$ and its corresponding variation is $\boldsymbol{v}(t) = \Phi_y(t,0)\boldsymbol{v}(0)$, $\boldsymbol{y}(t) = \boldsymbol{f}^{-1}(\varphi(t)) + \boldsymbol{v}(t)$. From the transformation (3), it follows that $\boldsymbol{y}(t) = \boldsymbol{f}^{-1}(\varphi(t)) + \boldsymbol{v}(t) = \boldsymbol{f}^{-1}(\varphi(t)+\boldsymbol{u}(t)) = \boldsymbol{f}^{-1}(\varphi(t)) + K(\varphi(t))\boldsymbol{u}(t)$, or $\boldsymbol{v}(t) = K(\varphi(t))\boldsymbol{u}(t)$, where $K(\varphi(t)) = \left(\dfrac{\partial \boldsymbol{f}^{-1}(\boldsymbol{x})}{\partial \boldsymbol{x}}\right)_{\boldsymbol{x}=\varphi(t)} = J^{-1}(\boldsymbol{f}^{-1}(\varphi(t)))$ is the Jacobian matrix of the inverse transformation. This leads to

$$\Phi_y(t,0) = K(\varphi(t))\Phi_x(t,0)K^{-1}(\varphi(0)) \,. \tag{13}$$

The eigenvalues are invariant (i.e., independent of the coordinates), if Eq. (13) is a similarity transformation. If the transformations are restricted to the linear group, then K is a constant matrix and Eq. (13) is a similarity transformation for any reference orbit. For a general transformation, Eq. (13) is a similarity transformation if $K(\varphi(t)) = K(\varphi(0))$ for some t. The equality holds for a reference orbit which is either an equilibrium point, i.e., $\varphi(t) = \varphi(0)$ for all t, or a periodic orbit, i.e., $\varphi(T) = \varphi(0)$ for some T. In the first case the eigenvalues of $\Phi(t,0)$ are invariant for all t, while in the second case the invariancy holds only for the monodromy matrix $\Phi(T,0)$.

Remarks

1. For an equilibrium point, $C(t)$ is a constant matrix, and $\Phi(t,0) = e^{Ct} = Pe^{St}P^{-1}$, where the columns of P are the basis for the generalized eigenspaces of C, and S is the real canonical form associated with the eigenvalues of C. For detail see Hirsch and Smale (1974).

2. For a periodic orbit, $C(t)$ is periodic, and by Floquet theory $\Phi(t,0) = P(t)e^{Jt}P^{-1}(0)$, where $P(t)$ is periodic. Here the columns of $P(0)$ are the basis for the generalized eigenspaces of the monodromy matrix $\Phi(T,0)$, and J is the characteristic exponent matrix associated with the eigenvalues of $\Phi(T,0)$. For details see Lefschetz (1962).

3. If A is symplectic, then $A^TA = I$, $AA = -I$, and $\dfrac{d}{dt}[\Phi^T(t,0)A\Phi(t,0)] = \dot{\Phi}^TA\Phi + \Phi^TA\dot{\Phi} = \Phi^TC^TA\Phi + \Phi^TAC\Phi = \Phi^T(HA^TA + AAH)\Phi = 0$. This leads to $\Phi^T(t,0)A\Phi(t,0) = A$, or $\Phi^{-T}(t,0) = A\Phi(t,0)A^{-1}$. Since $\Phi(t,0)$ and $\Phi^{-T}(t,0)$ are related by a similarity transformation, the set of eigenvalues of $\Phi(t,0)$ is preserved under inversion. This implies that the set of eigenvalues of $\Phi(t,0)$ is the same as the set of their reciprocals (i.e., the reciprocal of every eigenvalue is itself an eigenvalue). Therefore, for Hamiltonian systems, the eigenvalues of $\Phi(t,0)$ appear in reciprocal pairs. Notice that this property is not true for an arbitrary constant skew-symmetric matrix A.

ACKNOWLEDGEMENT

The author is indebted to Dr. Victor Szebehely for his encouragement and support. The contributions of NATO ASI and of R. B. Curran Chair at The University of Texas are gratefully acknowledged.

REFERENCES

Abraham, R. H. and Marsden, J. E., 1978, *Foundations of Mechanics*, Benjamin/Cummings, Reading, MA.

Arnold, V. I., 1988, *Dynamical Systems*, Vol. I–V., Springer-Verlag, New York.

Arnold, V. I.,1978, *Mathematical Methods of Classical Mechanics*, Springer-Verlag, New York.

Arnold, V. I., 1973, *Ordinary Differential Equations*, The MIT Press, Cambridge, MA.

Birkhoff, G. D., 1927, *Dynamical Systems*, American Mathematical Society, Providence, RI,.

Coddington, E. A., and Levinson, N., 1962, *Theory of Ordinary Differential Equations*, McGraw-Hill, New York.

Hirsch, M., and Smale, S., 1974, *Differential Equations, Dynamical Systems, and Linear Algebra*, Academic Press, New York.

Hori, G., 1966, "Theory of General Pertrurbations with Unspecified Canonical Variables," *Publ. Astr. Soc. Japan*, 18, 287–296.

Lefschetz, S., 1962, *Differential Equations: Geometric Theory*, Wiley (Interscience), New York.

Pars, L. A., 1965, *Treatise on Analytical Dynamics*, Heinemann Educational Books Ltd., London.

Poincaré, H., 1892, *Les Méthods Nouvelles de la Méchanique Céleste*, Vol. 1, 2, 3; Gauthier-Villars, Paris. Reprinted by Dover, New York, 1957.

Smale, S., 1980, *The Mathematics of Time Essays on Dynamical Systems, Economic Processes, and Related Topics*, Springer-Verlag, New York.

Whittaker, E. T., 1959, *A treatise on the Analytical Dynamics of Particles and Rigid Bodies*, 4th Ed., Cambridge University Press, Cambridge.

ANALYTIC HAMILTONIAN SYSTEMS, THE VICINITY OF A PERIODIC SOLUTION

C. Marchal and J.P. Issartel

D.E.S. ONERA
92320 Chatillon
France

ABSTRACT

After the presentation of ordinary analytic systems of differential equations, the successive simplifications allowing the study of a periodic solution and its vicinity are described.

The first-order study leads to classical results and remains inconclusive in the critical case.

The critical and regular analytic Hamiltonian systems are analysed in great detail and the final conclusion of either instability or all-order stability can be reached in almost all cases.

When there is all-order stability there are also, almost always, an infinite number of Arnold tori.

INTRODUCTION

The interest of periodic solutions and their vicinity was emphasised by Henri Poincaré when he wrote about the three-body problem[1]: "The periodic solutions seems to be our only real means to force this unapproachable fortress..." These solutions have indeed many advantages, not only in the three-body problem but also in many other complex problems:

A) Their existence can generally be rigorously demonstrated.

B) They can be computed with any given accuracy and a numerical solution given for one period remains valid for all time.

C) The continuous families of periodic solutions can be followed numerically step by step by the method of successive infinitesimal modifications of initial conditions.

[1] See Reference 1.

From Newton to Chaos, Edited by A.E. Roy
and B.A. Steves, Plenum Press, New York, 1995

D) In many problems the Poincaré conjecture is valid: "The periodic solutions are dense in the set of the bounded, or even recurrent, solutions". These periodic solutions are thus a very efficient means of exploration.

E) The vicinity of a periodic solution can be analysed accurately with many details both numerically and analytically as emphasised by the theoretical works of Floquet, Poincaré, Liapounov, Birkhoff, Kolmogorov, Arnold, Moser, Nekroroshev...[2]

ORDINARY DIFFERENTIAL EQUATIONS

We will consider at the same time the general case and the case of Hamiltonian systems. The state vector $\vec{X}$ collects all necessary parameters for the system of interest.

$$\vec{X} = (x_1, x_2, \ldots, x_s) \tag{1}$$

Its evolution in terms of the description parameter t (generally the time) is governed by a system of ordinary differential equations:

$$\frac{d\vec{X}}{dt} = F(\vec{X}, t) \tag{2}$$

that is:

$$\frac{dx_k}{dt} = f_k(\vec{X}, t); \quad k = \{1, 2, \ldots, s\} \tag{3}$$

We will assume that the equations of motion (2) and (3) are either autonomous (as in the usual n-body problem) or at least periodic in terms of t (as in the elliptic restricted three-body problem):

$$F(\vec{X}, t) \equiv F(\vec{X}, t + P) \tag{4}$$

The system (1) -(4) is Hamiltonian, with r degrees of freedom, if it can be written:

$$\vec{X} = (\vec{p}, \vec{q}) = (p_1, p_2, \ldots p_r, q_1, q_2 \ldots q_r) \tag{5}$$

with an Hamiltonian function $H(\vec{p}, \vec{q}, t)$ and with:

$$\frac{d\vec{p}}{dt} = -\frac{\partial H}{\partial \vec{q}}; \quad \frac{d\vec{q}}{dt} = \frac{\partial H}{\partial \vec{p}} \tag{6}$$

Most conservative problems can be written in Hamiltonian form and the condition (4) is equivalent to:

$$H(\vec{p}, \vec{q}, t) \equiv H(\vec{p}, \vec{q}, t + P) \tag{7}$$

We will especially consider analytic systems (the functions $f_k(\vec{X}, t)$ and $H(\vec{p}, \vec{q}, t)$ are analytic, or at least decomposable into converging power-Fourier series in a vicinity of the periodic solutions of interest).

THE CLASSICAL SIMPLIFICATIONS

Let $\vec{X}(t)$ be the periodic solution of interest.

$$\forall t \begin{cases} \frac{d\vec{X}(t)}{dt} = F[\vec{X}(t), t] \\ \vec{X}(t) \equiv \vec{X}(t + T) \end{cases} \tag{8}$$

[2]see References 1 to 10.

The period T is arbitrary in the autonomous case, but it must be a multiple of the period P of (4) and (7) in the periodic case.

The first simplification is the reduction to the origin; we will study the small values of $\vec{Y}$:

$$\vec{Y} = \vec{X} - \vec{X}(t) \tag{9}$$

The equation of motion of the small vector $\vec{Y}$ is then:

$$\frac{d\vec{Y}}{dt} = G(\vec{Y}, t) \tag{10}$$

with of course:

$$G(\vec{Y}, t) = F[\vec{X}(t) + \vec{Y}, t] - F[\vec{X}(t), t] \tag{11}$$

and with the properties:

$$G(\vec{0}, t) = 0; \quad G(\vec{Y}, t) \equiv G(\vec{Y}, t + T) = 0 \tag{12}$$

In the Hamiltonian case these conditions (8)-(12) become:

$$\left.\begin{array}{l}
\vec{X}(t) = [\vec{p}(t), \vec{q}(t)]; \quad \vec{p}(t) \equiv \vec{p}(t+T); \quad \vec{q}(t) \equiv \vec{q}(t+T) \\
\vec{p_a} = \vec{p} - \vec{p}(t); \quad \vec{q_a} = \vec{q} - \vec{q}(t) \\
H_a(\vec{p_a}, \vec{q_a}, t) = H(\vec{p}, \vec{q}, t) - \vec{p_a}\frac{d\vec{q}(t)}{dt} + \vec{q_a}\frac{d\vec{p}(t)}{dt} - H[\vec{p}(t), \vec{q}(t), t] \\
H_a(\vec{p_a}, \vec{q_a}, t) \equiv H_a(\vec{p_a}, \vec{q_a}, t + T) \\
H_a(\vec{0}, \vec{0}, t) = 0; \quad \frac{\partial H_a(\vec{0},\vec{0},t)}{\partial \vec{p_a}} = 0; \quad \frac{\partial H_a(\vec{0},\vec{0},t)}{\partial \vec{q_a}} = 0
\end{array}\right\} \tag{13}$$

The Hamiltonian function $H_a(\vec{p_a}, \vec{q_a}, t)$ has neither zero - order terms (in terms of $\vec{p_a}$ and $\vec{q_a}$) nor first-order terms; it can be decomposed into terms of degree n in $\vec{p_a}$, $\vec{q_a}$:

$$H_a(\vec{p_a}, \vec{q_a}, t) = \sum_{n=2}^{\infty} H_{an}(\vec{p_a}, \vec{q_a}, t) \tag{14}$$

The second simplification is a result of the Floquet first-order analysis: this first order analysis will become independent of the time.

To the first-order (10) is:

$$\frac{d\vec{Y}}{dt} = M(t)\vec{Y} + O\{Y^2\} \tag{15}$$

The matrix M(t) has the period T and the first-order analysis is:

$$\frac{d\vec{Y}}{dt} = M(t)\vec{Y} \tag{16}$$

The integration of (16) leads to

$$\vec{Y}(t) = C(t)\vec{Y}(0) \tag{17}$$

with, for the matrix $C(t)$:

$$C(0) = I = unit\ matrix; \quad \frac{dC}{dt} = M(t)C(t) \tag{18}$$

Let us call A the matrix $C(T)$, it is the "monodromy matric" or "transition matrix" or "Floquet matrix", its eigenvalues $\mu_1, \mu_2, \cdots, \mu_s$ are the "Floquet multipliers".

(18) and the periodicity of $M(t)$ imply:

$$\forall t \left\{ \begin{array}{l} C(t+T) = C(t).A \\ det\ C(t) = exp\left[\int_0^t [trace\ M(u)]\ du\right] > 0 \end{array} \right\} \forall m \in \mathbf{Z}: \quad C(mT) = A^m \tag{19}$$

Let $\vec{V_k}$ be an eigenvector of A:

$$A\vec{V_k} = \mu_k \vec{V_k} \tag{20}$$

Hence with (17) and (19) the first-order analysis gives:

$$\vec{Y}(0) = \varepsilon \vec{V_k} \text{ implies } \vec{Y}(mT) = \varepsilon \mu_k^m \vec{V_k} \tag{21}$$

and thus classically, with $\mu = sup|\mu_k|$; $1 \leq k \leq s$:

$$\left\{ \begin{array}{l} \mu > 1 \Rightarrow \text{ the periodic solution } \vec{X}(t) \text{ is "exponentially unstable"} \\ \mu < 1 \Rightarrow \text{ the periodic solution } \vec{X}(t) \text{ is "exponentially stable"} \\ \mu = 1 \text{ :limit or "critical" case, the stability depends on higher-order effects} \end{array} \right.$$

The importance of critical cases comes from conservative systems (and especially Hamiltonian systems) for which the product of all μ_k is one. The exponential stability is then impossible and the critical cases are very common.

The Floquet transformation is a linear and periodic transformation:

$$\vec{Z} = R(t)\vec{Y}; \quad R(t) \equiv R(t+T) \tag{22}$$

leading from the periodic equation $\frac{d\vec{Y}}{dt} = M(t)\vec{Y}$ written in (16) to the corresponding autonomous equation:

$$\frac{d\vec{Z}}{dt} = N\vec{Z} \tag{23}$$

The constant matrix N must verify:

$$\exp(NT) = R(0)\ A\ [R(0)]^{-1} \tag{24}$$

which gives many possible N and leads to:

$$R(t) = \exp(Nt).\ R(0).\ [C(t)]^{-1} \tag{25}$$

The eigenvalues λ_k of N verify $\exp(T\lambda_k) = \mu_k$, their real parts are the Liapunov characteristic exponents.

If the monodromy matrix A is diagonalizable the matrix N can be chosen diagonal and the complete equations (2), (3), (10), (15) become:

$$\frac{dz_k}{dt} = \lambda_k z_k + g_k(z_1, z_2, \cdots, z_s, t); \quad k = \{1, 2, \cdots, s\} \tag{26}$$

The analytic functions $g_k(\vec{Z}, t)$ contain the higher-order terms and are periodic in terms of t with the period T.

In the Hamiltonian case the linear and periodic Floquet transformation is canonical and "symplectic" :

$$\begin{pmatrix} \vec{p_b} \\ \vec{q_b} \end{pmatrix} = R(t) \begin{pmatrix} \vec{p_a} \\ \vec{q_a} \end{pmatrix} \; ; \; R(t) \equiv R(t+T) \; ; \; R(t) \text{ symplectic for all t} \tag{27}$$

The transformation $(H_a, \vec{p_a}, \vec{q_a}, t) \rightarrow (H_b, \vec{p_b}, \vec{q_b}, t)$ can be obtained by the method of Von Zeipel (Ref. 11) with a generatrix C_2-function S with mixed parameters:

$$\left. \begin{array}{l} S(\vec{p_a}, \vec{q_b}, t, H_b) = tH_b + S_1(\vec{p_a}, \vec{q_b}, t) \\ \vec{p_b} = \frac{\partial S}{\partial \vec{q_b}}; \quad \vec{q_a} = \frac{\partial S}{\partial \vec{p_a}}; \quad H_a = \frac{\partial S}{\partial t} \end{array} \right\} \tag{28}$$

This method gives canonical transformations and uses here an homogeneous function S_1 of degree two in $\vec{p_a}$ and $\vec{q_b}$ in order to obtain the linear transformation (27).

The Hamiltonian function H_b corresponding to (26) is then, with its decomposition in terms of degree n in $\vec{p_b}$, $\vec{q_b}$:

$$\left. \begin{array}{l} H_b(\vec{p_b}, \vec{q_b}, t) = \sum_{n=2}^{\infty} H_{bn}(\vec{p_b}, \vec{q_b}, t) \\ H_{b2} = \sum_{k=1}^{r} \lambda_k p_{bk} q_{bk} \\ n \geq 3 \;\; implies: \quad H_{bn}(\vec{p_b}, \vec{q_b}, t) = H_{an}(\vec{p_a}, \vec{q_a}, t) \\ H_b(\vec{p_b}, \vec{q_b}, t) \equiv H_b(\vec{p_b}, \vec{q_b}, t+T) \end{array} \right\} \tag{29}$$

Notice that λ_k is the eigenvalue corresponding to q_{bk} while that corresponding to p_{bk} is $(-\lambda_k)$. In Hamiltonian problems the eigenvalues are pairs of opposites.

If the monodromy matrix A is not diagonalisable (exceptional case) these expressions (26)-(29) are only slightly modified (Ref. 12, 13).

THIRD SIMPLIFICATION. THE NEAR RESONANCE THEOREM

The equations of motion (26) and the Hamiltonian function H_b of (29) are equivalent to the initial equations and Hamiltonian, but they are also much simpler after the suitable linear and periodic transformation from $\vec{X}$ to $\vec{Z}$ and from $\vec{p}, \vec{q}$ to $\vec{p_b}, \vec{q_b}$.

The near-resonance theorem uses the same idea for higher order terms: a suitable T-periodic transformation $\vec{Z} \rightarrow \vec{U}$ will allow us to simplify them considerably (Ref. 12, 13).

That theorem is valid not only for analytic Hamiltonian systems but also for all analytic systems and even for systems that are locally decomposable into converging power-Fourier series. Let us then use the expression (26) and let us decompose the functions $g_k(\vec{Z}, t)$ into power-Fourier series.

$$g_k(\vec{Z}, t) = \sum_{\vec{\alpha}, m} G_k(\vec{\alpha}, m) \vec{Z}^{\vec{\alpha}} \exp\{mi\Omega t\} \tag{30}$$

with:

$$\left. \begin{array}{l} \vec{\alpha} = (\alpha_1, \alpha_2, \alpha_3, \cdots, \alpha_s); \quad \alpha_1, \alpha_2, \cdots, \alpha_s \in \mathbf{N}; \\ \alpha_1 + \alpha_2 + \cdots + \alpha_s \geq 2; \quad \vec{Z}^{\vec{\alpha}} = z_1^{\alpha_1} z_2^{\alpha_2} \cdots z_s^{\alpha_s}; \\ m \in \mathbf{Z}; \quad G_k(\vec{\alpha}, m) \in \mathbf{C}; \quad \Omega = 2\Pi/T \end{array} \right\} \tag{31}$$

The near resonance theorem is then the following:

For any positive scalar ε, even very small, there exists a transformation $\vec{Z} \rightarrow \vec{U}$ locally decomposable into converging power-Fourier series and such that:

$$\left. \begin{array}{l} \vec{U} - \vec{Z} = O(Z^2): \;\; \vec{U} \; and \; \vec{Z} \text{ are equivalent} \\ \vec{U}(\vec{Z}, t) \equiv \vec{U}(\vec{Z}, t+T)\text{: the transformation has the period T} \end{array} \right\} \tag{32}$$

$$\left.\begin{array}{l}\frac{du_k}{dt} = \lambda_k u_k + j_k(\vec{U},t); \quad k=(1,2,\ldots s)\\ j_k(\vec{U},t) = \sum_{\vec{\alpha},m} J_k(\vec{\alpha},m).\vec{U}^{\vec{\alpha}}.\exp(mi\Omega t)\end{array}\right\} \quad (33)$$

The power-Fourier series $j_k(\vec{U},t)$ have only "near-resonant terms" i.e. terms for which:

$$\left|\sum_{q=1}^{s} \alpha_q \lambda_q + mi\Omega - \lambda_k\right| \leq \varepsilon(\alpha_1 + \alpha_2 + \ldots + \alpha_s) \quad (34)$$

The near-resonance theorem leads to many properties. For instance if ε is smaller than any positive $\mathcal{R}e(\lambda_k)$ the usual "unstable subset" of the periodic trajectory of interest is the subspace of all u_k with positive $\mathcal{R}e(\lambda_k)$. The integration of the equations of motion (32) in that subspace is very simple (and become of the type $u_k = P_k(t).\exp\{\lambda_k t\}$, with $P_k(t)$ a polynomial in terms of t, for sufficiently small ε).

The Hamiltonian version of this theorem leads to a periodic canonical transformation $(\vec{p_b},\vec{q_b},t) \to (\vec{p_u},\vec{q_u},t)$ and to a simplified Hamiltonian function:

$$H_u(\vec{p_u},\vec{q_u},t) = \sum_{k=1}^{r} \lambda_k p_{uk} q_{uk} + \sum_{\vec{\alpha},\vec{\beta},m} K(\vec{\alpha},\vec{\beta},m)\vec{p_u}^{\vec{\alpha}}\vec{q_u}^{\vec{\beta}} exp\{mi\Omega t\} \quad (35)$$

that contains only near resonant terms:

$$\left.\begin{array}{l}|mi\Omega + \sum_{k=1}^{r} \lambda_k(\beta_k - \alpha_k)| \leq \varepsilon D\\ \text{with D = degree of the term of interest} = \sum_{k=1}^{r}(\alpha_k + \beta_k) \geq 3\end{array}\right\} \quad (36)$$

THE TYPES OF STABILITY AND INSTABILITY. THE CRITICAL AND REGULAR HAMILTONIAN SYSTEMS

We have already met the "exponential stability" and the"exponential instability" given by the first-order analysis of the periodic solution of interest and its vicinity. With the possibilities given by the limit "critical case" we can list the six following main types for analytic systems, from the most unstable to the most stable.

A) The exponential instability, as in $\frac{dx}{dt} = \lambda x; \lambda > 0$
B) The power-n instability, as in $\frac{dx}{dt} = x^n; n \geq 2$
C) The Arnold diffusion: escape slower than any power of x (i.e. any power of the distance to the trajectory of interest).
D) The non-asymptotical stability.
E) The power-n stability, as in $\frac{dx}{dt} = -x^n$; n odd.
F) The exponential stability, as in $\frac{dx}{dt} = \lambda x; \lambda < 0$ (37)

For Hamiltonian systems the types E and F are impossible while the types C and D are called "all-order stability" even if the type C is theoretically unstable (but practically almost as good as the type D: what is the meaning of an instability requiring years, if not millions of years, to appear?)

We will now consider especially the critical and regular analytic Hamiltonian systems. They represent a large majority of not yet solved systems and, because of the difficulty of discrimination between the types C and D, the remaining discussion is essentially the choice between the "all-order stability" (cases C and D) and the "power-n instability" (case B). In the latter case it is also the determination of n.

Finally an analytic system is "regular" if its monodromy matrix A is diagonalisable (a usual case that includes all cases in which the Floquet multipliers μ_k are different). Then critical and regular analytic and periodic systems can be reduced (by an analytic and periodic transformation) to systems of the type (35)-(36) in which all eigenvalues λ_k are purely imaginary, and thus all Liapunov characteristic coefficients are zero.

Notice that, while $\vec{p_a}, \vec{q_a}, H_a$ used in (13) after the first simplification are real, this cannot be the case for $\vec{p_b}, \vec{q_b}, H_b$ of (27)-(29) and $\vec{p_u}, \vec{q_u}, H_u$ of (35)-(36) because of the complex character of the eigenvalues λ_k.

If we are interested in the reality of expressions the transformations (27) and (28) must be real. For critical and regular Hamiltonian systems the simplest possible expressions are then the following after the succession of simplifications and with the real quantities $\vec{p_N}, \vec{q_N}, H_N$:

$$\left.\begin{array}{l} \vec{p_N} = (p_{N1}, p_{N2}, \ldots, p_{Nr}); \quad \vec{q_N} = (q_{N1}, \ldots, q_{Nr}) \\ p_{Nk} + iq_{Nk} = z_{Nk}; \quad p_{Nk} - iq_{Nk} = \overline{z_{Nk}}; \quad k = \{1, 2, \ldots, r\} \\ (z_{N1}, z_{N2}, \ldots, z_{Nr}) = \vec{z_N}; \quad (\overline{z_{N1}}, \ldots, \overline{z_{Nr}}) = \vec{\overline{z_N}} \\ H_N = \sum_{k=1}^{r} \frac{\omega_k}{2}(p_{Nk}^2 + q_{Nk}^2) + \sum_{\vec{\alpha},\vec{\beta},m} K_N(\vec{\alpha}, \vec{\beta}, m).\vec{z_N}^{\vec{\alpha}}.\vec{\overline{z_N}}^{\vec{\beta}}.exp\{mi\Omega t\} \\ \text{with } \vec{\alpha} = (\alpha_1, \alpha_2, \ldots, \alpha_r); \quad \vec{\beta} = (\beta_1, \beta_2, \ldots, \beta_r) \\ \text{All } \alpha_k \text{ and } \beta_k \in \mathbf{N} \\ \vec{z_N}^{\vec{\alpha}} = z_{N1}^{\alpha_1}.z_{N2}^{\alpha_2}.\ldots.z_{Nr}^{\alpha_r}; \quad \vec{\overline{z_N}}^{\vec{\beta}} = \overline{z_{N1}}^{\beta_1}.\ldots.\overline{z_{Nr}}^{\beta_r} \end{array}\right\} \qquad (38)$$

The reality condition of H_N is obviously:

$$\forall \vec{\alpha}, \vec{\beta}, m: \quad K_N(\vec{\beta}, \vec{\alpha}, -m) = \overline{K_N}(\vec{\alpha}, \vec{\beta}, m) \qquad (39)$$

and the near-resonance condition (36) becomes

$$\left.\begin{array}{l} |m\Omega + \sum_{k=1}^{r} \omega_k(\alpha_k - \beta_k)| \leq \varepsilon D \\ \text{with: } \Omega = 2\pi/T = \text{ general pulsation} \\ D = \sum_{k=1}^{r}(\alpha_k + \beta_k) = \text{ degree of the term of interest } \geq 3 \end{array}\right\} \qquad (40)$$

The series of H_N is absolutely converging for all t in a small vicinity of the origin but unfortunately that vicinity is usually going to zero with ε.

The r real quantities ω_k are the usual "eigenpulsations", with a definite sign that is not at all indifferent. The sign of ω_k will precisely be defined by the expression of H_N in (38).

The quantities $\omega_k/2\pi$ are the usual "eigenfrequencies".

The relations between $\vec{p_u}, \vec{q_u}, H_u$ of (35)-(36) and $\vec{p_N}, \vec{q_N}, H_N$ of (38)-(40) for the same value of ε are then the following:

$$\left.\begin{array}{l} \text{The } (2r+1) \text{ ratios } (p_{uk}/\overline{z_{Nk}}),\ (q_{uk}/z_{Nk}) \text{ and } (H_u/H_N) \text{ are fixed.} \\ \text{The roles of } \vec{\alpha} \text{ and } \vec{\beta} \text{ are interchanged} \\ \lambda_k = i\omega_k \text{ for } k = (1, 2, \ldots, r) \end{array}\right\} \qquad (41)$$

The $2r$ eigenvalues are then the purely imaginary quantities $\pm i\omega_k$; for $k = (1, 2, \ldots, r)$.

It is of course possible to consider that the $2r$ conjugate parameters are $\overline{z_{Nk}}$, z_{Nk} instead of p_{Nk}, q_{Nk}. The Hamiltonian function becomes then $H_{Nz} = 2iH_N$ and is purely imaginary.

$$H_{Nz} = i\sum_{k=1}^{r} \omega_k z_{Nk}\overline{z_{Nk}} + 2i \sum_{\vec{\alpha},\vec{\beta},m} K_N(\vec{\alpha}, \vec{\beta}, m)\vec{z_N}^{\vec{\alpha}}\vec{\overline{z_N}}^{\vec{\beta}} exp\{mi\Omega t\} \qquad (42)$$

with Hamiltonian equations of motion that maintain the complex conjugacy:

$$\left.\begin{array}{l}\left.\begin{array}{l}\frac{dz_{Nk}}{dt} = \frac{\partial H_{Nz}}{\partial \overline{z_{Nk}}} \\ \frac{d\overline{z_{Nk}}}{dt} = -\frac{\partial H_{Nz}}{\partial z_{Nk}}\end{array}\right\} \\ \text{usual "formal partial derivatives" of complex conjugate quantities:} \\ \frac{\partial(z^\alpha \overline{z^\beta})}{\partial z} = \alpha z^{(\alpha-1)}\overline{z^\beta}\end{array}\right\} \qquad (43)$$

We can of course choose ε in (40) as small as desired and we can thus reduce at will the number of non strictly resonant terms of a given degree. **We will define the integer N of $\mathbf{H}_N, \vec{\mathbf{p}}_\mathbf{N}, \vec{\mathbf{q}}_\mathbf{N}, \mathbf{H}_{NZ}$ as the largest degree for which the near resonance condition (40) implies a strict resonance:**

$$\left.\begin{array}{l}D = \sum_{k=1}^r(\alpha_k + \beta_k) \leq N \ implies: \ m\Omega + \sum_{k=1}^r(\alpha_k - \beta_k)\omega_k = 0 \\ \text{and also then: } K_N(\vec{\alpha}, \vec{\beta}, m) \text{ is independent of N}\end{array}\right\} \qquad (44)$$

These $K_N(\vec{\alpha}, \vec{\beta}, m)$ for $N \geq D$ are the generalisation of "Birkhoff invariant coefficients". This definition of N implies also:

$$\left.\begin{array}{l}\left.\begin{array}{l}\vec{p_{N+1}} - \vec{p_N} \\ \vec{q_{N+1}} - \vec{q_N}\end{array}\right\} \text{ are of } N^{th} \text{ order} \\ H_{N+1} - H_N \text{ is of } (N+1)^{th} \text{ order} \\ H_N, H_{N+1}, H_{N+2}, \ldots \text{ have an identical expression up to the degree N}\end{array}\right\} \qquad (45)$$

THE FIRST "QUASI-INTEGRALS"

Assume that the relation of strict resonance:

$$m\Omega + \sum_{k=1}^{r}(\alpha_k - \beta_k)\omega_k = 0 \qquad (46)$$

implies

$$\alpha_5 - \beta_5 = 0 \qquad (47)$$

(that is: there is no rational relation between ω_5 and the r other pulsations $\Omega, \omega_1, \omega_2, \omega_3, \omega_4, \omega_6, \omega_7, \ldots, \omega_r$) it is then easy to verify that the quantities w_{N5} given by:

$$w_{N5} = p_{N5}^2 + q_{N5}^2 = z_{N5}\overline{z_{N5}} \geq 0; \quad N = \{3, 4, 5, 6, \ldots\} \qquad (48)$$

are such that:

$$\left.\begin{array}{l}\frac{dw_{N5}}{dt} = (N+1)^{th} \text{ order} \\ w_{N+1,5} - w_{N5} = (N+1)^{th} \text{ order}\end{array}\right\} \qquad (49)$$

A sequence as $w_{3,5}, w_{4,5}, w_{5,5} \ldots$ having the properties (49) will be called a "quasi-integral". If it converges toward a true integral of motion it would be extremely helpful, if not it has nevertheless a great interest: the quasi-integral w_{N5}; $N = \{3, 4, 5 \ldots\}$ shows that the parameters p_{N5} and q_{N5} have the "all-order stability".

More generally let us call $\vec{w_N}$ the following vector:

$$\left.\begin{array}{l}\vec{w_N} = (w_{N1}, w_{N2}, \ldots, w_{Nr}) \text{ with} \\ w_{Nk} = p_{Nk}^2 + q_{Nk}^2 = z_{Nk}\overline{z_{Nk}} \geq 0; \quad k = \{1, 2, \ldots, r\}\end{array}\right\} \qquad (50)$$

The vector $\vec{w_N}$ is a vector of the $\mathbf{R}^r$ space and even a vector of its "positive corner" $(\mathbf{R}_+)^r$.

Notice that a term as $K_N(\vec{\alpha}, \vec{\beta}, m) \vec{z_N}^{\vec{\alpha}} \vec{\bar{z}_N}^{\vec{\beta}} \exp(mi\Omega t)$ of the Hamiltonian H_N will move the vector $\vec{w_N}$ in a direction parallel to $(\vec{\alpha} - \vec{\beta})$ and also notice that the relation (46) of strict resonance can be written:

$$\left.\begin{array}{l} (\vec{\alpha} - \vec{\beta})\vec{\omega} = 0 \text{ modulo } \Omega \\ \text{with of course: } \vec{\omega} = (\omega_1, \omega_2, ..., \omega_r) \end{array}\right\} \quad (51)$$

Hence an analysis using only the $(r+1)$ pulsations $\Omega, \omega_1, \omega_2, ..., \omega_r$ leads to the following.

1) Let us look for vectors $\vec{V} = \vec{\alpha} - \vec{\beta}$, with integer coordinates and satisfying the strict resonance condition:

$$\vec{V}\vec{\omega} = 0 \text{ modulo } \Omega \quad (52)$$

The corresponding degree D will verify:

$$D = \sum_{k=1}^{r} (\alpha_k + \beta_k) \geq \sum_{k=1}^{r} |v_k|; \text{ with } \vec{V} = (v_1, v_2, ..., v_r) \quad (53)$$

2) As long as, for increasing degrees D, the possible vectors $\vec{V}$ and the corresponding vector-space will not intersect the "positive corners" $(\mathbf{R}_+)^r$ outside the origin it would be impossible to escape: this gives an obvious lower limit of n for a "power-n instability".

3) This property can sometimes be extended to terms of all degrees, there is then "all-order stability" and there exists then a vector $\vec{c}$ such that:

$$\left.\begin{array}{l} \text{All r coordinates of } \vec{c} \text{ are positive} \\ (\vec{\alpha} - \vec{\beta})\vec{\omega} = 0 \text{ modulo } \Omega \text{ implies } (\vec{\alpha} - \vec{\beta})\vec{c} = 0 \end{array}\right\} \quad (54)$$

The sequence of scalar products $\vec{c}\vec{w_N}$; $N = \{3, 4, 5, ...\}$ is then a quasi-integral the existence of which is indeed a sufficient condition of all-order stability (this sequence can be used in order to improve the results of the Nekroroshev theorem, Ref 8).

The condition (54) is just opposed to the existence of "positive resonances".

$$\left.\begin{array}{l} \vec{a}\vec{\omega} = \sum_{k=1}^{r} a_k \omega_k = 0 \text{ modulo } \Omega \\ \text{with } \vec{a} \neq 0; \text{ all } a_k \in \mathbf{N} \end{array}\right\} \quad (55)$$

Thus the "positive resonances" (55) are the only ones that are dangerous for the stability: in their absence there is "all-order stability".

This result is an extension of the classical Lejeune-Dirichlet theorem: "If the Hamiltonian H_N is autonomous it is also an integral of motion, if furthermore all ω_k of (38) are positive that integral of motion is minimum at the origin, which implies stability".

We find indeed then that the pulsation Ω is arbitrary, the strict resonance condition becomes $\vec{a}\vec{\omega} = 0$ and this forbids positive resonances with $\vec{a} \neq 0$, all $a_k \geq 0$, all $\omega_k > 0$.

Notice that we can obtain quasi-integrals of the type $\vec{c}\vec{w_N}$; $N = \{3, 4, 5, ...\}$ even if some coordinates of $\vec{c}$ are negative. It is sufficient to verify the second condition of (54):

$$(\vec{\alpha} - \vec{\beta})\vec{\omega} = 0 \text{ modulo } \Omega \text{ implies } (\vec{\alpha} - \vec{\beta})\vec{c} = 0 \quad (56)$$

These quasi-integrals don't give directly the "all-order stability" but they will be very helpful in the analysis of the next section.

THE FOURTH SIMPLIFICATION. THE "MAIN PROBLEM"

The dangerous "positive resonances" (55) are very common and it is thus necessary to go beyond the analysis of the general pulsation Ω and the eigenpulsations $\omega_1, \omega_2, ..., \omega_r$.

The following fourth simplification is then very useful.

Let us define the real scalars P_{Nk}, Q_{Nk} and the complex scalars $Z_{Nk}, \overline{Z_{Nk}}$ by:

$$\left.\begin{array}{l} Z_{Nk} = P_{Nk} + iQ_{Nk} = z_{Nk}\exp\{-i\omega_k t\} \\ \overline{Z_{Nk}} = P_{Nk} - iQ_{Nk} = \overline{z_{Nk}}\exp\{i\omega_k t\} \end{array}\right\} \tag{57}$$

The transformation $(\vec{p_N}, \vec{q_N}) \to (\vec{P_N}, \vec{Q_N})$, with of course $\vec{P_N} = (P_{N1}, P_{N2}, ..., P_{Nr})$; $\vec{Q_N} = (Q_{N1}, Q_{N2}, ..., Q_{Nr})$ is canonical and leads to the following Hamiltonian function $\mathcal{H}_N(\vec{P_N}, \vec{Q_N}, t)$.

$$\mathcal{H}_N(\vec{P_N}, \vec{Q_N}, t) = H_N(\vec{p_N}, \vec{q_N}, t) - \sum_{k=1}^{r} \frac{\omega_k}{2}(p_{Nk}^2 + q_{Nk}^2) \tag{58}$$

The Hamiltonian $\mathcal{H}_N(\vec{P_N}, \vec{Q_N}, t)$ is then real and can be developed into power-Fourier series that are converging in a small vicinity of the origin.

$$\mathcal{H}_N(\vec{P_N}, \vec{Q_N}, t) = \sum_{(\vec{\alpha},\vec{\beta},m)} K_N(\vec{\alpha}, \vec{\beta}, m).\vec{Z_N}^{\vec{\alpha}}.\vec{\overline{Z_N}}^{\vec{\beta}}.\exp\{[m\Omega + (\vec{\alpha} - \vec{\beta})\vec{\omega}]it\} \tag{59}$$

The real Hamiltonian $\mathcal{H}_N(\vec{P_N}, \vec{Q_N}, t)$ has no more terms of degree 0, 1 or 2, it is no more periodic, it is quasi-periodic and, because of (44), it is autonomous up to the degree N.

As for $\vec{\overline{z_N}}, \vec{z_N}$ we can consider that the 2r conjugate parameters are $\overline{Z_{Nk}}, Z_{Nk}, k = \{1, 2, ..., r\}$. The new Hamiltonian function $\mathcal{H}_{NZ}$ and the equations of motion are then:

$$\left.\begin{array}{l} \mathcal{H}_{NZ}(\vec{\overline{Z_N}}, \vec{Z_N}, t) = 2i\mathcal{H}_N\text{: purely imaginary} \\ \frac{d\overline{Z_{Nk}}}{dt} = -\frac{\partial \mathcal{H}_{NZ}}{\partial Z_{Nk}};\ \frac{dZ_{Nk}}{dt} = \frac{\partial \mathcal{H}_{NZ}}{\partial \overline{Z_{Nk}}};\ k = \{1, 2, ..., r\} \end{array}\right\} \tag{60}$$

$\mathcal{H}_N$ being autonomous up to the degree of N and autonomous Hamiltonians being integrals of motion we have to consider two possibilities.

I) Either all $\mathcal{H}_N$ have no terms up to the degree N.

The 2r sequences of parameters P_{Nk} and $Q_{Nk}, N = \{3, 4, ..\}$ are then quasi-integrals and there is "all-order stability".

II) Or for some N the Hamiltonian $\mathcal{H}_N$ has non-zero autonomous terms of degree smaller than or equal to N and, because of (44), these terms appear also in all subsequent $\mathcal{H}_N$.

The sequence of $\mathcal{H}_N, N = \{3, 4, ...\}$, is then a new quasi-integral and, with the quasi-integrals of the type $\vec{c}\vec{w_N}$, it can sometimes lead to the "all-order stability".

The main part of the Hamiltonian $\mathcal{H}_N$ for large N is a real autonomous and homogeneous polynomial of degree M (larger than 2) that we will call h_M.

The problem of the motion with the Hamiltonian $h_M(\vec{P}, \vec{Q})$ will be called "main problem", it is both simpler than the real problem and extremely close to it.

Is there some interest in the study of the main problem? What are the similarities between the real and the main problems?

The two following very interesting properties have been demonstrated (Ref 14).
First property:

$$\left\{\begin{array}{l}\text{If the main problem has a solution that is:}\\ \text{A) Asymptotic to the origin when } t \to -\infty.\\ \text{B) Escaping as fast as possible for a Hamiltonian of degree M, that}\\ \text{is as a power (M-1) instability.}\\ \text{Then the real problem has also the same type of asymptotic solution}\\ \text{and has thus a power (M-1) instability.}\end{array}\right.$$

Such asymptotic solutions with a power (M-1) escape are characterized by the following: when t approaches minus infinity the distance $\rho(t)$ to the origin verifies:

$$A(-t)^{\frac{-1}{M-2}} \leq \rho(t) \leq B(-t)^{\frac{-1}{M-2}}, \text{ with } 0 < A \leq B < +\infty \tag{61}$$

For instance

$$\left.\begin{array}{l} M \geq 3; r = 1; h_M = i\frac{(Z^M - \overline{Z}^M)}{2M(M-2)} \\ \frac{dZ}{dt} = 2i\frac{\partial h_M}{\partial \overline{Z}} = \frac{\overline{Z}^{(M-1)}}{(M-2)} \end{array}\right\} \tag{62}$$

which gives the "radial asymptotic solutions":

$$Z = (t_0 - t)^{[\frac{-1}{M-2}]} \tag{63}$$

We may think that to each favourable asmyptotic solution $\vec{Z_m}(t)$ of the main problem corresponds an asymptotic solution $\vec{Z_r}(t)$ of the real problem that is "tangent" to $\vec{Z_m}(t)$, that is a real solution for which:

$$\lim_{t \to -\infty} \frac{||\vec{Z_r}(t) - \vec{Z_m}(t)||}{||\vec{Z_m}(t)||} = 0 \tag{64}$$

However this is not always true. It happens that only a part of the favourable asymptotic solutions $\vec{Z_m}(t)$ are sufficiently regular and "robust", i.e. have a corresponding $\vec{Z_r}(t)$ satisfying (64).

Of course that part of robust solutions is always non-empty and we can use the "first property" as soon as we have found a favourable asymptotic solution.

Second Property:

The second property concerns the stability and not the instability.

Notice that in most cases the number M is equal to 4. Indeed terms in $Z_{Nk}^2 \overline{Z}_{Nk}^2$ and in $Z_{Nk}\overline{Z_{Nk}}Z_{NL}\overline{Z_{NL}}$ are always strictly resonant and we thus usually have with a fixed, real and symmetrical matrix **F**:

$$\left.\begin{array}{l} h_M = \vec{w_N}\mathbf{F}(^t\vec{w_N}) \text{ ; M} = 4 \\ \text{with as in (50):} \\ \vec{w_N} = (w_{N1}, w_{N2}, ..., w_{Nr}) \\ w_{Nk} = p_{Nk}^2 + q_{Nk}^2 = P_{Nk}^2 + Q_{Nk}^2 = Z_{Nk}\overline{Z_{Nk}}; k = \{1, ..., r\} \\ \text{and with: } (^t\vec{w_N}) = \text{ transposed of } \vec{w_N} \end{array}\right\} \tag{65}$$

The motions given by h_M are then simple: all w_{Nk} are constant and the motions of the Z_{Nk} are of the type "differential rotation".

These motions are stable and the main questions are then the following: In which cases, according to the matrix F, this stability is robust, i.e. is always transmitted to the real problem (under the form "all-order stability")? If not what limit can we give on n for a power-n instability?

The present results are the following (Ref 14).

Let us look for a possible power-n instability; it will require $n \geq 4$.

We will need:

A) The strictly resonant vectors $\vec{V}_j$ with integer coordinates $v_{1j}, v_{2j}, ..., v_{rj}$ and such that:

$$\vec{V}_j.\vec{\omega}_j = 0, \text{ modulus } \Omega \text{ (condition of strict resonance)} \tag{66}$$

$$\sum_{k=1}^{r} |v_{kj}| \leq 3n - 7 \tag{67}$$

B) A non-zero real vector $\vec{W}$ such that:

$$\vec{W} \in (\mathbf{R}_+)^r \text{ : the r components of } \vec{W} \text{ are positive or zero} \tag{68}$$

$$\vec{W} = \sum_j b_j \vec{V}_j \text{ : for some vectors } \vec{V}_j \text{ of (66) - (67)} \tag{69}$$

$$\vec{V}_j F({}^t\vec{W}) = 0 \text{ : for all } \vec{V}_j \text{ of the sum (69)} \tag{70}$$

Notice that (69) and (70) imply:

$$\vec{W} F({}^t\vec{W}) = 0 \tag{71}$$

If no such non-zero real vector $\vec{W}$ exists the periodic solution of interest has no power-n instability.

Of course if for all n no power-n instability exists the periodic solution of interest has the "all-order stability".

The demonstration is based on the construction of new quasi-integrals using rational fractions of the P_{Nk} and Q_{Nk} (with as denominators only polynomials of the w_{Nk}).

For almost all matrices F no $\vec{W}$ vector exists and there is "all-order stability". But if a vector $\vec{W}$ exists in (66)-(71) it is possible to build examples of Hamiltonians $\mathcal{H}_N = \vec{w_N} F({}^t\vec{w_N})$ + a finite number of higher order terms, such that the motion has a power-n instability with radial escapes of $\vec{w_N}$ in the direction of $\vec{W}$.

Finally let us notice that if det. F $\neq 0$ the conditions of application of the K.A.M. theorem are satisfied (Ref 15) and an infinity of Arnold tori exist in the vicinity of the periodic solution of interest: most neighbouring solutions will remain forever in a close vicinity.

Thus it is possible to build examples of systems with a power-n instability but also with stability of most neighbouring solutions.

CONCLUSION

The periodic solutions of analytic Hamiltonian systems and the neighbouring solutions can be analysed very accurately. This leads to many possible conclusions corresponding to very different possible types of motion.

The three main types of motion are the exponential instability, the power-n instability and the all-order stability. The first type appears with the classical first-order analysis while the other two types correspond to critical systems and require first an analysis of eigenfrequencies. The "positive resonances" are the only ones that are dangerous for the stability and that may lead to a power-n instability.

In this last case the influence of high-order terms become essential and our utmost simplification leads to a "main problem" very near to the real problem, but also much simpler.

When the main problem has its maximum possible instability the real problem has the same power-n instability and it seems that then the instability is general: almost all neighbouring solutions would escape. When the main problem is stable the real problem is very often all-order stable and simple corresponding conditions can be developed. These conditions also give informations on the presence of an infinite number of neighbouring Arnold tori.

However there remain several limit cases in which the analysis remains incomplete such as those of critical and non regular Hamiltonian systems. There remains also a major question: "The all-order stability corresponds either to a true non-asymptotic stability or to an extremely slow Arnold diffusion and, even if the physical differences between these two types of motion are mostly theoretical, it would be interesting to discriminate them."

REFERENCES

1. Poincaré, H. : 1957, "Les Méthodes nouvelles de la Mécanique Céleste", Dover Publication, New York, Vol. 1, p82.

2. Floquet, G. : 1883, "Sur les équations différentielles linéaires à coefficients périodiques", Annales de l'Ecole Normale Supérieure, Paris 12, p47 - 89.

3. Liapounov : 1907, "Problème général de la stabilité du mouvement", Annales de la Faculté des Sciences de l'Université de Toulouse, 9.

4. Birkhoff, G. D. : 1927, "Dynamical Systems", Am. Math. Soc. Colloq. Publ. IX, New York, Am. Math. Soc. VIII.

5. Kolmogorov, A. N. : 1957, "General theory of dynamical systems and classical mechanics", Proc. Int. Congr. Math. 1954 - Amsterdam 1 p315 - 333.

6. Arnold, V. I. : 1988, "Equations différentielles ordinaires", Editions Mir, Moscou.

7. Moser J. : 1980, "Various aspects of integrable Hamiltonian systems. Dynamical systems", C.I.M.E. Lectures, Bressanone 1978, Progress in Mathematics 8, p233 -290.

8. Nekroroshev : 1977, "An exponential estimate of the time of stability of near integrable Hamiltonian systems", Uspekhi Mat. Nauk. 32, 6 (1977), Russian Math. Surveys 32, 6.

9. Brjuno : 1967, "Formal stability of Hamiltonian systems", Math. Notes 1, p216 - 219.

10. Glimm : 1964, "Formal stability of Hamiltonian systems", Communications on Pure and Applied Math. 17.

11. Von Zeipel, H. : 1916 - 1917, "Recherches sur le mouvement des petites planètes", Ark Matematik Astronomi och Fysik 11(1), 12(9), 13(3).

12. Marchal, C. : 1988, "The near-resonance theorem: Analysis of the vicinity of periodic solutions of analytic differential systems" in "Long term dynamical behaviour of natural and artificial n-body systems", A. E. Roy editor, Nato Advanced Study Institute, Cortina d'Ampezzo, 2 - 13 August 1987, Kluwer Academic Publishers.

13. Marchal, C. : 1990, "The three body problem", Chapter 10.7.7 to 10.7.10, p174 - 247, Elsevier Science Publications B.V.

14. Issartel, J. P. : 1993, Systèmes dynamiques. Les résonances et les conditions de la stabilité", Thèse de Doctorat de l'Université Paris 6, 23 Septembre.

15. Arnold V.I.: 1988 "Dynamical systems III" Encyclopedia of Mathematical Sciences, Vol.3; R.V. Gamkrelidze ed. Springer-Verlag, Berlin Heidelberg, p182-199

COMPLEX INSTABILITY

H.Papadaki, G.Contopoulos
and C.Polymilis

Department of Astronomy
University of Athens
Panepistimiopolis
GR 157 83-Athens, Greece

ABSTRACT

We describe a linear theory that represents with great accuracy the orbits close to a complex unstable orbit in a system of three degrees of freedom. The complex unstable orbit has a stable and an unstable manifold of dimension 3 in the 5-dimensional phase space of constant energy. The successive consequents of orbits on a Poincare surface of section form a spiral in a 2-dimensional projection, or a helix in a 3-dimensional space. The agreement between theory and numerical integrations is very good if the distance of the consequents from the periodic orbit is small.

INTRODUCTION

In a Hamiltonian system a stable family of periodic orbits becomes unstable when the eigenvalues of the monodromy matrix of the corresponding linearized equations leave the unit circle. In Hamiltonian systems of 3 degrees of freedom the transition to instability may occur when one or two pairs of eigenvalues leave the unit circle. Complex instability occurs when two pairs of the eigenvalues collide on the unit circle at points not equal to ± 1 and leave the unit circle (Arnold 1978), moving outwards and inwards. This type of instability does not appear in systems of 2 degrees of freedom.

Complex instability has been observed in families of periodic orbits of a time independent 3-D Hamiltonian system

$$H = \frac{1}{2}(\dot{x}^2+\dot{y}^2+\dot{z}^2) + \frac{1}{2}(Ax^2+By^2+Cz^2) - \varepsilon xz^2 - \eta yz^2 = h \qquad (1)$$

representing 3 coupled harmonic oscillators (Magnenat 1982; Contopoulos 1983, 1986; Contopoulos and Magnenat 1985). This system may represent the central part of a deformed galaxy.

The purpose of this study is to investigate analytically the form of the asymptotic manifolds of a complex unstable periodic orbit very close to it. We found a linear theory that

From Newton to Chaos, Edited by A.E. Roy
and B.A. Steves, Plenum Press, New York, 1995

describes accurately these asymptotic manifolds close to the periodic orbit. Then we used the results of this theory to calculate numerically the form of the asymptotic manifolds far from the periodic orbit where the linear theory is no more valid.

DEFINITIONS

A set $S \subset R^n$ is defined as **invariant** for the flow $f_t(x)$, if for any x_o S we have $f_t(x)$ S for all t R (Guckenheimer and Holmes 1983; Wiggins 1990).

Let $A: R^n \longrightarrow R^n$ be an operator and $\lambda_1, \lambda_2, \ldots \lambda_n$ its eigenvalues. Suppose that λ_i, $i=1,2,\ldots,n$ are pairwise distinct; v of them are real and 2μ are complex conjugate $(2\mu+v=n)$. Then the linear space R^n decomposes into a direct sum of v one-dimensional and μ two-dimensional invariant subspaces with respect to A.

Each of the v one-dimensional subspaces is generated by one of the v eigenvectors that correspond to the v real eigenvalues of the operator and each of the μ two-dimensional subspaces is generated by the real and complex part of the complex conjugate eigenvectors which correspond to a pair of complex conjugate eigenvalues (Arnold 1987).

In a conservative Hamiltonian system of 3 degrees of freedom a complex unstable periodic orbit has four non trivial complex conjugate eigenvalues.

We call the direct sum of the invariant subspaces which correspond to eigenvalues with absolute value greater than one "**unstable asymptotic manifold**", while the direct sum of the invariant subspaces which correspond to eigenvalues with absolute value less than one "**stable asymptotic manifold**".

The four complex eigenvalues (Fig.1) are of the form:

$$
\begin{aligned}
\lambda_1 &= e^{\alpha+i\theta} = e^{\alpha}(\cos\theta+i\sin\theta) \\
\lambda_2 &= e^{\alpha-i\theta} = e^{\alpha}(\cos\theta-i\sin\theta) \\
\lambda_3 &= e^{-\alpha+i\theta} = e^{-\alpha}(\cos\theta+i\sin\theta) \\
\lambda_4 &= e^{-\alpha-i\theta} = e^{-\alpha}(\cos\theta-i\sin\theta)
\end{aligned} \tag{2}
$$

with $\alpha>0$, $0\le\theta<\pi$.

The eigenvalues are in complex conjugate pairs $[(\lambda_1,\lambda_2)$ and $(\lambda_3,\lambda_4)]$ and $|\lambda_1|=|\lambda_2|>1$, $|\lambda_3|=|\lambda_4|= 1/|\lambda_1|$.

The corresponding eigenvectors are

$$
\begin{aligned}
\underline{z_1} &= \underline{x_o} + i\underline{y_o} \\
\underline{z_2} &= \underline{x_o} - i\underline{y_o} \\
\underline{z_3} &= \underline{u_o} + i\underline{w_o} \\
\underline{z_4} &= \underline{u_o} - i\underline{w_o}
\end{aligned} \tag{3}
$$

We study these eigenvectors on a Poincare surface of section, which is 4-dimensional.

The vectors $\underline{x_o}, \underline{y_o}$ generate the unstable asymptotic invariant manifold and $\underline{u_o}, \underline{w_o}$ generate the stable manifold.

After n iterations

$$
\underline{z_i}^{(n)} = \lambda_i^n \underline{z_i}, \quad (i=1,2,3,4) \tag{4}
$$

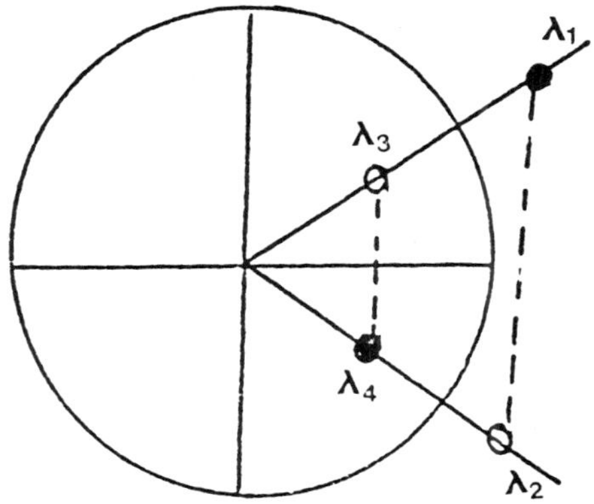

Figure 1. The position of the eigenvalues of a complex unstable orbit.

hence

$$\underline{x_o}^{(n)}=e^{n\alpha}(\cos(n\theta)\ \underline{x_o} - \sin(n\theta)\underline{y_o})$$

$$\underline{y_o}^{(n)}=e^{n\alpha}(\sin(n\theta)\ \underline{x_o} + \cos(n\theta)\underline{y_o})$$

$$^{(n)}=e^{-n\alpha}(\cos(n\theta)\ \underline{u_o} - \sin(n\theta)\underline{w_o}) \qquad (5)$$

$$\underline{w_o}^{(n)}=e^{-n\alpha}(\sin(n\theta)\ \underline{u_o} + \cos(n\theta)\underline{w_o})$$

STRUCTURE OF THE ASYMPTOTIC INVARIANT MANIFOLDS

Let us consider an arbitrary initial condition

$$\underline{\xi} = c_1\underline{x_o} + c_2\underline{y_o} \qquad (6)$$

on the unstable invariant manifold. After n iterations

$$\underline{\xi}^{(n)}=c_1\underline{x_o}^{(n)}+c_2\underline{y_o}^{(n)}=e^{n\alpha}[(c_1\cos(n\theta)+c_2\sin(n\theta))\underline{x_o}+ (c_2\cos(n\theta)-c_1\sin(n\theta))\underline{y_o}] \qquad (7)$$

The projection of $\underline{\xi}^{(n)}$ on the i-j plane (where i,j are any two directions on the Poincare 4-D surface of section) is described by the pair of equations

$$\xi_i^{(n)}=e^{n\alpha}[(c_1\cos(n\theta)+c_2\sin(n\theta))x_{oi}+ (c_2\cos(n\theta)-c_1\sin(n\theta))y_{oi}] \qquad (8)$$

$$\xi_j^{(n)}=e^{n\alpha}[(c_1\cos(n\theta)+c_2\sin(n\theta))x_{oj}+ (c_2\cos(n\theta)-c_1\sin(n\theta))y_{oj}] \qquad (9)$$

Eliminating $(n\theta)$ from the equations (8),(9) we obtain

$$\alpha_1(\xi_i^{(n)})^2-2\beta_1\xi_i^{(n)}\xi_j^{(n)}+\gamma_1(\xi_j^{(n)})^2=\delta, \qquad (10)$$

where

$$\alpha_1 = x_{oj}^2 + y_{oj}^2$$
$$\beta_1 = x_{oi}x_{oj} + y_{oi}y_{oj} \tag{11}$$
$$\gamma_1 = x_{oi}^2 + y_{oi}^2$$

and

$$\delta = -e^{2n\alpha}(c_1^2 + c_2^2)(y_{oi}x_{oj} - y_{oj}x_{oi})^2 \ . \tag{12}$$

Equation (10) represents an ellipse whose main axes grow with n. Thus Eq. (10) is the equation of a spiral curve if n takes all values, integer and non-integer. This curve is not the solution of the variational equations but a curve on the plane (i,j) which contains the projections of the Poincare consequents of the asymptotic orbit with initial condition $\underline{\xi}$. The successive consequents spiral outwards along the curve (10) (Fig. 2).

If the initial condition $\underline{\xi}$ is taken on the stable manifold

$$\underline{\xi} = c_3\underline{u}_o + c_4\underline{w}_o \tag{13}$$

we find a similar equation to (10). The only difference is that now

$$\delta = e^{-2n\alpha}(c_3^2 + c_4^2)(w_{oi}u_{oj} - w_{oj}u_{oi})^2, \tag{14}$$

therefore the spiral has decreasing main axes, and the successive Poincare consequents spiral inwards.

In Fig. 2 we have taken an initial condition on the unstable invariant manifold. Then we have found theoretically, using Equ. (10), a number of consequents which lie on a spiral. The theoretical consequents are marked by crosses. The circles are the Poincare consequents found numerically. We can see the perfect agreement between the numerical and theoretical results for at least 150 consequents. In this case α=0.092 and θ=60°8.

In the case of a larger α (Fig. 3, α=1.407897, θ=21°98) we see a good agreement between the numerical and theoretical results close to the periodic orbit, but larger deviations further away.

The deviations which can be observed away from that neighbourhood of the periodic orbit are due to the nonlinearity of the system.

In both Figs. 2 and 3 we have taken the axes i and j along x and $\dot{x}$ respectively on the Poincare surface of section $(x,\dot{x},y,\dot{y})$, defined by z=0 and a fixed value of the energy h. Similar figures are found if we take any two of the axes $(x,\dot{x},y,\dot{y})$.

If we change c_1 and c_2 in Eq. (6) continuously we have a continuous set of curves like those of Fig. 2, forming the unstable manifold of the complex unstable orbit. This is a 2-dimensional surface in the 4-dimensional Poincare surface of section. Similarly if we change c_3 and c_4 in Eq. (13) continuously we find the stable manifold, which is also a 2-dimensional is surface in the Poincare surface of section. In the 5-dimensional phase space of constant energy the unstable and stable asymptotic manifolds are 3-dimensional.

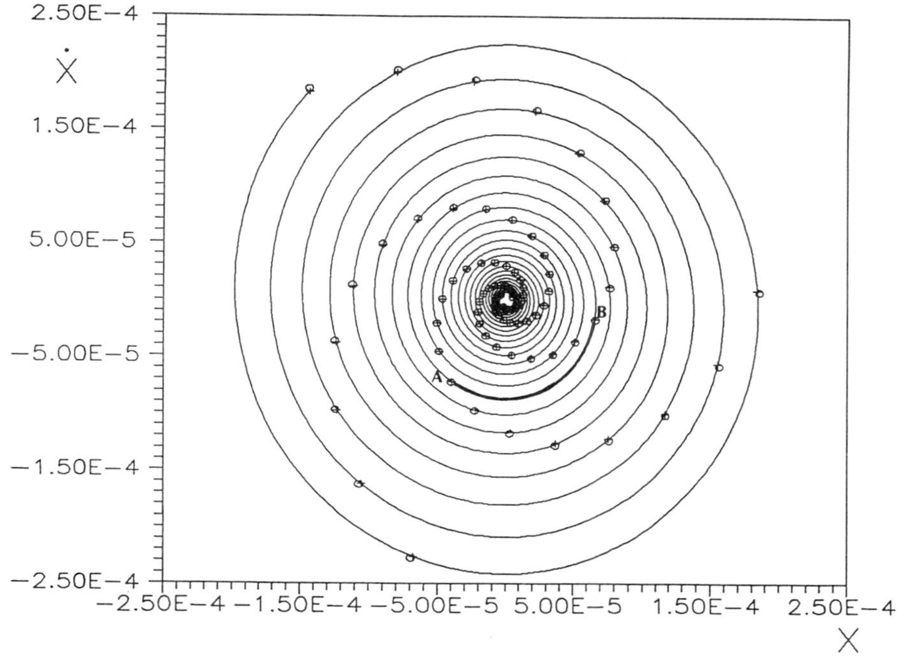

Figure. 2. The projections on the x-$\dot{x}$ plane of 150 Poincare consequents of an orbit initially located on the unstable eigendirection of a complex unstable periodic orbit. The consequents have been computed numerically (circles) and theoretically (crosses). The continuous curve is the spiral given by Eq. (10) (A=1.6, B=0.4, C=0.9, ε=0.08, η=0.01 and h=5). The points A,B are used in Fig. 6.

In the most general case we have all c_1, c_2, c_3, c_4 different from zero. Then the initial condition is

$$\underline{\xi} = c_1\underline{x_o}+c_2\underline{y_o}+c_3\underline{u_o}+c_4\underline{w_o}, \tag{15}$$

and after n iterations

$$\begin{aligned}\underline{\xi}^{(n)}&=c_1\underline{x_o}^{(n)}+c_2\underline{y_o}^{(n)}+c_3\underline{u_o}^{(n)}+c_4\underline{w_o}^{(n)} = \\ &e^{na}[(c_1\cos(n\theta)+c_2\sin(n\theta))\underline{x_o}+(c_2\cos(n\theta)-c_1\sin(n\theta))\underline{y_o}]+ \\ &e^{-na}[(c_3\cos(n\theta)+c_4\sin(n\theta))\underline{u_o}+(c_4\cos(n\theta)-c_3\sin(n\theta))\underline{w_o}]\end{aligned} \tag{16}$$

From this equation we derive that the terms with c_1, c_2 dominate over those with c_3, c_4 for large n. Therefore as n becomes large the consequents approach the unstable manifold

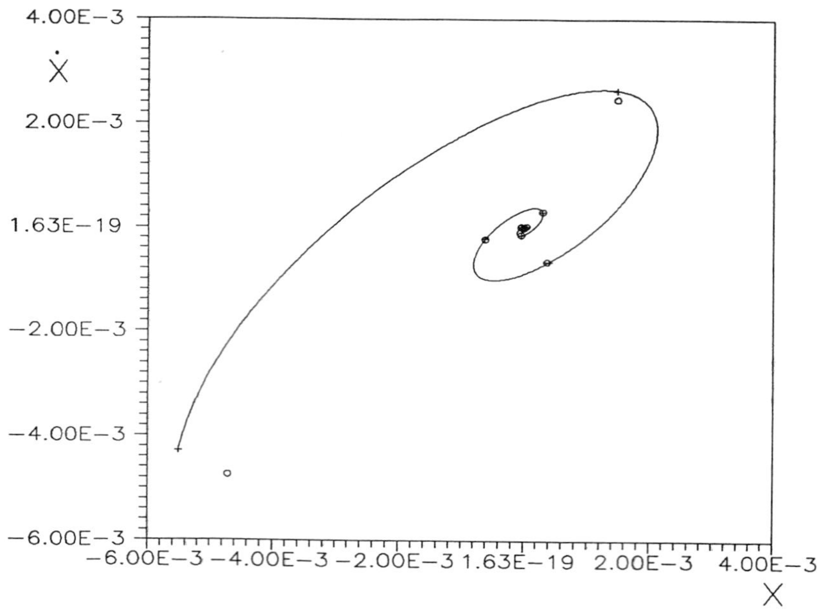

Figure. 3. The same as in Fig. 2 for the case A=0.9, B=0.4, C=0.225 ε=0.511, n=0.01 and h=0.0765. In this case the value of α is greater than in Fig. 2.

$(c_1^2+c_2^2\neq 0\,,\; c_3^2+c_4^2=0)$. Similarly for large negative n the consequents approach the stable manifold $(c_1^2+c_2^2=0,\; c_3^2+c_4^2\neq 0)$.

In Fig. 4 we have drawn the limiting stable and unstable manifolds corresponding to an orbit with initial $\underline{\xi}$ close to the stable manifold (small c_1,c_2, large c_3,c_4). The two manifolds seem to intersect themselves. However these curves are in fact the projections of curves in the 4-D Poincare surface of section, in which they do not intersect in general.

We notice that the successive consequents move initially inwards, close to the stable manifold. However, later on, the terms $c_1e^{n\alpha}$, $c_2e^{n\alpha}$ dominate over the terms $c_3e^{-n\alpha}$, $c_4e^{-n\alpha}$ and the consequents move outwards, close to the unstable manifold.

This behaviour is quite general. In fact it is very difficult to calculate numerically the stable manifold, because any small deviation from it will make the high order consequents to move further away from the stable manifold and spiral outwards, approaching the unstable manifold.

CONICAL SURFACE

In the general case (15) we can write

$$c_1 = R\cos\Phi\,,\quad c_2 = R\sin\Phi$$
$$c_3 = \tilde{R}\cos\tilde{\Phi}\,,\quad c_4 = \tilde{R}\sin\tilde{\Phi} \qquad (17)$$

where

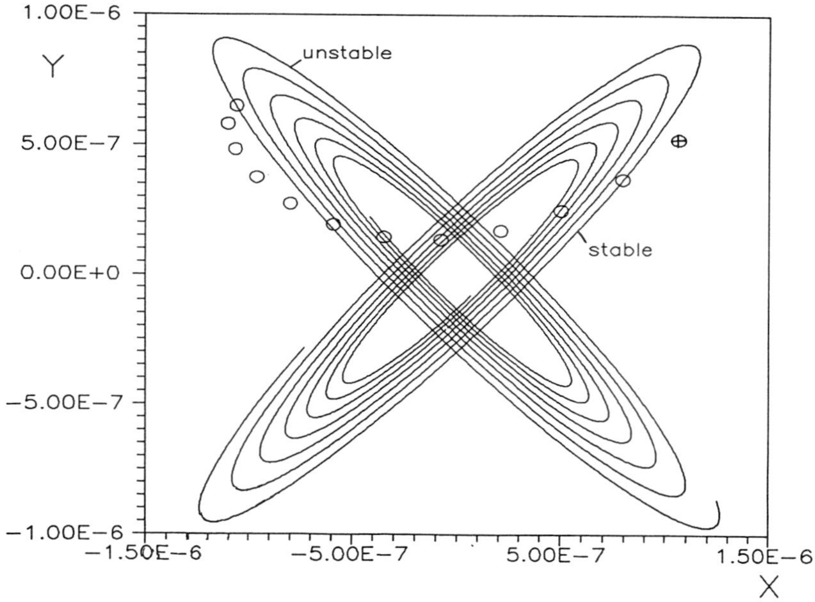

Figure 4. The projections of the consequents of an orbit on the $\dot{x}$-y plane with initial conditions close to the stable asymptotic manifold. The continuous curves are the two asymptotic manifolds. We mark every third consequent. The successive consequents move initially inwards, but later they move outwards. The initial condition has a cross.

$$R^2 = c_1^2+c_2^2 \ , \ \bar{R}^2 = c_3^2+c_4^2 \ . \tag{18}$$

After n iterations we find

$$\underline{\xi}^{(n)}=Re^{n\alpha}[\cos(\Phi-n\theta)\underline{x_o} + \sin(\Phi-n\theta)\underline{y_o}]+$$

$$+\tilde{R}e^{-n\alpha}[\cos(\tilde{\Phi}-n\theta)\underline{u_o} - \sin(\tilde{\Phi}-n\theta)\underline{w_o}]. \tag{19}$$

If we set

$$R^{(n)} = Re^{n\alpha} \ , \ \tilde{R}^{(n)} = \tilde{R}e^{-n\alpha} \tag{20}$$

we find

$$R^{(n)}\tilde{R}^{(n)} = R\tilde{R}=\text{const.} \qquad (n \in Z) \tag{21}$$

This equation represents a hyperbola in the plane $(R,\tilde{R})$.

If we set $b= \alpha/\theta$ then

$$e^{\alpha}=e^{b\theta} \ . \tag{22}$$

Therefore

$$\tilde{R}^{(n)} = \tilde{R}e^{-bn\theta} \tag{23}$$

The equations (19) show that the consequents of any asymptotic orbit on the Poincare surface of section lie on a helical curve, that spirals outwards on a conical surface in the space with coordinates $(x_o, y_o, \tilde{R})$ (Fig. 5), and at the same time the curve moves downwards as n increases, according to Eq. (23).

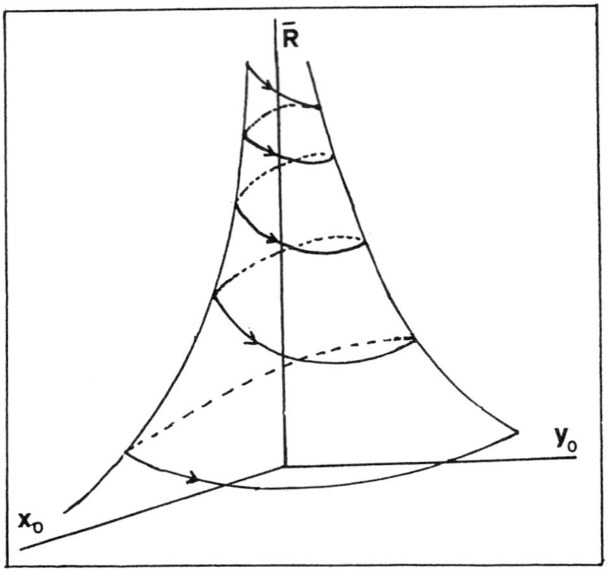

Figure 5. The Poincare consequents in the 3-D space $(x_o, y_o, \tilde{R})$ $(a>0,\ \theta<0)$ are on a helix.

LIMITING CASES

If α in Eq. (12) tends to 0, then δ tends to a constant value and the ellipse (10) tends to become fixed. In this case the orbit tends to become stable. In fact if $\alpha=0$ Eqs. (2) represent two pairs of colliding eigenvalues $\lambda=\cos\theta \pm i\sin\theta$. If a parameter of Eq. (1) (e.g. ε) varies further, the eigenvalues separate on the unit circle and we have two pairs of complex conjugate eigenvalues of measure one

$$\lambda_{\frac{1}{2}}=\cos\theta_1 \pm \sin\theta_1 \ , \ \lambda_{\frac{3}{4}}=\cos\theta_2 \pm i\sin\theta_2 \ . \tag{24}$$

On the other hand if $\theta \to 0$ the complex eigenvalues λ_i tend to become real. In the limiting case $\theta=0$ we have a doubly periodic orbit with $\lambda_1=\lambda_2=e^{\alpha}$ and $\lambda_3=\lambda_4=e^{-\alpha}$. As a parameter of Eq. (1) (e.g. ε) continues to vary the eigenvalues λ_1, λ_2 and λ_3, λ_4 separate and we have

$$\lambda_1=e^{\alpha_1} \ , \ \lambda_2=e^{\alpha_2} \ , \ \lambda_3=e^{-\alpha_1} \ , \ \lambda_4=e^{-\alpha_2} \ . \tag{25}$$

If θ becomes rational, e.g. $\theta=2\pi/3$, an orbit described 3 times has $\theta^{(3)}=0 \pmod{2\pi}$, therefore it is doubly unstable. However in this case a change of parameter (e.g. ε) of Eq. (1) does not generate a family of doubly periodic orbits. In fact a small change of ε produces a deviation of θ from $2\pi/3$ therefore $\theta^{(3)}$ is no more on the real axis $\theta^{(3)}=0$ (mod. 2π).

APPLICABILITY OF THE LINEAR THEORY

The linear thory is applicable when the deviation of an orbit from the periodic orbit is small. For example in Fig. 2 we have taken $c_1=c_2=10^{-5}$ and we found that the agreement between theory and numerical calculations is very good, as long as the radius $|\xi|$ is of the order of 10^{-3}. But if we extend our calculations for much longer times then the deviations are larger, and finally the linear theory is not even approximately correct.

In Fig. 6 we have calculated 300 consequents for 60 initial conditions on the unstable asymptotic manifold, continuing the curve of Fig. 2. We see that the asymptotic manifold extends up to a distance of order $|\xi|=1.5$ from the periodic orbit and then returns inwards.

The fact that the extent of the asymptotic curves reaches a maximum can be understood by an approximate nonlinear theory (Heggie 1985). However even a nonlinear theory is unable to describe the chaotic behaviour of the orbits in the neighbourhood of a complex unstable orbit. In fact the foldings of the unstable asymptotic curve of Fig. 6 indicate that after some time the orbits in the general neighbourhood of a complex periodic orbit behave in a chaotic way and they do not lie on simple integral surfaces. This means that both the linear and the nonlinear theory can describe the orbits only for a limited interval of time.

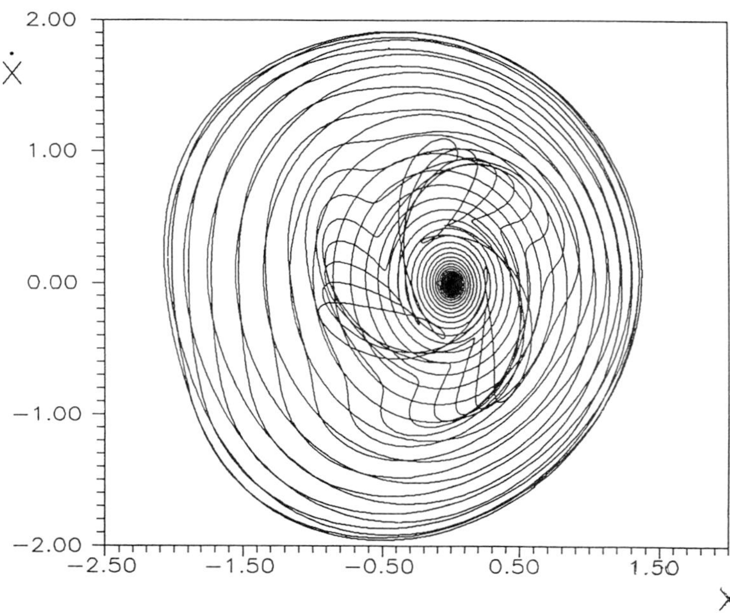

Figure 6. The projection of the unstable manifold on the x-$\dot{x}$ plane for the case A=1.6, B=0.4, C=0.9 ε=0.08 n=0.01, E=5. It is constructed of 60 initial conditions on the arc AB of Fig. 2 computed for about 300 consequents each.

CONCLUSIONS

We have studied the close neighbourhood of a complex periodic orbit on a Poincare surface of section in the case of a conservative Hamiltonian system of 3 degrees of freedom. Our main conclusions are:

1) The complex unstable periodic orbits have two asymptotic invariant manifolds, one stable and one unstable, of dimension 3 in the 5 dimensional phase space of constant energy. The consequents on the 4-D Poincare surface of section of every orbit with initial conditions on one of these manifolds lie on spirals. The projections of these consequents on a 2-D plane are spirals, while the projections on a 3-D space are helices.

2) Orbits with initial conditions close to the stable invariant manifolds have consequents that spiral initially inwards, but then spiral outwards.

3) If the eigenvalue $\lambda=e^{\alpha+i\theta}$ has an angle θ equal to a rational number $\frac{m}{n}$, times 2π, then the orbit described n times is doubly unstable. However such an orbit does not generate by bifurcation a family of doubly unstable orbits, except if $\theta=0$, or $\theta=\pi$.

4) The agreement between the theoretical consequents and the consequents calculated numerically is very good if their distance from the periodic orbit is small. However when the distance becomes large the agreement is worse.

5) The stable and unstable manifolds are 3-dimensional in the 5-dimensional phase space of constant energy. These manifolds reach a maximum distance from the periodic orbit and then turn inwards generating a chaotic behaviour.

REFERENCES

Arnold, V.I.: 1978, "Mathematical Methods of Classical Mechanics", Springer Verlag, New York.

Arnold, V.I.: 1987, "Ordinary differential equations". MIT Press, Cambridge, Mass.

Contopoulos, G.: 1983, Lett.Nuovo Cim. 38, 257.

Contopoulos, G., 1986, Celest. Mech. 38, 1.

Contopoulos, G.,and Magnenat P.,1985, Celest. Mech. 37, 387.

Guckenheimer, J. and Holmes, P.J., 1983, "Nonlinear Oscillations, Dynamical Systems and Bifurcations of Vector Fields", Springer Verlag, New York.

Heggie, D.C.: 1985, Celest. Mech. 35, 357.

Magnenat, P.: 1982, Celest.Mech., 28, 319.

Wiggins, S.: 1990, "Introduction to Applied Nonlinear Dynamical Systems and Chaos" Springer Verlag, New York.

INVERSE PROBLEMS LEADING TO INTEGRABLE SYSTEMS

F. Puel

Observatoire be Besançon
25000 France

INTRODUCTION

The inverse problem in dynamics is to find the forces from the knowledge, or from a partial knowledge, of the motions. Szebehely's equation (1974) is a first order linear partial differential equation that gives the force functions having these curves as orbits with the energy E. In fact, we obtain this equation by writing in the equations of motion along these curves, that the forces derive from a potential - U and that the kinetic energy is U + E. For three-dimensional motions, by considering a two-parameter family of orbits, Erdi (1982) obtained similarly a system of two partial differential equations. Szebehely's equation was used to know the families of plane orbits that are generated only in integrable systems (Puel, 1985; Ichtiaroglou & Meletlidou, 1990).

Our aim is to return to this problem of integrability, or more precisely of separability, by using Joukovsky's formula and a three-dimensional generalization we gave recently (Puel, 1992). After recalling some useful results concerning Liouville's and Stäckel's systems and Joukovsky's formula, we will study some types of families of two- and three-dimensional curves that can only be generated as orbits by separable systems.

CLASSICAL CASES OF SEPARABILITY (WHITTAKER, CHARLIER)

Notations: The dot $\dot{}$ denotes a derivation with respect to the time t.
T is the kinetic energy and U the force function.
The $q^i (i = 1, n)$ are generalized coordinates.
If we write, for instance, $a_r = a_r(q^r)$, it means that the function a_r depends only on the coordinate q^r.

a) Liouville's case:
$2T$ and U are of the following form:

$$2T = \left(\sum_{i=1}^{n} a_i(q^i)\right)\left(\sum_{i=1}^{n} b_i(q^i)(\dot{q}^i)^2\right) \tag{1a}$$

$$U = \frac{\sum_{i=1}^{n} u_i(q^i)}{\sum_{i=1}^{n} a_i(q^i)} \tag{1b}$$

b) Stäckel's case:
Let Φ be the determinant $\Phi = \det(\phi_{ij})$, with $\phi_{ij} = \phi_{ij}(q^j)$.
We note Φ^{kr} the first minor factor of ϕ_{kr} in the development of Φ. We have, of course, $\Phi = \sum_{i=1}^{n} \Phi^{li}\phi_{li}$. Let Ψ be defined by $\Psi = \sum_{i=1}^{n} \Phi^{li}\psi_{li}$ with $\psi_i = \psi_i(q^i)$.

$2T$ and U are of the form:

$$2T = \Phi \sum_{i=1}^{n} \frac{(\dot{q}^i)^2}{\Phi^{li}} \tag{2a}$$

$$U = \frac{\Psi}{\Phi} \tag{2b}$$

c) Relations and implications:
Liouville's case is a particular case of Stäckel's one. But in two dimensions they are equivalent.

d) Geometrical properties:
We will call "geometrical" the conditions about the kinetic energy because the metric is $ds^2 = 2T dt^2$.

For the planar motion of a point, geometrical separability occurs only in confocal coordinates and degenerate cases such as Cartesian coordinates, polar etc,... (Liouville, 1846).

For the motion of a point in a three-dimensional Euclidean space, geometrical separability occurs only if iso-coordinate surfaces are confocal quadric surfaces or degenerate cases (Eisenhart, 1934).

e) Incomplete separability:
We obtain an incomplete separability by generalization of Liouville's case with less strict conditions, i.e. with T and U of the form:

$$2T = [\alpha(q^1,q^2) + \beta(q^3)][b_1(q^1,q^2)(\dot{q}^1)^2 + 2b_{1,2}(q^1,q^2)\dot{q}^1\dot{q}^2 + b_2(q^1,q^2)(\dot{q}^2)^2 + b_3(q^3)(\dot{q}^3)^2] \tag{3a}$$

$$U = \frac{[u(q^1,q^2) + v(q^3)]}{\alpha(q^1,q^2) + \beta(q^3)} \tag{3b}$$

where α and u do not depend on q^3, β and v does not depend on q^1 and q^2.

More precisely, by using the change of variable defined by

$$dt = [\alpha(q^1,q^2) + \beta(q^3)]d\tau$$

we can obtain two first integrals of the motion.

JOUKOVSKY'S THEOREM

a) Two-dimensional case (Whittaker, Broucke & Lass, Molnar):
Let the orbits be implicitly given by:

$$f(x,y) = c \text{ (c = const)}$$

and let

$$g(x, y) = d \text{ (d = const)}$$

be the equation of a family of orthogonal curves.

We choose as coordinates the new variables

$$q^1 = f(x, y)\ ,\ q^2 = g(x, y)$$

With these new coordinates, the metric can be written

$$ds^2 = \frac{1}{2}[(h_1)^2(dq^1)^2 + (h_2)^2(dq^2)^2]$$

with the scale factors

$$h_1 = ||\vec{\mathbf{grad}}\ f||^{-1}\ ,\ h_2 = ||\vec{\mathbf{grad}}\ g||^{-1}$$

In this system Szebehely's equation can be written

$$h_2\frac{\partial U}{\partial q^1} + 2(U + E)\frac{\partial h_2}{\partial q^1} = 0$$

and Joukovsky's formula is

$$(h_2)^2 U = -\int_0^{q^1} E(q^1)\frac{\partial((h_2)^2)}{\partial q^1}dq^1 + A(q^2) \tag{4}$$

where A is an arbitrary function of q^2.

b) Three-dimensional case (Puel, 1992):

This generalization is not possible, unless the trajectories are orthogonal to a family of surfaces. We assume this property.

Let the orbits be implicitly given by

$$f(x, y, z) = c\ ,\ g(x, y, z) = d \text{ (c, d = const)}$$

According to the previous assumption, there exists a function $\phi(x, y, z)$ whose level surfaces are orthogonal to those trajectories.

We chose as coordinates the new variables

$$q^1 = f(x, y, z)\ ,\ q^2 = g(x, y, z)\ ,\ q^3 = \phi(x, y, z)$$

With these coordinates, the kinetic energy can be written as $T = \frac{1}{2}g_{ij}\dot{q}^i\dot{q}^j$ with the coefficients of the metric given by

$$g_{11} = (\vec{\mathbf{grad}}\ f)^{-2}\ ,\ g_{22} = (\vec{\mathbf{grad}}\ g)^{-2}\ ,\ g_{33} = (\vec{\mathbf{grad}}\ \phi)^{-2}$$

$$g_{12} = (\vec{\mathbf{grad}}\ f.\vec{\mathbf{grad}}\ g)^{-1}\ ,\ g_{13} = g_{23} = 0$$

We want to obtain the force functions $U(q^1, q^2, q^3)$ which can generate a two-parameter family of motions on the curves $q^1 = c, q^2 = d$ with the energy $E(q^1, q^2)$.

In the equations of motion we put the derivatives of q^1 and q^2 equal to zero and we write the velocity as given by the energy theorem.

We obtain Szebehely's equations:

$$\begin{cases} \frac{\partial}{\partial q^1}(g_{33}U) + E\frac{\partial g_{33}}{\partial q^1} = 0 \\ \frac{\partial}{\partial q^2}(g_{33}U) + E\frac{\partial g_{33}}{\partial q^2} = 0 \end{cases}$$

And the solution, Joukovsky's formula, is

$$U = \frac{1}{g_{33}}(A(q^3) - \int^{q^1,q^2} E(q^1,q^2)[\frac{\partial(g_{33})}{\partial q^1}dq^1 + \frac{\partial(g_{33})}{\partial q^2}dq^2]) \tag{5}$$

where A is an arbitrary function of q^3.

But we have to discuss the compatibility of Szebehely's equations:
Yet we have

$$\frac{\partial E}{\partial q^3} = 0 \tag{6}$$

The compatibility condition of Szebehely's equations is

$$\frac{\partial E}{\partial q^1}\frac{\partial g_{33}}{\partial q^2} - \frac{\partial E}{\partial q^2}\frac{\partial g_{33}}{\partial q^1} = 0 \tag{7}$$

In fact these two previous conditions involve *either*

$$\mathbf{g\vec{r}ad}\ E = \vec{\mathbf{0}}\ ,\ E = E_0\ ,\ \text{const.} \tag{8}$$

We said the family is "iso-energetic" *or*

$$\frac{\partial}{\partial q^3}(\frac{\partial g_{33}}{\partial q^1}/\frac{\partial g_{33}}{\partial q^2}) = 0 \tag{9}$$

This last equation implies that there exists a function $\alpha(q^1, q^2)$ such that g_{33} and E depend on q^1 and q^2 via α. Then g_{33} and E are of the form

$$g_{33}(q^1, q^2, q^3) = G(\alpha(q^1, q^2), q^3) \tag{10a}$$

$$E(q^1, q^2) = H(\alpha(q^1, q^2)) \tag{10b}$$

FAMILIES OF ORBITS AND SEPARABILITY

Now, we have only to replace the geometrical conditions of separability into Joukovsky's formula.

a) "Separable families" for the two-dimensional case (Puel, 1985): By replacing geometrical Liouville's condition (1;a) on $(h_2)^2$ in Joukovsky's formula, it is easy to see that the force function U is also of the Liouville form (1;b).

b) Three-dimensional case:

1) Liouville's case:
We recall geometrical Liouville's conditions (1;a), and particularly

$$g_{33} = [a_1(q^1) + a_2(q^2) + a_3(q^3)]b_3(q^3)$$

$$\frac{\partial g_{33}}{\partial q^1}dq^1 + \frac{\partial g_{33}}{\partial q^2}dq^2 = b_3(q^3)\ d[a_1(q^1) + a_2(q^2)]$$

Compatibility conditions (6) and (7) imply that $E(q^1, q^2)$ is a function of $a_1(q^1) + a_2(q^2)$, thus

$$E = H[a_1(q^1) + a_2(q^2)]$$

Therefore Joukovsky's solution (5) is

$$U = \frac{B(q^3) + \Gamma[a_1(q^1) + a_2(q^2)]}{a_1(q^1) + a_2(q^2) + a_3(q^3)} \tag{11}$$

where $B = A(q^3)/b_3(q^3)$ is an arbitrary function of q^3 and

$$\frac{\partial \Gamma}{\partial [a_1(q^1) + a_2(q^2)]} = H[a_1(q^1) + a_2(q^2)]$$

This solution for the force function U will be of the Liouville's form if

$$\Gamma = E_0.[a_1(q^1) + a_2(q^2)]$$

i.e. if the family is "iso-energetic".

Otherwise we have only incomplete separability (3;b) because we can't separate q^1 from q^2 in the force function U.

2) Less strict conditions:

In fact, we may weaken geometrical conditions and still have incomplete separability. We must have only separation between q^3 and the two other coordinates (q^1, q^2). Moreover the coordinate surfaces q^1 = const. and q^2 = const. don't need to be mutually orthogonal.

Let us consider the kinetic energy (3;a). We have

$$g_{33} = [\alpha(q^1, q^2) + \beta(q^3)]b_3(q^3)$$

Conditions (10;a;b) involve that $E(q^1, q^2)$ is of the form $E = H[\alpha(q^1, q^2)]$. Therefore Joukovsky's solution is

$$U = \frac{B(q^3) + \Gamma[\alpha(q^1, q^2)]}{\alpha(q^1, q^2) + \beta(q^3)} \tag{12}$$

where $B = A(q^3)/b_3(q^3)$ is an arbitrary function of q^3 and

$$\frac{\partial \Gamma}{\partial [\alpha(q^1, q^2)]} = H[\alpha(q^1, q^2)]$$

We recognize the partially separate form (3;b) for the force function U.

3) Stäckel's case:

We recall geometrical Stäckel's conditions (2;a).

We can develop g_{33} in order to separate the terms depending on q^3 from those depending on q^1 and q^2.

$$g_{33} = \frac{\Phi}{\Phi^{13}} = \phi_{13}(q^3) + \phi_{23}(q^3)\frac{\Phi^{23}}{\Phi^{13}}(q^1, q^2) + \phi_{33}(q^3)\frac{\Phi^{33}}{\Phi^{13}}(q^1, q^2) \tag{13}$$

(i) "Iso-energetic" families: $E = E_0$ const.

By replacing the previous expressions of g_{33} in Joukovsky's formula, after some manipulations, we obtain

$$U = \frac{\Psi}{\Phi}$$

with

$$\Psi = -E_0\phi_{11}(q^1)\Phi^{11} - E_0\phi_{12}(q^2)\Phi^{12} + A(q^3)\Phi^{13}$$

As for Liouville's case, geometrical separability implies the separability of the force function for "iso-energetic" families of orbits.

(ii) Non "iso-energetic" families

Generally, geometrical Stäckel's separability (2;a) does not imply the separability of the system, but we can hope to have at least a system partially separable.

Let us replace the compatibility condition (13) in the expression of g_{33}, we obtain the condition

$$[\frac{\partial}{\partial q^1}(\frac{\Phi^{23}}{\Phi^{13}})\frac{\partial}{\partial q^2}(\frac{\Phi^{33}}{\Phi^{13}}) - \frac{\partial}{\partial q^1}(\frac{\Phi^{33}}{\Phi^{13}})\frac{\partial}{\partial q^2}(\frac{\Phi^{23}}{\Phi^{13}})][\phi_{33}\frac{\partial\phi_{23}}{\partial q^3} - \phi_{23}\frac{\partial\phi_{33}}{\partial q^3}] = 0$$

So that we have *either* (2nd bracket null)

$$\frac{\phi_{23}(q^3)}{\phi_{33}(q^3)} = \lambda \ , \ \lambda = \text{ const.}$$

or (1st bracket null)

$$\frac{\Phi^{23}}{\Phi^{13}} \text{ and } \frac{\Phi^{33}}{\Phi^{13}} \text{ are functionally dependent.}$$

This is the case for instance if $\frac{\phi_{31}}{\phi_{21}} = \frac{\phi_{32}}{\phi_{22}} = -\mu$, $\mu =$ const., so that $\Phi^{23} = \mu\Phi^{33}$; but I am actually not sure there are not other cases of dependence of $\frac{\Phi^{23}}{\Phi^{13}}$ and $\frac{\Phi^{33}}{\Phi^{13}}$.

In both cases ($\frac{\phi_{23}(q^3)}{\phi_{33}(q^3)} =$ const. and $\frac{\phi_{31}}{\phi_{21}} = \frac{\phi_{32}}{\phi_{22}} =$ const.), we can recognize in $2T$ the geometrical incomplete separability (3;a), and from part b), 2) in the present section, we have incomplete separability for the force function U.

CONCLUSION

We have studied the potentials generating motions along curves that are orthogonal to the surfaces of a separable "á la Liouville" or "á la Stäckel" system of curvilinear coordinates (for the planar motions of a point, they are the nets of confocal conics or degenerate cases, for the motions in an Euclidean three-dimensional space, they are the families of confocal quadrics or degenerate cases.)

- For two-dimensional problems, these potentials are, also, always of the separable type.
- For three-dimensional problems, we must distinguish between two cases:

If all the motions of the family have the same total energy ("iso-energetic" family), then the potentials are of separable type.

If the family is not "iso-energetic", then the potentials are no longer separable; but by comparing the compatibility conditions of Szebehely's equations with the form of Liouville's or Stäckel's metric, we have seen there is still a partial separability in many cases (perhaps in all cases).

These results are resumed in the following table:

ORBITS	POTENTIALS
2-dim. separable	⇒ Separable
3-dim. separable	
"iso-energetic" family	⇒ Separable
non "iso-energetic" family	
Liouville's type	⇒ Partially Separable
Stäckel's type	
$\phi_{23}(q^3) = \lambda\phi_{33}(q^3)$	⇒ Partially Separable
$\frac{\phi_{31}}{\phi_{21}} = \frac{\phi_{32}}{\phi_{22}} = -\mu$	⇒ Partially Separable
other cases (?)	⇒ ?
Partially Separable	⇒ Partially Separable

REFERENCES

1. Broucke, R. and Lass, H.: 1977, Celest. Mech. 16, 215-225
2. Charlier, C.L.: 1902, *Die Mechanik des Himmels*, Liepzig
3. Eisenhart, L.P.: 1934, Ann. of Math. (2) 35, 284-305
4. Erdi, B.: 1982, Celest. Mech. 28, 209-218
5. Ichtiaroglou, S. and Meletlidou, E. : 1991, J. of Physics A; Math. Gen...
6. Molnar, S. : 1981, Celest. Mech. 25, 81-88
7. Puel, F. : 1985, in V.G. Szebehely (ed), "Stability of the Solar System and its minor natural and artificial bodies", 401.
8. Puel, F. : 1992, Celest. Mech. 53, 207-218
9. Szebehely, V. : 1974, in E. Proverbio (ed) Proc. Int. Mtg. on "The Earth's Rotation by Satellite Observations", Univ. of Cagliari, 31-35.
10. Whittaker, E.T. : 1937, *Analytical Dynamics*, Cambridge, 1937.

THE OBSERVER—NEW METHOD FOR NUMERICAL INTEGRATION OF DIFFERENTIAL EQUATIONS IN THE PRESENCE OF FIRST INTEGRALS

Andrzej J. Maciejewski

Institute of Astronomy, Nicolaus Copernicus University
87-100 Toruń, Chopina 12/18, Poland

INTRODUCTION

In celestial mechanics and dynamical astronomy the numerical integration of differential equations is one of the most popular and practical method of a study of dynamical systems. Recently a quite big progress was made in developing new more reliable methods of integration of differential equations. For Hamiltonian system different kinds of symplectic integrators were found (see Sanz-Serna 1992 for overview) and for more general Lie-Poisson system the Lie-Poisson integrators were developed (see Ge and Marsden 1988, Channell and Scovel 1991). This kind of integrators are especially important in celestial mechanics where most of dynamical systems are Hamiltonian. For general system of differential equations it seems that most promising recent achievements were done in the analytical continuations methods (see Gofen 1992 and references cited there).

The aim of this paper is to present, to apply and to test the observer method introduced in Busvelle *et al.* 1993. This method is a simple idea that can be used to improve reliable numerical integration of a system of differential equations in the presence of first integrals. In the cited paper the observer method was already tested on some important examples like the Kepler problem and the Gavrilov-Shilnikov system.

In this paper for testing of the observer method we chose four dynamical systems important for celestial mechanics. We continue tests of Busvelle *et al.* (1993) on the Kepler problem. Here we use very accurate integrator and apply the observer method for more integrals. Next we study the system of Célérier. This system is integrable but its orbit, in contrast to the Kepler problem, are very complicated. Last two examples are systems describing the rotational dynamics of a rigid satellite. The first one is the Beletskii equations describing planar oscillations of a satellite on an elliptic orbit. Here

From Newton to Chaos, Edited by A.E. Roy
and B.A. Steves, Plenum Press, New York, 1995

we show how it is possible to apply the observer method to time dependent system. The second system describe the rotational dynamics of a rigid satellite on circular orbit.

Let us note that the idea of improving precision of numerical integration by using first integrals has a long history. Here we refer to Nacozy (1971) paper where the idea of projection of numerical solution onto the level of constant integrals value was introduced. In two papers of Baumgarte (1973) and (1976) the level of constant integrals values where consider as a constrain of the system. The idea of stabilisation proposed by Baumgarte is similar but not equivalent to the observer method. Let us note that the observer method is much more general and simpler from Baumgarte method.

THE OBSERVER METHOD

Here we sketch the observer method. Details and proofs can be find in the paper of Busvelle *et al.* (1993).

Let us consider the initial value problem for a system of ordinary differential equations written in the vector form

$$\frac{dx}{dt} = F(x), \qquad (1)$$
$$x(0) = x_0$$

where $F \in C^1(U, \mathbb{R}^p)$, U is an open subset of $\mathbb{R}^p$ and $x_0 \in U$. Let us suppose that $H_1, \ldots, H_q \in C^1(U)$, $1 \leq q < p$, are the first integrals of the system (1) and let us denote

$$H = (H_1, \ldots, H_q)^T, \qquad H \in C^1(U, \mathbb{R}^q),$$

$$D_H(x) = \begin{pmatrix} \dfrac{\partial H_1}{\partial x_1}, & \ldots, & \dfrac{\partial H_1}{\partial x_p} \\ \ldots, & \ldots, & \ldots \\ \dfrac{\partial H_q}{\partial x_1}, & \ldots, & \dfrac{\partial H_q}{\partial x_p} \end{pmatrix},$$

where ^{T} denotes transposition. Let us consider $x_0 \in U$ and the level set of integrals

$$\Gamma = \{x \in U : H(x) = H_0\},$$

where $H_0 = H(x_0)$. We make two assumptions:

1. The set Γ is compact.
2. Integrals H are functionally independent on Γ.

From the implicit function theorem one easily deduces that for sufficiently small ϵ, $0 < \epsilon \leq \epsilon_0$, the following set

$$\Gamma_\epsilon = \{x \in U; \| H_0 - H(x) \| \leq \epsilon\}$$

is a compact subset of U and integrals $H_1, \ldots, H_q$ are functionally independent on it ($\| \cdot \|$ denotes the Euclidean norm).

The basic idea of the observer method consists of replacing the original initial value problem by the new one

$$\frac{dy}{dt} = F(y) + Ob(y), \qquad (2)$$
$$y(0) = y_0, \qquad (3)$$

where

$$Ob(y) = D_H^T(y)\left[D_H(y)D_H^T(y)\right]^{-1} K(y)\,(H_0 - H(y)),$$

and $K(y)$ is a diagonal matrix given by

$$K(y) = \begin{pmatrix} \theta_1 & 0 & \cdots & 0 \\ 0 & \theta_2 & & \vdots \\ \vdots & & \ddots & \\ 0 & \ldots & 0 & \theta_q \end{pmatrix},$$

where $\theta_1, \ldots, \theta_q$ are continuous functions on Γ_ϵ such that inequality

$$\eta = \min_{1 \leq i \leq q} \min_{y \in \Gamma_\epsilon}(\theta_i(y)) > 0$$

holds. The perturbation term $Ob(y)$ is called the observer. Its specific form and assumptions which we made, allow to prove following facts.

1. $\Gamma \subset U$ is an invariant submanifold with respect to the flow of the system (2).

2. Γ is the global attractor of the system (2) restricted to the Γ_ϵ with an exponential rate of convergence of orbits towards Γ, which is exactly controlled.

Moreover, when solving the initial value problem (2)–(3) by any standard one step explicit numerical method of first order, it is possible to prove, choosing appropriate values of the step-size, η and ϵ, that the error of integrals is bounded over infinite time interval. We again refer to the paper of Busvelle *et al.* (1993) for proofs and detailed discussion.

Finally, let us make two remarks concerning numerical integration of the initial value problem (2)–(3).

In the matrix $K(y)$ diagonal elements can be taken constant. Having the system of ODE, the initial condition and chosen integrator it is necessary to find the 'optimal' values of θ_i. Bad choice will not improve the precision of integration.

At first glance it seems that during numerical integration of the system (2) it is necessary to inverse matrix $A = D_H^T(y)D_H(y)$ once per integration step. However the observer can be rewritten as

$$Ob(y) = D_H^T z,$$

where $z = z(y)$ is the solution of the linear system

$$Az = K(y)(H_0 - H(y))$$

which can be solved numerically without an explicit matrix inversion.

THE KEPLER PROBLEM

The observer method was tested on the Kepler problem by Busvelle *et al.* (1993). Test show that for orbits with big eccentricity this method gives better results than standard RK and symplectic integration.

The main purpose of the presented test is to compare the observer method with high a precision integrator. For this purpose we chose procedure RA15 that implements the Everhard algorithm (see Everhard 1974). For test we chose an orbit with following elements: $a = 1$, $e = 0.8$, $i = \pi/4$, $\omega = \Omega = \pi/2$. The time of the pericenter passage was 0. The gravitational parameter was set 1 and we integrated equations up to the time of 10^6 revolutions.

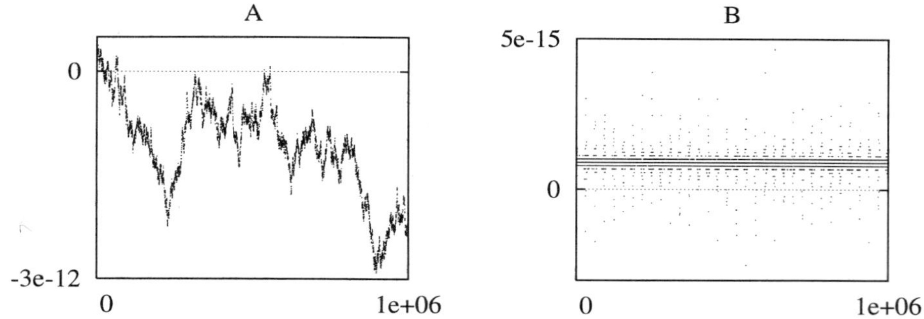

Figure 1. The absolute error in energy. A—integration without the observer term; B—integration with the observer.

We made two integrations with the local precision 10^{-12}. In the first one we integrated the equation of motion using the RA15 integrator. In the second test we use the same integrator but we modify the right hand sides of the equations of motion by the observer term. We calculate the observer term taking into account existence of four integral, namely the energy and three components of the angular momentum vector.

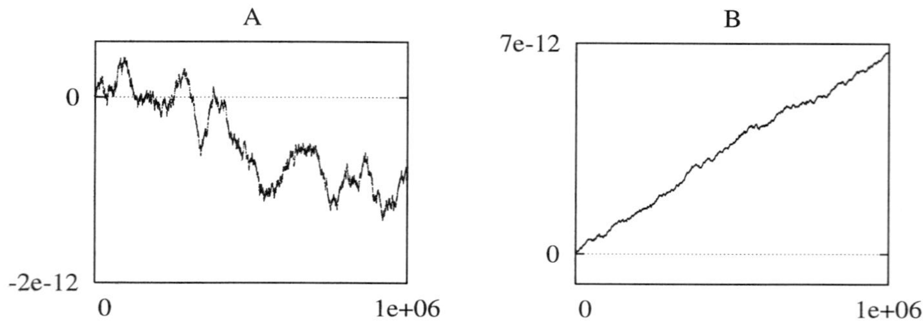

Figure 2. As in Figure 1 but for the second component of the Laplace vector.

Figure 1 shows how the observer improves the preservation of energy. In fact we have the same improvement in other observed integrals as well as in the first and the third component of the Laplace vector. Only in the case of the second component of the Laplace vector the observer method gives worse results. This is shown in Figure 2.

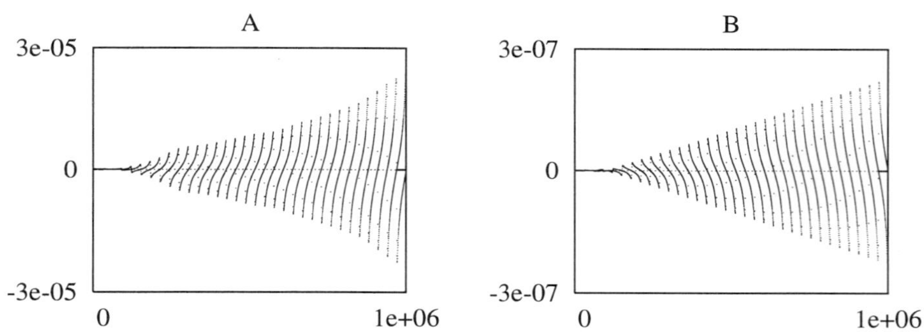

Figure 3. The absolute error in the first component of the radius vector. A—integration without the observer term; B—integration with the observer.

The most important result is shown in Figure 3 where we show the absolute error in x component of the position vector for both integration. It is evident that the observer method gives two orders small error but also that the growth of error in position for this method is linear. We have the same behaviour in other components of the radius vector as well as of the velocity.

THE BELETSKII SYSTEM

The well known Beletskii's equation (Beletskii, 1965):

$$(1 + e\cos(\nu))\frac{d^2\delta}{d\nu^2} - 2e\sin(\nu)\frac{d\delta}{d\nu} + n^2\sin(\delta) = 4e\sin(\nu) \tag{4}$$

describes planar oscillations of a rigid satellite on an elliptic orbit. One of the principal axes of the satellite, let us say the z-axis, is always perpendicular to the orbital plane ($\delta/2$ is angle between the radius vector and x-axis of the body, e is the eccentricity, ν is the true anomaly, $n^2 = (B - A)/C$, $A \leq B \leq C$ are principal moments of inertia). In (4) only second harmonic of the potential function was taken into account and the central body was assumed to be spherically symmetric.

The equation (4) is used as the standard model for spin-orbit coupling. It is equivalent to the Hamiltonian equations

$$\frac{dq}{d\nu} = \frac{\partial H}{\partial p}, \qquad \frac{dp}{d\nu} = -\frac{\partial H}{\partial q},$$

with time dependent Hamiltonian function

$$H = \frac{1}{2}\left\{\frac{p}{f(\nu)} - 2f(\nu)\right\}^2 - f(\nu)n^2\cos(q)$$

where

$$f(\nu) = 1 + e\cos(\nu).$$

For this system we cannot apply the observer method directly because it is time dependent and we do not know an integral of it. However, we make this system Hamiltonian and autonomous by introducing an additional degree of freedom.

The configuration space for our system is $M = S^1 = \mathbb{R}^1\{q\}/\mathrm{mod}\,2\pi$ and thus the phase space is the cylinder $T^*M = S^1 \times \mathbb{R}^1\{p\}$. In the extended phase space $T^*M \times \mathbb{R}^2\{E, t\}$, the equation of motion of our system can be written as

$$\dot{q} = \frac{\partial K}{\partial p}, \qquad \dot{p} = -\frac{\partial K}{\partial q}, \qquad \dot{E} = \frac{\partial K}{\partial t}, \qquad \dot{t} = -\frac{\partial K}{\partial E}, \tag{5}$$

where $K = H(q, p, t) - E$, $q \in M$, $p \in T_q^*M$. Thus, we introduced the energy as the generalised coordinate and the time as the momentum conjugated to it. The new system with two degrees of freedom has the first integral K.

The construction presented above can be applied for any time depended Hamiltonian system (see Arnold *et al.* 1988, Kozlov 1983).

For test we chose following values of parameters $n = 1$ and $e = 0.01$. Initial condition was $q(0) = 0$, $p(0) = 2.1$. For integration we used Burlish-Stoer extrapolation method with step-size control (see Hairer *et al.* 1985). Local precision of integration was 10^{-6}. We integrated equations (5) with and without observer term. Results are shown on Figure 4 where errors of integral K in logarithmic scale are shown.

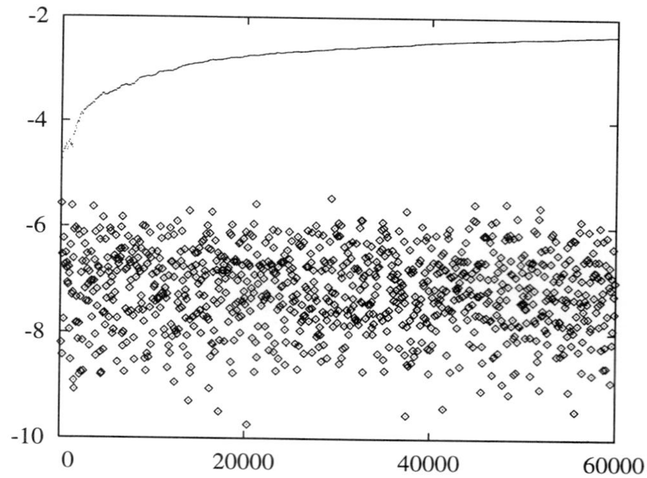

Figure 4. Error of the integral K of Beletskii system in logarithmic scale. Continuous line is for integration without the observer and scattered squares for integration with the observer method.

A RIGID SATELLITE ON A CIRCULAR ORBIT

Equations of rotational motion of a rigid body (satellite) on a circular orbit around a fixed gravitational center can be written in the following form

$$\begin{aligned}\frac{d}{dt}\mathbf{M} &= \mathbf{M}\times\boldsymbol{\omega} + 3\Omega\boldsymbol{\gamma}\times\mathbf{I}\boldsymbol{\gamma},\\ \frac{d}{dt}\boldsymbol{\gamma} &= \boldsymbol{\gamma}\times(\boldsymbol{\omega}-\Omega\mathbf{n}),\\ \frac{d}{dt}\mathbf{n} &= \mathbf{n}\times\boldsymbol{\omega},\end{aligned} \tag{6}$$

where:

$$\Omega^2 = \frac{Gm}{r^3}, \qquad \mathbf{M} = \mathbf{I}\boldsymbol{\omega},$$

$$\mathbf{I} = \begin{bmatrix} A & 0 & 0 \\ 0 & B & 0 \\ 0 & 0 & C \end{bmatrix},$$

G is the gravitational constant, m is the mass of the central body, r is the radius of the orbit and A, B, C are the principal moments of inertia of the body. $\mathbf{M}$, $\boldsymbol{\omega}$ are vectors of the angular momentum and the total angular velocity of the satellite, respectively; $\boldsymbol{\gamma}$ and $\mathbf{n}$ are unit vectors along the radius vector and normal to the orbital plane, respectively. Components of all vectors are taken here with respect to the principal axes reference frame.

The above equations where derived by Beletskii (1965). Their Lie-Poisson structure was described by Bogoyavlensky (1992).

Equations (6) possess four integrals, energy

$$H_1 = H(\mathbf{M},\boldsymbol{\gamma},\mathbf{n}) = \frac{1}{2}(\mathbf{M},\mathbf{I}^{-1}\mathbf{M}) + \frac{3}{2}\Omega^2(\boldsymbol{\gamma},\mathbf{I}\boldsymbol{\gamma}) - \Omega(\mathbf{M},\mathbf{n}),$$

and tree geometrical integrals

$$H_2 = (\boldsymbol{\gamma},\boldsymbol{\gamma}), \qquad H_3 = (\mathbf{n},\mathbf{n}), \qquad H_4 = (\boldsymbol{\gamma},\mathbf{n})$$

For real system we consider always $H_2 = H_3 = 1$ and $H_4 = 0$.

For this system we can apply the observer method in many different way. Here we present results obtained when the observer method was applied first to the integral H_1 and next to the integral defined as follow:

$$H_g = \frac{1}{2}(H_2 + H_3) + H_4.$$

For test we put $\Omega = 1$, $A = 1.1$, $B = 2.1$, $C = 2.5$ and initial conditions were $M_1 = -10$, $M_2 = 0.1$, $M_3 = 0.2$, $\gamma_1 = 0.1$, $\gamma_2 = -0.3$, $\gamma_3 \simeq 0.94898$, $n_1 = n_2 \simeq 0.6993786$, $n_3 \simeq 0.14744$.

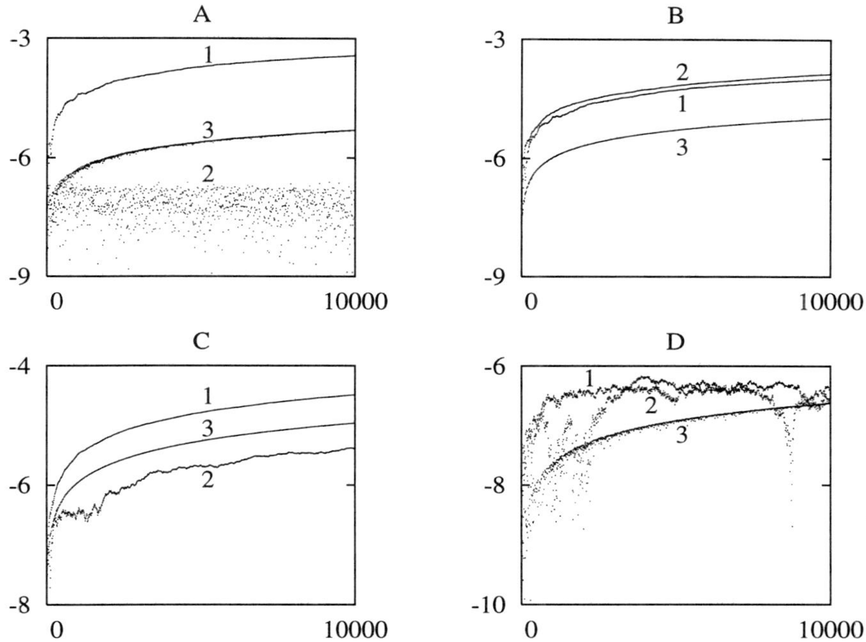

Figure 5. Logarithm of absolute error of integrals H_1, H_2, H_3 and H_4 (panels A, B, C and D, respectively) for three methods of integrations; 1—standard integration; 2—integration with the observer method applied to the integral H_1; 3—integration with the observer method applied to the integral H_g.

For integration we used the same procedure as in the previous example with local precision 10^{-8}. Equations of motion (6) were integrated three times: without the observer, with the observer applied to the integral H_1 and with the observer applied to integral H_g. Results are presented in Figure 5 where errors of integrals H_1, H_2, H_3 and H_4 are shown on logarithmic scale. Let us note the application of the observer method for different integrals gives different result. In Figure 5D curve label with '1' lies always higher than other twp curves.

THE CÉLÉRIER SYSTEM

Let us consider a point mass in the field of gravitational point center. Let us assume that, additionally to gravitational forces, a constant force acts on the body. In the appropriate Cartesian reference frame equations of motion of this system can be written in the following form

$$\ddot{x} = -\frac{x}{r^3} + f, \qquad \ddot{y} = -\frac{y}{r^3}, \qquad \ddot{z} = -\frac{z}{r^3}, \qquad r = \sqrt{x^2 + y^2 + z^2}. \tag{7}$$

In these equations we use the same convention as in the book of Beletskii (1977) where this system is described. We will called it the Célérier system.

System (7) is integrable. Except energy integral

$$H_1 = \frac{1}{2}(\dot{x}^2 + \dot{y}^2 + \dot{z}^2) - \frac{1}{r} - fx$$

it possesses two additional integrals, the x-component of the angular momentum

$$H_2 = z\dot{y} - y\dot{z}$$

and generalised Laplace integral

$$H_3 = \dot{x}r\dot{r} - \frac{x}{r} - \frac{3}{2}fx^2 - \frac{1}{2}fr^2 - 2H_1x.$$

As it was described in Beletskii (1977) book orbits of the Célérier system can be very complicated.

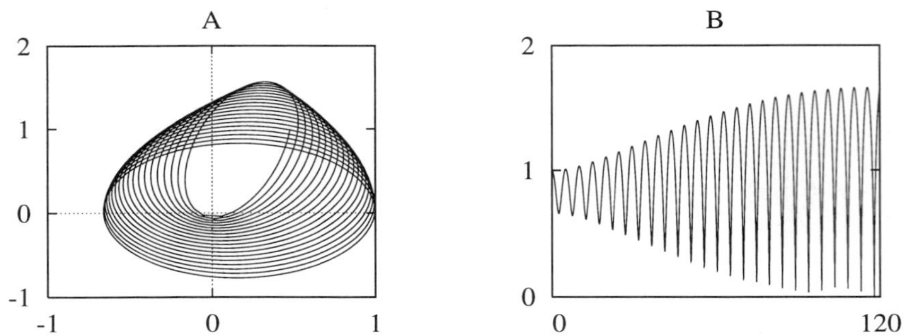

Figure 6. A—an arc of computed orbit on (x, y) plane corresponding to integration time 90; B—the radius vector as a function of time.

It is very difficult to integrate this system with the help of any fixed step-size method e.g. symplectic method. For wide range of initial conditions an orbit can pass near the origin where, for reliable integration, the step-size has to be very small.

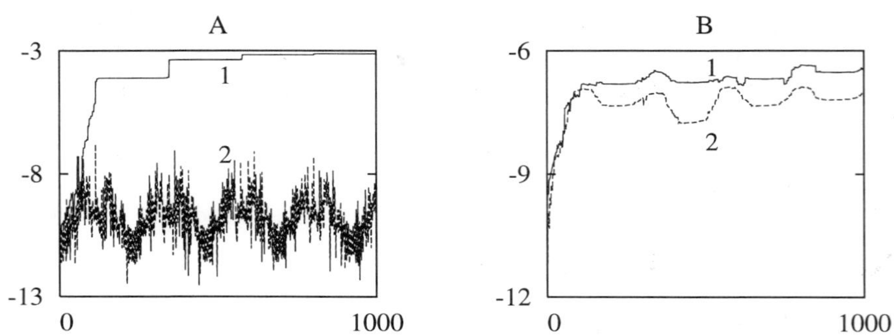

Figure 7. Errors in: A—energy integral H_1, B—integral H_3 for integration without and with the observer method (curves marked with '1' and '2', respectively).

For test we took planar Célérier system (we put $z = 0$) with $f = 0.01$. Initial condition was $x = 1, y = \dot{x} = 0$, $\dot{y} = 0.9$. First, we integrated equations of motion using ODEX procedure with local precision 10^{-8} and next, we applied the observer method for the energy integral only. We used normalised observer method (see Busvelle *et al.* 1993 for definition). Integration time was 1000. Figure 6 A shows the beginning arc of the computed orbit. On Figure 6B the radius vector as a function of time is shown. This figure shows that the orbit can pass very nearly to the gravitational center. Figure 7 shows errors (in logarithmic scale) in integrals for both integrations. In this example the observer method gives not only relatively big improvement of energy preservation but also errors in orbit are much more smaller. This is shown on Figure 8 relative error for both integrations is shown. The relative error is defined as

$$\frac{\|s(t) - s_{\text{num}}(t)\|}{\|s(t)\|},$$

where $s(t) = (x(t), y(t), \dot{x}(t), \dot{y}(t))$ and $s_{\text{num}}(t)$ denote analytical and numerical solutions, respectively. Here as the precise solution $s(t)$ we took numerically computed solution with local precision 10^{-16}. Let us not by the way that analytical solution for the Célérier system can be expressed by elliptic integrals.

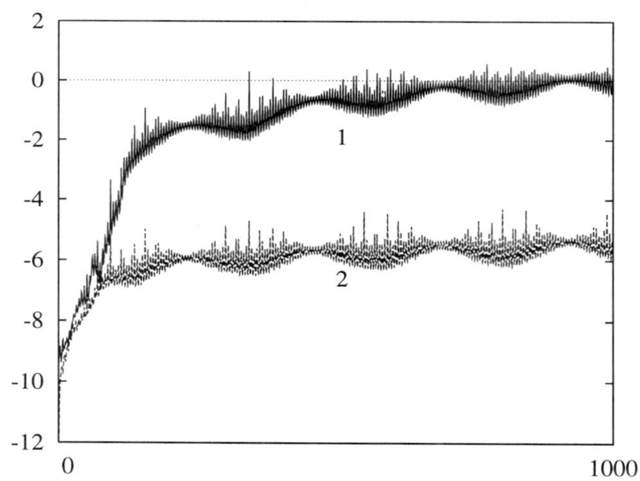

Figure 8. Relative error in computed orbit for integration without and with the observer method (curves marked with '1' and '2', respectively)

Let us underline that although we used variable step size procedure for integration without the observer errors in orbit grows rapidly and after time 500 it is of order 1.

It was practically impossible to integrate the orbit with prescribed initial condition for time of order 200 using a symplectic integrator. We tried to use for this purposes Yoshida (1991) fourth order symplectic integrator. Close approaches to origin cause rapid grow of errors. For reliable results the step-size must be chosen much smaller than 10^{-5}. With such a small step-size computation are very long and application of a symplectic integrator for study bounded orbit in this system is questionable.

ACKNOWLEDGEMENTS

I would like to thank warmly to J.-M. Strelcyn for many hours of common work and discussions.

I would like to thank J.-P. Marco from University Paris VI for his suggestion to study the Célérir system.

This work he was supported by the fellowship ERB3510PL9210479 from the European Community's Action for Cooperation in Science and Technology with Central and Eastern European Countries. The author acknowledges the Department of Mathematics of the Rouen University as well as URA CNRS 1378 for their hospitality and excellent working condition. In particular he thanks sincerely Gerard Grancher for his help and attention. Partially this work was supported also by grants KBN 22149203 and UMK 512A.

REFERENCES

Arnold V.I., Kozlov V.V. and Neishtadt A.I., 1988, Mathematical aspects of classical and celestial mechanics, *in Arnold V.I. (Edit.) Dynamical Systems III, Encyclopaedia of Mathematical Sciences*, **3**, (Berlin: Springer Verlag).

Baumgarte J., 1973, Stabilization of the differential equations of Keplerian motion, in Tapley B.D. and Szebehely V. (Eds) Recent Advances in Dynamical Astronomy, 38–44, *(Dodrecht: D. Reidel Publ Comp)*, .

Baumgarte J., 1976, Stabilization, manipulation and analytic step adaptation, in Szebehely V. and Tapley B.D. (Eds) Long Time Predictions in Dynamics, 153–163, *(Dodrecht: D. Reidel Publ Comp)*, .

Beletskii V.V., 1965, *The motion of an artificial satellite around the center of mass*, Moscow: Nauka, (in Russian).

Beletskii V.V., 1977, *Essays on motion of celestial bodies*, Moscow: Nauka, (in Russian)..

Bogoyavlensky O.I., 1992, Euler equations on finite-dimensional Lie coalgebras arising in problems of mathematical physics (in Russian), *Uspekhi Math. Nauk.*, **47 (1)**, 107–146; English translation in Russian Math. Surveys, **47 (1)**, 117–189.

Busvelle E., Kharab R., Maciejewski A.J. and Strelcyn J.-M., 1993, Numerical Integration of Differential Equations in the Presence of First Integrals: Observer Method, preprint, *Rouen University, March 1993.*

Channell P.J. and Scovel J.C., 1991 Integrators for Lie-Poisson dynamical systems, *Physica D*, **50**, 80–88.

Everhard E., 1974, Implicit Single Sequence Methods for integrating Orbits, *Celest. Mech.*, **10**, 35–55.

Ge Z.G. and Marsden J.E., 1988, Lie-Poisson-Hamilton-Jacobi theory and Lie-Poisson integrators, *Phys. Letters A*, **133(3)**, 134–139.

Gofen A.M., 1992, Taylor method integration of ODE's: the problem of steps and singularities, *Kossm.Issled.*, **30(6)**, 723–738.

Hairer E., Norsett S.P. and Wanner G., 1987, *Ordinary differential equations I, Nonstiff problem*, (Berlin: Springer Verlag).

Kozlov V.V., 1983, Integrability and nonintegrability in hamiltonian mechanics, *Russian Math. Survey*, **38 (1)**, 1–76.

Nacozy P.E., 1971 The use of integrals in numerical integrations of the n-body problem, *Astroph. and Space Science*, **14**, 40–51.

Sanz-Serna J.M., 1991, *Symplectic integrators for Hamiltonian systems: an overview, Acta Numerica, (annual volume published by Cambridge University Press)*, **1**, 243–286.

Yoshida H., 1990 Construction of higher order symplectic integrators, *Phys. Lett. A*, **150**, 262–268.

ANALYTICAL APPROXIMATIONS FOR SITNIKOV'S PROBLEM

Karl Wodnar

Institute of Astronomy
University of Vienna, Austria

INTRODUCTION

We treat the well known Sitnikov problem, which is characterized by two equally massive bodies in Keplerian (usually elliptic) orbits and an infinitesimally small mass, the motion of which is confined to the axis perpendicular to the primaries' orbital plane. In appropriate units Hamiltonian and equation of motion are respectively

$$H(t,z,\dot{z}) = \frac{1}{2}\dot{z}^2 - (r^2+z^2)^{-1/2}, \quad \ddot{z} + z(r^2+z^2)^{-3/2} = 0, \quad r(\varphi) = (1-\varepsilon^2)[2(1+\varepsilon\cos\varphi)]^{-1},$$

where $\varphi = \varphi(t)$ is one of the primaries' true anomaly, $r(\varphi)$ the primaries' distance from barycenter (O), dots are time derivatives, and ε is the primaries' eccentricity. The position of the small mass along the axis w.r.t. (O) is given by z.

The case of $\varepsilon = 0$ — usually referred to as the *circular case* — is completely integrable with H being constant and shows bounded periodic (elliptic) motion for $-2 \leq H < 0$ and unbounded motion for $H \geq 0$ (parabolic for $H = 0$, hyperbolic for $H > 0$). The limiting case $H \to -2$ is a zero amplitude linear oscillator with frequency $\omega = \sqrt{8}$, although the system in general is highly nonlinear.

In the *elliptic case* of $\varepsilon > 0$ — proceeding from small energies to large ones — we generally encounter the transition from quasiperiodic to chaotic motion, although, as has been shown by Martínez Alfaro and Chiralt (1993) there exists a neighbourhood of the origin $(z = 0, \dot{z} = 0)$ and there are tiny intervals within the range $\varepsilon \in]0,1[$, inside which there is *no* quasiperiodic motion for whatsoever small but positive amplitude.

For more introductory information see Sitnikov (1960), Alekseev (1968ab, 1969), Moser (1973), Wodnar (1991, 1994) and Dvorak et. al. (1994).

SOLUTION OF THE CIRCULAR CASE

Conventional Approaches

The solution of the circular case was expressed in terms of elliptic integrals in the Weierstraß form by Pavanini (1907) and in Legendre's form by Mac Millan (1913). Mac Millan gave

From Newton to Chaos, Edited by A.E. Roy
and B.A. Steves, Plenum Press, New York, 1995

in addition a Fourier series expansion with coefficients being essentially polynomials in the energy H. For small energies this solution shows satisfactory convergence, for large energies however the solution diverges, since, for one reason, a Taylor series in the energy giving the basic period of the system fails to describe the singularity at $H = 0$. The second problem arises from the singularity in amplitude entering the coefficients. Since usual numerical methods to represent elliptic integrals essentially suffer from the same shortcomings, the mentioned conventional approaches seem not to be useful for a representation of the circular case. The main interest in an easy to handle solution of this integrable case lies in its applicability to the generally nonintegrable elliptic case by means of perturbation techniques.

The Y-Transformation — using the Osculating Linear Oscillator

Expanding the Hamiltonian about $z = 0$ we obtain

$$H = \frac{1}{2}\dot{z}^2 - (1/4 + z^2)^{-\frac{1}{2}} = \frac{1}{2}\dot{z}^2 + 4z^2 - 2 + o(z^4). \tag{1}$$

Now let us introduce a set of new variables $(y, \overline{y})$ and require that inserting them into (1) instead of $(z, \dot{z})$ makes all the terms of order higher than z^2 vanish exactly. This corresponds to a horizontal projection of a trajectory onto that of the linear oscillator osculating at $z = 0$ in $(z, \dot{z})$ phase space geometry (see fig. 1), which uniquely defines a transformation of the circular system to a linear Hamiltonian system

$$R = \frac{1}{2}\overline{y}^2 + \frac{1}{2}\omega^2 y^2, \quad \omega = \sqrt{8}, \qquad \text{with} \quad H + 2 = R, \quad \dot{z} = \overline{y}, \quad \mathrm{sgn} z = \mathrm{sgn} y$$

being coupling conditions. From these one gets immediately

$$(1/4+z^2)^{-1/2} = 2(1-2y^2), \quad z = y(1-2y^2)^{-1}(1-y^2)^{1/2}, \quad dt = d\tau(1-2y^2)^{-2}(1-y^2)^{-1/2}, \tag{2}$$

where τ is the new time whose derivative is indicated as a bar. We shall call the above defined map $Y : (z(t), \dot{z}(t))_t \mapsto (y(\tau), \overline{y}(\tau))_\tau$ between trajectories in old and new variables the *Y-Transformation* for brevity. Using the energy-time parametrization

$$y(\tau) = \frac{1}{2}\sqrt{R}\cos\omega\tau, \qquad \overline{y}(\tau) = -\sqrt{2R}\sin\omega\tau \tag{3}$$

we have $t = t_0 + I$, with $I := \int_{\tau_0}^{\tau}(1-2y^2)^{-2}(1-y^2)^{-1/2}|_{y=y(\tau')}d\tau' = t_0 + \int_{\tau_0}^{\tau}(1-\frac{1}{2}R\cos^2\omega\tau')^{-2}(1-\frac{1}{4}R\cos^2\omega\tau')^{-1/2}d\tau'$ from (2), and note that the singularity at $|y| = \sqrt{2}/2$ does *not* occur within finite times, since it corresponds to infinite $|z|$ of the parabolic or hyperbolic case, but can be approached arbitrarily close. Hence always $|y| < \sqrt{2}/2$ and the singularity at $|y| = 1$ is thus not physically relevant.

One might linearize a system around any point; in the present case, the existence of a linear oscillator with well defined frequency $\omega = \sqrt{8}$ for vanishing amplitude suggested to choose $z = 0$ as a reference point, since it turned out that the system behaves nearly linear around this axis also for *arbitrary* (even infinite) amplitudes, with the osculating frequency ω being always the same. There are infinitely many ways to map the circular system to the harmonic oscillator osculating at $z = 0$ — set, for instance, $z/\dot{z} = y/\overline{y}$ — however, the choice $\dot{z} = \overline{y}$ is, as we shall see, well suited to the problem.

Approximation of the Quadrature Expression

Now we shall tackle expression I by formulating it in terms of

$$I = \int_{\tau_0}^{\tau} f\!f(y)\Big|_{y=y(\tau')} d\tau', \quad f\!f(y) := (1-2y^2)^{-2} f(y), \quad f(y) := (1-y^2)^{-1/2},$$

where I, $f\!f$ and f shall be approximated to $o[(1-2y^2)^N]$ as

$$I \approx \int_{\tau_0}^{\tau} gg(y)\Big|_{y=y(\tau')} d\tau', \quad f\!f(y) \approx gg(y) := (1-2y^2)^{-2} g(y), \quad f(y) \approx g(y) := \sum_{i=0}^{N} a_i(1-2y^2)^i, \tag{4}$$

and obtain, defining $u := 1 - 2y^2$, $F(u) := f(y)$, $FF(u) := ff(y)$, $G(u) := g(y)$, $GG(u) := gg(y)$, the relations

$$I = \int_{\tau_0}^{\tau} FF(u)\Big|_{u=u(y(\tau'))} d\tau' \approx \int_{\tau_0}^{\tau} GG(u)\Big|_{u=u(y(\tau'))} d\tau'$$

$$FF(u) = u^{-2}F(u), \quad GG(u) = u^{-2}G(u), \quad F(u) = \sqrt{2}(1+u)^{-1/2}, \quad G(u) := \sum_{i=0}^{N} a_i u^i. \quad (5)$$

The problem is thus reduced — apart from determining the a_i — to the elementary quadrature

$$I \approx \frac{1}{\omega}\sum_{i=0}^{N} a_i \underbrace{\int_{\Psi_0}^{\Psi}\left(1 - \frac{1}{2}R\cos^2\Psi'\right)^{i-2} d\Psi'}_{I_{i-2} :=}, \qquad \Psi := \omega\tau$$

of trigonometric rational functions ($i = 0, 1$) and polynomials ($i \geq 2$), resulting in

$$I_n = \tilde{I}_n\Big|_{\Psi_0}^{\Psi}, \qquad n = -2, -1, 0, 1, 2, \ldots, N-2$$

$$\tilde{I}_{-2}\begin{cases} R = 2 \;\Rightarrow\; \tilde{I}_{-2} = -\frac{1}{3}\left(\sin^{-2}\Psi + 2\right)\cot\Psi \\ R \neq 2 \;\Rightarrow\; \tilde{I}_{-2} = R\sin 2\Psi(2-R)^{-1}(4-R-R\cos 2\Psi)^{-1} \\ \qquad\qquad + (4-R)^{-1}(4-2R)^{-1}\tilde{I}_{-1} \end{cases}$$

$$\tilde{I}_{-1}\begin{cases} R = 0 \;\Rightarrow\; \tilde{I}_{-1} = \Psi \\ 0 < R < 2 \;\Rightarrow\; \tilde{I}_{-1} = (1-\frac{1}{2}R)^{-1/2}\Big\{\Psi + \arctan\Big[\Big(1-\sqrt{1-\frac{1}{2}R}\sin 2\Psi\Big) \\ \qquad\qquad \cdot\Big(1+\sqrt{1-\frac{1}{2}R} - \Big(1-\sqrt{1-\frac{1}{2}R}\Big)\cos 2\Psi\Big)^{-1}\Big]\Big\} \\ R = 2 \;\Rightarrow\; \tilde{I}_{-1} = -\cot\Psi \\ R > 2 \;\Rightarrow\; \tilde{I}_{-1} = \frac{1}{2}(\frac{1}{2}R-1)^{-1/2}\ln\Big[\Big(\sin\Psi - \sqrt{\frac{1}{2}R-1}\cos\Psi\Big) \\ \qquad\qquad \cdot\Big(\sin\Psi + \sqrt{\frac{1}{2}R-1}\cos\Psi\Big)^{-1}\Big] \end{cases}$$

$$\tilde{I}_{n\geq 0}\begin{cases} R \geq 0 \;\Rightarrow\; \tilde{I}_n = \left[\sum_{k=0}^{n}\binom{n}{k}\binom{2k}{k}\left(-\frac{R}{8}\right)^k\right]\Psi \\ \qquad\qquad + \sum_{l=1}^{n}\left[\sum_{k=l}^{n}\binom{n}{k}\binom{2k}{k-l}\left(-\frac{R}{8}\right)^k\right]\frac{\sin 2l\Psi}{l}. \end{cases}$$

Now the determination of the coefficients a_i in (4), (5) is left. Expanding $F(u)$ into a Taylor series would be most straightforward for this task, but would suffer from the same divergence problems as the conventional methods. Therefore let us split (5) into two parts

$$G(u) = G_{N_1,N_2}(u) := \sum_{i=0}^{N_1}\alpha_i u^i + \sum_{i=0}^{N_2}\beta_i^{N_1,N_2} u^{N_1+i+1},$$

and impose the conditions $\left(\frac{d}{du}\right)^i G(0) = \left(\frac{d}{du}\right)^i F(0)$ $(i = 0,\ldots,N_1)$, $\left(\frac{d}{du}\right)^j G(1) = \left(\frac{d}{du}\right)^j F(1)$ $(j = 0,\ldots,N_2)$, on the α_i, $\beta_i^{N_1,N_2}$. From these we obtain

$$\alpha_i = \sqrt{2}\binom{-1/2}{i}, \quad i = 0,\ldots,N_1, \qquad \vec{\beta} := \begin{pmatrix}\beta_0^{N_1,N_2} \\ \vdots \\ \beta_{N_2}^{N_1,N_2}\end{pmatrix} = D^{-1}\,\vec{V},$$

$$D := (D_{k,i})_{k,i=0,\dots,N_2} := \left(\prod_{j=0}^{k-1} (N_1 + i - j + 1) \right)_{k,i=0,\dots,N_2},$$

$$\vec{V} := (V_k)_{k=0,\dots,N_2} := \left((-2)^{-k} \prod_{l=0}^{k-1} (1/2 + l) - \sum_{i=k}^{N_1} \left[\prod_{j=0}^{k-1} (i - j) \right] \alpha_i \right)_{k=0,\dots,N_2},$$

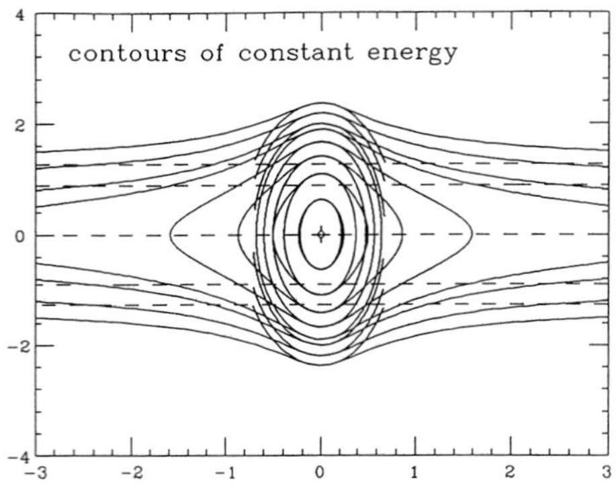

Figure 1. Horizontal phase space projection via the Y-transformation: $\dot{z}$ and $\overline{y}$ versus z and y.

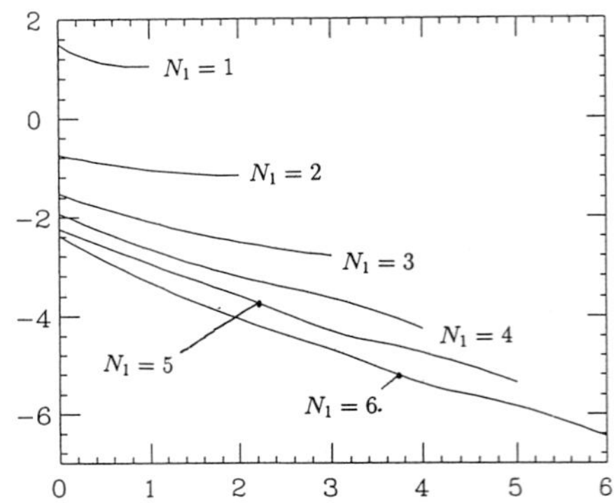

Figure 2. Maximum deviation Δff versus N_2, for various values of N_1.

where we have assumed $N_2 \leq N_1$ for technical reasons and notice the coincidence of the α_i with Taylor coefficients of an expansion about $u = 0$. We have thus obtained an unconventional series expansion insofar as the $\beta_i^{N_1,N_2}$ are dependent on the order of expansion, as indicated by superscripts, and, because derivatives of the original function up to a certain order have to coincide at *two* points ($u = 0, 1$), instead of only *one*. We shall call this kind of series *constrained powerseries*. This strategy is very rewarding, since we succeeded for instance in proving that this process shows even for fixed $N_2 = 0$ *uniform convergence* on $u \in]\delta_1, 1 - \delta_2]$ ($\delta_1, \delta_2 > 0$, arbitrarily small, δ_2 fixed), which is remarkable, because $u = 0$ corresponds to the singularity at $|y| = \sqrt{2}/2$. Thus we can make the *absolute* — not only the *relative* — error $|GG(u) - FF(u)|$, with $GG(u) := u^{-2} G_{N_1,0}$, arbitrarily small choosing N_1 sufficiently large for u approaching 0 as closely as desired, although $GG(u)$ and $FF(u)$ tend to infinity! The missing δ_2-neighbourhood around $u = 1$ seems — as numerical practice indicates — only a problem of a lacking proof, since this point is 'well behaved'; this is valid the more, if one admits to vary also N_2.

Summary of the Circular Solution

We are now able to formulate the $\varepsilon = 0$ solution

$$z = \frac{1}{2}\sqrt{R}\cos\Psi\sqrt{1 - \frac{1}{4}R\cos^2\Psi}\left(1 - \frac{1}{2}R\cos^2\Psi\right)^{-1}, \qquad \dot{z} = -\sqrt{2R}\sin\Psi$$

$$t = t_0 + \frac{1}{\sqrt{8}}\left[\sum_{i=0}^{N_1}\alpha_i I_{i-2} + \sum_{i=0}^{N_2}\beta_i^{N_1,N_2} I_{N_1+i-1}\right], \tag{6}$$

parametrized by $R \geq 0$ within the set of orbits, and by Ψ along an orbit, with initial conditions entering in the form[1] $R = H(z = z_0, \dot{z} = \dot{z}_0)|_{\varepsilon=0} + 2 = \frac{1}{2}\dot{z}_0^2 - (1/4 + z_0^2)^{-1/2} + 2$, where $\Psi_0 = -\arcsin(\dot{z}_0/\sqrt{2R})$, for $z_0 \geq 0$, and $\Psi_0 = \pi/2 + \arcsin(\dot{z}_0/\sqrt{2R})$, for $z_0 < 0$. Condition $|y| < \sqrt{2}/2$ for $R \geq 2$ restricts the range of physically relevant Ψ values to $\Psi_{\inf} := k\pi + \arccos\sqrt{2/R} < \Psi < \Psi_{\sup} := k\pi + \arccos\left(-\sqrt{2/R}\right)$, k integer.

This solution is able to cope with the mentioned singularity problem in practice through the *whole* range of energies, as can be seen from fig. 2, where the decadic logarithm of the maximum absolute error[2] $\Delta ff := \sup_{|y|<\sqrt{2}/2} |ff(y) - gg(y)|$ is plotted against N_2, for different N_1 values in the range $0 \leq N_2 \leq N_1 \leq 6$.

A Simple Application

Answering a question of C. Simo (1993), who asked for a proof different to his own one, of the fact that for bounded motions ($R < 2$) the period in the $\varepsilon = 0$ case increases monotonically with the amplitude, we considered the expression for the period $P = \frac{\sqrt{2}}{2}\int_0^\pi (1 - \frac{1}{2}R\cos^2\Psi)^{-2}(1 - \frac{1}{4}R\cos^2\Psi)^{-1/2} d\Psi$, which follows from the Y-transformation. Since $\frac{\partial}{\partial R}(1 - \frac{1}{2}R\cos^2\Psi)^{-2}(1 - \frac{1}{4}R\cos^2\Psi)^{-1/2}$ is positive for fixed Ψ, if $R\cos^2\Psi < 18/5$, we have proved the integrand increasing monotonically with R for any $\Psi \in [0, \pi]$, because of $R\cos^2\Psi < 2$, which follows from $|y| < \sqrt{2}/2$ and (3). Thus P also increases with R, but $R = 2 - (1/4 + z_{\max}^2)^{-1/2}$ evidently increases with amplitude $z_{\max}$ and the proof is finished.

PERTURBATIVE APPROACH TO THE ELLIPTIC CASE

The Y-Transformation for the Elliptic Case

The Y-transformation can be extended to the elliptic case keeping formally

$$
\begin{array}{rclr}
(t, H, z, \dot{z}), \qquad \dot{\ldots} := d/dt\ldots & \overset{Y}{\mapsto} & (\tau, K, y, \overline{y}), \qquad \overline{\ldots} := d/d\tau & \\
dt & = & d\tau(1 - 2y^2)^{-2}(1 - y^2)^{-1/2} & \\
H(t, z, \dot{z}) & = & K(t, y, \overline{y}) - 2 & (7) \\
z & = & y(1 - 2y^2)^{-1}(1 - y^2)^{1/2} & \\
\dot{z} & = & \overline{y} = dy/d\tau &
\end{array}
$$

where $K(t, y, \overline{y})$ is the expression resulting from transforming Hamiltonian[3] $H(t, z, \dot{z})$. Nevertheless, we need to check if Hamilton's equations are preserved under (7), before regarding K as the new Hamiltonian. Indeed, as is easily verified, this is the case for $\varepsilon = 0$. However, we have $1 = [z, \dot{z}]_{z,\dot{z}} \not\equiv [y, \overline{y}]_{z,\dot{z}} = (1 - 2y^2)^{-2}(1 - y^2)^{-1/2} = dt/d\tau$, where $[.,.]_{z,\dot{z}}$ denotes the Poisson-bracket w.r.t. the canonical set $(z, \dot{z})$ of variables. Thus we have found — against widespread believe — another example of a *noncanonical* transformation, which nevertheless preserves Hamilton's equation. Such thing ought to exist according to the footnote on p. 241 of Arnold (1978), who states that the confusion of defining a canonical transformation by just requiring preservation of Hamilton's equations appeared even in some high level mechanics text books.[4] For $\varepsilon > 0$, the situation is different. The pair of eqs. belonging to $(y, \overline{y})$ formally behaves canonical, but the extended phase space eqs. corresponding to $(t, -K)$ are only quasi-canonical, i.e., contain an additional factor. This factor cannot be removed by the usual change of the time variable, since this would affect the Hamiltonian character of the other pair

---

[1] Throughout this paper initial conditions will always be denoted by subscript 0.

[2] The solution is *exact* in $(z, \dot{z})$, the only error Δt occurs in t and can be estimated to be $\Delta t < \frac{2\pi}{\sqrt{8}}\Delta ff \approx 2.22\Delta ff$, per phase cycle $\Psi \in [0, 2\pi]$. $N_1, N_2 = 6$, for instance, implies a precision better than 10^{-6} per period of $z(t)$.

[3] Note the appearance of t in $K(t, y, \overline{y})$, which is due to the fact that t cannot be explicitely expressed without completely solving the system.

[4] Actually there are two different concepts in literature usually associated with the notion of a canonical transformation. Here we shall refer to the more stringent definition requiring preservation of symplectic structure *without* additional factors.

of equations. Interpreting $(\tau, -K)$ as canonical pair of variables results in *one* — consistent, but redundant — canonical eq. for τ and leaves the eq. for $-K$ quasi-canonical, but then the occurance of old time t in K as $\varphi(t)$ does not fit Hamiltonian formalism, since time dependence should have been removed by extension of phase space.

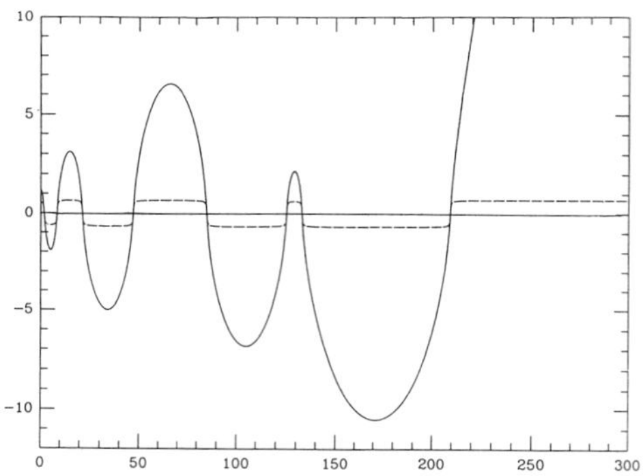

Figure 3. Numerically integrated chaotic orbit with $z(t)$ (solid line) and $y(t)$ (dashed line) plotted versus t.

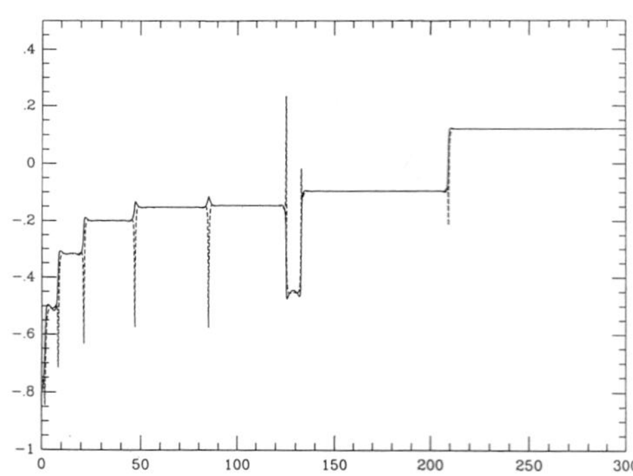

Figure 4. Energy $H(t,z,\dot{z})$ (solid line), and energy-like variable $H^0(z,\dot{z}) := H(t,z,\dot{z})|_{\varepsilon=0}$ (dashed line), with the orbit $(z(t),\dot{z}(t))_t$ of fig. 3 inserted.

Thus we are left now with a new 'pseudo'-Hamiltonian as well as equation of motion

$$K(t,y,\overline{y}) = \frac{\overline{y}^2}{2} + 4y^2 - 2(1-2y^2)\underbrace{\left\{[1+(4r^2-1)(1-2y^2)^2]^{-1/2} - 1\right\}}_{o(\varepsilon)}$$

$$\overline{\overline{y}} + 8y = -8y\left\{[1+(4r^2-1)(1-2y^2)^2]^{-3/2} - 1\right\} = o(\varepsilon) \tag{8}$$
$$r = (1-\varepsilon^2)[2(1+\varepsilon\cos\varphi)]^{-1},$$

which clearly define a nonlinearly perturbed linear oscillator with time dependent excitation. Figs. 3 and 4 give an idea of the effect of the transformation in the elliptic case.

Y-Transformation and Action-Angle-like Formalism

Using an analogous relation to (3), but now without assuming $\Psi = \omega\tau$ we transform (8) into the language of variables (τ, φ, R, Ψ)

$$\left.\begin{array}{rcl} y &=& \frac{1}{2}\sqrt{R}\cos\Psi \\ \overline{y} &=& -\sqrt{2R}\sin\Psi \end{array}\right\} \Rightarrow \left\{\begin{array}{rcl} \overline{\Psi} &=& \sqrt{8}\,[1+Q\cos^2\Psi] \\ \overline{R} &=& \sqrt{8}RQ\sin 2\Psi \\ \overline{\varphi} &=& (1-\varepsilon^2)^{-3/2}(1+\varepsilon\cos\varphi)^2 \\ && \cdot(1-\frac{1}{2}R\cos^2\Psi)^{-2}(1-\frac{1}{4}R\cos^2\Psi)^{-1/2} \\ Q &:=& [1+(4r^2(\varphi)-1)(1-\frac{1}{2}R\cos^2\Psi)^2]^{3/2} - 1 = o(\varepsilon). \end{array}\right. \tag{9}$$

For a perturbative treatment of eqs. of the latter form it always makes sense to collect as many as possible varying quantities on the left side (partial separation), thus we set $S := \ln R \Rightarrow \overline{S} = \overline{R}/R$. With a 1st order perturbation ansatz $\Psi = \Psi^0 + \varepsilon\Psi^1$, $S = S^0 + \varepsilon S^1 \Rightarrow R = R^0 e^{\varepsilon S^1} = R^0(R^1)^\varepsilon$, and $\varphi = \varphi^0 + \varepsilon\varphi^1$, we obtain to zero order the circular solution $\overline{\Psi^0} = \sqrt{8} \Rightarrow \Psi^0 = \sqrt{8}\tau$,

$\overline{S^0} = 0 \Rightarrow S^0 \equiv S^0_0$, and $\overline{\varphi^0} = \left(1 - \frac{1}{2}R^0 \cos^2 \Psi^0\right)^{-2} \left(1 - \frac{1}{4}R^0 \cos^2 \Psi^0\right)^{-1/2} \Rightarrow \varphi^0 = \varphi^0_0 + t - t_0$, and to first order

$$\begin{aligned} \overline{\Psi^1} &= 3\sqrt{8}\left(1 - \tfrac{1}{2}R^0 \cos^2 \Psi^0\right)^2 \cos^2 \Psi^0 \cos \varphi^0 \\ \overline{S^1} &= 3\sqrt{8}\left(1 - \tfrac{1}{2}R^0 \cos^2 \Psi^0\right)^2 \sin 2\Psi^0 \cos \varphi^0 \\ \overline{\varphi^1} &= 2\overline{\varphi^0} \cos \varphi^0 \quad \Rightarrow \quad \varphi^1 = 2 \sin \varphi^0 \end{aligned} \tag{10}$$

Because of its complicated structure and since it enters only the 1st order of the solution, we have exceptionally ignored an expession in the latter eq. for this first simple model.

Next we have to evaluate the quadratures

$$\Psi^1 = \Psi^1_0 + 3\sqrt{8} \int_{\tau_0}^{\tau} \left[1 - \frac{1}{2}R^0 \cos^2 \Psi^0(\tau')\right]^2 \cos^2 \Psi^0(\tau') \cos \varphi^0 d\tau' \tag{11}$$

$$S^1 = S^1_0 + 3\sqrt{8} \int_{\tau_0}^{\tau} \left[1 - \frac{1}{2}R^0 \cos^2 \Psi^0(\tau')\right]^2 \sin 2\Psi^0(\tau') \cos \varphi^0 d\tau' \tag{12}$$

The use of our precise circular solution at this stage would lead to extremly complicate integrands, thus we proceed with the expansions[5]

$$\overline{\varphi^0} = 1 + \frac{9}{2}(y^0)^2 + \frac{115}{8}(y^0)^4 + \ldots \Big|_{y^0 := \frac{1}{2}\sqrt{R^0}\cos\Psi^0} = \sum_{k=0}^{\infty} A_k(R^0) \cos 2k\Psi^0. \tag{13}$$

Now, in principle, (13) implies a representation of $\cos \varphi^0$ in the form

$$\cos \varphi^0 = \sum_{\Omega \in \Gamma} C_\Omega \cos(\Omega \Psi^0) + D_\Omega \sin(\Omega \Psi^0), \tag{14}$$

with Γ being an infinite set of discrete frequencies.[6] This form inserted into (11), (12) clearly results in easily integrable Fourier series as integrands. We shall demonstrate how to obtain (14) along with the simple case of considering (13) only up to $o[(y^0)^2]$. This gives — by means of integer order Bessel functions of the first kind —

$$\overline{\varphi^0} \approx \overline{\tilde{\varphi}^0} := 1 + \underbrace{\frac{9}{16}R^0}_{A_0} + \underbrace{\frac{9}{16}R^0}_{A_1} \cos 2\Psi^0 \tag{15}$$

$$\Rightarrow \quad \tilde{\varphi}^0 = \varphi^0_0 + \left(1 + \frac{9}{16}R^0\right)\tau + \frac{9}{32\sqrt{8}}R^0 \sin 2\Psi^0,$$

$$\begin{aligned} \cos \tilde{\varphi}^0 &= \cos\left[\varphi^0_0 + \left(1 + \frac{9}{16}R^0\right)\frac{\Psi^0}{\sqrt{8}}\right] \left\{ \mathcal{J}_0\left(\frac{9}{32\sqrt{8}}R^0\right) + 2\sum_{l=1}^{\infty} \mathcal{J}_{2l}\left(\frac{9}{32\sqrt{8}}R^0\right) \cos(4l\Psi^0) \right\} \\ &- \sin\left[\varphi^0_0 + \left(1 + \frac{9}{16}R^0\right)\frac{\Psi^0}{\sqrt{8}}\right] \left\{ 2\sum_{l=1}^{\infty} \mathcal{J}_{2l-1}\left(\frac{9}{32\sqrt{8}}R^0\right) \sin[2(2l-1)\Psi^0] \right\} \end{aligned}$$

and thus shows the idea behind (14).

In order to construct now the most straightforward solution — avoiding Bessel functions — within the foregoing framework, we even drop A_1 in (15) and integrate (11), (12) using the

[5] The first one is uniformely convergent for $|y^0| \leq \sqrt{2}/2 - \delta$, $\delta > 0$, arbitrarily small but fixed, and the second one is everywhere uniformly convergent in Ψ^0 for fixed $R^0 < 2$, with coefficients $A_k(R^0)$ being power series uniformly convergent for $R^0 \in [0, 2 - \delta]$, $\delta > 0$, arbitrarily small and fixed. We note that eq. (13) incidentally reproduces Mac Millan's (1913) series for the period of the circular problem, if we consider $P = \frac{\pi}{\sqrt{2}}A_0$, $A_0 = \frac{1}{2\pi}\int_0^{2\pi} \overline{\varphi^0} d\Psi^0 = 1 + \frac{9}{16}R^0 + \frac{345}{1024}(R^0)^2 + \ldots$.

[6] This series can be shown to be everywhere absolutely and uniformely convergent in Ψ^0, for fixed R^0 and a fixed finite number of terms of (13).

approximation $\varphi^0 \approx \tilde{\varphi}^0 := \varphi_0^0 + (1 + \frac{9}{16}R^0)\tau$, leading to

$$\begin{aligned}
\Psi^1 &= \Psi_0^1 + 3\sqrt{8}\Big\{[5(R^0)^2 - 24R^0 + 32]\frac{s_0}{36R^0 + 64} \\
&+ [15(R^0)^2 - 64R^0 + 64]\left[\frac{s_{-1}}{144R^0 - 512\sqrt{8} + 256} + \frac{s_1}{144R^0 + 512\sqrt{8} + 256}\right] \\
&+ R^0(3R^0 - 8)\left[\frac{s_{-2}}{72R^0 - 512\sqrt{8} + 128} + \frac{s_2}{72R^0 + 512\sqrt{8} + 128}\right] \\
&+ (R^0)^2\left[\frac{s_{-3}}{144R^0 - 1536\sqrt{8} + 256} + \frac{s_3}{144R^0 + 1536\sqrt{8} + 256}\right]\Big\}, \\
s_k &:= \sin(\tilde{\varphi}^0 + 2k\Psi^0)
\end{aligned}$$

$$\begin{aligned}
S^1 &= S_0^1 + 3\sqrt{8}\Big\{[5(R^0)^2 - 32R^0 + 64]\left[\frac{c_{-1}}{72R^0 - 256\sqrt{8} + 128} - \frac{c_1}{72R^0 + 256\sqrt{8} + 128}\right] \\
&+ R^0(R^0 - 4)\left[\frac{c_{-2}}{18R^0 - 128\sqrt{8} + 32} - \frac{c_2}{18R^0 + 128\sqrt{8} + 32}\right] \\
&+ (R^0)^2\left[\frac{c_{-3}}{72R^0 - 768\sqrt{8} + 128} - \frac{c_3}{72R^0 + 768\sqrt{8} + 128}\right]\Big\}, \\
c_k &:= \cos(\tilde{\varphi}^0 + 2k\Psi^0).
\end{aligned}$$

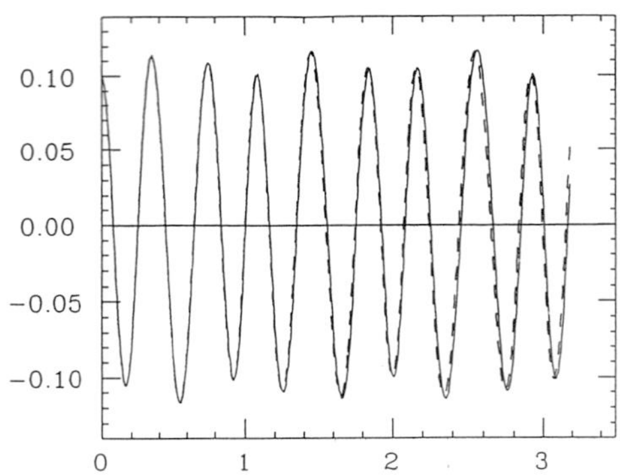

Figure 5. Quasiperiodic motion for $\varepsilon = 0.1$, $z_0 = 0.1$, and $\dot{z}_0, \varphi_0, t_0 = 0$.

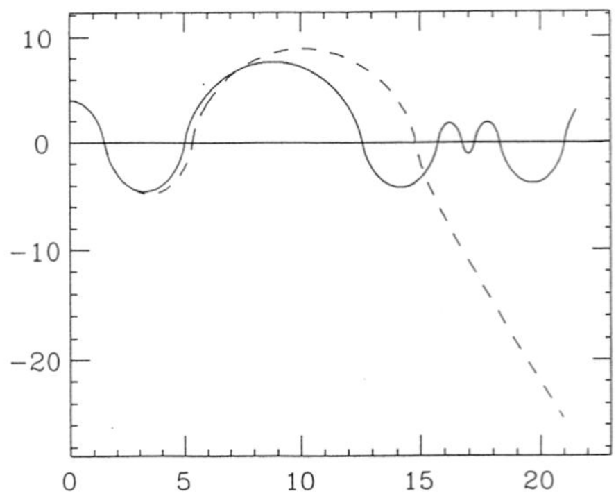

Figure 6. Chaotic orbit with $\varepsilon = 0.3$, $z_0 = 4$, and $\dot{z}_0, \varphi_0, t_0 = 0$.

Behaviour of the analytic solution (dashed line) compared to numerical integration (full line), in a $z(t)$ versus t diagram.

Now after having performed the quadrature, we return to our high precision circular solution (6) and replace $\tilde{\varphi}^0$ by $\varphi^0 = t$ in the above 1st order terms. We adjust initial conditions by means of the usual iteration process and translate the resulting 1st order solution back to the original variables. We thus dispose of an approximation valid for small values of $\varepsilon, z_{\max} < 0.1$. As can be seen from fig. 5, in spite of sufficient agreement with numerical integration[7] in the amplitude, the frequency behaviour of our solution is not yet satisfactory. This results from simplifications made, for instance, already in the computation of A_0, and is *not* a problem

[7]We used a Runge-Kutta-Fehlberg 5th order integrator with adaptive stepsize.

in principle. For larger values of $\varepsilon, z_{\max}$ the solution fails due to the expansion only up to $o(\varepsilon^1)$ as well as $o[(y^0)^2]$ in (13) and neglecting A_1. Nevertheless the case of large amplitudes is the most interesting one, since chaos appears. However, our solution is not only valid for small amplitude orbits, but also *each* time $|z|$ is small, i.e., near to each passage of the planet through barycenter (O), without regard to the maximum elongation. As can be seen from figs. 3, 4, within this small timespan the main changes in energy take place, so that we shall call it the *interaction phase*[8]; afterwards the energy remains nearly constant. But constant energy characterizes the circular solution, thus we decided to employ

A Piecewise Approach,

i.e., we switch between the 1st order perturbation solution and the 0th order solution, depending on the size of $|y^0|$. We switch on 1st order if $|y^0| < y_{\rm sw}$ and $\Psi \bmod \pi < \pi/2$ and change back to 0th order if $|y^0| \geq y_{\rm sw}$ and $\Psi \bmod \pi > \pi/2$. Empirically we found a threshold $y_{\rm sw} \approx .25$ to fit quite well, although we found extreme sensitivity w.r.t. this value for chaotic trajectories. An example of how this strategy works in a highly *chaotic* region, is given in fig. 6, where we reckognize our approximation (dashed line) follows the numerical integration (solid line) for some time but then quickly departs. This is quite clear, first, from the fact that we have used the most simple version of our model, and second, because we have exponential growth of errors present at any time. In forming an opinion about the quality of this solution, one should be aware of the fact that some 'quick and dirty' numerical integration methods show essentially the same divergence of initally identical orbits when changing slightly the stepsize, but take much more computing time than our solution. In addition this piecewise analytic solution qualitatively always reveals the character of the true curves quite closely and — unlike other models that are specialized to either quasiperiodic (e.g., Hagel 1992) or irregular motion (e.g., Liu and Sun 1990) — represents ordered as well as chaotic orbits in continous transition.

Our continous approximation can be evaluated at any point of the orbit, thus the construction of a succession of two alternatively applied mappings is obvious. Starting, say, at $|y^0| = y_{\rm sw}$, $\Psi \bmod \pi < \pi/2$, i.e., $z > 0$, $\dot{z} < 0$, we map using 1st order mapping M_1 through the interaction phase — passing (O) — to $y^0 = -y_{\rm sw}$, $\Psi \bmod \pi > \pi/2$, i.e., $z < 0$, $\dot{z} < 0$, then apply 0th order mapping M_0 to $y^0 = -y_{\rm sw}$, $\Psi \bmod \pi < \pi/2$, i.e., $z < 0$, $\dot{z} > 0$, and so forth. Since it may take a long time between successive zeroes of $z(t)$, (see, e.g., Sitnikov 1960, Wodnar 1994), our method saves a lot of computing time compared to numerical integration.

In addition to the minimum set of mapping points involved above, one may also — using an iteration — evaluate the solution expressions each time when $\Psi = \pi/2 + k\pi$, k integer, corresponding to $z = 0$. We then essentially dispose of the mapping for the Sitnikov problem theoretically investigated by Moser (1973), who maps speed and time from one zero of $z(t)$ to the next, which can be easily calculated from our variables S, φ. If we also evaluate at[9] $\Psi \approx \Psi^0 = k\pi$, corresponding to $z \approx z_{\max}$, we obtain the energy during the quiet regions between the interaction phases. Thus we achieve a mapping according to Petrosky's philosophy, who maps time from each center of interaction to the next, but — as a second mapping quantity — uses the well defined energy values *between* the fluctuative phases. Such a kind of mapping can be found, for instance applied to cometary motion, in Petrosky and Broucke (1988).

It still remains remarkable that we have to switch off our laboriously gained 1st order terms in some phase space regions to achieve better results. Let us therefore investigate this point a little closer. Consider the integrands of (11), (12), which are essentially $\overline{\Psi^1}$, $\overline{S^1}$. Especially concentrate on $\overline{S^1}$. This expression shows — for the moderate value $R^0 = 1$ — low frequency oscillations, when plotted against Ψ^0; thus over the whole range $[0, 2\pi]$ of phase angle Ψ^0 we have significant contributions to the value of integral (12) (see fig. 7). For a large energy

[8]At the very center $z = 0$ of the interaction phase there is *no* instantanous change of energy ($\overline{S} = 0$), as can be seen from (10), thus the time shortly before and after passage of (O) mainly determines the fate of the planet.

[9]This evaluation needs only to be made if $|y^0| = y_{\rm sw}$ does *not* occur within a half cycle $\Psi \in [0, \pi[$ or $\Psi \in [\pi, 2\pi[$, since otherwise we know already the energy value of the quiet phase from the switching evaluation.

value[10], say, $R^0 = 1.9$, we notice wild oscillations of $\cos\varphi^0$ near $\Psi^0 = k\pi$ (k integer), which are, however, damped by the factors $1 - \frac{1}{2}R^0\cos^2\Psi^0$ and $\sin 2\Psi^0$ at the same locations (see fig. 8). This concentration of high frequencies around phase angles corresponding to extremal elongation of z is due to the compensation of the nonlinearity in the dependent variables $(z, \dot{z})$ by the Y-transformation which thereby puts all the nonlinearity into the new independent variable $\tau = \Psi^0/\omega$. Thus we have significant contributions to the energy jump during the interaction phase only around $\Psi^0 = \pi/2 + 2k\pi$ corresponding to $z = 0$. The 'sidelobes' are not only insignificant because of their small absolute area, but also, because — as $\dot{z} = 0$ is approached — they contribute a quickly decaying sequence of areas entering the integral (12) with *alternating* sign, similiar to 'telescopic sums'.[11] Thus we are entitled to discard our 1st order perturbation far enough from $z = 0$ where it is of poor quality.[12] Intuitively it is also clear that for $|z|$ large enough the force field variations acting on the planet due to eccentric primary orbits are small. We note that the Y-transformation is useful in resolving the interesting interaction phase in detail.

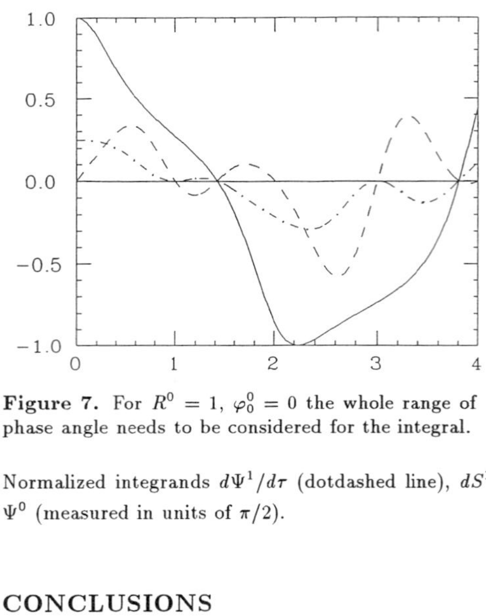

Figure 7. For $R^0 = 1$, $\varphi_0^0 = 0$ the whole range of phase angle needs to be considered for the integral.

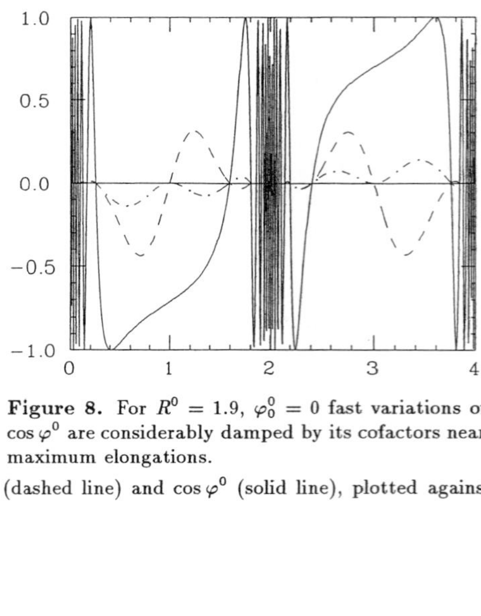

Figure 8. For $R^0 = 1.9$, $\varphi_0^0 = 0$ fast variations of $\cos\varphi^0$ are considerably damped by its cofactors near maximum elongations.

Normalized integrands $d\Psi^1/d\tau$ (dotdashed line), $dS^1/d\tau$ (dashed line) and $\cos\varphi^0$ (solid line), plotted against Ψ^0 (measured in units of $\pi/2$).

CONCLUSIONS

Results

Defining a *globally* valid transformation to linear oscillator normal form for the *circular* Sitnikov system ($\varepsilon = 0$) we constructed an easy to handle high precision analytical solution for this case. We extended our transformation to the *elliptic* case ($\varepsilon > 0$) resulting in a perturbed harmonic oscillator which we solved perturbatively to first order, obtaining an 'ad hoc' approximation valid for small z. By discarding $o(\varepsilon^1)$ terms each time $|z|$ exeeds a certain thres-

[10] Remember that for the circular case $R \in [0, 2[$ is the region of bounded motion, thus $R^0 < 2$, but close to 2 will lead to high amplitude chaotic motion near escape in the elliptic case; see, e.g., Moser (1973).

[11] This is also the principle of haed demagnetizers used for tape recorders.

[12] Similiar arguments hold for $\overline{\Psi^1}$, although the physical significance of $\Delta R := R(\Psi_0^0)\left\{ \exp\left[\varepsilon \int_{\Psi_0^0}^{\Psi_1^0} \overline{S}(\Psi')d\Psi'\right] - 1\right\}$, with $\Psi_0^0 = 0$, $\Psi_1^0 = \pi$ for $\dot{z} < 0$, and $\Psi_0^0 = \pi$, $\Psi_1^0 = 2\pi$ for $\dot{z} > 0$, as energy jump during the interaction phase — and thus essentially representing the Melnikov-Arnold integral — is more striking, than the deviation of Ψ from proportionality with τ.

hold,we obtain a qualitatively well fitting piecewise analytic solution for moderate eccentricities $\varepsilon < 0.3$. The solution expressions allow for the construction of various kinds of mappings, which are based on an alternating sequence of mappings M_0, M_1, the first of which can be shown to be symplectic to a high degree of accuracy, the second one — up to now — only in an approximative sense (on the average). Our derivation has been prepared for generalisations left to

Future Work,

which will be devoted to including terms containing Bessel functions etc. as indicated above. In addition M_0, M_1 should be improved to higher orders in ε, possibly using a generalisation of conventional perturbation theory, which relies on the principle of not just constructing a Taylor series, but matching higher derivatives w.r.t. the perturbation parameter at *more than one* point (*constrained powerseries*), as has already been successfully applied in the approximation of I in this paper, extending the usual radius of convergence.

Combination of the present analytical expressions with a formerly found transformation (Wodnar 1991) to achieve better results with low order perturbation theory seems also promising.

The construction of a new kind of area preserving twist mappings especially taylored for the Sitnikov problem containing as many free parameters as desired is in preparation. The present results and their extensions shall be used for determining these parameters thus resulting in symplectic mappings valid for *all* regions of phase space as well as for a wide range of eccentricities.

ACKNOWLEDGEMENTS

I express my gratitutde to R. Dvorak, Institute of Astronomy, Vienna, for continous support, especially when things sometimes seemed hopeless. For many important hints during our discussions I am indebted to Yi-Sui Sun, Center for Nonlinear Dynamical Systems, University of Nanjing, China, and to T.Y. Petrosky, Center for Studies in Statistical Mechanics and Complex Systems, University of Texas, Austin.

REFERENCES

Alekseev, V.M., 1968ab, 1969, Quasirandom dynamical systems I, II, III, *Math. USSR Sbornik* 5:73 (I, 1968a), 6:505 (II, 1968b), 7:1 (III, 1969).

Arnold, V.I., 1978, "Mathematical Methods of Classical Mechanics", Springer-Verlag, New York, Heidelberg and Berlin.

Dvorak, R., Vrabec, F., Wodnar, K., 1994, The Sitnikov problem: a short review, *in:* "Proceedings of the Primo Convegno Nazionale di Meccanica Celeste, l'Aquila, Italy", A. Celletti and E. Perozzi ed..

Hagel, J., 1992, A new analytic approach to the Sitnikov problem, *Cel. Mech.* 53:267.

Liu, J., Sun, Y.-S., On the Sitnikov problem, *Cel. Mech.* 49/3:285.

Mac Millan, W.D., 1913, An integrable case in the restricted problem of three bodies, *Astron. Jour.* 27:11.

Martínez Alfaro, J., Chiralt, C., 1993, Invariant rotational curves in Sitnikov's problem, *Cel. Mech.* 55/4:351.

Moser, J., 1973, "Stable and Random Motions in Dynamical Systems", *Annals of Mathematics Studies No. 77*, Princeton University Press and University of Tokio Press, Princeton, New Jersey.

Pavanini, G., 1907, Sopra una nuova categoria di soluzioni periodiche nel problema dei tre corpi, Annali di Mathematica, Serie III, Tomo XIII:179.

Petrosky, T.Y., Broucke, R., 1988, Area-preserving mappings and deterministic chaos for nearly parabolic motions, *Cel. Mech.* 42:53.

Simo, C., 1993, private communication.

Sitnikov, K., 1960, The existence of oscillatory motions in the three-body problem, *Dokl. Akad. Nauk USSR* 133/2:303.

Wodnar, K., 1991, New formulations of the Sitnikov problem, *in:* "Predictability, Stability, and Chaos in N-Body Dynamical Systems", *NATO ASI Series* B272:457, A.E. Roy, ed., Plenum Press, New York and London.

Wodnar, K., 1994, The original Sitnikov article — new insights, *in:* "Sitzungsber. der math.- nat. Kl. d. Österr. Akad. d. Wiss. Abt. II", 202/1–10:133, Springer-Verlag, Vienna and New York.

ELEMENTARY DERIVATION OF HOPF TYPE BIFURCATION FORMULAS

Franz Spirig

CH-9400 Rorchacherberg
Switzerland

ABSTRACT

In the case of a generalised Hopf bifurcation, several families of small periodic solutions may exist. An elementary procedure is presented for establishing those families as well as their stability behaviour, provided a certain non-degeneracy condition is satisfied. This seminar contribution gives a simplified version of [1], where further references may be found. The approach is based on the ideas of [2], [3]. The method is illustrated by deriving a well-known explicit bifurcation formula for the generic Hopf bifurcation.

Consider an autonomous differential equation depending on a small parameter ϵ:

$$\dot{u} = U(u, \epsilon) , \quad \cdot = \frac{d}{dt} , \quad u \in R^d , \quad |\epsilon| \leq \epsilon_o .$$

Suppose that there exists a family of equilibrium points $u(\epsilon)$ and that the matrix of the corresponding variational equation has two simple complex conjugate eigenvalues $\alpha(\epsilon) \pm i\beta(\epsilon)$, $\beta(\epsilon) > 0$. Without loss of generality, one may assume that the system has the form

$$\begin{aligned} \dot{x} &= \alpha(\epsilon)x - y + f(x, y, z, \epsilon) , \\ \dot{y} &= x + \alpha(\epsilon)y + g(x, y, z, \epsilon) , \\ \dot{z} &= \Gamma(\epsilon)z + h(x, y, z, \epsilon) , \end{aligned} \tag{1}$$

with $(x, y) \in R^2$, $z \in R^{d-2}$. The functions f, g and h represent terms of order $\mathcal{O}(x^2 + y^2 + |z|^2)$.

Assume that

$$\alpha(0) \approx \alpha^1(0) = \cdots = \alpha^{(m-1)}(0) = 0 , \quad \alpha^{(m)}(0) \neq 0 ,$$

From Newton to Chaos, Edited by A.E. Roy
and B.A. Steves, Plenum Press, New York, 1995

for some positive integer m. The eigenvalues $\gamma(\epsilon)$ of $\Gamma(\epsilon)$ are not integer multiples of the imaginary unit: $\gamma(0) \notin iZ$.

The aim is to establish the small, nearly 2π-periodic solutions of Eq. (1), as well as their stability. The following Lemma provides an *a priori* bound on small, nearly 2π-periodic solutions.

Lemma 1 *Let $u = (x, y, z)$ be a point on a small periodic orbit with a period T close to 2π. Then there is a constant K such that*

$$|z| \leq K(x^2 + y^2) \, .$$

The Lemma is proven by using the variation of constants formula and the implicit function theorem (see [1]).

Next, polar coordinates in the xy plane and a scaled vector are introduced:

$$x = r\cos\phi \, , \quad y = r\sin\phi \, , \quad z = r\zeta \, .$$

The scaling is justified by the Lemma, as will be seen below. Finally, ϕ is used as the new independent variable. Eq. (1) implies

$$t' = 1 + \theta(\phi, r, \zeta, \epsilon) \qquad (2)$$

$$\left. \begin{aligned} r' &= r(\,\alpha(\epsilon) + P(\phi, r, \zeta, \epsilon)\,) \\ \zeta' &= [\Gamma(\epsilon) - \alpha(\epsilon)I]\,\zeta + Q(\phi, r, \zeta, \epsilon) \end{aligned} \right\} \quad (3)$$

$' = \frac{d}{d\phi}$, where θ, P and Q are 2π-periodic with respect to ϕ and of order $\mathcal{O}(r)$.

Let $(t(\phi),\ r(\phi),\ \zeta(\phi))$ be a solution of Eqs. (2) and (3) corresponding to a small T-periodic solution of Eq. (1) with T close to 2π. Then $(r(\phi),\ \zeta(\phi))$ has to be a 2π-periodic solution of Eq. (3) with respect to ϕ. Therefore, from now on the range of ϕ is restricted to $0 \leq \phi \leq 2\pi$. The solutions under consideration satisfy

$$x^2 + y^2 < \delta^2 \, , \quad |z| \leq K(x^2 + y^2) \, ,$$

or, expressed in the new coordinates

$$|r| < \delta \, , \quad |\zeta| \leq Kr < K\delta \, ,$$

respectively. Hence the domain of definition for Eq. (3) is chosen as $\{(r, \zeta) \mid |r| < \delta \, , \ |\zeta| < K\delta\}$. The following theorem provides a representation of such solutions.

Theorem 1 *Any 2π-periodic solution of Eq. (3) has the form*

$$(r,\, \zeta) = (\, r(\phi,\, r_o,\, \epsilon),\, \zeta(\phi,\, r_o,\, \epsilon)\,) \, ,$$

with

$$r(0,\, r_o,\, \epsilon) = r_o \, , \quad \zeta(0,\, r_o,\, \epsilon) = \zeta(2\pi,\, r_o,\, \epsilon) \, ,$$

where r_o has to satisfy the bifurcation equation $r(2\pi,\, r_o,\, \epsilon) = r_o$. In addition, if all eigenvalues of $\Gamma(0)$ have negative real parts, and if $\frac{\partial r}{\partial r_o}(2\pi,\, r_o,\, \epsilon) < 1$ holds, then the solution is asymptotically stable.

Proof of the first part of the Theorem (for the stability condition see [1]): Let

$$(r(\phi, r_o, \zeta_o, \epsilon), \zeta(\phi, r_o, \zeta_o, \epsilon)),$$

be the solution of Eq. (3) with the initial value

$$(r(0, r_o, \zeta_o, \epsilon), \zeta(0, r_o, \zeta_o, \epsilon)) = (r_o, \zeta_o).$$

For $r_o = 0$ the solution is

$$r(\phi, 0, \zeta_o, \epsilon) = 0, \quad \zeta(\phi, 0, \zeta_o, \epsilon) = \exp[\phi(\Gamma(\epsilon) - \alpha(\epsilon)I)]\,\zeta_o.$$

A solution is 2π-periodic iff $r(2\pi, r_o, \zeta_o, \epsilon) = r_o$, $\zeta(2\pi, r_o, \zeta_o, \epsilon) = \zeta_o$. The function

$$F(r_o, \zeta_o, \epsilon) \equiv \zeta(2\pi, r_o, \zeta_o, \epsilon) - \zeta_o,$$

satisfies

$$F(0, 0, 0) = 0 \text{ and } \frac{\partial F}{\partial \zeta_o}(0, 0, 0) = e^{2\pi\Gamma(0)} - I.$$

Since $\frac{\partial F}{\partial \zeta_o}(0, 0, 0)$ is nonsingular, the implicit function theorem implies that ζ_o is a function of r_o and ϵ : $\zeta_o = Z(r_o, \epsilon)$.

The expansions of the 2π-periodic solution $(r(\phi, r_o, \epsilon), \zeta(\phi, r_o, \epsilon))$ and of the matrix $\Gamma(\epsilon)$ with respect to r_o and ϵ are of the form

$$r = \sum \epsilon^i \rho_{ij}(\phi) r_o^j, \quad \zeta = \sum \epsilon^i \eta_{ij}(\phi) r_o^j, \quad \Gamma = \sum \epsilon^i \Gamma_i.$$

These expansions and those of the right-hand side of Eq. (3) with respect to r, ζ and ϵ yield

$$\rho_{o1} = 1, \quad \rho_{i1} = 0, \quad \text{for } 1 < i < m, \quad \rho_{m1} = \alpha^{(m)}(0)\frac{\phi}{m!},$$

and

$$\begin{aligned}
\rho'_{ij+1} &= P_{ij}(\phi, \rho_{h2}, \cdots, \rho_{hj}, \eta_{h1}, \cdots, \eta_{hj-1}), \\
\eta'_{ij} &= \Gamma_o \eta_{ij} + \hat{q}_{ij}, \\
\hat{q}_{ij} &= \sum_{k<i} \Gamma_{i-k}\eta_{ki} + q_{ij}(\phi, \rho_{h2}, \cdots, \rho_{hj}, \eta_{h1}, \cdots, \eta_{hj-1}), \\
h &\le i, \quad \rho_{ij+1}(0) = 0, \quad \eta_{ij}(0) = \eta_{ij}(2\pi).
\end{aligned}$$

Finally, one obtains the bifurcation equation

$$r_o\left[\epsilon^m \rho_{m1}(2\pi) + \sum_{i<m} \epsilon^i \rho_{ij+1}(2\pi) r_o^j + \mathcal{O}(\epsilon^{m+1} + \epsilon^m r_o)\right] = 0.$$

As an example, the generic case of a Hopf bifurcation, i.e. $\alpha'(0) \neq 0$, is considered. Eq. (3) has the form

$$\begin{aligned}
r' &= \epsilon\alpha'(0)r + P_{10}r^2 + P_{11}r^2\zeta + P_{20}r^3 + \mathcal{O}(r^4 + r^3|\zeta| + r^2|\zeta|^2 + \epsilon r^2 + \epsilon^2 r), \\
\zeta' &= \Gamma_o\zeta + Q_{10}r + \mathcal{O}(r^2 + r|\zeta| + \epsilon r + \epsilon|\zeta|).
\end{aligned}$$

The expansions

$$\begin{aligned}
r &= (1 + \epsilon\alpha'(0)\phi)r_o + \rho_{02}r_o^2 + \rho_{03}r_o^3 + \mathcal{O}(r_o^4 + \epsilon r_o^2 + \epsilon^2 r_o), \\
\zeta &= \eta_{01}r_o + \mathcal{O}(r_o^2 + \epsilon r_o),
\end{aligned}$$

lead to

$$\begin{aligned}
\rho'_{02} &= P_{10}\,, \\
\rho'_{03} &= 2P_{10}\rho_{02} + P_{11}\eta_{01} + P_{20} = (\rho^2_{02})' + P_{11}\eta_{01} + P_{20}\,, \\
\eta'_{01} &= \Gamma_o\eta_{01} + Q_{10}\,.
\end{aligned}$$

It turns out that P_{10} is a homogeneous polynomial in $\sin\phi$, $\cos\phi$ of degree 3 (see below). Hence

$$\rho_{02}(2\pi) = 0\,, \quad \rho_{03}(2\pi) = \int_0^{2\pi}(P_{11}\eta_{01} + P_{20})d\phi\,.$$

Suppose that $\rho_{03}(2\pi) \neq 0$. Then one obtains the bifurcation equation (divided by r_o)

$$\epsilon 2\pi\alpha'(0) + \rho_{03}(2\pi)r_o^2 + \mathcal{O}(r_o^3 + \epsilon r_o + \epsilon^2) = 0\,,$$

or, with $\chi \equiv \frac{1}{2\pi}\rho_{03}(2\pi)$

$$\epsilon\alpha'(0)\,[1 + \mathcal{O}(r_o + \epsilon)] + \chi r_o^2\,[1 + \mathcal{O}(r_o)] = 0\,.$$

The only positive solution is found by an implicit function theorem argument

$$r_o = \sqrt{\frac{-\epsilon\alpha'(0)}{\chi}}\,[1 + \mathcal{O}(r_o + \epsilon)] = \sqrt{\frac{-\epsilon\alpha'(0)}{\chi}} + \mathcal{O}(|\epsilon|^{3/2})\,.$$

Whether the bifurcating solution exists for $\epsilon > 0$ or $\epsilon < 0$ depends on the signs of $\alpha'(0)$ and χ. It is stable if $\chi < 0$ and all eigenvalues of Γ_o have negative real parts. In this case the bifurcating solution exists if $\epsilon\alpha'(0) > 0$, i.e. if the equilibrium point $(r, \zeta) = (0, 0)$ is unstable. The crucial quantity χ can be easily expressed by the functions f, g and h of Eq. (1). The expansion of the function f has the form

$$\begin{aligned}
f(x, y, z, 0) &= \frac{1}{2}(F_{11}x^2 + 2F_{12}xy + F_{22}y^2) + xF_1z + yF_2z + \\
&\quad \frac{1}{6}(F_{111}x^3 + 3F_{112}x^2y + 3F_{122}xy^2 + F_{222}y^3) + \cdots\,.
\end{aligned}$$

Analogous expressions hold for g and h, where for h only the first three terms are needed. Using the abbreviations $s = \sin\phi$, $c = \cos\phi$ one obtains

$$\begin{aligned}
P_{10} &= \frac{1}{2}c^3F_{11} + c^2s(F_{12} + \frac{1}{2}G_{11}) + cs^2(\frac{1}{2}F_{22} + G_{12}) + \frac{1}{2}s^3G_{22}\,, \\
P_{11} &= c^2F_1 + cs(F_2 + G_1) + s^2G_2\,, \\
P_{20} &= P_{10}\left[-\frac{1}{2}c^3G_{11} + c^2s(\frac{1}{2}F_{11} - G_{12}) + cs^2(F_{12} - \frac{1}{2}G_{22}) + \frac{1}{2}s^3F_{22}\right] + \\
&\quad \frac{1}{6}\left[c^4F_{111} + c^3s(3F_{112} + G_{111}) + 3c^2s^2(F_{122} + G_{112}) + cs^3(F_{222} + 3G_{122}) + s^4G_{222}\right]\,, \\
Q_{10} &= \frac{1}{2}c^2H_{11} + csH_{12} + \frac{1}{2}s^2H_{22} = \frac{1}{4}(H_{11} + H_{22}) + \frac{1}{4}(H_{11} - H_{22})\cos 2\phi + \frac{1}{2}H_{12}\sin 2\phi\,.
\end{aligned}$$

The ansatz $\eta_{01} = X_o + X_2\cos 2\phi + Y_2\sin 2\phi$ yields

$$\begin{aligned}
X_o &= -\frac{1}{4}\Gamma_o^{-1}(H_{11} + H_{22})\,, \\
X_2 &= -\frac{1}{4}\Lambda\,[\Gamma_o(H_{11} - H_{22}) + 4H_{12}]\,, \\
Y_2 &= \frac{1}{2}\Lambda(H_{11} - H_{22} - \Gamma_oH_{12})\,,
\end{aligned}$$

with $\Lambda = (\Gamma_o^2 + 4I)^{-1}$ and finally

$$\begin{aligned} 16\chi &= (F_{11} + F_{22})F_{12} - (G_{11} + G_{22})G_{12} - F_{11}G_{11} + F_{22}G_{22} + F_{111} + F_{122} + G_{112} + \\ &\quad G_{222} - (F_1 - G_2)\Lambda\left[\Gamma_o(H_{11} - H_{22}) + 4H_{12}\right] - 2(F_1 + G_2)\Gamma_o^{-1}(H_{11} + H_{22}) + \\ &\quad 2(F_2 + G_1)\Lambda(H_{11} - H_{22} - \Gamma_o H_{12}) \, . \end{aligned}$$

This formula is given in [4], [5] with slightly different notations.

REFERENCES

1. F. Spirig, *An Elementary Approach to a Generalized Hopf Bifurcation*, Z.A.M.P., **44**, 6, (1993)
2. F. Spirig, *Bifurcation Equation For Planar Systems Of Differential Equations*, Z.A.M.P., **39**, 504-517, (1988)
3. F. Spirig, *Generic And Nongeneric Hopf Bifurcation.* In A. E. Roy (ed.), *Predictability, Stability and Chaos in n-body Dynamical Systems*, 565-572, Plenum Press, New York, (1991)
4. U. Kirchgraber, *On the Method of Averaging.* In V.Szebehely and B.D. Tapley (eds.), *Long-time Predictions in Dynamics*, 111-117, Reidel Publ. Comp., Dordrecht, (1976)
5. U. Kirchgraber and E. Stiefel, *Methoden der Analytischen Störungsrechnung und ihre Anwendungen*, 88-101, Teubner, Stuttgart, (1978)

CHAOTIC SCATTERING IN THE GAUSSIAN POTENTIAL

F. Casas[1] and J. Ros[2]

[1]Departament de Matemàtiques, Universitat Jaume I,
12071 Castelló, Spain
[2]Departament de Física Teòrica and IFIC,
Universitat de València, 46100 Burjassot, Spain

1. INTRODUCTION

It is well known that general classical Hamiltonian dynamical systems have as a rule chaotic behaviour. By such a term one usually understands a sensitive dependence on initial conditions which manifests itself in the topology of phase space. For the most studied case of bounded motions this behaviour is detected, for example, by analysing the Poincaré surfaces of section and by calculating Lyapunov characteristic exponents. The question then naturally arises of what are the effects of this complexity on the unbounded motions, i.e., on scattering phenomena. The signature of chaotic dynamics in these scattering regions of phase space has been the object of several papers appeared mainly in the last decade. Although it has been approached from different points of view it is true that both the number of case studies and the agreement over the quantitative characterization of the phenomenon is much less extensive than the corresponding situation for bounded motion.

Different techniques have been proposed to discuss the manifestations of these chaotic features on scattering phenomena. For example, the concept of Poincaré's surface of section, which rests on the recurrent character of bounded trajectories, has been extended (Jung, 1986) to the unbounded case. But by far the most used indicators of irregular scattering are the time delay and deflection angle as functions of the impact parameter: it is usually accepted that wild fluctuations of these functions betray chaotic behaviour. (Eckhardt, 1988) and (Smilansky, 1991) discuss applications in different physical situations of the idea of irregular scattering.

Most detailed analysis of the structure of the scattering process has been carried out (Bleher et al., 1989;1990) for non-relativistic Hamiltonians in two dimensions. The potentials used present several hills between which a set of unstable periodic orbits exists. A trajectory coming from infinity can be trapped in that region originating a singularity in the time delay or deflection angle as a function of initial conditions. In most cases studied so far the potential is given and the initial conditions are varied to cover different parts of phase space. Frequently the energy is used as the param-

From Newton to Chaos, Edited by A.E. Roy
and B.A. Steves, Plenum Press, New York, 1995

eter which sets the transition from regular to irregular behaviour. The set of initial conditions for which those singularities occur is shown to be fractal in nature.

In this contribution we present preliminary results for a family of simpler potentials which allows one to analyse, for a given set of initial conditions, the transition from a completely integrable situation to the general one. The potential has been used also to study relativistic effects and non conventional momentum dependences in chaotic Hamiltonian systems (Casas, 1989; 1992).

2. THE GAUSSIAN POTENTIAL

We are interested in the non-relativistic motion in two dimensions of a particle of unit mass under the action of the potential

$$V(x,y) = s\exp(-x^2 - qy^2), \tag{1}$$

where s distinguishes between attractive ($s = -1$) and repulsive ($s = 1$) cases. By variation of the q parameter this potential goes from the central case $q = 1$, which is integrable, to the non-central case ($q > 1$). This is a much simpler case than most of the potentials considered in the literature. Nevertheless, as we will show, it originates a very complex dynamics. The Hamiltonian of our problem will be

$$H = \frac{1}{2}(p_x^2 + p_y^2) + s\exp(-x^2 - qy^2), \tag{2}$$

where all the variables are dimensionless. For a wide spectrum of q values and initial conditions this system has shown to have all the characteristics of non-integrability for bounded motions (Casas, 1989), and here we will be interested in the study of the unbounded region of phase space. For that purpose we will have to integrate numerically the equations of motion. This will be done by using a Taylor series up to the seventh order, a scheme which has been proved to be very accurate and stable for this problem.

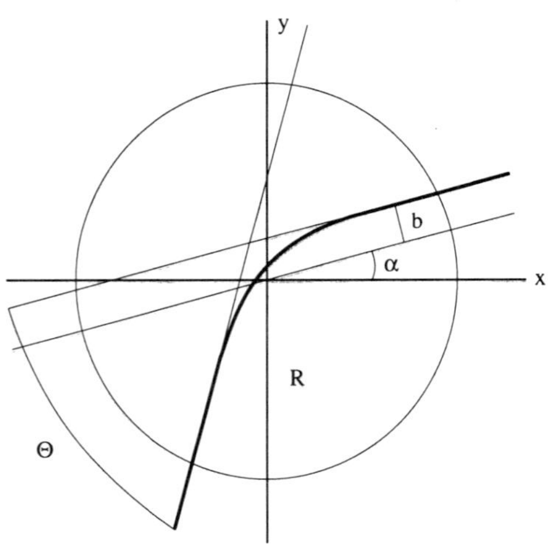

Figure 1. Schematic of a scattering trajectory.

A typical scattering trajectory is schematized in Fig. 1. A particle coming from infinity with energy $E > 0$, momentum $\vec{p}$ and angular momentum $\vec{l}$ enters a circle of radius R with incidence angle α. We take R large enough to ensure that outside that circle the particle is practically free. In our calculations we have taken $R^2 = 40$ and, due to symmetry reasons, we limit α to the range $[0, \pi/2]$. The impact parameter is defined by $b = |\vec{l}|/|\vec{p}|$. Each set of values of E, R, α and b determines uniquely a trajectory.

The magnitudes we are interested in are the deflection angle Θ and the time delay T, defined as the difference between the time it takes to the particle to traverse the region $x^2 + y^2 < R^2$ when the potential is on and the time taken when the particle is free.

The computational procedure to calculate $\Theta(b)$ and $T(b)$ consists in, keeping fixed the values of the energy E and the incidence angle α, dividing the interval $[b_{min} = -R, b_{max} = +R]$ in a number of points. For each b in this interval the equations of motion are integrated till the particle exits the circle of radius R, the scattering angle Θ being determined by the final momentum.

3. DISCUSSION OF THE RESULTS

Repulsive case

This corresponds to $s = 1$ and the form of the potential obviously does not allow the existence of periodic orbits, and so no irregular scattering is possible. Nevertheless, there is a peculiar phenomenon worth to be mentioned. In the central case ($q = 1$) a particle with $b = 0$ and energy $E = 1$, which is the top of the potential hill, asymptotically reaches the origin giving a divergence in the time delay function.

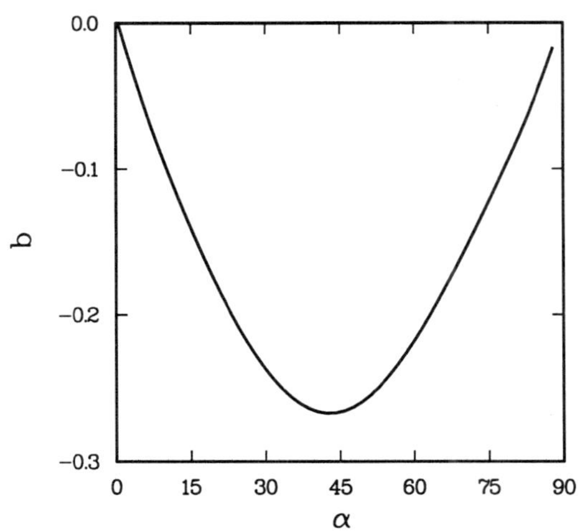

Figure 2. Distribution of points (b, α) for which a singularity appears in the functions $T(b)$ and $\Theta(b)$ for $q = 3$, $E = 1$ and $s = 1$.

For elliptical hill tops it has been argued (Bleher et al., 1989;1990) that, independently of the incidence angle α, the scattering angle Θ as a function of b presents a singularity in $b = 0$ for $E = 1$, and this behaviour is claimed to be a universal one for these potential shapes at the hill top energy. On the contrary, our results point in a different direction. In fact, we obtain that for $E = 1$, $\Theta(b)$ presents singularities at values of b which depend on the incidence angle α. In Fig. 2 we show the whole set of singularities b as a function of α for $E = 1$, $q = 3$. In particular, only for $\alpha = 0$ is $b = 0$ a singularity.

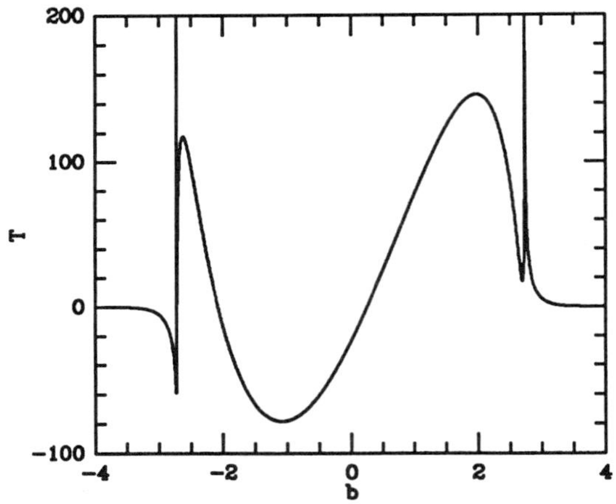

Figure 3. Time delay as a function of the impact parameter for $q = 1$, $E = 0.01$ and attractive potential ($s = -1$).

Attractive case

As an illustrative situation of this case we report the results obtained with $q = 1$ and $q = 3$ for the time delay and the scattering angle as functions of b for $E = 0.01$ and $\alpha = \pi/3$. Obviously, in the central potential there is no dependence on α.

Figure 3 presents the time delay $T(b)$ for $q = 1$. We can observe two discontinuities in that figure, which correspond to the well known phenomenon of orbiting (Newton, 1982). They appear symmetrically with respect to the origin, in agreement with the general theory. Furthermore, these particular values of b have been checked by an independent numerical procedure from the very definition of orbiting.

Figure 4 shows the corresponding results for $q = 3$. We observe a much intricate behaviour of the function $T(b)$ with a huge number of singularities distributed along an interval on the b-axis. To illustrate the possible fractal character of this set, a blowup of it is shown in Figure 5.

Instead of keeping fixed the energy and incidence angle and vary the impact parameter, we can fix b and study the dependence of T with E. Now the results show a 'resonance' structure similar to the quantum mechanical scattering problem. A typical illustration of this behaviour is given in Fig. 6.

Although the previous comments refer to the plots T vs. b, the same structure has been obtained for Θ vs. b.

4. CONCLUSIONS

A very simple potential with elliptic equipotential contours of varying ellipticity has been proposed to study the onset of chaotic scattering. Both repulsive and attractive cases have been considered. In the former case we have shown that for the hill top

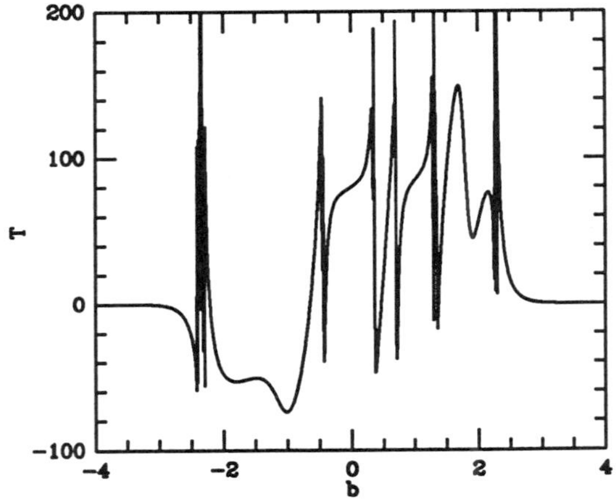

Figure 4. Time delay as a function of the impact parameter for the non-central case $q = 3$, $E = 0.01$, $\alpha = \pi/3$ and $s = -1$.

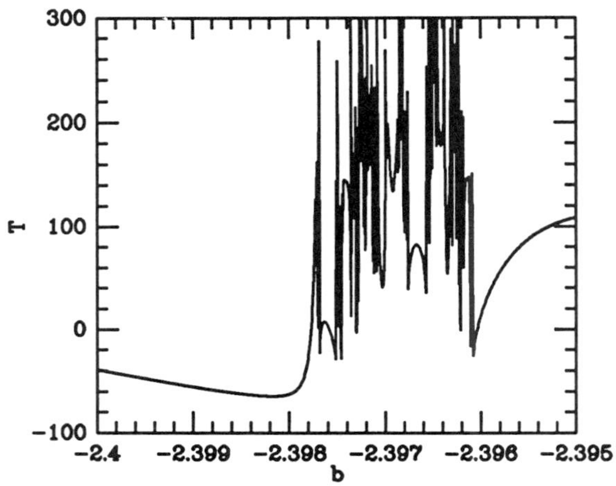

Figure 5. Blowup of the function $T(b)$ for $q = 3$, $E = 0.01$, $\alpha = \pi/3$ and $s = -1$.

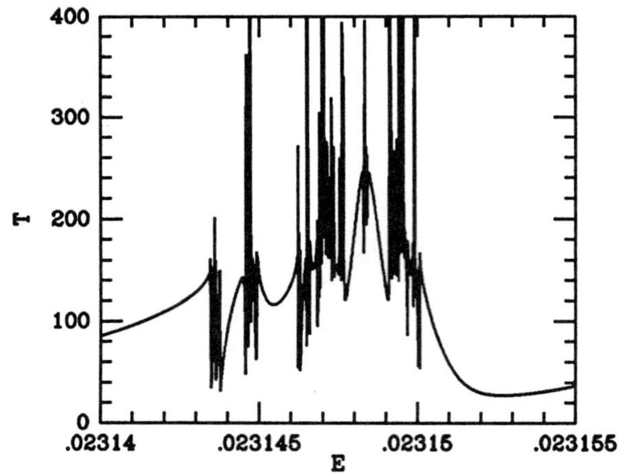

Figure 6. Time delay as a function of the energy for the non-central case $q = 3$, $\alpha = \pi/3$, $s = -1$ and impact parameter $b = 2$.

energy the scattering angle presents singularities which are movable with the incidence angle, at variance with what has been stated elsewhere. For the attractive potential we have observed the typical behaviour in the singularity structure of the time delay and deflection angle functions. Further work in this case to study the dependence of the fractal dimension of that set of singularities on the non-integrability parameter q is in progress.

Acknowledgments

JR thanks the spanish CICYT for partial financial support under grants number AEN 90-0049 and PB 92-0820. The work of FC has been partially supported by the collaboration programme UJI-Fundació Caixa Castelló 1993, grant number A42MA.

REFERENCES

Bleher, S., Ott, E., and Grebogi, C., 1989, Routes to chaotic scattering, *Phys. Rev. Lett.* **63**:919.
Bleher, S., Grebogi, C., and Ott, E., 1990, Bifurcation to chaotic scattering, *Physica D* **46**:87.
Casas, F., 1989, Master Thesis, Universitat de València.
Casas, F., 1992, Ph.D. Thesis, Universitat de València, Servei de Publicacions.
Eckhardt, B., 1988, Irregular scattering, *Physica D* **33**:89.
Jung, C., 1986, Poincaré map for scattering states, *J. Phys. A: Math. Gen.* **19**:1345.
Newton, R.G., 1982, "Scattering Theory of Waves and Particles," Springer-Verlag, New York.
Smilansky, U., 1991, The Classical and Quantum Theory of Chaotic Scattering, *in*: "Les Houches. Session LII. Chaos and Quantum Physics," M.-J. Giannoni, A. Voros and J. Zinn-Justin, eds., North-Holland, Amsterdam.

STOCHASTIC MOTION IN A CENTRAL FIELD WITH A WEAK NON–ROTATING BAR PERTURBATION

Pablo M. Cincotta, Josué A. Núñez and Juan C. Muzzio

Facultad de Ciencias Astronómicas y Geofísicas de la Universidad Nacional de La Plata and Programa de Fotometría y Estructura Galáctica del Consejo Nacional de Investigaciones Científicas y Técnicas de la República Argentina

INTRODUCTION

Chaotic motion was introduced in Stellar Dynamics three decades ago (e.g., Hénon and Heiles, 1964) and developed, among others, by Contopoulos who, in an old lecture (Contopoulos, 1973) wrote: "...It is perhaps dissapointing that most systems of interest are neither separable nor ergodic. Their behavior is quite complicated. Some orbits are periodic, or quasi–periodic, as in the separable systems, others behave in an ergodic way in a smaller or larger parts of the phase space. Thus a system is more or less near a separable system, or near an ergodic system...". It is surprising that it took nearly thirty years to accept that stochastic motion plays an essential role in, for example, the orbit of a star in a real galaxy (e.g., Udry and Pfenniger, 1988; Udry and Martinet, 1991).

In this work we present some results about the relevance of stochastic motion in the framework of the so–called Radial Orbit Instability (ROI) (Binney and Tremaine, 1987). Briefly, the ROI can be described in the following way: let us consider an N–Body spherical system where almost all particles have very low angular momentum, for example a collapse scenario; if we let it evolve under its own gravitation, the symmetry breaks down after a few crossing times and the spherical system becomes a bar. If the angular momentum of the particles is significant, however, the instability does not arise (Hénon, 1973). The ROI had been explained in terms of the classical Jeans Instability in the tangential direction by Barnes, Goodman and Hut (1986), but Merritt (1987) showed that this interpretation was inadequate. He suggested that the physical basis of the ROI lies rather on the transformation of the loop orbits of the spherical system into box orbits as a bar perturbation develops; these box orbits line up with the direction of the bar and reinforce its effect. Cincotta (1993) studied this

From Newton to Chaos, Edited by A.E. Roy
and B.A. Steves, Plenum Press, New York, 1995

transformation of loop into box orbits for increasing values of the bar perturbation and found that, as expected, loop orbits were turned into box orbits; nevertheless, if the perturbation was allowed to increase further, those box orbits gave way to chaotic orbits but, interestingly enough, only very weak bars were needed for the onset of chaos provided that the orbital angular momenta were low enough. Thus, in the present work we show some results on the relevance of these chaotic orbits.

THE DYNAMICAL MODEL

We will deal with plane orbits, and we adopt a central field represented by the isochrone potential:

$$\phi_0(r) = -\frac{GM}{b + \sqrt{b^2 + r^2}} \tag{1}$$

Our units are such that $G = 1, M = 1$ and $b = 1$. Now, let us consider a bar–like perturbation of the form

$$\alpha\phi_1(r, \varphi) = -\alpha\frac{r^2}{r^2 + a^2}\sin 2\varphi \tag{2}$$

Where α is the perturbation parameter and $a = 0.1$ throughout all the present paper. This election of ϕ_1 is similar to that of Udry and Pfenniger (1988), but slightly altered (by the softening parameter a) so that the force field does not diverge at the origin, also implying a finite value of the density at that point. Since we adopted a small value for a, the perturbation is nearly independent of r, i.e $\psi(r) \approx 1$ for $r \gg a$. Thus, we are now interested in the motion governed by the Hamiltonian:

$$H = H_0 + \alpha\phi_1 \tag{3}$$

Since the perturbation depends on the position on the orbit, we introduce the time–dependent effective perturbation parameter $\zeta(t)$, to estimate its actual value

$$\zeta(t) = 1 + \alpha\frac{\phi_1[\mathbf{r}(t)]}{\phi_0[r(t)]} \tag{4}$$

Thus, $\zeta(t)$ is a measure of the departure of the potential from spherical symmetry.

We will study the behavior of orbits in the Hamiltonian (3) solving numerically the Hamilton equations and obtaining the Poincaré Surface of Section (PSS hereafter) for a set of initial conditions. Then, using the algorithm discussed in the next section, we analyze the global stochasticity of this field.

THE METHOD

We are interested in the global orbital structure of the system. That is, we want to ascertain whether the stochastic regime is relevant or not in our model.

It is well known that regular motion is asociated with a well defined curve on the PSS (the invariant curve) and that the stochastic regime detroys this invariant set. Adopting a statistical point of view, we can recognize two diferent states of the system: ordered (regular) and disordered (stochastic). The quantity that gives us information on these states is the information entropy of the PSS, which is closely related to the entropy of statistical mechanics but different from the Kolmogorov entropy. The theorethical background and its aplication can be found in Eckmann y Ruelle (1985), Grassberg and Procaccia (1983) and Cohen and Procaccia (1985).

Let us denote with Σ a smooth manifold and μ a measure on Σ. We define the function $Z(x)$ in the interval $[0,1]$ as

$$Z(x) = -x \ln x \qquad 0 < x \leq 1, \qquad Z(0) = 0 \tag{5}$$

where Z has the following properties: is nonnegative, continuous and concave. Let $\chi = \{D_i\}_{i=1,\ldots,N}$ be mesurable a partition on Σ such that

$$\mu\left(\Sigma - \bigcup_{i=1}^{N} D_i\right) = 0, \qquad \mu(D_i \cap D_j) = 0 \qquad i \neq j \tag{6}$$

By definition, the entropy of the partition χ is

$$S(\chi) = \sum_{i=1}^{N} Z[\mu(D_i)] \tag{7}$$

Now, we apply this definition to the system of our interest. Let Σ be the sub–manifold of the phase space $\Gamma = \Re^{2N}$ where $dim(\Sigma) = 2N - 2$, defined by the equations: $q_1 = 0$, and $H(\mathbf{p}, \mathbf{q}) = h = cte$. In particular, we will consider $N = 2$ and we define the vector $\xi = (q, p)$, where $q \neq q_1$ and p is its conjugate momentum. If $\Sigma_1 = \{\xi_i : \xi_i = (q_{s_i}, p_{s_i}), i = 1, \ldots, M\}$ is the Poincaré Surface of Section of an orbit with initial condition $\mathbf{x_0}$, we can define a probability density $\rho(\xi)$ in Σ as

$$\rho(\xi) = \frac{1}{M} \sum_{i=1}^{M} \delta(\xi - \xi_i) \tag{8}$$

where δ is the *delta* function. From this definition follows

$$\int_{\Sigma} \rho(\xi) d^2\xi = 1 \tag{9}$$

Thus, we define the measure $\mu(D_i)$ as

$$\mu(D_i) = \int_{D_i} \rho(\xi) d^2\xi \tag{10}$$

Therefore, the entropy of the partition is given by (7).

The entropy may be interpreted as a measure of the amount of disorder in the system. For example, if the D_is are such that $\mu(D_i) = \mu(D_j), \forall (i,j)$ (i.e. a uniform distribution of points corresponding to an ergodic system), then the entropy reduces to $S = \ln N$, which can be shown to be its maximum value. Conversely, if $\mu(D_i) = 1$ for some i and $\mu(D_j) = 0, \forall j \neq i$ (a single periodic orbit), then $S = 0$.

Thus, we expect the entropy to have two regimes: a low–valued one for regular orbits and a high–valued one for stochastic orbits. It is known that the evaluation of the degree of chaoticity of a system is not easy and, up to date, it is an open subject. The errors of the numerical integrations grow, more or less, exponentially with time, so that it is very difficult to maintain the necessary accuracy when we need the phase–space coordinates after a long period. The more common indicator of chaoticity is the Liapunov exponent (see, e.g., Reichl, 1992) but, from its definition, the time must be infinite and to get a reasonable estimate one has to pursue the integration over long time intervals, so that the numerical errors may become very large. Thus, we preferred to use the PSS method where the time interval may be shorter (for excellent discussions of these problems see Heggie, 1991 and Szebehely, 1991).

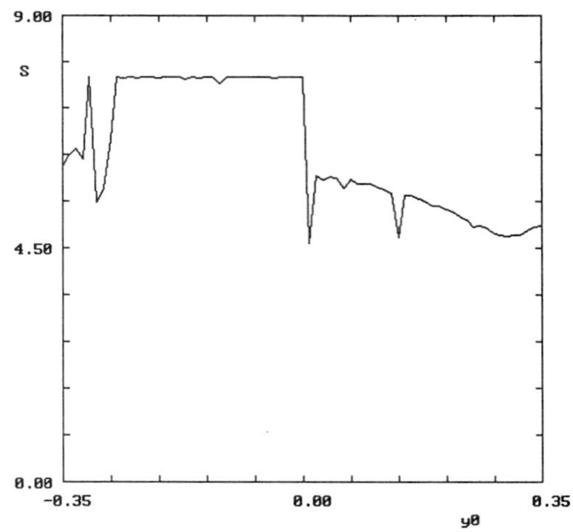

Fig. 1. Entropy vs. y_0 for the Hénon–Heiles potential

To check the proposed method we used the well known Hénon–Heiles potential (Hénon and Heiles, 1964). We obtained the PSSs with 2,500 points for 70 initial conditions, then we chose a particular partition $\beta = \{B_i\}_{i=1,\ldots,N}$ for the 70 PSSs, actually a grid of 2.5×10^5 squares covering all the section. Then we evaluated the measure $\mu(B_i)$ for each square, and finally the entropy. The set of initial conditions was $x_0 = 0, p_{y_0} = 0.02, E = 0.125$ and $-0.35 \leq y_0 \leq 0.35$, with $p_{x_0} > 0$. Fig. 1 shows the entropy S as a function of y_0. The two regimes can be clearly distinguished in the figure.

Benettin, Galgani and Strelcyn (1976, BGS hereafter) computed a quantity strongly related to the Kolmogorov entropy for some of the orbits considererd here. From Fig. 6 of BGS we see that the orbit with $y_0 = -0.15$ is stochastic while the one with $y_0 = 0.20$ is regular. Our Fig. 1 shows the same results. This agreement is not enough to conclude that our method is correct, so we obtained the PSSs for some of the 70 orbits considered and we compared them with the values of S given in Fig 1 for the same orbit (Cincotta, 1993). We found that the high and low values of S correspond to stochastic and regular motion, respectively, and that the presence of sharp valleys was asociated to stability islands around fixed points. Thus, we can be confident that the method reveals the global orbit structure of the Hénon–Heiles potential, and we can use it to investigate the Hamiltonian of our interest.

RESULTS

Cincotta (1993) investigated the ROI using the Hamiltonian given by eqs. (1)–(3). He concluded that the value of the perturbation parameter, α, must be of the order of 0.01 to obtain a time–scale adequate for the formation of the bar as predicted by N–body simulations. Therefore, in the present paper we investigated the behavior of orbits in the chosen potential, with $\alpha \geq 0.01$, using the method described in last section and with initial conditions $\{r_0 = 2, \pi/4 \leq \varphi_0 \leq 3\pi/4, p_{\varphi_0} = 0.01, E = -0.25\}$. The symmetry of the potential allowed us to consider only one quadrant, and we fixed the value of r because the perturbation is essentially independent of r, as we indicated in section II. Besides, we used a low value of p_{φ_0} to be consistent with the ROI conditions (Palmer and Papaloizou, 1987). Thus, we obtained the PSS with 5,000 points for 90 values of φ_0 and we chose a grid in the same way as we did for the Hénon–Heiles potential.

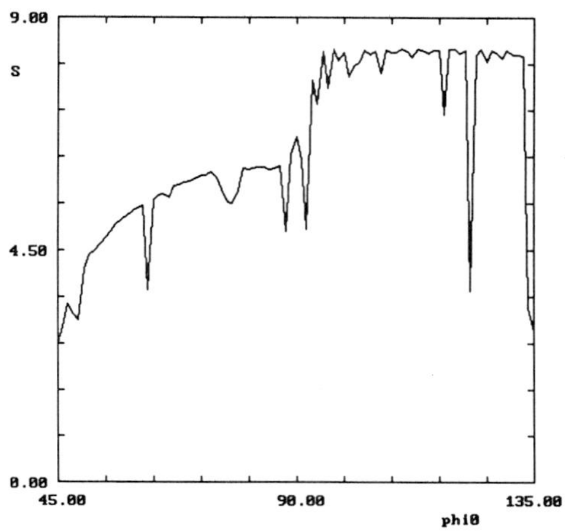

Fig. 2a. Entropy vs. φ_0 for $\alpha = 0.01$

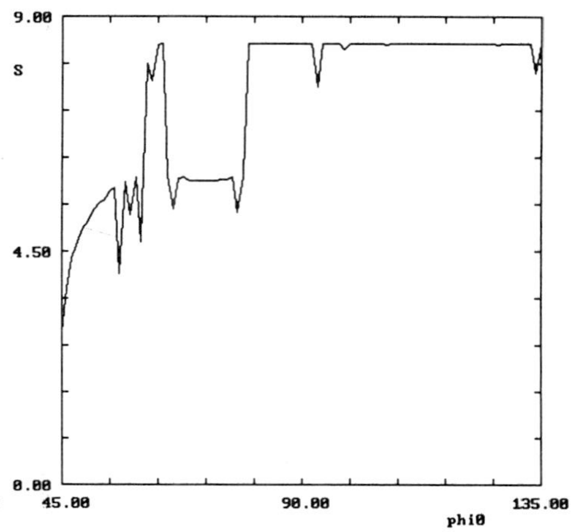

Fig. 2b. Entropy vs. φ_0 for $\alpha = 0.05$

Fig. 2 shows a plot of S as a function of φ_0 for two values of α : 0.01, and 0.05. We can notice, as in the Hénon–Heiles case, a very complicated structure. Regular behavior is closely related to a logarithmic dependence of the entropy on φ_0 and stability islands are present in this regime. Near $\varphi_0 = \pi/2$ an abrupt change in S takes place; the entropy remains essentially constant for larger φ_0 values, but the valley structure is still present. These valleys reveal the presence of stability islands embebed in the chaotic sea, besides being clear the presence of two regimes: a lower one, correponding to regular motion, and a higher one, corresponding to stochastic orbits. Notice that, for both α values, at least one half of the orbits are stochastic. This result is very important because, if we compute the effective perturbation parameter ζ, we obtain a maximum value of 1.04 for $\alpha = 0.01$ and 1.20 for $\alpha = 0.05$. This means that a large number of stochastic orbits exist with a departure of less than 4% from spherical symmetry (where all orbits are regular).

We also computed the value of S for $\alpha = 0.0001$ with the same initial conditions. It can be seen that in this case all the orbits must be regular. Fig. 3 shows this result, where only the low–valued regime is present. Although a change in the behavior of S at $\varphi_0 = \pi/2$ is present, it is not as abrupt as in the previous cases (the function S has a well defined derivative), and it rather indicates a change in the regular orbital structure, i.e., a transition from box orbits, aligned with the bar, to loop orbits.

Obviously, the value of S depends on the size of the partition and on the number of points in the PSS. We performed many other experiments (for example, with 10^4 points in the PSS and a grid of 10^6 squares) and, although the individual S values changed, its global behavior, which reveals the orbital structure of the Hamiltonian, remained the same.

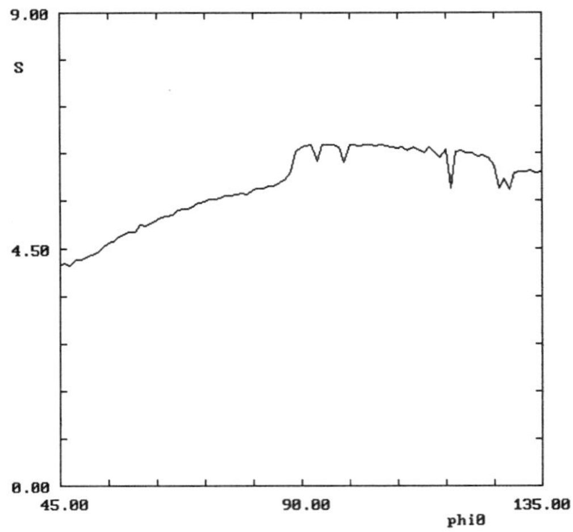

Fig. 3. Entropy vs. φ_0 for $\alpha = 0.0001$

CONCLUSIONS

We have used the information entropy for evaluating the global degree of stochasticity of orbits in a given potential. The low and high values of this entropy trace, in a quali–quantitative manner, the basins of initial conditions where the motion is regular or chaotic, respectively. The results obtained from its aplication to a bar–like perturbation to a central field allowed us to conclude that a chaotic regime is associated with at least 50% of the initial condition basin, although the perturbation was less than 4%. In this sense the importance of stochastic behavior for this kind of systems is very significant. In particular, chaotic motion is the normal behavior when the ROI is present.

AKNOWLEDGMENTS

The authors are very grateful to Daniel Carpintero for his advice on the use of computer software and codes and to Mariano Méndez for his useful comments. The technical assistance of S. D. Abal de Rocha, H. R. Viturro, M.C. Fanjul de Correbo and R.E. Martínez is gratefully acknowledged. This work was supported by grants from the Fundación Antorchas and from the Consejo Nacional de Investigaciones Científicas y Técnicas de la República Argentina, whose assistance is gratefully acknowledged.

REFERENCES

Barnes, J., Goodman, J y Hut, P. 1986. Ap. J. **187**, 112.

Benettin, G., Galgani, L. and Strelcyn J. 1976. Phys. Rev. **A 14**, 2338.

Binney, J. & Tremaine, S. 1987. Galactic Dynamics (Princeton:Princeton U.P.)

Cincotta, P. 1993. Thesis, La Plata University.

Cohen, A. and Procaccia, I. 1985. Phys. Rev. **A 31**, 1872.

Contopoulos, G. 1973. Topological Method in Stellar Dynamics, in Dynamical Structure and Evolution of Stellar Systems, Saas–Fee, p. 52, L. Martinet and M. Mayor eds.

Eckmann, J. and Rulle, D. 1985. Rev. Mod. Phys. **57**, N° 3, Part I, 617.

Grassberger, P. and Procaccia, I. 1983. Physica 9D, 189.

Heggie, D. 1991. Chaos in the N–Body Problem of Stellar Dynamics, in Predictability, Stability, and Chaos in N–Body Dynamical Systems, p. 47, A. Roy ed.

Hénon, M. 1973. Astron. & Astrophys. **24**, 229.

Hénon, M. and Heiles, C. 1964, Astron. J. **69**, 73.

Merrit, D. 1987. Stability of Elliptical Galaxies, Numerical Experiments, in Structure and Dynamics of Elliptical Galaxies, p. 315, T. de Zeeuw ed.

Palmer, P. and Papaloizou, J. 1987. Mon. Not. R. astr. Soc. **224**, 1043.

Reichl, L. 1992. The Transition to Chaos (New York: Springer–Verlag)

Szebehely, V. 1991. Chaos, Stability, and Predictability in Newtonian Dynamics, in Predictability, Stability, and Chaos in N–Body Dynamical Systems, p. 63, A. Roy ed.

Udry, S. and Pfenniger, D. 1988. Astron. Astrophys.**198**, 135.

Udry, S. and Martinet, L. 1991. Origin of Chaos and Orbital Behavior in Slowly rotating Models, in Predictability, Stability, and Chaos in N–Body Dynamical Systems, p. 115, A. Roy ed.

A NOTE ON THE CANONICAL CHARACTER OF THE STIEFEL–SCHEIFELE TIME ELEMENT

José M. Ferrándiz and M. Eugenia Sansaturio

Dep. de Matemática Aplicada a la Ingeniería
E.T.S. de Ingenieros Industriales
Universidad de Valladolid
47011 Valladolid, Spain

ABSTRACT

In this paper we present an easy way of transforming some standard canonical sets with the special feature that, in the new set, the time element by Stiefel and Scheifele appears as a canonical variable.

INTRODUCTION

Time elements are often used, specially when searching for the numerical solution of orbital problems, since the integration of the differential equation satisfied by the time element normally provides more accurate results than those of the equation which gives the relationship between the physical and the fictitious times. This fact is mainly due to the analytical control of the step–size, which is one of the most interesting effects involved when introducing a time element via a fictitious time.

There exists a wide number of references on time elements, but in this note it seems convenient to point out only the paper of Nacozy (1981), where he defines time elements associated to the ordinary anomalies of the elliptical motion, and that of Kwok and Nacozy (1981), which use Cartesian variables. On the other hand, KS transformation is always associated to an only element which, for the sake of brevity, will be designated here as Stiefel–Scheifele time element (Stiefel–Scheifele (1971)).

In general, all the deductions of the equations of the time elements are carried out in a non–canonical way and sometimes turn out to be quite long. Moreover, such equations are very similar, but not equivalent, depending on slight differences in the definitions. As far as we know, the only time elements introduced in a canonical way are those corresponding to the DS variables, either with the true anomaly as parameter or with the eccentric one (Scheifele and Graf (1974), Bond and Broucke (1980), Bond and Janin (1981)).

In this note we show how the Stiefel–Scheifele time element can become a coordinate

From Newton to Chaos, Edited by A.E. Roy
and B.A. Steves, Plenum Press, New York, 1995

of different canonical systems in a natural and easy way, without any reference to the KS and DS transformations. In this manner, the derivation of the differential equations which it satisfies is simpler than when using the usual direct procedures. In addition, it would also turn out to be possible and easy to calculate the analytical perturbations suffered by the element when the problem of the satellite is subjected to the action of canonical Lie transformations.

This paper is completed adding up some information about a canonical set of focal variables which incorporates the quoted time element and which is useful for the numerical integration, with a variable step-size code, of highly eccentric orbits under perturbations arising from tesseral harmonics or from a third body attraction. Likewise, we present a system of KS type in which the four coordinates have the character of anomalies, and the variable which gives an account of the size of the orbit is precisely the energy.

THE STIEFEL–SCHEIFELE TIME ELEMENT AS A CANONICAL VARIABLE

In this section we will adapt some standard canonical sets to include the Stiefel–Scheifele time element as a coordinate, without any reference to the KS transformation. In order to shorten the proofs we will start from the set of Hill variables $(r, \theta, h; R, \Theta, H)$, where r stands for the radial distance, θ is the polar angle of the particle in the orbital osculating plane reckoned from the ascending node, h is the longitude of such a node and R, Θ, H are their conjugate momenta, whose meaning is well known. We will extend this system to form a canonical set in the extended phase space by adding the pair of variables (t, p_0), t and p_0 being the physical time and the opposite of the energy, respectively. In what follows we will assume $p_0 > 0$, which corresponds to elliptical motion in the Keplerian case.

We have the following:

Proposition 1: The equations

$$q = r\sqrt{2p_0}, \qquad \tau = t + \frac{rR}{2p_0}, \qquad Q = \frac{R}{\sqrt{2p_0}}, \qquad p_0 = p_0 \tag{1}$$

define a canonical transformation on $p_0 > 0$. If the rest of the Hill variables remain unchanged, we obtain a new canonical set $(q, \theta, h, \tau; Q, \Theta, H, p_0)$, where the coordinate τ coincides with the Stiefel–Scheifele time element.

Proof: It suffices to show that the aforementioned transformation keeps the corresponding Poincaré–Cartan 2–form invariant, that is

$$dr \wedge dR + dt \wedge dp_0 = dq \wedge dQ + d\tau \wedge dp_0 .$$

Straightforward calculations lead to

$$\begin{aligned} dq &= \sqrt{2p_0}\, dr + \frac{r}{\sqrt{2p_0}}\, dp_0 , \\ dQ &= \frac{1}{\sqrt{2p_0}} - \frac{R}{(\sqrt{2p_0})^3}\, dp_0 , \\ d\tau &= dt + \frac{rdR + Rdr}{\sqrt{2p_0}} - \frac{rR}{(\sqrt{2p_0})^3}\, dp_0 , \end{aligned}$$

hence

$$dq \wedge dQ = dr \wedge dR - r\frac{dR \wedge dp_0}{\sqrt{2p_0}} - R\frac{dr \wedge dp_0}{\sqrt{2p_0}},$$

and the rest is immediate.

The fact that the variable τ coincides with the quoted time element is also easily verified by only transforming its defining equation in KS coordinates (Stiefel and Scheifele (1971), p.85) into polar variables. □

Let us recall that, in the unperturbed case, the element τ reduces to

$$\tau = t + \frac{\mu}{(\sqrt{2p_0})^3} e \sin E,$$

which is nothing more than the Kepler equation, μ being the reduced mass.

In the new set the Kepler problem admits the homogeneous Hamiltonian

$$H = \frac{1}{2}Q^2 2p_0 + \frac{1}{2}\frac{\Theta^2 2p_0}{q^2} - \frac{\mu\sqrt{2p_0}}{q} + p_0 = 0,$$

and carrying out the change of temporal independent variable from t to E defined by

$$dt = \frac{q}{2p_0} dE,$$

it becomes

$$K = \frac{1}{2}\left(Q^2 q + \frac{\Theta^2}{q}\right) - \frac{\mu}{\sqrt{2p_0}} + \frac{q}{2} = 0.$$

The equation of motion for τ is easily found to be

$$\frac{d\tau}{dE} = \frac{\mu}{(\sqrt{2p_0})^3}, \tag{2}$$

as expected.

In the perturbed case we only need to express the variables r and t in terms of the new canonical set by using the relationships

$$r = \frac{q}{\sqrt{2p_0}}, \qquad t = \tau - \frac{qQ}{2p_0}. \tag{3}$$

If the perturbation comes from a potential V, which in general will depend on all the coordinates and the momenta Θ and H, its contribution to the derivative of τ will be obtained by adding up the following term to the right hand side of (2)

$$\frac{q}{2p_0}\left[\frac{\partial V}{\partial r}\frac{\partial r}{\partial p_0} + \frac{\partial V}{\partial t}\frac{\partial t}{\partial p_0}\right],$$

with which it results

$$\frac{d\tau}{dE} = \frac{\mu}{(\sqrt{2p_0})^3} - \frac{q^2}{(\sqrt{2p_0})^5}\frac{\partial V}{\partial r} + \frac{q^2 Q}{4(p_0)^3}\frac{\partial V}{\partial t}, \tag{4}$$

the indicated partial derivatives being evaluated naturally in the new variables.

We find it interesting to remark the rapidity with which the equation of the time element is obtained in comparison with the traditional methods – see for instance Nacozy (1981).

In the rest of this section we will show other canonical sets in the extended phase space for which τ is a coordinate.

First, we state the following:

Proposition 2: Let $(X_i; P_i)$, $i = 1, 2, 3$, be the Cartesian coordinates and momenta and let us extend this set by adding the pair (t, p_0). The equations

$$\begin{aligned} q_i &= X_i\sqrt{2p_0}\,, \qquad \tau = t + \frac{\mathbf{X}\cdot\mathbf{P}}{2p_0}\,, \\ Q_i &= \frac{P_i}{\sqrt{2p_0}}\,, \qquad p_0 = p_0\,, \end{aligned} \tag{5}$$

where $\mathbf{X}\cdot\mathbf{P}$ stands for the dot product, give rise to a new canonical set $(q_i, \tau; Q_i, p_0)$.

The proof reduces to straightforward calculations so it will be omitted.

If we consider the extended set of focal variables $(x_i, \rho, t; p_i, p_\rho, p_0)$ used in Ferrándiz et al. (1992a) – where x_i $(i = 1, 2, 3)$ are the direction cosines and ρ is the inverse of the radial distance – proceeding as in the first case, the substitution of the new variables $(\zeta, \tau; p_\zeta, p_0)$ for the old ones $(\rho, t; p_\rho, p_0)$ via the relations

$$\begin{aligned} \rho &= \zeta\sqrt{2p_0}\,, \qquad \tau = t - \frac{\zeta p_\zeta}{2p_0}\,, \\ p_\rho &= \frac{p_\zeta}{\sqrt{2p_0}}\,, \qquad p_0 = p_0\,, \end{aligned} \tag{6}$$

provides again a canonical set which contains the time element τ as a coordinate.

Expressed in the new set, the homogeneous Hamiltonian of a perturbed two-body problem takes the form

$$H = \frac{\omega^2}{2}\left\{|\mathbf{x}|^2|\mathbf{p}|^2\zeta^2 + \zeta^4 p_\zeta^2\right\} - \mu\,\zeta\,\omega + 2p_0\zeta^2 W + p_0 = 0\,, \tag{7}$$

where, for the sake of conciseness in the notation, we have written $\omega = \sqrt{2p_0}$ and $W = \zeta^{-2}\omega^{-2}V$.

After performing the change of time $dt = (c\zeta^2\omega^2)^{-1}d\varphi$, where $c = |\mathbf{p}|.|\mathbf{x}|$ is the norm of the angular momentum, the equations of motion for $\mathbf{x}$ and its derivatives and for the angular momentum are simply obtained from those in Ferrándiz et al. (1992a) by replacing ρ with $\zeta\omega$. As for those corresponding to the new variables, we have the equation of a perturbed oscillator in the form:

$$\begin{aligned} \zeta' &= y + \frac{\zeta}{c\omega^2}\frac{\partial W}{\partial t}\,, \\ y' &= -\zeta + \frac{\mu}{c^2\omega} - \frac{c'}{c}y - \frac{1}{c^2}\frac{\partial(\zeta^2 W)}{\partial\zeta} + \frac{y}{c\omega^2}\frac{\partial W}{\partial t}\,, \end{aligned} \tag{8}$$

where $y = c^{-1}\zeta^2 p_\zeta$. This set of equations is completed by the following:

$$p_0' = -c^{-1}\frac{\partial W}{\partial t}\,, \qquad \tau' = \frac{\mu}{c\omega^3\zeta} - \frac{2\zeta'}{\zeta\omega^4}\frac{\partial W}{\partial t} + \frac{2}{c\omega^6}\left(\frac{\partial W}{\partial t}\right)^2 + \frac{1}{c\omega}\frac{\partial W}{\partial\omega}\,. \tag{9}$$

The application of these variables to the calculation of highly eccentric orbits can be found in Ferrándiz et al. (1992b).

A SPECIAL CANONICAL SET DERIVED FROM THE KS VARIABLES

Let us consider the set $(u_\alpha; U_\alpha)$, $(\alpha = 1, .., 4)$ of KS variables defined starting from the Cartesians $(X_i; P_i)$ by means of

$$\begin{pmatrix} \mathbf{X} \\ 0 \end{pmatrix} = L(\mathbf{u})\mathbf{u}, \qquad \begin{pmatrix} \mathbf{P} \\ B \end{pmatrix} = \frac{1}{2r} L(\mathbf{u})\mathbf{U},$$

where we have used vectorial notation, B stands for the bilinear relation – which vanishes along the trajectories of the KS transformed system – and $L(\mathbf{u})$ is the so-called KS matrix (Stiefel and Scheifele (1971), p.24).

Once the KS set has been extended to a 10-dimensional space by introducing the homogeneous formalism, we define new variables $(\mathbf{v}; \mathbf{V})$ by means of a slightly modified version of equations (5). More precisely we have:

Proposition 3: The transformation from $(\mathbf{u}, t; \mathbf{U}, p_0)$ to $(\mathbf{v}, \hat{\tau}; \mathbf{V}, p_0)$ defined by

$$\begin{aligned} \mathbf{v} &= \mathbf{u}\sqrt{\frac{2p_0}{\mu}}, & \hat{\tau} = t + \frac{\mathbf{u}\cdot\mathbf{U}}{2p_0}, \\ \mathbf{V} &= \mathbf{U}\sqrt{\frac{\mu}{2p_0}}, & p_0 = p_0 \end{aligned} \qquad (10)$$

is canonical on $p_0 > 0$.

Notice that such a transformation is inspired by the Keplerian relation $a = \mu/2p_0$, which provides the semimajor axis of an elliptic orbit. On the other hand, $\hat{\tau}$ does not coincide with the element τ any more, since $\mathbf{u}\cdot\mathbf{U} = 2\,\mathbf{X}\cdot\mathbf{P}$. The new variables turn out to be interesting due to the following:

Property: The coordinates v_α are adimensional, that is, they represent anomalies, and the momenta V_α have the dimension of an angular momentum.

Proof: It suffices to take into account that $[\mathbf{u}] = L$, $[\mathbf{U}] = L^{3/2}T^{-1}$ and with that it results $[\mathbf{v}] = 1$, $[\mathbf{U}] = L^2T^{-1}$. □

In the new variables the homogeneous Hamiltonian of a perturbed two-body problem can be written as

$$H = \frac{1}{2}\frac{|\mathbf{V}|^2}{|\mathbf{v}|^2}\frac{p_0^2}{\mu^2} - \frac{2p_0}{|\mathbf{v}|^2} + p_0 + V = 0\,.$$

The usual fictitious time corresponding to the eccentric anomaly in the unperturbed problem is introduced by means of

$$dt = \frac{|\mathbf{v}|^2\mu}{\omega^3}dE\,,$$

where we have denoted $\omega = \sqrt{2p_0}$, and, with respect to E, the variables $\mathbf{v}$ verify the equations of a perturbed oscillator with frequency 1/2. However, in contrast to the KS coordinates, they do not have information relative to the size of the orbit – which is contained in the variable p_0 – but only to the spatial direction. In fact, it is not difficult to realize that $\mathbf{v}$ is precisely a KS vector transformed through the scaling of the position vector of the particle when dividing by $a = \mu/2p_0$ to make it adimensional.

REFERENCES

Bond, V. and Broucke, R., 1980, Analytical satellite theory in extended phase space, *Celest. Mech.*, 21:357.

Bond, V. and Janin, G., 1981, Canonical orbital elements in terms of an arbitrary independent variable, *Celest. Mech.*, 23:159.

Ferrándiz, J. M., Sansaturio, M. E. and Pojman, J., 1992a, Increased accuracy of computations in the main satellite problem through linearization methods, *Celest. Mech.*, 53:347.

Ferrándiz, J. M., Sansaturio, M. E. and Vigo, J., 1992b, On the Accurate numerical Computation of Highly Eccentric Orbits, Paper AAS92-193, in "Advances in the Astronautical Sciences", Vol. 79, Part II: 1185:1204, R.E. Diehl, R.G. Schinnerer, W.E. Williamson and D.G. Boden eds., Univelt, San Diego.

Janin, G., 1979, Mission analysis for terrestrial satellites and planetary orbiters: software design and algorithm description, *ESA STM-208*, 47 pp., Darmstadt.

Kwok, J.H. and Nacozy, P., 1981, Time elements in rectangular coordinates, *Celest. Mech.*, 24:269.

Nacozy, P., 1981, Time elements in keplerian orbital elements, *Celest. Mech.*, 23:173.

Scheifele, G. and Graf, O., 1974, Analytical satellite theories based on a new set of canonical elements, *AIAA Paper 74-838*, 19 pp.

Stiefel, E. and Scheifele, G., 1971, "Linear and Regular Celestial Mechanics", Springer-Verlag, Berlin.

FRACTAL STRUCTURES IN SELF-GRAVITATING COSMOLOGICAL FLOWS

Giuseppe Murante[1] and Antonello Provenzale[2]

[1]Istituto di Cosmogeofisica del CNR, C.so Fiume 4, Torino, Italy
[2]Istituto di fisica Generale dell'Università di Torino

ABSTRACT

The galaxy and cluster distribution, as revealed by the observational data, displays a rich variety of nonlinear structures, such as filaments, walls, and a hierarchical organization at small scales. Here we discuss the results of different N-body numerical simulations which indicate that a fractal matter distribution at scales below 4–7 h^{-1} Mpc should be the natural outcome of the nonlinear evolution of a self-gravitating collisionless fluid in the Hubble flow. This is one of the few examples where an observed fractal distribution in physical space is dynamically related to the system evolution. We finally discuss the stability properties of the dynamically evolved fractal point distribution.

INTRODUCTION

The galaxy distribution at scales smaller than about 4–7 h^{-1} Mpc (where h is the Hubble constant in units of 100 km/sec Mpc^{-1}) appears to be strongly inhomogeneous [1], displaying evidence of scaling behavior characterized by a correlation dimension $D_2 \approx 1.2$. The analysis of different redshift samples has indicated a possible multifractal nature of the galaxy distribution, with $D_q \approx 1$ for $q > 2$ and $D_o \approx 2$, see e.g. Refs. 2,3.

The fractal behavior of the galaxy distribution at small scales requires a physical motivation. One possibility would be to accept a fractal nature of the initial conditions, a fact which however contradicts most accepted results on the properties of the density field at recombination epoch. An alternative approach is based on verifying whether a multifractal distribution may be naturally produced by the dynamical evolution of a self-gravitating "cosmological" collisionless fluid, i.e. a fluid embedded in an expanding frame of coordinates such as that provided by the Hubble expansion. This problem has been studied in the last two years by a series of cosmological N-body simulations based on different algorithms (PM and P^{3}M) [4–6]. The results of these simulations have confirmed that the strongly nonlinear clustering at scales smaller than 4–7 h^{-1} Mpc

From Newton to Chaos, Edited by A.E. Roy
and B.A. Steves, Plenum Press, New York, 1995

seems to naturally produce a multifractal matter distribution, whose general asymptotic properties turn out to be quite independent on the details of the initial conditions. In particular, the long–time state of the particle distribution is characterized by generalized dimension $D_q \approx 1$ for $q > 2$ and $D_q > 3$ for negative q's. The value of D_o is strongly dependent on bias and on the evolutive stage.

In the present work we review the main results of the N-Body simulations which have explored the scaling properties of the evolved density field. We also discuss the results of a numerical study on the stability of under gravitational evolution fractal distributions with given D_q spectra. The basic result in this framework is that a monofractal distribution is *not* stable under self-gravity; a diffuse background is in fact needed in order to stabilize the fractal properties of the matter distribution [7]. This implies that for the background it should be $D_o = 3$ even though we observe $D_2 \approx 1.2$ for the luminous matter.

MULTIFRACTAL FORMALISM

A number of different statistics are now available in order to study the clustering properties of the galaxy distribution. Here we are interested in the scaling properties of a point distribution, which may be quantitatively studied by the multifractal analysis (See e.g. Refs. 8,9 for a general discussion of this issue). A possible operative definition of the spectrum of the generalized fractal dimensions of a point distribution is based on the correlation integral method, which requires the evaluation of the partition function

$$Z(r,q) = \frac{1}{N}\sum_{i=1}^{N}(C_i(r))^{q-1} \tag{1}$$

where

$$C_i(r) = \frac{1}{N}\sum_{j=1}^{N}\Theta(r - |\vec{X}_i - \vec{X}_j|) \tag{2}$$

In the above formulas, N is the total number of objects in the distribution, $\vec{X}_i$ is the position of the i-th object, Θ is the Heaviside step function and q is the order of the moment. Vertical bars denote the norm of the vector and $C_i(r)$ measures the probability of finding another point in a sphere with radius r, centered on the i-th point of the distribution. For a scaling (scale-invariant) distribution, the partition function has a power-law behavior, i.e.

$$Z(r,q) \propto r^{(q-1)D_q} \tag{3}$$

where D_q is the generalized dimension of order q. For $q = 0$, the above formulas give the fractal (or box-counting) dimension D_o, which is insensitive to statistics and measures only geometrical properties. The behaviour of the correlation integral of order $q = 2$ is connected with the correlation function, which is the most commonly used statistic for characterizing the clustering properties of the galaxy distribution.

Monofractal objects are characterized by $D_q = D$ for all q's, in this case if $D < 3$ there are truly empty regions in space and high peaks have the same scaling properties of the underdense regions. By contrast, multifractal distributions are characterized by $D_q > D_{q'}$ if $q < q'$; in this case the large peaks are more strongly correlated than less dense regions and the background may be space-filling ($D_o = 3$) even though the peaks are strongly clustered. To practically determine the scaling properties of a distribution one has thus to evaluate the partition functions for various values of q as a function of r, to verify whether a power-law behavior is present and, if so, to evaluate the generalized dimensions D_q. To this end, one may consider the local generalized dimension $D_q(r)$ as given by a local least-square-fit of $\log Z$ versus $\log r$ on a small interval $(r - dr, r + dr)$. In this case, the constancy of $D_q(r)$ over a given range indicates scaling behavior on that range, and the value of $D_q(r)$ gives a direct estimate of D_q on those scales.

COSMOLOGICAL INITIAL CONDITIONS

A variety of numerical N-body simulations have been performed in order to compare different models of large–scale structure formation with the observational data on the galaxy distribution (for a review see e.g. Ref. 10). In particular, simulations with different cosmological initial conditions have been analyzed by Valdarnini, Borgani & Provenzale [4] (hereinafter VBP) in order to explore the possible presence of scaling behaviour at small scales. In VBP, a P^3M code has been used. The initial conditions have been chosen to be random Gaussian density perturbation fields with different power spectra; two scale–free spectra with spectral index $n = 1$ and $n = -2$ and a standard CDM spectrum have been considered.

The analysis of the above simulations has been conducted by using different generalized fractal dimension estimators, see e.g. Ref. 9 for a critical evaluation of the performances of the various methods. The analysis has been conducted on either different subsamples of the entire point distribution, corresponding to different biassing levels and ideally representing the galaxy distribution, as well as on the entire point set (representing the background matter distribution). The main conclusion of the analysis reported in VPB is that the strongly nonlinear gravitational evolution generates a well-defined scaling regime in the matter distribution at small scales. The value of the generalized fractal dimensions for $q > 2$ is $D_q \approx 1$; for $q < 0$, D_q becomes larger than three.

At intermediate evolutive stages, some differences may be observed between the different initial conditions; for example, the evolution seems to be slower for the scale-free spectrum with $n = -2$. Analogously, the precise value of the fractal dimension D_0 depends upon either the type of initial conditions and the biassing level. The results of the analysis do however suggest that the multifractal properties of the late (asymptotic) stage of the evolved matter distribution turn out to be almost independently on the details of the initial conditions. This result seems thus to indicate the existence of a statistical attractor for the strongly nonlinear gravitational evolution in expanding coordinates.

The results discussed in VBP have been independently confirmed by Colombi, Bouchet & Schaeffer [5] and by Yepes, Dominguez-Tenreiro & Couchman [6]. Both groups performed P^3M simulations on CDM initial conditions. At scales comparable with that of VBP, they found a “bifractal” scaling regime, characterized by $D_q \approx 1$ for large q and $D_q > 3$ when $q < 0$. The precise value quoted for the D_q's is not completely independent from the algorithm used to extract the analized point distribution: in particular Colombi, Bouchet & Schaeffer used the whole set of particles on the N-body simulation, roughly corresponding to the “background” of VBP, and found a value something higher for the high-q regime, while Yepes, Dominguez-Tenreiro & Couchman identified “galaxies” with an evolutive algorithm, obtaining more biased distribution and slightly lower values for the D_q's in the same range. A detailed study on how the scaling properties depend on the type of galaxy identification algorithm would thus be worthwhile. Apart from these minor differences, however, all these authors agree on the fact that a scaling regime is the natural outcome of gravitational evolution, for scales smaller than 4–7 h^{-1} Mpc.

EVOLUTION OF A FRACTAL DUST

The simulations discussed in the previous Section dealt with the evolution of an homogeneous, non–fractal initial state with small superposed density perturbations. A different question concerns the gravitational evolution of an initially fractal point distribution. This problem is related to the desire of verifying whether a given fractal configuration is statistically stable under gravitational evolution, as it should be expected for a true statistical attractor. Thus, the search for statistically stable fractal configuration may be taken as an attempt of determining the properties of the statistical attractors of the nonlinear gravitational evolution.

A study of the stability of the fractal properties of an ideally collisionless fluid, sampled with a finite number of massive particles, has been performed by Murante *et al* [7]. In that work, a tree-code algorithm has been used [11]; the various parameters of the code (e.g. smoothing length, time step) have been selected by the standard tests on collisionality and energy conservation. The initial conditions for these simulations have been chosen to be point distributions with known fractal properties; these were resembling the possible final states of the fully–developed gravitational evolution of the "cosmological" initial conditions discussed above. The initial fractal distributions were then embedded in the Hubble flow. Note that in this simulations the initial density field is already in a very nonlinear stage, since there are both overdense and void regions in a fractal distribution.

In these simulation, two different cases are of particular interest. First we consider an initially monofractal dust with generalized dimensions $D_q = D = 1$ at scales smaller than an appropriate homogeneity threshold L_h. This distribution ideally represents the distribution of visible matter; for this reason, a value $\Omega = 0.1$ has been chosen (Ω being the ratio of the average density to the critical density for a flat Universe). The multifractal analysis of the evolved distribution shows the the scaling properties are lost at early evolutive stages as the values of the local generalized dimensions do not stay constant on any range of scales. The distribution has lost its initial scale invariance and even at later times the scaling behaviour is not recovered, showing that a pure monofractal dust is *not* stable under self–gravitational evolution. Therefore, the asymptotic state of a self–gravitating collisionless particle distribution cannot be a monofractal distribution.

To further explore the statistical attractor of the gravitational evolution, it is then possible to consider the case of an initial multifractal dust generated by a superposition of a fractal point distribution (with $D = 1$) and of an homogeneus dust. Still the density contrast is very strong and the density field is in a nonlinear stage. The presence of a background suggest now to use a value $\Omega = 1$, consistent with cosmological dark matter simulations. In Ref. 7 the initially fractally distributed matter has been chosen to be the 10 per cent of the total matter, i.e., $\Omega_{fractal} = 0.1$ as in the previous simulation. At the initial time, $D_q \approx 1$ for $q > 2$ at small scales, since the large q moments measure the scaling properties of the strongly clustered (overdense) regions which are affected mainly by the monofractal component. Conversely, $D_0 = 3$ on the same scales. This implies that there is a space–filling background, generated by the fraction of matter which has been homogeneously distributed. At scales larger than L_h, the distribution becomes completely homogeneous.

For the above multifractal distribution, a very different scenario is observed: the homogeneous background now organizes itself around the initial overdensities of the $D = 1$ fractal component, and, as the simulation evolves, scaling is initially destroyed and then recovered, finding again $D_q \approx 1$ for $q > 2$. This evolved multifractal distri-

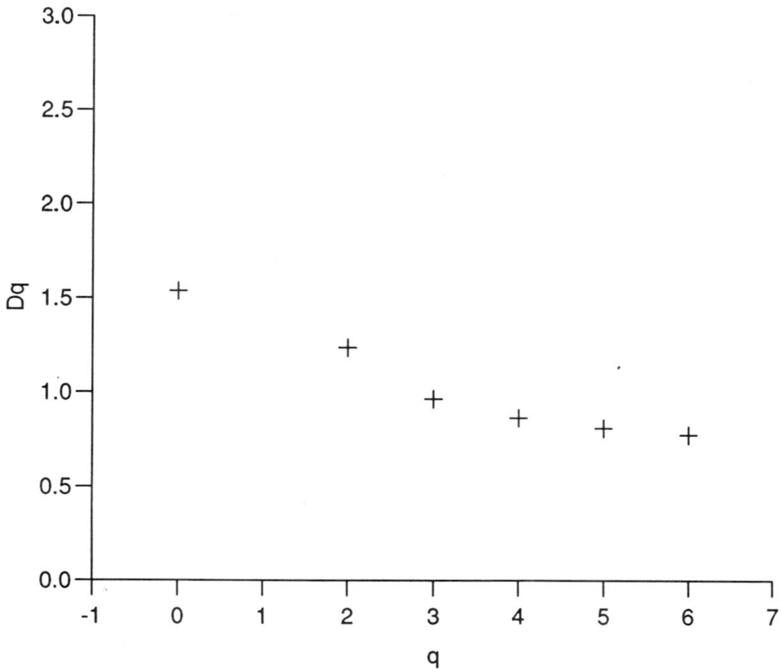

Figure 1. Spectrum of the generalised fractal dimensions D_q as a function of q, for non-negative q's for the multifractal distribution (see text). The value of the dimensions are obtained as a linear least-squares fit of log Z versus log r over one decade.

bution *does* now appear to be statistically stable under self-gravity. In Figure 1 the multifractal spectrum for $q \geq 0$ for the evolved distribution is shown; this appears to be very close to that obtained from the cosmological simulations discussed in the previous Section, even though the initial conditions are very different, confirming the existence of a statistical attractor.

CONCLUSIONS

The results of the numerical simulations discussed in Refs. 4-7 indicate that the strongly nonlinear gravitational dynamics at scales smaller than 4–7 h^{-1} Mpc seems to be capable of producing a multifractal matter distribution. This is one of the few examples where an observed fractal structure in physical space has been related to a well–defined dynamical origin. An analysis of the stability properties of fractal dusts indicate that a purely monofractal distribution with $D = 1$ (or with any other value of D) is *not* stable under nonlinear gravitational evolution. Conversely, a multifractal distribution composed by a fractal dust superposed onto a homogeneous background undergoes a rapid transient and then stabilizes as a distribution with $D_q \approx 1$ for $q > 2$ and $D_q > 3$ for negative q's; in general, $D_0 = 3$ for the background. This result is consistent with the previous studies on the evolution of different types of "standard" initial conditions [4–6]; in particular, it is consistent with the thermodynamical approach discussed in Ref. 12 and rephrased in terms of multifractal properties in Ref. 13. In general, these results seems to indicate the existence of a statistical attractor for the gravitational evolution.

From a cosmological point of view, the results presented here suggest that the fractal properties of the large scale distribution of matter in the Universe should be a natural product of strongly nonlinear gravitational dynamics. The results of the study on the stability properties also indicate that a fractal distribution of visible matter at small scales (say less than 4–7 h^{-1} Mpc) requires, in order to be stable, the existence of a diffuse background which is space-filling and much less clustered than the high density peaks.

REFERENCES

1. B. J. T. Jones 1992, *Observational and Physical Cosmology*, ed., (Cambridge Un. Press), in press.
2. V. J. Martinez, B. J. T. Jones, R. Dominguez-Tenreiro, & R. van de Weygaert, 1990, *Ap. J.,* **357**, 50.
3. A. Provenzale, 1991, *Applying Fractals in Astronomy,* ed. A. Heck & J. Perdang, Springer, Berlin, in press.
4. R. Valdarnini, S. Borgani, & A. Provenzale, 1992, *Ap. J.,* **394**, 422.
5. S. Colombi, F. R. Bouchet, & R. Scheffer, 1992, *Astr. Ap.,* **263**, 1.
6. G. Yepes, R. Dominguez-Tenreiro, & H. M. P. Couchman, 1992, *Ap. J.,* **401**, 40.
7. G. Murante, A. Provenzale, E. Branchini, & F. Governato, 1994, *sub judice.*
8. A. Provenzale, P. Galeotti, G. Murante, & B. Villone, 1992, *Ap. J.,* **401**, 455.
9. S. Borgani, G. Murante, A. Provenzale & R. Valdarnini, 1993, *Phys. Rev. E,* **47**, 3879.
10. C. S. Frenk, 1991, *Physica Scripta,* **T36**, 70.
11. J. Barnes, & P. Hut, 1986, *Nature,* **324**, 446.
12. W. C. Saslaw, & A. J. S. Hamilton, 1984, *Ap. J.,* **276**, 13.
13. S. Borgani, 1993, *M.N.R.A.S.,* **250**, 537.

QUATERNIONIC GENERALISATION OF THE MANDELBROT SET

Jagannathan Gomatam, John Doyle, and Bonnie Steves

Department of Mathematics
Glasgow Caledonian University
Glasgow, Scotland G4 0BA

ABSTRACT

The complex quadratic map leading to the celebrated Mandelbrot set is generalised in a natural way to R^4 using the framework of quaternions. Calculations are presented with the aid of Pauli spin matrices. Stability analyses of 3-cycles are summarised, with explicit expressions for generalised Mandelbrot domains in R^4. It is conjectured that similar expressions hold for stability domains of k-cycles.

1. INTRODUCTION

A wealth of both scientific and popular literature has been devoted to the Mandelbrot set which arises from the complex iterative map $Z \dashrightarrow Z^2 + C$ (Mandelbrot, 1980; Douaday and Hubbard, 1985; Peitgen 1986). It is logical to attempt to generalise the complex quadratic map to higher dimensions (Gomatam et al, 1993; Doyle et al, 1994). These higher dimensions could call for iterative maps of more general objects such as matrices of complex numbers or even more general rings, or quadratic maps involving generators of the groups SU(2), SU(3) or SU(n).

The first stage in this extension of the Mandelbrot set to higher dimensions was to four dimensions using W. R. Hamilton's creation, the quaternion (first proposed in 1884, Ashurst, 1982), but written in terms of complex matrices, the generators of SU(2). The four dimensional Mandelbrot set arises from the iteration of the quaternionic ring (Jacobson, 1974) using the quadratic map

$$Q_{n+1} = Q_n^2 + C$$

where Q_n and C are quaternions expressed in terms of 2x2 Pauli spin matrices (Goldstein, 1980).

After the presentation of the current investigation, we came across several publications on computer graphics associated with the quaternionic Mandelbrot and Julia sets (Norton, 1982, 1989; Pickover, 1988; Hart, Sandin and Kauffman, 1989; Hart, Kaufmann and Sandin, 1990; Ke and Panduranga, 1990; Griffith and Joshi, 1992). But to our knowledge, no systematic attempt has been made to analytically characterise the stability domains of the multiple cycles of the quaternionic quadratic map. Griffiths and Joshi (1993) include a discussion of the fixed point (1-cycle).

In this summary paper, we only briefly describe the analysis of stability for the first three cycles. The inequalities which characterise the quaternionic Mandelbrot domains in four dimensions are stated. A complete description of the analysis can be found in Gomatam et al (1993). Section 2 contains a description of the quaternionic quadratic map, while

From Newton to Chaos, Edited by A.E. Roy
and B.A. Steves, Plenum Press, New York, 1995

sections 3 to 5 are devoted to a discussion of the stability domains of the 1, 2 and 3 point cycles of the map, respectively.These results and future lines of research are discussed in Section 6.

2. THE QUATERNIONIC QUADRATIC MAP

Using 2x2 Pauli matrices, a quaternion can be written in matrix form as:

$$Q = X I + i \underline{\sigma} . \underline{V}$$

where $\underline{\sigma} = (\sigma_1, \sigma_2, \sigma_3)$ is the set of 2x2 Pauli matrices and $\underline{V} = (Y, Z, W)$. The quaternion components X, Y, Z and W span R^4. The special properties of quaternions facilitate the algebraic analysis of the stability of the multi-cycles.

Following the analogy of the complex quadratic map (Mandelbrot, 1980), the quaternionic quadratic map becomes

$$Q_{n+1} = Q_n^2 + C \tag{1}$$

where C is a quaternion constant defined by $C = a I + i \underline{\sigma} . \underline{K}$, $\underline{K} = (b, c, d)$ with $\{a, b, c, d\} \varepsilon R^4$. The iterative map (1) can be expressed in terms of its components $(X_n, \underline{V}_n)$:

$$X_{n+1} = X_n^2 - (Y_n^2 + Z_n^2 + W_n^2) + a$$

$$\underline{V}_{n+1} = 2 X_n \underline{V}_n + \underline{K}$$

Using the definitions

$$F^{(1)}(Q) = Q^2 + C \tag{2a}$$

$$F^{(2)}(Q) = F(F(Q)) \tag{2b}$$

and

$$F^{(3)}(Q) = F(F(F(Q))) \tag{2c}$$

the 1, 2 and 3 cycle points are found by setting $F^{(k)}(Q)=Q$ for k = 1, 2 or 3 and solving for Q, excluding the fixed points of the lower cycles. With the introduction of a perturbation about Q, a variational equation for the specific cycle is obtained. The stability matrix, written **formally**, becomes

$$\frac{\partial F^{(k)}}{\partial Q}(Q) \tag{3}$$

where k= 1, 2 or 3 according to the cycle chosen and $F^{(k)}(Q)$ is as defined in Eq (2). The dominant eigenvalue (ie the eigenvalue with the greatest modulus) of the stability matrix in Eq (3) is found to be

$$|\lambda|^2 = \det \frac{\partial F^{(k)}}{\partial Q}(Q) \qquad k = 1, 2 \text{ or } 3 \tag{4}$$

A detailed algebraic computation reveals that inside the determinantal operation in Eq (4), it is possible to apply the rules of the standard abelian calculus to the matrix functional $F^{(k)}(Q)$. It is our conjecture that this result is true for $k > 3$. The boundary surface of stability in {a, b, c, d} parameter space is then given by $|\lambda| = 1$.

3. THE FIXED POINT (1-CYCLE) AND ITS STABILITY

The fixed point or 1-cycle for Eq. (1) is defined by the equation

$$Q = Q^2 + C \tag{5}$$

Since it can be shown that $[Q, C] = QC - CQ = 0$, Eq. (5) can be treated as an ordinary polynomial in Q.

Solving for Q and using the expression for the square root of a quaternion (Gomatam et al, 1993), it can be shown that the fixed point is given by

$$X = \frac{1}{2} \pm \frac{1}{2\sqrt{2}}\left[1 - 4a + \sqrt{(1-4a)^2 + 16\underline{K}^2}\right]^{\frac{1}{2}} \tag{6a}$$

$$\underline{V} = \mp\frac{\underline{K}}{2\sqrt{2}\,|\underline{K}|}\left[\sqrt{(1-4a)^2 + 16\underline{K}^2} - (1 - 4a)\right]^{\frac{1}{2}} \tag{6b}$$

Using Eq (4), the fixed point is found to be stable if and only if

$$\det \frac{\partial F^{(1)}}{\partial Q}(Q) < 1$$

or after some algebra

$$\det 2Q < 1.$$

In terms of X and $\underline{V}$, the fixed point is stable if and only if

$$X^2 + \underline{V}^2 < \frac{1}{4}$$

Since X^2 must be less than 1/4 , the negative sign in Eq. (6a) must be chosen. The boundary surface in {a, b, c, d} parameter space then simplifies to

$$(1 - 4a)^2 + 16\underline{K}^2 = (4-8a) \pm 2\sqrt{3 - 8a} \tag{7}$$

In general, Eq. (7) describes a cardioidal surface in R^4. Its cusp is located at $(1/4, \underline{0})$, while its round 'nose' is situated at $(3/4, \underline{0})$. Projections onto any of the three planes involving the parameter 'a', ie {a,b}, {a, c} or {a, d}, exhibit the Mandelbrot geometry. Projections onto the other planes, where 'a' is taken to be a constant, display rings which result from spheres in the {b, c, d} parameter space. For example, at a = 0

$$\underline{K}^2 = \frac{\sqrt{3}\,(\sqrt{3} + 2)}{16}$$

which is a sphere of radius

$$\left(\frac{\sqrt{3}\,(\sqrt{3} + 2)}{16}\right)^{\frac{1}{2}}.$$

4. THE 2-CYCLE AND ITS STABILITY

The 2-cycle is defined by the solution of

$$Q = (Q^2 + C)^2 + C \tag{8}$$

It can be shown that $[Q,C] = 0$. Therefore Eq. (8) can be treated as a polynomial equation for Q, leading to the factorisation:

$$(Q^2 - Q + C)(Q^2 + Q + C + I) = 0 \tag{9}$$

Excluding the 1-cycle, the solutions to Eq. (9) are given by:

$$Q_{\pm} = \frac{-I \pm i\,(3I + 4\,C)^{\frac{1}{2}}}{2} \tag{10}$$

It can be shown that

$$|\lambda|^2_{max} = \det\,[4\,Q(\,Q^2 + C)] = \det\,(4Q_+\,Q_-)$$

where $Q_{\pm}$ are given by Eq (10). Thus

$$|\lambda|^2_{max} = 16 \det\,(\,I + C)$$

The boundary surface in {a,b,c,d} parameter space is defined by

$$(a + 1)^2 + \underline{K}^2 = \frac{1}{16} \tag{11}$$

Eq (11) describes a sphere of radius 1/4 in R^4,which touches the primary cardiac surface at $(-3/4, \underline{0})$.

5. THE 3-CYCLE AND ITS STABILITY

The 3-cycle is defined by the equation

$$Q = [\,(Q^2 + C)^2 + C]^2 + C \tag{12}$$

After lengthy algebra it follows that solutions of Eq (12) satisfy $[Q,C] = 0$. Exclusion of the 1-cycle from Eq (12) leads to a sixth degree polynomial in Q:

$$1+C+2C^2+C^3 + (1+C)^2Q + (1+3C+3C^2)Q^2 + (1+2C)Q^3 + (1+3C)Q^4 + Q^5 + Q^6 = 0 \tag{13}$$

Detailed algebraic calculations show that the 3-cycle is stable if

$$\det\,[8Q(Q^2 + C)\,\{(Q^2 + C)^2 + C\}] < 1 \tag{14}$$

To make further analytical headway, we note that the product of the roots $Q_{(j)}$ of Eq(13) is given by

$$\prod_{j=1}^{6} Q(j) = 1 + C + 2C^2 + C^3 \tag{15}$$

Exploiting the corresponding result for the 3-cycle Mandelbrot set (Peitgen and Richter, 1986), we have

$$\prod_{j=1}^{6} Q_{(j)} = (C + 1.755)(C + 0.123 + i\,0.745\,\underline{\sigma}\cdot\underline{m})\cdot$$

$$(C + 0.123 - i\,0.745\,\underline{\sigma}\cdot\underline{m}) \equiv G(C) \qquad (16)$$

where $\underline{m}$ is an arbitrary vector of unit magnitude. Since the 3-cycle is constrained by

$$Q_{(1)} = F(Q_{(2)})$$
$$Q_{(2)} = F(Q_{(3)})$$
$$Q_{(3)} = F(Q_{(1)})$$

the stability criterion Eq (14) can be reformulated as

$$\det\left[\frac{G(C)}{Q_{(4)}\,Q_{(5)}\,Q_{(6)}}\right] < \frac{1}{64} \qquad (17)$$

This reproduces the well documented Mandelbrot domain for the 3-cycle in the {a, b} plane.

6. CONCLUSIONS AND FUTURE WORK

Stability calculations for the 1, 2 and 3 cycles lead to explicit characterisations of surface boundaries for the generalised Mandelbrot set in {a, b, c, d} parameter space. Projection of the quaternionic Mandelbrot set onto {b, c, d} space, results in a series of spherical shells whose radii and thickness are functions of the parameter 'a'. Projection of the generalised set onto any plane involving the parameter 'a' produce the familiar complex Mandelbrot sets.

The above theoretical results were confirmed numerically through a series of 2-dimensional cross-sections of the quaternionic Mandelbrot set in R^4. All cross-sections involving two of the other parameters {b, c, or d} displayed circular rings. Further magnification of the rings reveals a Cantor like division in the radii of the shells. It follows that each symmetrical point on the (a, b), (a, c) or (a, d) Mandelbrot set is a spherical shell in (b, c, d) space.

Of particular note for the quadratic map is the possibility that the stability eigenvalue with the largest modulus can be found for any cycle from the determinant of the stability matrix derived using standard laws of abelian calculus.

The quaternionic map in R^4 and the stability analysis of its cycles provide a theoretical framework for extending the quadratic map to higher dimensions, with the possibility of generalising results to groups such as SU(n).

ACKNOWLEDGEMENTS

We (J. Gomatam and J. F. Doyle) would like to thank the Department of Mathematics, Glasgow Caledonian University and the Carnegie Foundation, respectively, for travel grants.

REFERENCES

Ashurst, G., 1982 "Founders of Modern Mathematics", Frederick Muller Ltd, London.

Douaday, A., and Hubbard, J.H., 1985, On the dynamics of polynomial-like mappings, *Ann. Scient. Ec. Norm. Sup., 4e série,* t.18:287.

Doyle, J.F., Gomatam, J. and Steves, B., 1994, Fractals, chaos and quaternions, *in* "From Newton to Chaos: Modern Techniques for Understanding and Coping with Chaos

in N-Body Dynamical Systems", A.E. Roy and B.A. Steves, eds., Plenum, London.
Goldstein, H., 1980, "Classical Mechanics", II Edition, Addison-Wesley Publishing Co., Reading, Massachusetts, 155.
Gomatam, J., Doyle, J.F. and Steves, B., 1993, Generalisation of the mandelbrot set: Quaternionic quadratic maps, "Technical Report: August 1993", Glasgow Caledonian University.
Griffith, C.J. and Joshi, G.C., 1992, Octonionic julia sets, *Chaos, Solitons and Fractals* 2:11.
Griffith, C.J. and Joshi, G.C., 1993, Transition points in octonionic julia sets, *Chaos, Solitons and Fractals* 3:67.
Hart, J.C., Kaufmann, L.H. and Sandin, D.J., 1990, Interactive visualisation of quaternionic julia sets, *in* "Visualisation 90: Proceedings of the First IEEE Conference on Visualisation", A. Kaufmann, ed., IEEE Computer Society Press, San Francisco, California, 209-218.
Hart, J.C., Sandin, D.J. and Kaufmann, L.H., 1989, Ray tracing deterministic 3-D fractals, *Computer Graphics* 23:289.
Jacobson, N., 1974, "Basic Algebra I", W.H. Freeman and Co., San Francisco.
Ke, Y. and Panduranga, E.S., 1990, A journey into the fourth dimension, in "Visualisation 90: Proceedings of the First IEEE Conference on Visualisation", A. Kaufmann, ed., IEEE Computer Society Press, San Francisco, California, 219-229.
Mandelbrot, B.B., 1980, Fractal aspects of the iteration of $z \rightarrow \lambda z(1-z)$ for complex λ and z, *Annals NY, Acad. Science* 357:249.
Norton, A., 1982, Generation and display of geometric fractals in 3-D, *Computer Graphics* 16:61.
Norton, A., 1989, Julia sets in the quaternions, *Computers and Graphics* 13:267.
Peitgen, H.O. and Richter, P.H., 1986, "The Beauty of Fractals: Images of Complex Dynamical Systems", Springer-Verlag, New York.
Pickover, C.A., 1988, Visualisation of quaternion slices, *Image and Vision Computing* 6:235.

PARTICIPANTS AND SPEAKERS

Baille, P.	CANADA	Mathematics Dept, R.M.C., Kingston, Ontario, K7K 5LO, CANADA
Barre, C. de la	USA	1016 Middle Ave #4, Menlo Park, CA, 94025, USA
Boudin, F.	FRANCE	Observatoire de la Cote d'Azur, C.E.R.G.A., Av. Nicolas Copernic, Grasse, F-06130, FRANCE
Broucke, R.	USA	Dept of Aerospace Engineering & Engineering Mechanics, The University of Texas at Austin, Austin, TX, 78712-1085, USA
Campos, B.	SPAIN	Departamento de Matematicas, Universitat Jaume I, Campus Penyeta Roja, Castello, 12071, SPAIN
Casas, F.	SPAIN	Departament de Matematiques, Universitat Jaume I, Campus de Penyeta Roja, Castellon, 12071, SPAIN
Celletti, A.	ITALY	Dipartimento di Matematica Pura ed Applicata, Universita degli Studi di L'Aquila, L'Aquila, 67100, ITALY
Cincotta, P.	ARGENTINA	Facultad de Ciencias Astonomicas Y Geofisicas, Universidad Nacional de La Plata, Paseo del Bosque, La Plata, 1900, ARGENTINA
Cipriani, P.	ITALY	Physics Dept. - Group G9, University of Rome "La Sapienza", P.le A. Moro, 2, Roma, 00185, ITALY

Contopoulos, G.	GREECE	Dept of Astronomy, University of Athens, Panepistimiopolis, Zografos Athens, GR - 157 63, GREECE
Conway, B.	USA	Dept of Aeronautical & Astronautical Engineering, University of Illinois at Urbana-Champaign, 306 Talbot Laboratory, 104 South Wright Street, Urbana, IL, 61801-2935, USA
Cors, J.	SPAIN	Departament de Matematica Aplicada III, Universitat Politecnica de Catalunya, Av. Bases de Manresa, 61-73, Manresa (Barcelona), 08240, SPAIN
Doyle, J.	UK	Dept of Mathematics, Glasgow Caledonian University, Cowcaddens Rd, Glasgow, G4 0BA, UK
Ferraz-Mello, S.	BRAZIL	Instituto Astronomico e Geofisico, Universidade de Sao Paulo, Caixa Postal 9638, Sao Paulo, 01065-970, BRAZIL
Floria, L.	SPAIN	Departamento de Matematica Aplicada a la Ingenieria, Universidad de Valladolid, Paseo del Cauce s/n, Valladolid, E-47011, SPAIN
Froeschlé, Cl.	FRANCE	Observatoire de la Cote d'Azur, B.P. 229, Nice Cedex 4, 06304, FRANCE
Ge, Y.	UK	Dept of Mathematics, Southampton University, Southampton, UK
Giorgilli, A.	ITALY	Dipartimento di Matematica, Universita di Milano, Via Saldini 50, Milano, 20113, ITALY
Gomatam, J.	UK	Dept of Mathematics, Glasgow Caledonian University, Cowcaddens Rd, Glasgow, G4 0BA, UK
Goudas, C.	GREECE	Department of Mathematics, University of Patras, Patras, 26110, GREECE
Hadjidemetriou, J.	GREECE	Dept of Theoretical Mechanics, University of Thessaloniki, Thessaloniki, GR- 540 06, GREECE

Irigoyen, M.	FRANCE	Universite Paris 2, 92 rue d'Assas, Paris, 75006, FRANCE
Issartel, J.	FRANCE	D.E.S. ONERA, B.P. 72, Chatillon Cedex, 92322, FRANCE
Kaya, D.	USA	Astrodynamics Division, USAF Space Command/CNY, Suite 4110, 150 Vandenberg Ave., Peterson AFB, CO, 80914-4110, USA
Kiseleva, L.	RUSSIA	Institute of Astronomy, University of Cambridge, Madingley Rd., Cambridge, CB3 0HA, U.K.
Labianca, A.	ITALY	Dipartimento di Matematica, Universita Degli Studi di Bari, Via E. Orabona, 4, Bari, 70125, ITALY
Laskar, J.	FRANCE	Bureau des Longitudes, 77, avenue Denfert-Rochereau, Paris, F-75014, FRANCE
Leftaki, M.	GREECE	Dept of Mathematics, University of Patras, Patras, 26110, GREECE
Lohinger, E.	AUSTRIA	Institut fur Astronomie, Universitat Wien, Turkenschanzstrasse 17, Wien, A-1180, AUSTRIA
Maciejewski, A.	POLAND	Institute of Astronomy, Nicolaus Copernicus University, Chopina 12/18, Torun, PL-87-100, POLAND
Marangio, L.	ITALY	Via Spinoza, 38, Roma, 00137, ITALY
Marchal, C.	FRANCE	D.E.S. ONERA, B.P. 72, Chatillon Cedex, 92322, FRANCE
Markellos, V.	GREECE	Dept of Engineering Science, University of Patras, Rion, Patras, GR 261 10, GREECE
Martínez Alfaro, J.	SPAIN	Departament de Matematica Aplicada i Astronomia, Universitat de Valencia, C. Doctor Moliner 50, Burjassot, Valencia, 46100, SPAIN

Message, P. J.	UK	Dept of Applied Mathematics & Theoretical Physics, University of Liverpool, Liverpool, L69 3BX, UK
Milani, A.	ITALY	Departimento di Matematica, Universita di Pisa, via Buonarotti 2, Pisa, 56127, ITALY
Murante, G.	ITALY	Istituto di Cosmo - Geofisica, Consiglio Nazionale delle Ricerche, Corso Fiume 4, Torino, 10133, ITALY
Neron_de_Surgy, O.	FRANCE	Equipe Astronomie et Systemes Dynamiques, Bureau des Longitudes, 77 avenue Denfert-Rochereau, Paris, 75014, FRANCE
Neslusan, L.	SLOVAKIA	Astronomical Institute, Slovak Academy of Sciences, Tatranska Lomnica, 059 60, SLOVAKIA
Nesvorný, D.	CZECH REPUBLIC	Astronomical Institute, Academy of Sciences of the Czech Republic, Bocni II 1401, Prague, 141 31, CZECH REPUBLIC
Niederman, L.	FRANCE	Bureau des Longitudes, 77 avenue Denfert-Rochereau, Paris, F-75014, FRANCE
Papadaki, H.	GREECE	Dept of Astronomy, University of Athens, Panepistimiopolis, Zografos Athens, GR - 157 63, GREECE
Papadakis, C.	GREECE	Dept of Engineering Science, University of Patras, Rion, Patras, GR 261 10, GREECE
Perdios, E.	GREECE	Dept of Engineering Science, University of Patras, Rion, Patras, GR 261 10, GREECE
Perozzi, E.	ITALY	Telespazio, Via Tiburtina 965, Roma, 00156, ITALY
Petrosky, T.	USA	Center for Studies in Statistical Mechanics and Complex Systems, The University of Texas, Austin, TX, 78712, USA

Pucacco, G.	ITALY	Physics Dept., University of Rome "Tor Vergata", V.le Ricerca Scientifica, 1, Roma, 00137, ITALY
Puel, F.	FRANCE	Observatoire de Besancon, Universite de Franche-Comte, B.P. 1615, 41 bis, Avenue de L'Observatoire, Besancon Cedex, 25010, FRANCE
Ries, J. Gyorgyey	USA	Dept of Astronomy, University of Texas at Austin, Austin, TX, 78712, USA
Robutel, P.	FRANCE	Astronomie et Systemes Dynamiques, Bureau des Longitudes, 77 av. Denfert-Rochereau, Paris, 75014, FRANCE
Roy, A. E.	UK	Dept of Physics & Astronomy, University of Glasgow, Glasgow, G12 8QQ, UK
Sansaturio, M. E.	SPAIN	Departamento de Matematica Aplicada a la Ingenieria, Universidad de Valladolid, Paseo del Cauce s/n, Valladolid, 47011, SPAIN
Sekiguchi, M.	JAPAN	Kisarazu National College of Technology, Kisarazu, Chiba 292, JAPAN
Simó, C.	SPAIN	Departament de Matematica Aplicada i Analisi, Universitat de Barcelona, Gran Via, 585, Barcelona, 08071, SPAIN
Snow, D.	USA	Astrodynamics Division, USAF Space Command, Suite 4110, 150 Vandenberg Ave., Peterson AFB, CO, 80914-4110, USA
Spirig, F.	SWITZERLAND	Wilenstrasse 10, Rorschacherberg, CH-9400, SWITZERLAND
Stavliotis, Ch.	GREECE	Dept of Physics, University of Patras, Patras, 26110, GREECE

Steves, B.	CANADA	Dept of Mathematics, Glasgow Caledonian University, Cowcaddens Rd, Glasgow, G4 0BA, UK
Szebehely, V.	USA	Dept of Aerospace Engineering WRW415, The University of Texas at Austin, Austin, TX, 78712-1085, USA
Valsecchi, G.	ITALY	I. A. S. - Planetologia, Viale dell' Universita, 11, Roma, 00185, ITALY
Varadi, F.	USA	Dept of Atmospheric Sciences, University of California, Los Angeles, CA 90024-1565, USA
Vicente, R.	PORTUGAL	R. Mestre Aviz, 30, R/C, Lisboa, 1495, PORTUGAL
Vindel, P.	SPAIN	Departamento de Matematicas, Universitat Jaume I, Campus Penyeta Roja, Castello, 12071, SPAIN
Vokrouhlický, D.	CZECH REPUBLIC	Astronomical Institute, Charles University, Svedska 8, Prague 5, CZ 15000, CZECH REPUBLIC
Waldvogel, J.	SWITZERLAND	Applied Mathematics, FLI, ETH - Zentrum, Zurich, CH-8092, SWITZERLAND
Williams, I.	UK	School of Mathematics, Queen Mary & Westfield College, Mile End Rd, London, E1 4NS, UK
Winter, O.	UK	School of Mathematical Sciences, Queen Mary and Westfield College, Mile End Rd, London, E1 4NS, UK
Wnuk, E.	POLAND	Astronomical Observatory, A. Mickiewicz University, ul. Sloneczna 36, Poznan, 60-286, POLAND
Wodnar, K.	AUSTRIA	Institut fur Astronomie, Universitat Wien, Turkenschanzstrasse 17, Wien, A-1180, AUSTRIA

Yokoyama, T.	BRAZIL	Dept. Estat. Mat. Aplic. Comp., Universidade Estadual Paulista - IGCE, Caixa Postal 178, Rio Claro, CEP 13500-970, BRAZIL
Zafiropoulos, B.	GREECE	Dept of Physics, University of Patras, Patras, 26110, GREECE
Zare, K.	USA	Dept of Aerospace Engineering & Engineering Mechanics, The University of Texas at Austin, Austin, Texas, 78712-1085, USA

NATO ADVANCED STUDY INSTITUTE ON FROM NEWTON TO CHAOS:
Modern Techniques for Understanding and Coping with
Chaos in N-Body Dynamical Systems
Held July 25 to August 6, 1993 in Cortina D'Ampezzo, Italy

AUTHOR INDEX

SUBJECT INDEX